THEORY OF
POPULATION GENETICS
AND
EVOLUTIONARY ECOLOGY:
AN INTRODUCTION

THEORY OF POPULATION GENETICS AND EVOLUTIONARY ECOLOGY: AN INTRODUCTION

JONATHAN ROUGHGARDEN

Stanford University

MACMILLAN PUBLISHING CO., INC.
New York

COLLIER MACMILLAN PUBLISHERS
London

MACMILLAN PUBLISHING CO., INC.
866 Third Avenue, New York, New York 10022
COLLIER MACMILLAN CANADA, LTD.

Library of Congress Cataloging in Publication Data

Roughgarden, Jonathan.
  Theory of population genetics and evolutionary ecology.

  Includes index.
  1. Population genetics. 2. Population biology.
3. Evolution. I. Title. [DNLM: 1. Genetics, Population. 2. Evolution. 3. Ecology. QH455 R856t]
QH455.R68   575.1   78-7245
ISBN 0-02-403180-1

Printing: 1 2 3 4 5 6 7 8   Year: 9 0 1 2 3 4 5

**To the memory of
Wolf Vishniac**

PREFACE

THE last 10 years have witnessed enormous progress in theoretical population genetics and population ecology. During this time several journals dealing solely with theoretical and mathematical papers in biology have been formed and have prospered. Also in this time, the traditional journals dealing with evolution and ecology have published a much higher fraction of theoretical papers than before. As a result, understanding evolution and ecology today involves knowing a much larger body of quantitative theory than was necessary 10 years ago. This situation has created a need for an introductory book in theoretical population genetics and population ecology, for a book whose ideas and motivations are biological but whose concentration is on the mathematical theory itself.

With this book, I am trying to make contemporary population theory accessible to as many people as I possibly can. Approximately one-third of the book is devoted to elementary and basic models. The elementary chapters require only one semester of calculus. The remaining chapters treat more advanced and specialized biological topics and also use more advanced mathematical methods. With this balance of elementary and advanced chapters, I hope that people of many and varied backgrounds can gain an entrée into theoretical population biology, a field that is one of the most active and exciting in science today.

The elementary chapters can be read separately without reading the advanced chapters. The elementary chapters are intended for undergraduates, beginning graduate students, and others who want an introduction to the very basic models of population biology. The advanced chapters are written for a somewhat different audience, one that includes graduate students, professional scientists, and others who want a survey of models on comparatively specialized topics. Some of the advanced chapters are monographic in character and represent my view on how the theory in these areas may be synthesized. The theory in the advanced chapters is often very recent and may also be synthesized in other ways with equal validity.

Theoretical population biology is not a new field although its current visibility is unprecedented. Both theoretical population genetics and theoretical population ecology are easily traced into the 1920s and 1930s with the works of R. A. Fisher, J. B. S. Haldane, and S. Wright in genetics; and of V. Kostitzin, A. J. Lotka, and V. Volterra in ecology, just to name a few. What is truly recent is the beginnings of a union of population genetic theory with the theory of population ecology. This unified theory is evolutionary population ecology. In this book I have tried to present both classical theoretical work and coverage of contemporary evolutionary population ecology.

I thank Paul Ehrlich, Marcus Feldman, Daniel Hartl, Richard Holm, Samuel Karlin, Robert May, George Oster, Ronald Pulliam, Michael Turelli, and Marcy Uyenoyama for encouragement and helpful comments on the manuscript. I also thank Woodrow Chapman and others in the Macmillan organization for their enthusiastic support of this project.

JONATHAN ROUGHGARDEN

Palo Alto, California

CONTENTS

*Elementary chapters—Only one semester of calculus needed.

ix

**THEORY OF
POPULATION GENETICS
AND
EVOLUTIONARY ECOLOGY:
AN INTRODUCTION**

Chapter 0
INTRODUCTORY SURVEY OF
POPULATION PHENOMENA

THIS book is an introduction to one of the most exciting and active areas of biology today, the biology of populations. In the next several paragraphs we shall mention some of the many phenomena that population biologists study. Also, we shall explain why it is natural for this area of biology to make more use of mathematics than other areas. This book may be your first experience in using simple mathematical models as a tool in learning about biology. Hence it is important to understand at the beginning that the use of mathematics is both natural and essential in this area.

What Is a Population? A population is a collection of organisms that we have lumped together because we believe they function together as a unit. A species is a particular kind of population, one in which all members are potentially able to interbreed with one another. (We must make due allowance, of course, for the fact that many species are split into two sexes within which interbreeding cannot occur.) But the populations we shall be discussing need not be a species by this definition. All we require is that the collection of organisms function together evolutionarily and ecologically as a unit. Thus a population can be part of a species, a so-called local population; or it can refer to groups of organisms that do not usually interbreed, like certain microorganisms and some plants.

The Genetic Structure Varies Through Time What sorts of phenomena do populations exhibit? To begin, there are phenomena involving the "genetic structure" of populations. The individuals in natural populations are never genetically identical. Indeed, the genetic variation among the individuals in a population is patterned in remarkable ways. Let us consider several examples. Many species of *Drosophila*, a genus of small fruit flies, have long been favorites of geneticists. One reason is that the salivary glands of these animals have chromosomes which permit especially clear cytological preparations. Because *Drosophila* chromosomes can be examined visually under the microscope, one of the first studies of the genetic structure of a population was carried out with *Drosophila* populations. Dobzhansky and Wright (1946) observed three forms of a chromosome in *Drosophila pseudo-obscura* at locations in California. These forms are labeled ST, CH, and AR. Figure 0.1 shows that the proportions of individuals in the population carrying any one of these chromosomes changes throughout the year. We see at once that there is a temporal pattern to the genetic structure. Moreover, if you reflect a moment, the changes in the genetic structure are occurring very rapidly. Until these observations it was felt that changes in the genetic composition of a population occurred only slowly over "evolutionary time." But instead, the genetic composition shows dramatic changes within a year. We shall be interested in the causes of this kind of phenomenon. Is natural selection responsible? How strong would natural selection have to be to produce changes this fast?

A particularly painful example of fast evolutionary change has been caused by industrial pollution in the vicinity of Birmingham, England. Before the pollution, a number of insect species had coloration that made

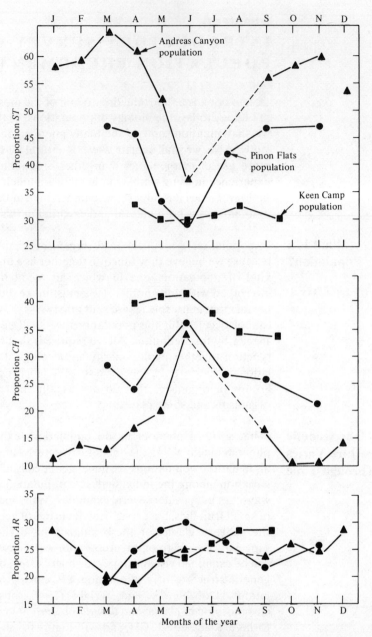

FIGURE 0.1. *Changes in the proportion of different types of chromosomes through time. There are three types of chromosomes named ST, CH, and AR. There are parallel observations from three populations of* Drosophila pseudo-obscura *in the San Jacinto mountains east of Los Angeles, California. Data from the populations are labeled separately with triangles, circles, and squares. Note that in two of the populations (the low-elevation populations) ST decreases during the summer while CH increases. The AR chromosome is comparatively constant through time. These data show that the genetic composition of a population can change dramatically within a year's time. [From T. Dobzhansky and S. Wright (1946), Genetics of natural populations XII, Experimental reproduction of some changes caused by natural selection in certain populations of* Drosophila pseudo-obscura, *Genetics* **31**: 125–156.]*

them very difficult to see against a background of bark covered with lichen. See Figure 0.2 (top). The pollution from industry at Birmingham, however, caused the trees to become black and the lichens to die. The original coloration was then very conspicuous to bird predators in this changed environment. Therefore, a "selection pressure" existed in favor of black insects, which would be less susceptible to predation. See Figure 0.2 (bottom). Kettlewell (1958) has studied the evolution of black coloration in several insect species. He has shown that the evolution of the black coloration occurred in only 40 years in some places. Thus we see that changes in the genetic structure of populations can definitely occur very rapidly. Again, we shall want to know how strong natural selection must be to cause evolution to be this rapid. However, we also want to understand the "equilibrium" conditions of the genetic structure of a population. If the genetic structure is not changing, why not? These are among the questions we want to answer concerning the temporal pattern in the genetic structure of populations.

The Genetic Structure Varies in Space

There is also a spatial pattern to the genetic structure of populations. Several plant species in California have ranges that extend from the moist Pacific coast, westward through the drier central valley of California and into the Sierra Nevadas (a mountain range with many locations at elevations over 10,000 feet). This topographic profile is illustrated in Figure 0.3. Clausen, Keck, and Heisy (1941) established study sites in the high Sierra, at a medium elevation in the Sierra, and at the Pacific coast. As Figure 0.4 shows, the plants from these locations differ greatly in appearance. A priori, these differences might be due to the influences of different environments on the physiology of germination and growth, or to genetic differences, or some combinations of both. A series of transplant experiments (see Figure 0.4) illustrates that both genetic and environmental factors are important. There is a spatial pattern both in the genetic structure of the populations and in the expression of the genotypes caused by the influence of the environment on the physiology and development of a plant.

The work described above documents a spatial pattern on a very large scale. There are also spatial patterns on a very small scale. The mineral zinc is toxic to most plants. However, some grasses that are commonly found in old fields were observed growing near abandoned zinc mines in England. Jain and Bradshaw (1966) established that the individuals from soils with high zinc content had a zinc tolerance that is inheritable while individuals from the usual soils did not. Figure 0.5 illustrates the spatial scale with which the genetic structure of this population changes. Note that there is almost a total change in the genetic structure over 20 meters. This abrupt change is maintained in the face of pollen and seed movement, which tend to smooth over this pattern. We want to understand this kind of phenomenon. To what extent is the spatial pattern of the genetic composition of a population a reflection of spatial pattern in the environment? Although Figure 0.5 shows a very fine scale of spatial pattern, could there be permanent patterns of a still finer scale?

A Metaphor

The common denominator in these temporal and spatial patterns is the presence of genetic variation in populations. Genetic and phenotypic variations among individuals of a population play a central role in population biology. Genetic variation, in particular, has always been of special interest

FIGURE 0.2. (Top) *Two individuals of the moth,* Biston betularia, *one with typical coloration and the other with black coloration. They are resting on an unpolluted lichen-covered tree trunk.* (Bottom) *Here the two individuals are resting on a soot-covered tree trunk. Notice that the typical coloration is difficult to see against the lichen-covered tree but stands out conspicuously against the soot-covered tree.* [Drawings by Anne Ehrlich from photographs by H. B. D. Kettlewell (1958), *Proc. 10th Int. Congr. Entomol.* **2**: 133, 134. From *The Process of Evolution* by P. Ehrlich, R. Holm, and D. Parnell, copyright 1975 by the McGraw-Hill Book Company, New York. Used with permission of the authors and of the McGraw-Hill Book Company.]

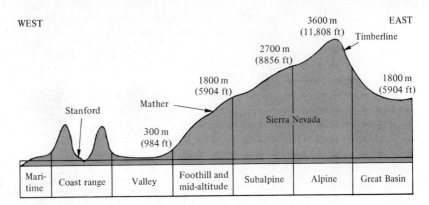

WEST

3600 m
(11,808 ft)   EAST
                               Timberline

2700 m
(8856 ft)

1800 m
(5904 ft)   1800 m
                                                   (5904 ft)

Stanford   Mather

                                        Sierra Nevada

                  300 m
                  (984 ft)

Mari-time	Coast range	Valley	Foothill and mid-altitude	Subalpine	Alpine	Great Basin

FIGURE 0.3. *Altitudinal profile of a transect from the Pacific Coast through the campus of Stanford University and across the Sierra Nevada Moutains. Note the elevations of the stations at Stanford, Mather, and Timberline.* [From J. Clausen, D. Keck, and W. Hiesey (1941), Regional differentiation in plant species, *Amer. Natur.* **75**: 231–250.]

to biologists because it is a prerequisite for evolution. The metaphor often used is that natural selection "works upon" genetic variation to produce *adaptations.* An adaptation is a trait that permits an organism to function well in its environment, a trait that endows an organism with capabilities especially appropriate in its particular environment. In this metaphor genetic variation is the "raw material" out of which selection has fashioned the diversity of living things. We shall be interested in the appropriateness of this metaphor in light of our contemporary account of the process of evolution.

We sometimes distinguish population genetics from evolution by the fact that population genetics is often more concerned with a detailed description of the genetic structure of a population and with changes over a shorter time scale than is evolution. Phenomena over long time scales include the radiation of groups of organisms, as exemplified by the radiation of lung fish in Figure 0.6. We want to understand phenomena of this sort. Why has the radiation been as extensive as it has been? Could certain forces have caused it to be more or less extensive; what controls the rate at which the radiation unfolds; and is there any causal connection between the radiation patterns in different groups? These are all important and fundamental questions, yet we cannot answer them very well. When research into population genetics was begun, it was assumed that these kinds of long-term evolutionary phenomena would be explained as a result. We need to reassess the relevance of population genetics to these kinds of evolutionary issues.

Ecology Population genetics and evolution are principally concerned with the genetic and phenotypic properties of the members of a population. In contrast, ecology is traditionally concerned with population size and distribution. Among typical ecological issues are the explanation of why some species are rare and others common or why some species are widespread and others restricted to local areas. Also, ecology tends to be more concerned with population interactions like competition and predation than is population genetics. All these areas of population biology are, of course, very

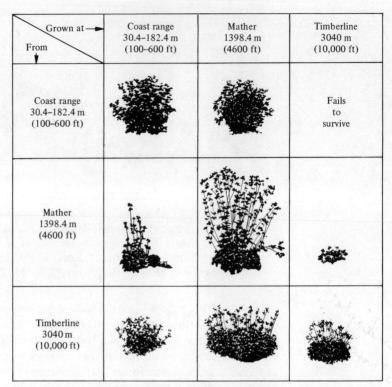

Grown at → From ↓	Coast range 30.4–182.4 m (100–600 ft)	Mather 1398.4 m (4600 ft)	Timberline 3040 m (10,000 ft)
Coast range 30.4–182.4 m (100–600 ft)			Fails to survive
Mather 1398.4 m (4600 ft)			
Timberline 3040 m (10,000 ft)			

FIGURE 0.4. *Pictorial chart illustrating transplant experiments with the herb,* Potentilla glandulosa. *Each row indicates the size of a plant from a certain elevation when grown at each of the three stations. Each column indicates the sizes of the plants from all elevations when grown at a given station. Note in the right column that the timberline station yields the smallest plants but that timberline plants grow relatively better at this station than plants from other elevations. Similarly, note in the middle column that all plants grow best at Mather but that plants originally from Mather grow relatively larger at Mather than plants originally from other elevations. Note also that Coast Range plants grow relatively better at stations in the Coast Range than the other plants. These data suggest the existence of both genetic and environmental influences on plant growth.* [Adapted from J. Clausen, D. Keck, and W. Hiesey (1941), Regional differentiation in plant species, *Amer. Natur.* **75**: 231–250.]

closely related, but by tradition and inclination most biologists tend to specialize more in one of these areas than another.

Populations Have Age Structure

We now mention some phenomena from the ecological side of population biology. Figure 0.7 illustrates graphs that indicate the percentage of the human population in different age classes for North and Central America. There are generally more young people than old people in both regions, but the relative proportions differ substantially. Moreover, the region with the higher rate of population growth contains relatively more young people. This is no coincidence, and we shall study the relationship between population growth and age structure.

We shall see that the age structure of a population, as mentioned previously with human populations, can be predicted from knowledge of the fertility and mortality rates for individuals of all the various age groups. The

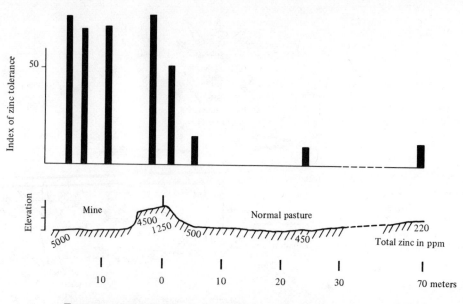

FIGURE 0.5. *The evolution of zinc resistance in grasses over a very fine spatial scale. The top graph illustrates the degree of zinc tolerance exhibited by plants collected from several places along a transect of approximately 100 meters in length. The lower graph illustrates the amount of zinc in the soil along the transect. Note the abrupt drop in zinc concentration at the boundary between the mine and the pasture.* [From S. K. Jain and A. D. Bradshaw (1966), Evolutionary divergence among adjacent plant populations I, *Heredity* **21**: 407–441.]

same considerations also apply to plant and animal populations. However, a new dimension arises with natural populations. The pattern of fertility and mortality across ages has evolved differently in different species. Some populations, like salmon and some squid, have evolved a pattern in which there is a burst of reproductive activity at a special time followed by rapid senescence. In other populations there is an essentially constant rate of reproductive activity and a more or less constant mortality rate for individuals of all ages. We want to understand how natural selection shapes the fertility and aging patterns for different species.

Populations Interact

Many questions in ecology concern the interactions between populations. The most commonly studied interactions are the competition interaction between two populations, usually for limiting resources, and the predator-prey interaction. Concerning competition, we want to know the impact of competition between two populations on the abundance and distribution of each other. For example, we want to know if competition can restrict the range of a species. Perhaps a species has the physiological capability of occurring in many more places than it actually does. Is competition from some other species a possible cause in excluding a species from places where it has the physiological ability to live? A type of geographical pattern, called zonation, refers to the placement of species ranges in a sequence along an environmental gradient. See Figure 0.8. Is competition between adjacent species along the gradient responsible for shaping the pattern of zonation?

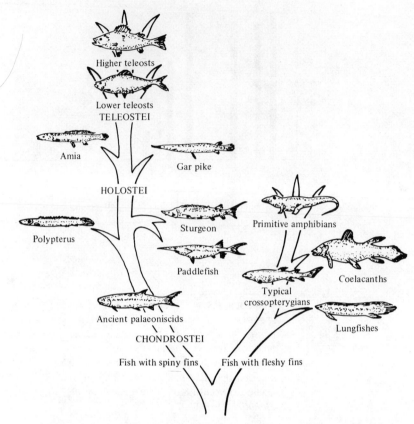

FIGURE 0.6. *A simplified family tree of the bony fishes, to show their relations to one another and to the amphibians. The top of the tree is more recent than the bottom of the tree.* [From A. S. Romer (1962), *The Vertebrate Body*, 3rd ed., W. B. Saunders Company, Philadelphia.]

Competition One of the most fascinating situations where the results of competition are expressed is on remote oceanic islands. Lack (1947) demonstrated that the finches on the Galapagos islands, originally studied by Darwin, show an intriguing size pattern. As Figure 0.9 shows, islands with three species contain species of greatly different sizes. Notice especially that *Geospiza fuliginosa* and *G. fortis* have a greatly different bill depth when they occur with one another but converge to a similar bill depth on islands where they do not occur with one another. There is evidence that a difference in bill shape between two species corresponds to a difference in the resources which those species use. If so, the patterns of bill size on the islands suggest that where *G. fuliginosa* and *G. fortis* co-occur they differ in their resource use, whereas they use more similar resources wherever they do not co-occur. Perhaps competition causes this pattern. Perhaps competition causes species to evolve differences in their resource use that do not evolve in the absence of competition.

Another example illustrating the apparent evolutionary effect of competition on oceanic islands is provided by the *Anolis* lizards of the islands throughout the Caribbean. *Anolis* lizards are the small arboreal insect-eating lizards that are also widespread in Florida and throughout the southeastern United States. They are often green and are sometimes called

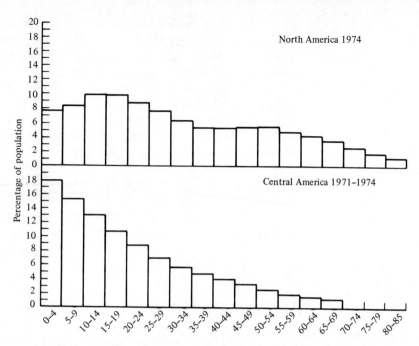

FIGURE 0.7. *The age structure in human populations from North America and Central America. Age in years is plotted on the horizontal axis; percentage of the population is plotted on the vertical axis. Note that there is a higher proportion of young people in the Central American population than in the North American population. The figure for North America combines the data for Canada and the United States. The figure for Central America combines the data for Costa Rica, El Salvador, Guatemala, Honduras, Mexico, Nicaragua, and Panama. The data were compiled from the* Demographic Yearbook, 1975, *published by the United Nations in 1976.*

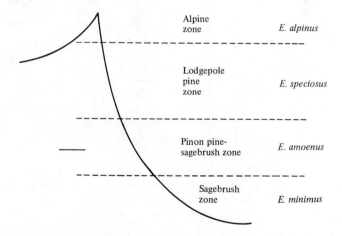

FIGURE 0.8. *Altitudinal zonation in chipmunks of the genus* Eutamias *on the East-facing slope of the Sierra Nevada.* [From Heller and Poulsen (1972), Altitudinal zonation of chipmunks (Eutamias); adaptation to temperature and high humidity, *American Mid. Nat.* **87**: 296–313.]

chameleons even though they are not related to the true chameleons of Africa. In Figure 0.10 Region A includes many islands. Each island has either one or two species of *Anolis* lizards. These islands are old and have never been submerged; the species on each island are native to it. As sketched in Figure 0.10 the size of lizards on islands with only one species lies between the sizes of the lizards from islands with two species. The size differences on the two-species islands are correlated with differences in resource use by the two species. Again, it appears that an evolutionary effect of competition is to cause the competing populations to evolve differences in their resource use that do not evolve in the absence of competition.

Region B in Figure 0.10 is the Bahama islands. These islands are very low in elevation and have often been submerged by the ocean. The *Anolis* lizards in these islands are not native to them but are recent arrivals from the nearby large islands of Cuba, Hispaniola, and Puerto Rico. Although the

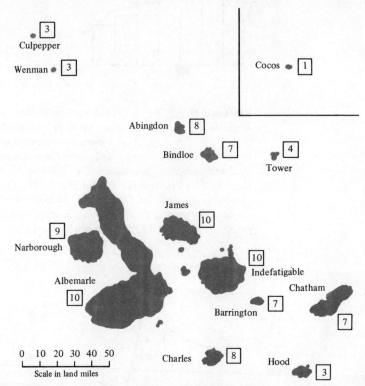

FIGURE 0.9a. *Map of the Galapagos islands with a tabulation of the number of finch species on each island.* [From D. Lack (1947), *Darwin's Finches*, Cambridge University Press, New York.]

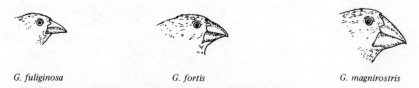

FIGURE 0.9b. *Profile of the bills of three species of finches of the genus* Geospiza *from the Galapagos Islands.* [From D. Lack (1947), *Darwin's Finches*, Cambridge University Press, New York.]

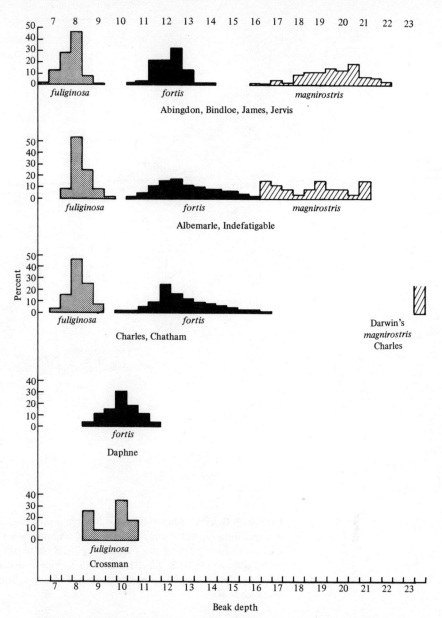

FIGURE 0.9c. *Histogram of beak depth in* Geospiza *species. Measurements in millimeters are placed horizontally, and the percentage of specimens of each size vertically. Note that* G. fortis *is larger where it co-occurs with* G. fuliginosa *than where it is solitary on an island. Also note that* G. fuliginosa *is smaller where it co-occurs with* G. fortis *than where it is solitary.* [From D. Lack (1947), *Darwin's Finches,* Cambridge University Press, New York.]

Bahama populations are new relative to the populations on the islands in Region A, they have been there a very long time, many thousands of years. Why have the Bahama populations not diverged from their parental populations in all this time? What sets the rate of evolutionary divergence between isolated populations and why is the rate so slow? We shall explore this question in Chapter 6.

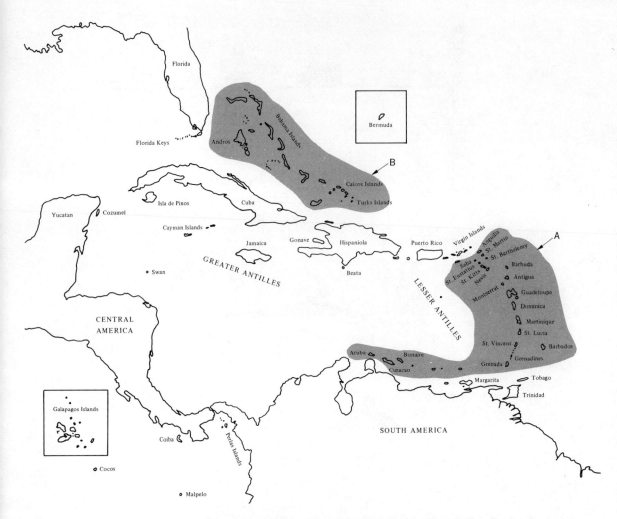

FIGURE 0.10a. *Map of the islands of the West Indies. Region A consists of comparatively old islands. Each island contains either one or two native species of lizards of the genus* Anolis. *Region B consists of the comparatively recent Bahama islands. The lizards on these islands are recently descended from populations on the nearby large islands of Cuba, Jamaica, Hispaniola, and Puerto Rico.*

Predation   The predator-prey interaction also leads to many interesting questions. A basic question concerns the impact of the predator population on the abundance of the prey population. Do predators significantly reduce the population size of their prey? This question has great theoretical interest and also great practical importance in connection with biological pest control. Other practical questions that arise in the predator-prey interaction concern the effects of human disturbances on a predator-prey balance. What happens to the balance if a generally harmful chemical like DDT is added? What happens to the balance if the resource base for the prey is increased, as it is when phosphate from dish washing detergents is added to natural streams and lakes? And finally, a famous issue involving the predator-prey interaction is the oscillation in population size shown by lynx in Canada. As illustrated in Figure 0.11, the population size of lynx has oscillated in a

FIGURE 0.10b. *A typical lizard of the genus* Anolis. (Anolis cristatel-lus, ♂, *from Puerto Rico.*)

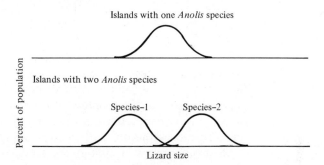

FIGURE 0.10c. *Sketch of size distributions for adult lizards in Region* A.

remarkably regular fashion for a very long time in Canada. We shall explore whether this phenomenon could be caused by a predator-prey interaction between the lynx and their prey.

Why Mathematics?   The paragraphs above illustrate some of the phenomena that biological populations exhibit. Even a very quick glance through this book reveals that there is a great deal of mathematics in the study of population biology. So the natural question to ask is "Why is mathematics needed to study population phenomena?" There are two simple reasons why we cannot avoid mathematics even if we wanted to.

The first reason why mathematics is necessary in population biology is that the questions which define the field of population biology are essentially quantitative to begin with. For example, we want to know how *fast* evolution occurs and how *strong* natural selection must be to cause a certain *speed* of evolution. This question is a quantitative question; it cannot even be stated

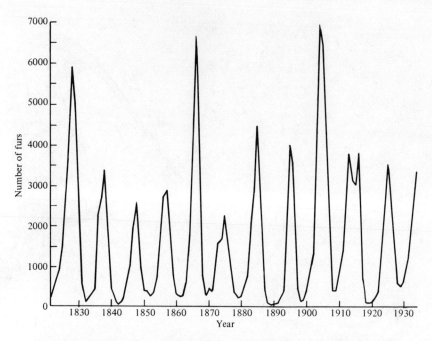

FIGURE 0.11. *The number of lynx captured by the Hudson Bay Company in the Mackenzie River District of northern Canada from the years 1821 through 1934. These data illustrate huge and rather regular fluctuations in the population size of lynx.* [From C. Elton and M. Nicholson (1942), The 10-year cycle in numbers of the lynx in Canada, *J. Anim. Ecol.* **11**: 215–244.]

precisely, much less answered without using a mathematical approach. Similarly, we want to know how competition and predation influence the *population size* and *distribution* of the interacting populations. Again this question is a quantitative question and again it cannot be stated precisely and answered without a mathematical approach.

Most other areas of biology do not ask quantitative questions. In particular, molecular biology today is preoccupied with the discovery of the mechanisms that bring about life processes. To answer a question about the mechanism of some life process is generally to identify the components of the mechanism and to establish the sequence of action of those components. It is rarely necessary to invoke mathematics either for stating a mechanistic question or as a help in providing the answer.

The second reason why mathematics is necessary in population biology is that a natural population is often an entity which exists on a larger scale than human experience. This fact implies that we often cannot directly see an entire population and we cannot directly manipulate them experimentally. For example, even if we know the mechanism of an interaction between the individuals of two populations, we cannot directly witness the results, *for the populations,* of that interaction. Models help us to infer results on the population level from information about individuals in those populations. Not only are populations often large relative to the scale of human experience, but also the time scale of population processes is often slow relative to the human life span. Again, mathematical models help us to infer what we cannot experience in our life time from what we can.

PART ONE
THE BASICS OF POPULATION GENETICS

Chapter 1
AN OVERVIEW OF POPULATION GENETICS

Population genetics is about the collection of genes belonging to all the members of a population taken together; it is about what is called the "gene pool" of a population. Two general kinds of questions are always at issue: What is the structure of the gene pool at any time, and what are the causal forces that change the structure of the gene pool from time to time? This field comprizes half of population biology, and in this section we shall mention in general terms what sorts of topics turn up in population genetics and why it is important for you to learn population genetics.

As discussed with examples in the previous chapter, a basic property of *all* populations is the presence of individuals with different phenotypes. For the most part, population genetics concerns the genetic aspects of the phenotypic variation of a population. It is the genetic part of a population that is passed from generation to generation and must be understood to appreciate the long-term properties of that population. Some phenotypic traits are passed on from generation to generation through teaching, learning, and other mechanisms of "cultural inheritance." However, the mathematical study of cultural evolution is in its infancy, and population genetics is principally concerned with the evolution of traits that are inherited in a traditional genetic manner.

Assume a Specific Genetic Mechanism Typically, the formulation of a problem in population genetics involves first assuming that some trait is the expression of a specific genetic mechanism. For example, as Mendel showed, the surface of a pea may be either wrinkled or smooth. This trait depends on one locus at which there are often just two alleles, with one being dominant. There are many traits, especially color characters, in both plants and animals which are known to be produced by a simple genetic mechanism. Yet, obviously, many other traits are the result of a very complicated genetic mechanism. However, most of population genetics to date assumes the very simplest sorts of genetic mechanisms. So first, we shall consider population genetics assuming that the trait under discussion is produced by only one locus with two alleles; later we shall branch out into more complicated mechanisms.

The Basic Questions Once having assumed a particular genetic mechanism, say one locus with two alleles, typically we want to determine whether the population can persist with genetic variation at the locus or whether one of the alleles will eventually disappear. If, in the end, only one allele remains and the other is lost, we say that the population has become *fixed* for the allele that remains. If both alleles persist, we say the population is *polymorphic* at the locus in question. In general, we want to know whether a locus will become fixed or polymorphic; if fixed, then for which allele, and if polymorphic, then for what proportion of each allele relative to the other. Moreover, we want to know low long it will take for the final result to occur from any initial condition. To answer these questions is to know the structure of the gene pool at the locus in question through time.

The Basic   Whether a locus becomes fixed or polymorphic depends on the "forces"
Forces   impinging on the individuals carrying different alleles for the locus. One of
the most elementary and basic results in population genetics, called the
Hardy–Weinberg law, is that in the absence of such forces, any genetic
variation initially present in the gene pool will persist without change. To
elicit a change, certain forces are needed. Typically, three kinds of forces are
distinguished: natural selection, mutation, and genetic drift. If carriers of
one allele have more offspring on the average than carriers of another allele,
then clearly the allele that confers higher reproductive output will increase
in the gene pool relative to the other. This process is called *natural selection*.
It can be a very powerful force, and we shall derive formulas to predict how
fast the results of natural selection occur and how selection can also lead to
polymorphism. If alleles differ in their mutation rate, then clearly the allele
that mutates least often or is formed most often from the mutations of other
alleles will increase in the gene pool while the others relatively decrease. In
this sense recurrent mutation is a force that must be taken into account, but it
is a weak and slow force when compared to selection. Mutation, however, is
the ultimate source of new or novel genes and it also prevents old alleles
from ever being *entirely* eliminated. *Genetic drift* is the change in the gene
pool structure due to error or "noise" that creeps into the transmission of the
gene pool from generation to generation. As a result, due to the "transmission noise," the proportions of the alleles in the gene pool can change,
sometimes erratically, from generation to generation. This force, too, is
weak when compared to selection, but it can be important in *small* populations. However, there may be cases where selection is almost absent and if
so, the gene pool properties are determined by the interaction of the two
weaker forces—mutation and genetic drift.

In the following sections we shall deal with these forces on the gene pool
one by one. Later in the book we shall also consider certain advanced topics
in population genetics including the extension of the theory to complicated
genetic mechanisms, gene pool change due specifically to spatial and
temporal variation in the environment, certain moderately general
theorems about how natural selection works, ideas about how the genetic
system itself evolves, and theories about the evolution of altruism.

Although the intrinsic interest of population genetics may be sufficient
reason to study the field, it is also relevant to mention why theoretical
population genetics was initiated and that the theory has several important
areas of practical application.

Some History   Traditional theoretical population genetics developed during the 1920s
from the efforts of Ronald Fisher, J. B. S. Haldane, and Sewall Wright. For
summaries, see Fisher (1958), Haldane (1966), and Wright (1969). These
scientists were interested in a rigorous genetic formulation of Darwin's
theory of evolution by natural selection. By evolution we mean phenotypic
change in a population over many generations. Darwin's theory contained
two new propositions. The first is that the phenomenon of organic evolution
actually exists; that all organisms today are descendants of organisms alive in
the past. By this proposition there is a lineage connecting even the highest
organisms alive today and some lower organisms of the remote past. This
proposition was initially disputed because it contradicts a literal reading of
Genesis. But the empirical evidence provided by fossils and other sources

has led to a general acceptance that evolution actually occurs. Darwin's second proposition was that the causal mechanism for evolution, its driving force, is natural selection. This proposition was disputed well into the 1900s by thoughtful scientists for a variety of reasons, which we shall discuss later. The initial motivation for what is now traditional population genetics was to consider evolution in a rigorous genetic context and to assess exactly how strong natural selection had to be in order to bring about evolution.

Today   Today traditional population genetics remains the cornerstone for our understanding of evolution. But it is not a perfect cornerstone. It is not perfect because it deals at the level of the genome whereas evolution is observed at the level of the phenotype. So the actual use of population genetics with evolutionary problems usually involves trying to translate phenotypic observations down into gene pool structure and then back up again. As a result of this difficulty, other approaches to evolutionary problems that deal directly with the phenotype have been developed recently. We shall consider these other approaches in addition to the traditional topics of population genetics. (See Chapter 9.)

Applications   Although evolutionary problems may have provided the inspiration for population genetics, population genetics today is relevant to other issues. First, the genetic structure of a population is of immense practical concern. Although genes are made of DNA, which is, of course, a chemical, genes cannot simply be stored in bottles on shelves. The only storage place for different genes at this time is in living organisms themselves. A reservoir of genetic variation is needed in any species to allow it to adapt to new environments. Currently, we are extremely dependent on many cultivated plants which all are to some extent genetically similar because they are all members of the same family (i.e., the grass family, Gramineae, which includes corn, rice, wheat, oat, barley, and sugar cane, among others). The wild relatives of some of these plants are gradually being eliminated by increasing land development, and also the process of domestication often produces genetic uniformity. Both factors are clearly reducing the reservoir of genetic variation in these species. It is important to know how much genetic variation would be good insurance for expected changes in the environment and how perhaps to carry out agriculture in ways that would maintain enough variation in crop plants to meet those expected changes in the environment. See Wilkes (1977) for more discussion on this point.

   Second, population genetics is sometimes applied in plant and animal breeding to produce a desired phenotype by artificial selection. The theory exists to predict roughly how much change in phenotype will occur in the next generation as a function of the intensity of artificial selection that is applied.

   Third, population genetics has numerous medical applications. Indeed, much current research in population genetics is done at medical schools and may be part of the medical curriculum [Cavalli-Sforza and Bodmer (1971)]. It is important for a medical doctor to understand that the set of patients he treats is a sample drawn from a natural population. His patients collectively will show many of the patterns first investigated in the more strictly biological world. Various blood types occur in patterns expected by the Hardy–Weinberg law, and certain conditions like sickle cell anemia occur as an evolutionary response of the local human population caused by natural selection in the local environment—in this case the presence of malaria. Moreover, varying disease susceptibility and drug effectiveness among individuals often reflect genetic differences. In fact, an M.D. operates closer to the level of the genome than most population biologists because of the mode of action of many drugs. Moreover, increasingly in medicine special care must be taken in the application of medicines to avoid causing the evolution of drug immunity among pathogens. To avoid a continual (and losing) arms race with the pathogens, it will, perhaps, be increasingly necessary to design disease treatment programs that take account of the selection forces produced on the pathogens.

Chapter 2
THE HARDY–WEINBERG LAW

$\mathbf{T}$HE path into population genetics starts with the Hardy–Weinberg law. Let us examine first what this theorem asserts and then discuss the derivation. In this way we can first decide that the theorem is interesting before becoming involved with the technical conditions under which it is valid.

Notation Consider a population containing genetic variation at one locus with two alleles, A and a. Then the possible genotypes are AA, Aa, and aa. We must introduce some notation as follows. The *genotype numbers*, denoted N_{AA}, N_{Aa}, N_{aa}, are the numbers of organisms of each genotype, respectively. Let the total population size be N; then

$$N = N_{AA} + N_{Aa} + N_{aa} \qquad (2.1)$$

We need still more notation. We define the *genotype frequencies* or fractions as

$$D = \frac{N_{AA}}{N}$$

$$H = \frac{N_{Aa}}{N} \qquad (2.2)$$

$$R = \frac{N_{aa}}{N}$$

The sum of D, H, and R must equal one. (We shall use this notation even if A is not dominant.) Finally, we define the *gene frequencies* as

$$p = \frac{2N_{AA} + N_{Aa}}{2N}$$

$$q = \frac{2N_{aa} + N_{Aa}}{2N} \qquad (2.3)$$

According to these formulas p is the fraction of A alleles in the gene pool and q is the fraction of a alleles. If there are N individuals, then there are a total of $2N$ genes in the gene pool at the A locus. Now each AA homozygote has two A alleles and each heterozygote has one A allele, so that p, the fraction of A alleles, must be $(2N_{AA} + N_{Aa})/(2N)$. The sum of p and q is one. It is important to understand the distinction between genotype frequencies and gene frequencies. The gene frequencies refer only to the genetic variation within the whole gene pool while the genotype frequencies also indicate how that variation is organized into genotypes.

The Hardy– The Hardy–Weinberg law is a relationship between the genotype fre-
Weinberg Law quencies and the gene frequencies. It asserts that any arbitrary initial D, H, and R will tend in one generation to the values $D = p^2$, $H = 2pq$, and $R = q^2$, and thereafter remain constant. In symbols, upon letting a subscript denote

time, we have

$$
\begin{array}{ccccc}
D_0 & & D_1 = p_0^2 & & D_t = p_0^2 \\
H_0 & \xrightarrow{\text{after one generation}} & H_1 = 2p_0q_0 & \xrightarrow{\text{in all generations after the first}} & H_t = 2p_0q_0 \quad (2.4)\\
R_0 & & R_1 = q_0^2 & & R_t = q_0^2
\end{array}
$$

where $t > 1$. Let us take an example. Suppose that some aquarium fish have a fin coloration character determined by one locus with two alleles such that AA is red, Aa is purple, and aa is blue. One supplier sells only red finned fish, the other only blue, and we stock the aquarium with 75 percent red and 25 percent blue. Then $D_o = \frac{3}{4}$, $H_0 = 0$, and $R_0 = \frac{1}{4}$. Also, clearly, $p_0 = \frac{3}{4}$ and $q_0 = \frac{1}{4}$. Then, by the Hardy–Weinberg law, in one generation the proportion of red fish is $\frac{9}{16}$, purple fish $\frac{6}{16}$, and blue fish $\frac{1}{16}$. Moreover, these ratios will show no further change in future generations. Note that the heterozygous fraction, H, increased at the expense of the homozygous fractions. Note also that the initial gene frequency does not change, that is, $p_t = p_0$. (In Problem 2.2, you should prove this for yourself.)

The Hardy–Weinberg law is a very strong statement. According to the law, the initial gene frequency, p_0, does not change and the genotype frequencies come to the equilibrium computed from p_0 in only one generation. Thus the Hardy–Weinberg law predicts both that an equilibrium exists and that the approach to that equilibrium is very rapid.

Some Assumptions The Hardy–Weinberg law depends on several assumptions, and the best way to see what these assumptions are is to derive the law. Consider a population with a "life history" as diagramed in Figure 2.1. This kind of life history has discrete generations, and the individuals are all synchronized in these activities. The whole population begins as a collection of zygotes, matures to adults, and breeds, leaving zygotes to begin the cycle again. The breeding occurs with *random union of gametes*. This kind of breeding method is often appropriate to organisms that release their gametes into the water, and to wind-pollinated plants. For example, sea urchins, as illustrated in Figure 2.2, shed gametes directly into the water at breeding season and are likely to show random union of gametes. Moreover, we assume here that generations are discrete—all the organisms alive at any time are the same

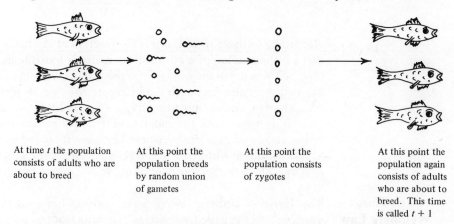

At time t the population consists of adults who are about to breed

At this point the population breeds by random union of gametes

At this point the population consists of zygotes

At this point the population again consists of adults who are about to breed. This time is called $t + 1$

FIGURE 2.1. *Diagram of a synchronized discrete generation life history. The census time is at the adult phase.*

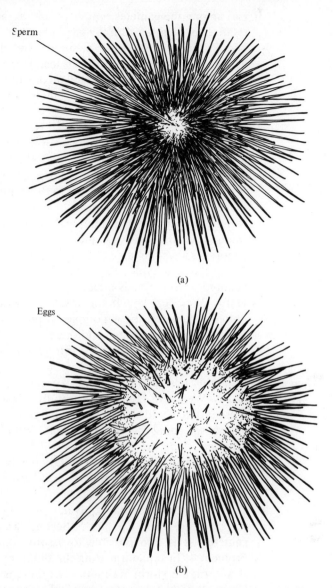

Sperm

(a)

Eggs

(b)

FIGURE 2.2. *Example of an organism that reproduces with a random union of gametes: the sea urchin. (a) A male sea urchin; (b) a female sea urchin. Sea urchins release gametes directly into the water. The gametes are released from pores in the top surface. The gametes may be collected and used to observe fertilization under the microscope.*

age and are members of the same generation. Annual plants may fit this assumption, provided that their seeds do not remain dormant in the soil through more than one generation, as do many insects and occasional vertebrates. We may choose to census the population at any point in its life cycle. Let us chose the adult phase as illustrated in Figure 2.1.

The Derivation These assumptions set the stage. Let us begin with arbitrary D_0, H_0, R_0, p_0 and q_0 measured in the adults at time t. (1) Assume that individuals of the different genotypes do not differ in their fertilities. If so, the gene frequencies among the gametes remain p_0 and q_0, since neither A nor a increases relative to the other. But the gametes are all haploid and, when the gamete pool is formed, information about the previous organization of these gametes into diploid genotypes is lost. Hence the initial relationship among D_0, H_0, and R_0 is lost due to the mixing of the gametes in the gamete pool. (2) Assuming random union of gametes, we must form zygotes. To form a zygote, we must draw two gametes at random from the gamete pool and consider them thereafter as fused. Now, the probability of drawing two gametes with the A allele is p_0^2. The probability of drawing an A allele and an a allele is $2p_0q_0$ because they can be drawn in either order. Similarly, the probability of drawing two a alleles is q_0^2. These probabilities assume that the gamete pool is large relative to the number of zygotes to be drawn and that each drawing of a gamete is independent of other drawings. Therefore, with random union of gametes and a large gamete pool the zygotic *genotype* frequencies are p_0^2, $2p_0q_0$, and q_0^2. (3) Assume that individuals of different genotypes have the same likelihood of surviving from zygotes to adulthood. If so, the genotypic frequencies will not change because none of the genotypes has any advantage over the others. This is the essence of the Hardy–Weinberg law. In the absence of natural selection and with random union of gametes, the genotypic proportions assume the values p_0^2, $2p_0q_0$, and q_0^2 in one generation, and thereafter do not change, and, also, the gene frequency remains unchanged throughout ($p_t = p_0$ for all t).

When Is It Valid? With this derivation in mind, we can consider the assumptions to the Hardy–Weinberg law more carefully. First, we assumed no forces were present that would change the gene frequency, p_0. Although the absence of natural selection, that is, differential fertility and/or survival, was *explicitly* assumed, there are other forces which are also implicitly assumed to be absent. Specifically, recurrent mutation between A and a, immigration from other populations, and "genetic drift" are all forces that also can alter the local gene frequencies and hence are implicitly assumed to be absent. The absence of genetic drift is equivalent to requiring a large population size, as will be discussed in Chapter 5. Second, we assumed the random union of gametes. It is also possible to assume random mating, as discussed in Problem 4. In random mating the different mating types (e.g., $AA \times AA$, $AA \times Aa$, etc.) form at random but there is no common *gamete* pool. The Hardy–Weinberg law obtains with this assumption too. Third, the derivation above ignores sexual differences. It can be shown, however, that if sex (i.e., two gamete classes) is considered, then the Hardy–Weinberg law, as stated, requires equal initial gene frequencies, p, in both sexes. Otherwise, the approach to Hardy–Weinberg equilibrium takes two generations instead of one. However, it is often reasonable to assume equal gene frequencies in both sexes. Fourth, the requirement of a life history with discrete generations is not necessary, since the Hardy–Weinberg law has also been derived for various other kinds of life histories. However, this assumption greatly simplifies the algebra and we shall use it often solely for this reason. It is often better to understand a result in a restricted context before approaching more general formulations.

Two Consequences In concluding this section, let us note two consequences of the Hardy–Weinberg law. First, this law shows that any genetic variation in the population is not lost solely as a result of the propagation of the population. Thus, in the absence of selection, mutation, genetic drift, and so forth, any genetic variation in the population is conserved through time. Or, in other words, there is no natural decay of genetic variation in the population *unless* the forces of selection, mutation, and drift are explicitly introduced into the situation. The second point concerns the mathematical models in population genetics. With one locus and two alleles, we potentially have two independent variables to describe the genotypic composition of the population, D, H, and R. (Remember $D + H + R = 1$.) The assumption of random union of gametes, as in the derivation of the Hardy–Weinberg law above, provides a relationship between D, H, and R, and p and q, which thereby reduces the specification of the system from two to one independent variable. As a result, we do not have to solve as many equations as would otherwise be necessary.

PROBLEMS

2.1. Become familiar with the following notation:
  Let genotypes be AA, Aa, and aa and the corresponding genotype frequencies be D, H, and R. Let p be the frequency of A and q for a.

(a) Prove

$$p = D + \tfrac{1}{2}H, \quad q = R + \tfrac{1}{2}H$$

(b) $p^2 + 2pq + q^2 = 1$.
(c) Did you use any biological assumptions to prove (a) and (b)?

2.2. Prove that if $D_t = p_0^2$, $H_t = 2p_0q_0$, and $R_t = q_0^2$ for all $t > 0$, then $p_t = p_0$ for all $t \geqslant 0$.

2.3. Prove that the Hardy–Weinberg law is also valid with random union of gametes if we had decided to census the population at the zygotic phase. In the derivation in the text we censused at the adult phase.

2.4. Derive the Hardy–Weinberg law based on random mating as distinct from random union of gametes, that is, show that $D_{t+1} = p_t^2$, $H_{t+1} = 2p_tq_t$, and $R_{t+1} = q_t^2$.

(a) Write down all possible mating types in a vertical column at the left of a page. For example, $AA \times AA$, $AA \times Aa$, and so forth, where $\times$ means "mates-with."
(b) Write down the probability of each mating type in the next column, for example, D_t^2, $2D_tH_t$, and so forth.
(c) Using Mendel's laws, write down the genotype of the progeny from each mating type.
(d) Sum up the progeny of each genotype. For example, the frequency of AA genotypes in the progeny should be $D_t^2 + D_tH_t + \tfrac{1}{4}H_t^2$ representing the progeny from the $AA \times AA$, $AA \times Aa$ and $Aa \times Aa$ matings.
(e) Simplify the totals using the identities in (1a) above.
  For example,

$$(D_t^2 + D_tH_t + \tfrac{1}{4}H_t^2) = (D_t + \tfrac{1}{2}H_t)^2 = p_t^2$$

(f) You are now finished, are you not?
(g) Did you use any biological assumptions? What were they?

Chapter 3
NATURAL SELECTION AND MUTATION AT ONE LOCUS WITH TWO ALLELES

$\mathbf{N}$ATURAL selection is the most important concept in population biology. It is one of the most important principles in science, and we shall begin to explore its properties in detail in this chapter.

Darwin in *The Origin of Species* proposed two controversial ideas. First, he proposed that evolution is a fact, that all populations today, however splendid, share a lineal relationship with all other populations, however lowly. A particular case is that human beings, as a population, share a lineal relationship with populations of animals, and that our nearest relatives are other primates. This proposition contradicts the teachings of some religious circles that man was created *de novo* without having evolved from "lower" forms. The public controversy surrounding Darwin's work was principally concerned with this proposition. But now it is generally accepted that all organisms alive today are descendants, perhaps greatly modified, of organisms that were alive in the past. Darwin's second proposition is that the mechanism, or driving force, of evolution is natural selection. This proposition has long been debated among scientists, all of whom agreed that evolution occurred but disagreed on how. We shall examine in this section how strong a force natural selection can be.

To begin, we must agree that genetic evolution involves changing gene frequencies. By definition, the appearance of a single mutant individual does not denote a significant amount of evolution, for we must have a major change in the *population's* characteristics, not simply the single one-time appearance of a new kind of individual. Remember, the fossil record is a sample drawn from *populations* of the past. Although diagrams of evolutionary trees (recall Figure 0.6) illustrate a sequence of single individuals through evolutionary time, each individual represents an entire population.

In genetic terms Darwin's second proposition is that natural selection is the major force changing gene frequencies. An alternative hypothesis is that *recurrent* mutation is the major force which alters gene frequencies. "Recurrent" is emphasized because, as noted above, one mutation alone does not produce evolution; we need enough mutations to change the gene frequencies. A priori it is not obvious that selection is stronger, or faster, than recurrent mutation. One of the principal uses of the model in this chapter is to allow us to assess how fast even weak selection can change gene frequencies relative to the speed with which recurrent mutation can change gene frequencies. This will help us to decide whether we should accept Darwin's second proposition that natural selection is the driving force of evolution.

The Basic Model For Natural Selection at One Locus with Two Alleles

We shall consider a simple model for natural selection based on the discrete generation life history that was introduced in the last section. Natural selection occurs because individuals of different genotypes differ in their abilities to survive and in their fertilities. An individual's ability to survive is called its *viability*. We shall derive a formula to predict the changes in gene frequency that result from natural selection. To do this, we shall derive an equation that predicts the gene frequency at time $t + 1$, given the

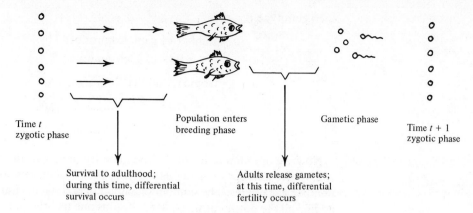

Time *t*
zygotic phase

Population enters
breeding phase

Gametic phase

Time *t* + 1
zygotic phase

Survival to adulthood;
during this time, differential
survival occurs

Adults release gametes;
at this time, differential
fertility occurs

FIGURE 3.1. *Diagram illustrating the points during the life history at which differential survival occurs and at which differential fertility occurs.*

gene frequency at time *t and* the measurements of the strength of natural selection.

Refer to Figure 3.1. We shall work through the derivation based on the diagram of the life history. Let us agree to census the population at the *zygotic* phase. (The choice of the census point is very important now. In contrast it was not important in the last chapter on the Hardy–Weinberg law.) At this time the zygotes have just been formed by random union of gametes. Therefore, the numbers of each genotype are in Hardy–Weinberg ratios as follows

$$\begin{bmatrix} \text{number of each genotype at} \\ \text{time } t \text{ among zygotes} \end{bmatrix}$$

$$AA: p_t^2 N_t \tag{3.1}$$

$$Aa: 2p_t q_t N_t$$

$$aa: q_t^2 N_t$$

Now let us proceed to the adult phase. Let the fraction of type AA individuals who survive to adulthood be denoted as l_{AA}, and similarly for other genotypes. Then, the number of each genotype among adults is

$$\begin{bmatrix} \text{number of each genotype} \\ \text{among adults} \end{bmatrix}$$

$$AA: l_{AA} p_t^2 N_t \tag{3.2}$$

$$Aa: l_{Aa} 2p_t q_t N_t$$

$$aa: l_{aa} q_t^2 N_t$$

Next, let us move to the gametic phase. Let the number of gametes shed by type AA individuals be denoted as $2m_{AA}$. (Thus, if an AA individual on the average leaves 140 gametes, then m_{AA} is 70.) Similarly, let m_{Aa} and m_{aa} refer to the fertilities of the other genotypes. Then the gametic output from each genotype is

$$\begin{bmatrix} \text{number of gametes produced by} \\ \text{adults of each genotype} \end{bmatrix}$$

from AA adults: $2m_{AA}l_{AA}p_t^2N_t$ (3.3)

      Aa $2m_{Aa}l_{Aa}2p_tq_tN_t$

      aa $2m_{aa}l_{aa}q_t^2N_t$

Now we are almost finished. The gene frequency in the gamete pool is simply the number of gametes carrying the A allele divided by the total number of gametes. The number of gametes carrying an A allele is found by adding all the output of the AA adults plus one half the output from the Aa adults. Accordingly, we have

$$\begin{bmatrix} \text{frequency of } A \\ \text{in gamete pool} \end{bmatrix} = \frac{2m_{AA}l_{AA}p_t^2N_t + \frac{1}{2}[2m_{Aa}l_{Aa}2p_tq_tN_t]}{[\text{total number of gametes}]} \qquad (3.4)$$

The total number of gametes is

$$\begin{bmatrix} \text{total number} \\ \text{of gametes} \end{bmatrix} = 2m_{AA}l_{AA}p_t^2N_t + 2m_{Aa}l_{Aa}2p_tq_tN_t + 2m_{aa}l_{aa}q_t^2N_t \qquad (3.5)$$

Finally, we move from the gametic phase to the new zygotic phase. If we assume that the gametes themselves survive equally well regardless of which gene is carried, then the gene frequency among the zygotes will be the same as that among the gametes

$$\begin{bmatrix} \text{frequency of } A \text{ among} \\ \text{zygotes, } p_{t+1} \end{bmatrix} = \begin{bmatrix} \text{frequency } A \text{ in} \\ \text{gamete pool} \end{bmatrix} \qquad (3.6)$$

Also, the population size among the zygotes will be simply half the number of gametes

$$\begin{bmatrix} \text{population size among} \\ \text{zygotes, } N_{t+1} \end{bmatrix} = \tfrac{1}{2}[\text{total number of gametes}] \qquad (3.7)$$

We are now finished. If we put all this together and divide out the quantities present in both the numerator and denominator of Equation (3.4), we get

$$p_{t+1} = \frac{m_{AA}l_{AA}p_t^2 + m_{Aa}l_{Aa}p_tq_t}{m_{AA}l_{AA}p_t^2 + m_{Aa}l_{Aa}2p_tq_t + m_{aa}l_{aa}q_t^2}$$

$$N_{t+1} = (m_{AA}l_{AA}p_t^2 + m_{Aa}l_{Aa}2p_tq_t + m_{aa}l_{aa}q_t^2)N_t \qquad (3.8)$$

Notice that the quantity $m_{ij}l_{ij}$ appears throughout the formulas for p_{t+1} and N_{t+1}. This fact suggests that we can simplify matters by lumping m's and l's into a single quantity. Let us then define

$$W_{AA} = m_{AA}l_{AA}$$

$$W_{Aa} = m_{Aa}l_{Aa} \qquad (3.9)$$

$$W_{aa} = m_{aa}l_{aa}$$

The W's are called *absolute selective values*. The *fitness* of an individual is defined as the contribution which that individual makes to the next generation's gene pool. An individual's fitness depends on both its fertility and viability. The W's defined above combine both the fertility and viability into a single overall measure of an individual's fitness†. Using these W's, the equations become

$$p_{t+1} = \frac{(p_t W_{AA} + q_t W_{Aa})p_t}{[p_t^2 W_{AA} + 2p_t q_t W_{Aa} + q_t^2 W_{aa}]} \tag{3.10a}$$

$$N_{t+1} = [p_t^2 W_{AA} + 2p_t q_t W_{Aa} + q_t^2 W_{aa}]N_t \tag{3.10b}$$

These are the basic equations to predict the gene frequency and population size at generation $t+1$, given the gene frequency and population size at time t together with the W's.

Some Notation Conventions
Now that we have these equations, what can we do with them? The answer is that we can, as a thought experiment, assign a different fertility and survival ability to each of the three genotypes and then use the equations to determine how the gene pool and population size change as a result. To do this, we should first simplify matters still further. Notice in the equation for p_{t+1} that the W's appear in both the numerator and denominator. This fact means that we can multiply all the W's by some constant, say C, and the constant will divide out. Therefore, in the equation for p_{t+1} the exact values of the W's are not important, but the relationship among them is. This fact, that only the relationship among the W's is important, allows us to use a simple convention. Let us agree to set one of the selective values equal to 1. For example, suppose that we wish to set the selective value of type AA equal to 1. Then we alter the selective values of the other types so that the relationship among them is preserved. This is achieved simply by dividing the W of each type by the W of the genotype that we have decided to set equal to 1. Thus we have

$$w_{AA} = \frac{W_{AA}}{W_{AA}} = 1$$

$$w_{Aa} = \frac{W_{Aa}}{W_{AA}} \tag{3.11}$$

$$w_{aa} = \frac{W_{aa}}{W_{AA}}$$

We call the selective values that have been scaled in this way the *relative selective values*; they are denoted with lower-case letters to distinguish them from the absolute selective values.

Let us consider a simple example. Suppose that $m_{AA} = 100$, $m_{Aa} = 50$, $m_{aa} = 25$, $l_{AA} = \frac{3}{4}$, $l_{Aa} = \frac{1}{2}$, and $l_{aa} = \frac{1}{5}$. Then the absolute selective values are

$$W_{AA} = (100)(\tfrac{3}{4}) = 75$$

$$W_{Aa} = (50)(\tfrac{1}{2}) = 25 \tag{3.12}$$

$$W_{aa} = (25)(\tfrac{1}{5}) = 5$$

† The W's may equal any number ≥ 0. Moreover, the W's need not be constants; they may depend on population size (density-dependent selection), or on the relative abundance of the other genotypes (frequency-dependent selection). In most cases we shall assume that the W's are constant, but this assumption is not inherent in Equations (3.10).

Now the relative selective values, provided that we want to make $w_{AA} = 1$, are

$$w_{AA} = \tfrac{75}{75} = 1$$
$$w_{Aa} = \tfrac{25}{75} = \tfrac{1}{3} \qquad (3.13)$$
$$w_{aa} = \tfrac{5}{75} = \tfrac{1}{15}$$

In this example we have let the largest w equal 1. This is the most common convention although there are occasional exceptions. The use of relative selective values in this way allows us to pool together many cases that would otherwise have to be treated separately. For example, the absolute selective values of $W_{AA} = 15$, $W_{Aa} = 5$, and $W_{aa} = 1$ also reduce to the same relative selective values in (3.13), as would any other multiple.

There is one more brief notational point. The *selection coefficients* are defined as follows

$$s_{AA} = 1 - w_{AA}$$
$$s_{Aa} = 1 - w_{Aa} \qquad (3.14)$$
$$s_{aa} = 1 - w_{aa}$$

Thus a selection coefficient is simply the difference between the relative selective value and 1. For example, the selection coefficients corresponding to the w's in (3.13) are

$$s_{AA} = 0$$
$$s_{Aa} = \tfrac{2}{3} \qquad (3.15)$$
$$s_{aa} = \tfrac{14}{15}$$

Thus the selection coefficients measure the strength of selection *against* some type while the relative selective values measure the strength of the selection *for* a type.

We can now begin our exploration of how natural selection changes gene frequencies in a population. We shall concentrate on the equation for p_{t+1}. Although we also have an equation for N_{t+1} we shall not use it until Chapter 17 in the ecology part of the book.

Case with All the Selective Values Equal

Suppose that the selective values are equal

$$w_{AA} = w_{Aa} = w_{aa} \qquad (3.16)$$

In this case the w's divide out of the numerator and denominator of the equation for p_{t+1} (3.10a) leaving $(p_t + q_t)p_t$ in the numerator and $(p_t^2 + 2p_t q_t + q_t^2)$ in the denominator. But both these expressions simply equal one so that (3.10a) reduces to saying that $p_{t+1} = p_t$ for any p_t. Perhaps you will now recognize this as the Hardy–Weinberg situation considered in the last chapter. If the w's are equal, there is no natural selection and therefore *any* initial gene frequency does not change and is perpetuated. When the w's are equal, the situation is sometimes called *selective neutrality* and the phenotype, as affected by the locus, is said to be neutral with respect to natural selection.

Selection Against a Dominant Allele

Suppose that A is dominant and that it is deleterious to its carriers. Since the phenotypes of AA and Aa individuals are then the same, they must have the same selective values. Hence the selective values for this case are

$$w_{AA} = w_{Aa} = 1 - s \qquad (0 < s \leq 1)$$
$$w_{aa} = 1$$

(3.17)

We wish to know how fast this natural selection changes the gene frequency as a function of the strength of the selection s, against the A allele. We can use (3.10a) to answer this question. We assume that the population begins with some initial gene frequency, p_0. Using (3.10a), we can compute p_1. Then, using (3.10a) again, we compute p_2, and so on. This process is called *iteration*. Suppose that we are interested in the time taken for natural selection to change p_0 to, say, p_z. Then we iterate (3.10a) starting with p_0 and simply observe how long it takes until p_z is reached. In this sense (3.10a) can tell us how fast natural selection changes gene frequencies from one given value to another.

Iteration can be done either numerically on a computer, or analytically. Technically, (3.10a) is a nonlinear recursion equation, and there is no general way to iterate it analytically, although special cases have been solved. So we shall use a computer to iterate (3.10a). However, later in this section, when we consider mutation, we shall be able to iterate analytically.

Figure 3.2 presents the results of computer iterations of (3.10a) for different strengths of selection against A. An s of 1 indicates that A is lethal to its carriers while s of .001 indicates very weak selection. An s of .001 represents a difference in relative selective values in the third decimal place and is the current limit to which we can detect any selection at all. We can see immediately that the time taken to change p from .9 to .1 varies with the strength of selection. If A is lethal, then p goes from .9 to zero in one generation because all the A's are immediately eliminated. At the opposite extreme if $s = .002$, then 6500 generations are needed to reach .1.

In addition, the curve of p through time (this curve is called a trajectory) tends to start out rather flat and to accelerate as the evolution progresses. This is because at the beginning, the aa phenotype is very rare. If q is .1, then R (the frequency of the aa type) at the zygotic phase is .01 (i.e., q^2) so that only one hundredth of the initial population is being favored by selection and all the rest are being selected against. At this initial stage in the evolution, most of the a alleles are in heterozygotes, where they are not expressed. Since the heterozygotes are phenotypically the same as AA homozygotes, the a alleles, which are tied up in heterozygotes, are not favored by selection but are selected against. Because only the aa homozygotes are favored and these are initially rare, evolution gets off to a slow start. But as the evolution proceeds, more and more aa homozygotes appear and the speed accelerates. Moreover, complete elimination of the A allele is approached rapidly because, being dominant, A is always expressed and hence always exposed to the selection against it.[†]

Selection Against a Recessive

Again let us assume that A is selected against and let A be recessive. (In genetics an upper case letter usually refers to a dominant allele, and lower

[†] The approach to $p = 0$ in this case is said to be "geometric" because for p near 0, p_t is given by $p_t \approx \lambda^t$ where $0 < \lambda < 1$.

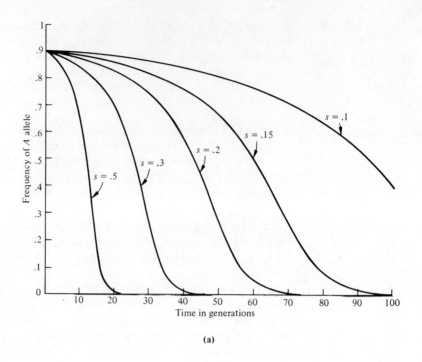

(a)

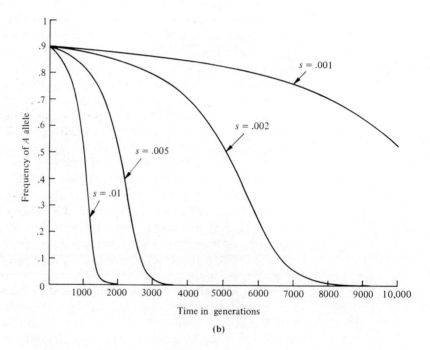

(b)

FIGURE 3.2. *Selection against a dominant allele*: (a) *Strong selection*; (b) *weak selection. The vertical axis is the frequency of the A allele; the horizontal axis is time measured in generations. For this figure* $w_{AA} = w_{Aa} = 1 - s$ *and* $w_{aa} = 1$. *Note that the time needed to eliminate the A allele depends on the strength of the selection against it. For very strong selection,* $s = .5$, *the A allele is essentially eliminated in 20 generations whereas with weak selection,* $s = .002$, *the time to elimination is about 8000 generations.*

case to a recessive. This convention is *not* being used here.) Since the *Aa* phenotype is now the same as the *aa* phenotype, the selective values for this case are

$$w_{AA} = 1 - s \qquad (0 < s \leqslant 1)$$

$$w_{Aa} = w_{aa} = 1$$

(3.18)

Computer iterations from (3.10a) for this case appear in Figure 3.3. We see

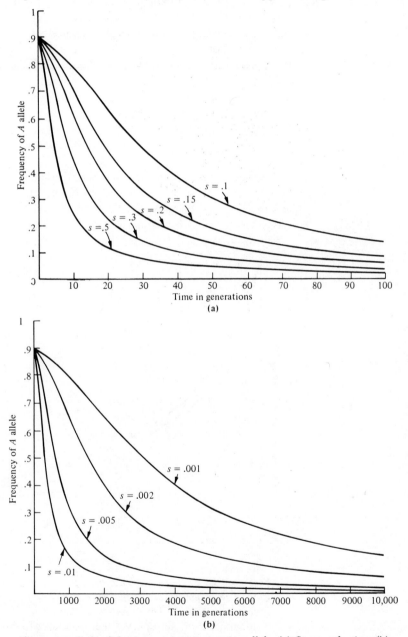

FIGURE 3.3. *Selection against a recessive allele*: (a) *Strong selection*; (b) *weak selection. For this figure* $w_{AA} = 1 - s$, $w_{Aa} = w_{aa} = 1$. *Note the rapid speed of evolution at the beginning and the slow rate at the end. Contrast these curves with those in Figure 3.2.*

immediately that these trajectories are very different from those in Figure 3.2. Indeed, they show properties opposite to Figure 3.2 in that the evolution is initially fast but becomes increasingly slower. Also, the time for p to go from .9 to .1 is much longer. If s is .5, the required time is 21 generations; if s is .002, then 7500 generations are needed.

The reason evolution is so slow in this case is that the selection against A becomes increasingly less effective as the frequency of A decreases. Since A is recessive, it is not expressed in heterozygotes and, as A becomes rarer, virtually no AA homozygotes exist. Hence A is increasingly "protected" from the selection against it as A becomes rarer.† On the other hand, the initial speed of evolution is rapid because, in contrast to the previous case, there is an initial abundance of the phenotype favored by selection. If q is .1, then the initial fraction of the population favored by selection is .19. (This is $2pq + q^2$.) Thus the initial fraction favored by selection is 19 percent of the population in this case, whereas in the previous case the initial fraction being favored was only 1 percent. Hence evolution gets off to a fast start because all the a alleles, no matter how rare initially, are exposed to the selection in their favor.

An experimental demonstration of this case has been provided by Wallace (1963) with *Drosophila*. In the experiment, the frequency of a recessive lethal gene was followed for 10 generations. The generations were kept separated to fulfill the conditions of the model. See Figure 3.4. One line

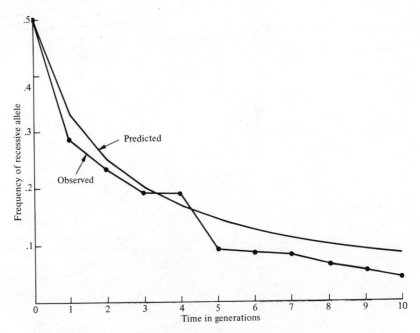

FIGURE 3.4. *Experiment illustrating selection against a recessive lethal gene. The frequency of the recessive allele is on the vertical axis, time in generations is on the horizontal axis.* [Data from B. Wallace (1963), The elimination of an autosomal lethal from an experimental population of *Drosophila melanogaster, Amer. Natur.* **97**: 65–66.]

† The approach to $p = 0$ in this case is said to be "algebraic" in that for p near zero, p_t is given by $p_t \approx 1/t$. This is a far slower rate of approach than the geometric rate.

represents the data; another line is the theoretical prediction obtained by iterating the equation (3.10a), for p_{t+1}, assuming $w_{AA} = 0$ and $w_{Aa} = w_{aa} = 1$. The data agree quite well with the predictions, and the standard errors of the points are large enough so that most deviations are not significant. But since the data indicate a somewhat faster loss of the allele than expected, it is surmised that the allele being eliminated by selection is not fully recessive and that, as a result, there is some selection against heterozygotes for the deleterious allele.

Incomplete Dominance In the previous two cases we assumed that A was either completely dominant or completely recessive. This assumption allowed us to set w_{Aa} equal to one of the homozygote selective values. In general, however, A need not be completely dominant or recessive, and if so, the w's are ranked as follows

$$w_{AA} < w_{Aa} < w_{aa}$$
$$w_{aa} = 1 \tag{3.19}$$

There are two special cases of incomplete dominance. Let e_A be some number that measures the "effect" of an A allele on the fitness of its carrier, and similarly for e_a. Then the *additive* scheme is defined as

$$
\begin{aligned}
w_{AA} &= e_A + e_A \\
w_{Aa} &= e_A + e_a \qquad (0 < e_A < \tfrac{1}{2}) \\
w_{aa} &= e_a + e_a = 1 \qquad (e_a = \tfrac{1}{2})
\end{aligned}
\tag{3.20}
$$

In this case the selective value of the heterozygote is the usual arithmetic average of the two homozygotes. The *multiplicative* scheme is defined as

$$
\begin{aligned}
w_{AA} &= v_A v_A \qquad (0 < v_A < 1) \\
w_{Aa} &= v_A v_a \qquad (v_a = 1) \\
w_{aa} &= v_a v_a = 1
\end{aligned}
\tag{3.21}
$$

Here w_{Aa} is the geometric mean of w_{AA} and w_{aa}.† There is little concrete biological justification for these schemes, but they are mathematically convenient. Nevertheless, they do illustrate the sorts of trajectories that arise with incomplete dominance.

Figure 3.5 illustrates the additive case, and Figure 3.6 the multiplicative case. As the figures show, they are essentially identical to one another. Both are qualitatively in between the two previous cases where full dominance or recessiveness was assumed. The multiplicative scheme, rather than the additive scheme, leads to a pattern of trajectories slightly closer to that where A is fully dominant because w_{Aa} is nearer to w_{AA} in the multiplicative scheme than in the additive scheme. Again note the times required for p to change from .9 to .1.

---

† The geometric mean of two numbers x_1 and x_2 is defined as $\sqrt{x_1 x_2}$. The geometric mean is always less than the arithmetic mean (unless $x_1 = x_2$).

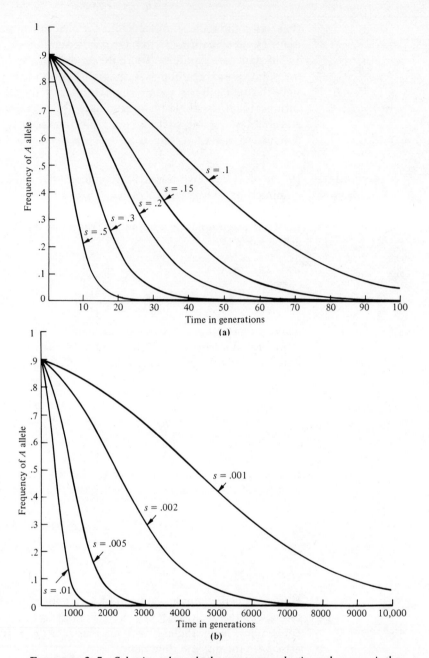

FIGURE 3.5. *Selection where the heterozygote selective value,* w_{Aa}, *is the arithmetic average of the selective values of the homozygotes.* (a) *Strong selection with additive fitnesses;* (b) *weak selection with additive fitnesses.*

Heterozygote Superiority Leads to Polymorphism

Suppose now that the heterozygote has a higher selective value than either homozygote,

$$w_{AA} = 1 - s_{AA}$$

$$w_{Aa} = 1 \qquad (0 < s_{AA}, s_{aa} < 1) \qquad (3.22)$$

$$w_{aa} = 1 - s_{aa}$$

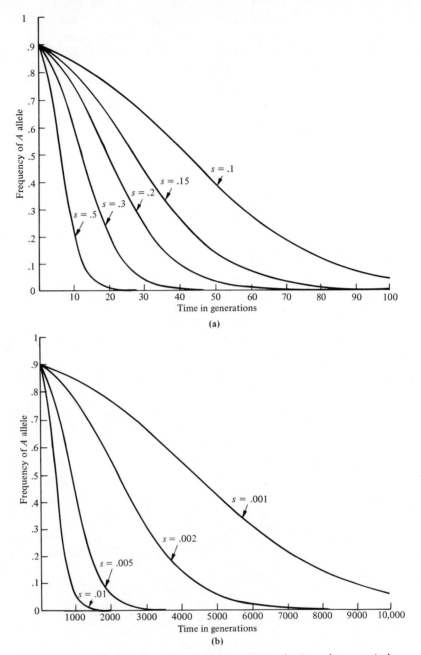

FIGURE 3.6. *Selection where the heterozygote selective value, w_{Aa}, is the geometric average of the selective values of the homozygotes.* (a) *Strong selection with multiplicative fitnesses*; (b) *weak selection with multiplicative fitnesses.*

A rationale for this assumption is that the A allele produces a protein that functions well in certain conditions, and the a allele produces a protein that functions well in other conditions. The heterozygote, in having both proteins, can perhaps survive under a wider range of conditions than either homozygote. If so, the heterozygote may have a higher selective value than either homozygote. This situation leads to a very different outcome than anything we have considered so far. Figures 3.7 and 3.8 show that trajectories starting from .9 or .1 both converge to a value of p in the interior of the graph. This means that neither A nor a is eliminated, but that the population comes to equilibrium with both alleles being maintained in the population. The reason for this result is simple. The heterozygote is favored

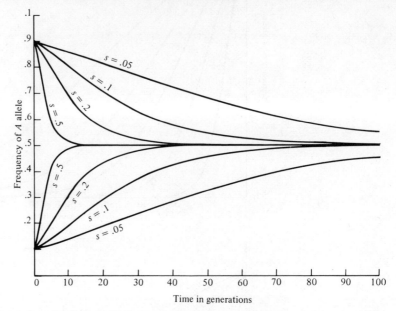

FIGURE 3.7. *Selection with heterozygote superiority. In these runs the selection coefficients against the homozygotes are equal, that is, $s_{AA} = s_{aa}$. Note that all trajectories approach an equilibrium at $\hat{p} = \frac{1}{2}$.*

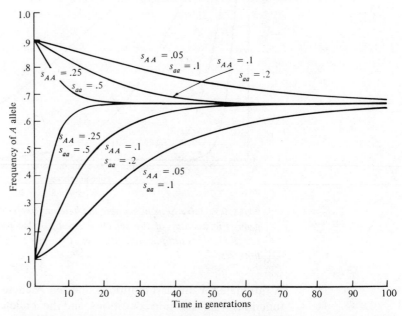

FIGURE 3.8. *Selection with heterozygote superiority. In these runs the selection against genotype aa is twice as strong as the selection against genotype AA, that is $s_{aa} = 2s_{AA}$. Note that all trajectories approach an equilibrium at $\hat{p} = \frac{2}{3}$.*

and every time a heterozygote survives, it carries both alleles with it. Therefore, favoring heterozygotes will maintain both alleles in the population.

A glance at Figures 3.7 and 3.8 shows that the homozygote selection coefficients, s_{AA} and s_{aa}, influence both the speed of the approach to equilibrium and where the equilibrium is. We can separate these effects as follows. First, we can easily derive a formula for the equilibrium gene frequency, which we denote as $\hat{p}$. The definition of an equilibrium is that $p_{t+1} = p_t$. Moreover, we are interested only in the "interior" equilibrium, that is, $\hat{p}$ which does not equal either 0 or 1. Now, to have $p_{t+1} = p_t$ in (3.10a), we need to have

$$pw_{AA} + qw_{Aa} = p^2 w_{AA} + 2pq w_{Aa} + q^2 w_{aa} \qquad (3.23)$$

Now this equality will certainly not be true in general, but only for a special value of p. This value is obtained by regarding p as an unknown in (3.23) and solving for p in terms of the w's. Let us denote the solution to (3.23) as $\hat{p}$. The solution to (3.23), $\hat{p}$, is an equilibrium because if $p_t = \hat{p}$, then p_{t+1} will also equal $\hat{p}$. The arithmetic is in the box. (Incidentally, we shall consider less tedious ways of obtaining $\hat{p}$ in the next section.) The solution is

$$\hat{p} = \frac{s_{aa}}{s_{AA} + s_{aa}} \qquad (3.24)$$

The frequency of A at equilibrium, $\hat{p}$, increases as the selection against its alternative, a, increases, as seems intuitive. Note that $\hat{p}$ depends on the relationship among the s's. For example, whenever both homozygotes are *equally* selected against ($s_{AA} = s_{aa}$), then $\hat{p}$ is $\frac{1}{2}$ regardless of the strength of selection. In Figure 3.8 $s_{AA} = \frac{1}{2} s_{aa}$ so $\hat{p}$ is $\frac{2}{3}$ regardless of the strength of selection. Thus the *relationship among the s's controls the location* of the equilibrium. *The magnitude of the s's controls the speed* with which the equilibrium is obtained.

Solve for $p \neq 0$ or 1 in the quadratic equation

$$pw_{AA} + qw_{Aa} = p^2 w_{AA} + 2pq w_{Aa} + q^2 w_{aa}.$$

Rewrite the right-hand side as

$$p(pw_{AA} + qw_{Aa}) + q(pw_{Aa} + qw_{aa}).$$

Then move the first term on the right over to the left leaving

$$(pw_{AA} + qw_{Aa}) - p(pw_{AA} + qw_{Aa}) = q(pw_{Aa} + qw_{aa}).$$

Rearrange the left-hand side leaving

$$(1 - p)(pw_{AA} + qw_{Aa}) = q(pw_{Aa} + qw_{aa})$$

Note that $(1 - p) = q$ so that the quadratic equation can be factored as

$$q[(pw_{AA} + qw_{Aa}) - (pw_{Aa} + qw_{aa})] = 0$$

therefore, one of the roots is $q = 0$, or equivalently $p = 1$. The other root is the one we want; it is found from the linear equation

$$pw_{AA} + qw_{Aa} = pw_{Aa} + qw_{aa}$$

Rearranging, we have

$$q(w_{Aa} - w_{aa}) = p(w_{Aa} - w_{AA})$$

Then substituting $q = 1 - p$ and solving for p, we obtain

$$\hat{p} = \frac{(w_{Aa} - w_{aa})}{(w_{Aa} - w_{AA}) + (w_{Aa} - w_{aa})}$$

If we use selection coefficients, $w_{AA} = 1 - s_{AA}$, $w_{Aa} = 1$, and $w_{aa} = 1 - s_{aa}$, then this formula converts to

$$\hat{p} = \frac{s_{aa}}{s_{AA} + s_{aa}}$$

The case of heterozygote superiority is very important in evolutionary theory. This is the simplest case where natural selection maintains genetic variation. In the other cases discussed in this chapter, the end result of natural selection is to fix one of the alleles.

Dobzhansky (1954) has produced an elegant experimental example of this case with *Drosophila pseudo-obscura*. Hence *CH* and *ST* are names of blocks of genes. Because of a special chromosomal feature called an *inversion*,[†] the genes within each block are held together. As a result, *CH* and *ST* are effectively two alleles at a single locus. Dobzhansky determined that heterozygotes for these blocks have higher fitness than either homozygotes and therefore a polymorphism should result. Figure 3.9 illustrates four replicate experiments in which the frequencies of *CH* and *ST* were allowed to change by natural selection. The solid line is the prediction obtained by iterating (3.10a) with selective values of $w_{ST,ST} = .90$, $w_{ST,CH} = 1$, and $w_{CH,CH} = .41$. The dotted lines are the 95 percent confidence limits associated with the experimental sampling technique. Note that the experiments are all within the 95 percent confidence limits of the prediction. This is strong support for the theory that predicts how natural selection influences gene frequencies. In these experiments the theory correctly predicted *both* the equilibrium frequency itself and the dynamics of the approach to equilibrium.

Heterozygote Inferiority

In this case the heterozygote is assumed to have a lower selective value than either homozygote

$$
\begin{aligned}
w_{AA} &= 1 + s_{AA} \\
w_{Aa} &= 1 \qquad\qquad (s_{AA} > 0, \; s_{aa} > 0) \\
w_{aa} &= 1 + s_{aa}
\end{aligned}
\qquad (3.25)
$$

In this case our convention is to scale the heterozygote to a relative selective value of one even though it is the smallest. A rationale for this kind of selection might be that the locus is responsible for a protein having two chains that assemble in the cytoplasm. One could imagine that a protein would be functional with two of the chains from the A allele, or two from the a allele, but that a protein with one chain from each (the "hybrid" protein)

† Chromosome inversions are discussed in more detail in Chapter 11.

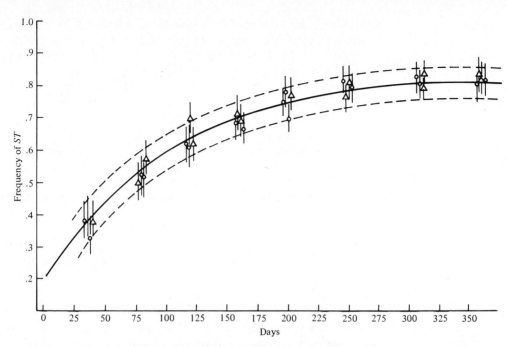

FIGURE 3.9. *Experiment illustrating selection with heterozygote superiority in* Drosophila pseudo-obscura. *The data are from four replicate experiments. Here ST and CH are the names of blocks of genes. The genotypes that these blocks form are ST, ST; ST, CH; and CH, CH. Heterozygotes for these blocks, that is, the genotype ST, CH, have the highest fitness. The vertical axis is the frequency of the ST block and the horizontal axis is time. The solid line is the prediction obtained by iterating equation (3.10a) with selective values of* $w_{ST,ST} = .90$, $w_{ST,CH} = 1$ *and* $w_{CH,CH} = .41$. *The dotted lines are the 95 percent confidence limits associated with the experimental sampling technique. Note that the experiments are all within the 95 percent confidence limits of the prediction.* [From T. Dobzhansky (1954), Evolution as a creative process, *Proc. of 9th Int. Cong. of Genetics*, in *Caryologia* 435–449.]

might not function well. If such a protein influenced the organism's ability to survive, then heterozygotes would be at a disadvantage.

Figure 3.10 shows that the pattern of trajectories that result from iterating (3.10) in this case is very different from any of the patterns considered so far. The trajectories lead either to $p = 0$ or $p = 1$ depending on where they begin. There is a demarcation point such that if a trajectory begins above this point, then it heads up to $p = 1$, and if it begins below this point, then it heads down to $p = 0$. Thus the outcome of natural selection in this case is always the elimination of one of the alleles, but which is eliminated depends on the initial condition.

There is a simple rationale for the result in Figure 3.10. Consider the example where the heterozygote is lethal and both homozygotes equally fit. An experimental analogy to this example is to draw beads two at a time from an urn containing two colors of beads. Whenever two beads of the same color are drawn, they are replaced back into the urn, and whenever two beads of different colors are drawn, they are both discarded. Then if the initial number of beads of each color is different, the color to be exhausted first will be the color with the numerical disadvantage to begin with. In the same sense, selection against a heterozygote eliminates one of each allele. Eventually then, the gene pool will lose the allele that began with a numerical disadvantage.

We can develop a formula to predict the demarcation point. Again consider the urn example above. Suppose that the two colors of beads are initially *exactly* equal in number. Then neither color can be exhausted before the other and the initial ratio between the number of each color will not change. This idea suggests that the demarcation point in our model may be an equilibrium point. If the initial p is *exactly* equal to the demarcation point, then the trajectory will neither go to $p = 0$ or $p = 1$ but will indicate no change at all in p. Indeed, if we set up our condition for an equilibrium

$$pw_{AA} + qw_{Aa} = p^2 w_{AA} + 2pqw_{Aa} + q^2 w_{aa} \qquad (3.26)$$

and substitute the w's for this case from (3.25), we do obtain a familiar solution. The demarcation point is (see the box for more data)

$$\hat{p} = \frac{s_{aa}}{s_{AA} + s_{aa}} \qquad (3.27)$$

Again we see that the location of the demarcation point depends on the relationship among the s's while the speed with which the trajectories move to $p = 0$ or $p = 1$ depends on the magnitudes of the s's.

In the preceding box we derived

$$\hat{p} = \frac{w_{Aa} - w_{aa}}{(w_{Aa} - w_{AA}) + (w_{Aa} - w_{aa})}$$

If we use selection coefficients of $w_{AA} = 1 + s_{AA}$, $w_{Aa} = 1$, $w_{aa} = 1 + s_{aa}$, we obtain

$$\hat{p} = \frac{-s_{aa}}{-s_{AA} - s_{aa}}$$

Upon dividing out -1, we obtain our familiar result.

Although the demarcation point is technically an equilibrium point, this terminology can be misleading. If p_0 is even infinitesimally above or below $\hat{p}$, then the trajectories lead to the top or bottom boundaries, respectively. The significance of $\hat{p}$ is not that it is an equilibrium point but that it is the demarcation point. Absolutely no real system can realize the equilibrium at $\hat{p}$; all real systems must move either to $p = 0$ or $p = 1$ depending on the initial condition. The demarcation point is usually called an *unstable equilibrium* to distinguish it from the kind of equilibrium that arose before with the case of heterozygote superiority. With the heterozygote superiority, $\hat{p}$ is called a *stable equilibrium* because trajectories actually lead up to it instead of away from it as in Figure 3.10.

It is difficult to find documented examples of this case of natural selection. A population in nature under this selection scheme would be found fixed for one of the alleles. But this information alone could just as easily suggest any of the first cases of selection we considered and there would be little cause to pursue the matter further. However, a clue to the presence of this scheme

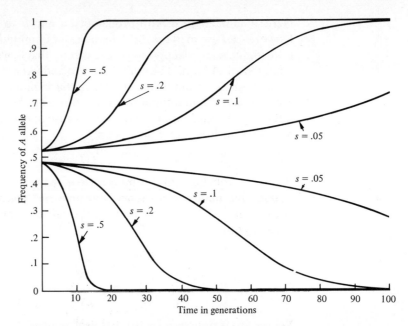

FIGURE 3.10. *Selection with heterozygote inferiority. Note that all trajectories beginning above p = .5 lead upward and culminate in fixing the A allele. All trajectories beginning below p = .5 lead downward and culminate in fixing the a allele.*

might be the presence of one allele in certain populations and the other allele in others so that the geographic pattern is a mosaic. If hybrids formed from parents of different populations have reduced fitness, then the hypothesis of heterozygote inferiority must be given serious consideration.

Now that we have examined all these different schemes of natural selection, we can better appreciate the importance of our simple model for natural selection. It is rich with possibilities. Simply by choosing different values for the *w*'s, we can represent different kinds of dominance relationships between the alleles and we can generate many different patterns of trajectories representing the evolution of a population. The derivation which produced our equation for p_{t+1} certainly did not suggest that such a wealth of biological predictions would emerge. But it was by systematically examining various mathematical cases and then interpreting each mathematical case in its biological context that we were led to discover the wealth of biological predictions.

One-Way Recurrent Mutation   Lest you have forgotten, one of the original motivations for developing a model for natural selection was to assess the speed with which selection changes gene frequencies. A glance back at Figures 3.2–3.3, 3.5, and 3.6 shows that the time needed for selection to reduce *p* from .9 to .1 varies from one generation to about 15,000 generations depending on the strength of selection and the dominance relationship between *A* and *a*. Let us now turn to recurrent mutation and ask the same question, how long would mutation take to reduce *p* from .9 to .1? In general there is both forward and backward mutation between the alleles

$$A \underset{v}{\overset{u}{\rightleftharpoons}} a \qquad\qquad (3.28)$$

Here A mutates to a with probability u and a to A with probability v. However, let us "load the dice" in the favor of mutation as a force changing gene frequencies by supposing that v is zero; that is, we shall assume one way mutation of A to a. The time taken to reduce p from .9 to .1 with this assumption is certainly faster than if back mutation is also taken into account. But if mutation at its fastest proves to be very slow, then we can safely reject recurrent mutation as the cause of evolution. The equation for p_{t+1} with one way mutation is simply

$$p_{t+1} = (1-u)p_t \qquad (3.29)$$

If u is the probability that A mutates to a, then $(1-u)$ is the probability that A does *not* mutate to a. Hence p at time $t+1$ is simply the fraction of A alleles from t that does not mutate. We can iterate Equation (3.29) analytically as follows:

$$p_1 = (1-u)p_0$$
$$p_2 = (1-u)p_1 = (1-u)^2 p_0 \qquad (3.30)$$
$$p_3 = (1-u)p_2 = (1-u)^3 p_0 \qquad \text{and so forth}$$

We see from Equation (3.30) that each generation simply raises $(1-u)$ to the next higher power, so in general

$$p_t = (1-u)^t p_0 \qquad (3.31)$$

We can solve this formula for the time needed to reduce p from .9 to .1. The time for p to move from p_0 to p_t is given by

$$t = \frac{\log (p_t/p_0)}{\log (1-u)} \qquad (3.32)$$

Solve for the time needed for p to move from p_0 to p_t. Begin with

$$p_t = (1-u)^t p_0$$

Rearrange as

$$\frac{p_t}{p_0} = (1-u)^t$$

Take the log of both sides, yielding

$$\log \left(\frac{p_t}{p_0}\right) = t \log (1-u)$$

Solve for t, giving

$$t = \frac{\log (p_t/p_0)}{\log (1-u)}$$

(See the box for details.) A typical mutation rate is 10^{-6}. Substituting for p_0 and p_t in Equation (3.32) gives

$$t = \frac{\log (.1/.9)}{\log (1-10^{-6})} = 2.2 \times 10^6 \text{ generations} \qquad (3.33)$$

Thus the time taken by recurrent mutation to reduce p from .9 to .1 is 2.2 million generations, which is much longer than required by even weak selection on a recessive allele in Figure 3.3. This process is illustrated in Figure 3.11. This simple calculation shows even weak selection to be so much faster than recurrent mutation in changing gene frequencies that we can usually dismiss mutation from further consideration as a force changing gene frequencies. If, however, the alleles at some locus are *really* neutral, then by default we must once again take account of mutation. But so long as even weak selection is present, $s \approx .001$ or more, then mutation can be disregarded.

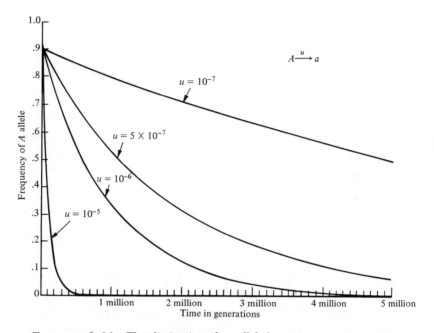

FIGURE 3.11. *The elimination of an allele by recurrent one-way mutation. Note how slowly evolution occurs by mutation as compared with natural selection in the preceding figures.*

Mutation Is a Source of Variation

If mutation is unimportant as a force changing gene frequencies, then what is its importance in evolution? Recurrent mutation is fundamental to the process of evolution because it helps to maintain a supply of genetic variation for selection to act on. Even if selection is tending to eliminate an allele, recurrent mutation tends to maintain its presence in the population in spite of the selection pressure. Should the environment change at some time in the future in such a way that the selection pressures reverse, then the allele that was previously opposed by selection will, in a large population, still be available for selection to act upon. Mutation is the ultimate source of genetic variation and thus plays a fundamental role in the process of evolution.

Mutation May Oppose Selection

It is interesting to explore the balance between selection and mutation. What is the level of genetic variation that is maintained when selection and mutation are in opposition? The algebraic details are given in a self-guiding problem set at the end of the chapter. The results are as follows. Suppose there is selection against A and that A is dominant; that is, the w's are $1-s$, $1-s$, and 1, and that v is the mutation rate of $a \to A$. Then the equilibrium frequency of A maintained by a balance between selection and mutation is

$$\hat{p} = \frac{v}{s} \qquad \text{(selection against a dominant)} \qquad (3.34)$$

If there is selection against A and A is recessive (the w's are $1-s$, 1, 1) then the equilibrium is at

$$\hat{p} = \sqrt{\frac{v}{s}} \qquad \text{(selection against a recessive)} \qquad (3.35)$$

There is an intuitive difference between these formulas. Suppose that the mutation rate, v, is 10^{-6} and let $s = .01$. Then if there is selection against a dominant $\hat{p} = 10^{-8}$ and if there is selection against a recessive, $\hat{p} = 10^{-4}$. Thus if A is recessive, mutation maintains a much higher frequency than if A is dominant. This is simply because if A is recessive, it is usually hidden from selection and hence mutation can build up the frequency to a higher equilibrium level.

Two-Way Recurrent Mutation   The equilibrium above is established by the joint action of selection and mutation. For completeness we might ask, what happens when there is recurrent mutation alone? If $A \to a$ at rate u and $a \to A$ at rate v, then the gene frequency at $t+1$ due to mutation alone is

$$p_{t+1} = (1-u)p_t + vq_t \qquad (3.36)$$

By setting $p_{t+1} = p_t$ and solving for the equilibrium, we obtain

$$\hat{p} = \frac{v}{u+v} \qquad (3.37)$$

This is the equilibrium maintained by recurrent mutation alone. But this equilibrium should not be taken too seriously. With even weak selection (e.g., $s = .001$) the formulas for $\hat{p}$ in terms of a mutation-selection balance [(3.34) and (3.35)] become important; or if there is heterozygote superiority, then $\hat{p}$ is effectively given by (3.24), assuming realistic mutation rates. Also, even in the complete absence of selection, the equilibrium given by (3.37) above is unlikely to be realized. The rate of approach to the above equilibrium is extremely slow and, as we shall see in the chapter on genetic drift, the population size must exceed 10^6 before mutation can "overpower" genetic drift. Because mutation rates are so low, any equilibrium set by the mutation rates alone is not realistic.

Other Issues   The theory above predicts the outcome of natural selection, given the knowledge that natural selection is occurring. But in the real world, how can we detect the presence of natural selection? In principle we could measure the fertility and viability of several genotypes and directly determine whether or not the w's are the same. But in practice this is quite inefficient. There have been several papers in recent years about the practical aspects of detecting natural selection. Interested readers should see Lewontin and Cockerham (1959), Prout (1965, 1969), Bundgaard and Christiansen (1972), Christiansen and Frydenberg (1973), and the symposium volume on this topic by Christiansen and Fenchel (1977).

There has been long-standing interest in models that include more complex assumptions than those in the model described above. Specifically, a model involving two sexes and allowing for selection in opposing directions on males and females is treated by Bodmer (1965) and Karlin (1972). A model for natural selection through differential fertility, assuming random mating (not random union of gametes), is discussed in Bodmer (1965) and Hadeler and Liberman (1975).

PROBLEMS

3.1. Write a computer program to iterate Equation (3.10). Use the program on a computer to verify the conclusions we obtained from Figures 3.2 through 3.10.

3.2. Suppose that $w_{AA} = \frac{3}{4}$, $w_{Aa} = 1$, and $w_{aa} = \frac{1}{2}$ what is $\hat{p}$?

3.3. Derive Equation (3.34). The selective values are $w_{AA} = 1 - s$, $w_{Aa} = 1 - s$, and $w_{aa} = 1$. The mutation rate of $a \to A$ is v. The equation for Δp under both mutation and selection is $\Delta p = \Delta p_{sel} + \Delta p_{mut}$, where Δp_{sel} is the change in p due to selection alone and Δp_{mut} is the change in p due to mutation alone.

(a) Show that

$$\Delta p_{sel} = \frac{pw_{AA} + (1-p)w_{Aa}}{p^2 w_{AA} + 2p(1-p)w_{Aa} + (1-p)^2 w_{aa}} p - p$$

(b) Convince yourself that if p is sufficiently small, then the quantity $[p^2 w_{AA} + 2p(1-p)w_{Aa} + (1-p)^2 w_{aa}]$ is effectively equal to one.

(c) Show that the assumption in (b) leads to

$$\Delta p_{sel} = (1 - s)p - p$$
$$= -sp$$

(d) Show that

$$\Delta p_{mut} = v(1 - p)$$

(e) Convince yourself that if p is sufficiently small, then $v(1 - p)$ is effectively equal to v.

(f) Then the equation for the change in p becomes $\Delta p = -sp + v$.

(g) Solve for the equilibrium. What was the main assumption?

3.4. Derive Equation (3.35). The selective values are $w_{AA} = 1 - s$, $w_{Aa} = 1$, and $w_{aa} = 1$. The rate of $a \to A$ is again v.

(a) Show that if p is sufficiently small, then

$$\Delta p_{sel} = (1 - sp)p - p$$
$$= -sp^2$$

(b) Show that Δp is then given by

$$\Delta p = -sp^2 + v.$$

(c) Solve for the equilibrium. What was the main assumption?

Chapter 4

THE FUNDAMENTAL THEOREM OF NATURAL SELECTION

THE detailed outcome of natural selection at one locus and two alleles depends critically upon the dominance relationship between the two alleles. Specifically, the detailed outcome depends on whether the allele favored by selection is dominant or recessive and on whether the heterozygote is more or less fit than either homozygote.

But is there any common denominator among the various possible outcomes? In spite of detailed differences in the results of natural selection, is there some general principle that subsumes all the particular cases? This chapter is a search for some general principles about natural selection. We shall examine several candidates for general principles, and see whether these principles do indeed subsume the detailed results we have derived in Chapter 3.

A Modest Proposal: The Adaptive Topography
A candidate for a general principle about natural selection is supplied by the notion of adaptation. The spirit of the Darwinian theory of natural selection is that natural selection uses the genetic variation in a population to produce individuals that are adapted to their environment, that is, can function particularly well in their environment. Perhaps a general result of natural selection is to increase, through time, the average level of adaptation of the members of a population. Let us explore this idea further.

Let us take the selective value of an organism, w, as a measure of its degree of adaptation. An organism's w is certainly a reasonable measure of adaptation, for if an organism does function well in its environment, we expect this to be reflected in its ability to survive and to reproduce there. To the extent that an organism's capabilities in an environment are not reflected in its fitness, then that capability is irrelevant to natural selection, regardless of how strange and intriguing the capability may seem to us. If we take an organism's w as a measure of degree of adaptation, then the average or mean level of adaptation in the population is $\bar{w}$, the mean selective value. If we consider one locus with two alleles, then the full expression for $\bar{w}$ is

$$\bar{w} = p^2 w_{AA} + 2pq w_{Aa} + q^2 w_{aa} \qquad (4.1)$$

Our hypothesis now becomes that selection increases $\bar{w}$ through time and that this is true regardless of the particular details of the dominance relationship among the alleles.

The first general connection between natural selection and $\bar{w}$ was established by Sewall Wright as follows. Recall our equation for p_{t+1} as

$$p_{t+1} = \left(\frac{p_t w_{AA} + q_t w_{Aa}}{\bar{w}} \right) p_t \qquad (4.2)$$

It is mathematically convenient in this chapter to consider the quantity Δp, which is simply the *change* in the gene frequency

$$\Delta p = p_{t+1} - p_t \qquad (4.3)$$

Since we know p_{t+1} from (4.2), we can write Δp as

$$\Delta p = \frac{(p_t w_{AA} + q_t w_{Aa})}{\bar{w}} p_t - p_t \tag{4.4}$$

Now we want to rearrange (4.4) into a more pleasant form (see the algebra in the box) and we obtain

$$\Delta p = \frac{pq}{\bar{w}} \{p(w_{AA} - w_{Aa}) - q(w_{aa} - w_{Aa})\} \tag{4.5}$$

We stress that (4.5) is simply a rearranged form of (4.2). There is no new biology involved and no new biological assumptions have been made.

---

1. On Wright's adaptive topography, $\bar{w}$.
   (a) Note that if w_{AA}, w_{Aa}, and w_{aa} are constants, then

   $$\frac{d\bar{w}}{dp} = \frac{d}{dp}(p^2 w_{AA} + 2pq w_{Aa} + q^2 w_{aa})$$

   $$= 2(p w_{AA} + q w_{Aa} - p w_{Aa} - q w_{aa})$$

   $$= 2[p(w_{AA} - w_{Aa}) - q(w_{aa} - w_{Aa})]$$

   Use the chain rule remembering that $q = 1 - p$ in the derivative above.
   (b) Rearrange the equation of Δp as follows:

   $$\Delta p = \frac{(p w_{AA} + q w_{Aa})}{\bar{w}} p - p$$

   $$= \frac{p(p w_{AA} + q w_{Aa} - \bar{w})}{\bar{w}}$$

   $$= \frac{p}{\bar{w}}(p w_{AA} + q w_{Aa} - p^2 w_{AA} - 2pq w_{Aa} - q^2 w_{aa})$$

   $$= \frac{p}{\bar{w}}(pq w_{AA} + q w_{Aa} - 2pq w_{Aa} - q^2 w_{aa})$$

   $$= \frac{pq}{\bar{w}}(p w_{AA} + w_{Aa} - 2p w_{Aa} - q w_{aa})$$

   $$= \frac{pq}{\bar{w}}(p w_{AA} + q w_{Aa} - p w_{Aa} - q w_{aa})$$

   $$= \frac{pq}{\bar{w}}[p(w_{AA} - w_{Aa}) - q(w_{aa} - w_{Aa})]$$

   (c) Substitute from (1a) assuming that the selective values are not functions of the gene frequencies and obtain

   $$\Delta p = \frac{pq}{\bar{w}}\left(\frac{1}{2}\right)\frac{d\bar{w}}{dp}$$

Now consider $\bar{w}$ again. If natural selection is to cause $\bar{w}$ to increase, we would expect that an equilibrium condition would refer to a situation where

$\bar{w}$ cannot be increased any further. Or, in other words, if p is at equilibrium, then $\bar{w}$ must be at its maximum value if we are to believe that selection in general acts to increase $\bar{w}$. With this idea in mind, we ought to inspect the derivative of $\bar{w}$ to see whether there is any connection between the slope of $\bar{w}$ and the equation for Δp. The function, $\bar{w}$, is a function of only one variable, p (remember that $q = 1 - p$), so the derivative of $\bar{w}$ with respect to p is found to be (see boxed material)

$$\frac{d\bar{w}}{dp} = 2\{p(w_{AA} - w_{Aa}) - q(w_{aa} - w_{Aa})\} \tag{4.6}$$

There *is* an assumption here. *It is assumed that the w's are independent of p.* If the w's do depend on p, then terms in $dw_{AA}(p)/dp$, and so forth, would have to be included in (4.6). We immediately observe that the expression in braces in (4.6) is identical to that in braces in (4.5). Hence, *assuming* that the w's are independent of p, we can rewrite (4.4) as

$$\Delta p = \frac{pq}{\bar{w}}\left(\frac{1}{2}\right)\frac{d\bar{w}}{dp} \tag{4.7}$$

This equation is fundamental in population genetics. It establishes a connection between changes in the gene frequency, Δp, and the slope of the function, $\bar{w}$. In particular, we do indeed observe that if $\bar{w}$ is a maximum with respect to p (so that its slope is zero), then Δp is zero; that is, the gene pool is at equilibrium. Also, if the slope of $\bar{w}$ is negative, then Δp is negative too; similarly, if the slope of $\bar{w}$ is positive, then Δp is positive so that generally Δp has the same sign as the slope of $\bar{w}$. The function $\bar{w}$ is sometimes called the *adaptive topography*. It represents the average degree of adaptation in the population as a function of the gene frequency, p.

Examples of the Adaptive Topography   Let us now discuss some of the cases considered in the last chapter.

1. Suppose that the heterozygote is the *most fit*. The selective values are then

$$w_{AA} = 1 - s_{AA}$$
$$w_{Aa} = 1 \tag{4.8}$$
$$w_{aa} = 1 - s_{aa}$$

and $\bar{w}$ is found by substituting these values into $\bar{w}$

$$\bar{w} = 1 - p^2 s_{AA} - q^2 s_{aa} \tag{4.9}$$

Figure 4.1 plots $\bar{w}$ in this case for the same values of s considered in Figure 3.8 of the last chapter. Note that the curves always have a maximum point. Since the slope of $\bar{w}$ at equilibrium must be zero, the equilibrium frequency $\hat{p}$ can be found by finding the p that maximizes $\bar{w}$. The algebra is set as a problem. If you do the problem, you will discover that this derivation of the formula for $\hat{p}$ is *much* shorter than that of the last chapter.

The stability of the equilibrium is also suggested by the $\bar{w}$ function. Suppose that p is displaced to the right of $\hat{p}$. The slope of $\bar{w}$ is negative to the right of $\hat{p}$ so that p will tend to return to $\hat{p}$. Similarly, if p is displaced to the left of p then the slope of $\bar{w}$ is positive and p will increase. Thus if $\bar{w}$ has a maximum, natural selection should produce a stable equilibrium.

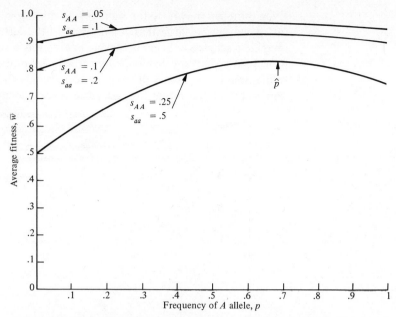

FIGURE 4.1. *The average fitness, $\bar{w}$, as a function of p for the case of heterozygote superiority. In this figure $s_{AA} = .5\, s_{aa}$ and the equilibrium gene frequency is at $\hat{p} = \frac{2}{3}$. Note that the peak to $\bar{w}$ coincides with $\hat{p}$.*

2. Figure 4.2 plots the $\bar{w}$ that results with the selective values for *heterozygote inferiority.*

$$w_{AA} = 1 + s_{AA}$$
$$w_{Aa} = 1 \qquad\qquad (4.10)$$
$$w_{aa} = 1 + s_{aa}$$

In this case $\bar{w}$ has a minimum value. To the right of the minimum the slope is positive so that p increases, and to the left the slope is negative so that p decreases. As we discovered in the last section, if the heterozygote is inferior, either A or a will be fixed depending on the initial condition. The adaptive topography analysis supports this result. Note too that fixation of either A or a can be viewed as selection increasing $\bar{w}$ from the given starting point.

3. Figures 4.3 and 4.4 illustrate the adaptive topographies corresponding to *selection against the A allele*, assuming that it is either recessive or dominant. Here $\bar{w}$ is maximized when A is fully eliminated, and $\bar{w}$ increases through time as A is progressively eliminated. The shapes of $\bar{w}$ depends on the dominance relationship, but the general result that $\bar{w}$ increases as evolution proceeds does not.

A Refinement   Thus the concept of the adaptive topography does provide a unifying theme for the diverse results possible from natural selection at one locus and two alleles. More recently, an important refinement of this theory has been added. To understand why this refinement is important, consider again the adaptive topography where the heterozygote is superior, Figure 4.1. As we have seen, if p is to the right of $\hat{p}$, then Δp is negative, while if p is to the left, Δp is positive. But this information alone is not sufficient to conclude that p will converge to $\hat{p}$. One could imagine that large overshoots might occur.

Even though p may be to the right of $\hat{p}$ so that Δp is negative, the magnitude of Δp may be so large that p is moved to the left *past* $\hat{p}$. One could then imagine a scheme of ever-increasing overshoots resulting in p being fixed at 0 or 1. We shall see behavior like this in several ecological models. The refinement to the theory of adaptive topography shows that this behavior is impossible in the model for selection at one locus, which we have been

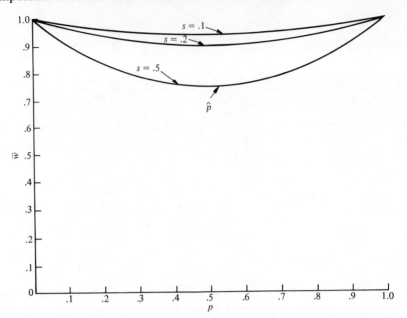

FIGURE 4.2. *Average selective value, $\bar{w}$, as a function of p for the case of heterozygote inferiority. In this figure $s_{AA} = s_{aa}$ and the demarcation point lies at $\hat{p} = \frac{1}{2}$. Note that the minimum of $\bar{w}$ coincides with the demarcation point.*

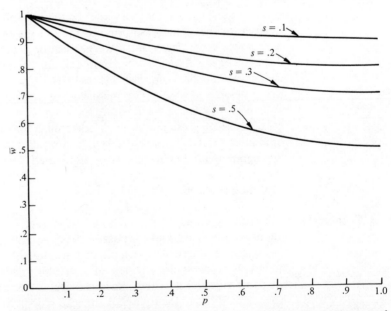

FIGURE 4.3. *Average selective value, $\bar{w}$, as a function of p for the case of selection against a dominant allele.*

considering. Specifically, we shall see in Chapter 7 that

$$\Delta \bar{w} \geq 0 \qquad (4.11)$$

This assertion is that $\bar{w}$ always increases from generation to generation to generation, except if p equals $\hat{p}$ in which case $\Delta \bar{w}$ must be zero. In particular, $\Delta \bar{w}$ is never negative so that the mean fitness does not decrease as a result of natural selection at one locus with two alleles. This fact rules out the possibility of overshoots that are so large as to prevent the convergence to $\hat{p}$. One important point is that the proof of (4.11) refers to one locus with an arbitrary number of alleles. It is not restricted to one locus with only two alleles. We shall discuss this result later when we consider the theory at one locus with multiple alleles in Chapter 7.†

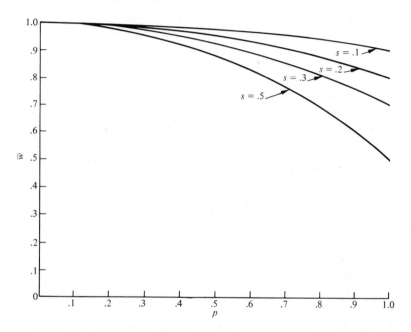

FIGURE 4.4. *Average selective value, $\bar{w}$, as a function of p for the case of selection against a recessive allele.*

Adaptive Topography with Frequency-Dependent Selection

The derivation of the equation connecting Δp with the slope of $\bar{w}$ assumes that the w's are constants. But suppose that the w's are not constants. Then what? In Chapter 17 we shall suppose that the w's are functions of population size and consider a natural extension of the theory of adaptive topographies to "density-dependent selection." We might also suppose that the w's depend on p and that the relative fitness of a type of individual depends on whether it is rare or common. This condition is "frequency-dependent selection." The adaptive topography idea applies to frequency-dependent selection in a "point-wise" fashion as follows. Consider a point p_0. Define a set of constants, $w_{0,\,AA}$, $w_{0,\,Aa}$, $w_{0,\,aa}$, which are the w's evaluated at p_0, for example, $w_{0,\,AA} = w_{AA}(p_0)$, and so forth. Finally,

† Since $\Delta \bar{w} \geq 0$, with the equality holding only at equilibrium, $\bar{w}$ is said to be a *Liapunov function*.

define the adaptive topography "as seen from" p_0 as $\bar{w}_0(p) = p^2 w_{0,\,AA} + 2pq w_{0,\,Aa} + q^2 w_{0,\,aa}$. Then Δp, given p_0, is connected to the slope of $\bar{w}_0$ by the usual formula (4.7). Specifically, given p_0 then

$$\Delta p = p_1 - p_0 = \frac{p_0 q_0}{2\bar{w}_0} \frac{d\bar{w}_0(p_0)}{dp} \tag{4.12}$$

Metaphorically, we might describe this condition by saying that Δp is governed at p_0 by the adaptive topography "as seen from" p_0. In this sense, as p changes from generation to generation, the adaptive topography changes too. Selection changes p at any particular time based only on the adaptive topography at that time and without regard to the future appearance of the adaptive topography. As a result, the actual mean fitness may show an overall decrease through time in frequency-dependent selection even though selection has at each time increased the proportion of the type with highest fitness at that time. Metaphorically, it is sometimes said that selection is "shortsighted" when $\bar{w}$ is caused to decrease in this way.

The Fundamental Theorem of Natural Selection for Asexual Populations

Another general theorem about natural selection is Fisher's "fundamental theorem of natural selection." We shall consider a special case of this theorem presented in Crow and Kimura (1970). It concerns a different model of population genetics, one that originates in a simple model for population growth. Some readers may wish to defer reading about Fisher's theorem until after reading Chapter 16 on exponential growth.

Assume that the organisms are asexual and growing exponentially with different growth rates, for example, a vat containing many genotypes of bacteria, each with a different growth rate. Let the growth rate of an individual of type i be labeled r_i, and $n_i(t)$ be the number of organisms of type i at time t. Then the rate of increases of $n_i(t)$ in the vat is

$$\frac{dn_i(t)}{dt} = r_i n_i(t) \tag{4.13}$$

This model differs from the preceding model in that reproduction (binary fission) is possible at all times, whereas previously we assumed discrete generations with all the reproduction occurring at one specific time. In this model we can view r_i, an individual's growth rate, as a measure of its fitness. (Note that the appropriate measure of fitness depends on the choice of model.) If r_i is the fitness of an individual of type i, and if there are $n_i(t)$ individuals of type i at time t, the mean fitness at time t must be

$$\bar{r}(t) = \sum_i r_i \left(\frac{n_i(t)}{N(t)} \right) \tag{4.14}$$

where $N(t)$ the total number of individuals in the population, $N(t) = \sum n_i(t)$. There is variation in the r's because the individuals of different types have different r's. A measure of the amount of variation in the r's at time t is the variance of the r's, $\sigma_r^2(t)$

$$\sigma_r^2(t) = \sum_i [r_i - \bar{r}(t)]^2 \left(\frac{n_i(t)}{N(t)} \right) \tag{4.15}$$

Fisher's theorem relates the rate of increase in the mean fitness to the variance in the r's. Specifically, the theorem is that

$$\frac{d\bar{r}(t)}{dt} = \sigma_r^2(t) \tag{4.16}$$

For asexual populations, as assumed here, the proof is simple and is in the box. This theorem differs from the adaptive topography theorem in that the

---

Consider a population of phenotypes reproducing clonally according to exponential growth. Then for each phenotype, i, we have

$$\frac{dN_i}{dt} = r_i N_i \tag{1}$$

where r_i is the rate of increase for phenotype i, assumed to be a measure of fitness. Each r_i is assumed to be constant. For the population as a whole we have

$$\frac{dN}{dt} = \sum r_i N_i = N \sum \frac{N_i}{N} r_i = \bar{r} N \tag{2}$$

where the summation is over all phenotypes in the population. The average fitness or rate of increase $\bar{r}$ is defined as

$$\frac{\sum r_i N_i}{N} \tag{3}$$

The population variance in the rate of increase is defined as

$$\sigma_r^2 = \sum \frac{N_i}{N}(r_i - \bar{r})^2 = \sum \frac{N_i}{N}(r_i)^2 - \bar{r}^2 \tag{4}$$

Then the rate of change in the average fitness is found, by differentiating using the quotient rule, as

$$\frac{d\bar{r}}{dt} = \frac{d}{dt}\left(\sum \frac{N_i r_i}{N}\right) = \frac{N \sum r_i (dN_i/dt) - (\sum N_i r_i)(dN/dt)}{N^2} \tag{5}$$

This can be simplified by substituting the definitions above:

$$\frac{d\bar{r}}{dt} = \frac{\sum r_i(r_i N_i)}{N} - \frac{(\sum N_i r_i)}{N} \cdot \frac{(\bar{r}N)}{N} = \sum \frac{N_i}{N}(r_i)^2 - \bar{r}^2 \tag{6}$$

Finally, using the definition of the variance in fitness (4) we have

$$\frac{d\bar{r}}{dt} = \sigma_r^2 \tag{7}$$

Thus the rate of increase of the average fitness in the population is equal to the population variance in fitness.

---

speed with which $\bar{r}$ increases is related to the amount of variation present in the population. An important metaphor for evolution is that selection operates on the variation within a population to improve the average fitness of the members of a population. We might expect that the rate at which selection can do this is related to the amount of variation that is present.

Fisher's theorem makes this relationship explicit and it concisely encapsulates the traditional view of the role of natural selection in evolution. In this sense it is the fundamental theorem of natural selection.

The extension of this theorem to diploid random mating populations requires the concepts introduced in Chapter 9.

PROBLEMS

4.1. Find the value of p that maximizes $\bar{w}$.

(a) Differentiate $\bar{w}$ with respect to p.
(b) Set $d\bar{w}/dp = 0$ and solve for p.
(c) Compare your answer with that which we obtained in the last chapter.

4.2. Seven individual bacteria are sampled from a bacterial infusion containing many bacterial phenotypes. By cloning the individuals who were sampled under the same culture conditions as the infusion, the following growth rates were found:

$$.3, .8, .6, .5, .5, .8, .4$$

What is your estimate, based on this sample, for the rate of change of the average fitness, $\bar{r}$, in the infusion at the time the sample was drawn?

Chapter 5
GENETIC DRIFT

$\mathbf{E}$VOLUTION is an affair of chance. Some gametes by chance are drawn from the gamete pool and incorporated into zygotes while others are washed to sea; some types of individuals may, on the average, leave more offspring than others but what actually happens depends in part on chance. Chance is an essential and natural part of the process of evolution, and this section is devoted to the business of chance.

The theory we have developed so far is what is called a *deterministic* theory. According to this theory, if we know the gene frequency at time t and the relative selective values, then we predict one specific value for the gene frequency at time $t+1$. In contrast, in a *stochastic theory* we would *not* predict one specific value for p, but instead would predict the probability that p is one of several possible values. More generally, in a stochastic theory we assume that it is possible for the system under study to be in many states, and we develop theory to predict the *probability* of the system's being in each possible state as a function of time.

Genetic drift is our entrée to a stochastic theory of evolution. There are two ways in which an element of chance may enter the evolutionary process. First, the chance effects may be of environmental origin. The w's, for example, might differ from generation to generation because the environment is not constant from year to year. Second, the chance effects may be internal to the population in the sense that they would still occur even in a perfectly constant environment. "Genetic drift" originally referred to all chance effects in evolution but has come to refer to a special source that is internal to the population. This internal source is "sampling error" by gametes from the zygote pool. This chapter concerns genetic drift in this narrow sense while Chapter 13 will treat the issue of evolution in fluctuating environments.

What Is Sampling Error? Let us begin with a simple graphic example of what sampling error means. Suppose that the adult population has a gene frequency, p, of $\frac{1}{2}$. Assume that these adults produce a large gamete pool and that zygotes are formed by random union of gametes. Suppose, moreover, that a population size of *only two zygotes* is to be formed from the gametes. Thus we are discussing a very small population and *only four gametes* are to be drawn at random from the pool. When these four gametes are drawn, it is possible that exactly two A gametes and two a gametes will be selected, but it is also possible, indeed fairly probable, that three A's and one a, or one A and three a's will be drawn instead. Suppose then that three A's and one a are, in fact, drawn. Then the gene frequency in the zygote pool is $\frac{3}{4}$. Thus p has changed from $\frac{1}{2}$ among the adults to $\frac{3}{4}$ among the zygotes without the intervention of selection or mutation but solely as an accident of chance. This phenomenon is called sampling error. In this example, if p is $\frac{1}{2}$ among the zygotes, then an accurate sample has been drawn while if p differs from $\frac{1}{2}$, then sampling error has occurred. The expected severity of the sampling error, that is, the average extent to which p in the zygotes differs from p in the gamete pool, depends on the population size. Intuitively, drawing 1000 gametes instead of only 4 will yield a p quite close to $\frac{1}{2}$. A larger population size ensures that the zygotes will be a more accurate representation of the gamete pool. Thus sampling error adds an element of chance to population genetics, in a way that becomes more important as the population size decreases. Genetic drift is the name for changes in gene frequencies caused by this sampling error.

One may think of sampling error in the terminology of communication theory. The gene frequency, p, is a "signal" that must be transmitted from one generation's gene pool to the next. The effect of sampling error is to introduce "noise" into the communication channel. Because of this noise, the signal that is received fluctuates from generation to generation.

The cumulative effect of sampling error may have important consequences for a population. Figure 5.1 illustrates some computer simulations of populations. In each generation, using a random number generator, gametes were drawn at random and fused to make zygotes. The results in Figure 5.1 show that eventually all the populations became fixed for one of the alleles. In the absence of mutation, $p = 0$ and $p = 1$ are said to be absorbing barriers because the gene pool then consists entirely of one allele or the other. If, by chance, a population should "wander" into either of these barriers, it cannot escape thereafter. Intuitively, given enough time all populations should eventually end up at $p = 0$ or 1. Thus we see that the effect of genetic drift is the eventual loss of any genetic variation initially present. Indeed, understanding if and how genetic variation is maintained in small populations poses a difficult problem.

Now that we understand intuitively what sampling error is and what its cumulative effect is, we can proceed to discuss drift more quantitatively. We want to know how "strong" a force genetic drift is, how fast it changes gene frequencies, and how to tell if genetic drift rather than some other force is the cause of an observed change of gene frequencies in nature.

Because any specific population shows a random amount of sampling error, we cannot hope to predict exactly what any given population will do. But we can combine the results of several populations and observe the properties of a group of populations. Figure 5.2 is a compilation of information on the trajectories of 50 populations including those seen before in Figure 5.1. There is a histogram of the fraction of populations with different gene frequencies through time. All the populations begin with $p = \frac{1}{2}$. If $N = 1$ the histogram rapidly becomes U-shaped as the populations fix at either $p = 0$ or 1. If $N = 8$, the histogram also eventually becomes U-shaped but changes much more slowly. The difference in speed reflects the fact that genetic drift is more important in smaller populations. There clearly appears to be a regular pattern when a group of populations is considered. Both groups eventually assumed U-shape distributions, and the rate at which this U-shape is attained can be interpreted as a measure of the "strength" of genetic drift. Perhaps we can develop a theory to predict the properties of a *group* of populations showing genetic drift even though we cannot predict exactly what any one population will do.

Theory of Genetic Drift for $N = 1$

To illustrate the equations for a group of populations, consider the special case of a *constant population size of only one individual*. Let us label the state of the gene pool (censused at the gametic phase) in terms of the number of A alleles. Thus the state of the gene pool is either 0, 1, or 2. Let $\rho(0)$ be the fraction of populations in state 0, and similarly for $\rho(1)$ and $\rho(2)$. Remember that states 0 and 2 are absorbing states. Then the equation for $\rho_{t+1}(0)$ is as

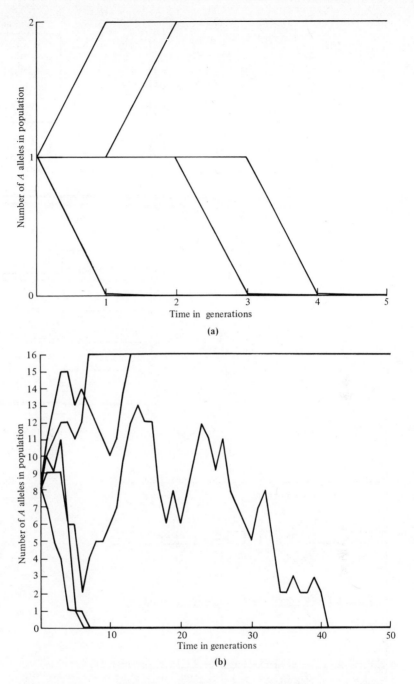

FIGURE 5.1. *Computer simulations of genetic drift. There are two alleles, A and a, in each population. The number of A alleles in each population is plotted on the vertical axis. Time in generations is plotted on the horizontal axis. All populations are started with an initial gene frequency of $\frac{1}{2}$. There are five replicate populations, each of which has a population size of one individual, and five replicate populations, each of which has a population size of eight individuals. Note that fixation has occurred after 4 generations for the runs with a population size of one individual and after 41 generations for the runs with a population size of eight individuals. (a) Population size = 1 diploid individual, 5 runs; (b) population size = 8 diploid individuals, 5 runs.*

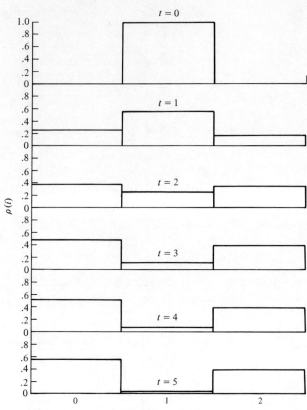

Number of A alleles in population, i

(a)

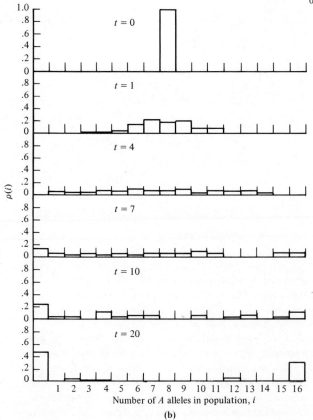

Number of A alleles in population, i

(b)

follows:

$$\rho_{t+1}(0) = 1\rho_t(0) + (\tfrac{1}{4})\rho_t(1) + 0\rho_t(2) \tag{5.1a}$$

Let us consider this formula term by term. First, all populations in state 0 at time t must remain there, so $\rho_{t+1}(0)$ includes all of those who were in state 0 at time t. Second, $\tfrac{1}{4}$ of the populations in state 1 at time t will move to state 0 at time $t+1$. To see this, remember that, for populations in state 1, generation $t+1$ is formed by drawing two gametes from a gamete pool in which the gene frequency is $\tfrac{1}{2}$. So the probability of drawing two A gametes is $\tfrac{1}{4}$, of drawing an A gamete and an a gamete is $\tfrac{1}{2}$ (i.e., $2pq$), and of drawing two a gametes is $\tfrac{1}{4}$. Therefore, $\tfrac{1}{4}$ of the populations in state 1 will move to state 0, $\tfrac{1}{2}$ will stay in state 1, and $\tfrac{1}{4}$ will move to state 2. The second term in our formula represents the fraction that enter state 0 from state 1. Third, since state 2 is an absorbing state, nothing can leave and hence there is a zero contribution from state 2 to state 0. The next two equations follow from the same kind of reasoning:

$$\rho_{t+1}(1) = 0\rho_t(0) + (\tfrac{1}{2})\rho_t(1) + 0\rho_t(2) \tag{5.1b}$$

$$\rho_{t+1}(2) = 0\rho_t(0) + (\tfrac{1}{4})\rho_t(1) + 1\rho_t(2) \tag{5.1c}$$

We can now iterate these equations and predict the fraction of the populations in each state through time. In particular, if all the populations start in state 1,

$$[\rho_0(0) = 0, \rho_0(1) = 1, \rho_0(2) = 0]$$

Then iterating the equations leads to the curves depicted in Figure 5.3. These theoretical predictions for the fractions in each state through time differ slightly from that observed in Figure 5.2. But Figure 5.2 does not involve a very large group of populations. In the limit of infinitely many populations, each with the same finite number of individuals in it, the theory would be exact.† Equations like (5.1) are called a *Markov chain* in mathematics.

Now that we have a theory to predict the properties of a group of populations, what use is it? One very important use is to test whether gene frequency changes in natural or experimental populations could have been caused by genetic drift. Consider three examples. Suppose that one isolated population of one individual is observed to change from $p = \tfrac{1}{2}$ to $p = 1$, in one generation. Can genetic drift be the cause? Perhaps, for as we have seen, the probability is $\tfrac{1}{2}$ of being fixed for either A or a in each generation, and thus one out of two populations should change this much in one generation. Now

---

† The infinite group of populations in physics is called an *ensemble* and $\rho_t(i)$ is the ensemble distribution. In mathematics and statistics, $\rho_t(i)$ is called a probability density function. This theory is a stochastic theory in the sense that what is predicted, $\rho_t(i)$, is the *probability* of the systems being in state i as a function of time.

[OPPOSITE] **FIGURE 5.2.** *Compilation of* 100 *computer runs including those illustrated in the preceding figure. There are* 50 *runs with a population size of one, and* 50 *runs with a population size of eight. All populations were started at a state corresponding to* $p = \tfrac{1}{2}$. *Note that the histograms gradually approach a* U-*shaped distribution with a speed that depends upon the population size.* (a) *Population size* = 1, *ensemble of* 50 *populations;* (b) *population size* = 8, *ensemble of* 50 *populations.*

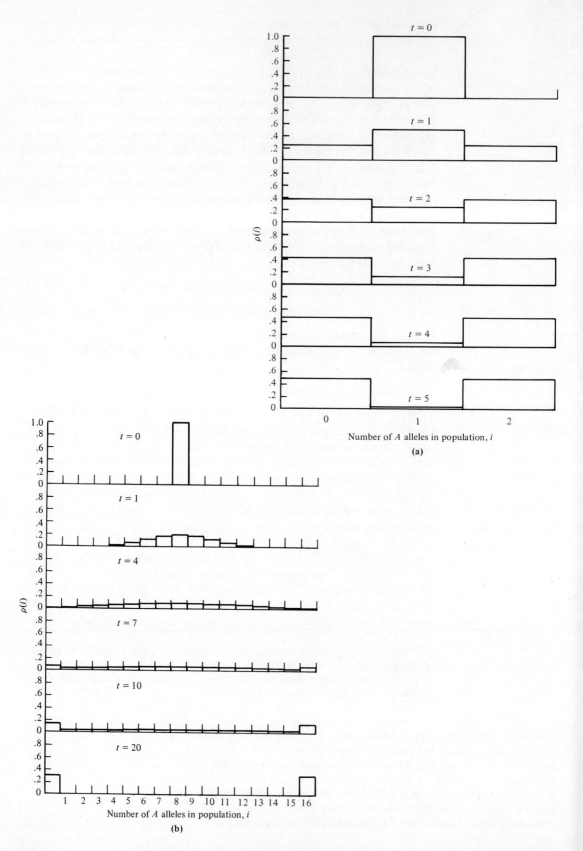

suppose that 12 isolated populations of one individual each change from $p = \frac{1}{2}$ to $p = 1$ in one generation. Again, can drift be the cause? Very unlikely. If only drift were occurring, the theory predicts the following results: three populations with $p = 0$, six populations with $p = \frac{1}{2}$, and three populations with $p = 1$. The χ^2 statistic for this discrepancy between what was observed and what is expected, assuming only drift is occurring, is 36.† With 2 degrees of freedom, the probability of a deviation this large is far less than .001. Hence some cause in addition to drift should be sought, and of course natural selection is a prime candidate. On the other hand, suppose that a population of only one individual persists with $p = \frac{1}{2}$ for seven generations. Is this consistent with drift? Again, this observation is very improbable if drift alone is occurring. The probability of a population of one individual remaining unchanged at $p = \frac{1}{2}$ for seven generations is $(\frac{1}{2})^7 = .0078$, which is less than .01. Hence some other force must be sought. With a little practice and imagination one can infer a great deal about what should and should not happen with genetic drift, using the stochastic theory we have just developed.

An Empirical Example   A clear empirical illustration of the action of genetic drift is provided in experiments of Kerr and Wright (1954) with *Drosophila melanogaster*. They established 96 populations each with four males and four females. Each new generation was produced by randomly choosing four individuals of each sex from the preceding generation. There are two alleles segregating, the wild type allele and a mutant allele. The mutant is the sex-linked recessive gene called "forked," which produces forked bristles. It was chosen because it is only slightly deleterious. Therefore, selection will not obscure the tendency for drift to cause fixation of either allele, depending on chance. Table 5.1 and Figure 5.4 present the data. Notice that the qualitative features of the experiment are exactly what is expected according to our discussion so far. The distribution of states for the populations is initially unimodal and gradually becomes U-shaped as various populations become fixed for one of the alleles. Note that there is a slightly higher tendency for the wild-type allele to fix more often than the mutant allele. This result suggests that there is some slight selection pressure against the mutant. Kerr and Wright compared these data with the detailed predictions from the mathematical theory in genetic drift. They found that the rate at which the populations in the experiment were becoming fixed was very slightly faster than that predicted from the theory. They interpreted this small discrepancy by suggesting that, on the average, one of the eight flies failed to reproduce. If so, the effective population size would be slightly lower and the loss of polymorphism would proceed correspondingly faster.

The topic of genetic drift involves some of the most advanced mathematics in theoretical biology, notwithstanding its humble origin in the sampling of gametes. (The remainder of this chapter is an introduction to this more advanced mathematics and may be skipped if necessary.) To introduce these more advanced methods, we shall first discuss the Markov chain for drift in

† $\chi^2 = [(12-3)^2/3] + [(0-6)^2/6] + [(0-3)^2/3] = 36$. The degrees of freedom are $3 - 1 = 2$.

[OPPOSITE] **FIGURE 5.3.** *Theoretical ensemble distributions for populations whose gene frequencies are changing due to genetic drift.* (a) *Population size* = 1; (b) *population size* = 8.

Table 5.1. Experiment Using 96 Populations of *Drosophila melanogaster* Each with 8 Individuals. From Kerr and Wright (1954).

Generation	Fraction of Populations Fixed for + Allele	Fraction of Populations Polymorphic	Fraction of Populations Fixed for "Forked" Allele
0	0	1.000	0
1	.010	.979	.010
2	.010	.958	.031
3	.021	.906	.073
4	.073	.823	.104
5	.104	.729	.167
6	.115	.688	.198
7	.167	.615	.219
8	.177	.583	.240
9	.208	.542	.250
10	.250	.490	.260
11	.302	.406	.292
12	.323	.385	.292
13	.354	.354	.292
14	.385	.313	.302
15	.396	.302	.302
16	.427	.271	.302

populations larger than only one individual and point out difficulties in solving this problem. Then we shall consider an equation called a diffusion equation, which approximates the Markov chain. This equation will also allow us to assess the relative strength of natural selection and genetic drift.

Theory of Genetic Drift for Arbitrary N

The equations for ρ_{t+1}, given ρ_t, which were derived previously, are easily generalized to larger population sizes. If we rewrite the equations [(5.1a), (5.1b), and (5.1c)] in terms of matrix multiplication, we obtain

$$[\rho_{t+1}(0), \rho_{t+1}(1), \rho_{t+1}(2)] = [\rho_t(0), \rho_t(1), \rho_t(2)]\begin{pmatrix} 1 & 0 & 0 \\ \frac{1}{4} & \frac{1}{2} & \frac{1}{4} \\ 0 & 0 & 1 \end{pmatrix} \quad (5.2)$$

This is simply a shorthand way of writing the three equations, using the notation of matrix multiplication; there is no new biology involved. If you are not familiar with this notation, you should refer now to Appendix III, where it is explained. From now on we let ρ_t denote a row vector whose elements are $\rho_t(i)$. Also we let P stand for the matrix. With these symbols we can write Equation (5.2) in a very compact form

$$\rho_{t+1} = \rho_t P \quad (5.3)$$

Now with this notation it becomes very easy to generalize the equations to any population size. If the population size is N, there are $2N$ genes total at a given locus. Then ρ_t contains $2N + 1$ elements and P is a $(2N + 1) \times (2N + 1)$ square matrix. Let us focus our attention on the P matrix.

Here P is called the matrix of transition probabilities; P_{ij}, that is, the element in the ith row and the jth column, is the probability that a population

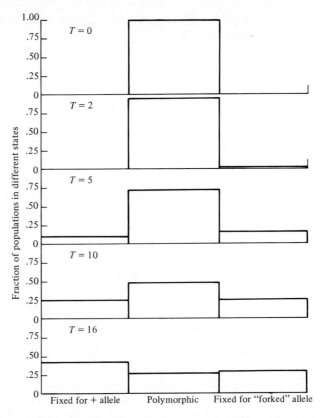

FIGURE 5.4. *Experimental demonstration of genetic drift in an ensemble of 96 laboratory populations of* Drosophila melanogaster. *Population size =* 8. [Data from W. Kerr and S. Wright (1954), Experimental studies of the distribution of gene frequencies in very small populations of *Drosophila melanogaster*: I. Forked, *Evolution* **8**: 172–177.]

in state i at time t moves to state j at time $t + 1$. In our simple example, where $N = 1$, we have the following transition probabilities:

$$
\begin{array}{cc}
 & \text{state at } t+1 \\
 & \begin{array}{ccc} 0 & 1 & 2 \end{array}
\end{array}
$$

$$
\text{state at time } t \;\; \begin{array}{c} 0 \\ 1 \\ 2 \end{array} \left|
\begin{array}{ccc}
1 & 0 & 0 \\
\frac{1}{4} & \frac{1}{2} & \frac{1}{4} \\
0 & 0 & 1
\end{array}
\right. \tag{5.4}
$$

You should verify that the entries really are the probabilities of a population moving from one state to another as we discussed before. Now, to generalize to large population sizes we must find the appropriate entries for larger transition matrices. The general formula for the entries is

$$
P_{ij} = \binom{2N}{j}\left(\frac{i}{2N}\right)^{j}\left(1 - \frac{i}{2N}\right)^{2N-j} \tag{5.5}
$$

This formula is simply the jth term in the binomial expansion of $(p + q)^{2N}$

where $p = \frac{i}{2N}$ and $q = 1 - (\frac{i}{2N})$. The symbol $\binom{2N}{j}$ is short-hand for the binomial coefficient

$$\binom{2N}{j} \equiv \frac{(2N)!}{(2N-j)!\,j!} \tag{5.6}$$

You should verify that $N = 1$ leads to the matrix (5.4). With this formula we can produce the transition matrix for any population size. Then, given any initial distribution of population states, ρ_0, we can simply iterate (5.3) to obtain the distribution at future times.

There are two important properties to this scheme of transition probabilities. (1) Suppose that the numbers of A alleles in a population at times t and $t+1$ are denoted as i_t and i_{t+1}. Then what is the average value of i_{t+1}, given i_t? The answer is that the average i_{t+1}, conditional on i_t, equals i_t. That is, there is no net tendency for drift to move the population up or down from its current value. The tendency for i to increase is balanced by that to decrease, leaving the average unchanged. (2) Next, what is the variance of i_{t+1}, given i_t? The variance of i_{t+1}, given i_t, is $2Np_t q_t$, where $p_t = \frac{i_t}{2N}$ and $q_t = 1 - \frac{i_t}{2N}$. Equivalently, if we write $p_{t+1} = i_{t+1}/(2N)$, then the variance of p_{t+1} given p_t is $\frac{p_t q_t}{2N}$. These properties are very important and will be used again later.

We could explore the equations for drift by iterating the equation for ρ_{t+1} on a computer. With this model, however, we can obtain some but not many, results analytically. The main question we want to investigate is how the population size influences the speed at which genetic variability is lost by drift. This question is answered by examining the eigenvalues of the transition matrix. What you need to know is that it is possible to express any arbitrary ρ_t as follows

$$\rho_t = c_1 \lambda_1^t \hat{\rho}^{(1)} + c_2 \lambda_2^t \hat{\rho}^{(2)} + \cdots c_{2N+1} \lambda_{2N+1}^t \hat{\rho}^{(2N+1)} \tag{5.7}$$

where λ_i is the ith eigenvalue of P (ranked from the largest to the smallest) and $\hat{\rho}^{(i)}$ is the eigenvector corresponding to λ_i. The eigenvalues and eigenvectors are quantities that can be calculated from P. What is important here is that all transition matrices have as their largest eigenvalue $\lambda_1 = 1$ and the remaining eigenvalues have a magnitude between 0 and 1. Some of the remaining eigenvalues may also equal 1, but none are larger than one. The consequence of this fact is that as time increases λ^t tends to zero for those λ which are less than one. Therefore, for t sufficiently large, ρ_t is determined only by the first two or three terms in (5.7); the remaining terms become negligible. For this reason, the behavior of the distribution as time becomes large is determined by the magnitude of the several top eigenvalues. Hence there is considerable biological interest in the eigenvalues of the transition matrix.

The eigenvalues of the transition matrix have been determined in general by Feller. For more information, see Ewens (1969, pp. 28–32). The first three are $\lambda_1 = 1$, $\lambda_2 = 1$, and $\lambda_3 = 1 - 1/(2N)$. The first two eigenvalues correspond to the absorption states where one of the alleles is fixed. The interesting eigenvalue is

$$\lambda_3 = 1 - \frac{1}{2N} \tag{5.8}$$

This is the eigenvalue that principally controls the rate at which genetic variability is lost. If λ_3 is near one, then the variability is lost very slowly because $\lambda_3^{\,t}$ tends to zero slowly with t. If λ_3 is near zero, then the variability is lost rapidly because $\lambda_3^{\,t}$ tends to zero rapidly with t. Now observe that λ_3 varies from $\frac{1}{2}$ to 1 as N varies from 1 to infinity. This expression provides the connection between population size and the loss of genetic variability. It is graphed in Figure 5.5. Note that with a population size of 100 to 1000, drift causes only a slight loss of variability each generation. The quantity $1-1/(2N)$ arises frequently in population genetics, and we shall meet it again in the study of inbreeding.

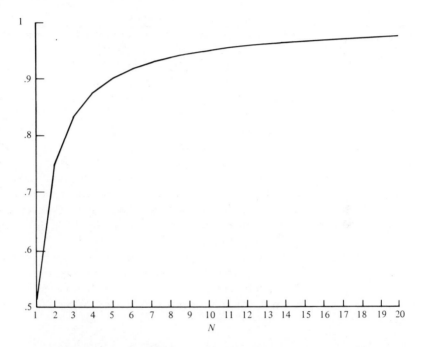

FIGURE 5.5. *The dominant eigenvalue of the transition matrix for genetic drift as a function of population size.*

Some Refinements The paragraphs above present the generalization of the traditional drift model to population sizes of N individuals. But still, much realism is omitted. Specifically, virtually all populations of interest have two sexes and show population sizes that vary in time. Moreover, as we saw in the example of drift with *Drosophila* populations, certain genes are sex-linked. Again, we want to know the connection between population size and the loss of genetic variability in these cases.

In each of these cases there is a quantity, N_e, that controls the rate of loss of variability in a manner analogous to the population size in the simple model discussed. Specifically, in these cases involving more realism, the interesting eigenvalue is given by

$$\lambda = 1 - \frac{1}{2N_e} \tag{5.9}$$

where N_e is a quantity that must be calculated by choosing a formula which depends on the assumptions. With two sexes N_e is given by

$$N_e = \frac{4N_\female N_\male}{N_\female + N_\male} \qquad (5.10)$$

With sex-linked loci, assuming that males are XY and females are XX, N_e is

$$N_e = \frac{9N_\female N_\male}{4N_\female + 2N_\male} \qquad (5.11)$$

where $N_\male$ is the number of males and $N_\female$ the number of females. The total population size is $N = N_\female + N_\male$. Also, N_e *is called the effective population size relative to genetic drift* because of its analogy to the role of the regular population size, N, in the elementary theory of genetic drift. If $N_\female = N_\male$ in (5.10), then N_e in fact equals N. And if $N_\female = N_\male$ in (5.11), then $N_e = \frac{3}{4}N$.

The formula for N_e with a sex-linked locus is relevant to the rate of loss of polymorphism in the *Drosophila* experiments quoted previously. In those experiments the loss should proceed at the rate that occurs in a population of $\frac{3}{4} \cdot 8 = 6$ individuals, where the loci are not sex-linked. In fact, N_e was determined to be 5.6; the slight discrepancy could easily be explained by one individual's occasionally failing to mate.

Equation (5.10), for two sexes at an autosomal locus, is of interest if the sex ratio is not near 50-50. If the sex ratio greatly favors one sex, say females, then the effective population size is closer to the number of the rarer sex, the males. Thus the simple theory above is qualitatively correct even with these more complicated assumptions, provided that the effective population size is used in place of the actual population size.

There is also an expression for N_e if the population size varies in a cyclic fashion. Suppose that the population size assumes the values $N_1, N_2, N_3 \cdots N_k$ in consecutive generations and then repeats the cycle over and over. Then it can be shown that N_e is approximately the harmonic mean of the N_i's. Specifically,

$$N_e = \frac{1}{(1/N_1) + (1/N_2) + (1/N_3) + \cdots (1/N_k)} \qquad (5.12)$$

Once again the effective size is determined by the smallest size that occurs during a cycle. A point of low population abundance is called a "bottle neck." This formula shows that these occasional periods of low abundance can have a large influence on the effective population size relative to drift.

The formulas for N_e have been quoted from Ewens (1969, pp. 32–36) and their derivations occur there. Still more formulas for N_e under other conditions are found in Crow and Kimura (1970), Crow (1954), and Karlin (1968a).

Genetic Drift with Natural Selection and Mutation   Our next question concerns the joint action of genetic drift with both natural selection and mutation. We want to know the comparative strengths of these forces. How small must the population size be for drift to "overpower" a given selection pressure, and conversely, how strong must the selection pressure be to "overpower" drift in a population with a given size? To answer these and other questions without resorting to computer simulation, we must develop some additional machinery. We shall introduce an equation, called a "diffusion equation," which approximates the Markov chain discussed above, and which will also allow a convenient introduction of selection and mutation into the model as well.

Your first formal encounter with the process of diffusion was probably in chemistry. You have seen how a drop of colored dye in a glass of water gradually spreads out and eventually imparts a uniform color throughout the water. Diffusion works because of the random motion, called Brownian motion, of the dye particles. Because of this random movement of the particles, the drop of dye spreads out. We now view a population as analogous to a particle, and its gene frequency as analogous to position. Then if a bunch of populations are placed at some initial p, we shall envision them as diffusing away from that point. See Figure 5.6. The populations spread out from the initial p in a way analogous to dye particles spreading away from their initial location. As we shall see, the use of the machinery originally developed to study the physical process of diffusion will allow us to obtain important biological conclusions about evolution. Moreover, this machinery is widely used throughout biology (and other fields) to study stochastic processes. We shall meet it again in the chapter on spatially varying selection pressures and in the chapter on population dynamics in stochastic environments.

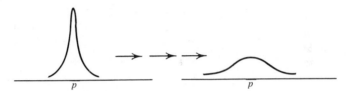

FIGURE 5.6. *Sketch of the result of diffusion.*

The next several paragraphs explain, in an intuitive way, the machinery used to describe any diffusion process. Then we shall return to the specific question of genetic drift and natural selection. There are two steps involved in describing diffusion. The first relates the change in concentration at any location to the flow of particles in the vicinity of that location. The second relates the flow to the forces influencing the movement of particles.

What Is a Diffusion Equation? Let $\rho(x, t)$ denote the density of particles at location x at time t. Thus the actual amount of particles in the interval between x and $x + \Delta x$ is $\rho(x, t)\,\Delta x$. See Figure 5.7. Let the flow across the surface at $x + \Delta x$ be denoted as $J(x + \Delta x, t)$; similarly, the flow across the surface at x is $J(x, t)$. If J is

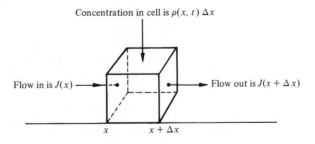

FIGURE 5.7. *Sketch of the setup in a transport equation.*

positive, the flow is to the right; if negative, then to the left. Now the idea is very simple—the change in the amount in the cell is simply the flow across x minus the flow across $x + \Delta x$, that is, $J(x, t) - J(x + \Delta x, t)$. Or in symbols we have

$$\frac{\partial}{\partial t}[\rho(x, t)\, \Delta x] = -[J(x + \Delta x, t) - J(x, t)] \tag{5.13}$$

Upon dividing both sides by Δx and then taking the limit as $\Delta x \to 0$, we have

$$\frac{\partial}{\partial t}\rho(x, t) = -\frac{\partial}{\partial x}J(x, t) \tag{5.14}$$

Thus the change in the density at location x is equal to the spatial derivative of the flow.† This result is really quite intuitive; the concentration at any place changes if the flows in the vicinity are not all the same. But if the flows in the vicinity of some location are all equal, then the input into that location equals the output and hence there is no net change in the amount.

Now let us consider the expression for the flow, $J(x)$. There are two causes for the movement of particles. First, the particles may be subjected to some force. For example, dye particles are usually quite large and are subject to the force of gravity. Or if the particles are charged, they could be subject to an electric field, and so forth. Second, the particles diffuse from regions of high concentration to regions of low concentration. This is due to the random motion of the particles as mentioned earlier. The expression for $J(x)$ contains two terms, one for the movement due to some external force and the other due to the diffusion,

$$J(x, t) = M(x)\rho(x, t) - \frac{1}{2}\frac{\partial}{\partial x}V(x)\rho(x, t) \tag{5.15}$$

The first term represents the movement due to some external forces. Consider a short time interval Δt. By definition, $M(x)\,\Delta t$ is the *average* distance traveled during the time interval Δt by a particle that started at x. Thus the average speed of a particle at x is $\frac{M(x)\,\Delta t}{\Delta t} = M(x)$. In chemistry some expression involving gravity or an electric field would be found for $M(x)$. Here, in population biology, $M(x)$ will involve mutation and natural selection. Usually, $M(x)$ is zero if these sorts of external forces are absent. The random diffusion does not *usually* contribute to $M(x)$ because the random motion occurs in both directions and usually cancels out when calculating the average distance traveled by a particle.

The diffusion is typically represented only in the second term. To illustrate the meaning of the diffusion term suppose, just briefly, that $V(x)$ is a constant and let $D = V/2$. In chemistry D would be called the diffusion constant and the second term would simply be proportional to $-\partial\rho/\partial x$. This term means that the flow due to diffusion depends on the presence of a gradient in the density of particles. If the density is the same everywhere, then $\partial\rho/\partial x$ is zero and there is no net flow attributable to diffusion. But if there is a concentration gradient, then the diffusion results in a net flow from regions of high concentration into regions of low concentration. For this reason the flow varies as $-\partial\rho/\partial x$. In chemistry the proportionality constant, D, depends on temperature and other factors, and it can usually be regarded as a constant, provided that the temperature is the same throughout the

† In physics this is called a continuity equation or a transport equation.

experimental apparatus. But, in general, the coefficient in the diffusion term is not a constant, and especially not in most biological applications.

The diffusion coefficient is related to the *variance* of the movements of the particles. Again consider a short time interval, Δt. By definition $V(x)\Delta t$ is the variance of the distances traveled during the time interval Δt by particles that started at x. Thus $V(x)$ is the variance of the distance traveled from x per unit time. Note that if $V(x)$ is a constant, then the diffusion constant, D, is seen to be proportional to the variance of the distances traveled by particles per unit of time regardless of their present position. If the particles are heated up, then their movement increases, thereby increasing the diffusion coefficient. So, to summarize, the diffusion term relates the flow, J, to the gradient in the concentration from place to place, and the coefficient in this relationship involves the variance of the distances traveled by particles from their initial positions during a small time interval.

We can now combine the formula for the flow, $J(x, t)$, with the formula for $\partial\rho/\partial t$ to obtain

$$
\begin{aligned}
\frac{\partial\rho(x, t)}{\partial t} &= -\frac{\partial}{\partial x}\left[M(x)\rho(x, t) - \frac{1}{2}\frac{\partial}{\partial x}V(x)\rho(x, t)\right] \\
&= -\frac{\partial}{\partial x}[M(x)\rho(x, t)] + \frac{1}{2}\frac{\partial^2}{\partial x^2}[V(x)\rho(x, t)]
\end{aligned}
\tag{5.16}
$$

This equation is called a diffusion equation. It predicts the density of particles at any place through time.† To use it, we need to specify how the external forces enter into $M(x)$ and how the random component in the movement of the particles enters into $V(x)$. Then the equation can be used to predict $\rho(x, t)$ through time, given an initial distribution, $\rho(x, 0)$, and the boundary conditions.

The Equilibrium Distribution of a Diffusion Process   In many applications, $\rho(x, t)$ tends to an equilibrium distribution, $\hat\rho(x)$. When there is an equilibrium distribution, its formula is given in terms of $M(x)$ and $V(x)$ by

$$
\hat\rho(x) = \frac{c}{V(x)}\exp\left\{2\int\left[\frac{M(x)}{V(x)}\right]dx\right\}
\tag{5.17}
$$

where c is the normalizing constant to ensure that the area under $\hat\rho(x)$ equals 1. The integral is simply the indefinite integral; in most cases the constant of integration is absorbed into the normalization constant, c. [The derivation of (5.17) from (5.16) is straightforward and is set as a self-guiding problem at the end of the chapter.] Thus, in practice, the use of the diffusion equation may be quite simple. If we have determined $M(x)$ and $V(x)$ and if we know there is an equilibrium distribution, $\hat\rho(x)$, all that must be done is to substitute for $M(x)$ and $V(x)$ in (5.17) to obtain the resulting equilibrium distribution.‡

† Equation (5.16) is also called the Chapman–Kolmogorov equation in mathematics, and the Fokker–Planck equation in physics. It is derived in hundreds of books from many points of view and degrees of rigor. For a physical interpretation see Morse and Feshback (1953), and for a probabilistic interpretation see Cox and Miller (1968).

‡ Diffusion theory may become much more complex if more than one dimension is involved, for example, selection with three or more alleles combined with mutation and drift. Also, it may be difficult to establish the behavior near boundaries in the system. In addition to the solution for $\hat\rho(x)$, diffusion theory provides machinery for studying the problem of extinction, including the expected time required for the loss of an allele by drift.

*The Diffusion
Approximation for
Genetic Drift
Combined with
Selection and
Mutation*

Now let us return to the study of selection and mutation combined with genetic drift. In population genetics the basic time interval is one generation, and in a finite population, the gene frequency, p, varies in discrete steps of $1/(2N)$. But diffusion theory is in continuous time and the gene frequency of a population—which is analogous to the position of a particle—is a continuous variable. Nonetheless, predictions from the diffusion equation will be very close to those of the exact treatment in discrete time if two conditions are met. First we measure time in units of $2N$ generations. Thus one generation becomes only $1/(2N)$th unit of time. For example, if N is 100 then one generation is $(\frac{1}{200}) = .005$ units of time. Moreover, if N is large then, although p varies in steps, the steps are very small. Therefore, in the limit as N becomes large the time and gene frequency steps become small and can be approximated with continuous variables. The second requirement is that the selection and mutation coefficients be "small" in the following sense: Let s be the selection coefficient against some genotype. Then we require that the product sN remain finite in the limit as $N \to \infty$. This requirement is satisfied if s is of the order $1/N$ or less because then the product, sN, will remain finite as $N \to \infty$. The requirement ensures that a population cannot "move too far" during an infinitesimal time increment. To summarize, the diffusion equation should approximate the Markov chain, provided that time is measured in units of $2N$ generations, N is large, and the selection and mutation parameters are sufficiently weak.

To use the diffusion formulas, we must find the appropriate expressions for $M(p)$ and $V(p)$. Consider first how mutation and selection can be combined to give an expression for $M(p)$. Let u be the mutation rate of A to a and v for a to A. Then Δp due to mutation is

$$\Delta p_{\text{mut}} = vq - up$$
$$= v - (u+v)p \tag{5.18}$$

Our familiar formula for Δp due to selection is

$$\Delta p_{\text{sel}} = \frac{pq[p(w_{AA} - w_{Aa}) + q(w_{Aa} - w_{aa})]}{\bar{w}} \tag{5.19}$$

For simplicity we shall assume that the total change in gene frequency over one generation is $\Delta p_{\text{mut}} + \Delta p_{\text{sel}}$. This is not exact because mutation is usually assumed to act after the selection within any generation, but with weak selection the error in this approximation is negligible. Next we shall assume that the selection is moderately weak so that $\bar{w}$ is effectively equal to 1. With these assumptions we have

$$\Delta p = v - (u+v)p + p(1-p)[p(w_{AA} - w_{Aa}) + q(w_{Aa} - w_{aa})] \tag{5.20}$$

This formula for Δp is based on the deterministic theory of the preceding chapters. How do we know that it represents the average Δp, given p_t, in the stochastic model with drift? Recall that if only drift is present, then the average p_{t+1}, given p_t, equals p_t. That is, drift produces no *net* tendency to move up or down from the current p_t. Hence any net change to p must come from the deterministic forces of mutation and selection. For this reason we are safe, in this case, in taking the Δp from deterministic theory as representing the average Δp starting from p in the stochastic theory. The final step is to convert to a time scale where $2N$ generations equals one unit of time. To convert Δp to the new time units, we scale as follows:

$$M(p) = \frac{\Delta p}{(1 \text{ generation})} \frac{(2N \text{ generations})}{1 \text{ time unit}}$$

Therefore, $M(p)$ is given by

$$M(p) = \frac{\left[\begin{array}{c} \text{average } \Delta p \\ \text{starting from } p \end{array}\right]}{\text{time unit}} = 2N\{v - (u + v)p + p(1 - p)[p(w_{AA} - w_{Aa})$$

$$+ q(w_{Aa} - w_{aa})]\}$$

$$= 2Nv - [2Nv + 2Nu]p + p(1 - p)[p2N(w_{AA} - w_{Aa})$$

$$+ q2N(w_{Aa} - w_{aa})] \tag{5.21}$$

This formula gives $M(p)$ in terms of mutation and selection. One can notice here why it is important that the selection and mutation coefficients are small. In the limit as $N \to \infty$, $M(p)$ blows up unless Nu, Nv, $N(w_{AA} - w_{Aa})$, and $N(w_{Aa} - w_{aa})$ remain finite. This fact imposes the requirement that the mutation and selection coefficients be small.

The variance, $V(p)$, is the variance of Δp per unit time for a population starting at p. Recall from our earlier discussion of the binomial transition probabilities that the variance of p_{t+1} starting from p is $p(1 - p)/(2N)$. This expression is also the variance of Δp over one generation for a population starting from p. To understand this, recall that the *average* p_{t+1} starting from p_t is simply p_t. Drift causes no net tendency to move up or down. By definition the variance of p_{t+1} is $E(p_{t+1} - \bar{p}_{t+1})^2$.[†] Then this is the same as $E(p_{t+1} - p_t)^2$, which in turn is the variance of Δp conditional on starting at p. Therefore, the variance of Δp over one generation for a population starting at p is $p(1 - p)/(2N)$. This formula for the variance of Δp is based on the binomial transition probabilities in the absence of mutation and drift. Again, if we assume moderately weak selection, then the difference between the variance predicted by this formula and the true variance becomes negligible. Finally, we convert to the appropriate time units as follows.

$$V(p) = \left(\frac{\text{variance of } \Delta p}{1 \text{ generation}}\right)\left(\frac{2N \text{ generations}}{1 \text{ unit time}}\right)$$

Therefore, $V(p)$ is given by

$$V(p) = \frac{\left[\begin{array}{c} \text{variance of } \Delta p \\ \text{starting from } p \end{array}\right]}{\text{time unit}} = 2N\left\{\frac{p(1 - p)}{2N}\right\} = p(1 - p) \tag{5.22}$$

Thus $V(p)$ is certainly not a constant. The diffusion coefficient varies with p and is highest at intermediate values of p and lowest near the boundaries of 0 and 1.

---

† The symbol $E(x)$ means that the average value of x is to be calculated.

The General
Equilibrium
Distribution of
Gene Frequency
with Drift,
Selection, and
Mutation

When genetic drift is combined with recurrent mutation, then the boundaries of 0 and 1 are not necessarily absorbing states. Mutation provides the opportunity for a population that is fixed for an allele to escape from this state and return to the polymorphic states. In the presence of mutation, the distribution, $\rho(p, t)$, is expected to approach an equilibrium distribution that represents a balance between mutation generating genetic variation and drift eliminating it. Moreover, in the presence of selection, the equilibrium distribution is shaped by the selection coefficients as well. To find this equilibrium distribution, we have to substitute the expressions for $M(p)$ and $V(p)$ into the formula for $\hat{\rho}(p)$. The computational details are left as a self-guiding problem at the end of the chapter. The result, assuming the w's are constants, is

$$\hat{\rho}(p) = c p^{4Nv-1} (1-p)^{4Nu-1}$$

$$\times \exp\{2N[w_{AA} + w_{aa} - 2w_{Aa}]\,p^2 + 4N[w_{Aa} - w_{aa}]p\} \quad (5.23)$$

where c is the normalization constant which, except in special cases, must be determined by numerically integrating (5.23) between 0 and 1 and then dividing by the result in order to obtain a distribution with unit area. The shape of $\hat{\rho}(p)$ is the topic of interest and it is determined, except for the constant multiplier, by (5.23). We now discuss how the shape of $\hat{\rho}(p)$ depends on the relative strengths of mutation, selection, and drift.†

*Special Cases
of the Equilibrium
Distribution*

First, let us examine the joint action of genetic drift and mutation in the absence of selection. With no selection the w's are equal and $\hat{\rho}(p)$ reduces to

$$\hat{\rho}(p) = c p^{4Nv-1} (1-p)^{4Nu-1} \quad (5.24)$$

This distribution can have several shapes.‡ (See Figure 5.8.) If both $4Nv \ll 1$ and $4Nu \ll 1$, then $\hat{\rho}(p)$ is approximately $1/[p(1-p)]$. In this case the distribution is U-shaped. This is the shape we have come to expect with genetic drift. Most populations are fixed for one or the other allele. In this case we could say that genetic drift has "overpowered" mutation. In contrast, if $4Nv \gg 1$ and $4Nu \gg 1$ the distribution has a mode at an intermediate gene frequency. Indeed, it can be shown (from the tabulated properties of beta distributions) that the average value of p is

$$\bar{p} = \frac{v}{u+v} \quad (5.25)$$

This is the equilibrium gene frequency from our deterministic theory of recurrent mutation. Hence we might say in this case that mutation has overpowered drift. Finally if, say, $4Nu \ll 1$ while $4Nv \gg 1$, then the distribution leans to the side favored by mutation. From these curves we learn that whether mutation overpowers drift depends on how the quantities $4Nu$ and $4Nv$ compare with 1. If both $4Nu$ and $4Nv$ are less than 1, then drift prevails, while if both $4Nu$ and $4Nv$ are greater than 1, then mutation prevails.

---

† In addition to determining the stationary distribution, it is also possible to classify the boundaries as discussed in Chapter 20.

‡ This distribution is called a beta distribution. In this case the normalization constant is given explicitly by $c = \Gamma(4Nu + 4Nv)/[\Gamma(4Nu)\Gamma(4Nv)]$, where $\Gamma(x)$ is the gamma function.

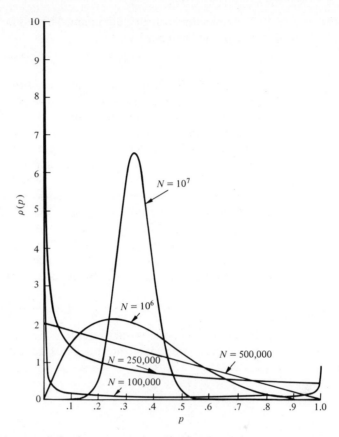

FIGURE 5.8. *Stationary ensemble distribution with recurrent two-way mutation, genetic drift, and no selection.* $A \xrightarrow{u} a$, $a \xrightarrow{v} A$, $u = 1 \times 10^{-6}$, $u = .5 \times 10^{-6}$, *and* $\bar{p} = \frac{1}{3}$.

The result at the end of the preceding paragraph is quite important, biologically. How large must a population be in order to observe an equilibrium polymorphism maintained by recurrent mutation? If u and v are of the order of 10^{-6}, then the population must exceed 1 million individuals. Obviously, many interesting species, not to mention separate populations within a species, fail to satisfy this criterion. Even in large populations drift and mutation can be forces of comparable strength. In the absence of selection, mutation and drift must always be considered together. In the absence of selection, however, it is never safe to consider only mutation while ignoring drift, even in large populations.

Next let us explore the comparative strengths of selection and genetic drift. For the sake of illustration, assume that the selection is *for A and against a*. The typical cases are

1. Selection against a recessive. The w's are $w_{AA} = 1$, $w_{Aa} = 1$, $w_{aa} = 1 - s$. In this case $\hat{\rho}(p)$ reduces to

$$\hat{\rho}(p) = cp^{4Nv-1}(1-p)^{4Nu-1} \exp\left[2Ns(2p-p^2)\right] \qquad (5.26)$$

2. Heterozygote intermediate. The w's are $w_{AA} = 1$, $w_{Aa} = 1 - s/2$, $w_{aa} = 1 - s$. So $\hat{\rho}(p)$ reduces to

$$\hat{\rho}(p) = cp^{4Nv-1}(1-p)^{4Nu-1} \exp(2Nsp) \qquad (5.27)$$

3. Selection against a dominant. The w's are $w_{AA} = 1$, $w_{Aa} = 1-s$, $w_{aa} = 1-s$. Then $\hat{\rho}(p)$ is

$$\hat{\rho}(p) = cp^{4Nv-1}(1-p)^{4Nu-1}\exp(2Nsp^2) \qquad (5.28)$$

Figure 5.9 illustrates some examples of the $\hat{\rho}(p)$ above. If $4Nu$ and $4Nv$ are both less than 1, then the curves have peaks near 0 and 1 as before. The role of selection is to increase the area under the peak corresponding to fixation of the favored allele. If $2Ns \ll 1$, then there is essentially no effect of selection, while if $2Ns \gg 1$ the influence of selection is conspicuous. For example, in the figures N is 1000, u and v are 10^{-6}, and there are curves with $s = .005$, $.002$, and $.001$. We see that whether selection is conspicuous in these cases depends on the relationship of $2Ns$ to 1.

Perhaps the most interesting case is that of heterozygote superiority. Here the effect of selection is to maintain genetic variation as a polymorphism while the effect of drift is to cause the loss of genetic variation. In this case the two forces are directly opposed to each other. The w's are $w_{AA} = 1 - s_A$, $w_{Aa} = 1$, $w_{aa} = 1 - s_a$. Let us also define $s = (s_A + s_a)/2$. Then $\hat{\rho}(p)$ reduces to

$$\hat{\rho}(p) = cp^{4Nv-1}(1-p)^{4Nu-1}\exp\left\{4Ns\left[\left(\frac{s_a}{s}\right)-p\right]p\right\} \qquad (5.29)$$

Figure 5.10 illustrates the curves that may result. If $4Ns \gg 1$, then there is a pronounced mode at an intermediate frequency, while if $4Ns \ll 1$ the curve is

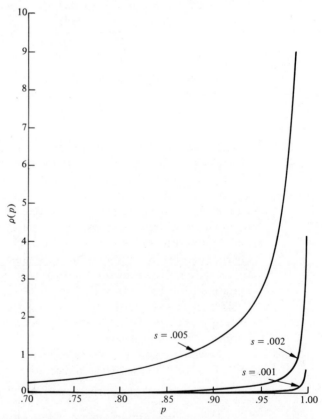

FIGURE 5.9a. *Stationary ensemble distribution with recurrent two-way mutation, genetic drift, and selection against a recessive allele. $N = 1000$, $u = v = 10^{-6}$, $w_{AA} = 1$, $w_{Aa} = 1$, and $w_{aa} = 1-s$.*

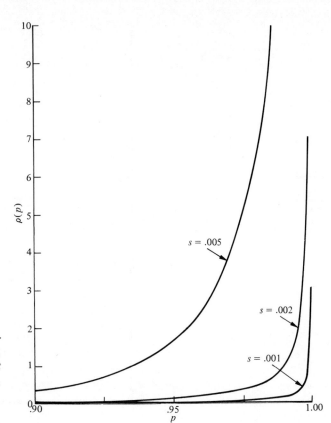

FIGURE 5.9b. *Stationary ensemble distribution with recurrent two-way mutation, genetic drift, and selection against a dominant allele.* $N = 1000$, $u = v = 10^{-6}$, $w_{AA} = 1$, $w_{Aa} = 1 - s$, and $w_{aa} = 1 - s$.

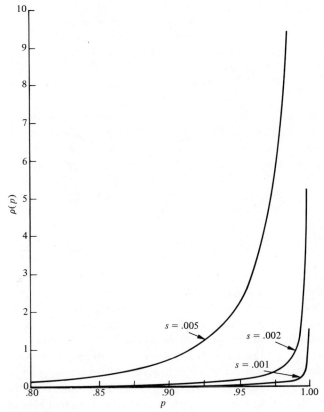

FIGURE 5.9c. *Stationary ensemble distribution with recurrent two-way mutation, genetic drift, and an intermediate heterozyte* $N = 1000$, $u = v = 10^{-6}$, $w_{AA} = 1$, $w_{Aa} = 1 - s/2$, and $w_{aa} = 1 - s$.

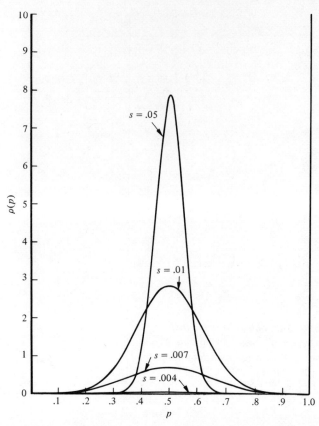

FIGURE 5.10. *Stationary ensemble distribution with recurrent two-way mutation, genetic drift, and heterozygote superiority. It is important to note that these curves are trimodal; there are also peaks at $p = 0$ and $p = 1$ although they are too thin to be graphed in the figure. $N = 1000$, $u = v = 10^{-6}$, $w_{AA} = 1 - s$, $w_{Aa} = 1$, $w_{aa} = 1 - s$, and $\hat{p} = \frac{1}{2}$.*

U-shaped. Thus again the quantity $4Ns$ proved to be important in determining the qualitative outcome. If $4Ns \ll 1$, then most populations possess little genetic variation and we might say that drift has overpowered selection. If $4Ns \gg 1$, then many populations are polymorphic as would be predicted from the deterministic theory. The location of the mode in $\hat{\rho}(p)$ caused by the selection is found by differentiating $(s_a/s - p)p$ and setting the derivative equal to zero, yielding

$$p_{\text{mode}} = \frac{s_a}{2s} = \frac{s_a}{s_A + s_a} \qquad (5.30)$$

Indeed, p_{mode} is exactly the polymorphism frequency predicted by the standard deterministic theory. Thus, if $4Ns \gg 1$, most populations attain a polymorphism frequency very near that predicted by the standard theory of natural selection. So if the population is sufficiently large and the selection sufficiently strong that this condition is met, then genetic drift can be safely ignored and the standard deterministic theory is sufficient for most purposes.

The discussion above summarizes much of the classical theory on genetic drift. It was developed principally by Wright (1931, 1945) and Kimura (1964). Further coverage of the theory of genetic drift is contained in Moran

(1962) and Ewens (1969). It should be noted that there is still another possible formulation of the process of genetic drift in terms of "branching processes." In this formulation the basic matrix of binomial transition probabilities is seen as a special case resulting from a Poisson offspring distribution. For an introduction to this formulation, consult Ewens (1969 pp. 39–41 and 79–89), and Karlin and McGregor (1964, 1968).

PROBLEMS

5.1. (a) Consider 12 isolated populations of one individual each. Initially, $p = \frac{1}{2}$ in all the populations. (We are considering one locus with two alleles.) After two generations $p = 0$ in eight of the populations, $p = \frac{1}{2}$ in three of the populations, and $p = 1$ in one of the populations. Is genetic drift an acceptable explanation for the changes in gene frequency?

      (b) Suppose that the genetic changes mentioned in Problem 1 occurred in one generation, not two. Is genetic drift still an adequate explanation?

5.2. Write a program to iterate Equation (5.3) for a population size of N individuals. Use it to work out an example similar to that in Problem (1) but use a larger population size.

5.3. Show that the equilibrium distribution of the diffusion equation, $\hat{\rho}(x)$ equals

$$\hat{\rho}(x) = \frac{c}{V(x)} \exp\left\{2 \int \left[\frac{M(x)}{V(x)}\right] dx\right\}$$

where c is a normalization constant.

      (a) From (5.14) observe that at equilibrium the flux, $J(x)$, must be constant with respect to x.

      (b) Convince yourself that the constant which $J(x)$ equals is, in fact, zero as follows. Suppose that the walls of the container are located at 0 and 1; thus the diffusion is confined to within 0 and 1. Suppose that the mass in the container is fixed and no material is being added or withdrawn. Therefore, $\int_1^0 \hat{\rho}(x)\, dx = m$, where m is the given total mass. The assumption of a fixed total mass means that $J(0) = J(1) = 0$, since there is no flow across the boundaries. But J is a constant, so that if it equals 0 at $x = 0$ and $x = 1$, then it must be zero everywhere else as well.

      (c) Show that $\hat{\rho}(x)$ must now satisfy the first-order ordinary differential equation

$$M(x)\hat{\rho}(x) - \frac{1}{2}\frac{d}{dx}V(x)\hat{\rho}(x) = 0$$

      (d) Introduce $f(x) \equiv V(x)\hat{\rho}(x)$. Show that $f(x)$ satisfies

$$\frac{M(x)}{V(x)} f(x) - \frac{1}{2}\frac{d}{dx} f(x) = 0$$

      (e) Using the fact that $(1/f(x))df(x) = d \ln(f(x))$, show that the above differential equation can be rearranged as

$$d \ln [f(x)] = 2\frac{M(x)}{V(x)} dx$$

(f) Integrate both sides to obtain

$$\ln\left[f(x)\right] = 2\int \frac{M(x)}{V(x)}\,dx + \text{const}$$

(g) Rearrange the above formula to

$$f(x) = c\exp\left\{2\int\left[\frac{M(x)}{V(x)}\right]dx\right\}$$

(h) Recover $\hat{\rho}(x)$ as

$$\hat{\rho}(x) = \frac{c}{V(x)}\exp\left\{2\int\left[\frac{M(x)}{V(x)}\right]dx\right\}$$

5.4. Find the equilibrium distribution $\hat{\rho}(p)$, where $M(p)$ and $V(p)$ are given by (5.21) and (5.22).

(a) Show that

$$\frac{M(p)}{V(p)} = \frac{2Nv}{p(1-p)} - \frac{2Nv+2Nu}{(1-p)} + 2N(w_{AA}-w_{Aa})p$$
$$+ 2N(w_{Aa}-w_{aa})(1-p)$$

(b) Using the standard formulas that

$$\int \frac{dx}{a+bx} = \frac{1}{b}\ln|a+bx| \qquad \text{and} \qquad \int \frac{dx}{x(a+bx)} = -\frac{1}{a}\ln\left|\frac{a+bx}{x}\right|$$

show that

$$2\int\left[\frac{M(p)}{V(p)}\right]dp =$$

$$-4Nv\,\ln\left[\frac{1-p}{p}\right] + (4Nv+4Nu)\ln(1-p)$$

$$+\frac{4N(w_{AA}-w_{Aa})p^2}{2} + 4N(w_{Aa}-w_{aa})p - \frac{4N(w_{Aa}-w_{aa})p^2}{2}$$

(c) Show that this can be rearranged as

$$\ln\left[p^{4Nv}\right] + \ln\left[(1-p)^{4Nu}\right] + 2N(w_{AA}+w_{aa}-2w_{Aa})p^2 + 4N(w_{Aa}-w_{aa})p$$

(d) Show that $\exp\left\{2\int\left[\frac{M(p)}{V(p)}\right]dp\right\}$ equals

$$p^{4Nv}(1-p)^{4Nu}\exp\left\{2N[w_{AA}+w_{aa}-2w_{Aa}]p^2 + 4N[w_{Aa}-w_{aa}]p\right\}$$

(e) Now show that $\hat{\rho}(p)$ is given by (5.23).

Chapter 6
THE NEUTRALITY CONTROVERSY

YOU have now learned the basic vocabulary of population genetics. You have explored the ideas of natural selection, genetic drift, adaptation, and fitness. Now you are in a position to apply these basic ideas to a topic that is currently under very active research.

We have discovered that the forces of natural selection, mutation, and genetic drift all *can* combine in shaping the genetic structure of a population. Therefore, the natural questions to ask at this point are (1) what is the actual genetic structure of natural populations, and (2) how have the basic evolutionary forces actually been involved in producing the observed genetic structure? The empirical examination of the genetic structure of populations has only been attempted in very recent years. The techniques have been borrowed from molecular biology, which itself is a rather recent field. Perhaps because this empirical knowledge is so recent, a currently unresolved controversy exists about how the structure has evolved. This chapter is an introduction to this topic. We discuss (1) the technique used to assay the genetic structure of a population, (2) the data that have been obtained for population of *Drosophila*, and (3) the controversy surrounding the interpretation of those data in terms of the basic evolutionary forces.

Detecting the Genetic Structure of a Population

When investigating the genetic structure of a population, we may have two very different purposes. First, we may be interested in the specific set of loci that influence a particular trait. Typically, the trait will be one that we know is relevant to natural selection, for example, color traits that allow an animal to blend with its background, or control its ability to thermoregulate, and so forth. Second, we may be interested in a random sample of the entire genome. We may want to know what percentage of the entire genome is polymorphic for different alleles and, if so, whether natural selection is responsible for maintaining the polymorphism or whether the polymorphism is caused by some mixture of genetic drift and mutation. Thus, if we want to make statements about the entire genome, we certainly want to avoid considering only those special loci that have already been implicated with natural selection. Most of the other chapters in this book are concerned with the special loci and traits that are believed, in advance, to be involved in natural selection, but this chapter focuses on data from a random sample of the genome.

To obtain a random sample of the genome, Lewontin and Hubby (1966) and Harris (1966) independently proposed exploring the enzymes found in natural populations. We know from molecular genetics that the linear sequence of amino acids in a protein corresponds to a linear sequence of nucleotides in DNA. Hence a protein provides information about the gene that produced it. Specifically, suppose that we find two slightly different forms of a given enzyme among the individuals in a population. For example, we may find two slightly different forms of the enzyme, glucose-6-phosphate dehydrogenase (6GPDH), among the individuals in a population. Then it is possible that the population is polymorphic at the 6GPDH locus. To establish whether the different forms of 6GPDH correspond to different alleles, we have to perform crosses and search for Mendelian segregation.

We would also have to establish biochemically that the two forms were not simply two conformations of the same molecule. After all this work we would be able to say whether the locus was polymorphic or not. Our general procedure, then, will be to seek out as many randomly chosen loci as practically possible and determine whether they are polymorphic or not.

Before we present the data from this approach, it is important to dwell in more detail on the biochemical techniques used for the analysis of proteins. Recall from your introductory biology that amino acids all have the general structure sketched in Figure 6.1—different amino acids have different residues but all have the NH_2 and COOH groups at the terminal carbon. These NH_2 and COOH groups are used in the peptide bonds that bind the polypeptide together, while the residues are what give the polypeptide its character. Some amino acids have positively charged residues, others are neutral, and still others are negative. Hence the entire polypeptide has a net charge that depends upon the residues of its component amino acids. Substitution of just one amino acid for another may easily alter the overall charge of the polypeptide. This fact is the basis of a technique called *electrophoresis* for separating polypeptides that differ by as little as one amino acid.

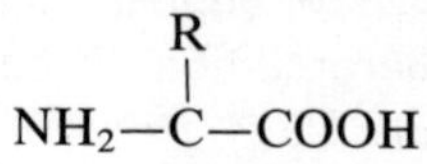

$$
\overset{\displaystyle R}{\underset{\displaystyle }{NH_2-\overset{|}{C}-COOH}}
$$

FIGURE 6.1. *An amino acid. All amino acids have an NH_2 and a COOH group attached to the terminal carbon atom. Amino acids differ from one another in the rest of the molecule, which is labeled as R in the figure.*

Electrophoresis works as follows: A solution containing the proteins is deposited at one end of a slab of jellylike material (usually starch or a plastic called acrylamide). The solutions from many individuals can be run at the same time. (See Figure 6.2.) The slab is then subjected to an electric field for several hours. If a protein has a net electrical charge, say negative, then it migrates to the positive pole. (See Figure 6.3.) After several hours the electric field is removed and the location of the proteins is detected with various stains and other assays. Specifically, the enzymes are supplied with a substrate and the stains combine with the *product* of the chemical reaction which the enzyme catalyzes. The stain thus identifies an enzyme by its function. In summary, electrophoresis measures the ability of a protein to move through the jellylike material in the presence of an electric field. The

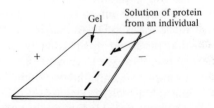

FIGURE 6.2. *Electrophoresis. A small amount of protein solution is placed on the gel. The gel is then placed in an electric field.*

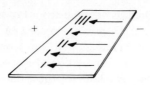

FIGURE 6.3. *After several hours the protein molecules with a net negative charge will have moved to the positive pole. If the organism possesses several kinds of protein molecules that differ slightly in charge, then the molecules will separate slightly into distinct bands.*

ability of a protein to move under these conditions is called the electrophoretic *mobility* of a protein.

The mobility of a protein depends both on its charge and on its shape. In the example given above, a protein with a more negative charge than another protein would travel farther through the slab in a given time. But also, two proteins of the same charge but with different sizes or shapes would also travel different distances in a given time. To separate these two effects, the proteins should be used with slabs of different porosity. Thus two proteins that appear identical in a slab of one porosity may be differentiated in a slab of another porosity [Johnson (1974 and 1977)]. In addition, two proteins that appear the same with respect to charge and shape at one pH may become differentiated at other pH's. Similarly, proteins may be differentiated by heat sensitivity, as seen in studies by Bernstein, Throckmorton, and Hubby (1973). When all these techniques and tools are combined, there would seem to be little chance for two different proteins to masquerade as one.

One caveat should be kept in mind. The biochemical techniques above are addressed only to soluble proteins, most of which are enzymes. The techniques are not generally used with insoluble proteins or with the products of regulatory genes. It is generally believed (or hoped) that this class of proteins is representative of the whole genome. But is is conceivable that this class of proteins is a biased sample of the genome after all, even though all the studies of this type are intended to be a random sample.

The Case of In order to discuss the issues that arise in trying to understand the genetic
Drosophila structure of populations, it is helpful to focus on one concrete example. The
Pseudo-obscura example for this section is condensed from Lewontin (1974).

Drosophila pseudo-obscura occurs in the western United States and Canada, and in Central America. There is a separate and disjunct population in the highlands of Colombia near Bogota. According to Lewontin (1974) the species is typically found in cool forests and is restricted to the higher elevations in the more southern or arid parts of the range. It is found at sea level in northern California but not below 5500 feet in Guatemala or below 7000 feet in Colombia. In the spring it can also be found around small oases in the Mojave Desert and even in Death Valley although the temporal continuity of these oasis populations is disputed. In general terms the population is almost continuously distributed during the spring and fall but may contract to isolated cool, moist refuges during the summer. Unfortunately, little is known of the ecology of this species.

Prakash, Lewontin, and Hubby (1969) and Prakash, Lewontin, and Crumpacker (unpublished) assembled the data presented in Table 6.1, which is adapted from Lewontin (1974). They examined 24 separate loci from *D. pseudo-obscura* in 12 localities. Ten of the localities are in the continental United States; one is in Guatemala; and one is in the range of the isolated population at Bogota, Colombia. First, let us consider some summary statistics for these data and then examine them in more detail.

Of the 24 loci examined, 11 were absolutely monomorphic and homogeneous over the entire species range, and 13 were polymorphic at least in some places. Let us focus attention on the localities in the continental United

Table 6.1. 13 Polymorphic Loci in *D. pseudo-obscura*. Adapted from Lewontin (1974).

Locus	Name of Each Allele	Strawberry Canyon, Calif.	Wild Rose, Calif.	Charleston, Nev.	Sheep Ranch, Nev.	Cerbat, Ariz.	Mesa Verde, Colo.	State Recreation, Colo.	Hardin Ranch, Colo.	Nelson Ranch, Colo.	Austin, Tex.	Guatemala	Bogotá, Colombia
Larval acid	0.93	—	—	—	—	—	—	—	—	—	.028	—	—
phosphatase-4	1.00	1.000	1.000	1.000	1.000	1.000	1.000	1.000	1.000	1.000	.860	1.000	1.000
	1.05	—	—	—	—	—	—	—	—	—	.112	—	—
Acetaldehyde	0.90	.010	—	—	—	—	—	—	—	—	—	—	—
oxidase-2	0.93	.030	—	—	—	—	—	—	—	—	—	—	.050
	1.00	.940	1.000	1.000	1.000	1.000	1.000	1.000	1.000	1.000	1.000	1.000	.830
	1.02	.020	—	—	—	—	—	—	—	—	—	—	.120
Malic	0.80	—	—	—	—	.011	—	.019	—	—	—	—	—
dehydrogenase	1.00	.969	1.000	.936	.929	.954	.948	.962	.882	.904	.957	.727	1.00
	1.20	.031	—	.064	.071	.034	.052	.019	.118	.096	.043	.273	—
Octanol	null	—	—	.014	—	—	—	—	—	—	—	—	—
dehydrogenase-1	0.75	—	0.26	—	—	—	—	—	—	—	—	—	—
	0.86	—	—	—	.056	.020	.013	—	—	—	—	—	—
	1.00	.977	.871	.951	.902	.939	.961	.965	.972	.885	1.000	1.000	1.000
	1.05	—	.077	.014	.028	.010	—	.035	—	.115	—	—	—
	1.22	.023	.026	.021	.014	.031	.026	—	.028	—	—	—	—
Leucine	0.83	—	.012	—	.012	.009	—	—	—	—	—	—	—
aminopeptidase	0.90	.008	—	.016	—	—	.025	—	—	.019	.043	.036	—
	0.95	.050	.024	.039	.024	.018	.008	—	—	—	.022	—	—
	1.00	.892	.916	.897	.893	.954	.940	.875	1.000	.923	.870	.964	.947
	1.10	.050	.048	.048	.071	.018	.025	.125	—	.058	.054	—	.054
	1.12	—	—	—	—	—	—	—	—	—	.011	—	—
Protein-7	0.68	.005	—	—	—	—	—	—	—	—	—	—	—
	0.73	.005	—	.021	.014		.009	.040	.025	—	.012	—	.050
	0.75	.954	.950	.979	.971	.987	.955	.960	.950	1.000	.966	1.000	.925
	0.77	.036	.050	—	.014	.013	.036	—	.025	—	.023	—	.025
Protein-8	0.80	.014	.025	.008	—	—	.009	.019	.027	—	.011	—	.870
	0.81	.472	.450	.600	.514	.473	.410	.539	.595	.480	.441	.625	.093
	0.83	.514	.525	.392	.472	.527	.576	.442	.378	.480	.512	.375	.037
	0.85	—	—	—	.014	—	—	—	—	.040	.035	—	—
Protein-13	1.23	.057	—	.082	—	.045	.025	.058	—	.022	.022	—	—
	1.30	.943	1.000	.918	1.000	.940	.975	.942	1.000	.935	.978	1.000	.725
	1.37	—	—	—	—	.015	—	—	—	.043	—	—	.275

Table 6.1. 13 Polymorphic Loci in *D. pseudo-obscura*. Adapted from Lewontin (1974) (*cont.*)

Locus	Name of Each Allele	Strawberry Canyon, Calif.	Wild Rose, Calif.	Charleston, Nev.	Sheep Ranch, Nev.	Cerbat, Ariz.	Mesa Verde, Colo.	State Recreation, Colo.	Hardin Ranch, Colo.	Nelson Ranch, Colo.	Austin, Tex.	Guatemala	Bogotá, Colombia
Esterase-5	null	—	—	—	—	.016	—	—	—	—	—	—	—
	0.85	—	.013	.007	.047	—	.035	—	—	—	—	—	—
	0.90	—	.027	.030	—	—	—	—	—	—	.015	—	—
	0.95	.123	.149	.096	.140	.114	.113	.237	.216	.114	.031	.158	.026
	0.97	—	.027	.015	—	—	—	—	—	—	.031	—	—
	1.00	.424	.460	.356	.419	.317	.365	.474	.486	.341	.292	.579	.974
	1.02	.014	.013	.022	—	—	.048	—	.108	.182	.108	.053	—
	1.03	.080	—	—	—	—	.039	—	—	—	—	—	—
	1.04	.004	.041	.193	.198	.211	.104	.017	.135	.045	.154	—	—
	1.07	.193	.243	.200	.174	.260	.196	.271	.054	.273	.262	.210	—
	1.09	.009	—	.007	—	—	—	—	—	—	—	—	—
	1.12	.132	.027	—	.023	.081	.100	—	—	.045	.046	—	—
	1.16	.019	—	.073	—	—	—	—	—	—	.062	—	—
Xanthine dehydrogenase	0.90	.053	—	.007	—	—	.016	.035	—	—	.018	—	—
	0.92	.074	—	.030	.040	—	.073	.089	.026	.020	.036	—	—
	0.97	—	.133	.098	.077	.012	—	—	—	—	—	—	—
	0.99	.263	.200	.188	.173	.131	.300	.286	.289	.220	.232	.278	—
	1.00	.600	.667	.647	.710	.857	.581	.555	.633	.720	.661	.722	1.000
	1.02	.010	—	.030	—	—	.030	.035	.052	.040	.053	—	—
Protein-10.	1.02	.005		.007	.014	.015	.022	—	—	—	.010	—	—
	1.04	.615	.898	.945	.943	.985	.970	.770	.694	.308	.935	—	—
	1.06	.380	.102	.048	.043	—	.008	.230	.306	.692	.054	1.000	1.000
Protein-12	1.18	.550	.736	.750	.792	.733	.940	1.000	1.000	.972	.900	1.000	1.000
	1.20	.450	.264	.250	.208	.267	.060	—	—	.028	.100	0	0
α-Amylase	0.74	.030	—	—									
	0.84	.290	.206	.090	.172	.194	.211	.380	.391	.548	.125	1.00	1.00
	1.00	.680	.794	.910	.828	.806	.789	.620	.609	.452	.875	—	—

States. Note that the first two loci are nearly monomorphic. Typically, if enough specimens are examined, some variation can be found at nearly every locus. Thus depending on whether we wish to regard the first two as effectively monomorphic or not, we can say that either 13 of 24 or 11 of 24 loci were polymorphic. This corresponds to percentages of 54 and 46 percent, respectively, so that roughly half the loci are polymorphic in the population.

The fact that a locus is polymorphic in the population, of course, does not mean that every individual is heterozygous at that locus. The fraction of the population that is heterozygous at any locus depends on both the gene frequencies at that locus and the number of alleles. For example, if there are two alleles, then the fraction of individuals who are heterozygous is simply

$$H = 2p_1p_2 \tag{6.1}$$

where p_1 and p_2 are frequencies of the alleles A_1 and A_2, respectively.

Similarly, with three alleles

$$H = 2p_1p_2 + 2p_1p_3 + 2p_2p_3 \qquad (6.2)$$

These formulas assume that the genotypes are approximately in Hardy–Weinberg ratios. Even with rather strong selection this is still approximately true, provided that there is random mating. In this way we can calculate the heterozygosity at each locus in *each* locality. Then, if we average the heterozygosity at each locus over the 10 continental United States localities, we obtain the results in Table 6.2, adapted from Lewontin (1974).

Table 6.2. Heterozygosity at Different Loci Averaged over 10 Locations. Adapted from Lewontin (1974).

11 monomorphic loci	0
Acetaldehyde oxidase-2	.012
Larval acid phosphatase	.025
Protein-7	.063
Protein-13	.070
Malic dehydrogenase	.102
Octanol dehydrogenase	.109
Leucine aminopeptidase	.155
Protein-10	.229
Protein-12	.234
α-Amylase-1	.353
Xanthine dehydrogenase	.492
Protein-8	.513
Esterase-5	.741

Note: Average overall 24 loci is .128. Average over the 13 polymorphic loci is .238.

Note that the locus with the highest heterozygosity is the locus with the largest number of alleles. Almost 75 percent of the individuals are heterozygous at this locus. In contrast only about 1 percent of the individuals are heterozygous for acetaldehyde oxidase-2 even though there are four alleles because the allele No. 1.00 is so very common. If we average the heterozygosity for the 13 polymorphic loci, we obtain .238. This figure means that roughly 24 percent of the individuals in a population are heterozygous at an average polymorphic locus. This conclusion assumes that the sample of 24 loci is representative of the entire genome.[†]

These data were among the very first obtained on the degree of heterozygosity and polymorphism in a random sample of the genome. These data are more or less typical of other populations as well. Table 6.3, adapted from Lewontin (1974) summarizes the analogous data for other populations. The species include *Homo sapiens*; several species of mice; several *Drosophila* species; and the horseshoe crab, *Limulus*, which represents an extremely primitive line of arthropods. In the 13 populations in Table 6.3, between 20 and 86 percent of the loci are polymorphic and the average percent

[†] If there is no linkage disequilibrium (see Chapter 8) between the loci, that is, all the pairwise and higher-order D's are zero, then this figure also means that *any individual* is heterozygous at 24 percent of the loci which are polymorphic in the population. In fact, the linkage disequilibrium usually is very low and this interpretation is acceptable.

Table 6.3. Summary Data for 13 Species. Adapted from Lewontin (1974).

Species	Number Populations	Number of Loci	Proportion of Loci Polymorphic	Proportion Heterozygous at the Polymorphic Loci
Homo sapiens	1	71	.28	.239
Mus musculus musculus	4	41	.29	.214
M. m. brevirostris	1	40	.30	.367
M. m. domesticus	2	41	.20	.280
Peromyscus polionotus	7	32	.23	.248
Drosophila pseudo-obscura	10	24	.54	.238
D. persimilis	1	24	.25	.424
D. obscura	3	30	.53	.204
D. subobscura	6	31	.47	.162
D. willistoni	2–21	28	.86	.214
D. melanogaster	1	19	.42	.283
D. simulans	1	18	.61	.262
Limulus polyphemus	4	25	.25	.244

heterozygosity at the polymorphic loci falls within the range of 16 to 42 percent. The median percentage of polymorphic loci is 30 percent and the median heterozygosity per polymorphic locus is 25 percent. These data may be summarized by saying that these *sexually reproducing species of animals are polymorphic for a third of their genes, and a quarter of the individuals are heterozygous at an average polymorphic locus.*

There are also several reports of populations with abnormally low heterozygosity. These reports include the cricket frog, *Acris crepitans*, discussed by Dessauer and Nevo (1969), two species of kangaroo rat, *Dipodomys*, studied by Johnson and Selander (1971), and the fossorial mole rat, *Spalax ehrenbergi*, described by Nevo and Shaw (1972). Thus, although generalizations can be made about the high levels of heterozygosity, exceptions can be found that may prove very important in the future.

The data in *D. pseudo-obscura* reveal some other rather curious points as well: (1) Except for the first two loci discussed above, the classification of a locus as polymorphic or not applies throughout the species range within the continental United States. Thus, there are no loci that are polymorphic at 50 percent of the locations and monomorphic at the other locations. Polymorphism versus monomorphism is almost an all-or-none phenomenon throughout the species range. (2) The actual gene frequencies are very much the same throughout the species range in the continental United States except for the last three loci, which will be mentioned below. The most common allele is the most common form just about everywhere. Even at the locus of protein-8, where the No. 81 and No. 83 alleles are more or less equally common, they are about equally common everywhere. (3) The isolated Colombian population near Bogota does not share continuity with the continental U.S. populations as mentioned in (1) and (2) above. The Bogota population is markedly less heterozygous; indeed, it is monomorphic or nearly so at the loci for esterase-5 and xanthine dehydrogenase, which are extremely polymorphic in the continental United States. Comparison of the Bogota population with that in Guatemala suggests that it is the isolation of the Bogota population rather than the mountain tropical environment which is responsible for the distinctive lack of polymorphism. Unfortunately, the Guatemala sample was very small and the comparison cannot be pursued in much detail. (4) The last three loci in the table are all polymorphic but show substantial variation in gene frequencies from place to place within the continental United States. These loci happen to be located on the third chromosome where *D. pseudo-obscura* is known to have a series of chromosome inversions. (See Chapter 11 for more detail on inversions.) It is possible to relate the gene frequency patterns at these loci to the known geographic pattern of chronosomal architecture in *D. pseudo-obscura.*

To summarize, we observe from these studies that about one-third of the genome is polymorphic and approximately one-fourth of the population is heterozygous at the polymorphic loci. Moreover, if the allelic content at any locus is not correlated with the content at other loci, we can also say that each individual is heterozygous at approximately 10 percent of its loci ($\frac{1}{3} \times \frac{1}{4} = \frac{1}{12} = \approx 10$ percent). In addition, a detailed inspection of the data reveals interesting geographical patterns in the gene frequencies. The bulk of the population shows very similar gene frequencies throughout its range, while considerable differentiation can arise in isolated populations that are peripheral to the main species range. The question we now face is to explain these data in terms of the basic evolutionary forces of selection, drift, and mutation.

The Hypotheses

At present there are two major hypotheses about how the basic evolutionary forces have combined to produce the observed genetic structure. The hypotheses differ in the kind of natural selection involved. In the first hypothesis it is assumed that a protein has certain amino acids which are especially critical to the protein's biochemical function and other amino acids which are not. It is assumed that substitution at the critical sites produces a nonfunctional protein, whereas substitution at the uncritical sites does not alter the function very much. If this assumption is correct, we would expect natural selection to eliminate mutations influencing the critical parts of the protein while ignoring mutations to the noncritical parts of the protein. If so, the *observed* variation at a locus represents alleles that are selectively neutral to one another—alleles that involve substitutions at the noncritical regions of the protein. Of course, the observed variation is not neutral relative to the mutants that are eliminated by selection. This hypothesis is called the *neutralist hypothesis*. It asserts that most of the remaining variation at a locus is neutral while the deleterious mutants have already been eliminated by natural selection. Selection against deleterious mutants is sometimes called *purifying selection*. It is analogous to a sieve, filtering out the deleterious mutants and leaving mutants that are neutral to one another. According to the neutralist hypothesis the kind of natural selection involved in producing the observed genetic structure is mostly purifying selection.

You should note that the neutralist hypothesis cannot be confirmed or refuted only with the techniques of enzymology. It may be possible to detect differences in enzyme kinetics and temperature optima of many of the allelic enzymes at a locus. But what is important is whether these enzymes lead to sufficiently different fitnesses for their carriers. Recall from the last chapter that we need a fitness difference, s, which is large enough so that $4Ns$ is greater than one, where N is the population size.

The other hypothesis assumes that the various *observed* allelic proteins *do* have an important effect on fitness. Indeed, it is assumed that at polymorphic loci the polymorphism is being "actively" maintained by natural selection. Typically, it is visualized that there is heterozygote superiority at each locus leading to a polymorphism at frequencies set by the selective values. This hypothesis, which accounts for the observed polymorphism, is called the *selectionist hypothesis*. According to the selectionist hypothesis, the observed variation is not neutral to selection and the kind of natural selection involved in producing the observed genetic structure is that which maintains a polymorphism.

Resolving which of these hypotheses is correct is an interesting question in its own right. But there are also many fascinating additional consequences of this dispute as summarized in Lewontin (1974). One particularly important consequence concerns what might be called the "rate-limiting step" in evolution. To illustrate this issue, let us refer again to the *Anolis* lizards throughout the islands of the Caribbean as presented in the introductory chapter. Recall that the Bahama islands are very low-lying islands that have frequently been submerged during times when the sea level was higher. In contrast, the islands of the Lesser Antilles are mostly of volcanic origin with tall, occasionally active volcanoes. These volcanic islands have never been fully submerged. For this reason we know that the lizards on the Bahama islands must have been there a shorter period of time than the lizards of the Lesser Antilles. It is easy to observe that the lizards on the Bahamas have colonized from Cuba, Hispaniola, and Puerto Rico and that they have not differentiated very much during their stay in the Bahamas. In contrast, the lizards on each of the Lesser Antillean islands (or island banks) have become very differentiated from one another and are all considered separate species. Of course, they have had a lot more time to evolve this differentiation. The question is, what controls the rate at which the differentiation occurs? The rate seems quite slow. Although the Bahama islands are recent in terms of geological time relative to the Lesser Antilles, there have nonetheless been thousands of generations during which the Bahama lizards have evolved only slight differentiation. Why is this rate of differentiation so slow?

According to the neutralist hypothesis, the rate-limiting step occurs in the production of favorable mutations, since most mutations are deleterious and favorable mutations are very rare. Thus the rate of evolutionary differentiation is limited by the lack of suitable genetic variation for selection to act upon. But according to the selectionist hypothesis, the rate-limiting step must be attributable to selection pressure itself. There is no lack of genetic variation for selection to act upon. Instead, perhaps the new environment in the Bahamas is not substantially different from the original habitat of a colonizing species so that the selection driving the evolutionary differentiation is weak. Thus explaining the speed with which evolution actually occurs poses a hard problem, and the neutralist and selectionist hypotheses suggest very different answers.

There have been two approaches to resolving these hypotheses. The first is to derive predictions from the neutrality hypothesis and to test these predictions against the data. The second is to search for direct evidence of the kind of natural selection which produces polymorphism. In the next several paragraphs we shall present a brief status report on what has been accomplished so far with these approaches.

Predictions of the Neutrality Hypothesis

The most complete set of predictions that have been derived from the neutrality hypothesis is contained in the "sampling theory of selectively neutral alleles" developed by Ewens (1972) based on a formulation of Crow and Kimura (1964). To express the neutrality hypothesis, we assume that the possible alleles at any locus are divided into two classes. One class consists of the viable alleles that are neutral with respect to one another; the other class consists of the alleles that are deleterious mutants. For simplicity we shall assume that the deleterious alleles are immediately eliminated by purifying selection. The number of possible alleles that are in each class depends on the enzymology of the locus. By assumption, loci coding for enzymes with a very critical amino acid composition have fewer alleles in the neutral class and more in the lethal class relative to loci coding for enzymes whose composition is not as critical. Next, we shall also assume that the number of *possible* neutral alleles at any locus is much larger than the number of neutral alleles actually at the locus at any one time. This assumption will allow us to view each mutation to a neutral allele as producing an allele that is not presently in the population. A mathematical model for the neutrality hypothesis using these assumptions is developed as follows. The population is supposed to consist of N diploid individuals, and hence $2N$ genes at every locus. To produce the generation at $t + 1$, we first draw $2N$ gametes from the gamete pool formed from the preceding generation, exactly as we did in the last chapter. Next we allow each gamete that we have drawn to mutate. We let u denote the probability of mutating to a neutral allele, and if this happens, the mutant is assumed to be different from any presently in the population. We let v denote the probability of mutating to a deleterious allele, and if this happens, the gamete is thrown away and another is drawn to replace it. Thus v never really enters into the theory because the purifying selection is assumed to act immediately. The constants that do enter into the theory are N, the population size, and u, the mutation rate to *neutral* alleles at the locus in question.†

When this process of drawing gametes, mutation, and purifying selection is carried out for a long time, say 1000 generations, the general characteristics of the gene pool come to an equilibrium, which we can compare with the observed genetic structure in a population. The techniques for making this comparison are provided by the sampling theory of Ewens (1972), and we now quote some of his results. In practice, the data are obtained from a sample of diploid individuals that were collected from a population. From this sample we observe the number of different alleles, k, and the frequency of each of these alleles $p_i (i = 1, 2 \ldots k)$ within the sample. Some results of Ewens follow.

1. The mean and variance of the number of different alleles observed per locus in a sample is predicted to depend on the sample size and the quantity $4Nu$ is a certain way. Here, N is the population size and u is the mutation rate to neutral mutations at the locus under study. The neutral mutation rate, u, will be very low for loci that produce proteins with a very critical

† This model is called the "infinite alleles" model because of the assumption that mutations produce genes not currently in the population.

amino acid composition, and comparatively high for loci that manufacture proteins whose composition is less critical. Figure 6.4 illustrates the predictions. It shows how the average number of different alleles at the locus is predicted to depend on the quantity 4 Nu and on sample size according to the "infinite alleles" model. Note that loci with a very critical amino acid composition, that is a very low u, will have on the average just one kind of allele and thus appear monomorphic. Also, loci with a low u have a low variance in the number of different alleles. In contrast, loci with a high u should support many different alleles, but there should also be a high variance in the number of alleles observed at loci with a high u.

2. The number of different alleles, k, in a sample of a given size is a "sufficient statistic" for estimating the quantity $4Nu$. That is, by observing k in any given sample one obtains all the information necessary to calculate the best estimate of $4Nu$ based on that given sample. In particular, the *pattern* of gene frequencies, p_i, adds *no* additional useful information relevant to estimating the quantity $4Nu$. Figure 6.5 graphs the best estimate of $4Nu$, given the sample size and the number of alleles in the sample, k.

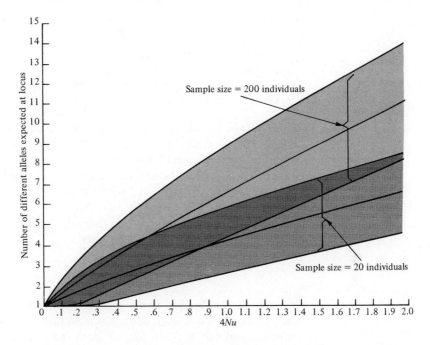

FIGURE 6.4. *The number of different alleles that should be found at a locus as a function of 4Nu at that locus according to the "infinite alleles" model. The band at the top refers to the sample size of 200 individuals, the band at the bottom to a sample size of 20 individuals. The line in the center of each band represents the average number of alleles expected at the locus. The top border of each band represents the average plus one standard error; the bottom represents the average minus one standard error, as predicted by the infinite alleles model. The width of the band therefore represents an approximate 65 percent confidence interval. That is, for a given 4Nu and a given sample size, approximately 65 percent of the samples will contain an allele number that lies within the interval, according to the infinite alleles model. Note that the average allele number and the width of the interval both increase with sample size. [Plotted from tables in Ewens (1972).]*

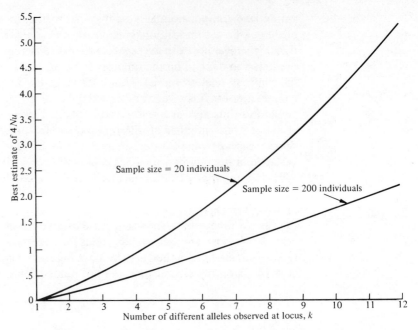

FIGURE 6.5. *The best estimate of* 4Nu *for a locus as a function of the number of alleles observed at that locus, based on the infinite alleles model. The best estimate means the maximum likelihood estimate.* [Plotted from a table in Ewens (1972).]

These results all imply a qualitative prediction from the neutrality hypothesis. If we rank the loci in a given population with respect to u, the mutation rate to neutral alleles, and if we also rank the loci with respect to the number of alleles at each locus, then the two rankings should match one another. That is, there should be a so-called "rank correlation" between u and k.

3. It is also possible to test whether the pattern of allele frequencies is consistent with neutrality. The quantity

$$I_k = -\sum_{i=1}^{k} p_i \ln p_i \qquad (6.3)$$

is widely used in biology as a measure of evenness. A low I_k indicates that one of the alleles is very abundant and the others rare while a high I_k indicates that the alleles all have about the same abundance. Ewens has defined an index, L_k, as follows

$$L_k = \frac{I_k - E(I_k)}{\sigma(I_k)} \qquad (6.4)$$

where $E(I_k)$ means the expected value of the quantity I_k, given that there are k-alleles in the sample. The term, $\sigma(I_k)$, is the standard deviation of I_k, given k-alleles. Both $E(I_k)$ and $\sigma(I_k)$ can be predicted from the neutrality hypothesis, and I_k itself is computed directly from the data on the gene frequencies, p_i. The value of L_k provides an approximate test of neutrality. If L_k is larger than two, then the allele frequencies are more even than expected by the neutrality hypothesis. If L_k is smaller than two, however, then one allele is more common and the others are more rare than expected by the neutrality hypothesis. Ewens has developed a simple computer program to calculate the L_k indices directly from the data. The complete listing of the program appears in Ewens (1974). An example by Ewens, based on hypothetical data, appears in Table 6.4. The gene frequencies in sample 1 are too even for the neutrality hypothesis; those in sample 2 are exactly what might be expected under neutrality; and those of sample 3 are too uneven for the neutrality hypothesis.

Table 6.4. Patterns of Allele Frequencies. Adapted from Ewens (1972).

$$N = 350, \; k = 4$$

Sample	Allele Frequencies				L
1	.35	.30	.20	.15	2.36
2	.83	.11	.04	.02	.02
3	.99	.005	.0025	.0025	−1.70

There are two preconditions that must be met for the use of these tests. First, nearly all of the alleles must be identified. Failure to distinguish different alleles that happen to produce proteins with the same electrical charge can lead to a spurious appearance of selection. In particular, the allele distribution will seem too even for the neutrality hypothesis even if the neutrality hypothesis is correct, according to Ewens and Gillespie (1974). Second, the population must be at equilibrium with respect to drift and mutation. (The purifying selection is assumed to be comparatively fast.) This equilibrium generally requires a long time because drift and mutation are very slow processes. Therefore, the population chosen for the study must have been in its present habitat and environment for, say, 1000 generations. Otherwise, an appearance of selection might simply represent a transient configuration of the gene pool. Interested readers are urged to consult the papers by Ewens (1972) and Ewens and Gillespie (1974) for more detail.

It should also be remembered that a failure to prove that the neutrality hypothesis is false does not necessarily mean that the neutrality hypothesis is true. It is conceivable that certain selection pressures would produce polymorphisms similar to those also produced under the neutrality hypothesis. At this time surprisingly few studies have taken advantage of Ewen's theory in order to test the neutrality hypothesis. One reason is the stringent requirement of the identification of alleles. Few studies present data where extensive effort has been placed into achieving full identification.

The Search for Natural Selection

The other major approach is to search directly for the selection pressures that are presumed to maintain the polymorphism at polymorphic loci. When this approach is successful, the results are very satisfying. Advocates of the selectionist hypothesis would much prefer to know what the selection pressure is in any particular case rather than merely that the neutrality hypothesis is false. But although this approach is potentially the most satisfying, it is by far the most difficult to carry out. There are two difficulties, one theoretical and the other practical. The theoretical problem, as we shall see, is that we do not know what kind of selection pressures we ought to be looking for and we can show that the *simplest* hypothesis about the selection pressure is *not* consistent with the data. The practical problem is that the detection of selection at any locus involves all of the labor of achieving identification of the alleles as before, plus obtaining data on survival and fertility of the various genotypes often under several environmental conditions. Thus, although most biologists instinctively favor the selectionist hypothesis, the case is not yet ready for the jury.

The simplest hypothesis about the kind of selection maintaining polymorphism is that there is heterozygote superiority at each of the polymorphic loci. But predictions from this hypothesis do not stand up well against the data. For the purposes of illustration, suppose that the (marginal) fitness of a heterozygote at a polymorphic locus is w_{het} and that of a homozygote at a polymorphic locus is w_{ho} ($w_{\text{het}} > w_{\text{ho}}$). The (marginal) fitness of a homozygote at a monomorphic locus is w_{mon}. If the loci are independent of one another, it is more or less reasonable to assume that the overall fitness of an individual is the *product* of the (marginal) selective values at each of the loci. Suppose that there are a total of T loci in the genome. Let P be the number of polymorphic loci in the population and let H be the number of loci at which the individual is heterozygous. Then the fitness of an individual for a given T, P, and H is

$$w_t = (w_{\text{mon}})^{T-P}(w_{\text{ho}})^{P-H}(w_{\text{het}})^{H} \tag{6.5}$$

To see whether this model of selection is consistent with the data, we experimentally produce an individual with a lower degree of heterozygosity and compare the fitnesses of individuals in the experimental stock with those in the wild stock. Let H' be the extent of heterozygosity in the experimental stock. Then the fitness of an individual is

$$w_e = (w_{\text{mon}})^{T-P}(w_{\text{ho}})^{P-H'}(w_{\text{het}})^{H'} \tag{6.6}$$

Then the ratio of the fitnesses of the experimental type to the wild type is

$$\frac{w_e}{w_t} = \left(\frac{w_{\text{ho}}}{w_{\text{het}}}\right)^{H-H'} \tag{6.7}$$

We can measure H and H' and also the reduction in fitness of the experimental type relative to the wild type and thus see whether the results are consistent with reasonable levels of heterozygote superiority at the polymorphic loci.

The experimental results have been obtained with *Drosophila melanogaster*. It is conservatively estimated that there are 10,000 loci in the *D. melanogaster* genome. (A low figure favors the selectionist hypothesis.) From Table 6.3 we observe that *D. melanogaster* is polymorphic at 42 percent of its loci and at these loci the average percentage of individuals who are heterozygous is 28 percent. Therefore, the number of polymorphic loci, P, is $.42 \times 10,000 = 4200$. The number of loci at which an average individual is heterozygous, H, is $.28 \times 4200 = 1175$. This value of H refers to wild-type stock. Now let us determine H', the number of loci that are heterozygous in an average experimental individual. The experimental technique consists of making individuals that are homozygous for an entire chromosome. In particular chromosome II comprises about 40 percent of the *D. melanogaster* genome. Hence this experimental stock will be heterozygous at the remaining 60 percent of the loci at which the wild-type stock is heterozygous. Therefore, $H' = .6 \times 1175 = 705$ so that $H - H' = 470$. With these data, by (6.7), we predict in Table 6.5 the following ratios of the fitness of experimental animals to wild-type animals as a function of the degree of heterozygote superiority at the polymorphic loci. Thus, the fitness of flies that are homozygous for chromosome II should be essentially zero if w_{ho}/w_{het} is .9, very small but perhaps measurable if w_{ho}/w_{het} is .99, and approximately half that of wild-type flies if w_{ho}/w_{het} is .999. As you can see, the degree of heterozygote superiority at the polymorphic loci has a tremendous consequence for the fitness of flies that are homozygous for one of their chromosomes.

Table 6.5. Ratios of Fitness of Experimental Animals to Wild-Type Animals.

w_{ho}/w_{het}	$s = 1 - w_{ho}/w_{het}$	w_e/w_t
.900	.100	3.12×10^{-22}
.990	.010	.0089
.999	.001	.6249

Sved and Ayala (1970) and Sved (1971) have succeeded in measuring the fitness of *D. melanogaster* flies homozygous for chromosome II relative to the wild type. (The experimental techniques of these researchers are very elegant, and interested readers are urged to consult the original papers and the summary in Lewontin (1974), pp. 47–66). They determined that the relative fitness of chromosome II homozygotes, w_e/w_t, is .17. From this datum we can infer that the average degree of heterozygote superiority at the polymorphic loci is

$$\frac{w_{ho}}{w_{het}} = (.17)^{1/470} = .996$$

Or, if the fitnesses are expressed in terms of selection coefficients such as $1-s$, 1, and $1-s$, then the average selection coefficient against the homozygotes at each locus is $s = .004$.

This calculation is not very comforting to advocates of the selectionist hypothesis. A selection coefficient of .004 indicates very weak selection. Selection this weak only slowly restores any accidental perturbation to gene frequencies. The fluctuation from year to year in the selection pressures may easily be larger than .004. With selection this weak, drift can be a very significant factor. We learned in Chapter 5 that for selection to "overpower" drift requires $4 Ns \gg 1$. With $s = .004$, the population size of the breeding component of the population must be substantially over 100, say near 1000 *every generation*. Although this requirement may often be met by *Drosophila* (though this, like other features of its ecology, is not known), it may not be met by many other populations including the rarer species of plants and vertebrates. The final rub is that a selection coefficient of .004 would be extremely difficult, almost impossible, to measure directly.

The calculation above may, of course, not be appropriate. The overall fitness of an individual may not be the product of the fitnesses at each of the loci. We do not, however, currently know of any good alternatives. One avenue of future work is to search for other models for the selection pressure that would still account for the polymorphism and yet not be subject to the above criticisms or to others that are equally bad. Typically, the search for selection pressures other than heterozygote superiority requires great insight into the population structure and ecology of the population. As we shall see in later chapters, polymorphism can also be maintained by the spatial and temporal pattern in the environment, and by more complex selection pressures that are founded in the forces of inter- and intraspecific competition for resources. The use of these more complex selection pressures, however, requires an intimate knowledge of the ecology of the populations in their natural environment.

The attempts to detect the selection pressures at specific loci so far have met with varying success. In an extensive and very carefully designed study of a fish, the eel pout, reviewed in Christiansen (1977), Christiansen and Frydenberg were able to detect a very weak selection pressure at an esterase locus. But the selection pressure changed somewhat from year to year and was not obviously of the type that could maintain the observed polymorphism.

An important beginning has been made in terms of developing a *Drosophila* system whose ecology and genetics can be studied simultaneously. The *Drosophila* species are those associated with desert cactus. See Fellows and Heed (1972), Johnson and Heed (1975), and Barker (1977).

Other studies have found suggestive correlations between certain alleles and environmental features. Koehn (1969), Koehn et al. (1971), and Meritt (1972) have established correlations between temperature optima for some esterase enzymes assayed *in vitro*, and gradients in the environment. Also, M. S. Johnson (1971) in a blenny, and Clegg and Allard (1972) and Hamrick and Allard (1972) in wild oats, have established a detailed correlation between certain alleles and features of the environment. More recent data in this vein, combining field and enzymological information on *Colias* butterflies, are found in Johnson (1976) and Watt (1977). Kojima, Gillespie, and Tobari (1970), as well, have pointed out a suggestive difference between the degree of polymorphism in enzymes whose substrates are presumed to be brought in as food from the outside environment versus enzymes whose substrates are presumably of internal origin. In the laboratory with *D. melanogaster*, Gibson (1970) has been able to show substantial differences in the activities of several allelic forms of alcohol dehydrogenase under different environmental conditions.

This direct examination of the selection at particular loci has begun to yield interesting findings and much research of this type is in progress. But a great many loci will have to be examined before it can be concluded that a major part of the polymorphism in the genome is maintained by selection. Also, people must report the well-designed attempts to detect selection that have failed as well as those that have succeeded in order to prevent the published literature from representing a biased sample of the studies.

As you can see, the controversy about how the basic forces of selection, drift, and mutation have shaped the genetic structure of natural populations is far from resolved. It is hoped that this chapter has given you food for thought and provided an opportunity to apply the ideas learned in the previous chapters to a real problem under current research. To pursue this topic further, consult the books by Hochachka and Somero (1973), Kimura and Ohta (1971), Lewontin (1974), and Nei (1975) and also recent review articles by Ayala et al. (1974), Ayala (1977), Johnson (1973 and 1974), Selander and Johnson (1973), Ewens (1977), Ewens and Feldman (1976), and A. C. Wilson et al. (1977).

PART TWO
COMPLEX GENETIC SYSTEMS

Chapter 7
NATURAL SELECTION WITH MULTIPLE ALLELES AT ONE LOCUS

THERE are often more than two alleles at one locus. Indeed, the recent surveys of enzyme polymorphism, mentioned in the last chapter, reveal many loci with 4 to 5, even 10 alleles. What, then, are the forces that cause so many alleles to occur at a locus? In particular, what kinds of selection pressures can cause a multiple allele polymorphism? To answer these questions, we need to extend our theory of population genetics to cover multiple alleles at one locus. We shall see that virtually all of the *techniques* of two-allele theory can be elegantly generalized to multiple alleles— however, the conclusions from these techniques can become complex. Also the theory in this chapter is a prerequisite to understanding multilocus population genetics and is a helpful mathematical analogy to ecological models involving many interacting species.

The Basic Equations　The equations for the gene frequencies at $t+1$ are obvious extensions of the two allele theory. Let $p_{i,t}$ be the frequency of allele A_i at time t, w_{ij} be the relative selective value of the phenotype produced by the A_iA_j genotype, and let n be the number of alleles. Then

$$p_{i,t+1} = \left[\frac{\sum\limits_{j=1}^{n} p_{j,t}w_{ij}}{\bar{w}_t} \right] p_{i,t} \qquad (i = 1, 2 \dots n) \qquad (7.1)$$

where, as before, the mean selective is

$$\bar{w}_t = \sum_{i=1}^{n} \sum_{j=1}^{n} p_{i,t}p_{j,t}w_{ij} \qquad (7.2)$$

Note that the matrix, w_{ij}, is a symmetric matrix. If n equals 2, these become the familiar equations for p and q. Our first task will be develop a method for finding the equilibrium gene frequencies, $\hat{p}_i$.

The Equilibrium Frequencies　The equilibrium frequencies are, as before, found by setting the expression in brackets in Equation (7.1) equal to 1, giving

$$\sum_{j=1}^{n} \hat{p}_j w_{ij} = \bar{w} \qquad (7.3)$$

This is a set of simultaneous equations; for example, if $n = 3$, we have

$$\hat{p}_1 w_{11} + \hat{p}_2 w_{12} + \hat{p}_3 w_{13} = \bar{w}$$

$$\hat{p}_1 w_{21} + \hat{p}_2 w_{22} + \hat{p}_3 w_{23} = \bar{w} \qquad (7.4)$$

$$\hat{p}_1 w_{31} + \hat{p}_2 w_{32} + \hat{p}_3 w_{33} = \bar{w}$$

We should solve these simultaneous equations for $\hat{p}_1$, $\hat{p}_2$, and $\hat{p}_3$, subject to the constraint, of course, that $\hat{p}_1 + \hat{p}_2 + \hat{p}_3 = 1$. There is a trick for solving these equations. The $\bar{w}$ on the right can temporarily be taken as a constant and absorbed into the p's. More specifically, let us define a new quantity, p_i^*

as

$$p_i^* \equiv \frac{\hat{p}_i}{\bar{w}} \tag{7.5}$$

The reason for using p_i^*'s is that we can easily solve for them and then regenerate the $\hat{p}$'s. For example, when n is three we have three simultaneous linear equations for the p_i^*'s,

$$p_1^* w_{11} + p_2^* w_{12} + p_3^* w_{13} = 1$$
$$p_1^* w_{21} + p_2^* w_{22} + p_3^* w_{23} = 1 \tag{7.6}$$
$$p_1^* w_{31} + p_2^* w_{32} + p_3^* w_{33} = 1$$

Then by Cramer's rule

$$p_i^* = \frac{\Delta_i}{\Delta} \tag{7.7}$$

where Δ denotes the determinant of the w_{ij} matrix and Δ_i denotes the determinant of a matrix made from the w_{ij} matrix by replacing the ith column with 1's.† This formula gives us the p_i^*'s, and all that is left is to regenerate the $\hat{p}$'s. We do this by first finding $\bar{w}$ as follows: The sum of the p_i^* is

$$\sum_{i=1}^{n} p_i^* = \frac{\left(\sum_{i=1}^{n} \hat{p}_i\right)}{\bar{w}} = \frac{1}{\bar{w}} \tag{7.8a}$$

so

$$\bar{w} = \frac{1}{\sum_{i=1}^{n} p_i^*} = \frac{\Delta}{\sum_{i=1}^{n} \Delta_i} \tag{7.8b}$$

Then, since $\hat{p}_i = \bar{w} p_i^*$, we have

$$\hat{p}_i = \bar{w} \frac{\Delta_i}{\Delta} = \left(\frac{\Delta}{\sum_{i=1}^{n} \Delta_i}\right)\left(\frac{\Delta_i}{\Delta}\right) = \left(\frac{\Delta_i}{\sum_{i=1}^{n} \Delta_i}\right) \tag{7.9}$$

This is the answer; the equilibrium frequency of A_i is given by Δ_i divided by the sum of all the Δ_i's. As the first problem you should confirm for two alleles that $\hat{p}$ computed from Equation (7.9) gives the same result that we obtained earlier in various ways.

Stability of the Equilibrium with All n Alleles Present

Once we have found an equilibrium by the method above, the problem arises of determining whether the equilibrium is stable or unstable. The way to find out whether an equilibrium is stable or not relies on the idea that natural selection maximizes the mean fitness, $\bar{w}$. Briefly, the way we shall determine whether an equilibrium is stable is to determine whether $\bar{w}$ is maximized at the equilibrium point.

There are several parts to this section: (1) We must show that equilibria in a one-locus n allele system correspond to critical points of the $\bar{w}$ function.

† We assume $\Delta \neq 0$, for otherwise two or more alleles would be indistinguishable.

(2) We show that fitness is always increasing except at equilibrium. With this result we then know that *stable* equilibria correspond to maxima of the $\bar{w}$ function and that unstable equilibria correspond to minima or saddlepoints of the $\bar{w}$ function. (3) Finally, we present some practical techniques to calculate whether $\bar{w}$ is maximized at an equilibrium point and (4) some miscellaneous results.

Equilibria Lie at
Critical Points
of $\bar{w}$

First, we consider equilibria in a one-locus n-allele system. We show that they lie at critical points of the $\bar{w}$ function subject to the constraint that $\sum p_i = 1$. To begin, let us inspect the equations for Δp_i. If we take each equation for $p_{i,t+1}$ and subtract $p_{i,t}$ from both sides, and then rearrange slightly, we obtain

$$\Delta p_i = \frac{p_i(w_i - \bar{w})}{\bar{w}} \qquad (i = 1 \cdots n) \tag{7.10a}$$

where

$$w_i = \sum_j p_j w_{ij} \tag{7.10b}$$

and w_i is called the marginal fitness of allele A_i. Now by inspection of (7.10), we observe that the necessary and sufficient condition for a polymorphic equilibrium point is that

$$w_i = \bar{w} \qquad \text{for all } i \tag{7.11}$$

Next we show that this condition is identical to that for a point to be a critical point of $\bar{w}$ subject to the constraint that $\sum p_i = 1$.

We now want to find critical points of the function

$$\bar{w} = \sum p_i w_i = \sum_i \sum_j p_i p_j w_{ij} \tag{7.12a}$$

subject to the constraint that

$$\sum p_i - 1 = 0 \tag{7.12b}$$

To do this, we introduce a Lagrange multiplier, λ, and form the function

$$\bar{w} - \lambda \left(\sum p_i - 1 \right) \tag{7.13}$$

The critical point $(\hat{p}_1 \cdots \hat{p}_n)$ together with the unknown value of λ are determined from the following $n + 1$ equations:

$$\frac{\partial \bar{w}}{\partial p_i} - \lambda = 0 \qquad (i = 1 \cdots n) \tag{7.14a}$$

$$\sum_i p_i = 1 \tag{7.14b}$$

Let us first find the value of λ from this system of equations. Note that each equation in (7.14a) can be written as

$$2\sum p_j w_{ij} - \lambda = 0 \qquad (i = 1 \cdots n) \tag{7.15}$$

Then multiply each of these equations by p_i yielding

$$2\sum_j p_i p_j w_{ij} - \lambda p_i = 0 \qquad (i = 1 \cdots n) \tag{7.16}$$

Now add all the n-equations of this form together, yielding

$$2\sum_i \sum_j p_i p_j w_{ij} - \lambda \sum_i p_i = 0 \qquad (7.17)$$

that is,

$$2\bar{w} - \lambda = 0 \qquad (7.18)$$

Hence the value of the Lagrange multiplier is

$$\lambda = 2\bar{w} \qquad (7.19)$$

Now we can proceed with finding the critical point. By (7.14a) the critical point satisfies

$$\frac{\partial \bar{w}}{\partial p_i} = \lambda = 2\bar{w} \qquad (i = 1 \cdots n) \qquad (7.20)$$

Since $\partial \bar{w}/\partial p_i = 2w_i$, we have determined that the critical point of $\bar{w}$, subject to the constraint $\sum p_i = 1$, satisfies

$$w_i = \bar{w} \qquad \text{for all } i \qquad (7.21)$$

Clearly, (7.21) is the same as (7.11). Thus we have demonstrated that polymorphic equilibria of the one-locus n-allele model correspond with critical points of the function $\bar{w}$ subject to the constraint that $\sum p_i = 1$.

The Mean Fitness Always Increases Except at Equilibrium
The mean fitness, $\bar{w}$, always increases from generation to generation in the one-locus n-allele model except at equilibrium, at which point the mean fitness remains constant. A shorthand way to state this result is that

$$\Delta \bar{w} \geqslant 0 \qquad (7.22)$$

with equality holding if and only if the system is at equilibrium. We prove this result by simply verifying that it is true. The proof follows Ewens (1969) and Kingman (1961b). The mean fitness at time $t+1$ is given by

$$\bar{w}_{t+1} \equiv \sum_i \sum_j w_{ij} p_{i,t+1} p_{j,t+1}$$

$$= \sum_i \sum_j w_{ij} \left[\left(\frac{\sum_k w_{ik} p_{k,t}}{\bar{w}_t} \right) p_{i,t} \right] \left[\left(\frac{\sum_l w_{jl} p_{l,t}}{\bar{w}_t} \right) p_{j,t} \right]$$

$$= \bar{w}_t^{-2} \sum_i \sum_j \sum_k \sum_l w_{ij} w_{ik} w_{jl} \, p_{i,t} p_{j,t} p_{k,t} p_{l,t} \qquad (7.23)$$

$$\Delta \bar{w} \equiv \bar{w}_{t+1} - \bar{w}_t$$

$$= \bar{w}^{-2} \left[\sum_i \sum_j \sum_k \sum_l w_{ij} w_{ik} w_{jl} p_i p_j p_k p_l - \bar{w}^3 \right] \qquad (7.24)$$

thus $\Delta \bar{w} \geqslant 0$ is equivalent to.

$$\sum_i \sum_j \sum_k \sum_l w_{ij} w_{ik} w_{jl} p_i p_j p_k p_l \geqslant \bar{w}^3 \qquad (7.25)$$

with equality holding only at equilibrium.

To show the inequality in (7.25), we use two other general inequalities. Assume that $p_i \geq 0$ and $\sum p_i = 1$. For any set of $b_i \geq 0$ and for any $\alpha \geq 1$ we have

$$\sum p_i b_i^{\alpha} \geq \left(\sum p_i b_i \right)^{\alpha} \tag{7.26}$$

This result states the convexity property of the function b^{α}. The second general inequality is the so-called arithmetic-geometric mean inequality,

$$\tfrac{1}{2}(x + y) \geq \sqrt{xy} \tag{7.27}$$

for nonnegative x and y.

We begin with some steps of rearrangement that yield

$$\sum_i \sum_j \sum_k \sum_l w_{ij} w_{ik} w_{lj} p_i p_j p_k p_l$$

$$= \sum_i \sum_j \sum_k w_{ij} w_{ik} (\tfrac{1}{2}) \left[\sum_l w_{lj} p_l + \sum_m w_{mk} p_m \right] p_i p_j p_k \tag{7.28}$$

Now by the arithmetic-geometric mean inequality (7.27) we have

$$\geq \sum_i \sum_j \sum_k w_{ij} w_{ik} \left[\sum_l w_{lj} p_l \sum_m w_{mk} p_m \right]^{1/2} p_i p_j p_k$$

$$= \sum_i p_i \left\{ \sum_j w_{ij} p_j \left(\sum_l w_{lj} p_l \right)^{1/2} \right\}^2 \tag{7.29}$$

Now we apply the convexity inequality (7.26) to get

$$\geq \left\{ \sum_i p_i \sum_j w_{ij} p_j \left(\sum_l w_{lj} p_l \right)^{1/2} \right\}^2$$

$$= \left[\sum_j p_j \left(\sum_l w_{lj} p_l \right)^{3/2} \right]^2 \tag{7.30}$$

Now applying the convexity inequality (7.26) again, we obtain

$$\geq \left[\sum_j \sum_l w_{lj} p_l p_j \right]^3 = \bar{w}^3 \tag{7.31}$$

Furthermore, equality obtains in each of the three key steps above if and only if $\sum_j w_{ij} p_j = $ const, where const is a constant independent of i. This condition is met if and only if $\sum w_{ij} p_j = \bar{w}$ for all i; that is, if and only if the system is at equilibrium.

The result above shows that the mean fitness increases from generation to generation until equilibrium is reached. Some implications of this result are: First, a polymorphic equilibrium is locally stable if and only if it is a local *maximum* of $\bar{w}$ subject to the constraint that $\sum p_i = 1$. Second, suppose that there exists a full polymorphic equilibrium that is a local maximum of $\bar{w}$ and hence is locally stable. We would also like to know if there is global convergence to that equilibrium point. The problem is that it is conceivable that some trajectories could lead to some other equilibrium points; equilibria at which one or more alleles are lost. However, it has also been shown that if a full polymorphic equilibrium is locally stable, then all of the other equilibria are unstable, that is, local stability of the full polymorphic equilibrium implies global convergence as studied by Kingman (1961a). Third, persistent oscillatory behavior cannot arise in the one-locus n-allele model; instead, there is always convergence to an equilibrium point. Hence, unlike ecological models, or even other genetic models to which frequency dependence and other complications are added, the standard one-locus n-allele model always shows convergence to an equilibrium point.

How to Determine If an Equilibrium Point Lies at the Maximum of $\bar{w}$

Next we consider several practical criteria for determining whether an equilibrium point maximizes $\bar{w}$ subject to the constraint that $\sum p_i = 1$. Armed with the result just stated, our problem of determining whether an equilibrium is stable reduces to determining whether the equilibrium point maximizes $\bar{w}$ subject to the constraint that $\sum p_i = 1$. Some general techniques for determining conditions for a constrained maximum can be found in Hancock (1960, pp. 115–116) and Bellman (1970, pp. 78–80). However, it has been shown that these general conditions can be simplified when applied to our specific problem. We now state, without proof, these simplified conditions, which are necessary and sufficient for an equilibrium point to be a constrained maximum to $\bar{w}$. The proofs themselves are purely technical and are found in the original papers.†

MANDEL'S DETERMINANTAL CRITERIA. Let

$$w\begin{pmatrix} 1, 2, 3 \cdots r \\ 1, 2, 3 \cdots r \end{pmatrix}$$

denote the rth successive principal minor of the w_{ij} matrix as follows

$$w\begin{pmatrix} 1, 2, 3 \cdots r \\ 1, 2, 3 \cdots r \end{pmatrix} = \begin{vmatrix} w_{11} & w_{12} & w_{13} & \cdots & w_{1r} \\ w_{21} & w_{22} & w_{23} & \cdots & w_{2r} \\ w_{31} & w_{32} & w_{33} & \cdots & w_{3r} \\ \vdots & \vdots & \vdots & & \vdots \\ w_{r1} & w_{r2} & w_{r3} & \cdots & w_{rr} \end{vmatrix} \tag{7.32}$$

The argument in the symbol

$$w\begin{pmatrix} 1, 2, 3 \cdots r \\ 1, 2, 3 \cdots r \end{pmatrix}$$

enumerates, on the top, the labels of all the rows and, on the bottom, the labels of all the columns from the full w_{ij} matrix that are used in the determinant. Mandel's (1959) criteria are that *a positive polymorphic equilibrium is stable if and only if*

$$(-1)^r w\begin{pmatrix} 1, 2, 3 \cdots r \\ 1, 2, 3 \cdots r \end{pmatrix} < 0 \qquad (r = 2, 3 \cdots n) \tag{7.33}$$

Thus successive principal minors of the w_{ij} matrix must alternate in sign, beginning with $w\begin{pmatrix} 1, 2 \\ 1, 2 \end{pmatrix}$ being negative.

KINGMAN'S EIGENVALUE CRITERION. Again consider the w_{ij} matrix. It is symmetric because the genotype A_iA_j is identical to A_jA_i, and therefore $w_{ij} = w_{ji}$. The eigenvalues of a symmetric matrix are real. Kingman's (1961a) criterion is that *a positive polymorphic equilibrium is stable if*

† Here we consider the generic situation wherein none of the minors and eigenvalues of the w_{ij} matrix is zero.

and only if the w_{ij} matrix has one positive eigenvalue and $(n-1)$ negative eigenvalues.

These criteria are completely equivalent to one another. They both provide conditions for the mean fitness, $\bar{w}$, to be a constrained maximum as discussed above. Indeed, Mandel's criteria follow immediately from Kingman's on application of Jacobi's theorem. (See Gantmacher, Vol. 1, p. 303.) However, it is usually easier to use Mandel's criteria than Kingman's unless the w_{ij} matrix has a special structure.

Some Miscellaneous Results

We record here some miscellaneous results on the one-locus n-allele model that are of general interest and are proved in the papers by Mandel and Kingman.

(a) A necessary (but by no means sufficient) condition for a polymorphism with n-alleles is that

$$w_{ii} < \hat{\bar{w}} \qquad \text{for all } i \qquad (7.34)$$

where w_{ii} is the fitness of a homozygote for A_iA_i and $\hat{\bar{w}}$ is the mean fitness at the polymorphic equilibrium. Mandel has interpreted this condition as an analogue to heterosis. Roughly speaking, it means that not too many of the heterozygotes can be less viable than any of the homozygotes. In the two-allele case this is, of course, the classic condition that the heterozygote is fitter than both homozygotes. In the three-allele case Mandel has shown explicitly that at most one heterozygote viability may fall below that of at most two homozygotes.

(b) If the w_{ij} matrix has more than one positive eigenvalue, then the full polymorphic solution, if it exists, cannot be stable. Kingman shows that if the w_{ij} matrix has k-positive eigenvalues and if all n-alleles are initially present, then *at least* $k-1$ genes must die out before a stable equilibrium can be reached.

The Boundary Equilibria

The equilibrium with all n alleles coexisting is often called an "interior equilibrium." The terminology arises from a geometrical representation of the gene frequencies. Figure 7.1 illustrates equilateral triangles with unit height. Any point in the middle can be used to represent (p_1, p_2, p_3). By elementary geometry one can prove that the sum of the perpendiculars to the sides equals one in an equilateral triangle with unit altitude. A point in the "interior" of the triangle represents a condition where none of the p_i equal zero, as illustrated in Figure 7.1a. A point on the "boundary" represents a condition where one allele is absent; for example, in Figure 7.1b the allele A_2 is absent. The vertices represent conditions where only one allele is present. This geometrical representation can be generalized to higher dimensions, although it rapidly becomes tedious to illustrate.

The complete characterization of the equilibria of a one-locus n-allele system is the enumeration of all the equilibria, both interior and boundary, and their stability properties. The analysis of a boundary equilibrium occurs in two parts. The first is simply the application of the preceding theory in a reduced system. Thus, to find the equilibrium on the A_1-A_3 boundary, simply solve for p_1 and p_3 using a w_{ij} matrix with the second row and column deleted. Similarly, to determine the stability of this equilibrium with respect to *perturbations along the boundary*, simply apply the criterion of the preceding section to the reduced w_{ij} matrix.

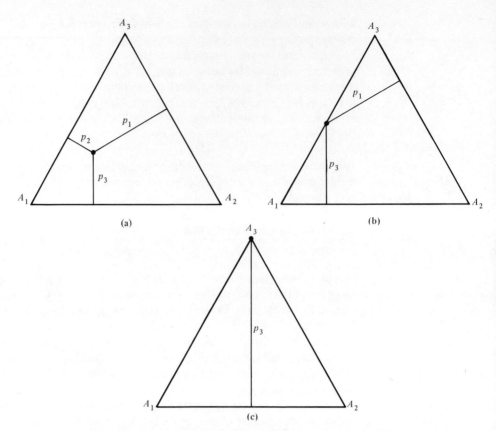

FIGURE 7.1. (a) *An interior equilibrium, p_1, p_2, $p_3 > 0$; (b) a boundary equilibrium, $p_1, p_3 > 0, p_2 = 0$; and (c) a corner or fixation equilibrium, $p_3 = 1, p_1 = p_2 = 0$.*

The second part of the boundary analysis is to determine stability with respect to *perturbations away from* the boundary. For example, an A_1-A_3 boundary equilibrium may be stable to perturbations to p_1 and p_3, provided that p_2 is kept equal to zero, but unstable if p_2 is perturbed. This problem is often synonymously stated as: Can A_2 increase when rare? If $\Delta p_2 > 0$ when evaluated at the boundary equilibrium, then the boundary equilibrium is unstable and A_2 can enter the system. If $\Delta p_2 < 0$, then a small introduction of A_2 will be counteracted and the system will return to the boundary equilibrium.

The criterion for increase when rare is very simple. Let A_k denote the allele about to be introduced. Then from Equation (7.10)

$$\Delta p_k = \left(\sum_{j \neq k}^{n} p_j w_{jk} - \bar{w} \right) \frac{p_k}{\bar{w}} \tag{7.35}$$

The criterion for $\Delta p_k > 0$ is simply

$$\sum_{j \neq k}^{n} \hat{p}_j w_{jk} > \bar{w} \tag{7.36}$$

where the $\hat{p}_j$ are computed using the reduced w_{ij} matrix obtained by deleting the kth row and column and $\bar{w}$ is the mean fitness at this boundary point. In words, this condition asserts that the average fitness of individuals that are heterozygous for the new allele, A_k, must exceed the average fitness in the population in the absence of A_k. Thus, for A_k to increase when rare, it must on the average convey an advantage to its heterozygotes relative to the rest of the population. This analysis does not include the case of full dominance. Thus the analysis of boundary equilibria is carried out with the preceding theory applied to a reduced w_{ij} matrix and is supplemented with the criterion above for increase when rare.

The Adaptive Topography

The idea of the adaptive topography can also be generalized to the model for one locus with n alleles. Recall that with two alleles we were able to arrange Δp in the form

$$\Delta p = \frac{p(1-p)}{2\bar{w}} \frac{d}{dp} \bar{w}(p) \qquad (n=2) \tag{7.37}$$

The problem in generalizing this idea to more than two alleles is that $\bar{w}$ becomes a function of several variables, $\bar{w}(p_1, p_2 \cdots p_n)$. So the analogue to (7.37) must involve some sort of partial derivative; that is, the slope of $\bar{w}$ must be specified *in some direction*. It is tempting to suggest that the usual partial derivative, $\partial\bar{w}/\partial p_i$ can be used in a formula for Δp_i, but in fact this suggestion does not even work with two alleles, since even then Δp depends on $d\bar{w}/dp$, not $\partial\bar{w}/\partial p$.

The answer to the problem is as follows: Consider the equation for Δp_k. Let all the other alleles p_i be considered together as a block; this block has frequency $(1-p_k)$. Let us consider this block of all the alleles except A_k as a unit such that the relative proportion of the alleles within the block is constant. Specifically, let the c_{ik} be the fraction of A_i within the block from which a_k is excluded. Then by definition

$$p_i = c_{ik}(1-p_k) \qquad (i \neq k) \tag{7.38}$$

because $(1-p_k)$ is the frequency of a non-A_k allele, and c_{ik} is the fraction of the non-A_k alleles which is A_i. Then let us regard the c_{ik} as fixed constants that describe the composition of the block of non-A_k alleles. The idea behind the generalization of adaptive topographies to n alleles is to consider the slope of $\bar{w}$ in a direction such that changes in the frequency of A_k are matched by equal and opposite changes in the frequency of the non-A_k block, thereby preserving the constraint that $p_k + (1-p_k) = 1$. It is understood that the c_{ik} are constants in any such change, that is, that the relative proportions of the different non-A_k alleles are fixed. Let the symbol, $\partial^*/\partial^* p_k$, denote a differentiation under this convention. In practice, to differentiate by this convention, substitute every occurrence of $p_i (i \neq k)$ with $c_{ik}(1-p_k)$ and then take the derivative with respect to p_k. Then by this convention Δp_k can be arranged as

$$\Delta p_k = \frac{p_k(1-p_k)}{2\bar{w}} \frac{\partial^*}{\partial^* p_k} \bar{w} \tag{7.39}$$

As exercises you should verify that, with two alleles, $\partial^*\bar{w}/\partial^* p$ is identical to $d\bar{w}/dp$, and also work an example of this differentiation convention with three alleles. This differentiation convention achieves the generalization of adaptive topographies to n alleles; it specifies in what direction the slope of $\bar{w}$ is to be taken so as to be relevant for controlling Δp_k.

In summary, we have seen that the one-locus two-allele theory of the preceding chapters can be extended to n alleles. This extension takes the form of a general machinery for determining the equilibrium behavior of selection models with n alleles. We shall explicitly use these results in the next chapter.

Chapter 8

POPULATION GENETICS WITH MULTIPLE LOCI

$\mathbf{T}$HE theory of population genetics at two and more loci has proved to be very different from single locus theory, and this theory is currently the subject of very active research. Two of the important issues in multiple locus theory concern independence and dependence, respectively. If two loci are independent in their functions, for example, one locus controls flower color and the other controls something entirely different, like leaf thickness, then can we use one locus theory separately for each locus? Of course, we would hope so, but can we show that, given functional independence of two loci, it follows that the population genetics at each locus is independent of the other? On the other hand, suppose that the two loci are simultaneously involved in the expression of a given character, for example, the loci may produce enzymes in the same biosynthetic pathway. Can selection on these two loci cause the aggregation of "packets" of alleles from different loci which function well together? These kinds of questions underlie the study of multilocus population genetics.

There are three large parts in this chapter. First, we shall carefully introduce the language used in the theory of two loci; second we shall derive an analogue of the Hardy–Weinberg law for two loci; and third, we shall review some of the results from models of natural selection at two loci.

The Language Let us agree to census the population at the gametic phase and assume that zygotes are formed by random union of the gametes. There are four kinds of gametes, and by convention we shall enumerate them in the following order:

$$
\begin{array}{ccc}
\text{gamete type} & & \text{frequency} \\
AB & \leftrightarrow & x_{1,t} \\
Ab & & x_{2,t} \\
aB & & x_{3,t} \\
ab & & x_{4,t}
\end{array}
\tag{8.1}
$$

The frequency of gamete type AB at time t is denoted as $x_{1,t}$, and similarly for the other types. (The sum of the x's is one, of course.) In addition, we have, again, the allele frequencies within each locus. Since we are censusing the gametes that are haploid, we have simply

$$
\begin{aligned}
p_{A,t} &\equiv x_{1,t} + x_{2,t} \\
p_{a,t} &\equiv x_{3,t} + x_{4,t} \\
p_{B,t} &\equiv x_{1,t} + x_{3,t} \\
p_{b,t} &\equiv x_{2,t} + x_{4,t}
\end{aligned}
\tag{8.2}
$$

(Of course, $p_A + p_a = 1$ and $p_B + p_b = 1$.) These symbols could be taken as completely describing the system, but it proves convenient to introduce still another quantity, D. This quantity is a measure of the statistical dependence between the two loci. Suppose that one draws a gamete from the gamete pool and determines which allele is at the A locus. Then, is this knowledge of the A locus relevant to predicting what is at the B locus of the same gamete? If not, the two loci are independent and if so, there is statistical dependence between them.† The measure of the statistical dependence, D, is defined as

$$D_t = x_{1,t}x_{4,t} - x_{2,t}x_{3,t} \tag{8.3}$$

Now let us see how a quantity defined in this curious way is a measure of statistical dependence. In the boxed material it is shown that the x's are related to the p's and D by the following identities

$$
\begin{aligned}
x_{1,t} &= p_{A,t}p_{B,t} + D_t \\
x_{2,t} &= p_{A,t}p_{b,t} - D_t \\
x_{3,t} &= p_{a,t}p_{B,t} - D_t \\
x_{4,t} &= p_{a,t}p_{b,t} + D_t
\end{aligned}
\tag{8.4}
$$

To see what these identities mean, consider first $x_{1,t}$. Suppose that the A and B loci are independent. Then the probability of a gamete containing *both* an A *and* a B should equal the product of the probabilities of drawing an A from the gene pool times the probability of drawing a B from the gene pool. Hence if the two loci are statistically independent, then x_1 must equal $p_A p_B$.

---

1. Prove that

$$x_1 \equiv p_A p_B + D$$

Substitute for p_A, p_B, and D yielding

$$
\begin{aligned}
x_1 &= (x_1 + x_2)(x_1 + x_3) + (x_1 x_4 - x_2 x_3) \\
&= x_1^2 + x_1 x_2 + x_2 x_3 + x_1 x_3 + x_1 x_4 - x_2 x_3 \\
&= x_1^2 + x_1 x_2 + x_1 x_3 + x_1 x_4 \\
&= x_1(x_1 + x_2 + x_3 + x_4) \\
&\equiv x_1
\end{aligned}
$$

2. Prove that

$$x_2 \equiv p_A p_b - D$$

Substitute for p_A, p_b, and D yielding

$$
\begin{aligned}
x_2 &= (x_1 + x_2)(x_2 + x_4) - (x_1 x_4 - x_2 x_3) \\
&= x_2 x_1 + x_1 x_4 + x_2^2 + x_2 x_4 - x_1 x_4 + x_2 x_3 \\
&= x_2 x_1 + x_2^2 + x_2 x_3 + x_2 x_4 \\
&= x_2(x_1 + x_2 + x_3 + x_4) \\
&\equiv x_2
\end{aligned}
$$

3. The identities for x_3 and x_4 are proved in a similar way.

---

†This idea is expressed more precisely by saying the two loci are independent if the conditional probability of B, given A, equals the unconditional probability of B.

Similarly, x_2 will equal $p_A p_b$, x_3 will be $p_a p_B$, and x_4 will be $p_a p_b$. If we substitute these particular values for the x's into the formula for D, we see that D is zero, thus confirming that D is zero if the loci are independent. However, if the loci are not independent, then D is not zero and D can be viewed as a "correction factor" in formula (8.4) above. The term D *indicates the extent to which the x's differ from the values they would have if the loci were independent.* Also, D is *defined* so that it can be used in this way. This is the sense in which D is a measure of statistical dependence between the loci. It is very important to understand that D measures only *statistical* dependence between loci and that there are many kinds of causes of such statistical dependence including causes that do not have anything to do with biology. For example, if we stock a fish tank with gametes of different origin such that 50 percent of the gametes are "AB" and the other 50 percent are "ab" then $x_1 = \frac{1}{2}$, $x_2 = 0$, $x_3 = 0$, and $x_4 = \frac{1}{2}$. With these values of x_i, D is found to be $\frac{1}{4}$. There is statistical dependency here simply because an experimenter set it up this way and not because of any underlying biological process. Thus D only *describes* statistical dependence; it does not in any way suggest the cause of the dependency.[†]

We shall use this language of x's, p's, and D to describe population genetics at two loci. For example, if D is positive, then we have the beginning of what might be described as gene complexes—both the A allele is found more often with the B allele, and the a allele more often with the b allele, than expected by chance if they were distributed independently. Similarly, if D is negative, the gene complexes are Ab and aB. To be considered a complex, however, the loci must not only have a nonzero value of D, but also must be located close to one another on the chromosome. Therefore, we shall want to seek the conditions under which selection causes D to stabilize at some nonzero value and perhaps thereby begin to understand whether gene complexes can form. On the other hand, if the two loci are independent in their functions, we would hope that D becomes zero. The key to understanding multilocus population genetics is to become adept at phrasing biological questions in terms of the special language in which the mathematics is carried out.

As a final note on terminology, D has many names and the most common is the "linkage disequilibrium coefficient." It is also commonly agreed that this name is misleading in that D need not have anything to do with linkage at all. Nonetheless, we shall follow common usage and refer to D as the linkage disequilibrium coefficient, but again, its importance is as a measure of statistical dependence between the loci.

The Hardy–Weinberg Situation

We begin our two-locus study with the Hardy–Weinberg situation, that is, by assuming that there is no natural selection, no mutation, no genetic drift, and that there is random union of gametes among the whole population. One new feature, however, is the introduction of a parameter, r, which characterizes the linkage between the loci. This parameter is called the *recombination fraction* and is related to the map distance between the loci. The

[†] The origin of D as a measure of statistical dependency is as follows. Let Y_1 be a random variable with value 1 if the A locus in a gamete contains an A allele, and value 0 if an a allele. Let Y_2 take value 1 for a B allele and 0 for a b allele. Then $E(Y_1) = p_A$ and $E(Y_2) = p_B$. The covariance of Y_1 with Y_2 is defined as $\mathrm{cov}(Y_1, Y_2) = E(Y_1 Y_2) - \bar{Y}_1 \bar{Y}_2$. Hence $\mathrm{cov}(Y_1, Y_2)$ equals $x_1 - p_A p_B$. Thus the coefficient, D, is simply the covariance between the two loci.

recombination fraction, r, is the map distance divided by 100 and is used to indicate the extent to which recombination occurs between the loci. For example, consider the gametes produced (by meiosis) from the dihybrid containing AB on one chromosome and ab on the other chromosome.

$$\frac{AB}{ab} \rightarrow \begin{cases} \frac{1}{2}(1-r)AB \\ \frac{1}{2}(1-r)ab \\ \frac{1}{2}rAb \\ \frac{1}{2}raB \end{cases} \tag{8.5}$$

If r approaches zero, then A and B are immediately adjacent on the chromosome and there should be negligible crossing over between them. Hence the gametic output from this dihybrid should be half AB gametes and half ab gametes. At the other extreme, if the A and B loci are on different chromosomes or at opposite ends of one long chromosome, then the gametic output will be $\frac{1}{4}AB, \frac{1}{4}Ab, \frac{1}{4}aB$, and $\frac{1}{4}ab$ because of independent segregation at meiosis. In this case, r is $\frac{1}{2}$. In general r lies between 0 and $\frac{1}{2}$ and if r is close to zero we say there is "tight" linkage and if near $\frac{1}{2}$ there is "loose" linkage.

Now let us examine how the gametic frequencies change from generation to generation as a function of r in the absence of selection, mutation and drift. The derivation is straightforward and is summarized in Table 8.1. First, the x's are censused among the gametes at time t. Then all 10 possible zygotic types are formed by random union of the gametes. Then gametes are in turn produced from these zygotes by meiosis. Note that the effect of crossing over can only be detected from the two kinds of dihybrids. Next the gametes of each type at time $t+1$ are summed up from the output of all 10 parental genotypes. Upon rearrangement we obtain

$$\begin{aligned} x_{1,t+1} &= x_{1,t} - rD_t \\ x_{2,t+1} &= x_{2,t} + rD_t \\ x_{3,t+1} &= x_{3,t}^* + rD_t \\ x_{4,t+1} &= x_{4,t} - rD_t \end{aligned} \tag{8.6}$$

And if we calculate the new value of D, we obtain

$$D_{t+1} = (1-r)D_t \tag{8.7}$$

These results are very important. In general terms they show how recombination between the loci, r, influences the statistical dependence between the loci, D. Specifically, let us first examine (8.7). By (8.7), after each generation, the statistical dependency, D, is reduced by a factor of $(1-r)$. In fact, we may iterate (8.7) starting with D_0 to obtain

$$D_t = (1-r)^t D_0 \tag{8.8}$$

Figure 8.1 illustrates D_t for several values of r. The figure shows that the initial statistical dependency, D_0, fades away rather slowly, and with a rate set by r. The loss of the statistical dependency between the loci is fastest with loose linkage, and slow with tight linkage. This result is surprising. The two loci in this Hardy-Weinberg situation may be functionally independent and may even be in different chromosomes. Yet if, by historical accident, there is

Table 8.1.

Gamete Pool at Time t			Zygote Pool			Gametic Output at Time $t+1$			
Frequency	Gamete Type		Frequency	Zygote Type		AB	Ab	aB	ab
$x_{1,t}$	AB		x_{1t}^2	$\dfrac{AB}{AB}$		1	0	0	0
$x_{2,t}$	Ab	$\rightarrow \rightarrow$ Random union of gametes	$2x_{1,t}x_{2,t}$	$\dfrac{AB}{Ab}$	$\rightarrow \rightarrow$ Production of new gametes through meiosis	$\frac{1}{2}$	$\frac{1}{2}$	0	0
$x_{3,t}$	aB		$2x_{1,t}x_{3,t}$	$\dfrac{AB}{aB}$		$\frac{1}{2}$	0	$\frac{1}{2}$	0
$x_{4,t}$	ab		$2x_{1,t}x_{4,t}$	$\dfrac{AB}{ab}$		$\frac{1}{2}(1-r)$	$\frac{1}{2}r$	$\frac{1}{2}r$	$\frac{1}{2}(1-r)$
			$x_{2,t}^2$	$\dfrac{Ab}{Ab}$		0	1	0	0
			$2x_{2,t}x_{3,t}$	$\dfrac{Ab}{aB}$		$\frac{1}{2}r$	$\frac{1}{2}(1-r)$	$\frac{1}{2}(1-r)$	$\frac{1}{2}r$
			$2x_{2,t}x_{4,t}$	$\dfrac{Ab}{ab}$		0	$\frac{1}{2}$	0	$\frac{1}{2}$
			$x_{3,t}^2$	$\dfrac{aB}{aB}$		0	0	1	0
			$2x_{3,t}x_{4,t}$	$\dfrac{aB}{ab}$		0	0	$\frac{1}{2}$	$\frac{1}{2}$
			$x_{4,t}^2$	$\dfrac{ab}{ab}$		0	0	0	1

Total each column to obtain
the frequency of each gamete type at
time $t+1$

$$\begin{aligned}
x_{1,t+1} &= x_{1,t}^2 + \tfrac{1}{2}(2x_{1,t}x_{2,t}) + \tfrac{1}{2}(2x_{1,t}x_{3,t}) + \tfrac{1}{2}(1-r)(2x_{1,t}x_{4,t}) + \tfrac{1}{2}r(2x_{2,t}x_{3,t}) \\
&= x_{1,t}(x_{1,t} + x_{2,t} + x_{3,t} + x_{4,t}) - r(x_{1,t}x_{4,t} - x_{2,t}x_{3,t}) \\
&= x_{1,t} - rD_t
\end{aligned}$$

$$\begin{aligned}
x_{2,t+1} &= \tfrac{1}{2}(2x_{1,t}x_{2,t}) + \tfrac{1}{2}r(2x_{1,t}x_{4,t}) + x_{2,t}^2 + \tfrac{1}{2}(1-r)(2x_{2,t}x_{3,t}) + \tfrac{1}{2}(2x_{2,t}x_{4,t}) \\
&= x_{2,t}(x_{1,t} + x_{2,t} + x_{3,t} + x_{4,t}) + r(x_{1,t}x_{4,t} - x_{2,t}x_{3,t}) \\
&= x_{2,t} + rD_t
\end{aligned}$$

Similarly, we obtain

$$x_{3,t+1} = x_{3,t} + rD_t$$

$$x_{4,t+1} = x_{4,t} - rD_t$$

some initial statistical dependency between the loci, this dependency is lost gradually through time. One might say that recombination through meiosis, even at its best, incompletely mixes the contents of the two loci. As a result, some time is needed for recombination to remove any initial statistical dependency between functionally independent loci. This result foreshadows

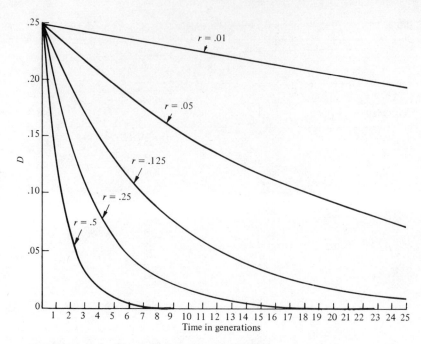

FIGURE 8.1. *The decay of linkage disequilibrium through time caused by recombination between loci. Time is measured in generations.*

the fact that the dynamics of natural selection will be much different at two loci than at only one, partly because meiotic recombination imparts a qualitatively new feature to the system.

Recall that the absence of selection caused the gene frequency to remain constant in our earlier one-locus study. This feature is still true here. Notice that $p_{A,t+1}$ is $x_{1,t+1} + x_{2,t+1}$. A glance at (8.6) shows that $p_{A,t+1}$ must therefore equal $x_{1,t} + x_{2,t}$ because the rD_t terms cancel out. Therefore, $p_{A,t+1} = p_{A,t}$, that is, the gene frequency does not change. Thus, as before, in the Hardy–Weinberg situation, the gene frequency within a locus remains constant. It is only the x's that are changing, not the p's.

Natural Selection
To understand the results of natural selection in multilocus genetic systems, it will be very helpful to consider a simple analogy. As you know, mutation is a source of genetic variability in a population. Recurrent mutation is also capable, in principle at least, of influencing gene frequencies. But since mutation rates are so low, we can usually (but not always) ignore the role of mutation in influencing gene frequencies and concentrate on the selection pressures instead. Recombination between loci is also a source of variation and is somewhat analogous to mutation. Recombination is *not* the source of new genes; it is the source of new *combinations* of genes. For example, if a population initially contains gametes only of the types AB and ab, then recombination in AB/ab heterozygotes results in the production of Ab and aB gametes. Thus recombination is a source of new gamete types. But, unlike mutation, recombination can be very fast. For example if A and B are in different chromosomes, then r is $\frac{1}{2}$. This is to be compared to mutation rates that are on the order of 10^{-6} to 10^{-9}. Recombination, like mutation, influences evolution but the strength of its effect is often comparable to the

selection pressures themselves. For this reason, we cannot generally ignore recombination when considering the action of natural selection in multilocus systems. To understand how recombination combines with natural selection to produce a *net* evolutionary outcome, we first explore, as an analogy, the joint action of natural selection and recurrent mutation.

Selection and Recurrent Mutation

Recall that selection at one locus with two alleles is governed by

$$\Delta p_{\text{sel}} = \frac{p(1-p)}{2\bar{w}(p)} \frac{d}{dp} \bar{w}(p) \tag{8.9}$$

where $\bar{w}(p) = p^2 w_{AA} + 2p(1-p)w_{Aa} + (1-p)^2 w_{aa}$. Suppose also that the mutation rate of $A \to a$ is u and of $a \to A$ is v.

$$A \overset{u}{\to} a$$

$$a \overset{v}{\to} A$$

Then the change in gene frequency due to recurrent mutation is

$$\Delta p_{\text{mut}} = -up + v(1-p) \tag{8.10}$$

The change in p caused by both forces together is

$$\Delta p = \frac{p(1-p)}{2\bar{w}(p)} \frac{d}{dp} \bar{w}(p) - up + v(1-p) \tag{8.11}$$

Now we can rewrite this equation for Δp in terms of a function that is maximized by the *joint* action of selection and mutation. Consider the function

$$L(p) = \bar{w}(p) p^{2v} (1-p)^{2u} \tag{8.12}$$

One can directly verify that Δp can be written in terms of $L(p)$ as

$$\Delta p = \frac{p(1-p)}{2L(p)} \frac{dL(p)}{dp} \tag{8.13}$$

If there is no mutation, then u and v are zero and $L(p)$ reduces to $\bar{w}(p)$. Also, if there is no selection ($w_{AA} = w_{Aa} = w_{aa} = 1$), then $L(p)$ reduces to $p^{2v}(1-p)^{2u}$. In the absence of mutation, the evolutionary equilibrium occurs where $\bar{w}(p)$ is maximized. And in the absence of selection, you will recall that there is an equilibrium caused by recurrent mutation at $p = v/(u+v)$. Notice that the function $p^{2v}(1-p)^{2u}$ is, in fact, maximized at $p = v/(u+v)$. Also, $L(p)$ combines the two separate maximization principles, one for selection and the other for mutation. The maximum point of $L(p)$ involves a compromise between maximizing $\bar{w}(p)$ and the function $p^{2v}(1-p)^{2u}$.

It is convenient to distinguish three cases. First, suppose that the heterozygote has the highest fitness. Then the action of selection is to maintain a polymorphism. The effect of recurrent mutations is also to maintain a polymorphism. See Figure 8.2. In this case the qualitative tendency of the two forces is the same and the actual equilibrium point is somewhere between the equilibrium that would result from selection alone and that from mutation alone. Moreover, it is possible that the pure selection

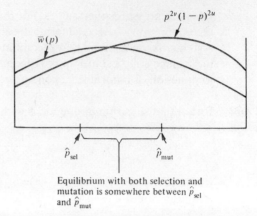

Equilibrium with both selection and mutation is somewhere between $\hat{p}_{sel}$ and $\hat{p}_{mut}$

FIGURE 8.2. *Heterozygote superiority and recurrent mutation. Both selection alone and mutation alone act to produce a polymorphism. There continues to be a stable polymorphism if they act jointly.*

equilibrium coincides with the pure mutation equilibrium. For example, both equilibria coincide at $\hat{p} = \frac{1}{2}$ if there is symmetry both in the selection ($w_{AA} = w_{aa}$) and in the mutation ($u = v$). Thus, in this case, the overall impact of mutation is quite small—both the existence of the equilibrium and perhaps also its approximate position could be predicted from knowledge of the selection regime alone, even if mutation rates are moderately large.

Second, suppose that selection favors one allele ($w_{AA} < w_{Aa} < w_{aa}$). Then selection and recurrent mutation are qualitatively in opposition. Selection is leading to fixation while mutation is resisting fixation. See Figure 8.3. The selection-mutation balance leads to a polymorphism. But the *existence* of this polymorphism could not be predicted from the selection regime alone. Moreover, the location of the equilibrium depends critically on the strength of mutation relative to selection. Hence in this case both the existence and position of the equilibrium can be predicted only by considering both mutation and selection.

In the third case selection is against the heterozygote. If the selection is sufficiently stronger than the mutation, then there will be two stable polymorphic equilibria, near $p = 0$ and $p = 1$, and one unstable equilibrium in between the two stable ones. See Figure 8.4a. However, if the mutation is

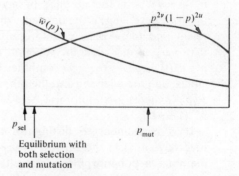

Equilibrium with both selection and mutation

FIGURE 8.3. *Directional selection against A and recurrent mutation. Selection alone leads to fixation of A while mutation alone leads to polymorphism. With both mutation and directional selection, there is a polymorphism very near the boundary.*

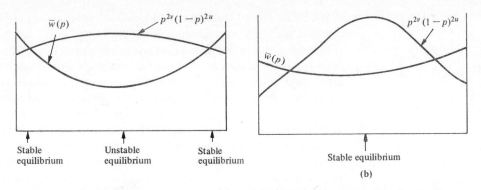

FIGURE 8.4. (a) *Strong heterozygote inferiority and weak recurrent mutation. There is a stable equilibrium on each side of an unstable equilibrium*; (b) *weak heterozygote inferiority and strong recurrent mutation. There is one stable equilibrium.*

sufficiently stronger than the selection, then there will be just one polymorphic equilibrium even though $\bar{w}$ will be nearly minimized at this point. See Figure 8.4b. This selection process is similar to the second case because mutation and selection are again opposing forces but it differs from the second instance in that this resolution of opposing forces often arises in symmetric fitness schemes. In particular, suppose that $w_{AA} = w_{aa}$ and that the mutation is strong. Then there is a transition from case 1 to the second form of case 3 (i.e., Figure 8.4b) as the magnitude of w_{Aa} is decreased. The second case, where selection clearly favors one allele, can usually be immediately recognized from the pattern in the selection coefficients. But especially with multiple alleles, the multiallele analogue of the second form of the third case is often hard to distinguish from the first case without detailed mathematical analysis of the w's. In both the first case and second form of the third case, there is one stable equilibrium, but in the first case $\bar{w}$ is nearly maximized while in the second form of the third case $\bar{w}$ is nearly minimized at the equilibrium.

In all these cases it is helpful to keep in mind that two forces are operating and that the result is a compromise. The equilibrium always satisfies $dL(p)/dp = 0$, and by the product rule, this derivative is given by

$$\frac{dL(p)}{dp} = \bar{w}(p)\frac{d}{dp}[p^{2v}(1-p)^{2u}] + [p^{2v}(1-p)^{2u}]\frac{d}{dp}\bar{w}(p) = 0 \quad (8.14)$$

This formula brings out that an equilibrium requires slopes of opposite sign for $\bar{w}(\hat{p})$ and $\hat{p}^{2v}(1-\hat{p})^{2u}$. In this sense the system is being pulled by selection in one direction, since $d\bar{w}(\hat{p})/dp$ is, say, negative and in the opposite direction by mutation, since $d\hat{p}^{2v}(1-\hat{p})^{2u}/dp$ would be positive.

Another general point is that the equilibrium is seldom exactly that which maximizes the average fitness, $\bar{w}(p)$. Unless the pure selection and pure mutation equilibria happen to coincide, there is a lower average fitness at equilibrium than would occur if selection alone were acting. Presumably, the genetic variation supplied by recurrent mutation is critical to a population's response to environmental changes. Nevertheless, at least in the short term, the typical effect of recurrent mutation is to displace the equilibrium away from the point that maximizes $\bar{w}(p)$.

We shall see phenomena analogous to all these examples when we combine selection with recombination later.

There is a caveat that must be added before proceeding. Our approach has been to view the system of selection combined with mutation by conceptually adding together the behavior of each of these processes separately. However, there is a possible hazard in this approach. If an equilibrium point is stable in each system separately and if it still exists as an equilibrium point in the combined system, it does not *automatically* follow that it is still stable in the combined system. One reason that is important in discrete time systems is that the combined force which pulls the system to the equilibrium may be so strong as to cause overshoots. In the mutation-selection example a necessary condition for divergent overshoots is extremely high mutation rates; specifically, it is necessary that $u + v$ exceed one. Obviously, since u and v are usually on the order of 10^{-6}, this issue would never arise. Furthermore, in multidimensional systems the characteristic directions of the forces typically will not coincide and the mismatch may lead to instability. Nonetheless, our approach of seeking to understand a combined system by conceptually synthesizing their separate tendencies aids the development of intuition. The validity of the results obtained by this approach always has to be established rigorously.

The Two-Locus Model with Selection

The basic model for natural selection with multiple loci has two parts—one part for the natural selection itself, and the other the effect of recombination. We shall review the effect of each force separately and then put them together.

The single, most important fact to remember in multilocus genetics is that every gamete type is analogous to an allele at one locus. Consider the following four gamete types: AB, Ab, aB, and ab. Assume that the recombination, r, between the A and B loci is zero. Then each of these gamete types is effectively an allele at one locus. If r is zero, we have a one-locus four-allele system. Hence, in the limit as r tends to zero, the two-loci-two-allele model must reduce to a one-locus four-allele model. When r is not zero, the system is roughly analogous to a one-locus four-allele model with recurrent mutation. The "alleles" are the gamete types and the "mutation" is the recombination that converts some gamete types into others.

In the preceding discussion of the two locus analogue of the Hardy–Weinberg law, we reduced the set of four gamete frequencies, x_1, x_2, x_3, and x_4, into one variable, D, the linkage disequilibrium coefficient. But from now on, we shall want to keep track of all the variables. The variable, D, is still very important, but it is not the whole story.

Let us consider the effect of recombination in the absence of selection. As we learned from Equation (8.6),

$$\Delta x_1 = -r(x_1 x_4 - x_2 x_3)$$
$$\Delta x_2 = r(x_1 x_4 - x_2 x_3)$$
$$\Delta x_3 = r(x_1 x_4 - x_2 x_3)$$
$$\Delta x_4 = -r(x_1 x_4 - x_2 x_3)$$

(8.15)

where, as before, $D = x_1 x_4 - x_2 x_3$. As you know, D tends to zero monotonically in the absence of selection.

It is important to understand that D itself is a function of the four variables, x_i. We could use the notation $D(x_1, x_2, x_3, x_4)$ to make this fact explicit. Since D is a function of several variables, then $D = 0$ is not a point but an entire surface. Recombination in the absence of selection moves the population to a nearby place on the $D = 0$ surface. The actual equilibrium point $(\hat{x}_1, \hat{x}_2, \hat{x}_3, \hat{x}_4)$ depends on the initial position. But all equilibrium points lie on the $D = 0$ surface. The fact that $D = 0$ is a surface and not a single point entails that there are many ways in which a pure recombination equilibrium and a pure selection equilibrium may coincide. Specifically, let us consider several of the points on the $D = 0$ surface. The "central point" $(\frac{1}{4}, \frac{1}{4}, \frac{1}{4}, \frac{1}{4})$ has $D = 0$. Also, the points corresponding to fixation at both loci $(1, 0, 0, 0)$, and so forth, have $D = 0$. And the points corresponding to fixation at one locus do, as well; for example, $(x_1, x_2, 0, 0)$ where $x_1 + x_2 = 1$.

As a rule, if a stable equilibrium due to selection in the absence of recombination happens to coincide with the $D = 0$ surface, then in the presence of recombination these points will continue to be stable equilibria [Karlin (1975)].

The points that do *not* lie on $D = 0$ surface typically are those representing polymorphism at *both* loci together with an excess of one pair of gamete types over the other. For example, $(x_1, 0, 0, x_4)$, where $x_1 + x_4 = 1$ is *not* on the $D = 0$ surface. Hence, if the stable equilibrium due to selection happens to lead to this sort of point, then adding recombination is certain to change the picture.

Now let us review the effect of selection in the absence of recombination. We are dealing now with a system of four alleles at one locus, and the equations are

$$\Delta x_i = \frac{x_i(w_i - \bar{w})}{\bar{w}} \tag{8.16}$$

where w_i is the "marginal fitness" of the ith gamete type defined as

$$w_i = \sum_{j=1}^{4} x_j w_{ij} \tag{8.17}$$

and $\bar{w}$ is the average fitness

$$\bar{w} = \sum_i \sum_j x_i x_j w_{ij} \tag{8.18}$$

We learned in the last chapter that this system enjoys some extremely convenient properties. First, the $\bar{w}$ function is always increasing, that is, $\bar{w}(t+1) \geq \bar{w}(t)$ with equality obtaining only at equilibrium. Second, there is at most one "interior" equilibrium, that is, $\hat{x}_i > 0$ for all i, and it is obtained by solving a set of linear equations. Third, the interior equilibrium is stable if and only if $\bar{w}$ is a maximum at the equilibrium. The equilibrium is unstable when the equilibrium point is a saddle point or a minimum of $\bar{w}$. There is a choice of two techniques to determine whether the equilibrium is stable or not; one involves determinants of the w_{ij} matrix and the other involves the eigenvalues of the w_{ij} matrix. With this machinery we are always able to describe fully the effect of selection on the gamete types in the absence of recombination.

Now we can simply combine the equations for Δx_i due to selection and to recombination. The only new point here is to remember that recombination occurs only in the double heterozygotes. Therefore, we have to weight the recombination term by the fraction of the population which, among the adults after selection, are double heterozygotes. Let w_H denote the double heterozygote fitness, that is, $w_H = w_{14} = w_{23}$. Then the weighting factor is $w_H/\bar{w}$. Putting all this together, we have for Δx_1

$$\Delta x_1 = \frac{x_1(w_1 - \bar{w})}{\bar{w}} - \frac{w_H}{\bar{w}} rD \tag{8.19}$$

Similarly, we have for the other gamete frequencies

$$\Delta x_2 = x_2 \frac{(w_2 - \bar{w})}{\bar{w}} + \frac{w_H}{\bar{w}} rD \tag{8.20}$$

$$\Delta x_3 = x_3 \frac{(w_3 - \bar{w})}{\bar{w}} + \frac{w_H}{\bar{w}} rD \tag{8.21}$$

$$\Delta x_4 = x_4 \frac{(w_4 - \bar{w})}{\bar{w}} - \frac{w_H}{\bar{w}} rD \tag{8.22}$$

These are the basic equations for the change in the gamete frequencies in the presence of both selection and recombination.

The equations for Δx_i involve two forces as discussed previously. The term $x_i(w_i - \bar{w})/\bar{w}$ represents selection pulling the system to the maximum of $\bar{w}$. The term with rD represents recombination pulling the system to the $D = 0$ surface. The equilibrium qualitatively represents some compromise between these two forces. To illustrate the kinds of evolutionary results that emerge from this model, we shall examine three examples. The first two examples are very straightforward and the third is complex.

Selection for Fixation   In our first example, let us suppose that there is selection for the A and B alleles over the a and b alleles. The AB/AB genotype is assumed to have the highest fitness. The w_{ij} matrix takes the form

		G_1 AB	G_2 Ab	G_3 aB	G_4 ab
G_1	AB	1	$1 - s_1$	$1 - s_1$	$1 - s_2$
G_2	Ab	$1 - s_1$	$1 - s_3$	$1 - s_2$	$1 - s_4$
G_3	aB	$1 - s_1$	$1 - s_2$	$1 - s_3$	$1 - s_4$
G_4	ab	$1 - s_2$	$1 - s_4$	$1 - s_4$	$1 - s_5$

where $s_1 < s_2, s_3 < s_4 < s_5$. This scheme represents a situation where fitness increases with the number of A and B genes carried by an individual. To assess the role of selection, consider the one-locus four-allele system. Let the alleles be G_1, G_2, G_3, and G_4. Clearly, with this w_{ij} matrix there is only one stable outcome: fixation of the G_1 allele. The action of selection alone leads to an equilibrium at $(1, 0, 0, 0)$. Next, observe that the point $(1, 0, 0, 0)$ is also on the $D = 0$ surface. Therefore, the pure-selection equilibrium coincides with one of the possible pure-recombination equilibria. Hence with both recombination and selection, the point $(1, 0, 0, 0)$ remains the stable equilibrium.

There are two important features to this example. First, since the equilibrium in the presence of recombination is the same as that resulting from selection alone, $\bar{w}$ is maximized at the equilibrium. In this important case, recombination does not displace the equilibrium away from the maximum of $\bar{w}$. Second, the fact that $\bar{w}$ is maximized, even with recombination, highlights a basic difference between mutation and recombination. Recombination can be viewed as a frequency-dependent mutation rate. The rate at which recombination produces new gamete types depends on the frequency of double heterozygotes. As an allele approaches fixation, the frequency of double heterozygotes drops to zero. Therefore, as an allele approaches fixation, the effect of recombination also drops to zero. For this reason, no polymorphism is maintained by a balance between recombination versus selection whenever the selection is leading to fixation of one of the alleles. In the important cases where there is fixation at one or both loci, the situation is dominated by the selection regime.

Coadaptation Between Loci Without Heterosis

As we shall discuss in more detail in Chapter 11, there is considerable interest in knowing whether the alleles at loci that are close together on the chromosome are functionally coadapted. One can imagine, for example, that neighboring loci produce enzymes that function in the same biochemical pathway. If so, then there is often a set of enzyme kinetic coefficients for the two enzymes that produces the best functioning of the pathway as a whole. Let us explore the consequences of this idea.

As our second example of equilibrium situations between selection and recombination, let us suppose that A and B are coadapted alleles in the sense that the products of these alleles function well together. Suppose also that a and b are equally coadapted. Finally, A and b, and a and B, are not coadapted. Then the highest fitness is attributed to genotypes with two coadapted pairs, an intermediate fitness to genotypes with one coadapted pair, and the lowest fitness to types with no coadapted pairs. Then the w_{ij} matrix is

		G_1	G_2	G_3	G_4
		AB	Ab	aB	ab
G_1	AB	1	$1-s_1$	$1-s_1$	1
G_2	Ab	$1-s_1$	$1-s_2$	1	$1-s_1$
G_3	aB	$1-s_1$	1	$1-s_2$	$1-s_1$
G_4	ab	1	$1-s_1$	$1-s_1$	1

where $s_1 < s_2$; s_i is the selection coefficient against a type with i nonadapted pairs.

We begin by considering the four gamete types as four alleles at one locus, G_1, G_2, G_3, G_4. This w_{ij} matrix leads to fixation of either G_1 or G_4. There is no polymorphism among the G's which is maintained by selection alone. Next we observe that the points $(1, 0, 0, 0)$ and $(0, 0, 0, 1)$ are on the $D = 0$ surface. These equilibria continue to be the stable equilibria when both selection and recombination occur. Again, the pure-selection equilibria coincide with pure-recombination equilibria.

This example is similar to the previous one. It is interesting because it shows that coadaptation alone does not lead to polymorphism. As we shall see in Chapter 11, there are many examples of polymorphisms between complexes of genes. It is often claimed that the alleles within any complex are coadapted with one another. Although it may be true that the alleles within the observed complexes are coadapted (no one really knows), it is nonetheless clear that coadaptation between loci is not sufficient to explain the occurrence of a *polymorphism* of gene complexes. Some additional factors, presumably heterosis among the alleles *within* a locus, must be involved to explain the existence of the polymorphism.

Heterosis at
the Two Loci

In our last example of equilibrium situations between recombination and selection, we consider a selection pressure that tends to maintain a polymorphism at both loci. As we shall see, there are many possible evolutionary results in this situation, but the general outline of the results is quite intuitive.

The natural selection is represented by the following w_{ij} matrix:

		G_1 AB	G_2 Ab	G_3 aB	G_4 ab
G_1	AB	$1-s_2$	$1-s_1$	$1-s_1$	1
G_2	Ab	$1-s_1$	$1-s_2$	1	$1-s_1$
G_3	aB	$1-s_1$	1	$1-s_2$	$1-s_1$
G_4	ab	1	$1-s_1$	$1-s_1$	$1-s_2$

In this scheme the double heterozygotes are the most fit. The selection coefficient against individuals with one homozygous locus is s_1 and that against individuals with two homozygous loci is s_2. If $s_1 < s_2$, there is said to be *ordered overdominance* between the loci. If $s_1 > s_2$, there is *between-locus underdominance*. See Figure 8.5 for a sketch of these relations. This model has been extensively studied in the literature. Key references about this topic include Lewontin and Kojima (1960), Bodmer and Felsenstein (1967), Ewens (1968), and Karlin and Feldman (1970). This model is often referred to as the *Lewontin–Kojima model*.

As always, we begin by considering the corresponding one-locus four-allele model. We cannot deduce the outcome of this w_{ij} matrix simply by inspection. Instead we shall use the one-locus machinery of the last chapter.

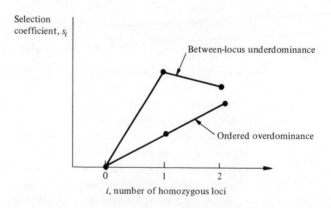

FIGURE 8.5. *Two possible schemes relating fitness to the number of homozygous loci.*

First, let us find the location of the four-allele polymorphism. By Equation (7.9) of the last chapter, we have

$$\hat{x}_1 = \frac{\Delta_1}{\sum \Delta_i}, \qquad \hat{x}_2 = \frac{\Delta_2}{\sum \Delta_i}, \qquad \hat{x}_3 = \frac{\Delta_3}{\sum \Delta_i}, \qquad \hat{x}_4 = \frac{\Delta_4}{\sum \Delta_i} \qquad (8.23)$$

where Δ_i is the determinant of the w_{ij} matrix with the ith column replaced by a column of 1's. It is now easy to verify that $\hat{x}_1 = \hat{x}_2 = \hat{x}_3 = \hat{x}_4 = \frac{1}{4}$.† So the single equilibrium point at which all $x_i > 0$ is at $(\frac{1}{4}, \frac{1}{4}, \frac{1}{4}, \frac{1}{4})$.

Next let us determine whether the equilibrium is stable. According to Mandel's criteria for stability, we require the successive principal minors of the w_{ij} matrix to alternate in sign,

$$\begin{vmatrix} 1-s_2 & 1-s_1 \\ 1-s_1 & 1-s_2 \end{vmatrix} < 0 \qquad (8.24a)$$

$$\begin{vmatrix} 1-s_2 & 1-s_1 & 1-s_1 \\ 1-s_1 & 1-s_2 & 1 \\ 1-s_1 & 1 & 1-s_2 \end{vmatrix} > 0 \qquad (8.24b)$$

and

$$\det (w_{ij}) < 0 \qquad (8.24c)$$

Upon evaluating the determinants, we find that these conditions reduce to

$$s_1 < s_2 \qquad (8.25a)$$

$$s_1 < 1 - \sqrt{\tfrac{1}{2}(1-s_2)(2-s_2)} \qquad (8.25b)$$

$$s_1 < \frac{s_2}{2} \qquad (8.25c)$$

(8.25a) is satisfied if (8.25c) is satisfied. And we can verify that (8.25b) is also satisfied whenever (8.25c) is satisfied.‡ Therefore, the key condition for *stability* of the polymorphism at $(\frac{1}{4}, \frac{1}{4}, \frac{1}{4}, \frac{1}{4})$ in the absence of recombination is that $s_1 < s_2/2$.

If the internal equilibrium is unstable, then we might expect that trajectories will converge to one or more of the boundary equilibria, that is, equilibria in which one or more of the alleles are absent. By inspection of the w_{ij} matrix, we can conclude that there is no equilibrium with only one allele absent, which is stable. For example, in a system with G_1, G_2, and G_3 the allele G_1 is certain to be lost. However, there are two reasonable equilibria with two alleles absent. These are $(\frac{1}{2}, 0, 0, \frac{1}{2})$ and $(0, \frac{1}{2}, \frac{1}{2}, 0)$. Clearly, a system of one locus with two alleles, G_1 and G_4, will attain the polymorphism of $\frac{1}{2}, \frac{1}{2}$ because this situation is the standard case of heterozygote superiority at one locus with two alleles, and similarly with the pair G_2 and G_3. Our question now becomes whether these boundary equilibria are stable with respect to perturbations away from the boundary into the interior region where all the four alleles are present. Consider the equilibrium $(\frac{1}{2}, 0, 0, \frac{1}{2})$ and let us

† One can verify that $\hat{x}_i = \frac{1}{4}$ by direct calculation of all determinants. But the easiest way is to show by elementary row and column operations that $\Delta_1 = \Delta_4$, $\Delta_2 = \Delta_3$, and then that $\Delta_1 = \Delta_2$. From this it follows that the equilibrium must be $(\frac{1}{4}, \frac{1}{4}, \frac{1}{4}, \frac{1}{4})$.

‡ To verify this, show that $s_2/2 < 1 - \sqrt{\tfrac{1}{2}(1-s_2)(2-s_2)}$.

introduce a little bit of G_2. Then (see last chapter) the condition for x_2 to increase when rare, that is, the condition for the boundary to the *unstable* relative to perturbations into the interior, is

$$\hat{x}_1 w_{21} + \hat{x}_4 w_{24} > \hat{x}_1^2 w_{11} + 2\hat{x}_1 \hat{x}_4 w_{14} + \hat{x}_4^2 w_{44} \qquad (8.26)$$

This inequality reduces to

$$s_1 < \frac{s_2}{2} \qquad (8.27)$$

Therefore, these boundary equilibria are unstable when the interior equilibrium is stable, and vice versa.

To summarize, in the absence of recombination if $s_1 < s_2/2$, then all trajectories converge to $(\frac{1}{4}, \frac{1}{4}, \frac{1}{4}, \frac{1}{4})$, which is the maximum of $\bar{w}$. But if $s_1 > s_2/2$, then all trajectories converge either at $(\frac{1}{2}, 0, 0, \frac{1}{2})$ or at $(0, \frac{1}{2}, \frac{1}{2}, 0)$ depending on where the trajectory starts. In this instance, the highest points of $\bar{w}$ are at these boundary equilibria and the point $(\frac{1}{4}, \frac{1}{4}, \frac{1}{4}, \frac{1}{4})$ is *not* the maximum of $\bar{w}$. This entire situation is analogous to the transition between heterozygote superiority and heterozygote inferiority at one locus with two alleles. When $w_{Aa} > w_{AA} = w_{aa}$, then there is a stable interior equilibrium at $(\frac{1}{2}, \frac{1}{2})$. But if $w_{Aa} < w_{AA} = w_{aa}$, then the interior point $(\frac{1}{2}, \frac{1}{2})$ is an unstable equilibrium and the boundary equilibria $(1, 0)$ and $(0, 1)$ are stable. Thus we have determined that the full w_{ij} matrix maintains a polymorphism among the four alleles if $s_1 < s_2/2$, and does not if $s_1 > s_2/2$.

As in the previous examples, the four-allele analysis allows us to perceive the general qualitative behavior of the two-locus two-allele model. If $s_1 < s_2/2$, then selection is tending to move the system to the point $(\frac{1}{4}, \frac{1}{4}, \frac{1}{4}, \frac{1}{4})$. This point is on the $D = 0$ surface. If $s_1 < s_1/2$, then the pure-selection equilibrium coincides with a pure-recombination equilibrium. Hence with two loci and two alleles the point $(\frac{1}{4}, \frac{1}{4}, \frac{1}{4}, \frac{1}{4})$ is the single stable equilibrium for any map distance between A and B, provided that $s_1 < s_2/2$. At this equilibrium, $\bar{w}$ is maximized. In this example the pure selection and pure recombination equilibria happen to coincide exactly because of the symmetry in the w_{ij} matrix. But, in general, if the w_{ij} matrix leads to a stable pure-selection polymorphism, it is then producing a result that is qualitatively the same as that with pure recombination. The two forces are working in concert. Hence there will exist one equilibrium point that is located between the pure-selection equilibrium and the $D = 0$ surface. The *existence* of this equilibrium will not depend on the map distance between the A and B loci, although its detailed position will shift closer to the $D = 0$ surface as r is increased. Typically, at this equilibrium $\bar{w}$ will be near the maximum value because the equilibrium will be close to the pure-selection equilibrium.

In contrast, if $s_1 > s_2/2$, then selection is driving the system to one of the boundary equilibria, but these boundary equilibria are *not* located on the $D = 0$ surface. (Note the contrast with the first and second examples we discussed under this topic.) Since selection is pushing the system to the points $(\frac{1}{2}, 0, 0, \frac{1}{2})$ and $(0, \frac{1}{2}, \frac{1}{2}, 0)$ while recombination is pulling it to the $D = 0$ surface, we obtain one or more stable equilibria involving all four gamete types. These equilibria are maintained by the balance of recombination versus selection. These equilibria are typically very sensitive to the details of the selection coefficients and to the linkage between A and B. Moreover, at these equilibria $\bar{w}$ is certainly not maximized and can be significantly lower than the maximum if r is large.

The condition for selection to maintain a polymorphism among the four gamete types, $s_1 < s_2/2$, means that there must be strong, ordered overdominance between the loci. Figure 8.6 sketches the sorts of between-locus relationships that lead to a stable pure-selection polymorphism at $(\frac{1}{4}, \frac{1}{4}, \frac{1}{4}, \frac{1}{4})$.

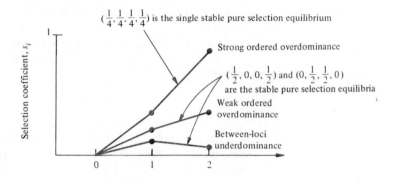

FIGURE 8.6. *Sketch of the between-locus relationships that lead to a single, stable, pure-selection equilibrium and to a pair of stable, pure-selection equilibria.*

Failure to consider selection at the level of the entire genotype has led to a great deal of misunderstanding in recent years. One popular model is to assume that the fitness of a two-locus genotype is obtained by taking the product of single-locus fitnesses. For example, suppose that the fitness at a single locus is $w, 1, w$ ($w < 1$). Then the fitness of a double heterozygote is taken as 1, the fitness with one heterozygous locus as w, and the fitness with two homozygous loci as w^2. Hence $s_1 = 1 - w$ and $s_2 = 1 - w^2$. Because s_1 is less than s_2, there is ordered overdominance between loci, *but* it is not strong. In this multiplicative scheme $s_1 > s_2/2$, therefore, the selection at the level of the whole genotype does not tend to maintain a full polymorphism among the gamete types. Hence this scheme leads to equilibria where $\bar{w}$ is not maximized, and where the number and position of the equilibria depend critically on the linkage between the loci as discussed previously. These results were felt to be surprising because the selection seemed well-behaved at the single-locus level (what could be nicer than fitnesses of $w, 1, w$?) and yet so strangely behaved at the two-locus level. But the results are not at all surprising when viewed in the context of the selection on the entire genotype. This multiplicative example is discussed further in Chapter 11 on the evolution of linkage.

It is important to understand that much of the complexity of multilocus models is inherent in the w_{ij} matrix itself and has nothing to do with linkage and recombination. This complexity is contained in the corresponding one-locus n-allele model. For this reason it is critical to understand the implication of the selection matrix itself before bringing recombination into the picture.

We are now ready to examine the two-locus model further. As we have seen, if $s_1 > s_2/2$, then selection and recombination are opposing forces.

What sort of balance results from these forces? What characterizes a polymorphism maintained by a selection-recombination balance? These questions cannot be answered in general. There is no general machinery to handle multilocus models analogous to that of the one-locus n-allele model. Instead, each case must be tediously analyzed. The detailed algebra for the symmetric two-loci model with heterosis appears in the papers of Bodmer and Felsenstein (1967) and Karlin and Feldman (1970). It is sufficient to summarize the results here. Interested readers are urged to consult the original papers.

Three classes of equilibria have been found in the equations for Δx_i.

The Central Equilibrium This point is

$$\hat{x}_1 = \hat{x}_2 = \hat{x}_3 = \hat{x}_4 = \tfrac{1}{4} \tag{8.28}$$

This equilibrium point exists for all s_1, s_2, and r. At this point $D = 0$. As discussed before, it is stable and is the maximum point of $\bar{w}$ if $s_1 < s_2/2$. It is also stable but is *not* the maximum of $\bar{w}$ if $s_1 > s_2/2$ and if the linkage is sufficiently loose.

$$r > \tfrac{1}{4}(2s_1 - s_2) \tag{8.29}$$

If $s_1 > s_2/2$ and if r satisfies (8.29), then the central point is stable because the recombination is overpowering the selection. This situation is analogous to the case with one locus and two alleles discussed earlier, where sufficiently strong mutation could overcome heterozygote inferiority and maintain a polymorphism at $p = \tfrac{1}{2}$.

The Symmetric Equilibria A pair of symmetric equilibria are given by

$$\hat{x}_1 = \hat{x}_4 = \frac{1}{4} + \frac{1}{4}\sqrt{1 - \frac{4r}{2s_1 - s_2}}$$

$$\hat{x}_2 = \hat{x}_3 = \frac{1}{4} - \frac{1}{4}\sqrt{1 - \frac{4r}{2s_1 - s_2}} \tag{8.30}$$

and

$$\hat{x}_1 = \hat{x}_4 = \frac{1}{4} - \frac{1}{4}\sqrt{1 - \frac{4r}{2s_1 - s_2}}$$

$$\hat{x}_2 = \hat{x}_3 = \frac{1}{4} + \frac{1}{4}\sqrt{1 - \frac{4r}{2s_1 - s_2}} \tag{8.31}$$

These equilibria *exist*, provided that $s_1 > s_2/2$ and

$$r < \tfrac{1}{4}(2s_1 - s_2) \tag{8.32}$$

Thus, whenever these symmetric equilibria exist, the central equilibrium is unstable. At these equilibria $D \neq 0$. In fact,

$$D = \pm\frac{1}{4}\sqrt{1 - \frac{4r}{2s_1 - s_2}} \tag{8.33}$$

These equilibria correspond to the boundary equilibria in the corresponding one-locus four-allele model. Indeed when $r = 0$ the first equilibrium becomes $(\frac{1}{2}, 0, 0, \frac{1}{2})$ and the second equilibrium becomes $(0, \frac{1}{2}, \frac{1}{2}, 0)$. These equilibria are *stable*, provided that

$$r^2 + r\left\{\frac{(s_2 - s_1)^2}{(2s_1 - s_2)} - s_1\right\} + s_2\frac{(2s_1 - s_2)}{4} > 0 \tag{8.34}$$

for r between 0 and $(2s_1 - s_2)/4$. This condition is a simple quadratic in r. It is sketched in Figure 8.7.

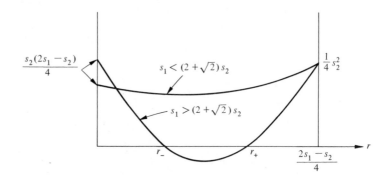

FIGURE 8.7. *Sketch of Equation (8.34) as a function of r.*

Ewens (1968) observed that if there is weak between-loci under-dominance, or ordered overdominance between the loci then (8.34) is satisfied for all r between 0 and $(2s_1 - s_2)/4$. Hence under these conditions whenever the symmetric equilibria exist, they are stable. But Ewens also pointed out that if there is very strong between-locus underdominance, specifically, $s_1 > (2 + \sqrt{2})s_2$, then (8.34) is not always satisfied. Under this condition the equilibria are stable for r between 0 and r_-, are unstable for r between r_- and r_+, and are stable for r between r_+ and $(2s_1 - s_2)/4$. The end points of these intervals are

$$= s_1 - \frac{(s_2 - s_1)^2}{2s_1 - s_2} \pm s_2\sqrt{\left(\frac{s_1}{s_2} - 2\right)^2 - 2} \tag{8.35}$$

So if there is strong between-loci underdominance, and an intermediate value of r, then there may be *no* stable interior polymorphism. (Remember that the central point is unstable whenever the symmetric equilibria merely exist, regardless of whether they are stable.) Therefore, no polymorphism is maintained and both loci become fixed. Thus even though there is heterozygote superiority within each locus, there is no polymorphism at all.

Note that regardless of the selection coefficients, provided that $s_1 > s_2/2$, then the symmetric equilibria are stable for r when it is sufficiently small. This is because these equilibria correspond to the boundary equilibria in the four-allele system. When $s_1 > s_2/2$ these boundary equilibria are the only stable equilibria in the four-allele system. As $r \to 0$, the two-locus two-allele system converges to the four-allele system and therefore these equilibria must be stable for r sufficiently small.

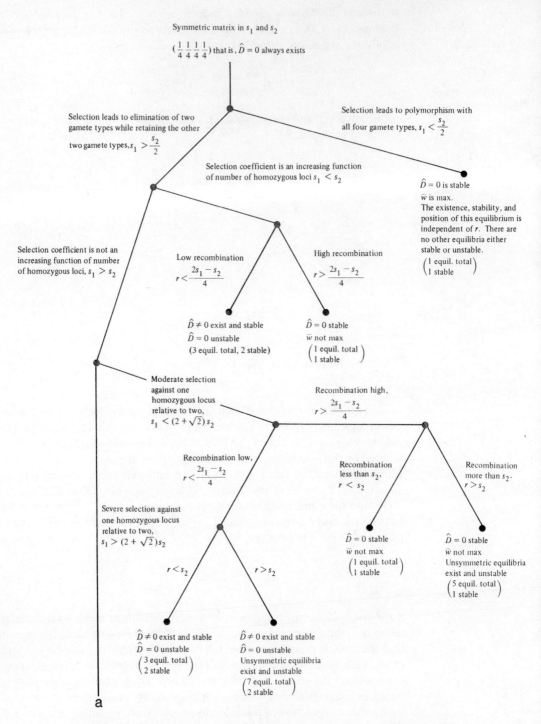

Symmetric matrix in s_1 and s_2

$(\frac{1}{4}\ \frac{1}{4}\ \frac{1}{4}\ \frac{1}{4})$ that is, $\hat{D} = 0$ always exists

Selection leads to elimination of two gamete types while retaining the other two gamete types, $s_1 > \frac{s_2}{2}$

Selection leads to polymorphism with all four gamete types, $s_1 < \frac{s_2}{2}$

Selection coefficient is an increasing function of number of homozygous loci $s_1 < s_2$

$\hat{D} = 0$ is stable
$\bar{w}$ is max.
The existence, stability, and position of this equilibrium is independent of r. There are no other equilibria either stable or unstable.
$\begin{pmatrix} 1 \text{ equil. total} \\ 1 \text{ stable} \end{pmatrix}$

Selection coefficient is not an increasing function of number of homozygous loci, $s_1 > s_2$

Low recombination

$r < \frac{2s_1 - s_2}{4}$

High recombination

$r > \frac{2s_1 - s_2}{4}$

$\hat{D} \neq 0$ exist and stable
$\hat{D} = 0$ unstable
(3 equil. total, 2 stable)

$\hat{D} = 0$ stable
$\bar{w}$ not max
$\begin{pmatrix} 1 \text{ equil. total} \\ 1 \text{ stable} \end{pmatrix}$

Moderate selection against one homozygous locus relative to two, $s_1 < (2 + \sqrt{2})s_2$

Recombination high,

$r > \frac{2s_1 - s_2}{4}$

Recombination low,

$r < \frac{2s_1 - s_2}{4}$

Recombination less than s_2, $r < s_2$

Recombination more than s_2. $r > s_2$

Severe selection against one homozygous locus relative to two, $s_1 > (2 + \sqrt{2})s_2$

$r < s_2$

$r > s_2$

$\hat{D} = 0$ stable
$\bar{w}$ not max
$\begin{pmatrix} 1 \text{ equil. total} \\ 1 \text{ stable} \end{pmatrix}$

$\hat{D} = 0$ stable
$\bar{w}$ not max
Unsymmetric equilibria exist and unstable
$\begin{pmatrix} 5 \text{ equil. total} \\ 1 \text{ stable} \end{pmatrix}$

$\hat{D} \neq 0$ exist and stable
$\hat{D} = 0$ unstable
$\begin{pmatrix} 3 \text{ equil. total} \\ 2 \text{ stable} \end{pmatrix}$

$\hat{D} \neq 0$ exist and stable
$\hat{D} = 0$ unstable
Unsymmetric equilibria exist and unstable
$\begin{pmatrix} 7 \text{ equil. total} \\ 2 \text{ stable} \end{pmatrix}$

a

FIGURE 8.8. *Flow chart summarizing all the interior equilibria in the example of heterosis at two loci; s_1 is the selection coefficient against genotypes with one homozygous locus and s_2 is the selection coefficient against genotypes with two homozygous loci. The full system also contains boundary and corner equilibria that are not included in this chart.*

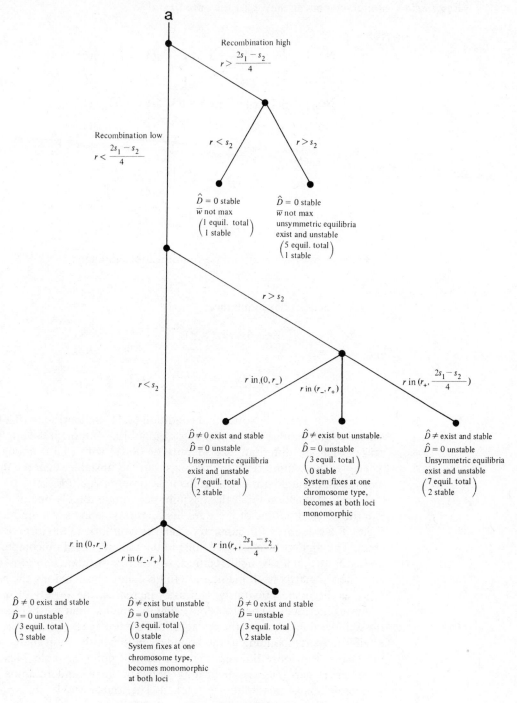

a

Recombination high

$$r > \frac{2s_1 - s_2}{4}$$

Recombination low

$$r < \frac{2s_1 - s_2}{4}$$

$r < s_2$

$\hat{D} = 0$ stable
$\bar{w}$ not max
$\left(\begin{array}{l} 1 \text{ equil. total} \\ 1 \text{ stable} \end{array}\right)$

$r > s_2$

$\hat{D} = 0$ stable
$\bar{w}$ not max
unsymmetric equilibria
exist and unstable
$\left(\begin{array}{l} 5 \text{ equil. total} \\ 1 \text{ stable} \end{array}\right)$

$r > s_2$

$r < s_2$

r in $(0, r_-)$

r in (r_-, r_+)

r in $(r_+, \frac{2s_1 - s_2}{4})$

$\hat{D} \neq 0$ exist and stable
$\hat{D} = 0$ unstable
Unsymmetric equilibria
exist and unstable
$\left(\begin{array}{l} 7 \text{ equil. total} \\ 2 \text{ stable} \end{array}\right)$

$\hat{D} \neq$ exist but unstable.
$\hat{D} = 0$ unstable
$\left(\begin{array}{l} 3 \text{ equil. total} \\ 0 \text{ stable} \end{array}\right)$
System fixes at one
chromosome type,
becomes at both loci
monomorphic

$\hat{D} \neq$ exist and stable
$\hat{D} = 0$ unstable
Unsymmetric equilibria
exist and unstable
$\left(\begin{array}{l} 7 \text{ equil. total} \\ 2 \text{ stable} \end{array}\right)$

r in $(0, r_-)$

r in (r_-, r_+)

r in $(r_+, \frac{2s_1 - s_2}{4})$

$\hat{D} \neq 0$ exist and stable
$\hat{D} = 0$ unstable
$\left(\begin{array}{l} 3 \text{ equil. total} \\ 2 \text{ stable} \end{array}\right)$

$\hat{D} \neq$ exist but unstable
$\hat{D} = 0$ unstable
$\left(\begin{array}{l} 3 \text{ equil. total} \\ 0 \text{ stable} \end{array}\right)$
System fixes at one
chromosome type,
becomes monomorphic
at both loci

$\hat{D} \neq 0$ exist and stable
$\hat{D} =$ unstable
$\left(\begin{array}{l} 3 \text{ equil. total} \\ 2 \text{ stable} \end{array}\right)$

FIGURE 8.8 [*continued*].

Karlin and Feldman (1970) discovered still another class of equilibria. In this model, whenever they exist they are *unstable*.

$$
\begin{array}{ccccc}
 & \hat{x}_1 & \hat{x}_2 & \hat{x}_3 & \hat{x}_4 \\
\text{No. 1} & \tfrac{1}{2}(\xi+\sqrt{R}) & \tfrac{1}{2}(1-\xi) & \tfrac{1}{2}(1-\xi) & \tfrac{1}{2}(\xi-\sqrt{R}) \\
\text{No. 2} & \tfrac{1}{2}(\xi-\sqrt{R}) & \tfrac{1}{2}(1-\xi) & \tfrac{1}{2}(1-\xi) & \tfrac{1}{2}(\xi+\sqrt{R}) \\
\text{No. 3} & \tfrac{1}{2}(1-\xi) & \tfrac{1}{2}(\xi+\sqrt{R}) & \tfrac{1}{2}(\xi-\sqrt{R}) & \tfrac{1}{2}(1-\xi) \\
\text{No. 4} & \tfrac{1}{2}(1-\xi) & \tfrac{1}{2}(\xi-\sqrt{R}) & \tfrac{1}{2}(\xi+\sqrt{R}) & \tfrac{1}{2}(1-\xi)
\end{array}
\tag{8.36}
$$

where

$$
\xi = \frac{s_1 + \{s_2^2/[2(r-s_2)]\}}{2s_1 - s_2 + \{s_2^2/[2(r-s_2)]\}}
\tag{8.37}
$$

$$
R = \xi^2 - \frac{r(1-\xi)^2}{(r-s_2)}
\tag{8.38}
$$

The existence of these equilibria requires loose linkage,

$$
r > s_2
\tag{8.39}
$$

between-locus underdominance,

$$
s_1 > s_2
\tag{8.40}
$$

and

$$
r^2 + r\left\{ \frac{(s_2-s_1)^2}{2s_1 - s_2} - s_1 \right\} + s_2\frac{(2s_1 - s_2)}{4} > 0
\tag{8.41}
$$

for r between s_2 and $\tfrac{1}{2}$. Notice that condition (8.41) is identical to (8.34), the condition for stability of the symmetric equilibria—although the applicable ranges of r are different. The identity of (8.41) with (8.34) implies that whenever the symmetric equilibria exist but are unstable, then the unsymmetric equilibria do not exist. At the unsymmetric equilibria $\hat{D} \neq 0$.

Your first reaction to all these equilibria is undoubtedly one of dismay. But the general outline of these results is intuitive and easy to understand. Figure 8.8 presents a full summary of all the equilibria in the form of a flow chart. The highest level of branching in the table is based on comparing s_1 and s_2. Note that if s_1 is small relative to s_2, there is only one solution—in this case the w_{ij} matrix itself maintains a full polymorphism among the gamete types as discussed previously. But as s_1 increases, the variety of possible solutions increases. When s_1 is only slightly smaller than s_2, there are two possibilities depending on the linkage. For tight linkage the symmetric $D \neq 0$ equilibria (corresponding to the boundary points in the four-allele system) are stable. For loose linkage then, the central point is stable. Next as s_1 becomes larger than s_2, indicating between locus underdominance, a pandora's box of possibilities is opened. The four unstable nonsymmetric equilibria may enter, depending on the linkage. Also, for severe between-locus underdominance, the complex stability criteria for the symmetric equilibria arise. Thus, in general terms, we see that between-locus underdominance has a "destabilizing" effect on the system. The system is simplest with strong, ordered overdominance and becomes very tricky with severe between-locus underdominance.

Some Final Remarks   The literature on multilocus models has become very extensive in recent years. Additional work on the two-locus model includes Karlin (1975) and Karlin and Carmelli (1975a). A symmetric three-locus system is analyzed in Feldman, Franklin, and Thomson (1974). Other studies include Feldman (1971) on the two-locus haploid model, and Karlin and Carmelli (1975b) on a two-locus model combining artificial and natural selection. Karlin (1968b) also includes cases of two-locus models with certain schemes of nonrandom mating. Studies involving large numbers of loci include numerical simulations for the multiplicative model for 36 loci (Franklin and Lewontin 1970) and an analysis by Slatkin (1972).

The topic of genetic drift in multilocus situations has been studied by Sved (1967), Ohta and Kimura (1969), and Hill (1974) among others. For a more complete set of references on these and other theoretical topics, see Karlin (1975).

Very recently, Karlin and his collaborators have extended the theory introduced in this chapter from two loci to N loci, where N is three or more. This extension includes the use of new ways to parameterize the multilocus recombination process and new ways to formulate the full genotypic selection pressure on the basis of the effects of the constituent loci. This new work is expected to appear in 1979.

Chapter 9
NATURAL SELECTION AND QUANTITATIVE INHERITANCE

THE genetic basis of many traits is unknown. Indeed, the most conspicuous characteristics of any organism—its size and shape—are influenced by many loci. A detailed genetic analysis of these characteristics is difficult, if not impossible, to obtain. What, then, can be done about population genetics and evolution of traits like size and shape? This section presents some theory for these kinds of traits. Although the early work by Fisher, Haldane, and Wright on population genetics was intended to explain macroscopic evolutionary phenomena, the recent orientation in population genetics has shifted to the details of the genetic structure of populations. Recent population genetics has focused on a locus by locus specification of gene frequencies; and on the influence of linkage, mating systems, and migration patterns on gene frequencies. Nonetheless, there remain many macroscopic phenomena revealed by the fossil record and observable in the field in terms of *phenotypic* patterns in natural populations. These patterns usually involve the size, color, and behavior of organisms. Such traits are precisely those whose genetic basis spans many loci and for which detailed genetic analysis is impractical. Moreover, it is relatively easy to observe organisms actually using these traits and to conjecture about the selection pressures on them. For these reasons a theory that avoids detailed reference to genetics and allows a rich specification of the selection pressures is appropriate.

This chapter relates to the need in evolutionary biology for a theory concerned principally with the phenotype and with natural selection. It is one of the most important theoretical topics in this book and will be covered in detail. First, we shall investigate a model for evolution that is defined purely at the level of the phenotype—with no reference to genetics whatsoever. Second, we shall introduce a purely genetic model that answers the same questions as the purely phenotypic model, but which assumes that a complete genetic analysis of the trait is available. Finally, we shall translate the purely phenotypic model into the purely genetic model, and vice versa. This translation will allow us to understand the genetic meaning of parameters that are originally defined only on the phenotypic level. Also the translation will allow us to determine the conditions under which a purely phenotypic model for evolution is consistent with a basic genetic account of the same evolutionary process.

Phenotype Model Based on the Segregation Kernel

The Segregation Kernel

A trait such as an organism's height is called a *quantitative character*. It shows continuous variation over a range of values within the population. A quantitative character is distinguished from a qualitative character. For example, a qualitative character is the presence or absence of a wrinkled coat in peas, or of a yellow versus green leaf. Often qualitative characters can be resolved into a simple genetic basis whereas quantitative characters involve a great many loci. The study of how quantitative characters are inherited between parents and offspring is called *quantitative inheritance*.

The inheritance of traits whose genetic basis is known can be inferred from Mendel's laws. But since we have admitted that we are unlikely to know the genetic basis of quantitative characters, we cannot appeal to Mendel's laws to tell us how quantitative characters are inherited. Therefore, we must start from scratch concerning quantitative inheritance. We must perform crosses and look at the progeny and see if any general pattern emerges.

In 1889 Galton published a classic work on the inheritance of quantitative characters. One of his studies concerns the inheritance of height in humans. (The height of females is multiplied by 1.08 to eliminate the effect of sexual dimorphism.) Galton determined the distribution of height among the offspring from parents of different sizes. Table 9.1 presents Galton's data. These data illustrate two principles.

The first principle which emerges from the data is that *there is a straight-line relationship between the size of offspring and the average size of their parents.* See Figure 9.1. This relationship is expressed in a special way as follows. We consider each individual's height in terms of its *deviation* from the average size in the population from which the parents were drawn. The average height in the population is 68.1 in. This value must be subtracted from everyone's height in order to express each height in terms of the deviation from the average. Figure 9.1 is a plot of the height of offspring

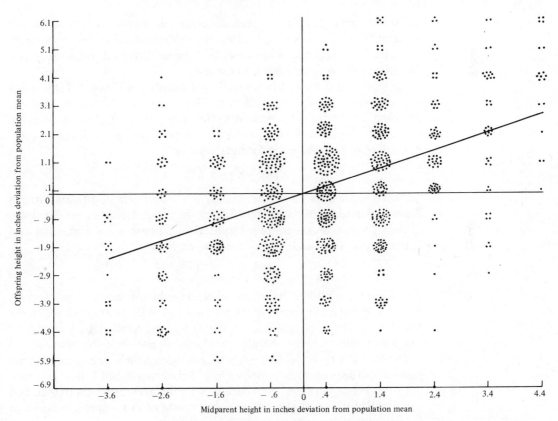

FIGURE 9.1. *Scatter diagram of offspring heights versus midparent heights. The regression line is drawn through the points. The slope of the regression line, h^2, is called the* heritability *and its value is* 0.65 *for these data. The clusters indicate data points that occur with a multiplicity. The degree of the multiplicity is indicated by the number of dots in the clusters.* [Data from Galton (1889) as summarized in Table 9.1.]

Table 9.1. Data on Inheritance of Height in Humans. All Female Heights Multiplied by 1.08. [From Galton (1889).]

| Height of Midparent | | Number of Adult Children of Various Height Classes | | | | | | | | | | | | | |
| | | 61.2 | 62.2 | 63.2 | 64.2 | 65.2 | 66.2 | 67.2 | 68.2 | 69.2 | 70.2 | 71.2 | 72.2 | 73.2 | 74.2 |
Height in Inches	Height in Inches of Deviation from Population Mean	-6.9	-5.9	-4.9	-3.9	-2.9	-1.9	-.9	.1	1.1	2.1	3.1	4.1	5.1	6.1
72.5	4.4								1	2	1	2	7	2	4
71.5	3.4					1	3	4	3	5	10	4	9	2	2
70.5	2.4	1		1		1	1	3	12	18	14	7	4	3	3
69.5	1.4			1	16	4	17	27	20	33	25	20	11	4	5
68.5	.4	1		7	11	16	25	31	34	48	21	18	4	3	
67.5	-.6		3	5	14	15	36	38	28	38	19	11	4		
66.5	-1.6		3	3	5	2	17	17	14	13	4				
65.5	-2.6	1		9	5	7	11	11	7	7	5	2	1		
64.5	-3.6	1	1	4	4	1	5	5		2					

against the average height of their parents. The average height of the two parents, that is $(X_♀ + X_♂)/2$, is called the *midparent* value. Then a straight line is fitted through the points. Note that the line passes through the origin. The slope can be calculated from the data.

Let us obtain the formula for the line illustrated in Figure 9.1, assuming that we have already determined its slope. Let X_o denote the average offspring height from parents with midparent value X_p, let the average height in the population be $\bar{X}$, and let the slope of the line be denoted by h^2. According to the figure, the formula must be

$$(X_o - \bar{X}) = h^2(X_p - \bar{X}). \qquad (9.1)$$

$(X_o - \bar{X})$ represents the vertical axis; it is the average offspring height expressed in terms of the deviation from the population average. Similarly, $(X_p - \bar{X})$ represents the average height of the parents on the horizontal axis. Furthermore, if we add $\bar{X}$ to both sides, we have

$$X_o = h^2 X_p + (1 - h^2)\bar{X} \qquad (9.2)$$

This formula describes the empirical relationship of the average offspring height to the average height of the parents and to the average height in the population. The slope of the line in Figure 9.1, h^2, is called the *heritability*.† The heritability of human height from Galton's data is found to be .65.

The heritability, h^2, indicates whether the character as represented in the parents also appears in their offspring; h^2 lies between 0 and 1. If h^2 is 1, then the average height of the offspring is exactly the average height of their two parents. But if h^2 is 0, then the average height of the offspring of any two

† A plot of points of y against x as in Figure 9.1 is called a *scatter diagram*. The straight line fitted to these points is called the *regression line* and the slope of the line is called the *regression coefficient*. Thus, in this terminology, the heritability, h^2, is the regression coefficient of offspring against midparent. The detailed calculations that are used to find the regression coefficient are discussed in all elementary textbooks in statistics. See, for example, Sokal and Rohlf (1969).

parents has nothing to do with the size of the parents. Instead, the average height of the offspring is simply the population average. Thus if h^2 is 0, the trait is not inherited in the sense that any deviation by the parents from the population mean is not, on the average, reflected in their offspring. It is usually true that h^2 is somewhere between 0 and 1, indicating partial heritability. It cannot be overemphasized that the measurement of heritability depends on the specific environmental conditions in which the breeding is done. For example, changing the nutritional regime can considerably alter the offspring-parent relationship. Furthermore, any particular estimate of a heritability is specific to the population studied. Galton studied people from England; had he studied people from somewhere else, he would probably have obtained data leading to a different estimate of h^2.

It also cannot be overemphasized that the magnitude of h^2 does not tell one whether any given trait is produced by a deterministic biochemical pathway during development. What it tells is whether any phenotype *deviations* from the population average are, on the average, transmitted to offspring. Moreover, the mechanism of inheritance need not be genetic to produce high h^2. Parents can communicate certain skills and aptitudes to offspring causing high parent-offspring resemblance without any genetics whatsoever. In short, h^2 is simply an empirical constant that indicates the extent to which parental phenotypic deviations from the population mean are transmitted to their offspring.

The second principle that emerges from the data concerns the nature of the variation in height among offspring from a given set of parents. The data suggest that *the variation among offspring height within families is nearly the same in all families regardless of the height of the parents*. Thus the average offspring height within a family depends on the size of the parents as we have already stated, but the variance of offspring height within a family is comparatively constant. Figure 9.2 illustrates this idea. Note that the shape of the offspring distribution is essentially the same throughout but that the location of the distribution depends on the midparent. Moreover, the distributions are approximately normal. The common variance of offspring height in the distributions illustrated in Figure 9.2 is denoted as σ_L^2. For Galton's data σ_L^2 is estimated to equal 5.2. The quantity, σ_L^2, is called the *segregation variance*.

We may summarize these two principles suggested by Galton's data with a formula to describe the inheritance of a quantitative character. Let $L(X)$ be a probability density function with mean of zero and variance σ_L^2. It describes the shape illustrated by the distributions in Figure 9.2. Typically, $L(X)$ is taken to be Gaussian. The *segregation kernel* (also called the *offspring phenotype distribution*) is the function that describes the offspring from any given midparent. It is obtained by locating $L(X)$ in the proper place corresponding to the given midparent. As we have seen, the mean offspring size is given by (9.2), so that to shift $L(X)$ to the appropriate location for a given midparent, we subtract the quantity, $h^2 X_p + (1 - h^2)\bar{X}$, from every X, giving

$$L\{X - [h^2 X_p + (1 - h^2)\bar{X}]\} \tag{9.3}$$

This is the segregation kernel. It is a function of one variable and has a mean of $h^2 X_p + (1 - h^2)\bar{X}$ and a variance of σ_L^2. For example, a Gaussian

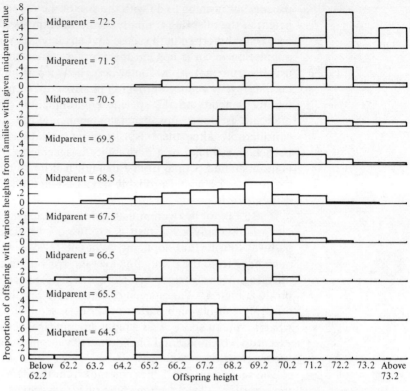

FIGURE 9.2. *Distribution of offspring heights from different midparents. Each histogram represents the offspring from many families with the same midparent value. Each histogram illustrates approximately the same variance in offspring height, but the mean offspring height depends on the midparent value. The common variance of the histograms, σ_L^2, is estimated to equal 5.2.* [Data from Galton (1899), *Natural Inheritance*, Macmillan Publishing Co., Inc., New York.]

segregation kernel is

$$L(X, X_p, \bar{X}) = \exp \frac{-\frac{1}{2}\{X - [h^2 X_p + (1 - h^2)\bar{X}]\}^2}{\sigma_L^2} \qquad (9.4)$$

We emphasize that the segregation kernel is an empirical quantity—it is not derived from underlying genetic assumptions. Instead, the segregation kernel summarizes the outcome of breeding experiments.†

Model for Evolution of a Quantitative Character

Now let us use our knowledge of the inheritance of a quantitative character to study the evolution of a quantitative character in a population. We no longer deal with gene frequencies. Instead, we consider the function, $p(X)$,

† The L observed in breeding experiments depends on the mating scheme in the population from which the parents are drawn. If L is to be used to predict evolution in a random mating population, then the parents for the breeding experiments used to measure L must also be drawn from a random mating population. In particular, the parents should not be from inbred lines because they would then be more homozygous on the average than phenotypically similar individuals in a random mating population. Hence the σ_L^2 measured from crosses among inbred parents might be quite different from that with crosses among parents from a random mating population.

which is the fraction of the population with a height of X.† We want to predict the distribution of height at time $t+1$, that is, $p_{t+1}(X)$, given the selection pressure and the height distribution at time t. The life history scheme is diagramed in Figure 9.3. We census at the zygotic phase. The organisms then survive to adults. We let $l(X)$ denote the probability that a zygote of phenotype X will survive to adulthood. Then there is random mating. It is assumed that every mating produces on the average $2m$ offspring. Thus the natural selection is through differential survival only and not differential fertility. The phenotypes of the progeny from any mating are then given by the segregation kernel and the next census is taken at the zygotic phase of the next generation. ‡

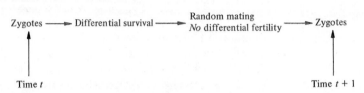

FIGURE 9.3. *The life history scheme underlying the segregation kernel model for natural selection on a quantitative character.*

Let us proceed through the life history step by step. We are given the distribution of the character among zygotes at time t as $p_t(X)$. To obtain the distribution after these zygotes have grown to adults, we must multiply by the probability of survival and divide by the necessary normalization constant giving

$$p_{w,t}(X) = \frac{l(X)p_t(X)}{\int l(X)p_t(X)\,dX} \tag{9.5}$$

We divide by $\int l(X)p_t(X)\,dX$ to ensure that the area under $p_{w,t}(X)$ sums to one; $p_{w,t}(X)$ is the distribution of the character among adults *after* selection has acted. Then these adults mate at random. The probability of an adult of type X mating with one of type Y, in that order, is the product, $p_{w,t}(X)p_{w,t}(Y)$. The midparent value for this mating is simply $(X+Y)/2$. So the offspring of this mating have phenotypes given by the segregation kernel, L, based on a midparent value of $(X+Y)/2$ and the population mean at time t, $\bar{X}_t$. Then the distribution of the character among the zygotes at $t+1$ is obtained by summing over the output from each kind of mating, that is, over all values of X and Y.

Putting all this together, we have

$$p_{t+1}(Z) = \int\int L\left\{Z - \left[h^2\frac{(X+Y)}{2} + (1-h^2)\bar{X}_t\right]\right\} p_{w,t}(X)p_{w,t}(Y)\,dX\,dY \tag{9.6}$$

This may look formidable but remember that all the biological assumptions in it are simple and straightforward. It says that the offspring distribution at

† Technically, $p(X)\,dX$ is the fraction of the population with a character value between X and $X+dX$.

‡ In using the segregation kernel to predict the progeny distribution after selection, it is implicitly assumed that the progeny distribution from the selected parents does not differ significantly from the progeny distribution estimated from breeding experiments conducted with a random sample of the population before selection.

$t + 1$ is found by summing up the output from all possible pairs of parents. The offspring from any pairs of parents is given by the segregation kernel, L. The parents mate at random, and selection through differential survival acts between the zygotic and adult phase. By combining the product, $ml(X)$, into a single measure of fitness, the selection pressure can be expressed in terms of the familiar relative selective value, $w(X)$. For example, $w(X)$ is the relative fitness of an individual with height X. Then substituting (9.5) into (9.6) and using the $w(X)$ gives an alternative form for (9.6)

$$p_{t+1}(Z) = \frac{1}{\bar{w}_t^2} \int\int L\left\{ Z - \left[h^2 \frac{(X+Y)}{2} + (1-h^2)\bar{X}_t \right] \right\}$$

$$\times w(X)p_t(X)w(Y)p_t(Y)\,dX\,dY \qquad (9.7)$$

where $\bar{w}_t = \int w(X)p_t(X)\,dX$. This expression for the population genetics of a quantitative character with natural selection is the analogue of that for p_{t+1} that we considered for one locus and two alleles with natural selection.

This equation for $p_{t+1}(X)$ can be iterated to predict the course of evolution of a quantitative character in a manner analogous to our iteration of the equation for p_{t+1} with one locus and two alleles. But for quantitative traits there is one very important additional assumption underlying the iteration of (9.7). It is assumed that the segregation kernel, L, remains the same from generation to generation. In particular, the heritability h^2, and the segregation variance, σ_L^2 are assumed to remain constant. This assumption may be valid in some circumstances but not in others, as will be discussed later, and therefore, some caution is appropriate. Because the one-locus two-allele equation of earlier sections is based on Mendel's laws, no proviso of this sort was needed. Nonetheless, in various experiments both h^2 and σ_L^2 do indeed remain approximately constant over 10 generations of strong selection. Continued selection for increased height, discussed in Falconer (1960, Chapter 12), is one example. Thus the use of (9.7) for fairly short term predictions can often be justified.

Kinds of Selection Pressures on Quantitative Characters

Now let us use (9.7) to infer something about the evolution of quantitative characters. It is easy to imagine many kinds of selection pressures on quantitative characters. Figure 9.4 illustrates three common schemes of selection. A selection pressure, $w(X)$, acting to move the population mean up or down from its current value is called *directional selection*. Figure 9.4a illustrates some forms of $w(X)$ that produce directional selection. If $w(X)$ is a step function, it is called *truncation* selection because it represents a situation where some individuals with a character value less than X_0 are prevented from breeding while all above X_0 do breed. Truncation selection is a special form of directional selection. Figure 9.4b illustrates a $w(X)$ that represents selection against individuals in the tails of the population. This $w(X)$ is called *stabilizing selection*; its effect is to reduce the population variance. Figure 9.4c illustrates a $w(X)$ that selects against a particular kind of individual, X_0, and in favor of those whose phenotype is either above or below X_0. This $w(X)$ represents *disruptive selection*; its effects can differ considerably. If the location of the notch in $w(X)$ in Figure 9.4c is constant through time, the population will be, in effect, directionally selected either up or down, depending upon the initial location of the population mean. However, if the location of the notch moves around so that the most

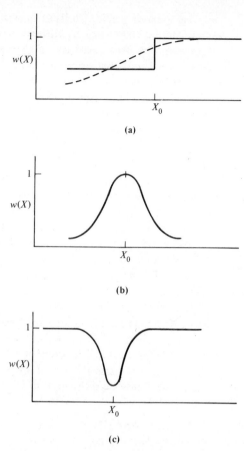

FIGURE 9.4. *Several types of selection function for a quantitative character. (a) Two kinds of directional selection; (b) stabilizing selection; and (c) disruptive selection.*

common type of individual is always selected against, then the effect is to increase the population variance. Many other schemes could be imagined, but these are the most commonly discussed.

Effect of Directional Selection on the Population Mean   Let us consider first the effect of *directional selection* on the population mean. Upon multiplying both sides of (9.7) by Z and integrating over Z, we obtain

$$\bar{X}_{t+1} = \bar{X}_t + h^2(\bar{X}_{w,t} - \bar{X}_t) \tag{9.8}$$

where $\bar{X}_{w,t}$ is the mean of $p_{w,t}(X)$, that is, the mean after selection. This formula predicts the average value of the trait, for example, the average height, in the population at time $t+1$. Given the average at time t, the formula represents a process whereby certain individuals from the original zygotic pool of time t are selected to become parents. The mean among these selected parents is $\bar{X}_{w,t}$. These parents as a group deviate from the population mean by $\bar{X}_{w,t} - \bar{X}_t$. Then $h^2(\bar{X}_{w,t} - \bar{X}_t)$ is the amount of this deviation that is transmitted to their offspring. The new population mean equals the old mean plus the amount inherited from the selected parents. The formula above for $\bar{X}_{t+1}$ is a classic formula in quantitative population genetics.

The formula for $\bar{X}_{t+1}$ tells us that the speed of evolution of the mean of a quantitative character depends on both the heritability and the *strength of selection*. The strength of selection, for directional selection, is often defined as $(\bar{X}_{w,t} - \bar{X}_t)$. If the individuals allowed to breed deviate greatly from the population mean, we say the selection is strong. The quantity $\bar{X}_{t+1} - \bar{X}_t$ is defined as the response to selection. By subtracting $\bar{X}_t$ from both sides of (9.8), we obtain the following relation

$$\binom{\text{response to}}{\text{selection}} = h^2 \binom{\text{strength of}}{\text{selection}} \tag{9.9}$$

The relation makes explicit the fact that fast evolution, that is, a fast change in the population mean of a quantitative character, requires both a high heritability and strong selection. The relation also brings out the fact that h^2 cannot in practice remain constant indefinitely. In practice, for example, one cannot select for increased height indefinitely. No selection program will produce corn 30 feet tall, or cows weighing 2 tons and so forth. There simply is not sufficient genetic variation in the population to produce such monstrosities regardless of the strength of selection. So after the mean has been shifted, say, three to five standard deviations by directional selection, no more response can be obtained and the heritability has become zero. Thus as the directional selection proceeds, the genetic variation for additional change is lost and the heritability tends to zero. But for a reasonably short time, h^2 can be treated as constant and equation (9.9) can be used to predict the population mean through time.

Effect of Stabilizing Selection on the Population Variance

Now let us consider the effect of *stabilizing selection* on the population variance. Multiplying each side of (9.7) by $(Z - \bar{X}_{t+1})^2$ and integrating over Z gives

$$\sigma_{t+1}^2 = \sigma_L^2 + \frac{(h^2)^2}{2} \sigma_{w,t}^2 \tag{9.10}$$

where $\sigma_{w,t}^2$ is the variance of $p_{w,t}(X)$. According to this formula, the population variance of the character at $t+1$ equals the segregation variance, σ_L^2, plus a constant times the variance in the population after selection. This constant is the heritability squared over two.[†]

We can use this formula to predict the effect of stabilizing selection on the population variance. The first point to keep in mind is that the population genetics of stabilizing selection involves some counteracting forces. There is selection against the tails, which tends to reduce the population variance. But the population variance cannot be reduced all the way to zero because there is variation that always resurfaces after any mating because of the segregation variance, σ_L^2. Even if, as an experiment, one were to cause the population to have zero variance after selection, the variance would

[†] This formula is also intuitive like that for $\bar{X}_{t+1}$. By the parent-offspring regression in Figure 9.1, the total variance among all offspring equals the variance predicted by the regression against midparent value plus the unexplained variance. The predicted variance is $(h^2)^2$ times the variance of the midparent value and the unexplained variance is σ_L^2. Since the midparent value is $\frac{1}{2}X_\circ + \frac{1}{2}X_\sigma$, the variance of the midparent value is $\frac{1}{4}\sigma_{w,t}^2 + \frac{1}{4}\sigma_{w,t}^2 = \frac{1}{2}\sigma_{w,t}^2$ assuming the distribution of both $X_\circ$ and X_σ is $p_{w,t}(X)$. Hence the variance among offspring as predicted by the regression against midparent is $(h^2)^2(\frac{1}{2})\sigma_{w,t}^2$. Adding this quantity to the unexplained variance, σ_L^2, yields (9.10).

rebound to σ_L^2 after random mating. Remember σ_L^2 is the variance among offspring from any mating, even from phenotypically identical parents. Thus, regardless of how far the variance is reduced during the selection phase, the variance cannot be less than σ_L^2 when censused among the zygotes after mating. Thus σ_L^2 provides a lower bound to the effect of stabilizing selection in the population variance.

Can we also find an upper bound? Yes. First let us define the "strength" of stabilizing selection as referring to the variance of the selection function $w(X)$ relative to the population variance, σ^2. If Var$[w(X)]$ is much less than σ^2, then there is strong selection. If Var$[w(X)]$ is much more than σ^2, there is weak stabilizing selection. Now consider weak selection. In the limit as the strength of stabilizing selection becomes very weak (i.e., Var$[w(X)] \to \infty$), the population variance after the selection will be the same as that before selection, that is

$$\sigma_{w,t}^2 \approx \sigma_t^2 \qquad \text{(weak selection)}$$

In the limit of weak selection Equation (9.10) becomes

$$\sigma_{t+1}^2 \approx \sigma_L^2 + \frac{(h^2)^2}{2}\sigma_t^2 \tag{9.11}$$

This expression indicates that σ^2 tends to an equilibrium under weak selection. Putting $\sigma_{t+1}^2 = \sigma_t^2 \equiv \sigma_u^2$ and solving gives the equilibrium,

$$\sigma_u^2 = \frac{\sigma_L^2}{1 - [(h^2)]^2/2} \tag{9.12}$$

Thus there is an upper bound to the population variance under weak stabilizing selection.

These results provide both an upper and lower bound to the population variance as summarized in Figure 9.5. Even the strongest stabilizing selection cannot reduce the variance below σ_L^2, and even the weakest does not cause it to spread beyond $\sigma_L^2/[1 - (h^2)^2/2]$. In particular with $h^2 = 1$, the upper bound is $2\sigma_L^2$ and with $h^2 = 1/2$, it is $(8/7)\sigma_L^2$. Figure 9.6 illustrates that the upper bound to the population variance is quite close to σ_L^2 unless h^2 is very near one. Thus the overall prediction from this model is that the population variance is confined, for any strength of stabilizing selection, to a

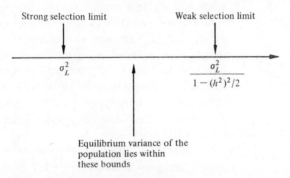

FIGURE 9.5. *The bounds to the population variance in a quantitative character with stabilizing selection. Predicted by the segregation kernel model for the evolution of a quantitative character.*

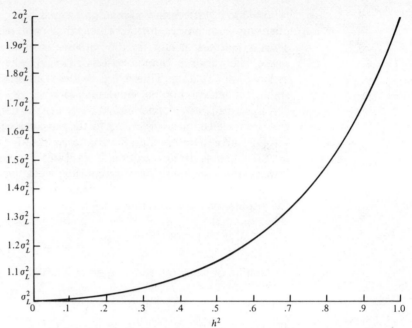

FIGURE 9.6. *The upper bound to the population variance as a function of the heritability,* h^2; $\sigma_L^2/\{1-(h^2)^2/2\}$.

rather narrow interval that depends principally on the segregation variance, σ_L^2.

These predictions of narrow bounds for the population variance bear on an important issue in evolution, namely, the adaptive significance of the *variance* of a character in a population. All quantitative characters show population variation, for example, variation in height, weight, bill length, jaw size, time of emergence in insects, flowering time, and so forth. A general question concerning this variation is whether it is somehow adaptive. Perhaps the population variation represents many individuals, each adapted to some part of a variable environment. For example, variation among individuals in flowering time could represent adaptations to the various sorts of microhabitats in which the plants are found. Alternatively, there may be only one optimal strategy in the environment, with the variation representing the segregation of suboptimal phenotypes. Now according to the model above, the selection pressure cannot *directly* influence the population variance to a significant degree—its bounds are set by σ_L^2. So any attempt to explain the population variance in terms of selection in a variable environment must go the route of showing that selection will alter σ_L^2 in order to produce the appropriate population variance. Conversely, if the variation is claimed to represent the segregation of suboptimal phenotypes, then it must be shown that there is no selection to reduce σ_L^2 to zero, thereby reducing the proportion of suboptimal phenotypes in the population.

To summarize, the phenotype model based on the segregation kernel has shown two general features of the evolution of a quantitative character. First, the response to directional selection depends both on the heritability, h^2, and on the strength of selection. Second, the population variance under stabilizing selection is bounded in a narrow interval set by the segregation variance, σ_L^2. This section on the phenotype model for the evolution of a quantitative character has been based on the papers by Bossert (1963), Slatkin (1970), Roughgarden (1972), and Slatkin and Lande (1976).

We now delve explicitly into some genetics, for two reasons. First, we want to know to what extent the assumptions of constant h^2 and σ_L^2 are consistent with Mendelian genetics. Since we have made no reference to genetics in the theory discussed so far, we may inadvertently have developed a theory that conflicts with Mendel's laws. We, therefore, need to explore the connection between the phenotype model above and models based more closely on genetic assumptions. Second, if we assume an underlying Mendelian basis for a quantitative character, we can derive alternative methods for measuring h^2. As discussed above, h^2 is obtained from plotting offspring against the average of the two parents. If underlying Mendelian laws are assumed, however, we can derive alternative methods for use when only one parent is known and we can use methods based on data about the resemblance among related individuals.

Gene Frequency Model at One Locus and Two Alleles We can develop a direct and simple approach to the evolution of quantitative characters if the genetic mechanism is known. To illustrate, let us pretend that some quantitative character is determined by one locus with two alleles. However, since the character is quantitative, not qualitative, it might show a range of values even among individuals with the same genotype. For example, we must allow the possibility that two individuals with genotype A_1A_1 may differ slightly in height because of random chance effects during organismal development. To allow such a possibility, we define the phenotype distribution for each genotype, $p(X|A_iA_j)$. This function is, in principle, measured for a given genotype by making a histogram of the different phenotypes produced by the given genotype. Figure 9.7 illustrates the sequence of phenotype distributions that would be needed to describe a one-locus two-allele system. We shall assume that the phenotype distributions $p(X|A_iA_j)$ are all normal with a common variance, V_e. Also, we shall label the mean phenotype for genotype A_iA_j as X_{ij},

$$X_{ij} = \int Xp(X|A_iA_j)\,dX \qquad (9.13)$$

The variance of the distributions, V_e, will be called the *environmental variance*. It represents the effect of the environment—both internal and external—upon the phenotype that is actually realized from the genotype. It is important that the variance, V_e, arises mostly from environmental effects and *not* from the influence of genes at other loci, say B, C, and so forth, on the expression of the A locus. If much of the variance originated from the effects of other loci we could not, in practice, view the trait as determined by

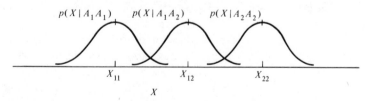

FIGURE 9.7. *Phenotype distributions for each of the genotypes. The distributions are normal and with the same variance, V_e. The mean of the distribution for genotype $A_i A_j$ is denoted as X_{ij}.*

one locus, as we are assuming, but instead would have to include the effects of selection on the other loci as well.

The natural selection is, as before, expressed in terms of the selection function $w(X)$. For example, $w(X)$ is the relative selective value of an individual with height X. We can use the phenotype distributions together with $w(X)$ to compute the relative selective values for the *genotypes*, w_{ij}. Then these genotypic selective values can be used in the formula for p_{t+1}, which we studied in great detail in earlier chapters. The formula for w_{ij} is

$$w_{ij} = \int p(X|A_iA_j)w(X)\,dX \qquad (9.14)$$

The term w_{ij} is simply the average fitness of individuals with genotype A_iA_j. The quantity, w_{ij}, is used in the familiar formula based on Mendelian inheritance

$$p_{t+1} = \left(\frac{p_t w_{11} + q_t w_{12}}{\bar{w}}\right)p_t$$

where $\bar{w} = p_t^2 w_{11} + 2p_t q_t w_{12} + q_t^2 w_{22}$ and p_t is the frequency of the A_1 allele at time t censused among the zygotes. Then the average value of X in the population at time t is found from

$$\bar{X}_t = p_t^2 X_{11} + 2p_t q_t X_{12} + q_t^2 X_{22} \qquad (9.15)$$

and the variance of X in the population is found from†

$$\sigma_t^2 = p_t^2(X_{11} - \bar{X}_t)^2 + 2p_t q_t(X_{12} - \bar{X}_t)^2 + q_t^2(X_{22} - \bar{X}_t)^2 + V_e \qquad (9.16)$$

These equations form a complete model for the evolution of a quantitative character. To use a theory of this sort, both the genetic basis of the character and the phenotype distribution for each genotype must be determined. Then the phenotypic selection function, $w(X)$, is converted into genotypic selective values by Equation (9.14). Next, the iteration through time is carried out in terms of gene frequencies and uses the genotypic selective values. The description of the mean and variance of the quantitative character in the population is then generated from the gene frequency information by (9.15) and (9.16).

It is important to understand the contrast between this model and the phenotype model considered before. In that model the iteration was carried out directly at the level of the phenotype. In contrast, in this model the description of the problem is translated into genetic terms, the iteration carried out at the level of gene frequencies, and then the genetic description translated back into phenotypic terms.

Influence of V_e on Evolution We can use this genetic model to illustrate the influence of the environmental variance, V_e, on the course of natural selection. We might expect that a given selection function, $w(X)$, can only be effective if V_e is small. For if V_e is large, the phenotype distributions produced by the different genotypes will greatly overlap. As a result, there will be little differential survival at the genotypic level even though there may be apparently strong natural selection at the phenotypic level. To examine this idea, consider the selec-

---

† This formula asserts that σ_t^2 is the sum of the between-group variance and the average within-group variance where a group is defined as a collection of individuals with the same genotype.

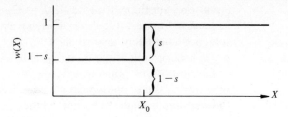

FIGURE 9.8. *The selection function for truncation selection.*

tion function illustrated in Figure 9.8. This $w(X)$ represents truncation selection.

$$w(X) = 1 \qquad X \geq X_0$$
$$w(X) = 1 - s \qquad X < X_0 \tag{9.17}$$

If s is 1, there is full truncation selection; otherwise there is partial selection. If we now calculate w_{ij} from Equation (9.14), using this $w(X)$, we obtain

$$w_{ij} = (1-s) + \frac{s}{2}\left[1 - \mathrm{erf}\left(\frac{X_0 - X_{ij}}{\sqrt{2 V_e}} \right) \right] \tag{9.18}$$

The function, erf (X), is called the *error function* and its value is tabulated in many tables. Figure 9.9 gives a sketch of erf (X). It is a sigmoid curve that tends from -1 to $+1$ as X tends from $-\infty$ to ∞. Then by inspection of Equation (9.18), we see that

$$\lim_{V_e \to \infty} w_{ij} = (1-s) + \frac{s}{2}[1 - \mathrm{erf}\,(0)] = 1 - \frac{s}{2} \tag{9.19}$$

That is, as the environmental variance, V_e, tends to ∞ all the w_{ij}'s come to

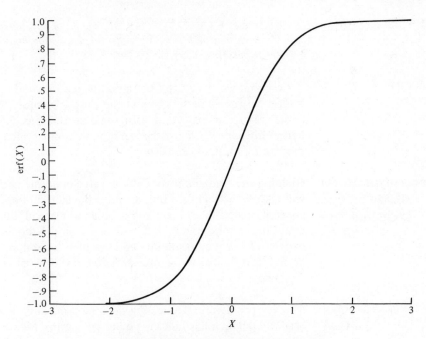

FIGURE 9.9. *The error function,* erf (X).

equal one another at the value $1-s/2$. This is illustrated in Figure 9.10. Indeed, in Figure 9.11, V_e does not have to be very large before the effective selection coefficient against A_1A_1 is only .3, while full truncation selection, $s = 1$, is occurring at the phenotypic level.

The important point to emerge from this analysis of w_{ij} is that strong selection at the phenotypic level need not translate into strong selection at the genotypic level—whether it does or not depends on V_e. This result provides a clue that V_e is closely related to the idea of heritability discussed earlier.

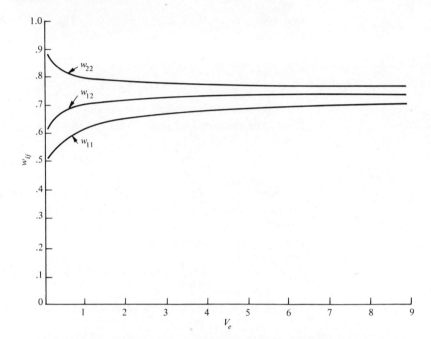

FIGURE 9.10. *The selective values of the three genotypes as a function of V_e. Note that increasing V_e tends to equalize the fitness among the three genotypes. Here $X_{11} = 1.0$, $X_{12} = 1.5$, $X_{22} = 2.0$, $X_0 = 1.75$, and $s = .5$.*

Now that we have explored a purely genetic model, let us try to build a bridge between this model and the purely phenotypic model considered earlier. We can expect to gain some further insight into the notion of heritability and to learn the conditions under which these two kinds of models are mutually consistent.

Phenotype Model Based on One Locus and Two Alleles

Building the bridge between the purely genotypic and phenotypic models will take several steps. First, we must develop an important new technical concept, called the "effect" of a gene, which will allow us to understand quantitative inheritance in Mendelian terms. Second, we shall derive an expression for the heritability and examine several methods for measuring heritability. Third, we shall also derive an expression for the segregation variance, σ_L^2.

The Effect of a Gene

The key problem in connecting the genotypic and phenotypic picture is to cast Mendelian inheritance in phenotypic terms. Mendelian genetics tells us how *genes* are inherited, but here we need to know how the *phenotype* is

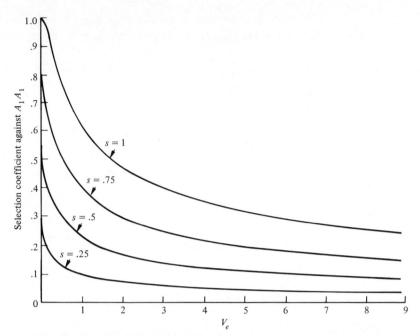

FIGURE 9.11. *Selection coefficient against the A_1A_1 homozygote as a function of V_e for various degrees of phenotypic selection, s. The vertical axis is the quantity $(1 - w_{11}/w_{22})$; $X_{11} = 1.0$, $X_{12} = 1.5$, $X_{22} = 2.0$, and $X_0 = 1.75$.*

inherited. We shall see that the phenotype can be viewed abstractly as being produced by the sum of two kinds of terms, and that it is the first kind, called the "additive effect of a gene," which is the part of the phenotype that is inherited, assuming underlying Mendelian segregation.

To develop the concepts we shall need, consider the following hypothetical problem. You are given certain information and asked to make a prediction. First, you are given the three average phenotypes X_{ij} produced by the genotypes A_iA_j. Second, you are given the mating scheme—in our case random union of gametes. Then you are asked to predict the average phenotype for any genotype based on *two* variables, one for each allele. Specifically, let χ_1 and χ_2 denote two variables whose values are as yet unknown. The first corresponds to the A_1 allele and the second to the A_2 allele. You are asked to *find* the values of χ_1 and χ_2 that allow you to predict X_{11} as the sum of $\chi_1 + \chi_1$, X_{12} as $\chi_1 + \chi_2$ and X_{22} as $\chi_2 + \chi_2$ as closely as possible. Obviously, if one is trying to collapse a set of three variables, X_{11}, X_{12}, and X_{22}, into two variables, χ_1 and χ_2, some error will usually result, but the values of χ_1 and χ_2 should be chosen to make the error as small as possible. Now the error in the prediction of X_{ij} from $\chi_i + \chi_j$ for any individual is

$$\left(\begin{array}{c}\text{prediction error}\\ \text{per individual}\end{array}\right) = X_{ij} - (\chi_i + \chi_j) \qquad (9.20)$$

Since we want to minimize the total prediction error, whether positive or negative, we should square the error so that + and − errors do not cancel out,

$$\left(\begin{array}{c}\text{prediction error}\\ \text{per individual}\end{array}\right)^2 = [X_{ij} - (\chi_i + \chi_j)]^2 \qquad (9.21)$$

Then the prediction error averaged over the whole population, assuming Hardy–Weinberg frequencies because of the random union of gametes, is

$$\begin{bmatrix} \text{average prediction} \\ \text{error}^2 \text{ for population} \end{bmatrix} = p_t^2[X_{11}-(\chi_1+\chi_1)]^2 + 2p_t q_t[X_{12}-(\chi_1+\chi_2)]^2$$

$$+ q_t^2[X_{22}-(\chi_2+\chi_2)]^2 = V_D \qquad (9.22)$$

We let V_D denote this average of the prediction error squared. Now it is easy to find the values of χ_1 and χ_2 that minimize V_D. Taking $\partial V_D/\partial \chi_1$ and $\partial V_D/\partial \chi_2$ and setting equal to zero gives two simultaneous linear equations for χ_1 and χ_2. (See the boxed material.) Solving gives

---

Derivation of formulas for the additive effects of alleles.

$$\frac{\partial V_D}{\partial \chi_1} = (-4)\{p^2[X_{11}-2\chi_1] + pq[X_{12}-(\chi_1+\chi_2)]\}$$

$$= (4p)\{(p+1)\chi_1 + q\chi_2 - (pX_{11}+qX_{12})\}$$

Similarly,

$$\frac{\partial V_D}{\partial \chi_2} = (4q)\{p\chi_1 + (q+1)\chi_2 - (pX_{12}+qX_{22})\}$$

Putting $\partial V_D/\partial \chi_1 = 0$ and $\partial V_D/\partial \chi_2 = 0$ gives

$$(p+1)\chi_1 + q\chi_2 = pX_{11}+qX_{12}$$

$$p\chi_1 + (q+1)\chi_2 = pX_{12}+qX_{22}$$

Solving for χ_1 and χ_2 in terms of p, q, and X_{ij} yields Equation (9.23).

$$\chi_1 = p_t X_{11} + q_t X_{12} - \tfrac{1}{2}\bar{X}_t$$
$$\chi_2 = p_t X_{12} + q_t X_{22} - \tfrac{1}{2}\bar{X}_t \qquad (9.23)$$

These formulas answer the prediction problem. To predict the average phenotype for the genotype, $A_i A_j$, based on the formula of $\chi_i + \chi_j$, the least average error results if χ_1 and χ_2 have the values given by (9.23). The variables, χ_1 and χ_2, are called the *additive effects* of alleles A_1 and A_2, respectively. The name arises because they are used in an additive formula to predict the phenotype X_{ij}.

As a numerical example, suppose that X_{11} is known to represent a height of 5 feet; X_{12} is 7 feet; and X_{22} is 10 feet. Suppose that p is $\tfrac{1}{4}$. From these data $\bar{X}$ is $\tfrac{1}{16}(5)+\tfrac{6}{16}(7)+\tfrac{9}{16}(10) = 8.56$. Then the additive effect of A_1 is $\chi_1 = [\tfrac{1}{4}(5)+\tfrac{3}{4}(7)] - \tfrac{1}{2}(8.56) = 2.22$ and of A_2 is $\chi_2 = [\tfrac{1}{4}(7)+\tfrac{3}{4}(10)] - \tfrac{1}{2}(8.56) = 4.97$. So the predicted phenotypes based on the additive effects are, for $A_1A_1, 2\chi_1 = 2(2.22) = 4.44$; for $A_1A_2, \chi_1 + \chi_2 = 2.22 + 4.97 = 7.19$; and for $A_2A_2, 2\chi_2 = 2(4.97) = 9.94$. These predictions commit the smallest average error throughout the population.

The formulas for χ_1 and χ_2 depend on the gene frequency, p. This dependence is natural because we are choosing the values of χ_1 and χ_2 to make the best prediction when averaged over the population. If p is high, then most of the population will consist of A_1A_1 and A_1A_2 individuals. Hence we would want χ_1 and χ_2 to yield accurate predictions of X_{11} and X_{12}

even if accuracy in predicting X_{22} were sacrificed. The dependence of χ_1 and χ_2 on p accomplishes this end.

If the heterozygote is exactly intermediate between both homozygotes, then there is no dominance; otherwise, one or the other allele is somewhat dominant. Thus A_1 is slightly dominant over A_2 in the numerical example above. It is the dominance interaction between the alleles that is the cause of any prediction error incurred when using $\chi_i + \chi_j$ to predict a phenotype. For this reason it is customary to view the actual phenotype as being produced by additive effects plus *deviations due to dominance*, d_{ij}. Thus the phenotype X_{ij} can be formally represented as

$$X_{ij} = \chi_i + \chi_j + d_{ij} \qquad (9.24)$$

where d_{ij} is the dominance deviation produced in genotype A_iA_j. In the numerical example above $d_{11} = .56$, $d_{12} = -.19$, and $d_{22} = .06$. The d_{ij} are also functions of p. The d_{ij} are *not* to be thought of as measures of the amount of interaction at the biochemical level between the biochemical products of the alleles. They are not physical measures of dominance. Instead, they simply document the extent to which the purely additive formula, $\chi_i + \chi_j$, fails to predict the phenotype X_{ij} based on the current p_t.

The formula above asserts that a phenotype can be viewed abstractly as resulting from the sum of two kinds of terms, terms for the additive effects of the individual alleles and a term for the dominance deviation. As we shall now see, it is the additive effects that are important for the inheritance of the phenotype.

Average Offspring Phenotypes Let us now consider inheritance in phenotypic terms. The problem is to infer the average phenotype of the *offspring* of any given individual from Mendelian laws. To solve this problem is to understand the extent to which the phenotype is on the average transmitted from parent to offspring. To solve this problem, we assume a mating scheme of random union of gametes and the absence of selection. Consider first the average offspring phenotype produced by an A_1A_1 individual. This individual can participate in three types of crosses as enumerated in Table 9.2. By the random union of gametes assumption, the probability of the given A_1A_1 individual crossing, for example, with type A_1A_2 is $2pq$ because $2pq$ is the frequency of A_1A_2 in the population. The frequency among the adults is the same as that among offspring (i.e., Hardy–Weinberg frequencies) because no selection is assumed. Based on this table, the average offspring phenotype from an individual of type A_1A_1 mating by random union of gametes without selection is

$$\binom{\text{average offspring phenotype}}{\text{for } A_1A_1} = p^2 X_{11} + 2pq\tfrac{1}{2}(X_{11} + X_{12}) + q^2 X_{12}$$

$$= pX_{11} + qX_{12} \qquad (9.25)$$

This expression can be stated in terms of the additive effect by substituting from Equation (9.23),

$$[\text{average offspring phenotype from } A_1A_1] = \chi_1 + \frac{\bar{X}}{2} \qquad (9.26)$$

Table 9.2. Offspring Table for Individuals of Each Genotype

Type of Cross	Frequency	Offspring Genotype	Average Offspring Phenotype in a Given Cross
		Offspring Table for A_1A_1 Individuals	
$A_1A_1 \times A_1A_1$	p^2	A_1A_1	X_{11}
A_1A_2	$2pq$	$\frac{1}{2}A_1A_1 + \frac{1}{2}A_1A_2$	$\frac{1}{2}(X_{11} + X_{12})$
A_2A_2	q^2	A_1A_2	X_{12}
		Offspring Table for A_1A_2 Individuals	
$A_1A_2 \times A_1A_1$	p^2	$\frac{1}{2}A_1A_1 + \frac{1}{2}A_1A_2$	$\frac{1}{2}(X_{11} + X_{12})$
A_1A_2	$2pq$	$\frac{1}{4}A_1A_1 + \frac{1}{2}A_1A_2 + \frac{1}{4}A_2A_2$	$\frac{1}{4}X_{11} + \frac{1}{2}X_{12} + \frac{1}{4}X_{22}$
A_2A_2	q^2	$\frac{1}{2}A_1A_2 + \frac{1}{2}A_2A_2$	$\frac{1}{2}(X_{12} + X_{22})$
		Offspring Table for A_2A_2 Individuals	
$A_2A_2 \times A_1A_1$	p^2	A_1A_2	X_{12}
A_1A_2	$2pq$	$\frac{1}{2}A_1A_2 + \frac{1}{2}A_2A_2$	$\frac{1}{2}(X_{12} + X_{22})$
A_2A_2	q^2	A_2A_2	X_{22}

By identical arguments we obtain

$$[\text{average offspring phenotype from } A_1A_2] = \frac{1}{2}(\chi_1 + \chi_2) + \frac{\bar{X}}{2}$$

$$[\text{average offspring phenotype from } A_2A_2] = \chi_2 + \frac{\bar{X}}{2}$$

(9.27)

And in general we see that the average offspring phenotype for a parent of genotype A_iA_j is $(\frac{1}{2})(\chi_i + \chi_j) + \bar{X}/2$. This is a remarkable result and yet intuitive. Recall that the average additive effect of alleles drawn at random from the gene pool is $p\chi_1 + q\chi_2 = \bar{X}/2$. To check this, use Equation 9.23. The offspring phenotype is, itself, produced by two additive contributions, the average additive effect from the known parent, $\frac{1}{2}(\chi_i + \chi_j)$, plus the average additive effect of an allele drawn at random from the gene pool, $\bar{X}/2$. In this sense it is the *additive effects* of the alleles that are transmitted to the offspring. The dominance effects d_{ij} are not relevant to the prediction of the average offspring phenotypes. This result is the reason why it is important to decompose the average phenotype, X_{ij}, of any genotype, A_iA_j, into the additive and dominance components. Only the additive component of the phenotype is used in the inheritance of the phenotype.

The derivation above accomplishes the description of inheritance in phenotypic terms as based on Mendelian laws. However, we also had to assume random union of gametes and no selection. We are now ready to calculate the heritability, h^2, in genetic terms under these assumptions.

The Heritability h^2

We can anticipate some properties of the formula for the heritability. As we learned from the purely genotypic model, h^2 must tend to zero as the environmental variance, V_e, tends to infinity. Recall that V_e controls how selection at the phenotypic level is translated into selection at the genotypic level. A large V_e causes the genotypes to have greatly overlapping phenotype distributions so that strong phenotypic selection does not result in strong genotypic selection. Also, as we just learned above, the additive component of the phenotype is the component that is transmitted to the offspring; the dominance deviations are not. Thus we would expect that if phenotypic variation in the population were caused mainly by dominance deviations, then h^2 would be low because the phenotypic variation would not be inherited. If the variation were mainly attributable to additive effects of the alleles, then h^2 would be high. Even strong phenotypic selection upon variation produced by dominance deviations should yield no response whereas the same selection on variation produced by additive effects should yield a definite response. Thus we conjecture that the formula for h^2 will depend both on whether the phenotypic variation exposed to the selection is caused by the additive effects of the alleles and will depend on the magnitude of V_e.

Before proceeding, let us formulate this conjecture more precisely. We can very simply quantify the extent to which the phenotypic variation that is exposed to selection is produced by different causes. The population variance is, as before in Equation (9.16).

$$\sigma^2(p) \equiv V_T(p) \equiv [p^2(X_{11} - \bar{X})^2 + 2pq(X_{12} - \bar{X})^2 + q^2(X_{22} - \bar{X}^2)] + V_e \tag{9.28}$$

The term in brackets is the component of the population variance caused by differing genotypes, each with its own average phenotype X_{ij} and V_e represents the variance among individuals with the same genotype. By convention, the population variance, σ^2, is often called the *total variance* and denoted, V_T. The terms in brackets are called the *genotypic variance*, V_G.

$$V_G(p) = p^2(X_{11} - \bar{X})^2 + 2pq(X_{12} - \bar{X})^2 + q^2(X_{22} - \bar{X})^2 \tag{9.29}$$

so that the total variance is the sum of V_G and V_e. Now as we have seen, the average phenotype produced by genotype A_iA_j can be decomposed into the components produced by the additive effects $(\chi_i + \chi_j)$ and the dominance effects, d_{ij}. We define the *additive genetic variance*, V_A, as the variance among the phenotypes based on predicting the phenotypes, X_{ij}, from the additive effects as $\chi_i + \chi_j$,

$$V_A(p) = p^2(2\chi_1 - \bar{X})^2 + 2pq(\chi_1 + \chi_2 - \bar{X})^2 + q^2(2\chi_2 - \bar{X})^2 \tag{9.30}$$

We also define the *dominance variance*, V_D, as the variance of the d_{ij},

$$V_D(p) = p^2 d_{11}^2 + 2pqd_{12}^2 + q^2 d_{22}^2 \tag{9.31}$$

This formula is especially simple because the average of the deviations, $\bar{d}$ is zero. Upon comparing Equation (9.31) with Equation (9.22) you will recognize $V_D(p)$ as the average prediction error^2, since $d_{ij} = X_{ij} - (\chi_i + \chi_j)$. You will recall that this is the quantity we minimized when finding the formula for χ_1 and χ_2. Now it can be shown that, for any p, the sum of the additive genetic variance and the dominance variance is identical to the genotypic variance†

$$V_G(p) \equiv V_A(p) + V_D(p) \tag{9.32}$$

† This result is easily though tediously demonstrated by adding $V_A(p)$ and $V_D(p)$, rearranging, and obtaining $V_G(p)$, thus proving that Equation (9.32) expresses an identity. This identity follows from viewing the prediction of X_{ij} in terms of $(\chi_i + \chi_j)$ as a regression. Here V_A is the variance explained by the regression and V_D is the unexplained variance. Their sum must be the variance of the X_{ij}.

As a result we can view the total phenotypic variance in the population in terms of three components

$$\sigma^2(p) \equiv V_T(p) = V_A(p) + V_D(p) + V_e \tag{9.33}$$

To illustrate, consider our previous numerical example; X_{11}, X_{12}, and X_{22} are 5, 7, and 10; p is $\frac{1}{4}$. Then $\bar{X}$ is 8.56, χ_1 is 2.22, and χ_2 is 4.97 as found previously. Then the genotypic variance is

$$V_G = \tfrac{1}{16}(5 - 8.56)^2 + \tfrac{6}{16}(7 - 8.56)^2 + \tfrac{9}{16}(10 - 8.56)^2 = 2.87$$

and the additive genetic variance is

$$V_A = \tfrac{1}{16}[2(2.22) - 8.56]^2 + \tfrac{6}{16}[2.22 + 4.97 - 8.56]^2 + \tfrac{9}{16}[2(4.97) - 8.56]^2$$

$$= 2.84$$

The dominance variance is

$$V_D = \tfrac{1}{16}(.56)^2 + \tfrac{6}{16}(-.19)^2 + \tfrac{9}{16}(.06)^2 = .04$$

Note that $V_A + V_D$ equals V_G as it should.

With the several possible causes of phenotypic variation separated in this way, we can pursue our conjecture about the formula for h^2. We might expect the quantity

$$\left[\begin{array}{l} \text{fraction of total phenotypic} \\ \text{variance due to additive effects} \end{array} \right] = \frac{V_A(p)}{V_A(p) + V_D(p) + V_e} \tag{9.34}$$

to be closely related to the heritability. We would expect a large response for a given selection intensity as the fraction of the phenotypic variation caused by additive effects tends to one. Also, as we have mentioned, we would expect h^2, the heritability factor, to tend to zero as V_e tends to infinity. Thus the quantity above behaves mathematically in the way we expect h^2 to behave. Let us derive the formula for h^2 to see if our conjecture is correct.

In statistics it is known that the general formula for the regression coefficient of y against x is

$$b_{y,x} = \frac{\text{Cov}(y, x)}{\text{Var}(x)} \tag{9.35}$$

To derive the formula for h^2, we must compute the covariance of the average offspring phenotype with the midparent and the variance of the midparent, and form the quotient. These quantities are calculated in the margin, resulting in

$$\text{Cov (average offspring, midparent phenotype)} = \tfrac{1}{2} V_A(p)$$

$$\text{Var (midparent)} = \tfrac{1}{2} V_T(p) \tag{9.36}$$

So

$$h^2(p) = \frac{(\tfrac{1}{2}) V_A(p)}{(\tfrac{1}{2}) V_T(p)} = \frac{V_A(p)}{V_A(p) + V_D(p) + V_e} \tag{9.37}$$

Thus the heritability, that is, the regression coefficient of offspring phenotype against midparent, equals the fraction of the phenotypic variance in the population attributable to the additive effects of the alleles. This formula is the bridge between pure phenotypic and genotypic models for the evolution of a quantitative character. This formula connects the regression coefficient of offspring against midparent phenotype in a random mating population with a *genetic* description of the trait in terms of gene frequencies, p, the average phenotypes X_{ij} produced by genotypes $A_i A_j$, and the environmental variance V_e.

Derivation of the formula for the heritability.

1. Calculate Cov (average offspring, midparent)

   Let O denote the offspring phenotype. Let M denote the midparent phenotype, and let $M = (P_\male + P_\female)/2$ where $P_\male$ and $P_\female$ are the phenotypes of the male and female parents, respectively.

   (a) It is an identity that

   $$\text{Cov}(O, M) \equiv \tfrac{1}{2}\text{Cov}(O, P_\male) + \tfrac{1}{2}\text{Cov}(O, P_\female)$$

   This identity is proved as follows:

   $$\text{Cov}(O, M) \equiv E\{[(O - \bar{O})(M - \bar{M})]\}$$

   $$= E\left\{(O - \bar{O})\left[\frac{P_\male + P_\female}{2} - \frac{\bar{P}_\male + \bar{P}_\female}{2}\right]\right\}$$

   $$= E\{(O - \bar{O})[\tfrac{1}{2}(P_\male - \bar{P}_\male) - \tfrac{1}{2}(P_\female - \bar{P}_\female)]\}$$

   $$= \tfrac{1}{2}E\{(O - \bar{O})(P_\male - \bar{P}_\male)\} + \tfrac{1}{2}E\{(O - \bar{O})(P_\female - \bar{P}_\female)\}$$

   $$\equiv \tfrac{1}{2}\text{Cov}(O, P_\male) + \tfrac{1}{2}\text{Cov}(O, P_\female).$$

   (b) If the sexes are identical with respect to the trait being considered, then $\text{Cov}(O, P_\male) = \text{Cov}(O, P_\female)$. Let P denote a single parent where the other parent is chosen at random from the population. Then, if the sexes are identical, we have

   $$\text{Cov}(O, M) \equiv \tfrac{1}{2}\text{Cov}(O, P_\male) + \tfrac{1}{2}\text{Cov}(O, P_\male) = \text{Cov}(O, P)$$

   (c) The problem of calculating $\text{Cov}(O, M)$ is now reduced to calculating the covariance of offspring with a single parent, provided the other parent is chosen at random. Equations (9.26) and (9.27) give the average offspring phenotype from parents of each genotype, provided the other parent is chosen at random. Using (9.26) and (9.27), we have

   $$\text{Cov}(O, P) = p^2(X_{11} - \bar{X})\left[\left(\chi_1 + \frac{\bar{X}}{2}\right) - \bar{X}\right]$$

   $$+ 2pq(X_{12} - \bar{X})\left[\left(\frac{\chi_1 + \chi_2}{2} + \frac{\bar{X}}{2}\right) - \bar{X}\right]$$

   $$+ q^2(X_{22} - \bar{X})\left[\left(\chi_2 + \frac{\bar{X}}{2}\right) - \bar{X}\right]$$

   $$= p^2(X_{11} - \bar{X})\left(\chi_1 - \frac{\bar{X}}{2}\right) + pq(X_{12} - \bar{X})\left(\chi_1 - \frac{\bar{X}}{2}\right)$$

   $$+ pq(X_{12} - \bar{X})\left(\chi_2 - \frac{\bar{X}}{2}\right) + q^2(X_{22} - \bar{X})\left(\chi_2 - \frac{\bar{X}}{2}\right)$$

$$= p[p(X_{11}-\bar{X})+q(X_{12}-\bar{X})]\left(\chi_1-\frac{\bar{X}}{2}\right)$$

$$+ q[p(X_{12}-\bar{X})+q(X_{22}-\bar{X})]\left(\chi_2-\frac{\bar{X}}{2}\right)$$

$$= p\left(\chi_1-\frac{\bar{X}}{2}\right)^2+q\left(\chi_2-\frac{\bar{X}}{2}\right)^2$$

(d) Recall that the additive genetic variance is defined in Equation (30) as

$$V_A = p^2(2\chi_1-\bar{X})^2+2pq(\chi_1+\chi_2-\bar{X})^2+q^2(2\chi_2-\bar{X})^2$$

This may be rearranged as

$$V_A = 2\left[p\left(\chi_1-\frac{\bar{X}}{2}\right)^2+q\left(\chi_2-\frac{\bar{X}}{2}\right)^2\right]$$

Hence $\mathrm{Cov}\,(O, P)=\frac{1}{2}V_A$

(e) In summary,

$$\mathrm{Cov}\,(O, M)\equiv\tfrac{1}{2}\mathrm{Cov}\,(O, P_\sigma)+\tfrac{1}{2}\mathrm{Cov}\,(O, P_\circ)$$

$$= \mathrm{Cov}\,(O, P)$$

$$= \tfrac{1}{2}V_A$$

2. Calculate Var (midparent)

$$\mathrm{Var}\,(M)\equiv\mathrm{Var}\left(\frac{P_\sigma+P_\circ}{2}\right)$$

(a) If the mating is independent with respect to the trait under consideration, we have

$$\mathrm{Var}\,(M)=\tfrac{1}{4}\mathrm{Var}\,(P_\sigma)+\tfrac{1}{4}\mathrm{Var}\,(P_\circ)$$

(b) If the sexes are identical to the trait under consideration as assumed in (1b) above, then $\mathrm{Var}\,(P_\sigma)=\mathrm{Var}\,(P_\circ)=\mathrm{Var}\,(P)$. The variance of the trait in the population is simply the total variance V_T. Hence

$$\mathrm{Var}\,(M)=\tfrac{1}{2}V_T$$

3. Therefore, the heritability, defined as the regression of offspring against midparent, is found as

$$h^2\equiv\frac{\mathrm{Cov}\,(O, M)}{\mathrm{Var}\,(M)}=\frac{V_A}{V_T}$$

Consider the numerical example again; X_{11}, X_{12}, and X_{22} are 5, 7, and 10; p is $\frac{1}{4}$. Suppose also that the variance in heights produced by any genotype, V_e, is 4. We previously computed $V_A = 2.84$ and $V_D = .04$. Therefore, h^2 as predicted by Equation (9.37) is

$$h^2=\frac{2.84}{2.84+.04+4.00}=.41$$

This result means that if we obtained a large random sample of animals from a random mating population with X_{ij}, p, and V_e as above, plotted a scatter diagram of offspring height against the average height of the two parents, and fitted a straight line through the data points, then the slope of this line would equal .41.†

Special Cases of $h^2(p)$ The formula predicting h^2, the heritability factor, from genetic considerations allows rich and interesting interpretation, and we shall consider some particularly illustrative cases. Also we should examine whether there is any case where $h^2(p)$ is effectively constant. Recall that we assumed that h^2 (and σ_L^2 too) were constant in the pure phenotypic model developed in the beginning of this chapter. We should examine, now, whether this assumption is warranted.

First, it should be noted that there are less cumbersome formulas for V_A and V_D than the equations that originally define these quantities. Simply, but tediously, rearranging Equations (9.30) and (9.31) leads to the following expressions

$$V_A = 2\left[p\left(\chi_1 - \frac{\bar{X}}{2}\right)^2 + q\left(\chi_2 - \frac{\bar{X}}{2}\right)^2\right] \tag{9.38a}$$

$$V_A = 2pq[p(X_{11} - X_{12}) + q(X_{12} - X_{22})]^2 \tag{9.38b}$$

$$V_D = p^2 q^2 [(X_{11} - X_{12}) - (X_{12} - X_{22})]^2 \tag{9.38c}$$

In particular, V_A and V_D from (b) and (c) are most useful in computing h^2 because they do not require a separate intermediate step of calculating the additive effects, χ_1 and χ_2.

INTERMEDIATE HETEROZYGOTE. In this case the additive effects exactly predict the phenotype so the dominance deviations, d_{ij}, are all zero. Therefore, the dominance variance V_D is also zero. Substituting $X_{12} = \frac{1}{2}(X_{11} + X_{22})$ into V_A yields

$$V_A = \tfrac{1}{2}pq(X_{11} - X_{22})^2 \tag{9.39}$$

Therefore, h^2 is

$$h^2(p) = \frac{\tfrac{1}{2}pq(X_{11} - X_{22})^2}{\tfrac{1}{2}pq(X_{11} - X_{22})^2 + V_e} \tag{9.40}$$

or upon rearranging

$$h^2(p) = \frac{1}{1 + [2/(pq)]\tilde{V}_e}$$

$$\tilde{V}_e = \frac{V_e}{(X_{11} - X_{22})^2} \tag{9.41}$$

This equation give $h^2(p)$ for the important case of the intermediate heterozygote; $\tilde{V}_e$ is a dimensionless measure of the environmental variance. It measures the phenotypic variance *within* a genotype relative to the difference *between* genotypes. Figure 9.12 presents the graph of $h^2(p)$ for several $\tilde{V}_e$. The heritability factor, h^2, tends to zero near fixation of either

---

† The intercept of the regression line is $(1 - h^2)\bar{X}$. The intercept, a, of a regression is found by solving $\bar{Y} = b_{yx}\bar{X} + a$ for a. So $a = \bar{Y} - b_{yx}\bar{X}$. If we obtain the data before selection occurs, then $\bar{X}$ is equal among both parents and offspring. Hence the intercept $= \bar{X} - h^2\bar{X} = (1 - h^2)\bar{X}$.

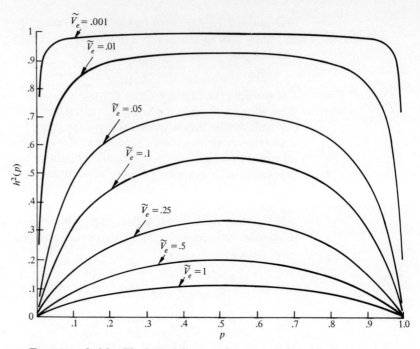

FIGURE 9.12. *The heritability as a function of allele frequency, assuming an intermediate heterozygote.*

allele. The maximum h^2 occurs at $p = \frac{1}{2}$ and depends inversely on $\tilde{V}_e$,

$$h^2_{\max} = \frac{1}{1 + 8\tilde{V}_e} \tag{9.42}$$

The term $h^2_{\max}$ is plotted in Figure 9.13. Because of the factor of 8, $h^2_{\max}$ is very sensitive to changes in $\tilde{V}_e$ when $\tilde{V}_e$ is small, but comparatively insensitive when $\tilde{V}_e$ is large. In Figure 9.12 we observe that h^2 is effectively constant for p from .3 to .7. Throughout this interval, h^2 nearly equals $h^2_{\max}$.

Study of the case of no dominance between alleles provides the conditions in which h^2 is effectively constant. As Figure 9.12 shows, h^2 is effectively constant, provided that the alleles act additively and neither allele is too rare. This result is particularly comforting because "no dominance" and "ample genetic variation" are precisely the situations for which the original purely phenotypic approach was intended. We expect traits like height, size, and so forth, to be those traits whose genetic basis allows h^2 to be regarded as constant. The error in treating h^2 as constant for p between .3 and .7 may be less than that arising from the assumptions in the derivation of the formula for h^2 (particularly the assumption of Hardy–Weinberg frequencies among adults).

FULL DOMINANCE. If V_D is not zero, then the variation of $h^2(p)$ with p is not confined only to the borders where p is near 0 or 1. Instead, $h^2(p)$ varies in a way that is essentially a restatement in phenotypic terms of the basic results of the one-locus two-allele theory of Chapter 3.

Let A_1 be fully dominant so that $X_{11} = X_{12}$. Then

$$\begin{aligned} V_A &= 2pq^3(X_{11} - X_{22})^2 \\ V_D &= p^2q^2(X_{11} - X_{22})^2 \end{aligned} \tag{9.43}$$

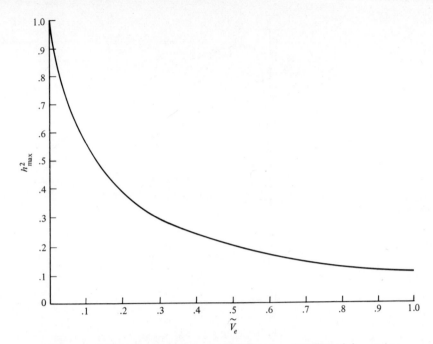

FIGURE 9.13. *The maximum heritability as a function of the environmental variance $\tilde{V}_e$, assuming an intermediate heterozygote.*

Then the heritability is

$$h^2(p) = \frac{2q^2}{q(q+1)+[1/(pq)]\tilde{V}_e}$$

$$\tilde{V}_e = \frac{V_e}{(X_{11}-X_{22})^2}$$

(9.44)

Figure 9.14 presents some plots of $h^2(p)$, assuming that A_1 is dominant. As before, h^2 drops to zero at the borders. But, in addition, h^2 decreases with increasing p in the central part of the graph. As p increases, the frequency of the recessive allele, A_2, decreases so that A_2 is increasingly locked up in heterozygotes where it is not expressed. As a result, there is less response to a given selection pressure as A_2 becomes rare. The decreasing character of $h^2(p)$ expresses this fact.

HETEROZYGOTE SUPERIORITY. If the heterozygote has a larger height than either homozygote, $h^2(p)$ becomes a very curious function. Suppose that $X_{11} = X_{22}$ and that X_{12} is the largest. Then the variance components are

$$V_A = 2pq(p-q)^2(X_{12}-X_{11})^2$$

$$V_D = 4p^2q^2(X_{12}-X_{11})^2$$

(9.45)

Hence h^2 is

$$h^2(p) = \frac{(p-q)^2}{1-2pq+[1/(2pq)]\tilde{V}_e}$$

$$\tilde{V}_e = \frac{V_e}{(X_{12}-X_{11})^2}$$

(9.46)

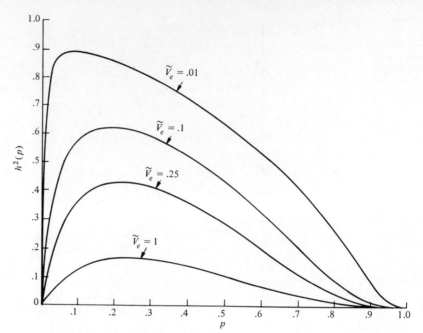

FIGURE 9.14. *The heritability as a function of allele frequency, assuming that A_1 is fully dominant.*

Figure 9.15 illustrates $h^2(p)$ in this case. The unusual feature is that h^2 is zero if $p = \frac{1}{2}$. As you know, selection for increased height would produce polymorphism at $p = \frac{1}{2}$. Thereafter, any selection would yield no further response so that h^2 must be zero at this point. In phenotypic terms, there would be no correlation between offspring phenotype and midparent phenotype within a random mating population when $X_{11} = X_{22}$ and p is $\frac{1}{2}$.

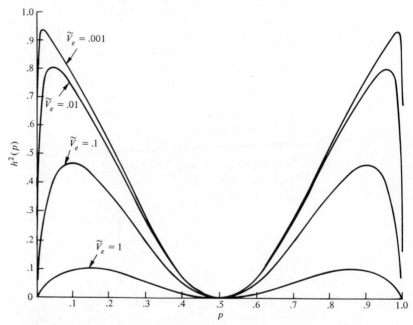

FIGURE 9.15. *The heritability as a function of allele frequency, assuming heterozygote superiority.*

This case further illustrates how the function $h^2(p)$ in effect restates the results of one-locus two-allele theory in phenotypic terms. The magnitude of h^2 indicates how fast evolution occurs for a given selection strength, and equilibrium frequencies can occur under directional selection at values of p where $h^2(p) = 0$, or equivalently where $V_A(p) = 0$.

This case also illustrates another important point. The additive genetic variance, V_A, is not necessarily a measure of genetic variation. In this example directional selection for increased height would lead to a stable polymorphism at $p = \frac{1}{2}$ because the heterozygote is assumed to have the maximum height. A genetic polymorphism obviously represents genetic variation, and yet V_A is zero at this polymorphism. This is because all the *phenotypic* variation that results when $p = \frac{1}{2}$ is explained by the dominance deviations, d_{ij}, while the phenotypes as predicted by the additive effects $(\chi_i + \chi_j)$ are identical and so explain none of the phenotypic variance. Thus V_A is a very technical concept, and the details of its definition must be clearly understood. The variance V_A is only a measure of *genetic* variation when there is negligible dominance among the alleles. In this case, the alleles *act* additively so that the additive genetic variance also becomes a measure of actual genetic variation.

Final Remarks Concerning $h^2(p)$

If we assume an underlying Mendelian basis for the quantitative character, then alternative methods for measuring h^2 are possible. None is as good, in practice, as the regression of offspring against midparent, but alternative methods may often be necessary. First, if only one parent is known, it is only possible to regress the offspring against one parent. This regression proves to equal one half the heritability so that

$$h^2 = 2\left(\begin{array}{c}\text{regression of offspring}\\\text{against one parent}\end{array}\right) \qquad (9.47)$$

The derivation of this result is very easy and left as a problem. If possible, the male parent should be used to avoid phenotypic resemblance acquired by maternal effects. Another method involves the variance among "half sibs." A half-sib group is a group of offspring having at least one parent in common. A strictly half-sib group is a group having *only* one parent in common. For example, assume that one male mates k females selected at random from the population. Then the collection of the first offspring from each female is a strictly half-sib group of k individuals. The group mean is the average X among these k individuals who share one parent in common but whose other parent is drawn at random from the population. If it is possible to obtain several of these half-sib groups, then h^2 can be estimated as

$$h^2 = 4 \frac{\left(\begin{array}{c}\text{variance among group means from}\\\text{groups of strictly half sibs}\end{array}\right)}{\text{population variance}} \qquad (9.48)$$

This derivation is also left as a problem. It is best if each group is the same size, that is, k is the same for each group. If the group is not of strictly half sibs, then a more complicated approach must be used. More data on this can be found in Falconer (1960, Chapter 10). Although this calculation based on the variance of group means is rather tedious, its virtue is that the parent need not actually be measured. All the quantitative information is measured among offspring. Information about the parents is used only to group the offspring.

There are important restrictions associated with the derivation of h^2 as the ratio, V_A/V_T. First, only one locus was assumed in the derivation. If other loci are involved, then the regression of offspring against midparent equals the ratio V_A/V_T plus a small correction term. At this juncture people differ in their definitions of heritability. Some prefer to *define* the ratio V_A/V_T as the heritability. If so, the offspring-midparent regression, which is the quantity actually used to predict the results of selection, must be called something else. Others, as in this book, define the heritability as the offspring-midparent regression on the understanding that it equals V_A/V_T with or without additional terms depending on the genetic basis of the character. For more information on the extension of this kind of theory to multiple loci, consult Kempthorne (1957).

The second restrictive assumption is that of weak selection so that the parental genotype frequencies are given by the Hardy–Weinberg proportions. If strong selection is present, predicting h^2 from the formula, V_A/V_T can involve significant error, with the nature of the error depending on the details of the selection function. For example, if there is upward directional viability selection, then at the time of random mating there will be a deficit, relative to HW proportions, of the small phenotypes. Consequently, the average offspring size from any parent mated at random will be shifted upward. This effect will influence both the slope, h^2, and the intercept of the offspring-midparent regression line.

The Segregation Variance, σ_L^2

The pure phenotype model for the evolution of quantitative characters uses two key parameters, the heritability, h^2, and the segregation variance, σ_L^2. Having analyzed h^2 in terms of an underlying genetic mechanism, we turn to σ_L^2. In the pure phenotype model we assumed that σ_L^2 is a constant. One task is to assess the plausability of this assumption; the other task is to examine the role of stabilizing selection in determining the population variance. Recall that we concluded, for any strength of stabilizing selection, that the population variance is bounded in a narrow interval that is determined principally by σ_L^2. We shall examine whether a qualitatively similar conclusion emerges upon assuming Mendelian segregation.

There is a quantity closely related to the segregation variance, σ_L^2, which is to be called the average within-cross variance, denoted $\overline{\sigma_c^2}$. Each type of cross, say $A_1A_1 \times A_1A_2$, produces a characteristic variance of offspring phenotypes. Table 9.3 lists the offspring variances for each type of cross *assuming an intermediate heterozygote*. The average within-cross variance, $\overline{\sigma_c^2}$, is simply the population average of the variance for each type of cross. This quantity is calculated in Table 9.3 as

$$\overline{\sigma_c^2}(p) = \tfrac{1}{2}V_A(p) + V_e = \tfrac{1}{4}pq(X_{11} - X_{22})^2 + V_e \qquad (9.49)$$

In practice $\overline{\sigma_c^2}$ would be measured as the average within-family variance in the population; that is, the phenotypic variance among the offspring within a given family is measured in many families and the average of these within-family variances is $\overline{\sigma_c^2}$.

The variance $\overline{\sigma_c^2}$ is, in general, not the same as σ_L^2; σ_L^2 is the variance around the regression line of offspring against midparent. Recall that σ_L^2 is

Table 9.3. Calculation of Average Within-Cross Variance.

Type of Cross	Frequency	Average Offspring with Intermediate Heterozygote	Variance of Offspring Phenotypes
$A_1A_1 \times A_1A_1$	p^4	X_{11}	V_e
A_1A_2	$2p^2 2pq$	$\frac{3}{4}X_{11} + \frac{1}{4}X_{22}$	$\frac{1}{16}(X_{11} - X_{22})^2 + V_e$
A_2A_2	$2p^2 q^2$	$\frac{1}{2}X_{11} + \frac{1}{2}X_{22}$	V_e
$A_1A_2 \times A_1A_2$	$2pq\,2pq$	$\frac{1}{2}X_{11} + \frac{1}{2}X_{22}$	$\frac{1}{8}(X_{11} - X_{22})^2 + V_e$
A_2A_2	$2 \cdot 2pqq^2$	$\frac{1}{4}X_{11} + \frac{3}{4}X_{22}$	$\frac{1}{16}(X_{11} - X_{22})^2 + V_e$
$A_2A_2 \times A_2A_2$	q^4	X_{22}	V_e

$$\overline{\sigma_c^2} = (4p^3 q)\tfrac{1}{16}(X_{11} - X_{22})^2 + (4p^2 q^2)\tfrac{1}{8}(X_{11} - X_{22})^2 + (4pq^3)\tfrac{1}{16}(X_{11} - X_{22})^2 + V_e$$

$$= \tfrac{1}{4}pq(X_{11} - X_{22})^2 + V_e$$

$$= \tfrac{1}{2}V_A(p) + V_e$$

used in connection with predicting the distribution of progeny phenotypes from given parental phenotypes. The average progeny from parents, $X_♀$ and $X_♂$, is $h^2(X_♀ + X_♂)/2 + (1 - h^2)\bar{X}$. Also, the variance around this average is σ_L^2. To find a formula for σ_L^2, we must find an expression for the variance around the regression line of progeny against midparent. The term σ_L^2 is the variance of the dependent variable (offspring phenotype) in excess of that predicted by the regression against the independent variable (midparent value). Using this definition, we calculate σ_L^2, assuming an intermediate heterozygote. Note $\sigma_L^2 \geqslant \overline{\sigma_c^2}$ with equality only when h^2 is one. See the boxed derivation.

$$\sigma_L^2(p) = \tfrac{1}{2}[1 - h^2(p)]V_A(p) + \overline{\sigma_c^2}(p) \tag{9.50}$$

Derivation of $\sigma_L^2(p)$

In the absence of selection, the regression of offspring against midparents leads to

$$V_T = (h^2)^2 \frac{V_T}{2} + \sigma_L^2$$

$$\sigma_L^2 = \left[1 - \frac{(h^2)^2}{2}\right] V_T$$

$$= \left(1 - \frac{1}{2}\frac{V_A^2}{V_T^2}\right) V_T$$

$$= V_T - \frac{1}{2}\frac{V_A^2}{V_T}$$

$$= V_T - \frac{1}{2}h^2 V_A$$

With intermediate heterozygote, V_T is $V_A + V_e$. Hence

$$\sigma_L^2 = V_A + V_e - \tfrac{1}{2}h^2 V_A$$

$$= \tfrac{1}{2}(1 - h^2)V_A + \tfrac{1}{2}V_A + V_e$$

$$= \tfrac{1}{2}(1 - h^2)V_A + \overline{\sigma_c^2}$$

The variance σ_L^2 is larger than $\overline{\sigma_c^2}$ when the heritability is less than one because the variance in progeny for a mating between X_δ and $X_\circ$ would include the average within-cross variance and also variance due to the variability in genotypes producing the same parental phenotypes. If $V_e > 0$, then different genotypes produce the same phenotype, implying that matings between a given X_δ and $X_\circ$ often represent different genetic crosses. Therefore, the variance among progeny represents both the average within-cross variance plus the variance caused by pooling different genetic crosses into the same phenotypic mating.[†]

Equation (9.50) shows that σ_L^2 equals $\overline{\sigma_c^2}$ when h^2 is one. At the other extreme, if h_2 is zero σ_L^2 equals $\overline{\sigma_c^2} + V_A/2 = V_A + V_e = V_T$. Also, if h^2 is zero then $V_A = 0$. Thus σ_L^2 equals the total population variance, V_T, if h^2 is zero. This result is intuitive because if h^2 is zero, the regression against midparent explains none of the population variance. Consequently, the unexplained variance, σ_L^2, equals the total, V_T.

Figure 9.16 illustrates $\overline{\sigma_c^2}(p)$ and $\sigma_L^2(p)$. The formulas for these variances are an important addition to those for $h^2(p)$. Using these formulas, one can predict in terms of Mendelian genetics both the offspring-midparent regression coefficient and variance around the regression line. These parameters in turn determine the evolution of the mean and variance of a quantitative character in a population. Note also in Figure 9.16 that $\sigma_L^2(p)$ is rather flat for p between .3 and .7 so that for some purposes $\sigma_L^2(p)$ may be taken as a constant in this interval.

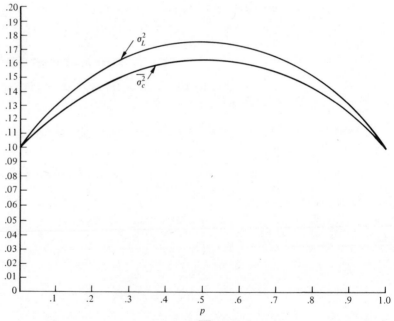

FIGURE 9.16. *Variances σ_L^2 and $\overline{\sigma_c^2}$ as a function of allele frequency, assuming an intermediate heterozygote. Here $V_e = .1$ and $(X_{11} - X_{22})^2 = 1$.*

† Galton termed a collection of progeny from parents of the same midparent value as a *cofraternity*. Thus a cofraternity is a pooling of several families each with the same midparent value. The variance σ_L^2 might also be called the "average within-cofraternity variance." Equation (9.50) then relates the average within-cofraternity variance to the average within-family variance.

Constraints on the Population Variance with Stabilizing Selection

We learned from the pure phenotype model that stabilizing selection has a rather limited effect on the population variance. We obtained bounds to the population variance under weak and strong selection. The model here leads to the same limit under weak selection. Specifically, the limit is (see the boxed material)

$$\sigma^2(p) = \frac{\sigma_L^2(p)}{1 - \frac{1}{2}[h^2(p)]^2} \tag{9.51}$$

Relation between population variance and σ_L^2:

$$\sigma^2 = (h^2)^2 \frac{\sigma^2}{2} + \sigma_L^2$$

$$\sigma^2 = \frac{\sigma_L^2}{1 - \frac{1}{2}(h^2)^2}$$

Relation between population variance and $\overline{\sigma_c^2}$ assuming intermediate heterozygote:

$$\frac{\overline{\sigma_c^2}}{\sigma^2} = \frac{\frac{1}{2}V_A + V_e}{V_A + V_e} = \frac{V_A + V_e - \frac{1}{2}V_A}{V_A + V_e}$$

$$= 1 - \frac{1}{2}\frac{V_A}{V_A + V_e}$$

$$= (1 - \frac{1}{2}h^2)$$

so that

$$\sigma^2 = \frac{\overline{\sigma_c^2}}{1 - \frac{1}{2}h^2}$$

This limit can also be expressed in terms of $\overline{\sigma_c^2}$ (see the boxed material), giving

$$\sigma^2(p) = \frac{\overline{\sigma_c^2}(p)}{1 - \frac{1}{2}h^2(p)} \tag{9.52}$$

Thus under weak stabilizing selection, the population variance does not simply expand and instead is at equilibrium at the limit given by Equation (9.51).

The weak selection limit on σ^2 in this one-locus two allele model is the same as in the pure phenotype model. But the strong selection limit is different. Indeed, it will emerge that the strength of stabilizing selection has no effect at all on the population variance if this simple genetic mechanism is assumed to underlie the quantitative character. That is, the relations (9.51) and (9.52) apply regardless of the strength of the stabilizing selection.

Consider then a set of selection functions $w(X)$, all with a mode at X_0 but differing in widths as sketched in Figure 9.17. The wide $w(X)$ indicate weaker stabilizing selection than the narrow $w(X)$. We continue assuming an intermediate heterozygote. Then any selection function with a mode at X_0 will, by ordinary directional selection, tend to cause $\bar{X}$ to evolve to equal

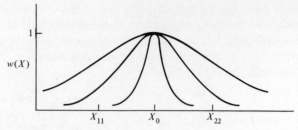

FIGURE 9.17. *Different strengths of stabilizing selection with the same optimum phenotype, X_0.*

X_0. Provided that X_0 is between X_{11} and X_{22}, there is a polymorphism at p_0 obtained from

$$X_0 = p_0^2(X_{11}) + 2p_0q_0\tfrac{1}{2}(X_{11} + X_{22}) + q_0^2 X_{22}$$
$$= p_0 X_{11} + q_0 X_{22} \tag{9.53}$$

Therefore,

$$p_0 = \frac{X_0 - X_{22}}{X_{11} - X_{22}} \tag{9.54}$$

Thus X_0, the value of the optimum phenotype, uniquely sets p_0, the polymorphism frequency. We are, as always, *censusing the zygotic phase after random union of gametes so that the genotype frequencies here are in Hardy–Weinberg proportions* regardless of the strength of the stabilizing selection. Therefore, the population variance is given by

$$\sigma^2 = p_0^2(X_{11} - X_0)^2 + 2p_0q_0(\tfrac{1}{2}(X_{11} + X_{22}) - X_0)^2 + q_0^2(X_{22} - X_0)^2 + V_e \tag{9.55}$$

It is important to notice that this expression is fully determined once X_0 is given because p_0 is computed from X_0 and the rest of the quantities are given parameters. Therefore, any $w(X)$ with mode, X_0, will produce the same population variance. In particular, all the $w(X)$ sketched in Figure 9.17 will produce the same σ^2. Thus the population variance, for this genetic basis, is completely *insensitive* to the strength of the stabilizing selection as measured by the width of $w(X)$.

Consequently, both the pure phenotype model and the model developed here predict that the population variance of a character is principally the expression of the segregation mechanism for the character and not, at least directly, of the selection function. But, as discussed earlier, there may be selection to modify both V_e and the dominance relationships, thereby changing the segregation variance and, by this indirect route, bringing the population variance into better accord with the selection regime.

The concepts of the additive effect of an allele and of the additive genetic variance, and the formula expressing the heritability as the ratio of V_A to V_T were derived by Fisher (1918). The formulas for σ_L^2 and $\overline{\sigma_c^2}$ in terms of h^2, V_A, and V_e are a new, though obvious, extension of Fisher's results.

The Fundamental Theorem of Natural Selection

A convenient spin-off of the theory developed in this chapter is that we can extend Fisher's fundamental theorem of natural selection to random mating diploid populations. Recall from Chapter 4 that we derived a relation between the rate of increase in mean fitness and the variance in fitness among the members of the population. Specifically, we found that

$$\frac{d\bar{r}(t)}{dt} = \sigma_r^2(t) \tag{9.56}$$

We derived this result assuming that each phenotype, say i, grows exponentially at rate r_i. This assumption would be appropriate to a population of bacteria with many phenotypes in it, each growing asexually. Obviously, we would like to generalize this result if possible to diploid sexually reproducing populations.

With the machinery of this chapter the generalization is surprisingly easy. The trick is to view the fitness of an organism as a quantitative trait, itself. If we do this, we can speak of the population variance in fitness, of heritability of fitness, of the additive variance in fitness and so forth. That is, we can apply all of the concepts of this chapter to a special example of a quantitative trait, namely the fitness. In particular, the additive component of the population's variance in fitness is, from (9.38b),

$$V_A = 2pq[p(w_{11} - w_{12}) + q(w_{12} - w_{22})]^2 \tag{9.57}$$

In this formula w_{ij} is, as you may expect, the average fitness of carriers of the $A_i A_j$ genotype. The mean fitness in the whole population is $\bar{w}$.

Our problem is to find an expression relating $\Delta\bar{w}$, that is, the change in the mean fitness, to the variance in fitness. If the selection is sufficiently weak, we may write

$$\Delta\bar{w} = \frac{d\bar{w}}{dp} \Delta p \tag{9.58}$$

Now

$$\frac{d\bar{w}}{dp} = 2\left[p(w_{11} - w_{12}) + q(w_{12} - w_{22})\right] \tag{9.59}$$

and

$$\Delta p = \frac{pq}{\bar{w}}\left[p(w_{11} - w_{12}) + q(w_{12} - w_{22})\right] \tag{9.60}$$

So substituting, and remembering V_A as defined above, we obtain

$$\Delta\bar{w} = \frac{1}{\bar{w}} V_A \tag{9.61}$$

Furthermore, since we have assumed weak selection in (9.58) we may safely assume that $\bar{w} \approx 1$ so that, for weak selection, we have

$$\Delta\bar{w} \approx V_A \tag{9.62}$$

This result asserts that the change in the mean fitness in the population approximately equals the additive variance in fitness. Thus the speed of evolution is connected to the amount of variability in fitness upon which the selection is acting. As we have seen, the additive component of the total variance in fitness is the component that is inherited, and hence this is the component that governs the speed of evolution.

An interesting implication of Equation (9.62) is that as natural selection leads to evolutionary equilibrium V_A approaches zero (since $\Delta\bar{w}$ is approaching zero). This fact, in turn, means that the heritability of any variation in fitness in the population approaches zero. Thus, in an equilibrium population, there is no heritability to whatever variation in fitness occurs among individuals.

Although the fundamental theorem is a famous result in evolutionary biology, be sure to remember that there is no frequency or density-dependence in the selection pressures, and that the one-locus genetic mechanism allows fitness to be maximized.

Chapter 10
NONRANDOM MATING

NONRANDOM mating covers a multitude of topics. Random mating is the occurrence of all possible mating combinations in frequencies expected by chance. For example, with random mating the frequency of matings between AA and AA individuals is simply D^2, where D is the frequency of the AA individual in the population. Nonrandom mating refers to the occurrence of the mating combinations in *any* other set of frequencies. Thus if $AA \times AA$ matings do not occur with frequency D^2, whatever the reason, then nonrandom mating is happening. The topic of nonrandom mating lumps together all kinds of mating patterns, and we shall investigate some of these in this chapter.

There are two general categories of nonrandom mating schemes. *Inbreeding in the broad sense* is the occurrence of matings between *relatives* either more or less often than expected by chance. Even in a random mating population, brothers and sisters will occasionally mate by chance, but inbreeding refers to the condition where this is either more or less common than occurs in a random mating population. *Inbreeding in the strict sense* refers to *more* mating between relatives than expected by chance, and *outbreeding (or outcrossing)* refers to less mating between relatives than expected by chance. We shall use inbreeding in the strict sense unless otherwise indicated.

The other general category of nonrandom mating is *assortative mating*. Assortative mating in the broad sense refers to matings between individuals of the same *phenotype* either more or less often than occurs in a random mating population. Assortative mating in the strict sense, sometimes termed *positive assortative mating*, is mating between individuals of the same phenotype more often than in a random mating population; and *negative assortative mating* is where it is less often.

We begin this chapter with some examples of nonrandom mating in plants and animals. We investigate the theory on these topics afterward.

Some Examples of Nonrandom Mating Systems in Plants and Animals

Inbreeding in Plants

Plant species show amazing variation in their mating systems. Some species have the most extreme form of inbreeding called *selfing*. In these plants pollen from any given flower often pollinates the very same flower in which it was produced. This form of "mating" can be recognized by flowers that never really open up, in which the anthers grow into physical contact with the stigmas. An intermediate degree of inbreeding occurs in "self-compatible" plants in which a plant is pollinated by both its own pollen and the pollen of other plants. Self-compatible plants have a mixture of selfing with random mating. At the other extreme there are self-incompatible plants which, through various mechanisms, never produce fruit from self-pollination. Different species in the same genus can cover the entire spectrum, as in *Leavenworthia* discussed in Solbrig (1972) and Rollins (1963).

There are curious patterns in the occurrence of different mating systems. In particular, there is a correlation between whether a plant is an inbreeder and whether it is an annual or perennial. Table 10.1 is adapted from Stebbins (1950, p. 166).

Table 10.1. Type of Fertilization and Growth Habit of Certain Grasses.

Type of Fertilization	Perennials with Extensive Root Systems	"Bunch-Grass" Perennials	Annuals
Self-incompatible and cross-fertilized	13	13	0
Self-compatible but mixed with outcrossing	1	39	3
Selfers	0	5	27

Clearly, the annuals tend to be selfers while the perennials either mix selfing with random mating, or are self-incompatible.

The existence of various mating systems in plants suggests two basic questions. (1) How does the organization of the genotypic variation change with nonrandom mating systems? Recall that with random mating the different genotypes occur in the ratios of p^2, $2pq$, and q^2. But with non-random mating the genotypic variation is no longer organized according to the Hardy–Weinberg frequencies, so what is its organization? (2) Why have different mating systems evolved? What is the advantage, if any, to one mating system over another in some environment? This second question—on the evolution of the mating system themselves—is very hard to answer and will be mentioned briefly in the next chapter. Here we shall concentrate on the first question. With a *given* mating system, what are the consequences for the organization of genetic variation into genotypic variation?

Outcrossing in Plants Another topic concerns the genetic mechanisms plants use to achieve outcrossing. One technique involves the production of flowers that are polymorphic for the length of the stalks bearing the anthers and stigma. This phenomenon is called *heterostyly*; if only two morphs are involved, it is called *distyly* as illustrated in Figure 10.1. The flower petals are arranged so that the pollinating insect lands in a fixed position in every flower and thereby pollinates only long stigmas with pollen from long anthers and short stigmas with pollen from short anthers. The genetic bases for heterostyly is sometimes surprisingly simple.

Another method very similar to distyly is to have plants with separate sexes, as is usually true in animals. This technique requires that flowers on some individuals contain only anthers while flowers on other individuals contain only stigma. Still another technique involves biochemical incompatibilities between pollen and stigma such that pollen of a given plant simply is not "activated" on the stigma from flowers of that same plant. Other techniques involve spatial or temporal separation between anthers and stigma of the same plant.

All these mechanisms have implications for the population. We would like to know in the case of distyly, for example, what determines the proportion in the population of individuals whose flowers have long or short styles. Does this system of outcrossing eventually lead to an equilibrium proportion for the two flower types? Or, why must there be only one equilibrium? Indeed, perhaps any initial proportion of the two kinds of flowers will continue unchanged, by analogy to the Hardy–Weinberg situation where any initial gene frequency is perpetuated? So, another general topic in the

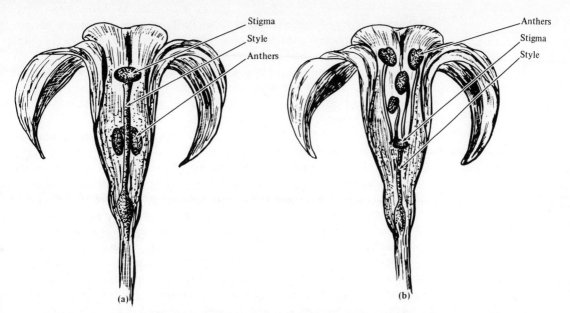

FIGURE 10.1. *Diagram illustrating the two types of primrose* (Primula polyantha) *flowers: In* (a) *called "pin," the anthers are located inside the flower tube below the stigma; in* (b) *called "thrum," the positions are reversed.*

population genetics of nonrandom mating is to predict the population consequences of various mechanisms for self-incompatibility. Important reviews of mating systems with special reference to plants include Allard, Jain and Workman (1968), Allard (1975), and Jain (1976).

Assortative Mating in Animals

Although plants provide the most dramatic illustrations of various inbreeding and outcrossing schemes, animals are the dramatists of assortative mating. Many examples in humans are common knowledge—there is a positive correlation in height between marriage partners, and so forth. Other examples involve offspring who prefer to mate with individuals having the same phenotypes as their parents. Behaviorists use the term *imprinting* for the development of a very specific recognition pattern in young animals—for example, newly hatched chicks imprint upon the mother and thereafter follow her around. By exposing a chick to a foster parent within a short time after hatching, the foster parent is imprinted as though it were the mother and the chick will follow the activities of the foster parent. It is often conjectured that early imprinting also influences mating preferences at maturity. If so, there would be a tendency for similar phenotypes to mate more often than in a random mating population.

Regardless of whether a mechanism of early imprinting is the cause, it has been shown that assortative mating occurs with blue and white geese, *Chen caerulescens*, discussed in Couch and Beardmore (1959), and with pale and dark phases of the Artic Skua, *Stercorarius parasiticus*, analyzed by O'Donald (1959). Another possible example involves bird song dialects. Some birds, like sparrows, have a song that is divided into two parts. The first part is variable among the individuals at the same locality. The second part is not variable among individuals at the same locality but is variable among localities. The geographical variants in bird song are sometimes termed *dialects*. It has been conjectured that these dialects influence courtship and result in assortative mating, according to Nottebohm (1969, and 1975). Thus there are many behavior patterns that can lead to assortative mating in animals and, as with inbreeding, we want to know what the population consequences are.

Breeding Experiments

Much of the motivation for the theory on nonrandom mating is based on breeding experiments, especially with economically important animals. Suppose that a set of organisms is taken from a random mating natural population and inbred in the laboratory. Several results invariably occur: (1) There is a reduction in the average fitness of the inbred animals. This is called the *inbreeding depression*. (2) There is an increase in the population variance—not simply for one or two traits but for almost any character. (3) Although there is an increase in the population variance, the variance among relatives decreases; that is, brothers, for example, come to resemble one another more closely as compared with the beginning of the breeding experiment. The problem then is to predict these results with quantitative accuracy. Can we use Mendelian genetics to predict how the population mean and variance change, and how the resemblance among relatives changes through time for any given mating system? The theory for this topic is not covered here, it is an extension of the theory on quantitative inheritance discussed in the last chapter. Interested readers should consult Crow and Kimura (1970).

The Concept of Genetic Relatedness

A fundamental topic in the study of nonrandom mating is the concept of genetic relatedness. Two individuals may possess the same genes as one another because of a common ancestor. For example, a brother and sister must share at least some genes because they have two parents in common. But they might conceivably share even more genes because their grandparents might also be related to one another. A basic problem is to calculate how similar, genetically, two individuals are to each other on the basis of known ancestry. This topic has long been of intrinsic interest to biologists. But it has also happened that the theory of genetic relatedness provides an important tool for answering other questions including the role of different mating systems in organizing the genotypic variation, and the analysis of phenotypic resemblance between relatives. Furthermore, this theory is fundamental to understanding the evolution of altruism as discussed in Chapter 14.

With this introduction you can see that the general subject of nonrandom mating covers many topics and issues. As a result, there is no single unified theory on this subject; instead there are a variety of approaches and techniques. Much of the theory of this chapter is condensed from Karlin (1968b); interested readers should turn to Karlin (1968b) for more details.

Simple Inbreeding Systems Without Natural Selection

How is the genotypic variation organized with inbreeding? In this section we answer this question for three simple inbreeding systems: pure selfing, selfing mixed with random mating, and brother-sister mating.

Pure Selfing

The consequences of selfing are very easy to understand. When *AA* individuals are selfed, all the progeny are of course *AA*. Similarly, selfing *aa* individuals yields only aa progeny. Selfing Aa individuals yields progeny that are $\frac{1}{4} AA$, $\frac{1}{2} Aa$, and $\frac{1}{4} aa$ according to Mendel's laws. Now consider a population and let D_t, H_t, and R_t be the frequencies of the AA, Aa, and aa genotypes. Then the frequencies at $t+1$ must be given by

$$D_{t+1} = D_t + \tfrac{1}{4}H_t \qquad (10.1a)$$

$$H_{t+1} = \tfrac{1}{2}H_t \qquad (10.1b)$$

$$R_{t+1} = R_t + \tfrac{1}{4}H_t \qquad (10.1c)$$

At each generation the frequency of heterozygotes drops by a factor of one half because only half of the progeny of heterozygotes are also heterozygous. Therefore, the effect of selfing is the eventual loss of heterozygosity.

It is important to note that the *gene* frequency, p, does not change as the following shows

$$p_{t+1} = D_{t+1} + (\tfrac{1}{2})H_{t+1} = (D_t + \tfrac{1}{4}H_t) + \tfrac{1}{2}(\tfrac{1}{2}H_t)$$

$$= D_t + (\tfrac{1}{2})H_t = p_t \qquad (10.2)$$

Therefore, selfing does not change the genetic variation, but it does change how the variation is organized into genotypes.

In summary, selfing does not change p, but it does change D, H, and R and in particular it causes H to approach zero by a factor of $\frac{1}{2}$ each generation. Selfing is the most extreme form of inbreeding and it causes the fastest loss of heterozygosity.

Selfing with Random Mating Now let us consider selfing mixed with random mating. Let α be the fraction of the population which self and $(1-\alpha)$ be the fraction that mates at random. Then the genotype frequencies at $t+1$ are

$$D_{t+1} = \alpha[D_t + \tfrac{1}{4}H_t] + (1-\alpha)p_t^2$$

$$H_{t+1} = \alpha[\tfrac{1}{2}H_t] + (1-\alpha)2p_tq_t \qquad (10.3)$$

$$R_{t+1} = \alpha[R_t + \tfrac{1}{4}H_t] + (1-\alpha)q_t^2$$

When α equals one, these equations reduce to the equations above for pure selfing, and if α equals zero, they reduce to the Hardy–Weinberg situation. Since p does not change in either the pure selfing or the Hardy–Weinberg case, we would not expect it to change here either. The algebra to prove this is easily worked out. Therefore, we can regard p and q as constants; only D, H, and R are variables changing through time.

We would also suspect that the heterozygotes cannot be completely eliminated if some random mating is present because whenever any AA and aa homozygotes mate with each other, heterozygotes are formed. Intuitively then, there seem to be two counteracting forces—selfing which tends to eliminate heterozygotes, and random mating which replenishes the supply of heterozygotes. These two forces lead to an equilibrium. The equilibrium, $\hat{H}$, is found by setting $H_{t+1} = H_t$. The equilibrium heterozygote frequency is

$$\hat{H} = \frac{(1-\alpha)}{(1-\alpha/2)} 2p_0q_0 \qquad (10.4)$$

It is set by α and the initial p and q. Notice that $\hat{H}$ is always between $2p_0q_0$

and zero; the endpoints refer to full random mating and full selfing, respectively. Knowing $\hat{H}$, we can also solve for the equilibrium $\hat{D}$ and $\hat{R}$ giving

$$\hat{D} = \frac{1}{2} \frac{\alpha}{(1-\alpha/2)} p_0 q_0 + p_0^2$$

$$\hat{R} = \frac{1}{2} \frac{\alpha}{(1-\alpha/2)} p_0 q_0 + q_0^2 \qquad (10.5)$$

In summary, selfing mixed with random mating involves two counteracting effects—the selfing reduces the heterozygotes while the random mating replenishes the supply. These effects lead to an equilibrium. The equilibrium has more homozygotes and fewer heterozygotes than would occur with purely random mating. The extension to mixed selfing, with random mating using a two-locus genetic system, appears in Weir and Cockerham (1973).

Sib Mating   Brother-sister matings are the most extreme form of inbreeding that can occur in most animals. Even hermaphroditic forms like many snails and worms cannot self-fertilize. Selfing is a process reserved principally for plants. Brother-sister mating is not nearly as common among animals in nature as is selfing among plants. The main use of the theory of brother-sister mating is with experimental breeding programs.

In determining the effect of selfing on the organization of the genetic variation in the preceding paragraphs, we developed equations for D, H, and R, the genotype frequencies. But we cannot easily write down analogous equations for D, H, and R with brother-sister mating. Instead we shall develop a new approach which is based on censusing not only the genotypes, but also the mating combinations into which the genotypes enter. Visualize a population at the time of mating. We shall census the population at this time and examine each mating pair to determine the genotype of both partners. There are six possible combinations:

$$AA \times AA, \quad AA \times Aa, \quad AA \times aa, \quad Aa \times Aa, \quad Aa \times aa, \quad aa \times aa$$

We let P_t, Q_t, R_t, S_t, T_t, U_t denote the fraction of all matings that are of type $AA \times AA$, $AA \times Aa$ $\cdots$, respectively. It is probably not obvious to you why one would want to census the mating combinations, but as you will see, this formulation turns out to be very convenient.

We can develop simple formulas to predict P_{t+1}, Q_{t+1} $\cdots$ from P_t, Q_t, $\cdots$, based on the assumption that all mating occurs between brothers and sisters. To develop these formulas for p_{t+1}, we need to summarize some information in the form of a table, in which the mating types that can occur among the offspring from a given parental mating combination are described. For example, consider,

	Kinds of Offspring Crosses					
	$AA \times AA$	$AA \times Aa$	$AA \times aa$	$Aa \times Aa$	$Aa \times aa$	$aa \times aa$
Row 1: parental cross is $AA \times AA$	1	0	0	0	0	0

This row is simple because the offspring of $AA \times AA$ matings are all AA so these offspring can obviously only form $AA \times AA$ crosses among themselves. The second row is

	Offspring Crosses					
	$AA \times AA$	$AA \times Aa$	$AA \times aa$	$Aa \times Aa$	$Aa \times aa$	$aa \times aa$
Row 2: parental cross is $AA \times Aa$	$\frac{1}{4}$	$\frac{1}{2}$	0	$\frac{1}{4}$	0	0

This row is also quite simple. The offspring of an $AA \times Aa$ parental cross are half AA and half Aa. Therefore, when these offspring mate among themselves (remember they must mate among themselves because there is brother-sister mating), they will form $AA \times AA$, $AA \times Aa$, and $Aa \times Aa$ combinations in the ratios of $\frac{1}{4}$, $\frac{1}{2}$ and $\frac{1}{4}$. The most complicated row is for crosses among the offspring of $Aa \times Aa$ parents. The offspring themselves are $\frac{1}{4}AA$, $\frac{1}{2}Aa$ and $\frac{1}{4}aa$. Thus they will form with one another all the possible combinations in the following ratios:

	Offspring Crosses					
	$AA \times AA$	$AA \times Aa$	$AA \times aa$	$Aa \times Aa$	$Aa \times aa$	$aa \times aa$
Row 4: parental cross is $Aa \times Aa$	$\frac{1}{16}$	$\frac{1}{4}$	$\frac{1}{8}$	$\frac{1}{4}$	$\frac{1}{4}$	$\frac{1}{16}$

The previous data are summarized in Table 10.2.

Table 10.2.

	Offspring Crosses					
	$AA \times AA$	$AA \times Aa$	$AA \times aa$	$Aa \times Aa$	$Aa \times aa$	$aa \times aa$
Parental cross:						
$AA \times AA$	1	0	0	0	0	0
$AA \times Aa$	$\frac{1}{4}$	$\frac{1}{2}$	0	$\frac{1}{4}$	0	0
$AA \times aa$	0	0	0	1	0	0
$Aa \times Aa$	$\frac{1}{16}$	$\frac{1}{4}$	$\frac{1}{8}$	$\frac{1}{4}$	$\frac{1}{4}$	$\frac{1}{16}$
$Aa \times aa$	0	0	0	$\frac{1}{4}$	$\frac{1}{2}$	$\frac{1}{4}$
$aa \times aa$	0	0	0	0	0	1

With this table we can immediately write down the equations for P_{t+1}, $Q_{t+1} \cdots$ by summing down each column. Thus, from the first column,

$$P_{t+1} = P_t + \tfrac{1}{4}Q_t + \tfrac{1}{16}S_t$$

This formula predicts the fraction of matings between AA and AA at time $t+1$ given the fraction of the various matings at time t. Similarly, by summing down the rest of the columns we obtain

$$Q_{t+1} = \tfrac{1}{2}Q_t + \tfrac{1}{4}S_t$$

$$R_{t+1} = \tfrac{1}{8}S_t$$

$$S_{t+1} = \tfrac{1}{4}Q_t + R_t + \tfrac{1}{4}S_t + \tfrac{1}{4}T_t \qquad (10.6)$$

$$T_{t+1} = \tfrac{1}{4}S_t + \tfrac{1}{2}T_t$$

$$U_{t+1} = \tfrac{1}{16}S_t + \tfrac{1}{4}T_t + U_t$$

These are the equations we have wanted; they predict the fractions of the different mating combinations at $t+1$ given the fraction at time t. Although the derivation is rather long, each step by itself is simple to understand.

We can always determine $\boldsymbol{D}$, $\boldsymbol{H}$, and $\boldsymbol{R}$, the genotype frequencies, given the mating-type frequencies P, Q, and R, ... (of course, we cannot do the reverse). Then $\boldsymbol{D}$, $\boldsymbol{H}$, and $\boldsymbol{R}$ are found as follows:

$$\boldsymbol{D}_{t+1} = P_t + \tfrac{1}{2}Q_t + \tfrac{1}{4}S_t \tag{10.7a}$$

The frequency of AA at $t+1$ equals all the offspring from $AA \times AA$ mating plus half the offspring of $AA \times Aa$ mating plus one quarter of the offspring from $Aa \times Aa$ matings. Similarly,

$$\boldsymbol{H}_{t+1} = \tfrac{1}{2}Q_t + R_t + \tfrac{1}{2}S_t + \tfrac{1}{2}T_t \tag{10.7b}$$

$$\boldsymbol{R}_{t+1} = \tfrac{1}{4}S_t + \tfrac{1}{2}T_t + U_t \tag{10.7c}$$

Therefore, as we iterate the equations for the mating-type frequencies, we can also keep a record of the genotype frequencies. This record can then be compared with that from selfing.

Figure 10.2 illustrates the comparison. With brother-sister mating, as with the selfing model treated here, it can be shown that p stays constant through time and, as the figure shows, there is a loss of heterozygosity with corresponding increase in the proportion of homozygotes. But brother-sister matings produce a much slower loss of the heterozygosity than pure selfing. It must be emphasized that in all these systems p remains unchanged; it is how the variability is organized into genotypes that change. In this sense, the

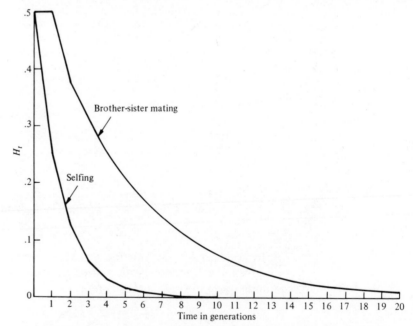

FIGURE 10.2. *The loss of heterozygosity through time caused by inbreeding. The loss is faster with selfing than with brother-sister mating. The selfing curve is obtained by iterating (10.1b) beginning with $H_0 = .5$. The brother–sister curve is obtained from (10.6) and (10.7b) beginning with $P_0 = \tfrac{1}{16}$, $Q_0 = \tfrac{1}{4}$, $R_0 = \tfrac{1}{8}$, $S_0 = \tfrac{1}{4}$, $T_0 = \tfrac{1}{4}$, and $U_0 = \tfrac{1}{16}$. This initial condition entails that $H_0 = .5$ to enable comparison with the results of selfing.*

mating system does not influence the amount of genetic variability in the gene pool but controls its expression and organization.

Some Remarks
About Inbreeding

Our results so far explain some basic observations about inbreeding. When a sample from a random mating natural population is inbred in a laboratory, the average fitness decreases for two reasons. First, heterozygotes often have higher fitness than homozygotes and so the loss of heterozygotes because of inbreeding lowers the mean fitness. Second, the increase in homozygotes resulting from inbreeding includes a more frequent appearance of the homozygote recessives. Although deleterious recessive genes are always present, they are not often expressed in a random mating population. But upon inbreeding, these genes occur in homozygotes more often and become exposed. Thus inbreeding cuts the average fitness in the population in two ways: It destroys the occurrence of heterozygote superiority and exposes deleterious recessives.

The phenotypic variability increases because of the loss of heterozygotes. If heterozygotes are intermediate between homozygotes for traits like height, size, and so forth, then the loss of heterozygotes produces a population with more individuals of extreme heights and sizes than in a random mating population.

Since inbreeding seems to have some undesirable consequences, one might wonder why anyone bothers to inbreed animals. The reason is that inbreeding places the genetic variation under a breeder's control. It exposes any unwanted recessive alleles so that they can be eliminated and reveals the presence of phenotypes which may be desired, but which would not often arise with random mating. After the different strains with desired alleles have been selected, they may be intercrossed to regain the advantages of heterozygosity.

The Study
of Genetic
Relatedness
and Its Uses

The Concept
of Identity
by Descent

The basic concept in the study of genetic relatedness is the concept of *identity by descent*. To say that two alleles are identical by descent means that both are copies of an allele present in their common ancestry. For example, Figure 10.3 illustrates a cross between parents with genotypes *Aa* and *aa*. Assume that these parents are drawn from an infinite population so that we can also assume that there are no common ancestors between the parents. Hence, by assumption, none of the parental genes are identical by descent with one another. The figure also illustrates their two children, both of which happen to be homozygous *aa*. The arrows show which parental alleles have

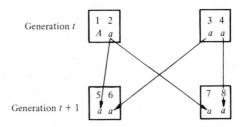

FIGURE 10.3. *The concept of identity by descent. Each number is a label for an allele. Arrows running from alleles in generation t to generation t + 1 indicate descent. According to the figure, alleles 5 and 7 are identical by descent to one another because they are both copies of allele 2.*

happened to pass to the offspring. Each individual allele is labeled with a number so that we can specifically refer to each of them. Gametes carrying copies of allele 2 happen to be incorporated into both zygotes, whereas the gametes from the other parent happen to be copies of different alleles. By definition, allele 5 is identical by descent with allele 7 because both are copies of allele 2. These are the only alleles among the progeny that are identical by descent. Alleles 6 and 8, for example, are identical in type but *not* identical by descent.

Figure 10.3 illustrates an example where two alleles are identical by descent because of a common parent. In principle, two alleles may be identical by descent because of a more remote common ancestor, say among grandparents or great grandparents. Of course, the more remote the common ancestor, then the more unlikely it is that a randomly chosen pair of alleles will be identical by descent.

States of Genetic Relationship Between Individuals

From the concept of identity by descent, we can build up to concepts of genetic relationship between individuals. Consider two individuals labeled I and J. If we say that an individual I is *related* to J, we mean that I and J share one allele that is identical by descent. If we say that I is *strongly related* to J, we mean that I and J share two alleles that are identical by descent. If we say an individual, I, is *inbred*, we mean that both of its alleles are identical by descent. Individuals can be both related and inbred. Table 10.3 illustrates all the possibilities. Each square represents an individual and the large dots represent alleles. The bar connecting the dots indicate identity by descent. There are nine possibilities. In the first column neither party is inbred, in the second column one party is inbred, and in the third column both parties are inbred.

Table 10.3. The States of Relationship Between Two Diploid Individuals.

Indices of Genetic Relatedness Between Individuals

Clearly, there are many possible ways in which two individuals can be related to one another. This fact suggests that it might be useful to find some index that boils all these possibilities into one number. There are a variety of indices of genetic relatedness in the literature, and we shall discuss two here.

The first is useful for the study of kin selection in Chapter 14. The index is to be used in the following context: Suppose that the ancestry of both I and J is known but that neither the actual genotypes of I and J nor the details of which alleles are identical by descent are known. If I and J share a common ancestor, then I and J are likely to be related somehow. But we cannot find out exactly how they are related. All we know is their ancestry. Now, as we shall illustrate, we can calculate from knowledge of the ancestry what the probabilities are for the different kinds of relationships. Thus we can actually calculate the following:

$$\text{prob}\,\{I :: J\} \quad \text{prob}\,\{I \,||\, J\} \quad \text{prob}\,\{I \mathrel{\vdash} J\}$$

$$\text{prob}\,\{I \mathrel{|:} J\} \quad \text{prob}\,\{I \doteq J\} \quad \text{prob}\,\{I \mathrel{\neg} J\}$$

$$\text{prob}\,\{I :| J\} \quad \text{prob}\,\{I = J\} \quad \text{prob}\,\{I \,\square\, J\}$$

Then a natural index of J's relationship to I, based on the known ancestry, is the *expected number of alleles carried by J that are identical by descent with alleles in I*. We call this quantity the *kinship coefficient*, $K_{I,J}$. It is computed as as

$$K_{I,J} = (1)\,\text{prob}\,\{I \doteq J\} + (2)\,\text{prob}\,\{I = J\}$$
$$+ (1)\,\text{prob}\,\{I \mathrel{\vdash} J\} + (2)\,\text{prob}\,\{I \mathrel{\neg} J\} + (2)\,\text{prob}\,\{I \,\square\, J\} \qquad (10.8)$$

The coefficient $K_{I,J}$ can vary between 0 and 2. (It may, on occasion, be appropriate to normalize this index. If $K_{I,J}$ is divided by the number of alleles in I, two for diploids, then this quantity is interpreted as the expected proportion of I's genome that is carried by J.) We calculate some examples of the kinship coefficient below.

But first it is interesting to inquire whether genetic kinship is necessarily symmetric. In other words, if I "views" J as a close genetic kin is it necessary that J view I as an equally close kin? To answer this question, we must compare $K_{I,J}$ and $K_{J,I}$. From the definitions of the different kinds of relatedness, there are certain identities connecting the relationship of I to J and J to I:

$$\text{prob}\,\{I \doteq J\} = \text{prob}\,\{J \doteq I\} = a$$
$$\text{prob}\,\{I = J\} = \text{prob}\,\{J = I\} = b$$
$$\text{prob}\,\{I \mathrel{\vdash} J\} = \text{prob}\,\{J \mathrel{\neg} I\} = c \qquad (10.9)$$
$$\text{prob}\,\{I \mathrel{\neg} J\} = \text{prob}\,\{J \mathrel{\vdash} I\} = d$$
$$\text{prob}\,\{I \,\square\, J\} = \text{prob}\,\{J \,\square\, I\} = e$$

Using these identities, we have

$$K_{I,J} = 1 \cdot a + 2 \cdot b + 1 \cdot c + 2 \cdot d + 2 \cdot e$$
$$K_{J,I} = 1 \cdot a + 2 \cdot b + 2 \cdot c + 1 \cdot d + 2 \cdot e \qquad (10.10)$$

Clearly, $K_{I,J} = K_{J,I}$ if and only if $c = d$. In fact, for almost all regular sorts of ancestries both c and d are zero, thereby ensuring symmetric kinship relations. But kinship relations are not *necessarily* symmetric, and one extremely important case of asymmetry will be considered in connection with the evolution of social behavior in the bees, wasps, and ants in Chapter 14.

The second index was introduced by Malecot and is called the *coefficient of coancestry*, $f_{I,J}$. The coefficient, $f_{I,J}$ is the *probability that an allele drawn at random from I is identical by descent with an allele drawn at random from J.*

$$f_{I,J} = \begin{array}{l} \text{prob \{an allele drawn at random from } I \text{ is identical} \\ \text{by descent with an allele drawn at random from } J \} \end{array}$$

This coefficient can be expressed in terms of the probabilities for the nine possible relations as follows:

$$f_{I,J} = \tfrac{1}{4} \text{prob} \{I \doteq J\} + \tfrac{1}{2} \text{prob} \{I = J\}$$
$$+ \tfrac{1}{2} \text{prob} \{I \urcorner J\} + \tfrac{1}{2} \text{prob} \{I \ulcorner J\} + \text{prob} \{I \square J\} \qquad (10.11)$$

The coefficients $\tfrac{1}{4}$, $\tfrac{1}{2}$, and so forth, in this formula are the probabilities that two alleles which are identical by descent will be drawn, given each relationship. Thus if $I \doteq J$, then there is only a $\tfrac{1}{4}$ chance that the allele drawn from I will be identical by descent with that drawn from J. Whereas if $I \square J$, then the probability is one that the allele from I will be identical by descent with that from J, and so forth.

Examples As examples of these concepts, we shall investigate the genetic kinship between, say, mother and daughter and between sister and sister. Figure 10.4 (left) illustrates the ancestry diagram (called a pedigree) for the parent-offspring calculation. We assume that the parents are both unrelated and not inbred, that is, none of the parental genes are identical by descent to one another. The offspring is produced by union of one gamete from each parent. Therefore, A_5 is identical by descent with either A_1 or A_2, and A_6 is identical by descent to either A_3 or A_4. Therefore,

$$\text{prob} \{M \doteq I\} = 1 \qquad \text{other prob} \{M \text{ versus } I\} = 0 \qquad (10.12)$$

Only one of the conceivable possibilities actually occurs because the mechanics of zygote formation ensure that exactly one gene from M will be identical by descent with exactly one gene in I. All the other conceivable possibilities have zero probability. Therefore, the average number of alleles in I that are identical by descent to alleles in M is, from (10.7),

$$K_{I,M} = 1 \cdot \text{prob} \{M \doteq I\} = 1 \qquad (10.13)$$

Since prob $\{M \ulcorner I\} = $ prob $\{M \urcorner I\}$, the relationship is symmetric. So there is, on the average, one allele in a daughter that is identical by descent to an allele of the mother. In fact, in this special case there is no variance—there is always exactly one allele in the daughter which is identical by descent with an allele of the mother.

The coefficient of coancestry in the mother-daughter case is

$$f_{M,I} = \tfrac{1}{4} \text{prob} \{M \doteq I\} = \tfrac{1}{4}$$

This means that the probability of an allele drawn at random from the mother being identical with one drawn from the daughter is $\tfrac{1}{4}$.

Next consider the relationship between two sisters. Figure 10.4 (right) illustrates the pedigree. Again we assume that the parents are unrelated. The possible offspring genotypes are A_1A_3, A_1A_4, A_2A_3, and A_2A_4. Thus all 16 possible pairs of offspring are tabulated as in Table 10.4.

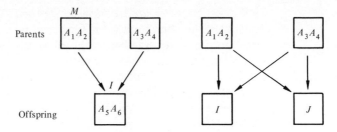

FIGURE 10.4. *Pedigree diagrams. Use the figure at the left to aid in calculating the parent-offspring kinship coefficient. Use the figure at the right to aid in calculating the brother-sister kinship coefficient.*

The number at each location in the table is the number of alleles in J that are identical by descent with alleles in I. From this table we infer that

$$\text{prob}\{I :: J\} = \tfrac{4}{16} = \tfrac{1}{4} \qquad \text{other prob}\{I \text{ versus } J\} = 0$$

$$\text{prob}\{I \div J\} = \tfrac{8}{16} = \tfrac{1}{2} \tag{10.14}$$

$$\text{prob}\{I = J\} = \tfrac{4}{16} = \tfrac{1}{4}$$

Since $\text{prob}\{I \vdash J\} = \text{prob}\{I \neg J\}$, the relation is symmetric. Therefore, the average number of alleles in J that are identical by descent to alleles in I is, from (10.7),

$$K_{I,J} = 1 \cdot \tfrac{1}{2} + 2 \cdot \tfrac{1}{4} = 1 \tag{10.15}$$

Thus, on the average, sisters share one allele that is identical by descent. Note that there is a nonzero variance here, unlike that in the mother-daughter relationship. We conclude that, *on the average*, mothers are related to daughters as strongly as sisters are to one another. We shall examine an example in Chapter 14, where this is not true. The probabilities in (10.14) also determine $f_{I,J}$ between sisters. The sister-sister relationship equals that between mother and daughter with this index too,

$$f_{i,j} = \tfrac{1}{4} \cdot \tfrac{1}{2} + \tfrac{1}{2} \cdot \tfrac{1}{4} = \tfrac{1}{4} \tag{10.16}$$

Geneticists have calculated the coefficients of relationship for many pedigrees. For more on this topic see Kempthorne (1957) and Karlin (1968b).

Table 10.4. Number of Genes Identical by Descent Between I and J

Genotype of I	Genotype of J			
	A_1A_3	A_1A_4	A_2A_3	A_2A_4
A_1A_3	2	1	1	0
A_1A_4	1	2	0	1
A_2A_3	1	0	2	1
A_2A_4	0	1	1	2

The Inbreeding Coefficient, f_I   The concepts above describe how an individual is related to another individual. A similar concept describes how an individual is related to itself. When we say an individual is *inbred* at some locus, we mean that *both of an* *individual's alleles are identical by descent.* As before, we cannot readily establish whether an individual is inbred because we would have to determine its genotype and whether the alleles are identical by descent by tracing them back to some common ancestor. Instead, the only information we are given is the ancestry of the individual. But what we can do with pedigree information is to calculate the *probability* that an individual is inbred. The techniques to do this are the same as those we used previously for the relationship between two different individuals.

There is a very important identity connecting the inbreeding coefficient with the coancestry coefficient. The inbreeding coefficient of some individual, K, is identical to the coancestry coefficient, $f_{I,J}$, of its parents, I and J. This identity occurs because the gametes that fuse to make K are drawn at random, one from I and the other from J. Therefore, the probability that the two alleles in K are identical by descent is the same as the probability that an allele drawn from I is identical by descent with an allele drawn from J.

There is also another connection between the inbreeding and coancestry coefficients. If the *mating* is by the random union of gametes, then the probability that any two gametes are identical by descent is independent of whether they happened to fuse into one zygote or whether they were incorporated into separate zygotes. In this case $f_{I,J}$ for any pair of individuals in the population equals the f_I of any individual. Nonrandom mating causes f_I to differ from $f_{I,J}$. Inbreeding systems, for example, cause f_I to increase while the $f_{I,J}$ for individuals between lines decreases and $f_{I,J}$ for individuals within lines increases.

As an example we calculate the inbreeding coefficient for the offspring of a brother-sister mating. (See Figure 10.5.) We have already computed the coancestry coefficient, $f_{I,J}$, between brother and sister (the same as sister-sister). Therefore, if K is their offspring, then f_K must equal $f_{I,J} = \frac{1}{4}$. Thus the probability that the two alleles in K are identical by descent is $\frac{1}{4}$. This simple example illustrates how the coancestry and inbreeding coefficients are often used together.

There are general procedures for calculating inbreeding coefficients. A useful formula exists for any pedigree that begins with unrelated common ancestors as sketched in Figure 10.6. The common ancestors may be inbred themselves, but they must be unrelated. Let $O_1, O_2, O_3 \cdots$ be the labels of

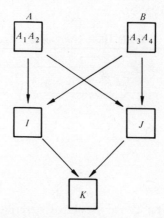

FIGURE 10.5. *Pedigree diagram for calculating the inbreeding coefficient for the offspring of a brother-sister mating.*

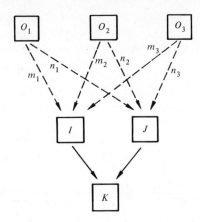

FIGURE 10.6. *Diagram of a pedigree that begins with unrelated, although possibly inbred, individuals.*

these original unrelated common ancestors of I and J. Let m_i be the number of generations separating I and O_i, and n_i be the number of generations separating J and O_i. Then it can be shown that K's inbreeding coefficient is

$$f_K = \sum_i (\tfrac{1}{2})^{m_i+n_i+1}(1+f_i) \tag{10.17}$$

As an example, this formula applied to the brother-sister mating treated above gives

$$f_K = (\tfrac{1}{2})^{1+1+1}(1+0)+(\tfrac{1}{2})^{1+1+1}(1+0)=\tfrac{1}{4} \tag{10.18}$$

For more on this theory, again, consult Kempthorne (1957) and Karlin (1968b).

The Inbreeding Coefficient as a Tool

The inbreeding coefficient would be an important concept even if its only use were to predict the likelihood of an individual's being inbred based on its pedigree. But the coefficient has also proven to be a useful tool in studying the role of an inbreeding system in organizing the genetic variation in a population. We discussed the influence of inbreeding upon the pattern of genotypic variation for selfing and brother-sister mating in the last section. Now we shall tie these two sections together. We shall show how the concepts of genetic relationship of this section, particularly the inbreeding coefficient, relate to the pattern of genotypic variability discussed in the last section.

Consider a population in which everyone in the population has the same inbreeding coefficient, f. That is, every member of the population has the same probability of being inbred, as occurs with the random union of gametes, for example. Then the gene and genotype frequencies may be connected to f by some relations. The basic assumptions are these: With probability f an individual is inbred. If it is inbred, then with probability p it is AA and with probability q it is aa. With probability $1-f$, it is not inbred. If not inbred, then with probability p^2, it is AA; with probability $2pq$, it is Aa; and with probability q^2, it is aa. Therefore, the total frequency of AA in the population must be the sum of the AA individuals that are inbred plus those that are not inbred.

$$D = fp + (1-f)p^2$$

Similarly,

$$H = (1-f)2pq \qquad (10.19)$$

$$R = fq + (1-f)q^2$$

Provided that the assumptions apply, these relations connect the inbreeding coefficient with the pattern of genotypic variation.

In the relations above, f can be viewed as measuring the deviation of the genotype frequencies from the Hardy–Weinberg proportions. If f is 1, everyone is completely inbred, and hence homozygous. Thus the genotype frequencies are p, 0, and q, whereas if f is 0, there is random mating and the genotype frequencies are p^2, $2pq$, and q^2. If f is between zero and one, the genotype frequencies can be viewed as a mixture of the pure inbreeding frequencies, $(p, 0, q)$, with the pure random mating frequencies $(p^2, 2pq, q^2)$. The inbreeding coefficient, f, controls the ratio of the ingredients in the mix.

Selfing with Random Mating Revisited We can use relations like (10.19) in either direction, either to compute f, given D, H, and R or to compute D, H, and R, given p, q and f. As an example of computing f from D, H, and R, recall that when selfing is mixed with random mating, an equilibrium D, H, and R is attained. As before, let α be the fraction of the population that selfs and $(1-\alpha)$ the fraction that mates at random. The equilibrium D, H, and R are given in (10.5). Rearranging these equations in the form of (10.19), we obtain

$$D = \frac{\alpha}{2-\alpha}p + \left(1 - \frac{\alpha}{2-\alpha}\right)p^2$$

$$H = \left(1 - \frac{\alpha}{2-\alpha}\right)2pq \qquad (10.20)$$

$$R = \frac{\alpha}{2-\alpha}q + \left(1 - \frac{\alpha}{2-\alpha}\right)q^2$$

Therefore, we can identify

$$f = \frac{\alpha}{2-\alpha} \qquad (10.21)$$

Thus at the equilibrium maintained by selfing and random mating, the inbreeding coefficient is $\alpha/(2-\alpha)$. Hence at this equilibrium, each individual has a probability of $\alpha/(2-\alpha)$ of having both of its alleles identical by descent. Equivalently, if we consider many loci, an individual is identical by descent (and hence homozygous) at a fraction $\alpha/(2-\alpha)$ of its loci.

Sib Mating Revisited As an example of computing D, H, and R from f, let us reconsider the brother-sister mating process. To analyze this breeding system, we previously used six simultaneous equations. Using f, we can develop a much simpler theory. Visualize a large population consisting of many lines each of which is being perpetuated by brother-sister mating. Suppose that each line was begun at the same time with unrelated individuals drawn from a random mating population. Suppose also that the generations are synchronized. Hence the inbreeding coefficient will be the same for all individuals and will be a function of time. Now recall that brother-sister mating does not change the allele frequencies, p and q; it influences only D, H, and R. Therefore, given the initial p and q, we can calculate D, H, and R, at any time from knowing f at that time. Thus the discussion of the process can be carried out with only one variable, f. We shall seek a recursion equation to predict f through time.

To develop a recursion equation for f, we first develop two simultaneous equations, which will then be boiled down to one equation. Figure 10.7 displays the now-familiar pedigree of brother-sister mating. Let $f_I(t)$ be the inbreeding coefficient of an arbitrary individual in generation t. Let $f_{I,J}(t)$ be the coancestry coefficient between sibs within a line in generation t. Recall that $f_{I,J}$ is the probability that an allele drawn at random from I is identical by descent with an allele drawn at random from J,

$$f_I(t) = f_{I,J}(t-1) \tag{10.22}$$

Now $f_{I,J}(t)$ depends on $f_I(t-1)$ and $f_{I,J}(t-1)$ as follows. With probability $\frac{1}{2}$ the two alleles drawn from I and J come from the same parent. This is

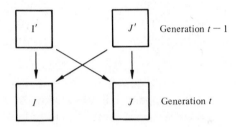

FIGURE 10.7. *The pedigree of a brother-sister mating.*

because the probability is $\frac{1}{4}$ that both are from I'; it is $\frac{1}{2}$ that one is from I' and the other from J'; and it is $\frac{1}{4}$ that both are from J'. Hence the probability is $\frac{1}{2}$ that both are from the same parent, either I' or J'. Now, given that they are from the same parent, the probability is $\frac{1}{2}$ that they are copies of the same allele, and with probability $\frac{1}{2}$ that they are copies of homologous alleles. If they are copies of the same allele, they are obviously identical by descent. If they are copies of homologous alleles, the probability is $f_I(t-1)$ that they are identical by descent. Thus if the two alleles drawn from I and J happen to be from the same parent, the probability of their being identical by descent is $\frac{1}{2} \cdot 1 + \frac{1}{2} \cdot f_I(t-1)$. Meanwhile, if the two alleles drawn from I and J are not from the same parent, then by definition the probability of their being identical by descent is $f_{I,J}(t-1)$. Putting all this together gives

$$f_{I,J}(t) = \frac{1}{2}[\frac{1}{2} \cdot 1 + \frac{1}{2} f_I(t-1)] + \frac{1}{2}[f_{I,J}(t-1)] \tag{10.23}$$

Equations (10.22) and (10.23) are a pair of linear recursion equations for $f_I(t)$ and $f_{I,J}(t)$ through time. We can combine these equations to obtain

$$f_I(t) = \frac{1}{2} f_I(t-1) + \frac{1}{4} f_I(t-2) + \frac{1}{4} \tag{10.24}$$

This is the recursion we sought, which predicts the inbreeding coefficient through time for brother-sister mating.

To explicitly illustrate the use of $f_I(t)$ in determining D, H, and R, we shall develop an equation for H_t based on assuming that the population is initially in Hardy–Weinberg equilibrium. By assumption $H_0 = 2p_0q_0$. The relations (10.19) apply and hence we have $H_t = (1-f_t)2p_0q_0$. Therefore, with initial

Hardy–Weinberg frequencies, we have

$$H_t = (1 - f_t)H_0 \qquad \text{and} \qquad f_t = \frac{H_0 - H_t}{H_0} \tag{10.25}$$

Substituting f_t in terms of H_t into the recursion equation gives (H_0 drops out),

$$H_t = \tfrac{1}{2}H_{t-1} + \tfrac{1}{4}H_{t-2} \tag{10.26}$$

This is an explicit equation to predict the frequency of the heterozygous genotype through time.† It is obviously a much simpler expression than the set of six simultaneous equations considered earlier. But its derivation has required a lot of underlying theoretical machinery. It can be shown that, for large t (say $t > 5$), H_t is approximately given by

$$H_t \approx (.809)H_{t-1} \qquad (t > \sim 5) \tag{10.27}$$

This is to be compared with selfing where $H_t = (.5)H_{t-1}$. Brother-sister mating is much slower in producing homozygosity than selfing. In fact, one generation of selfing is equivalent to about three generations of brother-sister mating.

Genetic Drift Revisited The concept of the inbreeding coefficient has led to a discovery about the role of the finiteness of a population in causing inbreeding. Intuitively, it would appear that all individuals in a finite population must be somehow related to one another. As residents of any small village can testify, if one searches hard enough, a common ancestor can almost always be found between two people. But this idea depends on the finiteness of the population. In an infinite population it is obviously possible for two individuals to have absolutely no ancestors in common. We can use the inbreeding coefficient to explore this issue. We shall develop a simple recursion equation for f. Let the population size be fixed at N and suppose that each generation is produced by a random union of $2N$ gametes. Because of the random union of gametes the inbreeding coefficient is the same for all individuals at any such time. Moreover, f_I always equals $f_{I,J}$ because, whether two alleles are together in one individual or are in separate individuals, is purely random. The probability that two gametes drawn from the gamete pool are copies of the same gene is $1/(2N)$. The probability that they are not copies of the same gene must be $1 - 1/(2N)$. If they are not copies of the same gene, then the probability that they are identical by descent is $f_I(t-1)$. This is so because $f_I(t-1)$ is the probability that any two alleles at $t-1$ are identical by descent, whether or not they reside in the same or in different individuals. Putting this together, we have

$$f_I(t) = \frac{1}{2N} \cdot 1 + \left(1 - \frac{1}{2N}\right) \cdot f_I(t-1) \tag{10.28}$$

Equation 10.28 predicts the inbreeding coefficient through time. It is conveneint to consider the quantity $[1 - f_I(t)]$, which is the probability that two alleles are *not* identical by descent. Equation (10.28) can be

---

† For those who enjoy trivia, if $H_0 = H_1 = \tfrac{1}{2}$ this recursion equation generates a sequence called the Fibonacci sequence: $\tfrac{1}{2}, \tfrac{2}{4}, \tfrac{3}{8}, \tfrac{5}{16}, \tfrac{8}{32}, \tfrac{13}{64}, \ldots$. This is the sequence that also results from Equation 10.6 as was illustrated in Figure 10.2.

arranged as

$$[1 - f_I(t)] = \left(1 - \frac{1}{2N}\right)[1 - f_I(t-1)] \qquad (10.29)$$

Thus the probability that two alleles are *not* identical by descent decreases by a factor of $1 - 1/(2N)$ every generation. Thus, as $t \to \infty$, all alleles tend to become identical by descent. This is a very strong result. We learned previously that genetic drift in finite populations leads to homozygosity. This result says not only that homozygosity is attained, but also that every individual's genes are copies of the same gene present in some common ancestor. Therefore a finite population (in the absence of mutation, selection, etc.) leads not only to homozygosity but also to completely inbred individuals.

There has been research into the extension of measures of descent beyond one-locus genetics to multiple loci. See Cockerham and Weir (1973) for results on this problem.

Assortative Mating

Assortative mating occurs wherever animals exercise preferences in their choice of mates. The preferences result in individuals with specific traits mating more or less often with one another than would occur in a random mating population. There is a seemingly endless variety of courtship and reproductive behavior patterns among the vertebrates and invertebrates. Few, if any, biological features show more variation among species than reproductive behavior. In some groups there is a very limited courtship while in others it may be very elaborate; in some groups the females exercise the choice in mating while in others it is the male and in still others both. In some groups, there is a pair bond between couples that may last a season or for life while in others the mating may be promiscuous. Also, in some social systems, the majority of males are prevented from mating by a few who maintain harems. Each scheme of reproductive and social behavior leads to assortative mating for those characters involved in the choice of mates. The *details* in each scheme of behavior can be very important in formulating a model for the assortative mating. For example, different models are appropriate if one or both sexes exercise preference, if all types of individuals have preferences or if only some do, if pair bonding is permanent or not, and if the origin of the preferences is due to imprinting on the parental phenotype or not. Because there are so many possible formulations for assortative mating, the following discussion will just touch upon some of the major points. For an important recent review of assortative mating models specifically related to sexual selection, see O'Donald (1977).

In the very simplest models for assortative mating the gene frequency, p, does not change and the heterozygote class decreases while the homozygote classes increase. This kind of model is rather similar to the simple inbreeding models considered earlier.

A Simple Model

As an example of a simple assortative mating model consider three different phenotypes, AA, Aa, and aa. Suppose that individuals of each phenotype try first to mate with individuals of their own type. Suppose, on the average, that a fraction, a_1, of the AA individuals mate with other AA individuals, that a fraction a_2 of Aa mate with other Aa, and similarly for a_3. That leaves a total of $(1 - a_1 D - a_2 H - a_3 R)$ individuals left over after the assortment phase. These remaining individuals mate at random. Thus the frequency of $AA \times AA$ matings that occur at this second phase is $(1 - a_1)D(1 - a_1)D/[1 - a_1D - a_2H - a_3R]$; similarly, the frequency of the $AA \times Aa$ matings is $2(1 - a_1)D(1 - a_2)H/[1 - a_1D - a_2H - a_3R]$. Thus the frequency of the different mating types is as summarized in Table 10.5.

Table 10.5.

Mating Type	Contribution from Assorting Phase		Contribution from Random Mating Phase
$AA \times AA$	a_1D	$+$	$(1 - a_1)^2D^2/X$
$AA \times Aa$			$2(1 - a_1)(1 - a_2)DH/X$
$AA \times aa$			$2(1 - a_1)(1 - a_3)DR/X$
$Aa \times Aa$	a_2H	$+$	$(1 - a_2)^2H^2/X$
$Aa \times aa$			$2(1 - a_2)(1 - a_3)HR/X$
$aa \times aa$	a_3R	$+$	$(1 - a_3)^2R^2/X$
	$X = [1 - a_1D - a_2H - a_3R]$		

This table of mating frequencies leads to equations for D_{t+1}, H_{t+1}, and R_{t+1} in the usual way. (Use Mendelian segregation with each mating type. Then sum up the output from each mating type, and so forth.) The equations for the genotype frequencies entail the following results, as discussed in Karlin and Scudo (1969). The gene frequency, p, does not change under this scheme of assortative mating. If all the a's equal one, that is, complete assortative mating, the heterozygotes are eliminated and only the homozygote classes remain. The model in this case is identical to the pure selfing model. But if the a's are less than one, indicating assortative mating mixed with random mating, H tends to an equilibrium. A new feature is that the inbreeding coefficient at equilibrium is itself a function of the gene frequency. Contrast this result with that found in Equation (10.21) for mixed selfing-random mating, where $f = \alpha/(2 - \alpha)$ regardless of the initial p.

A key assumption in this model is that there is no disadvantage to the failure to find the right mate. In the model above, those who fail to assort proceed happily with random mating among themselves and enjoy the same fertility as if they had mated assortatively. This assumption does not seem particularly realistic, for otherwise there would be no advantage to trying to mate assortatively to begin with. But incorporating a disadvantage for those who mate in the second phase is introducing a selection pressure into the model. With this additional assumption, one would not expect p to remain constant. Indeed, it does not; for the details of this extension, see Scudo and Karlin (1969), as well as Karlin and Scudo (1969). The extension of assortative mating models to encompass the inherent natural selection is very important because almost all patterns of reproductive and social behavior incorporate an inherent selective pressure.

Imprinting   The assortative mating models above all lead to equations for D, H, and R. O'Donald (1960) introduced a model for assortative mating based on imprinting that leads to equations for the freqencies of mating combinations, $(P, Q, R \cdots)$ analogous to the model for brother-sister mating considered earlier. The assumptions in the model are

1. The offspring from *like* homozygote × homozygote matings choose to mate only with the parental phenotype.

2. The offspring from homozygote × heterozygote matings choose to mate with either parental phenotype in equal ratios.
3. The offspring from heterozygote × heterozygote matings choose to mate only with the parental phenotype.
4. The offspring of *unlike* homozygote × homozygote matings choose to mate with either parental phenotype in equal ratios.

These assumptions lead to Table 10.6, which relates the parental crosses to offspring crosses.

Table 10.6.

		Offspring Crosses					
		$AA \times AA$	$AA \times Aa$	$AA \times aa$	$Aa \times Aa$	$Aa \times aa$	$aa \times aa$
Parental cross	$AA \times AA$	1					
	$AA \times Aa$	$\frac{1}{4}$	$\frac{1}{2}$		$\frac{1}{4}$		
	$AA \times aa$		$\frac{1}{2}$			$\frac{1}{2}$	
	$Aa \times Aa$	$\frac{1}{4}$	$\frac{1}{2}$		$\frac{1}{2}$	$\frac{1}{4}$	
	$Aa \times aa$				$\frac{1}{4}$	$\frac{1}{2}$	$\frac{1}{4}$
	$aa \times aa$						1

Table 10.6 leads to equations for P_{t+1}, Q_{t+1}, R_{t+1}, $S_{t+1} \cdots$ by the same procedure used with the brother-sister mating model earlier. It can be shown that the gene frequency, p, does not change through time with this model, and also that the heterozygotes are lost and the homozygote classes increase correspondingly. Thus this model is qualitatively the same for the given locus as an inbreeding model. Indeed the rate of approach to homozygosity is very close to that with brother-sister mating. Recall that for large t with brother-sister mating $H_{t+1} \approx .809\ H_t$. In this model of imprinting, for large t, it can be shown that $H_{t+1} \approx .853\ H_t$. So heterozygosity is lost at a slightly slower rate with this imprinting model than with brother-sister mating. An equivalent way of expressing this is to say that about three generations of brother-sister mating are equivalent to one of selfing, and that about four generations of imprinting according to this model, are equivalent to one of selfing.

This kind of model for imprinting has been extended to include mixed random mating, simple viability selection, and more. See Karlin (1968b).

Self-Sterility Mechanisms and Negative Assortative Mating

Self-sterility mechanisms refer to various techniques that plants employ to prevent self-pollination. There are many kinds of self-sterility mechanisms and there are different models for many of these. Again, we can only touch upon some of the major features of these models.

Diploid-Haploid Interaction

The first kind of model involves an *incompatibility between the diploid genotype of the plant bearing the ova and the haploid genotype of the pollen*. This model refers to the type of mechanism underlying the incompatibility system in tobacco, *Nicotiana alata*. East and Mangelsdorf (1925) discovered that there is one locus with three alleles in tobacco and that a pollen grain never functions on the style of a plant whose diploid genotype contains the same allele as itself. For example, a plant with genotype A_1A_2 can only be pollinated by a grain carrying A_3. A grain carrying either A_1 or A_2 simply does not function. Such a condition necessarily implies that homozygotes cannot be produced so that only the genotypes A_1A_2, A_1A_3, and A_2A_3 are possible. This scheme leads to Table 10.7, which lists possible crosses.

Table 10.7.

Row	Genotype of Plant Producing Ova	Genotype of Plant Producing Pollen	Offspring
1	A_1A_2	A_2A_3	$\frac{1}{2}A_1A_3 + \frac{1}{2}A_2A_3$
2	A_2A_3	A_1A_2	$\frac{1}{2}A_1A_2 + \frac{1}{2}A_1A_3$
3	A_1A_2	A_1A_3	$\frac{1}{2}A_1A_3 + \frac{1}{2}A_2A_3$
4	A_1A_3	A_1A_2	$\frac{1}{2}A_1A_2 + \frac{1}{2}A_2A_3$
5	A_2A_3	A_1A_3	$\frac{1}{2}A_1A_2 + \frac{1}{2}A_1A_3$
6	A_1A_3	A_2A_3	$\frac{1}{2}A_1A_2 + \frac{1}{2}A_2A_3$

There are two unusual facts about this scheme. First, the offspring from reciprocal crosses are not identical. Rows 1 and 2, for example, show different offspring. Second, the type of offspring produced from a given "maternal" genotype is independent of the "paternal" genotype. Rows 1 and 3, for instance, have the same offspring even though the genotype of the pollen-producing parent differs. With these assumptions it can be shown that there is a stable polymorphism with all three genotypes at a frequency of $\frac{1}{3}$, as discussed in Moran (1962).

There has been some attempt to investigate this kind of incompatibility system with more than three alleles. In nature there are usually many alleles; Darlington and Mather (1949), for example, quote 35 alleles among 500 plants of *Oenothera organensis*. But the extension has not proved tractable so far. It has been conjectured that the equilibria always involve all the types present in equal frequency as above. But Moran (1962) has pointed out that there are many boundary equilibria that must also be considered. For example, with four alleles the equilibria include the presence of all four with frequencies of $\frac{1}{4}$ and the presence of any three of the four in frequencies of $\frac{1}{3}$, $\frac{1}{3}$, $\frac{1}{3}$, and 0. But Moran has also pointed out that the boundary equilibria are probably unstable because any mutant is at a selective advantage when rare. Suppose that some new allele, M, is introduced. It can easily spread because it is not rejected by any type of ova. But as it becomes more common, the opportunity arises for its rejection on the stigma of plants already containing M. Moran conjectures that the presence of many alleles in an incompatibility system like this depends on this feature of advantage when rare.

Diploid-Diploid
Interaction

The second kind of incompatibility model involves the relation between the full diploid genotypes of the plants producing the ova and pollen. Typically, a set of genotypes are grouped into phenotypic classes such that breeding can occur between classes but not within. There is a further subdivision of diploid-diploid incompatibility mechanisms into those with "zygote elimination" and those with "pollen elimination." With zygote elimination, all the ova are pollinated by pollen from *all* the different phenotypes in proportion to their frequencies. Then the zygotes from incompatible types fail to mature. This results in a seed output that is a selected sample from the full zygotic array. With pollen elimination, zygotes never form between incompatible types and all ova are eventually pollinated by the appropriate pollen. The array of phenotypes among the seed output with pollen elimination is different from that with zygote elimination because with pollen elimination no "maternal" genotypes are lost because of accidental pollination by incompatible pollen. Moran (1962) provides a numerical example of the differences between these two techniques.

An example of the pollen elimination type of diploid-diploid incompatibility mechanism is distyly. In some species of *Primula* it is known that the long styles are homozygous, *aa*, and short styles are heterozygous, *Aa*. The pollinating insect delivers the pollen from the *Aa* plants to the *AA* plants, resulting in offspring that are $\frac{1}{2}Aa$ and $\frac{1}{2}AA$. Similarly, the insect delivers pollen from the *AA* plants to the *Aa* plants and again the progeny are $\frac{1}{2}AA$ and $\frac{1}{2}Aa$. Thus the system perpetuates with a dimorphism between short and long styles in a 50:50 ratio.

Karlin (1968b) presents a generalization that includes the possibility of selfing by the *aa* plants and differential survival of the two morphs. A basic assumption, however, is that the *Aa* plants cannot self and must cross with *aa*. One should note that this system is formally similar to *XY* sex determination where *XY* individuals always mate with *XX* individuals. (The *YY* genotype is not viable.) For a discussion of more aspects of incompatibilities in the primrose, see Bodmer (1960). A review of further theory on this topic can be found in Moran (1962).

There has been recent interest in models for negative assortative mating. In these models certain mating combinations are considered prohibited. Workman (1964) was interested in the fact that models involving (a) exclusively negative assortative mating, (b) mixed selfing and negative assortative mating, and (c) mixed random mating and negative assortative mating could explain the maintenance of polymorphism in the absence of heterozygote advantage. Karlin and Feldman (1968) have explored these models further. In general terms the specification of prohibited mating combinations involves an inherent selection pressure. Hence the mating system does not simply "organize" the genetic variation as with simple inbreeding models but causes the gene frequency to change. The dynamics that result can sometimes be very exotic; the references should be consulted for more detail.

Natural Selection in a Nonrandom Mating Population

There is very little theory on the topic of natural selection in nonrandom mating situations. By natural selection we refer here to the explicit use of fertility and viability coefficients for the different genotypes. There has been some recent work on the inherent selection in detailed assortative and negative assortative mating models as mentioned previously. But there has been rather little interest in the role of viability and fertility selection in nonrandom mating systems.

What interest there has been in natural selection mainly concerns the power of selection to maintain variation in spite of the reduction of heterozygosity caused by many mating systems. A representative example is selfing. As we have seen, the frequency of heterozygotes, H_t, decreases by a factor of $\frac{1}{2}$ each generation due to selfing. Intuitively then, to prevent H from tending to zero we would have to multiply each homozygote frequency by a factor less than $\frac{1}{2}$ so that the heterozygotes would still be better off by comparison. And indeed in the case of selfing with symmetric fitnesses, w, 1, w, it can be shown that if $w < \frac{1}{2}$, heterozygosity is retained, whereas if

$w > \frac{1}{2}$, then eventual homozygosity results. When w is less than $\frac{1}{2}$, the equilibrium is independent of the initial gene frequencies and is given by

$$\hat{D} = \hat{R} = \frac{w}{2(1-w)}$$

$$\hat{H} = \frac{1-2w}{1-w}$$

(10.30)

The full analysis of selfing with selection for nonsymmetric fertilities and viabilities appears in Karlin (1968b). Unlike selection with random mating, selection with selfing leads to linear equations whose full, time-dependent solution is easily obtained.

Karlin (1968b) also covers separate models for viability selection with brother-sister mating, parent-offspring mating, imprinting, and assortative mating. In general terms, if the mating leads to a slower loss of heterozygosity relative to selfing, then the selection against the homozygotes need not be as strong as with selfing in order to preserve heterozygosity.

PART THREE
SPECIAL TOPICS IN EVOLUTION

Chapter 11
EVOLUTION OF THE GENETIC SYSTEM

THERE are two levels in the study of evolution. At one level we want to understand how a trait evolves—given assumptions about its genetic basis, the mating system, the selection pressures, and so forth. But a deeper and certainly more difficult problem is to understand why a trait has a certain genetic basis to begin with. We want to understand the evolution of those quantities which we have taken as given constants until now. We want to know why, for example, one allele is dominant over another, why some trait is governed by two loci on different chromosomes while another is governed by two closely linked loci on the same chromosome, why one plant is self-incompatible while a closely related species is a selfer, why the mutation rate of one species is higher than that of another. We refer to the collection of genetic parameters that are necessary to determine the evolution of some trait as the *genetic system* of that trait. This chapter is an inquiry into the evolution of the genetic system itself. The topics in this chapter comprise some of the most important and interesting *unsolved* problems in the study of evolution.

Dominance

How Can Dominance Be Changed?

Dominance refers to the appearance of heterozygotes relative to homozygotes. If the phenotype of the heterozygote is not distinguished from that of a homozygote, we say there is full dominance, while if the heterozygote is more similar to one homozygote than the other, there is partial dominance. If the heterozygote is exactly intermediate between the homozygotes, there is no dominance.

It is sometimes helpful to think of dominance in biochemical terms. Suppose A_1 produces a pigment while A_2 is a nonfunctional alternative allele. Then the heterozygote A_1A_2 could conceivably be phenotypically identical to the homozygote A_1A_1 if the one A_1 allele in the heterozygote can produce enough of the pigment. But if one A_1 allele does not produce enough pigment, then the heterozygote will be somewhat intermediate between the colorless A_2A_2 and the fully pigmented A_1A_1.

Now suppose that A_1A_2 is *partially* pigmented. Then we may also imagine that another allele could arise, A_3, which makes more intense pigment than A_1 so that A_3A_2 appears fully pigmented. This allele, A_3, differs from A_1 in its degree of dominance. If there is selection in favor of pigmented individuals, we might expect that A_3 will replace A_1. If so, the genetic system determining pigmentation will have evolved a higher dominance in the direction of the pigmented phenotype. This model involves *selection on the primary locus, A.*

Another possibility is to consider a separate locus, M, which could cause the heterozygotes, A_1A_2, to become more pigmented. A different locus could slow the normal breakdown of the pigment and thereby cause a larger supply of pigment to accumulate. Suppose that allele m is initially present at this other locus and that an allele M arises which slows the breakdown of the pigment. If there is selection for pigmentation, we might expect M to evolve? If so, the effective dominance of A_1 will have increased. This model involves *selection on a modifying locus.*

Thus although the dominance relationships between any two given alleles may have a simple biochemical basis, we must also ask why those particular alleles are present in the population instead of others with different biochemical properties. It is important to understand that the elucidation of a biochemical mechanism is not itself the explanation for why that biochemical mechanism has evolved.

The theory in the evolution of dominance has been motivated by the generalization that the most common allele, the so-called "wild-type" allele, is usually dominant. This is obviously not necessarily true because, as we learned in Chapter 3, selection for either a dominant or recessive will eventually lead to fixation of the selected allele whether it is dominant or not—all that differs is the speed of the evolution. Given the preceding information, do you think the correlation between commonness and dominance is simply coincidence?

The Advantage The analysis of the evolution of dominance in the literature has principally
to Dominance concerned the evolution of dominance via a separate modifier locus. The substitution of a more dominant allele at the primary locus involves the same qualitative results. More specifically, consider an allele A that is favored by natural selection and another allele, a, that is opposed by selection. Suppose *initially* that A is recessive. Therefore, the initial selective values are $w_{AA} = 1$, $w_{Aa} = 1 - s$, and $w_{aa} = 1 - s$. Suppose also that a is being maintained in the population by one-way recurrent mutation from A to a with rate μ. Therefore, there is a selection-mutation balance. Recall from Chapter 3 that the equilibrium between mutation and selection against a dominant allele is $\hat{q} = \mu/s$ and $\hat{p} = 1 - \mu/s$. The mean fitness $\bar{w}$ in this situation is $\bar{w} = 1 - (2\hat{p}\hat{q} + \hat{q}^2) s$. Substituting for $\hat{p}$ and $\hat{q}$ gives $\bar{w} = 1 - 2\mu$ where terms in μ^2 are dropped. A population in this situation has $2\hat{p}\hat{q}$ heterozygotes, and they are opposed by selection because a is dominant. It might seem advantageous for A to evolve dominance because the heterozygotes would no longer be opposed by selection. Hence let us compare this situation with one where A is dominant.

If A is dominant, the selective values are $w_{AA} = 1$, $w_{Aa} = 1$, and $w_{aa} = 1 - s$. Recall that the equilibrium between mutation and selection against a recessive is $\hat{q} = \sqrt{\mu/s}$ and $\hat{p} = 1 - \sqrt{\mu/s}$. Therefore, if A is dominant there are *more* of the alleles, a, opposed by selection. There are more a alleles because many of them are now tied up in heterozygotes where they are not expressed. Thus an inherent disadvantage to the evolution of dominance is an increase in the equilibrium frequency of the allele being opposed by selection. But the advantage, as mentioned above, is that the heterozygotes are no longer selected against. Let us now examine whether the evolution of dominance would be a *net* advantage. The mean fitness at this equilibrium is $\bar{w} = 1 - \hat{q}^2 s$. Substituting for $\hat{q}$ gives $\bar{w} = 1 - \mu$. It is easy to check that if A is dominant, $\bar{w}$ is higher than if A is recessive, provided that $\mu < s$. Since the mutation rate is generally much smaller than the selection coefficient, it is indeed a net advantage for A to evolve dominance. But it is not a big advantage. If A is dominant, $\bar{w} = 1 - \mu$, whereas if A is recessive, $\bar{w} = 1 - 2\mu$ approximately. If $\mu = 10^{-6}$, for example, the difference is very slight.

A Model Now it may be advantageous for dominance to evolve, but this fact alone does not entail that it will evolve by natural selection. To investigate this, we need a two-locus model. As before, the A locus is the primary locus. There is selection against a, balanced by mutation from A to a. There are two alleles at the modifier locus. In individuals with mm at the modifier locus, A is recessive to a. In individuals with MM at the modifier, locus A is dominant to a. In individuals with Mm at the modifier locus, A is partially dominant. This setup is summarized in Table 11.1.

Table 11.1. Fitness Table.

Genotype at Modifier Locus	Genotype at Primary Locus		
	AA	Aa	aa
mm	1	$1-s$	$1-s$	
mM	1	$1-ks$	$1-s$	$(0<k<1)$
MM	1	1	$1-s$	

There is a surprisingly long history of analysis on this simple model in Fisher (1929), Wright (1929), Haldane (1930), Ewens (1965), and Feldman and Karlin (1971), even though the general features of the problem have been known for a long time. The equations are the typical expressions from the two-locus theory of Chapter 8 supplemented with the assumption of mutation from A to a. The results in the most recent treatment, presented in Feldman and Karlin (1971), are: The M-allele that causes A to be dominant does increase if initially rare and culminates in fixation. That is, M replaces m and, as a result, A evolves from recessive to dominant. When M is rare, its frequency increases by a factor of roughly $(1+c_1\mu)$ each generation where c_1 is a constant that depends on the parameters. As M approaches fixation, the frequency of its alternative, m, decreases by a factor of approximately $(1-c_2\sqrt{\mu})$ each generation where c_2 is a constant that depends on the parameters. These results on the rate of evolution show two points. First, the evolution of dominance begins more slowly than it ends. To see this, compare $1+c_1\mu$ with $1-c_2\sqrt{\mu}$. If μ is 10^{-6}, then the initial increase of M is slower than the final decrease of m (provided that c_1 and c_2 are of the same order, as they indeed are). Second, the evolution is at best very slow. For example at the beginning, with $\mu = 10^{-6}$, the frequency of M increases by a factor of about 1.000001 each generation. Clearly, this is quite slow.

A Controversy To summarize the theory so far, it can be shown that dominance is advantageous, but only slightly, and that dominance will evolve, but slowly. This qualitative conclusion was the basis for the original argument between Wright and Fisher on the evolution of dominance. Fisher suggested that the selection pressure for the evolution of dominance was, in spite of its weakness, an important factor in explaining the generalization that wild-type alleles are usually dominant. Wright suggested that such a weak selection pressure could not be important in practice for three reasons. First, the spontaneous mutation between m and M (which was not taken into account in the model above) produces an evolutionary force as strong as the selection pressure itself. Second, a very large population size is required for such a weak selection pressure to prevail in the face of genetic drift. For example with $\mu = 10^{-6}$, N must be at least of the order of 10^6. Third, the modifier locus itself is unlikely to be *absolutely neutral* in its effect on fitness, apart from its modification of the A locus in heterozygotes. It is unlikely that the fitness of AA-MM individuals is exactly the same as AA-mm individuals, as assumed in the model. But if the modifying locus is not selectively neutral, then the evolution of dominance at the A-locus is coupled with and perhaps overpowered by the evolution of other aspects of the phenotype.

At this time the question of why wild-type alleles are usually dominant is still open. O'Donald (1968) has proposed models for the evolution of dominance where the modifier locus is allowed a selective pressure of its own. Reviews of experimental work and its relation to the theory include Wallace (1968, pp. 328–331) and Sved and Mayo (1970).

An Example   An example of the evolution of dominance is provided by Kettlewell's (1965) studies of industrial melanism in moths in England. There have been successive waves of different black morphs during this evolution. The current black form, *carbonaria*, of *Biston betularia*, has been preceded in most locations by a number of intermediate forms collectively referred to as *insulara*. One of the *insularia* forms has been shown to be allelomorphic to *carbonaria*. The *carbonaria* heterozygotes are almost fully black, indicating dominance. But the earlier dark morphs were not as dominant as the *carbonaria*. This result suggests that alleles with higher dominance have been substituted for the original alleles. This interpretation is complicated, however, by the fact that the homozygous *carbonaria* are darker than homozygous *insularia* so it may be that the increased darkness, per se, of *carbonaria* is the basis for its increase rather than its increased dominance. Also, both factors could be involved.

In addition there appear to be dominance modifiers in the genome. When *carbonaria* moths from England are crossed with the cryptic forms of a closely related moth from Canada, the heterozygotes are more intermediate and *carbonaria* is no longer as dominant. These results show that the dominance of *carbonaria* depends on the *genetic background*, that is, on the genetic composition at other loci throughout the genome. Modifier genes appear to have evolved in the British moths that make *carbonaria* dominant there.

Examples like this show the reality of the evolution of dominance but do not necessarily provide evidence that the simple model treated here is the explanation. In particular, the observed rate of evolution of dominance seems much faster than that predicted by a model of neutral modifier genes.

Recombination   Recombination is the mixing of parental genes in the progeny. In eukaryotic organisms the basis of recombination is the production of gametes by meiosis and their subsequent union to form zygotes. Sex implies recombination. Without it, cells would simply duplicate by mitosis. With it, cells are made from a mix of parental genes. In prokaryotic organisms there are a variety of other mechanisms that cause recombination, but we shall be concerned mostly with eukaryotes.

The evolution of recombination actually involves two questions. First, why is there recombination at all, that is, why has sex evolved? Second, given the existence of sex, how much mixing should occur between any two loci? This second question concerns the evolution of chromosome structure and linkage. We shall discuss these questions separately.

Important monographs on the evolution of sex have very recently been written by Williams (1975) and Maynard Smith (1978).

Why Has Sex Evolved?

Sex and the Speed of Evolution

Fisher (1958) suggested that the evolutionary advantage to recombination is that it leads to the faster appearance of individuals containing several favorable mutants. Suppose that A^1 and B^1 are mutant alleles at different loci. Then, in an asexual population, the appearance of an individual containing both A^1 and B^1 must await both mutations within one line of descendents. In a sexual population A^1 and B^1 could arise in separate lines and might be brought together by mating between the lines. It might seem then that sex speeds the first appearance of the double mutant.

But, although recombination might speed the initial appearance of A^1B^1 individuals, it also breaks up such combinations with subsequent matings. Recombination stirs the gene pool; if it brings alleles together it also moves them apart. Nonetheless, if A^1B^1 is a particularly favorable combination, then its spread may be limited by the waiting time to its first appearance. Once formed, A^1B^1 may spread very rapidly by means of selection.

Because recombination both forms and destroys favorable gene combinations, it is not obvious that sexual populations have an advantage over asexual populations in this regard. A series of papers intended to resolve whether recombination is advantageous when both the pro and con are taken into account include Muller (1932), Crow and Kimura (1965 and 1969), Maynard Smith (1968), Eshel and Feldman (1970), and Karlin (1973). To briefly indicate the nature of these studies, we summarize a model discussed by many authors.

Consider a *haploid* population with two loci. Suppose that individuals with genotype AB are the most fit and that there is one-way recurrent mutation from a to A and from b to B. In sexual populations there is assumed to be recombination between the loci. Eventually, the genotype AB will be fixed because both the selection and mutation pressures are in the same direction. The question is whether AB evolves faster in some sense, in sexual or asexual populations.

We mention two cases studied in the literature. First, suppose that the contributions from each locus to an individual's fitness are independent in a multiplicative way. For example, let u_1 and u_2 be the contribution from the A and a, and v_1 and v_2 from B and b. Then the fitness of genotype Ab is u_1v_2, and so forth. Let $sex(n)$ and $asex(n)$ be the frequency of the AB type in the sexual and asexual populations, respectively, at generation n. We assume that the initial frequencies of the various genotypes are the same in both populations and then compare $sex(n)$ with $asex(n)$ through time. Using this model and assuming multiplicative fitnesses, it can be shown (Karlin, 1973) that whether the sexual or asexual population evolves faster depends solely on the initial condition, specifically on whether A and B are statistically associated or disassociated to begin with. The results are given in Table 11.2.

Table 11.2.

	Initial Condition	Result (all $n \geqslant 1$)
$D_0 > 0$:	A associated with B a associated with b	$Sex(n) < asex(n)$
$D_0 = 0$:	No association between loci	$Sex(n) = asex(n)$
$D_0 < 0$:	A associated with b a associated with B	$Sex(n) > asex(n)$

These results show that if there is no initial association between the A and B loci (the initial linkage disequilibrium coefficient is zero), then the sexual and asexual populations evolve at the same rate—see, Maynard Smith (1968). But if A is initially associated with B, then the sexual population evolves more slowly. The reason is that the recombination breaks up the initial association between A and B and this effect opposes the selection pressure in favor of AB. In contrast, if A is initially associated with b, then the sexual population evolves faster. Here recombination breaks up the initial association between A and b and thereby reinforces the selection pressure. Thus, in this case, which mode of reproduction is faster depends on the initial condition of the gene pool.

The second case introduced by Eshel and Feldman (1970) assumes that the AB type has a fitness higher than the product of the fitness of A and b. By this assumption A and B are especially "coadapted" to each other and are envisioned as working well together. In this case, as before, the sexual population is slower if $D_0 > 0$. But one new feature is that the sexual population is slower if $D_0 = 0$. The action of recombination in breaking up the AB types actually slows the evolution relative to the asexual population. Moreover, if $D_0 < 0$ the sexual population *may* be faster, as before, but not necessarily so. Both the recombination and mutation rate must be sufficiently small for the sexual population to be faster with $D_0 < 0$.

These results establish that a crucial factor in the discussion of the supposed advantages to recomination is the initial state of association between the new mutants destined to spread by selection. Crow and Kimura (1969) point out that an initial $D < 0$, that is, an initial disassociation between A and B, is very probable because A and B will probably originate in separate individuals. If there are N individuals of type ab, then after the mutations there will be $N - 2$ of type ab and 1 of Ab and another of aB. This condition leads to an initial $D_0 < 0$. [$D = x_1 x_4 - x_2 x_3$ so that $D_0 = 0 \cdot (N-2)/N - (1/N) \cdot (1/N) = -1/N^2$.] With an initial disassociation between A and B, the sexual population evolves faster than the asexual population as shown, provided that the other parameters are appropriate.

The issue of whether recombination is advantageous is logically distinct from that of how recombination has evolved. It is a separate and even more difficult problem to show how recombination could have evolved even if it is advantageous in the ways discussed above.

Sex and a Fluctuating Environment

Another hypothesis for an advantage to sexuality involves a fluctuating environment. It is envisioned that the environment changes from generation to generation and that phenotypes adapted to the conditions in one generation may not be adapted to those in the next generation. In this context a good "strategy" is to produce different kinds of offspring in the expectation that at least some of them will be adapted to future conditions. This hypothesis has substantial anecdotal evidence in its favor. Many organisms, for example aphids, can reproduce either sexually or asexually. Typically, rapid asexual proliferation occurs at the beginning of the growing season while sexual reproduction occurs only at the end of the season when conditions are clearly destined to change.

This hypothesis has yet to be developed in detail. Clearly, it depends on the predictability of the environment. If there is no correlation at all between conditions in consecutive generations, then perhaps it is a good strategy to produce many types of offspring; on the other hand, perhaps it is best to produce only one type of offspring that is the best type for average conditions. But if there is some correlation between the conditions at consecutive times, it is still not obvious what the best strategy is. Moreover, even with no correlation between consecutive conditions, the optimum strategy may depend on the distribution of possible conditions. Suppose, for example, that two environmental conditions are possible, E_1 and E_2, and that there is no correlation between the states at times t and $t+1$. But suppose also that the probability of state E_1 is very high and very low for E_2. Then perhaps the expected return in terms of fitness is higher if all the offspring are adapted to E_1 rather than a mixture of offspring with some adapted to E_1 and others to E_2. For more on this topic see Williams and Mitten (1973), and Williams (1975).

Our discussion has concentrated on why sexuality might be regarded as advantageous. But there is also reason to regard sexuality as disadvantageous based on "strategy" considerations, as discussed further in Maynard Smith (1971a and b). Suppose, as is not uncommon, that females are responsible for raising and providing for the young and that the males play no role in producing offspring beyond the original fertilization. If so, there is a certain expected number of offspring that the female parent will successfully rear. But only half of the genes in the offspring are hers. Therefore, she could contribute twice as many genes to the next generation if she reproduced asexually (parthenogenetically) than if she reproduced sexually. In terms of strategy rationale, the invitation to sexuality is very costly, for her parental investment in rearing the young puts half as many of her genes in the next generation with sexual reproduction as with asexual reproduction. Therefore, if the overall strategy of sexual reproduction is to be advantageous, from her point of view, then the mixing of her genome with that of the male must more than double the expected survival of her progeny. Sexual reproduction may not, in fact, be advantageous to an *individual's* fitness in this sense. But if it is, the implication is that the benefit from recombination must be substantial to offset the dilution factor of sexuality.

Indeed, there are many cases of very successful asexual populations. Among plants, asexual reproduction is called *apomixy*. Common apomicts include the dandelions (*Taraxacum*) and fruit trees of the genus *Citrus*. Most apomict populations differ in the details of seed formation but often the embryo in the seed is derived by mitosis from the maternal tissue without recombination or pollination. Among animals, examples are rapidly accumulating of fully parthenogenetic species including some whiptail lizards (*Cnemidophorus*) and many fish. It is often said that asexuality is an evolutionary dead end, and it is true that among higher plants and animals there is no asexual lineage in the fossil record that has endured for a long time or has led to an impressive phyletic radiation. However, apomictic populations are often extremely abundant and must be regarded as successful by this criterion.

There is little discussion, much less theory, on other topics related to sex. Why are there only two sexes? In *Paramecium bursaria*, for example, there are several mating types. Why is the egg invariably larger than the sperm? Again, in protozoan conjugation the two combining cells are the same size. As you can see, much remains to be learned about the evolution of sex.

Chromosome Structure

Biologists have amassed a tremendous quantity of information on the chromosomes of species. In botany and many areas of zoology it has almost become standard practice to obtain the chromosome configuration† of species while devising its taxonomic description. The chromosomal configuration of related species is often used as a basis for conjecture on the lineage relationships among the species. This body of information documents some remarkable and surprising patterns among the genetic systems of different species.

Concept of the Coadapted Gene Complex

The principal concept used to interpret patterns of chromosome structure is the *coadapted gene complex*. This concept is usually introduced on the basis of experiments comparing the outcome of crosses between individuals collected from geographically separated locations with the outcome of crosses between individuals from the same location. Typically, the first generation hybrids from the different locations are robust but in subsequent generations decline markedly in fitness. In contrast, the offspring from parents collected from the same location do not exhibit a continued decline in fitness over many generations under the same experimental conditions. Hence the decline in fitness in the hybrid stock under these conditions is not simply inbreeding depression resulting from a buildup of homozygosity. Instead, the decline in fitness in the hybrids must be due to a mixing of the genomes from different locations. These experimental results are customarily interpreted to mean that the genes at any location are *coadapted* to work with one another, and that the mixing of the genome in hybrids breaks up the coadapted sets into sets of genes that do not function well together. Wallace (1968, Chapter 18) summarizes the results of many experiments of this sort.

Interpopulational crosses provide evidence that has been interpreted in terms of coadaptation between genes at different loci. But it is *also* visualized that the coadapted genes are close to one another in some sense. As an example, recall the distyly polymorphism in the primrose, *Primula*, which was briefly mentioned in the last chapter. Recall that one morph (called thrum) had a short style and anthers placed near the flower opening while the other morph (called pin) had a long style and anthers placed well inside the flower. These morphs also differ in the nature of the papillae on the surface of the stigma, the surface of the pollen grains, and certain immunological characteristics that favor pollen tube growth in between-morph crosses and hinder it in within-morph crosses. Recall that these two morphs were treated as though determined by one locus with two alleles. But large-scale genetic experiments have shown that at least three very closely linked loci are involved. This is an example of a coadapted gene *complex*, a packet of alleles, one from each locus, which are physically close and function together.

Other possible examples of coadapted gene complexes include the "operons" described in *E. coli* and *Salmonella*. An operon is a section of the bacterial chromosome that includes regulatory genes together with the structural genes for the enzymes in some metabolic pathway. The histidine region of *Salmonella* is an especially dramatic example. However, the

---

† The chromosomal configuration of a species is called its *karyotype*. The karyotype is often described simply by a picture, taken through a microscope, of all the chromosomes contained in a cell.

structural genes for many biosynthetic pathways, including genes regulated by the same operator, are not closely linked together; for example, the genes controlling the arginine pathway in *E. coli* are not adjacent. This fact suggests that coadaptation between genes may not be sufficient to cause the formation of a *complex*.

We see then that a set of genes each from different loci is called a coadapted gene complex if three criteria are satisfied. First, the genes must function well together. Second, the genes that function well together must often occur together. Third, the loci must be closely linked.

It is usually assumed that selection will cause coadapted genes to bunch together into a coadapted gene complex. But it is a fundamental question to ask whether this is really true. Is the presence of coadaptation between genes either necessary or sufficient to cause the evolution of tighter linkage between their loci?

Model for the Evolution of Gene Complexes

There has been recent theoretical work on the evolution of gene complexes. Nei (1967) proposed a model which has been widely used to study this problem. Consider three loci. The first two are the primary loci, with alleles *A* and *a* and *B* and *b*, respectively. The fitness of an individual is a function of its genotype at these two loci only. The third locus is a modifying locus which controls the rate of recombination between the two primary loci. An individual's fitness (its viability) does not depend on its genotype at this modifying locus. This model is used as follows: The primary loci are assumed to be at equilibrium under some given selection scheme and with a given initial recombination rate. Then a new modifier allele that reduces the recombination between the primary loci is introduced into the population. If this new modifier allele spreads and becomes fixed, the recombination between the primary loci will have been reduced. Such an outcome could be interpreted, for example, as the evolution of closer positions of the primary loci on the chromosome.

This model differs in an important way from the model for the modification of dominance. Recall that the fitness of an individual with genotype *Aa* equals 1, $1 - ks$, or $1 - s$ depending on the genotype at the modifying locus. In the modification of dominance, the modifying locus *does* have a direct effect on fitness of certain individuals. But in the modification of recombination, the modifier locus is selectively neutral in this direct sense. The fitness of *any* individual depends *only* on its genotype at the primary loci and not on the genotype at the modifying locus. Nonetheless, the frequency of an allele at the modifying locus may change because of its statistical association with alleles from the primary loci. If this occurs, there is said to be *secondary selection* on the modifying locus.

With this model we can ask, does coadaptation between genes at the primary loci lead to the spread of a modifier allele that tightens the linkage between the primary loci? If so, we could say that coadaptation between genes at different loci leads to the formation of a *complex* of those genes. However, analysis of this model shows that coadaptation is neither necessary nor sufficient for the formation of a complex. Instead, what is required is a *selection pressure at the primary loci that maintains a stable polymorphism at both loci and also produces a nonzero disequilibrium coefficient,* $\hat{D}$. See Nei (1967), Feldman (1972), and Karlin and McGregor (1974) for further information. If this condition is met, then any new modifier allele that tightens the linkage will spread. Let us explore the meaning of this condition with two examples.

Coadaptation Is Not Sufficient to Cause Linkage Tightening

First, consider the model for pure coadaptation discussed in the second example of Chapter 8 on multilocus theory. In this model, A is coadapted with B and a with b. This assumption leads to the table of fitness illustrated in Figure 11.1. The variable, s_1, is the loss of fitness with only one pair of coadapted genes, and s_2 is the loss with no pairs of coadapted genes ($s_2 > s_1$). Recall that there is no stable polymorphism in this model. The population fixes for either A and B or a and b, depending on the initial condition. In this example, tighter linkage cannot evolve because the selection pressure does not maintain a polymorphism. This example shows that *coadaptation is not sufficient to produce a gene complex because coadaptation alone does not ensure a polymorphism.*

	AB	Ab	aB	ab
AB	1	$1-s_1$	$1-s_1$	1
Ab	$1-s_1$	$1-s_2$	1	$1-s_1$
aB	$1-s_1$	1	$1-s_2$	$1-s_1$
ab	1	$1-s_1$	$1-s_1$	1

FIGURE 11.1. *Fitness matrix for pure coadaption between alleles at different loci, assuming no heterosis between alleles at the same locus. In this example, A and B are coadapted with one another and a and b are coadapted with one another; s_i is the selection coefficient against an individual with i noncoadapted pairs. This matrix was discussed in the second example of Chapter 8.*

Coadaptation Is Not Necessary to Cause Linkage Tightening

Second, consider the multiplicative model for *independence* discussed in the third example of Chapter 8. Recall that in this model the fitness of a two-locus genotype is obtained by taking the product of the single-locus fitness. Suppose that the fitness of the three genotypes at a single locus are $w, 1$, and w where $w < 1$. Hence the fitness of a double heterozygote is 1; the fitness of a type containing one homozygous locus and one heterozygous locus is w; and the fitness of a double homozygote is w^2. In the terminology of the third example in Chapter 8, $s_1 = 1 - w$ and $s_2 = 1 - w^2$, where s_i is the selection coefficient against a type with i homozygous loci. These assumptions are expressed in the fitness matrix in Figure 11.2.

	AB	Ab	aB	ab
AB	w^2	w	w	1
Ab	w	w^2	1	w
aB	w	1	w^2	w
ab	1	w	w	w^2

FIGURE 11.2. *Fitness matrix for the multiplicative model of independence between loci, assuming heterozygote superiority within each locus. This is a special case of the matrix treated in the third example of Chapter 8. In comparing with this example, note that $s_1 = 1 - w$ and $s_2 = 1 - w^2$, where s_i is the selection coefficient against an individual with i homozygous loci.*

Recall that if $s_1 = 1 - w$ and $s_2 = 1 - w^2$, we have weak ordered over-dominance between the loci. To locate this case in Figure 8.8, note that $s_1 > s_2/2$ and $s_1 < s_2$. Hence there may be two possible outcomes. If the initial linkage between the loci is loose, $r > (2s_1 - s_2)/4$. Then there is one stable equilibrium at $\hat{D} = 0$. Thus, if the linkage is sufficiently loose to begin with, a modifier gene that tightens the linkage will not increase. On the other hand, if the initial linkage between the two loci is fairly tight to begin with, $r < (2s_1 - s_2)/4$, then there are two stable equilibria with $\hat{D} \neq 0$. [These are the so-called symmetric equilibria given by (8.30) and (8.31).] Since there are stable equilibria with $\hat{D} \neq 0$, then a modifier gene that tightens the linkage still further will increase. Thus, in summary, if there is sufficiently tight linkage to begin with, a modifier gene that further tightens the linkage will increase. This example is important because we could not say that there was coadaptation between the genes at different loci in view of the multiplicatively independent contribution which each locus makes to the fitness of a genotype. Therefore, this example shows that *coadaptation between loci is not necessary for the formation of a complex.*

A New Principle   Since the concept of coadaptation between the genes at different loci is not a decisive concept in the evolution of genes complexes, we might wonder whether there is some other principle that applies. In this regard, Lewontin (1971) and Karlin and McGregor (1974) observed that the result of the evolution of linkage tightening is often to increase the equilibrium mean fitness in the population. We can illustrate this fact by returning to the multiplicative model discussed in the previous paragraph. As noted, the multiplicative model leads to a fitness matrix that does not maintain gametic polymorphism with all four gametic types; instead the selection is tending to drive the system to the boundary equilibria of $(0, \frac{1}{2}, \frac{1}{2}, 0)$ or $(\frac{1}{2}, 0, 0, \frac{1}{2})$. At those points the mean fitness $\bar{w}$ is higher than at any of the interior points. But as we have seen, recombination is tending to move the system away from these boundary points and towards the $D = 0$ surface. And if the recombination is high enough, then it can force the stability to the central point, $(\frac{1}{4}, \frac{1}{4}, \frac{1}{4}, \frac{1}{4})$, even though this point is quite far away from the place where the mean fitness is maximized. Now the intuitive idea behind the fact that linkage tightening leads to an increase in $\hat{\bar{w}}$ is that tightening linkage reduces the amount of recombination and allows the system to come to equilibrium closer to the boundaries of $(0, \frac{1}{2}, \frac{1}{2}, 0)$ or $(\frac{1}{2}, 0, 0, \frac{1}{2})$. And the closer the equilibrium is to these boundaries, the higher the mean fitness. Thus reducing r leads to an increase in the mean fitness because it allows the equilibria to be closer to the boundaries $(0, \frac{1}{2}, \frac{1}{2}, 0)$ and $(\frac{1}{2}, 0, 0, \frac{1}{2})$.

We can explicitly illustrate this idea as follows. From Equation 8.30 we know that a representative stable equilibrium is given by

$$\hat{x}_1 = \hat{x}_4 = \tfrac{1}{4}(1 + \sqrt{Q})$$
$$\hat{x}_2 = \hat{x}_3 = \tfrac{1}{4}(1 - \sqrt{Q}) \tag{11.1}$$

where $Q = 1 - 4r/(1 - w)^2$, provided that $r < \frac{1}{4}(1 - w)^2$.

At this equilibrium $\hat{D} \neq 0$. At $r = \frac{1}{4}(1 - w)^2$ the equilibrium above merges into the central point, $(\frac{1}{4}, \frac{1}{4}, \frac{1}{4}, \frac{1}{4})$, which is then the only stable equilibrium. For $r > \frac{1}{2}(1 - w)^2$ the central point continues to be the only stable equilibrium. The mean fitness at equilibrium is itself a function of r because $\hat{x}_1$, $\hat{x}_2$, $\hat{x}_3$, and $\hat{x}_4$ are themselves functions of r. Thus

$$\hat{\bar{w}}(r) = \bar{w}(\hat{x}_1(r), \hat{x}_2(r), \hat{x}_3(r), \hat{x}_4(r)) \tag{11.2}$$

Then, substituting for $\hat{x}_i$ from above, and simplifying yields

$$\begin{aligned} \hat{\bar{w}}(r) &= \tfrac{1}{2}(1+w^2) - r && \text{for } r < \tfrac{1}{4}(1-w)^2 \\ &= \tfrac{1}{4}(1+w)^2 && \text{for } r \geqslant \tfrac{1}{4}(1-w)^2 \end{aligned} \tag{11.3}$$

This formula shows that $\hat{\bar{w}}(r)$ decreases monotonically in r so long as the stable equilibria are such that $\hat{D} \neq 0$. But $\hat{\bar{w}}(r)$ is flat for r such that the central point is stable. Figure 11.3 illustrates $\hat{\bar{w}}(r)$. Clearly, the effect of the evolution of tighter linkage is to increase the equilibrium mean fitness. At present this principle is the only qualitative concept we know of that seems to underlie the evolution of gene complexes.

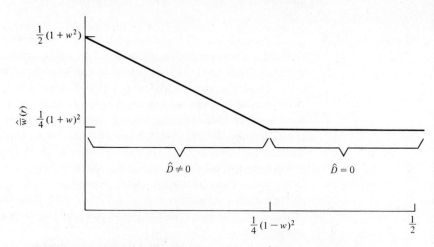

FIGURE 11.3. *Equilibrium mean fitness, $\hat{\bar{w}}$, as a function of r in the symmetric multiplicative model.*

Chromosomal Mechanisms

Inversion

When a complex of genes does evolve, there are a variety of chromosomal mechanisms that ensure tight linkage between loci. Tight linkage between a small number of loci is achieved simply by placing them next to one another along the chromosome. But to achieve linkage among a larger number of loci more complicated mechanisms exist. The first is a chromosome *inversion*. An inversion is a piece of chromosome that has its sequence of loci reversed from the usual arrangement. See Figure 11.4. An *inversion heterozygote* is an individual that has one regular chromosome and one inversion chromosome. An inversion binds together the loci contained within the inversion. To see how this works, we shall consider what happens during meiosis in an inversion heterozygote. When the regular chromosome and inversion chromosome attempt to pair during meiosis, a curious loop is formed. See Figure 11.5. By forming this loop, it is possible for all the loci to match up. Now suppose that there is a crossover within the inversion—say

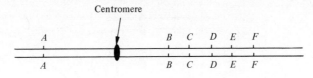

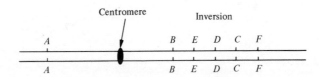

FIGURE 11.4. *Diagram of an inversion. Each chromosome has two strands. If the centromere lies outside the inversion, as in this example, the inversion is called a paracentric inversion. Most inversions are paracentric.*

between the *C* and *D* loci as illustrated in Figure 11.6. Then after the crossover, the chromosomes separate from one another and what they look like is illustrated in Figure 11.7. Note that the top chromosome lacks the *F* and *G* loci while having two of the *A* and *B* loci and two centromeres. The next chromosome lacks the *A* and *B* loci and a centromere and has two of the *F* locus. The gametes containing these chromosomes will be essentially useless and cannot produce viable offspring. Thus the products of a crossover within an inversion are eliminated. For this reason an inversion is often called a *crossover suppressor*. In this sense an inversion causes tight linkage among the loci contained in the inversion.

Because of the ease with which *Drosophila* salivary gland chromosomes can be inspected under the microscope, many inversions have been discovered in fruit flies. See Dobzhansky (1951) and White (1973) for many examples.

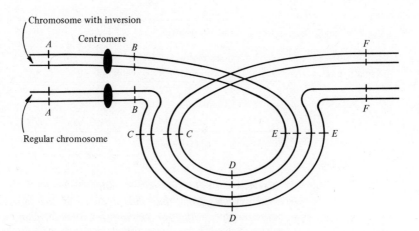

FIGURE 11.5. *Diagram of the meiotic pairing between homologous chromosomes in an inversion heterozygote.*

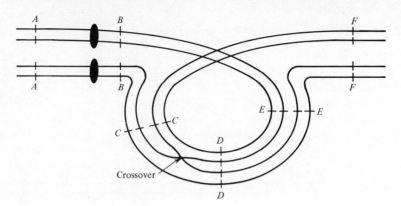

FIGURE 11.6. *A crossover between two of the strands in the four-strand stage of meiosis in an inversion heterozygote.*

Translocation The second chromosomal mechanism leading to increased linkage among many genes is *reciprocal translocation*. This mechanism causes two different (nonhomologous) chromosomes to function genetically as though they were one chromosome. The loci on the different chromosomes *can* recombine *but* no more often than they would if they were on the same chromosome. Figure 11.8 illustrates a reciprocal translocation. Consider two different chromosomes, one with loci *A* through *F* and the other loci *T* through *Z*. A reciprocal translocation occurs where the two chromosomes swap their end pieces. In the figure the *YZ* and *EF* loci have swapped chromosomes. A *reciprocal translocation heterozygote* is an individual who has one of each regular chromosome and one of each translocated chromosome. See Figure 11.9. When a translocation heterozygote undergoes meiosis, at the time

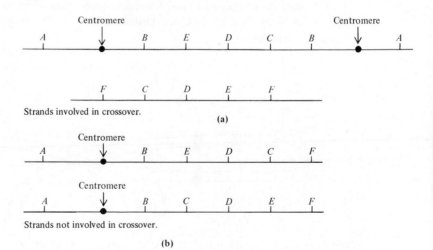

FIGURE 11.7. *Segregation products following a crossover in an inversion heterozygote. Note that the products of the crossover (a) include a chromosome with two centromeres and a chromosome without any centromeres. These chromosomes also lack, and duplicate, certain loci. In contrast, the strands (b) not involved in a crossover lead to fully functional chromosomes.*

A B C D E F
Centromere
T U V W X Y Z
A B C D Y Z
T U V W X E F

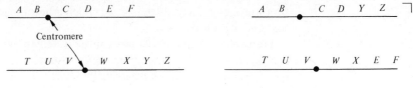

(a) Two nonhomologous chromosomes. (b) Reciprocal translocation.

FIGURE 11.8. *Diagram of a reciprocal translocation between nonhomologous chromosomes. (a) The chromosomes on the left represent the normal form; (b) the chromosomes on the right exhibit a translocation; the YZ loci have swapped places with the EF loci. (For convenience, the chromosomes are indicated as single lines although each is actually in a duplicated condition and consists of two chromatids.)*

when pairing usually occurs between homologous pairs, an aggregation of all four chromosomes forms. Then two of the chromosomes migrate to one pole and two to the other pole of the cell. All the possible gametes resulting from this process are tabulated in Figure 11.10† Observe that most gamete types

No. 1
A B C D E F
No. 2
A B C D Y Z
T U V W X Y Z
No. 3
T U V W X E F
No. 4

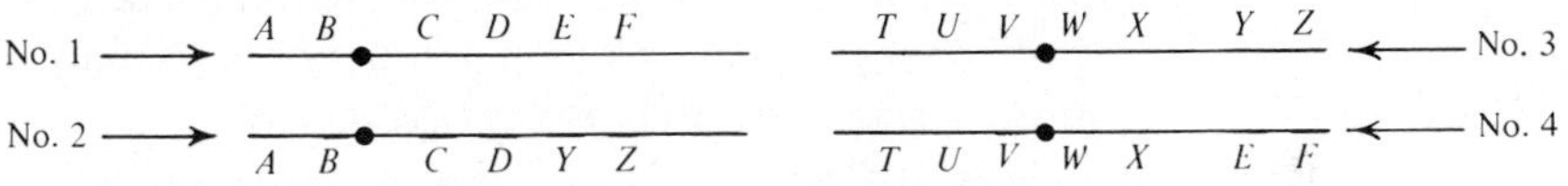

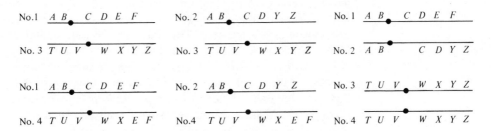

FIGURE 11.9. *Diagram of the chromosomes in a translocation heterozygote.*

have some loci duplicated and other loci absent. These gametes are inviable. In fact, the only gametes containing a full set of loci are the gamete types with chromosomes one and three and with two and four. Thus, among the viable gametes, chromosomes one and three always occur together while chromosomes two and four occur together. Any other pair of chromosomes results in inviable gametes. This fact entails that the genes in chromosome one are genetically linked to those in chromosome three and those in chromosome two are linked to those in chromosome four. One and three

No.1 A B C D E F
No. 3 T U V W X Y Z
No. 2 A B C D Y Z
No. 3 T U V W X Y Z
No. 1 A B C D E F
No. 2 A B C D Y Z
No.1 A B C D E F
No. 4 T U V W X E F
No. 2 A B C D Y Z
No.4 T U V W X E F
No. 3 T U V W X Y Z
No. 4 T U V W X Y Z

FIGURE 11.10. *Diagram of all possible gamete types produced by assortment in a translocation heterozygote.*

† The ratios in which these different gamete types are formed depends on the locations of the centromeres in the chromosomes. Minimum segregation distortion results if the centromeres are in the middle of the chromosomes.

become in effect one chromosome which is homologous to chromosome two together with four.

Translocation heterozygotes can be detected cytologically by observing the aggregation of the four chromosomes. The only way all four chromosomes can match up locus by locus is illustrated in Figure 11.11.

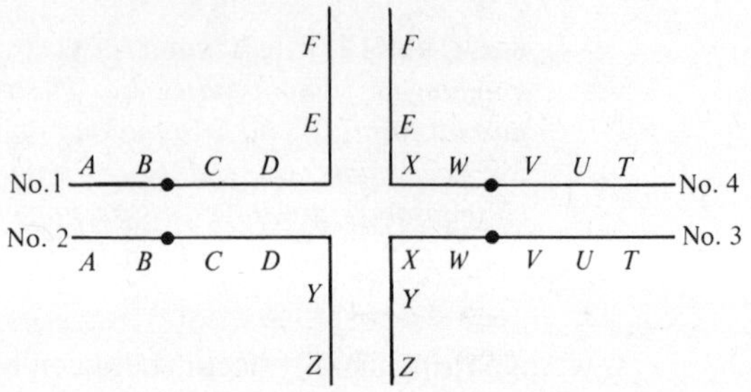

FIGURE 11.11. *Sketch of a translocation heterozygote at meiotic pairing.*

It is observed that as these chromosomes begin to move to opposite poles, they first detach near their centromeres and detach last at the end points. For a while they appear as a loop of four chromosomes attached to one another at their ends. This loop is very distinctive.

Note from Figure 11.11 that a crossover between chromosomes *does* result in perfectly viable gamete types. Crossovers products are not suppressed with translocation heterozygotes in contrast to inversion heterozygotes. Instead, it is the products of independent assortment among the chromosomes that are suppressed in translocation heterozygotes. For this reason translocation may be called an *assortment suppressor.* When two chromosomes cannot independently assort, they become one linkage group. Loci on the two different chromosomes can recombine but not more often than they would if they were on the same chromosome.

Renner Complex in Oenothera

A fascinating example of a system of translocations is found in the evening primroses, *Oenothera.* The genetic system in this genus combines translocations with recessive lethals as follows. Suppose, in the example above, that there is a recessive lethal on either chromosome one or three and another recessive lethal at a different locus on either chromosome two or four. Then individuals who are "homozygous" for either pair of chromosomes are inviable. The only chromosome configuration which is viable is that of the translocation heterozygote, that is, the type containing chromosomes one and three matched against two and four. In this system, because of the translocations, chromosomes one and three function as one linkage group, and chromosomes two and four function as another linkage group. Also, because of the recessive lethals an organism must always be "heterozygous" for the two groups. A linkage group like this, composed of several chromosomes containing translocations and recessive lethals, is called a "Renner complex." This system in addition to causing extensive linkage between loci in different chromosomes also allows permanent heterozygosity within any locus. The genetic system in *Oenothera* consists of two Renner complexes, and every individual is heterozygous for these complexes. A Renner complex presumably binds together coadapted genes from different loci; the two complexes are presumably coadapted with one another in a sense analogous to heterosis.

The genus *Oenothera* possesses 14 chromosomes. The populations east of the Rocky Mountains are primarily weedy and self-pollinating. All the chromosomes are involved in these Renner complexes. At meiosis a giant ring of 14 chromosomes can be observed. West of the Rockies there are outcrossing self-incompatible nonweedy species with no translocation and recessive lethals. These western species of *Oenothera* have the usual genetic system found in higher plants. There are also intermediate species whose Renner complexes involve fewer than 14 chromosomes with the remaining chromosomes functioning in the usual way. Thus the genus possesses all the intergrades between the typical plant genetic system and the exotic system of Renner complexes. Consult Cleland (1962) for more information.

At this time there simply is not any satisfactory theory that explains the occurrence of chromosomal phenomena like inversions and translocations. We cannot in principle, much less in practice, predict when an inversion should evolve to enclose loci given the nature of the selection on the loci, whether there is some maximum number of loci that can be effectively enclosed within an inversion, and whether an inversion is "easily" broken up by natural selection when the environment changes. Similarly, we cannot satisfactorily explain why some *Oenothera* populations have evolved the system of Renner complexes and why others have not.

The discussion above on inversions and translocations introduces the general problem of *chromosomal architecture*. What are the evolutionary forces shaping the number and appearance of species' chromosomes? Stebbins (1971) reviews the many patterns that emerge from the data on the chromosomes of many species of higher plants. In general terms the rationale offered to explain patterns in chromosomal architecture among species involves the concept of coadapted gene complexes. But as we have seen, the importance of this concept receives little theoretical support.

Mutation Rate

The Mutation Rate as a Trait

The process of gene mutation might, at first, seem to be basic and unalterable physical process like the Brownian motion of molecules. A naïve physicist might suppose that mutation rates are primarily determined by temperature and radiation. But gene mutation has proved to be a much more subtle and interesting process. In fact, the mutation rate of genes is controlled by other genes. This fact implies that mutation rates like other biological traits, may evolve by natural selection.

There are three factors influencing the mutation rate of a gene. First, there may be intrinsic differences in the mutability of different genes. A classic case concerns a locus for eye color in *Drosophila*, discussed in Timofeeff–Rossovsky (1937). The dominant "wild-type" allele in American strains mutates more often than that in a Russian strain. Moreover, the American allele changed to eosin eyes and white eyes with about equal frequency while the Russian allele changed mostly to white. Timofeeff–Rossovsky proved that the difference in the behavior of the Russian and the American strains was attributable to different mutability of the white gene itself and not to modifying genes at other loci. Similar observations have been made in maize in studies by Stadler (1946, 1948, and 1949).

Second, there may be genes that modify the general mutation rate throughout the genome. In one remarkable case the biochemical basis of the effect has been determined. The mutation rate in the phage T-4 is controlled by the structure of its DNA polymerase. The polymerase from phage with an allele causing high mutation rates produces more errors during DNA replication than does wild type polymerase; and polymerase from phage with an allele causing low mutation rates produces fewer errors during DNA replication, a phenomenon revealed by Muzyczba, Poland, and Bessman (1972) and Hershfield and Nossal (1973).

Third, there may be genes that specifically modify the mutation rate at a particular locus, or even of a particular allele. An example of a specific mutation modifier concerns a locus producing purple anthocyanin pigment in maize described by Rhoades (1941). There are several alleles known at the anthocyanin locus including A_1, which produces pigment, and a_1, which is colorless. Another locus in a different chromosome controls the mutation rate of a_1 to A_1. The effect of this modifier locus is highly specific. In the presence of the allele D_t (at the modifier locus) the mutability of a_1 to A_1 is greatly increased while that of the other alleles at the anthocyanin locus and at other loci is unaffected. Moreover, only the mutation rate of a_1 to A_1 and not the reverse is affected. This mutation modifier works both in somatic and germ tissues and can produce a mosaic individual.

Constant Environments As with other aspects of the genetic system, our ability to explain the evolution of mutation rates is poorly developed. But some progress has been made. It is clear that in a constant environment mutation is disadvantageous. A mutation-selection balance involves a mutation pressure counteracting selection. In this situation modifier genes which reduce mutation rate will increase. But Karlin and McGregor (1974) and others have calculated that the speed with which mutation modifiers spread is very slow, slower than the return to equilibrium following a perturbation at the primary locus. The primary locus itself evolves very slowly in the neighborhood of a mutation-selection equilibrium because the frequency of the mutant allele is so low. In any event, in a constant environment mutation ought to be as low as physiologically possible, or in the process of slowly evolving to that state.

Changing Environments It is also clear that mutation can be advantageous in a changing environment. But how much mutation is advantageous? The answer is not obvious. Current theory on the evolution of mutation rates offers several hypotheses about the "optimum" mutation rate and also some discussion on whether the optimum rate can evolve by natural selection.

Kimura (1960) was the first to explore the concept of an optimum mutation rate. He envisioned that a population is continually evolving through time in the following sense: Any allele that is currently favored by selection will eventually cease to be favored and some allele that is currently not favored will come to take its place. After a while this new allele will also be replaced by another as evolution continues.

Haldane (1957) had previously introduced the concept of the *cost of evolution*. During the time that an allele is in the process of evolving from an initially low frequency to its final frequency the average fitness in the population, $\bar{w}$, is less than the value it will finally have. Let $\bar{w}_t$ denote the average fitness at time t and let $\bar{w}_\infty$ denote the final value of the average fitness. At any t the difference between the current average fitness and the final average fitness is $(\bar{w}_\infty - \bar{w}_t)$. Haldane suggested that the cost of evolution can be measured by the *cumulative* difference between the current $\bar{w}$ and the final $\bar{w}$. This quantity has since become known as the *cost of a gene substitution*.

$$\begin{matrix} \text{cost of gene} \\ \text{substitution} \end{matrix} = \int_0^\infty (\bar{w}_\infty - \bar{w}_t)\, dt \qquad (11.4)$$

Figure 11.12 illustrates this idea. Haldane evaluated this expression using a model with one locus and two alleles. He discovered that the cost of substitution depends mostly on the initial gene frequency and on the dominance but only slightly on the selection coefficient. In particular, increasing the initial gene frequency, p_o, lowers the cost of substitution.

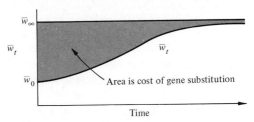

FIGURE 11.12. *Concept of the cost of a gene substitution.*

In a sense, there is also a cost to maintaining mutants in the population against the force of selection. Let $\hat{p}$ denote the equilibrium gene frequency at the mutation-selection balance. Let $\bar{w}(\hat{p})$ denote the average fitness at this equilibrium. Let $\bar{w}(0)$ denote the average fitness if there were no mutants. Then before the mutant becomes favored by selection, the difference between the average fitness without mutants and that with mutants is $[\bar{w}(0) - \bar{w}(\hat{p})]$. This quantity can be taken as a measure of the cost of maintaining mutants in the population at a frequency of $\hat{p}$. The mutation cost increases as $\hat{p}$ increases.

Kimura suggested that the optimum mutation rate represents the best compromise between the cost of maintaining mutants in the face of selection and the cost of allele substitution. In return for carrying deleterious mutants now, the cost of future evolution is reduced. This idea is sketched in Figure 11.13. There is an optimum frequency of mutants, $\hat{p}_{opt}$, and the optimum

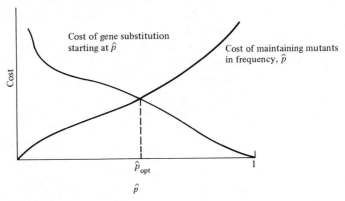

FIGURE 11.13. *Kimura's model for the optimum equilibrium frequency of a mutant allele that is deleterious in present circumstances but may be favored in future circumstances.*

mutation rate, μ_{opt}, is that which gives a mutation-selection equilibrium of $\hat{p} = \hat{p}_{opt}$. Kimura conjectured that populations that possessed the optimum mutation rate would often succeed where populations with other mutation rates would often fail; thus ultimately the optimum mutation rate would be established.

Levins (1967) has explored the importance of mutation in a different sort of changing environment. Consider an environment that changes from generation to generation but where it is assumed that the *average* environment is constant through time. Thus from generation to generation the environment fluctuates about a constant mean. Let $p_{opt,t}$ denote the *optimum* gene frequency at time t in the following sense. Suppose that the environment were fixed in the state which it happened to be in at time t, then $p_{opt,t}$ denotes the equilibrium gene frequency which would be attained in that environment. Let p_t denote the actual gene frequency at time t. The quantity, $(p_t - p_{opt,t})^2$, is a measure of the current deviation of the gene frequency from the optimum gene frequency. We might expect that it is to the advantage of the population to minimize this quantity over the long run.

If it is advantageous to minimize $(p_t - p_{opt,t})^2$, we should inquire how this is to be accomplished. The gene frequency p_t is an expression of the history of environments up to $t - 1$. This is because p_t is a function of the w's at time $t - 1$ and p_{t-1}; these in turn are functions of $t - 2$ and so forth. Thus the p_t that minimizes $(p_t - p_{opt,t})^2$ can be viewed as the best *prediction of* $p_{opt,t}$, based on the history of environments up through $t - 1$. Levins has pointed out that the mutation rate controls the "memory" of the genetic system. The presence of mutation destroys the correlation between the current gene frequency and environments of the distant past. Now, depending on the nature of the predictability of the environment, it may be important to consider ancient environments in developing the best prediction for p_t, or it may be irrelevant. But, if p_t is always correlated with the states of ancient environments, then its ability to respond rapidly to recent changes in the environment is constrained. In other words, the genetic system can not have it both ways—either it retains a long memory and is comparatively unresponsive to recent changes in the environment, or it has a short memory and does respond to recent changes. Levins envisions that an optimum mutation rate may be that which strikes the best compromise between the retention of a long memory and the responsiveness to recent environmental changes.

This hypothesis for an optimum mutation rate has yet to be developed rigorously. Indeed, there seems to be some disagreement concerning the qualitative prediction that follows from these considerations. Some seem to feel intuitively that higher mutation rates would be advantageous in unpredicatble environments. However, in an unpredictable environment there is little advantage in responding to recent environmental changes because the next environmental state is just as likely to reverse the previous change as it is to reinforce it. Perhaps the best strategy in an unpredictable environment is to let p be determined from a running average of as many of the previous environmental states as possible. In contrast, in a predictable environment, with positive serial correlation between consecutive environmental states, the chance of rapid reverses is much less and it becomes advantageous to respond to changes in the recent environment. If so, the mutation rate should be highest in predictable environments rather than in unpredicatable environments.

Levins (1968) has advanced an additional interesting hypothesis for an optimum mutation rate. Consider a locus with heterozygote superiority. Let the selective values be $w_{AA} = 0$, $w_{Aa} = 1$, and $w_{aa} = 1 - s_t$. Suppose that s_t varies randomly in time around a mean value of $\bar{s}(1 > \bar{s} > 0)$. Then there is usually heterozygote superiority. In a constant environment with $s = \bar{s}$ the equilibrium frequency of A is $\bar{s}/(1 + \bar{s})$. We might at first expect that in a fluctuating environment the average gene frequency $\bar{p}$ would also equal $\bar{s}/(1 + \bar{s})$. But this is not true. The average gene frequency $\bar{p}$ is less than $\bar{s}/(1 + \bar{s})$. Surprisingly, there are too few of the lethal allele, A. The reason is that perturbations to p that place p any given amount below $\bar{s}/(1 + \bar{s})$ are restored more slowly than equal but opposite perturbations that place p above $\bar{s}/(1 + \bar{s})$. Therefore, a fluctuating environment causes p, on the average, to occur below $\bar{s}/(1 + \bar{s})$. Nonetheless, the average fitness $\bar{w}$ would be maximized over the long run if $\bar{p}$ were equal to $\bar{s}/(1 + \bar{s})$. Therefore, Levins suggested that it would be advantageous to increase the mutation rate to lethals in order to move $\bar{p}$ closer to $\bar{s}/(1 + \bar{s})$. Levins notes that if a genome's loci are heterotic in this fashion, the optimum mutation rate would be greater than zero and less than 10^{-4}, which would accord with known mutation rates. By this hypothesis the mutation rate to lethals should be highest in the most fluctuating environments. Note that this hypothesis concerns the intensity of the fluctuation, that is, the variance of $s(t)$ while the preceding hypothesis concerns the predictability of the fluctuation, that is, the serial correlation of $s(t)$.

In view of all this theory, one might ask whether there is any basis in fact to believe that mutation rates do evolve by natural selection, even in asexual populations. Cox and Gibson (1974) have recently published results of a very elegant experiment using mutator strains of *E. coli* grown in chemostats. (A chemostat is a device for growing bacterial populations under steady state conditions.) In their competition experiments, a mutator strain of *E. coli* was more fit than the wild-type strain. The increased fitness was explained by the appearance of new mutants better adapted to the chemostat environment.

Other Features of the Genetic System

There are many more features of the genetic system than dominance, recombination, and the mutation rate. Some other basic aspects of the genetic system follow.

The Sex Ratio

A topic that has received much attention over the years is the evolution of the sex ratio in natural populations. Many, but not all, populations exhibit a nearly 50:50 ratio of males to females. Furthermore, whether this is true or not often depends on when the population is censused. Theoretical studies, beginning with Fisher (1958), indicated that natural selection may lead to a 50:50 sex ratio in the population when censused at the zygotic phase. Fisher's original argument is somewhat vague, however, and it is clear that the conclusion depends on many features of the ecology and genetics of the species. Important references on this topic include Shaw (1958), Bodmer and Edwards (1960), Leigh (1960), MacArthur (1965), Hamilton (1967), Eshel (1975), Leigh, Charnov, and Warner (1976), and Charnov (1978).

| *The Dispersal* | Little is known about dispersal strategies. Many seeds and larvae of marine |
| *Regime* | invertebrates, for example, are obviously designed for long distance dispersal, while others are destined to remain near their site of production. |

The Dispersal Regime

Little is known about dispersal strategies. Many seeds and larvae of marine invertebrates, for example, are obviously designed for long distance dispersal, while others are destined to remain near their site of production. How far, on the average, should a seed travel? It appears that dispersal in space is a strategy which expresses the expectation of temporal change. Seeds should not travel anywhere if the location at which they were produced could be relied upon to remain suitable for them. Seeds travel with the expectation of a better life elsewhere, but how far they should travel, on the average, remains an unsolved problem in biological strategy reasoning. A recent study on the evolution of dispersal appears in Hamilton and May (1977).

The Mating System

As we learned in the last chapter, plants show amazing variation in their mating systems. There are two situations in which plants evolve self-compatibility that are quite easy to understand. First, weedy species that regularly colonize newly opened patches of ground are usually self-compatible. Selfing is an adaptation to rapid proliferation in uncrowded conditions. Second, self-compatibility can evolve in peripheral regions of a species range where the principal pollenating insect is absent. But many species have a mixture of outcrossing and selfing. We lack any theory to explain the optimum mix. There are similar unanswered questions concerning the evolution of assortative mating in animals.

The Heritability, h^2, and Segregation Variance, σ_L^2, of a Quantitative Character

In Chapter 9 we discovered that the heritability and segregation variance of a quantitative character were influenced by V_e, the phenotypic variance of a given genotype. Recall that V_e measured the variety of phenotypes expressed by a given genotype. The term *phenotypic plasticity* may be used to describe a genotype with a large V_e.

There are interesting patterns among species in their extent of plasticity. Among plants, some of the most dramatic examples occur in aquatic species that produce one kind of leaf where the plant is submerged and another kind where it is above the water. This phenomenon is called *heterophylly*. But different species in the same genus often possess differing capabilities for heterophylly. This sort of observation suggests that the degree of plasticity is itself under genetic control. If so, what are the evolutionary forces determining the degree of plasticity? In a review of plasticity in plants Bradshaw (1965) suggests that plasticity is advantageous when environmental fluctuations occur very rapidly in time (within a life span) or in space. He also suggests that plasticity is analogous to dominance in allowing different genotypes to assume the same phenotype.

There are other considerations which may be relevant too. Recall that the response to selection is governed by h^2. As mentioned before, in an unpredictable environment, the best strategy may be to be unresponsive to fluctuations because the state of the environment is likely to reverse at consecutive times. But in a predictable environment, with serial correlation, reverses are not as likely and it becomes advantageous to respond to changes in the environment. If so, there should be selection for a low h^2 and high V_e in unpredictable environments. Thus the degree of plasticity should be correlated with the degree of unpredictability.

Recall also that the population variance for a quantitative character is largely determined by the segregation variance, σ_L^2, for any selection strength. Suppose that the selection pressure in a population is such that $\bar{w}$ is maximized with a specific population variance. If so, we might expect that modifier genes will spread that cause σ_L^2 to equal that value which yields the population variance that maximizes $\bar{w}$. Slatkin and Lande (1976) have explored a model for the modification of σ_L^2 in this sense.

There are similarities between phenotypic plasticity in plants and what is roughly termed "learned behavior" in animals. It is known, for example, that some birds have completely "stereotyped" songs that cannot be altered by subjecting the young to different environments. Other birds must "learn" the entire song and still other birds have songs parts of which are stereo-typed and other parts are learned. It is interesting to think of the evolution of the *aptitude* for learning in animals in terms similar to the evolution of phenotypic plasticity in plants.

Life Cycles   Most plants have a life cycle based on an alternation of haploid and diploid phases. In familiar vascular land plants and some algae, the haploid genera-tion is called the gametophyte because these individuals produce sperm and egg by mitosis. The diploid generation is called the sporophyte because these individuals produce spores by meiosis. In flowering plants the diploid sporophyte phase is prominent while the haploid gametophyte phase is reduced to a few cells. But among nonflowering terrestrial plants and the seaweeds, there is a spectrum of life cycles. There are plants that have a reduced sporophyte and large gametophyte (*Cutleria*, *Nemalion*, and other red algae; and also the mosses). There are plants with morphologically indistinguishable sporophyte and gametophyte (e.g., *Ulva*, the common "sea lettuce"). Also, there are plants which, like the flowering plants, have a large sporophyte and reduced gametophyte (e.g., the kelps and other brown algae).

Among the hydrozoan coelenterates (hydroids and jellyfish), there is also an alternation of phases. Both phases are diploid but one reproduces asexually and the other sexually. Also, one phase is sessile and the other is freely floating. Individuals in the sessile hydroid phase produce, by mitotic budding, individuals that grow into medusae. Freely floating medusoids produce, by meiosis, gametes that fuse into larvae, which then settle and develop into the hydroid individuals. There are species with a reduced hydroid phase and prominent medusoid phase, those with an equal allo-cation to each phase, and species with a prominent hydroid phase and reduced medusoid phase.

As you can readily guess, there has been little theoretical attention to the evolution of life cycles or of developmental sequences in general. This is still another topic in the study of the evolution of genetic systems on which we can expect progress in coming years.

Chapter 12
EVOLUTION IN SPATIALLY VARYING ENVIRONMENTS

MANY species occupy a rather broad area and they may also be subdivided into several more or less separate populations. Very few species, however, are so localized that all members can interbreed at random with one another and are confronted with the same selection pressures. Yet until now we have only considered populations where all individuals are subjected to the same rules of mating and selection. Thus in this chapter we shall explore what happens when the rules change from place to place—that is, what are the evolutionary implications of the spatial distribution of a population.

The first set of issues to explore are: (1) If the total population is subdivided into separate subpopulations, what are the consequences for the pattern of genetic variation? Clearly, population subdivision prevents random mating from occurring. Then some non-Hardy–Weinberg pattern of genetic organization must result and we shall discover what it is. (2) Even if the total population is large, the local subpopulations may be quite small. If so, we would expect genetic drift to operate. We shall explore how much migration between subpopulations is enough to make them function as one population, so far as genetic drift is concerned. (3) Because individuals travel finite distances from their place of birth, individuals who are related to one another tend to live near one another. We explore how the relatedness between individuals depends on the spatial distance between them.

This first set of issues arises simply from the existence of a spatial dimension. The environment may be identical for every subpopulation. Nonetheless, the existence of a spatial dimension itself imparts structure to the gene pool of the population.

The second set of issues deals specifically with spatial variation in the selection pressures. We explore three models: (1) Suppose that there is selection in opposite directions in subpopulations from different places. Is this assumption sufficient to guarantee a polymorphism? If not, what are the additional requirements? (2) Suppose that the population in any one place occupies several "microhabitats." That is, suppose that there is a small part of the environment where each genotype is most fit. Suppose also that all individuals mate at random after growth and maturity in the various microhabitats. Under what conditions can this situation maintain a polymorphism? (3) Suppose that the selection pressure varies continuously in space from one location to another. Suppose also that organisms regularly disperse away from their place of birth. How effectively can a population respond to a spatially varying selection pressure if the organisms in it are moving about? How severely will the flow of individuals into any place destroy the evolution of local adaptation in that place?

Wahlund's Effect: Organization of Genetic Variation in Space

Consider a species that is subdivided into separate subpopulations. Suppose, for any reason, that the gene frequencies differ from place to place. Then what is the effect of subdivision on the pattern of genetic variation realized in the entire species? To answer this question, we introduce a very simple model devised by Wahlund (1928). Let T be the number of subpopulations,

p_i be the frequency of the A allele (assuming one locus with two alleles) in the ith subpopulation. There is random mating *within* each subpopulation so that the genotype frequencies within the ith subpopulation are $p_i^2, 2p_iq_i, q_i^2$ when censused at the zygotic phase. For simplicity we assume the subpopulations are each of the same size although the result is easily generalized to subpopulations with different sizes. With this information we can calculate the average gene frequency in the whole species, $\bar{p}$, and the variance of the gene frequency among the subpopulations, σ^2 as

$$\bar{p} = \frac{1}{T} \sum_{i=1}^{T} p_i \tag{12.1}$$

$$\sigma^2 = \frac{1}{T} \sum_{i=1}^{T} p_i^2 - \bar{p}^2 \tag{12.2}$$

We can also calculate the average genotype frequencies in the whole population as

$$\bar{D} = \frac{1}{T} \sum_{i=1}^{T} p_i^2 \tag{12.3a}$$

$$\bar{H} = \frac{1}{T} \sum_{i=1}^{T} 2p_iq_i \tag{12.3b}$$

$$\bar{R} = \frac{1}{T} \sum_{i=1}^{T} q_i^2 \tag{12.3c}$$

Now we can express $\bar{D}$, $\bar{H}$, and $\bar{R}$ in terms of $\bar{p}$ and σ^2. For example, by adding $(-\bar{p}^2 + \bar{p}^2)$ to Equation (12.3a), we obtain

$$\bar{D} = \sigma^2 + \bar{p}^2 \tag{12.4a}$$

Similarly, we can arrange the formulas for $\bar{H}$ and $\bar{R}$ as

$$\bar{H} = -2\sigma^2 + 2\bar{p}\bar{q} \tag{12.4b}$$

$$R = \sigma^2 + \bar{q}^2 \tag{12.4c}$$

Note that the presence of variation among the subpopulations as measured by σ^2 leads to an increase in the homozygote classes, $\bar{D}$ and $\bar{R}$, at the expense of the heterozygote class. This effect is the same as that with inbreeding—an increase in homozygosity and loss of heterozygosity.

The Inbreeding Coefficient and Population Subdivision To further the analogy between population subdivision and inbreeding, we can calculate the inbreeding coefficient, f, produced by the subdivision. Recall that the inbreeding coefficient, f, controls the mixture of D, H, and R in terms of the pure random mating component $(p^2, 2pq, q^2)$ and the pure selfing component $(p, 0, q)$,

$$(D, H, R) = (1-f)(p^2, 2pq, q^2) + f(p, 0, q) \tag{12.5}$$

Then setting $\sigma^2 + \bar{p}^2$ equal to $(1-f)\bar{p}^2 + f\bar{p}$ and solving for f gives

$$f = \frac{\sigma^2}{\bar{p}\bar{q}} \tag{12.6}$$

This is the inbreeding coefficient that results from the subdivision of the population.

Thus we see that the presence of variation among the subpopulations leads to an increase in homozygosity and loss of heterozygosity as compared with the random mating frequencies that would result if the entire population were consolidated into one population. This result is called *Wahlund's effect* (Wahlund, 1928).

An example of the Wahlund effect in practice is offered by museum and herbarium collections. Typically, the collections of a species are a pooled sample of specimens obtained from many places in the species range. Such collections should show a deviation from Hardy–Weinberg frequencies based on the average gene frequencies in the pooled sample. This deviation from the Hardy–Weinberg proportions is in fact a measure of the amount of variation among the subpopulations. Variation from place to place is called *geographical variation*. The formula for f in terms of σ^2 relates the extent of the deviation from Hardy–Weinberg proportions to the amount of geographical variation in the population†.

The Wahlund effect depends on the existence of geographical variation among the subpopulations. What, we might ask, could cause such geographical variation in the first place? In principle either genetic drift or natural selection could cause these differences. We now discover, however, that even a tiny amount of migration between the subpopulations effectively prevents genetic drift from producing much geographical variation in gene frequencies.

Migration Versus Drift

Most populations from the same species are not totally isolated from one another. Instead, most subpopulations are "connected" to one another through a recurrent interchange of migrants. But we may ask, how much migration is needed to cause two populations to function evolutionarily as one population, or conversely, how rare must migration be so that the populations are effectively isolated from one another? There is a different answer to this question for each evolutionary force. For example, since drift is usually a weaker force than selection, we might expect that less migration is required to make two populations function as though one for drift than for selection. In the next several paragraphs we shall answer this question for the evolutionary force of genetic drift.

Some Definitions

We must be precise about what it means to function as a single or as two separate populations. As background, consider a *single* population with N *haploid* individuals. Each generation is formed by choosing N gametes from an effectively infinite gamete pool. Recall from Chapter 10 that this sampling process leads eventually to a state where all the alleles in the population are identical by descent. Recall also that if $h_t = (1 - f_t)$ is the probability that two alleles are *not* identical by descent, then

$$h_t = \left(1 - \frac{1}{N}\right)^t \tag{12.7}$$

Thus in a single population of N individuals, h_t, the probability that two alleles are not identical by descent tends to zero as $(1 - 1/N)^t$. Now consider a larger single population with $2N$ individuals. Obviously, h_t for this population tends to zero more slowly, indeed as $[1 - 1/(2N)]^t$.

† In practice, measuring f requires detecting significant deviations from the Hardy–Weinberg proportions. This requires large sample sizes.

Figure 12.1 illustrates the setup for our study of migration versus drift. There are two populations each with N haploid individuals. With probability m an individual migrates from one population to another and with probability $(1-m)$ an individual stays put. Let $f_{1,t}$ be the probability that two genes chosen at random from population 1 at time t are identical by descent, and similarly for $f_{2,t}$. Let $f_{12,t}$ be the probability that a gene drawn from population 1 is identical by descent with a gene drawn from population 2 at time t. (The variables, $f_{1,t}$ and $f_{2,t}$, are inbreeding coefficients, and $f_{12,t}$ is a coancestry coefficient.) Suppose initially that none of the genes is identical by descent, $f_{1,0}=f_{2,0}=f_{12,0}=0$. Then if there is no migration, each population will tend to a state of complete inbreeding as $h_t = (1-1/N)^t$, as mentioned above. Let $d_t = (1-f_{12,t})$ be a measure of the differentiation between the two populations. With no migration, $f_{12,t}$ must remain zero so that d_t remains equal to one. Therefore, we define

$$[\text{two isolated populations}] = \begin{cases} h_t = \left(1-\dfrac{1}{N}\right)^t \\ d_t = 1 \end{cases} \tag{12.8}$$

On the other hand, if there is complete mixing with $m = \frac{1}{2}$ then both populations will tend toward a state of complete inbreeding, as $h_t = [1-1/(2N)]^t$. Moreover, with complete mixing, two genes drawn from different populations have the same probability of being identical by descent as if they were drawn from the same population. Therefore, with complete mixing, $f_{1,t}=f_{12,t}=f_{2,t}$. Thus we define

$$[\text{single population}] = \begin{cases} h_t = \left[1-\dfrac{1}{2N}\right]^t \\ d_t = \left[1-\dfrac{1}{2N}\right]^t \end{cases} \tag{12.9}$$

With these definitions we can proceed to discover how small m must be to have two isolated populations each of size N, and how large m must be to have one population of size $2N$.

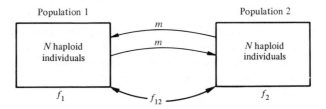

FIGURE 12.1. *Setup for the study of genetic drift in two populations coupled with migration. Coefficients f_1 and f_2 are the inbreeding coefficients in populations 1 and 2, respectively; f_{12} is the coancestry coefficient between individuals from the different populations. During each generation, a fraction, m, of the genes in each population migrates to the other population. The model includes only migration and drift. There is no natural selection in the model.*

The Model　The basic equations for this model are simple. The equations predict $f_{1,t+1}$, $f_{2,t+1}$, and $f_{12,t+1}$ from their values at time t. The first equation is

$$f_{1,t+1} = (1-m)^2 \left[\frac{1}{N_1} + \left(1 - \frac{1}{N_1}\right) f_{1,t} \right]$$

$$+ 2m(1-m)f_{12,t} + m^2 \left[\frac{1}{N_2} + \left(1 - \frac{1}{N_2}\right) f_{2,t} \right] \qquad (12.10)$$

The term, $f_{1,t+1}$, is the probability that two genes drawn at time $t+1$ from population 1 are identical by descent. The first term refers to two genes that were both in population 1 at time t. If they were, with probability $1/N_1$, they are copies of the same gene; with probability $(1 - 1/N_1)$, they are copies of different genes. But if they are copies of different genes, they still have probability $f_{1,t}$ of being identical by descent. The second term refers to a pair of genes one of which has migrated from the other population. By definition, $f_{12,t}$ is the probability that two genes from different populations are identical by descent. The third term refers to a pair of genes, both of which have migrated from population 2. Similarly, we can deduce that

$$f_{12,t+1} = (1-m)^2 f_{12,t} + m(1-m)\left[\frac{1}{N_1} + \left(1 - \frac{1}{N_1}\right) f_{1,t} \right]$$

$$+ m(1-m)\left[\frac{1}{N_2} + \left(1 - \frac{1}{N_2}\right) f_{2,t} \right] + m^2 f_{12,t} \qquad (12.11)$$

The term, $f_{12,t+1}$, is the probability that a gene drawn at $t+1$ from population 1 is identical with one drawn from population 2. The first term refers to two genes that have not migrated. The second term refers to genes that were both in population 1 at time t but one of which migrated to population 2. The third term refers to genes that both came from population 2. The fourth term refers to genes that both migrated (they interchanged populations). The equation for $f_{2,t+1}$ is symmetric to that for $f_{1,t+1}$.

Our interest is in the situation where $N_1 = N_2 \equiv N$ and where $f_{1,0} = f_{2,0}$. If $f_{1,t} = f_{2,t} \equiv f_t$, we need only consider two equations. Substituting $h_t = (1 - f_t)$ and $d_t = (1 - f_{12,t})$, we obtain

$$h_{t+1} = (1-x)\left(1 - \frac{1}{N}\right) h_t + x\, d_t \qquad (12.12a)$$

$$d_{t+1} = x\left(1 - \frac{1}{N}\right) h_t + (1-x)\, d_t \qquad (12.12b)$$

where $x = 2m(1-m)$. These equations predict the approach to a state of inbreeding in each population ($h \to 0$) and the loss of differentiation between populations ($d \to 0$) through time. The parameters are the migration rate m and the population size, N.

The Results　The equations for h_t and d_t are a pair of linear recursion equations which are easily solved analytically. The solution is explored in Problem 1 at the end of the chapter. The results are (1) If m is small and if $4mN \gg 1$, then

$$h_t \approx \left(1 - \frac{1}{2N}\right)^t$$

$$\qquad (12.13)$$

$$d_t \approx \left(1 - \frac{1}{2N}\right)^t \qquad (4mN \gg 1)$$

Thus if $4mN$ is much greater than one, the two populations function effectively as a single population with size $2N$. (2) If m is small and if $4mN \ll 1$, then

$$h_t \approx \left(1 - \frac{1}{N}\right)^t$$

$$d_t \approx 1 \qquad (4mN \ll 1) \qquad (12.14)$$

Thus if $4mN$ is much less than one, the populations are separate and effectively isolated from one another. If m is large or if $4mN$ is close to one, then the populations show intermediate behavior as described by the exact formulas discussed in Problem 1.

Surprising Consequences The model shows that whether two populations are effectively fused or separate depends on whether $4mN$ is much larger or smaller than one. This result has some surprising consequences. First, the amount of migration necessary to fuse two populations with respect to drift is surprisingly small. If only one individual from each population migrates to the other population, then $m = 1/N$, if two migrate $m = 2/N$, and so forth. If only two individuals migrate, then $4mN$ equals 8, and if three migrate then $4mN$ equals 12. These values are sufficiently above one that the populations are effectively fused. Thus a recurrent interchange of only two or three individuals is sufficient to fuse the populations with respect to genetic drift. Second, the amount of migration necessary to keep two populations separate is essentially zero. If only one individual migrates, $m = 1/N$ so that $4mN$ equals 4. Therefore, to make $4mN$ much less than one requires that less than one individual migrate. This is effectively zero migration. Thus we see that fusion versus separation turns in practice on whether two or three versus much less than one individuals migrate each generation.

A third consequence of the $4mN$ criterion is that the migration probability m, required to fuse large populations is smaller than that required to fuse small populations. This idea is intuitive—drift becomes weaker as the population size increases so that less migration should be needed to fuse large populations than small populations.

The results of this model have been particularly important in discussions of the data on enzyme polymorphisms that have been gathered by many in connection with the selection-neutrality controversy discussed in Chapter 6. It is sometimes observed that the same gene frequencies are found in two separate locations. If so, it is naïvely argued that selection must be responsible because drift would cause the populations to diverge. But, as we have seen, a mere trickle of migration is enough to prevent divergence between the populations due to drift. So uniform gene frequencies among locations can be taken as evidence for selection *only* if extremely accurate information on migration is available—information so accurate that a migration of only one individual per generation can be ruled out.

The conclusions here are based on analyzing the approach to homozygosity caused by drift. This is the approach taken by Maruyama (1970) in examining the more general problem of many subpopulations connected by migration between adjacent subpopulations. Another model, analyzed by Wright and Kimura, leads to a similar criterion based on $2mN$ where N is the population size of diploid organisms. Consider migration from a fixed infinite source population into a finite local population of size N. With diffusion techniques, these authors established that the probability distribution for the gene frequency in the local population is U-shaped if $2mN \ll 1$ and bell-shaped if $2mN \gg 1$. The local population tends to fixation if $2mN \ll 1$, showing that under this condition drift is stronger than migration. The local population tends to have the same gene frequency as the source region if $2mN \gg 1$, showing that under this condition drift is weaker than migration. For details of this treatment, see Crow and Kimura (1970).

Genetic Relatedness and Spatial Distance

A basic feature of the spatial structure of populations is that related individuals tend to live near one another. There is a large literature intended to predict the degree of relationship between two individuals as a function of the spatial distance separating them. We shall briefly quote some of the results of the theory, after discussing Malécot's (1969) treatment of this spatial relationship.

One Dimension

Consider a set of subpopulations each with N diploid individuals. Let the location of each subpopulation be labeled as x ($x = .. -2, -1, 0, 1, 2 ..$); see Figure 12.2. The distance between the locations is constant. Now as we discovered before with two populations, genetic drift combined with migration leads to a state of complete inbreeding within each subpopulation and a loss of differentiation between populations. (This occurs relatively fast if $4mN \gg 1$ but the eventual loss of differentiation would occur much more slowly if $4mN \ll 1$.) This same result would also occur with an infinite set of populations, provided that the migration is sufficiently local. That is, let the probability of migrating from x to $x + h$ be m_h ($h = .. -2, -1, 0, 1, 2 ..$). We require that the sum of these probabilities equal one, $\sum_{h=-\infty}^{\infty} m_h = 1$. For this series to converge, the probability of traveling long distances must be sufficiently less than traveling short distances. In this sense, then, the migration is local in that very distant locations are not strongly connected to one another. For sufficiently local migration, the entire system moves toward fixation, with all populations being entirely inbred and undifferentiated from each other. But we are interested in a rather different situation now, one where there is a regular pattern in the degree of relatedness between individuals of different colonies. This condition requires that there be some force preventing the eventual total loss of differentiation between colonies.

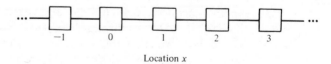

Location x

FIGURE 12.2. *Sketch of setup for model of genetic relatedness as a function of distance.*

One force that retains the genetic variability is mutation. Let u and v be recurrent mutation rates of A to a, and vice versa, and let k be their sum ($k = u + v$). With this additional assumption, the set of subpopulations tends to an equilibrium state where a permanent pattern of differentiation between the populations is maintained by the joint action of migration, mutation and drift. This pattern is described as follows. Let $f(h)$ be shorthand for $f_{x,x+h}$, that is, the coancestry coefficient between individuals from colonies separated by h units of distance. Let the mean and variance of the migration regime be

$$\bar{h} = \sum_{h=-\infty}^{\infty} h m_h$$

$$\sigma^2 = \sum_{h=-\infty}^{\infty} h^2 m_h - \bar{h}^2$$

(12.15)

$\bar{h}$ is zero for symmetrical migration. Then Malécot (1969, p. 84) has shown that at equilibrium

$$f(h) \approx \frac{\exp\left[-\sqrt{2k}(h/\sigma)\right]}{1 + 4N\sigma\sqrt{2k}}$$

(12.16)

provided that k is small (which, of course, it is). By this formula the coancestry coefficient decreases exponentially with the distance separating the colonies. Note also the influence of the expected dispersal distance σ, and the population size, N, as illustrated in Figure 12.3. Note also that the correlation extends over very long distances relative to the migration distance σ.

More Dimensions Equation (12.16) for the coancestry coefficient as a function of distance refers to a one-dimensional situation—all the colonies are orderable along one dimension as might occur with sand dune and coastal organisms, for example. Malécot and Kimura and Weiss (1964) have also studied two- and three-dimensional models. Also, they have investigated populations distributed *continuously* in space rather than localized into identifiable subpopulations. The formulas in continuous space from Malécot (1969) are in terms of r, the distance between two organisms,

$$f(r) \approx \exp\left[-\sqrt{2k}\left(\frac{r}{\sigma}\right)\right] \qquad \text{(one-dimensional continuum)} \qquad (12.17a)$$

$$f(r) \approx \frac{1}{\sqrt{r}} \exp\left[-\sqrt{2k}\left(\frac{r}{\sigma}\right)\right] \qquad \text{(two-dimensional continuum)} \qquad (12.17b)$$

$$f(r) \approx \frac{1}{r} \exp\left[-\sqrt{2k}\left(\frac{r}{\sigma}\right)\right] \qquad \text{(three-dimensional continuum)} \qquad (12.17c)$$

These formulas assume that $r \gg \sigma$. Notice that there are substantial differences between these formulas; thus the genetic relationship as a function of distance depends on the number of dimensions involved. In general, increasing the dimensionality lowers the expected relationship between two individuals separated by a given r. Crow and Kimura (1970) discuss this topic further.

Natural Selection with Small Migration Between Subpopulations In this section we begin our discussion of natural selection in spatially varying environments. Suppose selection favors one allele in one subpopulation and another allele in another subpopulation. Then, if there is migration between these subpopulations, will polymorphism result? That is, will each population contain both alleles in some stable proportion because of the migration?

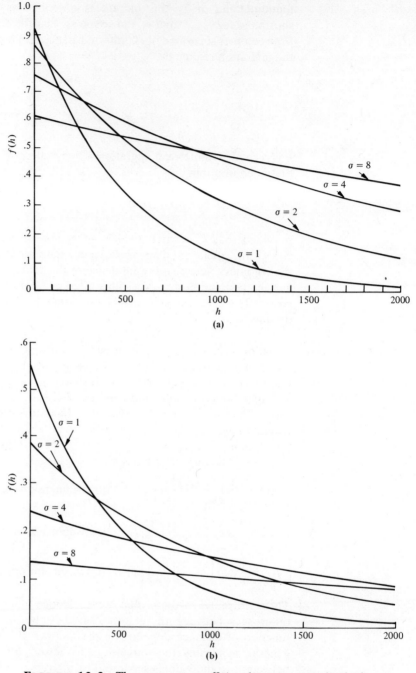

FIGURE 12.3. *The coancestry coefficient between two individuals as a function of the distance separating their colonies. From Equation* (12.16). (*a*) $N = 10$, $k = 2 \times 10^{-6}$; (*b*) $N = 100$, $k = 2 \times 10^{-6}$.

This problem has been studied in a very general way by Karlin and McGregor (1972a, b), provided that the populations are connected by a small amount of migration (weak coupling). We illustrate their results with a simple example involving two subpopulations. Let the dynamics of selection in population 1, in the absence of migration, be represented as

$$p_{1,t+1} = f_1(p_{1,t}) \tag{12.18}$$

where $p_{1,t}$ is the frequency of the A allele in population 1 at time t. We could assume the familiar formula

$$f_1(p_1) = (p_1 w_{1AA} + q_1 w_{1Aa}) \frac{p_1}{\bar{w}_1} \tag{12.19}$$

where w_{1AA} is the fitness of AA in population 1, and so forth. However, this theory applies to many models, not only the usual selection model.† Similarly, in population 2, without migration, we have

$$p_{2,t+1} = f_2(p_{2,t}) \tag{12.20}$$

We shall next be interested in this system when migration couples both subpopulations as follows,

$$p_{1,t+1} = (1-m)f_1(p_{1,t}) + mf_2(p_{2,t})$$
$$p_{2,t+1} = mf_1(p_{1,t}) + (1-m)f(p_{2,t}) \tag{12.21}$$

To use the technique of Karlin and McGregor, we must visualize the uncoupled system as a special case of Equation 12.21, namely for $m = 0$. Equation 12.21 is assumed to be understood for some particular value of m (i.e., $m = 0$) and then m is changed slightly. Next, the technique is to predict the properties of the system for this slightly different m from the properties of the system at the original value of m. Thus the technique is said to involve "small perturbations" to the parameters.

Rules for
Reasoning with
Weakly Coupled
Subsystems
To illustrate the rules of this technique, consider an example in which, in the absence of migration, A is fixed in population 1 and a in population 2. (See Figure 12.4.) Thus in the isolated population 1 there are two equilibria, $\hat{p}_1 = 0$ and $\hat{p}_1 = 1$; $\hat{p}_1 = 0$ is unstable and $\hat{p}_1 = 1$ is stable. Similarly, in the isolated population 2 the equilibria are $\hat{p}_2 = 0$ and $\hat{p}_2 = 1$, with the former stable and the latter unstable. For our example the dynamics are given by (12.19). The technique involves the following five rules: (A sufficient condition for these rules to be valid is also given.)

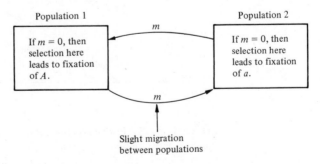

FIGURE 12.4. *Setup for a model of natural selection in two populations, together with a small amount of migration between the populations.*

† Here $f(p)$ is assumed to be a continuous mapping of p in $[0, 1]$ into $[0, 1]$, and $f'(p)$ is also assumed to be continuous in $[0, 1]$.

1. *Enumerate all the equilibria of the system with m = 0; see Table 12.1.* (See also Figure 12.5.)

Table 12.1.

$(\hat{p}_1, \hat{p}_2)$	Description
(0, 0)	*a* fixed in both subpopulations
(0, 1)	*a* fixed in 1 and *A* fixed in 2
(1, 0)	*A* fixed in 1 and *a* fixed in 2
(1, 1)	*A* fixed in both

2. *Classify the equilibria for m = 0 with respect to the stability characteristics of the components*; see Table 12.2.

3. *If an equilibrium is stable with respect to both components when m = 0, then with small migration, m > 0, there continues to be a unique equilibrium point nearby in the correct domain and the system continues to be locally stable at that point.* In our example, for *m* sufficiently small, there *is* a stable equilibrium near (1, 0). Since this equilibrium must be near (1, 0) it means

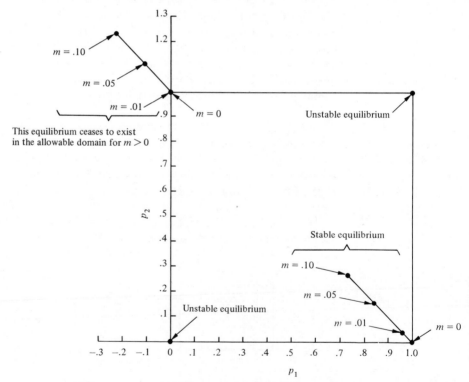

FIGURE 12.5. *Positions of the equilibria in the two-population system as a function of the amount of migration, m. Note that the equilibria in the lower left and upper right do not change positions as m is varied. However, as m increases, the stable equilibrium in the lower right moves from the corner into the interior of the unit square; the unstable equilibrium in the upper left moves out of the allowable domain and ceases to exist in the system for m > 0. Also, $w_{1,AA} = 1$; $w_{1,Aa} = .75$; $w_{1,aa} = .5$; $w_{2AA} = .5$; $w_{2,Aa} = .75$; and $w_{2,aa} = 1$. The dynamics are given by Equation (12.19).*

Table 12.2.

Equilibria $(\hat{p}_1, \hat{p}_2)$	Stability of $\hat{p}_1$ when $m = 0$	Stability of $\hat{p}_2$ when $m = 0$
$(0, 0)$	Unstable	Stable
$(0, 1)$	Unstable	Unstable
$(1, 0)$	Stable	Stable
$(1, 1)$	Stable	Unstable

that A is nearly fixed in population 1 and a is nearly fixed in population 2. Note in Figure 12.5 that the point moves from the corner into the interior of the square.

4. *If an equilibrium is unstable with respect to one or more of the components, then the equilibrium point may not continue to exist in the coupled system.* It may not exist because its position may shift outside the domain of the equations; that is, it may move outside the square where both p_1 and p_2 are in $[0, 1]$. *But if it does remain within the domain, then it continues to be unstable.* In our example, $(0, 0)$ and $(1, 1)$ continue to be unstable equilibria in the coupled system. In fact their positions do not change at all. The point $(0, 1)$, however, ceases to correspond to an equilibrium. As Figure 12.5 shows, it moves outside the domain.

5. *All the possible equilibrium points for small m are accounted for in this way. For sufficiently small m, no new equilibrium points arise that do not correspond to one of the equilibria present in the system with m = 0.* In our example, for small m, there are no more than four equilibria. There may be less than four if, as mentioned above, an unstable equilibrium moves outside the domain of the model. Thus for small m all trajectories converge to the stable equilibrium point near $(1, 0)$, where A predominates in population 1 and a is dominant in population 2.

When Are the Rules Valid? These simple rules allow us to deduce the qualitative properties of the evolution of populations connected by a small amount of migration based on knowledge of how the populations would evolve if isolated. Karlin and McGregor discuss the conditions under which the rules are valid in some detail. We state here perhaps the most useful of the various sufficient conditions for the rules to be valid. Consider $f_i(p_i)$, which describes the dynamics in region i. Let $\hat{p}_i$ denote an equilibrium gene frequency in region i when $m = 0$. The sufficient condition for the rules to apply is that, for all i, (a) $f_i'(\hat{p}_i)$ lie between -1 and 1 if $\hat{p}_i$ is stable *or* (b) $f_i'(\hat{p}_i)$ lie outside the interval $[-1, 1]$ if $\hat{p}_i$ is unstable. That is, a sufficient condition for the rules to apply is that $f_i'(\hat{p}_i)$ not equal exactly -1 or $+1$ in any region. If $f_i'(\hat{p}_i)$ does equal -1 or $+1$ in one of the regions, then the rules may apply or they may not and a deeper analysis is required to find out. The assumption that $|f'(\hat{p})| \neq 1$ entails that convergence or divergence from the equilibrium point $\hat{p}$ is at a geometric rate. A sequence p_t is said to converge geometrically fast to $\hat{p}$ at rate λ ($0 < \lambda < 1$) if the distance between p_t and $\hat{p}$ is of order λ^t. Thus geometric convergence or divergence from the $\hat{p}_i$ in each isolated system is a sufficient condition for the rules above to be valid.

If $f'(\hat{p})$ does equal one, then there are two possible meanings. First, it may mean that the convergence or divergence from the equilibrium is at a rate slower than geometric. In particular, with full dominance, the elimination of a deleterious recessive occurs at a slow rate called an algebraic rate. Algebraic convergence means an order of difference between p_t and $\hat{p}$ of $1/t^\alpha$ for some positive α, usually $\alpha = 1$. Second, it may mean that the equilibrium is "neutrally stable," that is, a perturbation away from the equilibrium is neither counteracted nor reinforced.

As an example, suppose that all the w's in population 2 are equal, thus indicating no selection in population 2 when it is isolated. Then if there *is* selection in population 1, even the slightest coupling qualitatively changes the stability properties of population 2. Thus neutrally stable equilibria of isolated systems are not preserved in the presence of weak coupling to other systems.

The theory above is based on the assumption of small migration. Suppose instead that migration is large, then what? It is hard to make any general statements on this. Typically, one system will dominate the other. For example, if the selection against a in population 1 is sufficiently stronger than that against A in population 2, then we would expect a to be eliminated from both populations. The condition for this would depend on the nature of the dominance and would require large migration. The answer can usually be found only by direct analysis of the coupled model. For example, see Karlin and McGregor (1972b) on a selection model with $m = \frac{1}{2}$, and Levin (1974) on an ecological model.

A Theorem on Structural Stability

The rules concerning small migration are special results of a theorem on *structural stability* advanced by Karlin and McGregor. The notion of structural stability concerns stability under perturbations, not to the variables in a model, but to the model itself. We want to know whether the qualitative predictions of a model change if the model itself is perturbed. Obviously, this point is of great practical importance, for we need to know whether the qualitative predictions of a model are still valid even though some parameters are not empirically measurable with perfect accuracy. On this topic we quote one of the theorems due to Karlin and McGregor.

THEOREM

Let x_t be a vector of k variables at time t and m be a vector of parameters. Consider a model of the form

$$x_{t+1} = f(x_t, m) \tag{12.22}$$

where $f(x, m)$ is a vector-valued function defined for a certain domain Ω. [That is, $f(x, m)$ maps Ω into itself.] Let us denote the matrix of partial derivatives $\partial f_i/\partial x_j$ $(i, j = 1, 2 \cdots k)$ as $f'(x)$. Let $\lambda_i(x)$ be the eigenvalues of $f'(x)$ evaluated at x. The theorem is

Suppose that $f(x)$ has exactly ν equilibrium points $x_1, x_2, \ldots x_\nu$, in Ω for $m = \tilde{m}$.

Assume that

1. *$f_i(x, m)$ is continuous in x and m.*
2. *$\partial f_i(x, m)/\partial x_j$ is continuous in x and m.*
3. *No $\lambda_i(x)$ of $f'(x)$ equals 1 or -1 at any equilibrium point, $x = x_\alpha$ $(\alpha = 1, 2 \cdots \nu)$.*

If m is sufficiently close to $\tilde{m}$, then

1. *There are no more than ν equilibrium points in Ω.*
2. *Let $x_\alpha(\tilde{m})$ denote a stable equilibrium point for $\tilde{m}$. There is exactly one*

corresponding equilibrium point for m, $x_\alpha(m)$, and it is stable, close to $x_\alpha(\tilde{m})$ and in Ω.

3. *Let $x_\alpha(\tilde{m})$ denote an unstable equilibrium point for $\tilde{m}$. There is exactly one corresponding equilibrium point for m, $x_\alpha(m)$ but it may not be in Ω. But if it is in Ω, then it is close to $x_\alpha(\tilde{m})$ and is unstable.*

The interesting qualitative conclusion from this theorem is that locally stable equilibria are structurally stable predictions from models while unstable and especially neutrally stable equilibria may not be structurally stable predictions.

The Multiple Niche Polymorphism

The ranges of most populations encompass many microhabitats or "niches." Microhabitats are little regions in the environment where physical conditions are different from adjacent regions. For example, microhabitats include patches of moist or dry ground, crotches at the branches of trees, and so forth. By definition, a niche is a characteristic manner of obtaining resources. For example, morphs who specialize on different types of food are said to occupy different niches.† Let us imagine that each genotype is most fit in a particular microhabitat or niche and that there is a certain amount of each microhabitat in the environment or "room" for each morph with a specialized niche. Moreover, we envision that at the time of mating, there is random mating in the whole population. The organisms emerge, as it were, from their microhabitats to mate at random. We would like to know whether this situation can maintain polymorphism.

This problem is at the opposite extreme of that in the last section. Here, complete mixing at mating among members of different microhabitats is assumed, whereas in the last section near zero mixing was assumed.

Constant-Zygote- Number Formulation

Models for a "multiple niche" polymorphism have been formulated in two ways. First, suppose that the fraction of *zygotes* which settle in any type of microhabitat is proportional to the amount of that microhabitat in the environment. Visualize, for example, a cloud of wind dispersed seeds which introduces seeds into patches of different soil types in proportion to the area of the different soil types in the environment. If C_h is the fraction of area of soil with type h, then we assume the fraction of seeds to land in soil type h is also C_h. Suppose that there are T different soil types. Let $w_{ij,h}$ denote the fitness of genotype $A_i A_j$ in microhabitat h. Then the average fitness of genotype $A_i A_j$, w_{ij}, is

$$w_{ij} = \sum_{h=1}^{T} C_h w_{ij,h} \tag{12.23}$$

After the seeds settle and mature, there is random mating among the survivors and the gene frequency in the next generation is the familiar

$$p_{t+1} = \frac{p_t w_{11} + q_t w_{12}}{\bar{w}} p_t \tag{12.24}$$

where w_{ij} is the average fitness of genotype $A_i A_j$ from Equation (12.23). The dynamics of this system are familiar. With two alleles, for example,

---

† This distinction between a habitat, or microhabitat, and niche is often put as: A habitat is an address; a niche is an occupation.

polymorphism requires $w_{12} > w_{11}, w_{22}$. Thus the condition for a multiple niche polymorphism in this formulation is

$$\frac{w_{11}}{w_{12}} = \frac{\sum C_h w_{11,h}}{\sum C_h w_{12,h}} < 1 \qquad (12.25a)$$

$$\frac{w_{22}}{w_{12}} = \frac{\sum C_h w_{22,h}}{\sum C_h w_{12,h}} < 1 \qquad (12.25b)$$

Constant-Fertile-Adult-Number Formulation
In the second formulation introduced by Levene (1953), we suppose that the fraction of *adults* that come from any type of microhabitat is proportional to the amount of that microhabitat in the environment. Visualize, for example, that a huge quantity of seeds are produced so that every patch of soil is completely saturated with seeds, more seeds than could possibly grow in the available space. In this situation we can assume that there is a fixed number of plants which could possibly occupy any patch and therefore the number of *adults* from soil type h is proportional to the fraction of area with soil type h, C_h. (We also assume that the seeds do not survive two or more seasons so that the generations are distinct.) There is random union of gametes among adults, and each patch begins with the same Hardy–Weinberg frequencies. As before, let $w_{ij,h}$ be the fitness of genotype $A_i A_j$ in microhabitat type h. Let $p_{t+1,h}$ be the gene frequency among the adults in microhabitat h after (viability) selection. Then p_{t+1} in the whole population will be

$$p_{t+1} = \sum_{h=1}^{T} C_h p_{t+1,h} = \sum_{h=1}^{T} C_h \frac{(p_t w_{11,h} + q_t w_{12,h})p_t}{p_t^2 w_{11,h} + 2 p_t q_t w_{12,h} + q_t^2 w_{22,h}} \qquad (12.26)$$

This model does not reduce to the familiar one-locus two-allele equation.†

Concept of a Protected Polymorphism
It is difficult with this model to solve directly for the polymorphism frequency and then to examine the conditions for its stability. Instead, a somewhat indirect approach has been used. We define a *protected polymorphism* as the condition where both alleles can increase when rare. If A can increase when rare, then A is said to be "protected." A polymorphic condition must result if both A and a are protected. But the existence of a protected polymorphism does not entail that there is one stable equilibrium *point*. There may, in principle, be several equilibrium points, a continuous curve of equilibrium points, or the frequencies may oscillate—all these possibilities are consistent with the idea of a protected polymorphism. Indeed, some of these unusual possibilities are known to occur with this model.

We now examine conditions for a protected polymorphism with Equation (12.26). The conditions will be sufficient but not necessary for a protected polymorphism. Consider $p_{t+1} = f(p_t)$ where $f(p)$ is the right-hand side of Equation (12.26). For A to increase when rare, we want $p = 0$ to be un-

† The distinction between these two formulations was pointed out by Dempster (1955), who termed the first as the "constant-zygote-number" hypothesis and the second as the "constant-fertile-adult-number" hypothesis. Wallace (1968) and Christiansen (1975) have termed the first as "hard" and the second as "soft" selection. The idea is that both population size and composition are influenced in the first formulation (hard selection), while only the genetic composition is influenced in the second (soft selection).

stable; for a to increase when rare, we want $p=1$ to be unstable. For these equilibria to be unstable, it is sufficient that $f'(0)>1$ and $f'(1)>1$. Of course, the equilibria could conceivably also be unstable if $f'=1$ and therefore our conditions are only sufficient, not necessary. Let $f_h(p)$ represent $p_{t+1,h}$ in microhabitat h according to the usual formula.

$$f_h(p)=\frac{p^2w_{11,h}+pqw_{12,h}}{p^2w_{11,h}+2pqw_{12,h}+q^2w_{22,h}} \tag{12.27}$$

Then, by definition,

$$f(p)=\sum_{i=1}^{T}C_h f_h(p) \tag{12.28}$$

Differentiating, we have

$$f'(p)=\sum C_h f'_h(p) \tag{12.29}$$

It can be verified that $f'_h(0)=w_{12,h}/w_{22,h}$ and $f'_h(1)=w_{12,h}/w_{11,h}$. Therefore, protection occurs if

$$f'(0)=\sum C_h w_{12,h}/w_{22,h}>1 \quad (A \text{ protected})$$
$$f'(1)=\sum C_h w_{12,h}/w_{11,h}>1 \quad (a \text{ protected}) \tag{12.30}$$

These conditions are often written in a slightly rearranged version. We can in each $f_h(p)$ divide numerator and denominator by a constant so as to scale $w_{12,h}=1$. Using this convention (all $w_{12,h}=1$) and rearranging gives

$$\frac{1}{\sum C_h\left(\dfrac{1}{w_{22,h}}\right)}<1 \quad (A \text{ protected})$$

$$\frac{1}{\sum C_h\left(\dfrac{1}{w_{11,h}}\right)}<1 \quad (a \text{ protected}) \tag{12.31}$$

In this form we see that a sufficient condition for a protected polymorphism is that the harmonic average of each homozygote fitness be less than that of the heterozygote (which equals 1 by convention).

Contrast Between Formulations   It is interesting to contrast the polymorphism conditions from the two formulations. In the earlier "constant-zygote-number" formulation the usual arithmetic average fitness for the homozygotes had to be less than that of the heterozygote, whereas in the "constant-fertile-adult-number" formulation, the harmonic average fitness of the homozygotes must be less than that of the heterozygote. It is "easier" to meet the condition with the harmonic average than with the arithmetic average as Figure 12.6 illustrates. Figure 12.6 illustrates three kinds of averages, where C_i is the probability of $x_i(C_1+C_2=1)$,

$$\text{(arithmetic)} \quad \bar{x}_a=C_1x_1+C_2x_2$$
$$\text{(geometric)} \quad \bar{x}_g=x_1^{C_1}x_2^{C_2} \tag{12.32}$$

$$\text{(harmonic)} \quad \bar{x}_h=\frac{1}{C_1(1/x_1)+C_2(1/x_2)}$$

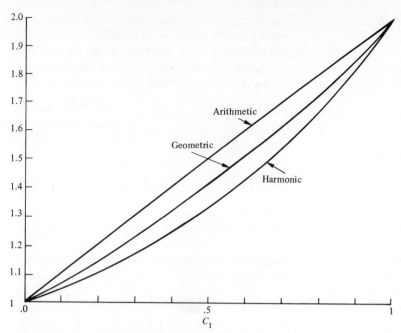

FIGURE 12.6. *The difference among three common types of means. The arithmetic mean is generally higher than the geometric mean, which, itself, is generally higher than the harmonic mean. Here $x_1 = 2$; $x_2 = 1$.*

Note that, for any given C_1 and C_2, the order is $\bar{x}_a > \bar{x}_g > \bar{x}_h$. For this reason there are fitness values which lead to polymorphism under the harmonic average criterion but which fail to produce polymorphism under the arithmetic average criterion. As a numerical example with two niches, if $w_{AA,1} = 1.75$ and $w_{AA,2} = .50$, and if $C_1 = C_2 = \frac{1}{2}$, then the arithmetic average is 1.125 while the harmonic average is .778 so polymorphism could not occur in the first case while it could in the second. Thus there is some quantitative difference in the "ease" with which polymorphism can be achieved depending on whether it is the number of adults or zygotes which are set by the availability of the microhabitats in the environment.

Maximization Principles for Selection in Multiple Niches

The above model for selection in multiple niches proposed by Levene has led to many interesting extensions and generalizations. One extension is the derivation of a maximization principle for selection in multiple niches.

In the first formulation—the "constant-zygote-number hypothesis"—it is easy to observe that the arithmetic mean fitness, $A = \sum C_h \bar{w}_h$, is maximized because the system is equivalent to the familiar one-locus multiallele model.

In the second formulation, that introduced by Levene, Cannings (1971) has shown that the geometric mean fitness is maximized. The geometric mean fitness is defined as

$$A = \prod_{h=1}^{T} \bar{w}_h^{C_h} \qquad (\textstyle\sum C_h = 1) \qquad (12.33)$$

For example, with two niches $A = \bar{w}_1^{C_1} \bar{w}_2^{C_2}$. To establish this result, we must show that equilibria correspond to critical points of A; that is, that $\Delta p = 0 \Leftrightarrow dA/dp = 0$. Next we must show that $A_{t+1} \geqslant A_t$, that is, that the equation for p_{t+1} causes A to increase steadily until equilibrium is attained.

For simplicity consider two alleles. Then with the constant-fertile-adult hypothesis

$$\Delta p = \sum C_h \, \Delta p_h \qquad (12.34)$$

where

$$\Delta p_h = \frac{pq}{2\bar{w}_h} \frac{d\bar{w}_h}{dp} \qquad (12.35)$$

Then Δp in full is given by

$$\Delta p = \frac{pq}{2} \sum \frac{C_h}{\bar{w}_h} \frac{d\bar{w}_h}{dp} \qquad (12.36)$$

Now if $A(p)$ is given by Equation (12.33), its derivative is found to be

$$\frac{dA(p)}{dp} = A(p) \sum \frac{C_h}{\bar{w}_h} \frac{d\bar{w}_h}{dp} \qquad (12.37)$$

So the equation for Δp can be written as

$$\Delta p = \frac{pq}{2A} \frac{dA(p)}{dp} \qquad (12.38)$$

where $A(p)$ is the geometric average over the niches of the mean fitness in each niche, Equation (12.33). Clearly, $\Delta p = 0$ is equivalent to $dA(p)/dp = 0$ provided $p, q \neq 0, 1$. Thus equilibrium corresponds to critical points of the $A(p)$ function as required. Moreover, Cannings (1971) has noted that the equation for p_{t+1} can be arranged in a form where a theorem of Baum and Eagen (1967) guarantees that $A(p_{t+1}) \geq A(p_t)$ as required. Thus it is the geometric mean fitness which is maximized under the constant-fertile-adult-number hypothesis of Levene.

Cannings (1971) also points out that with three alleles there may be a curve of stable equilibria. If so, the polymorphism frequency at equilibrium will depend on the initial conditions. Since $A(p)$ is maximized at equilibrium there must be a *curve* along which $A(p)$ is at a maximum. This may be visualized as the "rim of a volcano." Cannings provides numerical examples of the unusual behavior. Thus the existence of a protected polymorphism should certainly not be construed to imply convergence to a single unique interior equilibrium point as usually occurs in the typical single-locus multiallele model.

The multiple niche selection model of Levene has been extended in a series of papers by Deakin (1966 and 1968), Prout (1968), Maynard Smith (1970), Bulmer (1972), Strobeck (1974), and Christiansen (1974) to include various schemes of migration between the niches. The papers are principally concerned with the conditions for a protected polymorphism and rather general techniques have been developed for this problem. Important recent extensions to the theory above appear in Karlin (1976, 1977a, 1977b) and Karlin and Richter–Dyn (1976).

Clines and the Spatial Pattern of Gene Frequencies

Some Examples

Most populations have intriguing spatial patterns in gene frequencies. The simplest pattern is called a *cline*. A cline is a usually gradual change in gene frequency or phenotype from one place to another. Several sorts of clines have received special names. A *morph ratio* cline is a cline in a polymorphism frequency as in Figure 12.7. A *phenotypic* cline is a cline in some trait like size or shape which is a quantitative character, and a *step* cline is an

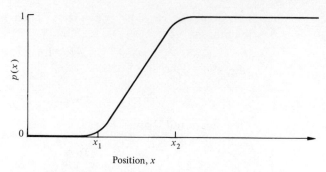

FIGURE 12.7. *Sketch of a gene frequency cline.*

abrupt, seemingly discontinuous, change in gene frequency in space as sketched in Figure 12.8.

To develop some appreciation for the phenomenon of clines, it will be helpful to cite some examples. Among the best studied are clines involving coastal and inland forms of *Peromyscus* in Florida by Sumner (1932) and melanic and nonmelanic forms of a moth in the Shetland Islands of Britain by Kettlewell (1961a, b) and Kettlewell and Berry (1961 and 1969). Figure 12.9, taken from Kettlewell and Berry (1961) illustrates the morph ratio cline in the melanic form. Note that there is nearly total fixation of the melanic type in the north and almost total fixation of the nonmelanic type in the south. Also, there is a valley, the Tingwall Valley, which runs across the Main Island and regularly has large winds through it. Kettlewell and Berry's (1969) experiments have shown that this valley is a barrier to the dispersal of moths. Note that the region where the cline is steepest appears to be near the Tingwall Valley. This example suggests two questions. What spatial pattern of selection pressures produces the pattern of gene frequency and what effect does a barrier to dispersal have on the slope of the cline?

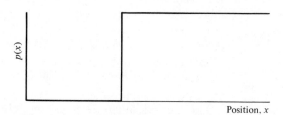

FIGURE 12.8. *Sketch of a step cline in gene frequency.*

Another example of a cline is provided by the spatial pattern of zinc tolerance near mines studied by Jain and Bradshaw (1966). Figure 0.5 in Chapter 0 illustrates a cline over a very short distance—measured in meters. The selection pressure is the presence of zinc in the soil; the trait is zinc tolerance. Notice that the amount of zinc abruptly drops off at the edge of the pasture while the cline in the grass tolerance to zinc is more gradual. The movement of seeds and pollen presumably smoothes over the abrupt change in the selection pressure and produces the gradual cline. The example suggests that the slope of the cline involves two considerations—the strength of the selection pressure and the average extent of dispersal. Using a model for the evolution of a cline, we can see how these two factors jointly determine the slope for a cline.

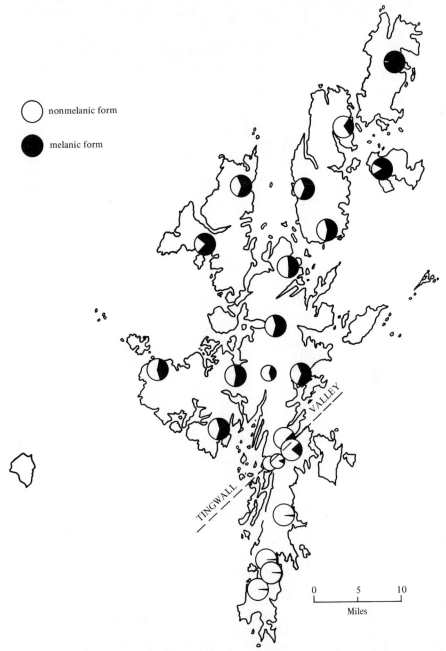

FIGURE 12.9. *Map illustrating the morph ratio cline in the melanic form of the moth,* Amathes glareosa. *Each circle has shading within it to indicate the relative proportions of the two types.* [From H. B. D. Kettlewell and R. J. Berry (1961), The study of a cline, *Amathes glareosa* Esp. and its melanic *F. edda* Staud. (Lep) in Shetland, *Heredity* **16**: 403–414.]

Still another example is what is often called a *hybrid zone*. Figure 12.10 illustrates a cline between two populations of birds that are considered to be separate species. See Sibley and Short (1959). One typical feature of such phenotype clines is the increased phenotypic variance at locations in the middle of the cline. This example poses several questions. Why is there an

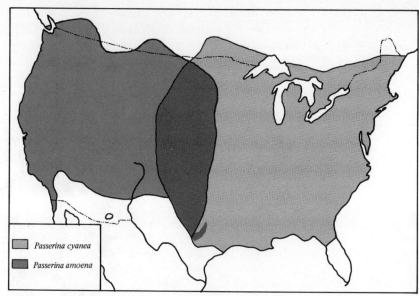

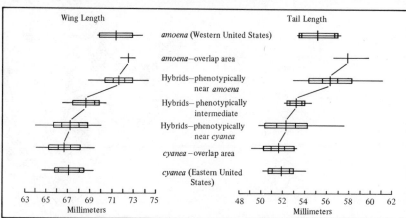

FIGURE 12.10. (*Top*) *Breeding ranges of the buntings,* Passerina cyanea *and* P. amoena, *showing the area of overlap.* (*Bottom*) *Statistical analysis of wing length and tail length of* P. cyanea, P. amoena *and hybrids. Horizontal lines represent the range; rectangles indicate one standard deviation from the mean and the solid black marks delineate twice the standard error of the mean. The means are indicated by vertical lines.* [From C. G. Sibley and L. L. Short (1959), Hybridization in the buntings, *Passerina*, of the Great Plains, the *Auk* **76**: 443–463.]

increased variance at locations in the middle of the cline, is it a mixing of migrants who originated from the ends of the cline, or is it a response to the local selection in the middle of the cline, for example, heterozygote superiority in the center? Alternatively, if hybrids between populations from ends of the cline have reduced fitness what is the effect on the slope of the cline?

A species may have many clines in different parts of its range resulting in a complicated spatial pattern of gene frequencies. British and European land snails in the genus *Cepaea* provide an instance of a complex spatial pattern extensively studied by Cain and Curray (1963), Goodhardt (1963), Ford

(1964), Clarke (1966), and Studies on *Cepaea* (1968). The snail species may contain many local populations, each with characteristic color banding patterns on their shells. The transition between regions with different color banding patterns is often extremely abrupt. Morph frequencies may change substantially over distances of 200 meters or less and often without any ready interpretation in terms of changes in the environment. These phenomena in *Cepaea* are referred to collectively as "area effects." The area effects in *Cepaea* raise two of the most difficult problems in the study of clines, why are the clines often so steep, and why are the clines found at places which do not correspond to any known sharp change in the environment? Indeed, these two questions arise in virtually every instance of extensively studied clines. The cline in the Shetland moths mentioned above has been shown to have its steepest part not at the Tingwall Valley as originally conjectured, but somewhat north of the valley in a region not associated with an obviously abrupt change in the environment. See Kettlewell and Berry (1969). Similarly, the steepest change in coloration in *Peromyscus* studied by Sumner occurred not near the coast but beyond twenty miles from the coast where there was no obvious gradient in soil color or other features of the environment that could explain the cline. Moreover, these examples of seemingly "misplaced" clines are not unusual and almost any naturalist can relate similar observations from his or her group of organisms.

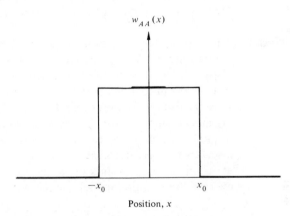

FIGURE 12.11. *Sketch of an environmental "patch."*

An important problem related to that of clines is the issue of the minimum "patch" size to which the population can respond. Figure 12.11 sketches a spatially varying selective pressure in which the A allele is favored between $-x_0$ and x_0 while the other allele is favored everywhere else. We might guess that if the patch is too small, then the influence of dispersal into the patch will swamp local evolution. If the patch is big enough, however, there is opportunity for local evolution of the favored allele. We shall be able to examine this question with the same model that was developed to predict the slopes of clines.

Finally, the theory of clines is relevant to a theory of the process of speciation. Generally, a sexually reproducing species is considered to be a population which does not interbreed with other populations. Moreover, it is understood that the failure of two populations to interbreed must be attributed to inherited traits in each population, like different courtship displays or failure of proper development of the hybrid zygote. The failure of two populations to interbreed because of geographical separation alone is not sufficient grounds to be considered separate species. By this definition the formation of a species is a problem in biological evolution. One original interbreeding population must somehow split into two populations and each of these must evolve *traits* that ensure reproductive isolation between them. A variety of mechanisms can accomplish this result in plants and animals. One mechanism that is traditionally believed to be the most common, geographical speciation, involves population divergence in different parts of the species range. It is also assumed that there are physical barriers to dispersal from one part of the range to another and that the barriers aid the evolution of divergence. Presumably, if the biological divergence becomes large enough while the dispersal is restricted between the diverging populations, then when dispersal is no longer restricted, modifier genes will evolve in both populations to prevent their interbreeding and to complete the speciation process. The theory of clines is critical to the first step in this and other speciation schemes. The theory predicts the conditions under which divergence can occur; and if so, how much divergence occurs as a function of the strength of selection, the amount of dispersal and the effectiveness of a barrier. The theory does not yet address the second step in which modifier genes evolve which prevent hybridization even though there is no longer any physical barrier to dispersal.

The Model The model for selection in a spatially varying environment should incorporate the familiar equation for selection at each place, together with dispersal. Let $M(u)$ be the probability that an individual disperses a distance u. Thus the probability that an individual disperses from y to x is $M(x - y)$, since the distance between x and y is simply $x - y$. Let us also assume that the organisms survive from zygotes to adulthood in some place and then disperse as adults before mating. This assumption is appropriate to may insects which, as caterpillars, eat plants in a local region and then as adults fly about in search of plants on which to oviposit. If the population density everywhere is approximately equal, we can calculate $p_{t+1}(x)$ as follows

$$p_{t+1}(x) = \int M(x - y) \left[\frac{p_t(y)w_{AA}(y) + q_t(y)w_{Aa}(y)}{\bar{w}(y)} \right] p_t(y)\, dy \qquad (12.39)$$

This formula asserts that natural selection occurs at each place, y, according to the usual formula where $w_{ij}(y)$ is the selective value of type A_iA_j at location y, and then migration occurs. This is the basic model for selection in a spatially varying environment. It is written here for one dimension, for example, a coast or valley, but also can be written for two dimensions with a double integral.

The dispersal regime $M(u)$ has been measured in a variety of organisms. Figure 12.12 illustrates an $M(u)$ adapted from Blair's study on the rusty lizard, *Sceloporus olivaceus*, from Austin, Texas. Blair's original figure lumps all individuals with a dispersal distance, u, regardless of direction. In drawing Figure 12.12, it was assumed that half of the individuals dispersed to the right and the other half to the left. Since there is no net drift of individuals to the right or left, the mean of $M(u)$ is zero. The standard deviation, l, of $M(u)$ is a measure of the average dispersal distance. For

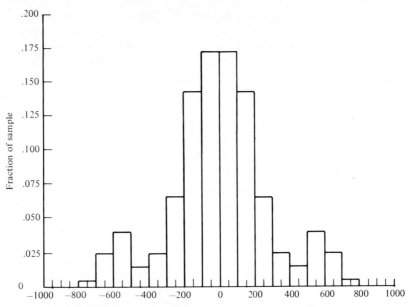

FIGURE 12.12. *The dispersal regime, M(u), for the ♂ rusty lizard* Sceloporus olivaceus *in the vicinity of Austin, Texas. The standard deviation of M(u) is 340 ft.* [Data from W. F. Blair (1960), *The Rusty Lizard*, University of Texas Press, Austin, Tex.]

males l is 339 ft. Note that $M(u)$ is approximately a pair of back to back exponential decay curves. This general appearance of $M(u)$ is commonly observed. See, for example, Cavalli-Sforza and Bodmer (1971, pp. 433–440), and Richardson (1970). $M(u)$ is rarely or never Gaussian as would be expected if the movement of individuals were solely like that of the Brownian motion of molecules. The absence of the rounded and broad peak of a Gaussian distribution may simply indicate that dispersing organisms do not reverse their direction as often as do particles in Brownian motion.

The Diffusion Approximation The basic equation above has not been solved with much success to date. Instead another equation which approximates it has been studied in its place. This new equation is called a diffusion equation and has the following layout:

$$\begin{bmatrix} \text{total} \\ \text{change in } p \\ \text{at location } x \end{bmatrix} = \begin{bmatrix} \text{change in } p \\ \text{due to migration} \end{bmatrix} + \begin{bmatrix} \text{change in } p \text{ due} \\ \text{to local natural selection} \end{bmatrix}$$

(12.40)

We have met diffusion equations before in Chapter 5 on genetic drift. The gene frequency at location x at time t is denoted as $p(x, t)$ and it is understood that time is now continuous, and not discrete. Therefore, the change in p at location x is denoted

$$\begin{bmatrix} \text{total change in } p \\ \text{at location } x \end{bmatrix} \equiv \frac{\partial p(x, t)}{\partial t}$$

(12.41)

The change in p due to migration is obtained by analogizing the movement of individuals to the movement of particles in the process of diffusion.

Briefly, it is assumed that the variance of the function describing the probability of a particle moving a distance u during a time interval Δt is of the form $l^2 \Delta t$. By this assumption, the larger the time interval, Δt, then the wider the dispersal function $M(u)$; and l^2 is, in effect, the variance per unit dispersal time for this function. It can then be shown that the change in p due to diffusion is

$$\left[\begin{array}{c} \text{change in } p \\ \text{due to migration} \end{array}\right] \equiv \frac{l^2}{2} \frac{\partial^2}{\partial x^2} p(x, t) \tag{12.42}$$

The rationale behind the second derivative is as follows. Any change in p at location x must arise from a differential *flow* of individuals from different directions. If we let $J(x)$ denote the flow of individuals across x, we must have $\partial p/\partial t \sim -\partial J(x)/\partial x$. Furthermore, the flow of individuals is itself proportional to a concentration gradient. Thus $J(x) \sim -\partial p/\partial x$. Combining these facts, we have $\partial p/\partial t \sim \partial^2 p/\partial x^2$. The constant of proportionality turns out to be simply $l^2/2$.

To find the change in p resulting from selection, we recall that

$$\Delta p = \frac{pq}{\bar{w}} [p(w_{AA} - w_{Aa}) + q(w_{Aa} - w_{aa})] \tag{12.43}$$

We introduce selective values with the form:

$$w_{AA} = 1 + sg(x)$$

$$w_{Aa} = 1 + hsg(x) \tag{12.44}$$

$$w_{aa} = 1 - sg(x)$$

The s measures the strength of selection and h measures the dominance. By convention, $g(x)$ varies between -1 and $+1$ and its functional form describes the spatial pattern of the selection pressure. For example, a step cline in the selection pressure is represented by a $g(x)$ which equals -1 for x below some x_o and which equals $+1$ for x above x_o. More gradual clines in the selection pressure would be represented by a $g(x)$ which changes continuously from -1 to $+1$ over a region. Substituting these selective values into the formula for Δp gives

$$\Delta p = \frac{pq}{\bar{w}} sg(x)[1 + h(1 - 2p)] \tag{12.45}$$

Now if s is sufficiently small that $\bar{w} \approx 1$, so for weak selection

$$\left[\begin{array}{c} \text{change in } p \text{ due} \\ \text{to selection} \end{array}\right] = sg(x)p(x, t)[1 - p(x, t)]\{1 + h[1 - 2p(x, t)]\} \tag{12.46}$$

Putting all this together, we obtain a differential equation for $p(x, t)$ which, as will be seen, approximates the original model:

$$\frac{\partial p(x, t)}{\partial t} = \frac{l^2}{2} \frac{\partial^2}{\partial x^2} p(x, t) + sg(x)p(x, t)[1 - p(x, t)]\{1 + h[1 - 2p(x, t)]\} \tag{12.47}$$

This equation has been extensively studied in the population genetic literature and was originally introduced by Fisher in 1937. Two kinds of questions have been treated. (1) The *equilibrium pattern* of gene frequencies is found by setting $\partial p/\partial t = 0$ whereupon we obtain an ordinary differential equation for the equilibrium $p(x)$,

$$\frac{l^2}{2}\frac{d^2}{dx^2}p + sg(x)p(1-p)[1+h(1-2p)] = 0 \qquad (12.48)$$

The solution to this equation for a given $g(x)$ and specified boundary conditions yields the equilibrium pattern of gene frequencies. (2) If a new favorable mutation arises at some point in the species range, it will spread from that point into the rest of the population. The movement of the new allele can be viewed as a *traveling wave* of alleles analogous, for example, to the wave in a pond caused by tossing a pebble. The wave of new alleles will move at a velocity set by the parameters in the system. The velocity is found by seeking a solution of the form $p(x - vt)$ where $p(x)$ denotes the profile of the wave. The term $-vt$ in the argument of $p(x)$ indicates that the coordinate frame is moving at a velocity v so that the location of the origin at time t is at vt in terms of the initial coordinate frame. We shall discuss several examples of solutions for the equilibrium pattern and at the end mention some work on the traveling wave problem.

The Step Cline in a Doubly Infinite Environment

As our first example, we consider a step cline in the selection pressure and assume the environmental region is effectively infinite in both directions. The function $g(x)$ then has the form (see Figure 12.13):

$$g(x) = -1 \qquad x < 0$$
$$g(x) = +1 \qquad x > 0 \qquad (12.49)$$

This case was fully solved by Haldane (1948) for arbitrary dominance, h. We present here the special case where $h = 0$. If we substitute $g(x)$ into the equation for the equilibrium pattern (12.48) and set $h = 0$, we obtain

$$\frac{d^2p}{dx^2} = \frac{-2s}{l^2}p(1-p) \qquad \text{for } x > 0$$
$$\frac{d^2p}{dx^2} = \frac{2s}{l^2}p(1-p) \qquad \text{for } x < 0 \qquad (12.50)$$

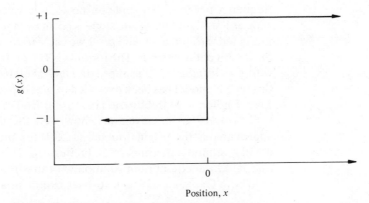

FIGURE 12.13. *Sketch of the spatial pattern of a selection pressure leading to a cline in a doubly infinite environment.*

We seek a solution to this system subject to the boundary conditions

(a) as $x \rightarrow \infty$, then $p \rightarrow 1$ and $\dfrac{dp}{dx} \rightarrow 0$

(b) as $x \rightarrow -\infty$, then $p \rightarrow 0$ and $\dfrac{dp}{dx} \rightarrow 0$ (12.51)

(c) at $x = 0$, $p = \dfrac{1}{2}$ and $\lim\limits_{x \rightarrow 0^+} \dfrac{dp}{dx} = \lim\limits_{x \rightarrow 0^-} \dfrac{dp}{dx}$

The first two conditions mean that the favored allele must approach fixation at locations very distant from the step cline in the selection pressure, and moreover, the slope of the gene frequency curve must be flat at regions very far away from $x = 0$. The third condition specifies how the curves from each side must join up. The solutions on the right and left must meet at $p = \frac{1}{2}$ and the curve and its derivative must be continuous at the junction. The continuity requirement is ensured if the derivative of $p(x)$, when approached from the right, equals the derivative when approached from the left. The solution is straightforward and a step-by-step guide is provided in Problem 2 at the end of this chapter. The solution is[†]

$$p(x) = \frac{-1}{2} + \left(\frac{3}{2}\right)\left[\tanh\left(\frac{\sqrt{s}}{l\sqrt{2}}x + c\right)\right]^2 \qquad (x > 0)$$

$$p(x) = \frac{3}{2} - \left(\frac{3}{2}\right)\left[\tanh\left(\frac{-\sqrt{s}}{l\sqrt{2}}x + c\right)\right]^2 \qquad (x < 0)$$

    (12.52)

where $c = \text{arctanh}\,(\sqrt{2/3}) = 1.14622$. Also, the slope dp/dx at $x = 0$ is the maximum value of the slope and it is

$$\left.\frac{dp}{dx}\right|_{x=0} = \frac{\sqrt{s}}{l\sqrt{3}}$$

    (12.53)

The solution is graphed in Figure 12.14 for several values of s.

    This solution allows us to determine how the steepness of a cline relates to both migration and selection. Suppose the selection coefficient, s, is .1 and the standard deviation of the dispersal function, l, is 339 ft as in male rusty lizards. Then the maximum slope of the cline is 5.39×10^{-4} units of gene frequency per foot. For some purposes, it is acceptable to approximate the cline as a straight line whose slope is given by Equation (12.53). If this line is projected through $p = 0$ and $p = 1$ we can derive an approximate formula for the length of the cline, L. The formula is $L \approx l\sqrt{3}/\sqrt{s}$. So the length of a cline with $s = .1$ and $l = 339$ is approximately 1860 ft or $\approx \frac{1}{2}$ km. The analysis of this type of model has been extended to other spatial patterns of selection by May, Endler, and McMurtrie (1975) and Endler (1977).

    Slatkin (1973) has examined how well the diffusion equation (12.48) approximates the original model (12.39) for the case of a step cline in a doubly infinite environment. He iterated (12.39) on a computer and compared the equilibrium solution with that predicted by (12.52). Slatkin used both Gaussian and back-to-back exponentials for the dispersal regime,

---

[†] tanh (x) is called the hyperbolic tangent. By definition $\tanh(x) = (e^x - e^{-x})/(e^x + e^{-x})$. The inverse hyperbolic tangent is denoted arctanh (x). It is computed as arctanh $(x) = (\frac{1}{2}) \ln[(1+x)/(1-x)]$, provided that $x^2 < 1$.

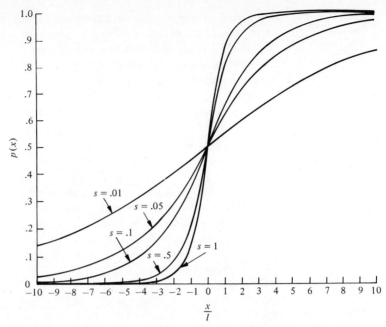

FIGURE 12.14. *The gene frequency cline in a doubly infinite environment produced by the step function selection pressure illustrated in Figure 12.13.*

$M(u)$. The results are recorded in Table 12.3. Notice that the solution from the diffusion equation is extremely accurate for weak selection with a Gaussian dispersal regime. The disparity between the diffusion model and the original model increases with the strength of selection and with the fourth moment of the dispersal function. A high fourth moment indicates a

Table 12.3. Computer Solutions from Slatkin (1973).

x/l	Formula (12.52)	Computer-Gaussian $M(u)$	Computer-Exponential $M(u)$
		$s = .05$	
0	.500	.500	.500
.5	.561	.557	.544
1.5	.666	.660	.624
2.5	.749	.743	.691
4.5	.861	.856	.795
6.5	.924	.921	.865
8.5	.959	.957	.913
10.5	.978	.977	.943
		$s = .1$	
0	.500	.500	.500
.5	.585	.576	.561
1.5	.720	.708	.665
2.5	.814	.804	.747
4.5	.921	.945	.895
6.5	.967	.990	.968
8.5	.987	.998	.990

dispersal function with a comparatively thick tail and narrow peak (for a given variance) as exemplified by the back-to-back exponential form. Slatkin (1973) has also provided a derivation of the diffusion equation from the original model which explicitly highlights the assumption of weak selection and Gaussian dispersal.

The Step Cline in a Finite-Infinite Environment

The next example of a cline is for an environment that is finite on the left and infinite on the right. For this case $g(x)$ is sketched in Figure 12.15. The function $g(x)$ equals 1 from x equal to 0 to b, and then equals -1 from b to ∞. This case has been solved by Nagylaki (1975) and is very interesting for two reasons. First, we may expect that if b is small enough and if the dispersal distance, l, is large enough, then the response to selection within the region $(0, b)$ will be swamped by migrants from outside. If so, perhaps no cline will exist and the a allele will be fixed throughout the species range. Second, the solution to this case is in fact also the solution to the problem of the environmental pocket posed earlier. Figure 12.16 sketches a $g(x)$ for a pocket. The solution for the finite-infinite cline of Figure 12.15 can be joined with its mirror image to produce the solution for the pocket of Figure 12.16. Therefore, the conditions for the existence of a cline for Figure 12.15 also

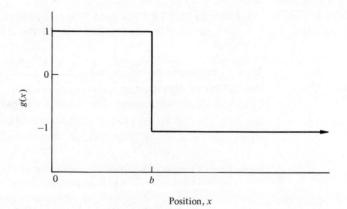

FIGURE 12.15. *Sketch of the spatial pattern of selection pressure leading a cline in an environment that is finite on the left and infinite on the right.*

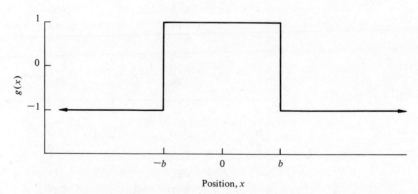

FIGURE 12.16. *Sketch of the selection pressure for an environmental patch obtained by combining the pressure illustrated in Figure 12.15, with its reflection around the vertical axis.*

provide the condition for the minimum pocket size that can be "tracked" by the gene pool of a population.

To analyze the cline of Figure 12.15, consider a new variable, $z = x/b$. In terms of z, the switch in the selection pressure is at $z = 1$, as sketched in Figure 12.17. Since $dz = dx/b$, we have to solve

$$\frac{d^2 p(z)}{dz^2} + \frac{2sb^2}{l^2} p(z)[1 - p(z)] = 0 \qquad z \text{ in } [0, 1]$$

$$\frac{d^2 p(z)}{dz^2} - \frac{2sb^2}{l^2} p(z)[1 - p(z)] = 0 \qquad z > 1 \tag{12.54}$$

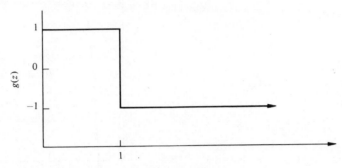

Scaled position, z

FIGURE 12.17. *Sketch of the spatial pattern of the selection pressure illustrated in Figure 12.15. However, for this figure, the length of the finite region is scaled to unit length.*

The solution to this system must have the general appearance of the sketch in Figure 12.18. Specifically, the boundary conditions are

(a) $p(z)$ must be flat at $z = 0$

(b) $p(z)$ and dp/dz must be continuous at $z = 1$ (12.55)

(c) $p(z)$ must tend to zero and be flat as $z \to \infty$

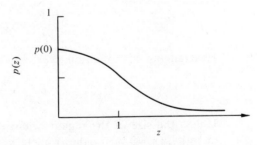

FIGURE 12.18. *Sketch of the gene frequency cline produced by the selection pressure illustrated in Figure 12.17.*

The solution to this problem has been studied by Nagylaki (1975). Figure 12.19 presents the value of $p(0)$ as a function of $\sqrt{2s}\, b/l$. As noted in Figure 12.18, $p(0)$ is the value of the gene frequency at $z = 0$. Note that the frequency of the A allele where it is highest, at $x = 0$, drops precipitously as the quantity $\sqrt{2s}\, b/l$ becomes small.

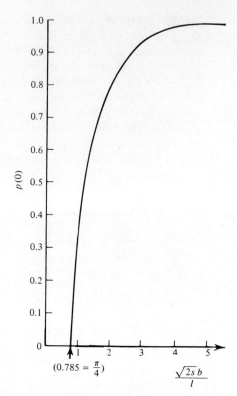

$(0.785 = \frac{\pi}{4})$ $\frac{\sqrt{2s}\,b}{l}$

FIGURE 12.19. *Maximum frequency of the A allele in the finite part of the environment as a function of the length of the finite part of the environment. Note that if the length is sufficiently small, then the A allele is completely* absent *from the region of the environment where it is favored.* [From T. Nagylaki (1975), Conditions for the existence of clines, *Genetics* **80**: 595–615.]

Nagylaki has shown that the condition for the existence of a cline in this case is that

$$\sqrt{2s}\frac{b}{l} > \arctan(1) = \frac{\pi}{4} \tag{12.56}$$

Rearranging this condition to express its implication for b, we have

$$b > \left(\frac{\pi}{\sqrt{32}}\right)\frac{l}{\sqrt{s}} \tag{12.57}$$

Unless the size of the region where the A allele is favored exceeds the quantity, $(\pi/\sqrt{32})l/\sqrt{s}$, the A allele will be completely swamped by migrations from the rest of the population and the a allele will be fixed throughout. Similarly, consider a pocket of length $c = 2b$. Then the pocket length c must exceed $(2\pi/\sqrt{32})l/\sqrt{s}$ if it is to produce any local response in the gene frequency pattern. For example, if l is 339 ft, as with male rusty lizards, and if $s = .1$, then the threshold pocket size is ~1200 ft or ~360 m. If some finite region in the infinite environment is smaller than this, then with $s = .1$, the A-allele will not be represented in the spatial pattern of gene frequencies

As the final example, we consider a cline in a finite environment as sketched in Figure 12.20, where A is favored to the left of the region and a is favored to the right. This case is also especially important because it applies to other situations as well. If we join the solution to this case with its mirror image, we obtain the solution for a pocket of length $2b$ contained in an environment of total length $2c + 2b$ as sketched in Figure 12.21. Moreover, if we repeatedly join solutions of the problem sketched in Figure 12.21, we obtain solutions for a periodic environment as sketched in Figure 12.22, where the region favoring the A allele is of length $2b$ and that favoring a is of length $2c$. The periodic environment model may be particularly useful in

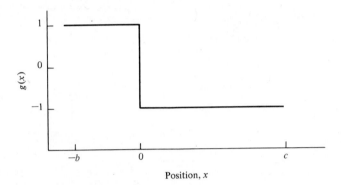

FIGURE 12.20. *Sketch of the spatial pattern of a selection pressure leading to a cline in a finite environment. The allele A is favored to the left and allele a is favored to the right.*

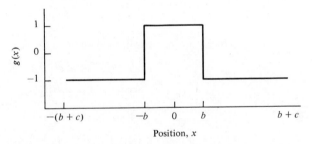

FIGURE 12.21. *Sketch of the spatial pattern of a pocket of length 2b located in an environment of total length 2c + 2b.*

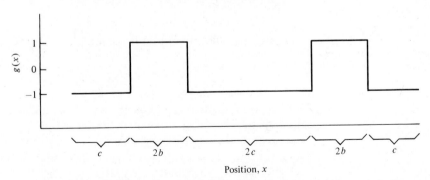

FIGURE 12.22. *The spatial pattern of the selection pressure in a periodic environment.*

explaining the relation between the spatial pattern of dorsal coloration in small mammals and lizards relative to the spatial pattern of substrate color in the environment. As we shall see, there are conditions on the width of the zones, b and c, which allow the dorsal coloration pattern to "track" the environmental pattern. If these conditions are violated the population will fix on whichever allele is adaptive in the most abundant substrate type.

For this problem, we introduce $z = x/b$ so that the left boundary is at -1 and the right boundary is at c/b as sketched in Figure 12.23. The equations for the cline then are

$$\frac{d^2 p(z)}{dz^2} = \frac{-2sb^2}{l^2} g(z) p(z)[1 - p(z)] \qquad (12.58)$$

where $g(z) = 1$ for $-1 \leqslant z < 0$ and $g(z) = -1$ for $0 < z \leqslant c/b$.

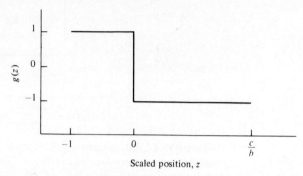

FIGURE 12.23. *The spatial pattern of the selection pressure from Figure 12.20 drawn to a more convenient scale.*

The boundary conditions once again are

(a) $p(z)$ is flat as $z \to -1$

(b) $p(z)$ and dp/dz are continuous at $z = 0$ $\qquad (12.59)$

(c) $p(z)$ is flat as $z \to c/b$

To solve this problem, Nagylaki (1975) applied some recent theorems of Fleming (1975) on finite clines. We shall quote these theorems without proof and then see how they can be used in this context following Nagylaki's treatment.

Fleming's Consider the quantity G that is the integral of $g(x)$ over the entire habitat.
Theorems

$$G = \int_{-b}^{c} g(x)\, dx \qquad (12.60)$$

If G is positive, there is a net selective advantage over the entire habitat to the A allele, while if G is negative, the net advantage is to the a allele. If G is exactly zero, then neither allele has a net selective advantage. In our case, G simply equals $b - c$. Now we might expect that if G is positive (i.e., net selection for A), then sufficiently large migration would lead to the fixation of A while weak migration would allow a polymorphic pattern to persist. Fleming's theorems are a proof of this conjecture. Consider the equation for

$p(x)$ at equilibrium in the form

$$\frac{d^2}{dx^2}p(x)+\lambda g(x)F(p(x))=0 \tag{12.61}$$

where $F(p)=p(1-p)[1+h(1-2p)]$. In our case, λ is $2sb^2/l^2$ and $F(p)$ is $p(1-p)$, since h is assumed to be zero. The quantity, λ, is the parameter that lumps together the selection, migration, and patch size characteristics. Fleming's results may be stated as follows:

1. If there is no net advantage in favor of either allele $(G=0)$, then complete fixation of either allele is an unstable solution. Instead, there is at least one stable polymorphic pattern of gene frequencies. Moreover, suppose that the problem is symmetric in the sense that $g(-x)=-g(x)$, $g(x)\neq 0$ unless $x=0$, and $h=0$ (no dominance). Now consider an equilibrium which is symmetric in the sense that $p(-x)=1-p(x)$. If $g(x)$ is symmetric, then there is only one symmetric equilibrium solution.

2. If there is a net selective advantage in favor of the A allele $(G>0)$, then complete fixation of the other allele, a, is an unstable solution. There is a threshold value of λ, called λ_0 (remember, $\lambda \sim s/l^2$) such that if $\lambda < \lambda_0$ (i.e., selection is weak and/or migration strong), then complete fixation of A is a stable solution. But if $\lambda > \lambda_0$, that is, selection is strong and/or migration is weak, then complete fixation of A is unstable and, instead, there is at least one stable polymorphic pattern of gene frequencies. Suppose that $\lambda > \lambda_0$ so that at least one polymorphic pattern does exist. Then as λ decreases to λ_0, there is a polymorphic pattern which becomes increasingly similar to the state of complete fixation of A. [It is known that there may be more than one stable polymorphic solution for special choices of $g(x)$ and h.]

3. The threshold, λ_0, is found by analyzing a differential equation obtained by "linearizing" around the solution whose stability is under examination. Total fixation of A means a solution where $\hat{p}(x)=1$ for all x. The linearized equation around some given equilibrium solution, $\hat{p}(x)$, is by definition

$$\frac{d^2p(x)}{dx^2}+\lambda g(x)F'(\hat{p}(x))p(x)=0 \tag{12.62}$$

The symbol, $F'[\hat{p}(x)]$, means the derivative of $F(p)$ is evaluated using the given function $\hat{p}(x)$. If $F(p)=p(1-p)$ and $\hat{p}(x)=1$, then $F'(\hat{p}(x))=-1$. In this case we consider

$$\frac{d^2p(x)}{dx^2}-\lambda g(x)p(x)=0 \tag{12.63}$$

There are only certain values of λ for which this equation has a valid solution subject to the boundary conditions that $dp/dx=0$ at the edges of the environment. The smallest positive λ for which a solution exists is the threshold λ_0.† If the actual λ is less than λ_0, then total fixation of A is

---

† In the terminology of differential equations, a value of λ, such that a solution exists to (12.61) subject to the given boundary conditions is called an eigenvalue. For example, the system $y''+\lambda y=0$ subject to $y(0)=0$ and $y(1)=0$ has a solution for $\lambda = \pi^2, 2^2\pi^2, 3^2\pi^2$, and so forth. The smallest positive eigenvalue is π^2. It is also known that the smallest λ may be computed in terms of Raleigh quotients and this technique is perhaps the most useful in practice for obtaining λ_0.

stable; whereas if the actual λ is greater than λ_0, then the fixation of A is unstable and instead there is a stable polymorphic solution.

4. If there is a net selective advantage in favor of a $(G < 0)$, then items 2 and 3 above are modified by interchanging A with a. In particular the threshold λ_0 is found from examining the linearized equation around $\hat{p}(x) = 0$ for all x. Thus when $F(p) = p(1-p)$ then $F'[\hat{p}(x)] = +1$ and λ_0 is then found from

$$\frac{d^2p(x)}{dx^2} + \lambda g(x)p(x) = 0 \tag{12.64}$$

Application to the Step Cline in a Finite Environment It is easy to apply these results to the case of a step cline in a finite environment. Recall that $G = b - c$. Let us examine the situation where $c > b$ so that the net selective advantage over the environment favors the a allele $(G < 0)$. To see whether fixation of a is stable or whether there will be a cline, we seek the smallest λ that allows a solution of (12.62). (Remember that $\lambda = 2sb^2/l^2$.) The function $g(z)$ is sketched in Figure 12.23 so that we obtain

$$\frac{d^2p(z)}{dz^2} + \lambda p(z) = 0 \qquad -1 \le z < 0$$

$$\frac{d^2p(z)}{dz^2} - \lambda p(z) = 0 \qquad 0 < z \le c/b \tag{12.65}$$

subject to the conditions at

(a) $dp/dz = 0$ at $z = -1$

(b) $p(z)$ and dp/dz continuous at $z = 0$ $\qquad$ (12.66)

(c) $dp/dz = 0$ at $z = b/c$

It is a straightforward task to find the smallest λ that allows a solution to (12.65) satisfying the conditions (12.66), and a step-by-step guide is provided in a problem set. Then λ_0 is found as the root of

$$\tan(\sqrt{\lambda_0}) = \tanh\left(\frac{c}{b}\sqrt{\lambda_0}\right) \tag{12.67}$$

Figure 12.24 is a plot of λ_0 as a function of c/b.

Notice that λ_0 very rapidly plateaus at a value which is in fact familiar to us. In the limit as $c/b \to \infty$, $\tanh(c/b\sqrt{\lambda_0})$ tends to 1; λ_0 is then found from

$$\tan(\sqrt{\lambda_0}) \approx 1 \qquad \left(\frac{c}{b} \to \infty\right) \tag{12.68}$$

so

$$\lambda_0 \approx [\arctan(1)]^2 = \left(\frac{\pi}{4}\right)^2 = .617 \tag{12.69}$$

Therefore, in this limit, a polymorphic cline exists if $\lambda > \lambda_0$, that is,

$$2s\frac{b^2}{l^2} > \left(\frac{\pi}{4}\right)^2 \tag{12.70}$$

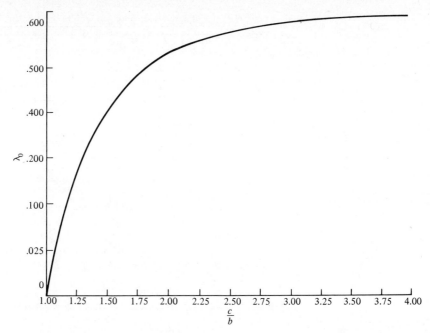

FIGURE 12.24. *The threshold, λ_0, as a function of c/b.*

which in turn reduces to our earlier condition of

$$\sqrt{2}s\frac{b}{l}>\frac{\pi}{4} \tag{12.71}$$

Recall that this is the condition for a cline in a finite-infinite environment discussed earlier (12.56). The rapid rise to a plateau in Figure 12.24 shows that c does not have to be much larger than b in order to treat the environment as effectively infinite.

An Example   These results are very important for predictions of whether a polymorphic pattern of gene frequencies will form in a spatially varying environment or instead, whether the population fixes for the allele which is found in the most common habitat type. As an example, consider the relationship between dorsal color and substrate color. Suppose that there are valleys in a desert where black lava strips alternate with a sandy region. In one valley the pattern is periodic, with $c = b$ as sketched in Figure 12.25. In another valley, $c = 1.5b$, and a third valley $c = 2b$. Which valley will support a spatially polymorphic population and which will fix on a dorsal coloration that blends with the sandy background? From Figure 12.24 we observe that $\lambda_0(1) = 0, \lambda_0(1.5) = .41 \; \lambda_0(2) = .54$. So for a polymorphic pattern, we require

$$2s\frac{b^2}{l^2}>0 \qquad (c=b)$$

$$2s\frac{b^2}{l^2}>.41 \qquad (c=1.5b) \tag{12.72}$$

$$2s\frac{b^2}{l^2}>.54 \qquad (c=2b)$$

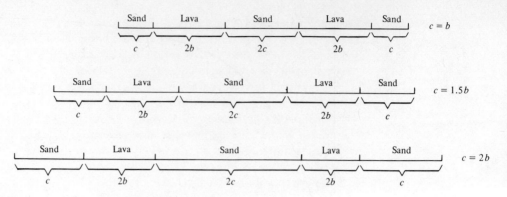

FIGURE 12.25. *Some hypothetical examples of environmental patterns in selection pressure. Suppose that b is 500 ft. According to the theory on clines, a population with the dispersal regime illustrated for ♂ rusty lizards in Figure 12.12 would genetically track the pattern if c = b and if c = 1.5b and would not genetically track the pattern of c = 2b.*

Rearranging these conditions in terms of b, we obtain

$$b > 0 \qquad (c = b)$$

$$b > \sqrt{.41} \frac{l}{\sqrt{2s}} \qquad (c = 1.5b)$$

$$b > \sqrt{.54} \frac{l}{\sqrt{2s}} \qquad (c = 2b)$$

Now suppose, for example, that $s = .1$, $l = 339$ ft as with male rusty lizards, and b, the width of a lava strip, is 500 ft. Then since $(\sqrt{.41})(339)/\sqrt{.2} = 485$ and $(\sqrt{.54})(339)/\sqrt{.2} = 557$, it is clear that a polymorphic pattern will evolve where $c = b$ and $c = 1.5b$, but will not evolve where $c = 2b$. This example is particularly interesting because the absolute size of the lava strips is the same 500 ft in each valley. All that is varying is the width of the sandy zone between the lava strips. Thus the evolution of a spatial pattern of gene frequencies can depend on both the absolute and relative sizes of the regions where the different selection pressures act.

Some Other Points on Clines

We could proceed with more examples of using the diffusion equation (12.48) to predict the population's response to a spatially varying environment. Obviously, many examples could be posed using different $g(x)$ curves, dominance assumptions, and so forth—but there simply is not space here for more. However, it is worth noting certain other conclusions that have been obtained.

1. Increasing the heterozygote fitness flattens and lengthens the cline while decreasing it steepens and shortens the cline, Slatkin revealed (1973). This fact is relevant to explaining very long clines, like that with melanism in lepidoptera of the Shetland Islands discussed earlier where some heterosis is believed present. It is also relevant to explaining why hybrid zones of clines between two quite differentiated populations, are often strangely steep. If hybrids have lower fitness, then the cline should be steeper than otherwise expected.

2. The presence of a barrier to dispersal leads to a discontinuity in the gene frequency curve, Slatkin noted (1973). Other than this result, little theorizing has been done to date on the role of barriers.

3. Increasing the dimensionality of the environment from one to two dimensions increases the ease with which a patch can be swamped by migration, Nagylaki observed (1975).

4. Because the dominance relationship between alleles influences the steepness of a cline, the evolution of dominance also influences the spatial pattern of gene frequencies. Clarke (1966) has explicitly analyzed the evolution of modifier alleles with a view toward their effects on clines. He has found conditions in which modifier genes cause a steepening of clines and has applied this idea to the sharp clines found in the area effects of *Cepaea* snails.

5. All the discussion above is directed to explaining a population's spatial pattern based on a correspondence between the spatial pattern in the population and the spatial pattern in the selection pressure. Although the relationship depends on many factors, there always was, in the discussion above, *some* explanation of the population's pattern in terms of a corresponding environmental pattern. But it is also conceivable that a permanent spatial pattern exists in the population, which is *not* explainable in terms of a corresponding environmental pattern. To see how this possibility arises, recall the analysis earlier in the chapter of selection among weakly coupled subpopulations. Suppose there is an *identical* regime of disruptive selection in every subpopulation. In particular, if there are two subpopulations and two alleles with heterozygote inferiority, then there are two possible polymorphic equilibria. In one of the equilibria, *A* predominates in subpopulation I and *a* in subpopulation II; in the other equilibrium state *a* predominates in subpopulation I and *A* in subpopulation II. Thus the selection pressure is the same everywhere, but the population shows a mosaic spatial pattern of gene frequencies. This scheme requires organisms with little migration and disruptive selection. It may be significant that bird predation on snails exerts a disruptive selection on coloration. This possibility appears relevant to the *Cepaea* area effects.

The Traveling Wave of Alleles   In addition to predicting the equilibrium pattern of gene frequencies, the diffusion model also can predict the spread of alleles in a population. Suppose that a new favorable mutant originates at some point in the population. It will spread through the population in a fashion analogous to a traveling wave produced by dropping a pebble into a pond. The velocity of the wavefront depends on the parameters in the model. If a solution of the form $p(x - vt)$ is substituted into the diffusion equation, conditions on v are obtained in terms of the parameters l and s. The algebra in this topic is developed in a problem at the end of the chapter. It is easy to show that the minimum velocity is given by

$$v_{min} = \sqrt{2sl}$$

It appears that if the alleles are initially localized in space, then the velocity of spread is actually v_{min} and not some greater velocity. If so, the velocity of the wavefront with $s = .1$ and $l = 339$, for example, would be about 150 ft per generation.

This problem is formally identical to the growth of a population in space. References on this topic include Fisher (1937), Kendall (1948), Skellam (1951), and Moran (1962, pp. 178–182).

PROBLEMS

12.1. Migration versus drift.

(a) Show that eigenvalues for Equation (12.12) are

$$\lambda = (1-x)\left(1 - \frac{1}{2N}\right) \pm \frac{1}{2N}\sqrt{4x^2N^2(1-1/N) + (1-x)^2}$$

(b) Show that if $m \ll 1$, then $2xN \gg 1-x$ is equivalent to $4mN \gg 1$ and that $2xN \ll 1-x$ is equivalent to $4mN \ll 1$.

(c) If $4mN \gg 1$, show that $\lambda_1 = 1 - (1-2m)/(2N)$ and that $\lambda_2 = (1-4m) - (1-2m)/(2N)$.

(d) If $h_0 = 1$ and $d_0 = 1$, show that $h_1 = 1 - (1-2m)/N$ and $d_1 = 1 - 2m/N$.

(e) With $h_t = c_1\lambda_1^t + c_2\lambda_2^t$ and if $h_0 = d_0 = 1$, show that $c_1 = 1 - (1-2m)/(8mN)$ and $c_2 = (1-2m)/(8mN)$.

(f) If $4mN \gg 1$, show that

$$h_t = \left[1 - \frac{(1-2m)}{(2N)}\right]^t - (1-2m)\frac{[\lambda_1^t - \lambda_2^t]}{(8mN)}$$

(g) Show that if $4mN \gg 1$, then

$$d_t = \left[1 - \frac{(1-2m)}{(2N)}\right]^t + \frac{[\lambda_1^t - \lambda_2^t]}{(8mN)}$$

(h) If $4mN \ll 1$, show that $\lambda_1 = 1 - 2m(1 - 1/(2N))$ and $\lambda_2 = (1 - 1/N) - m(2 - 1/N)$.

(i) Show that if $4mN \ll 1$, then

$$h_t = \left[1 - \frac{1}{N} - m\left(2 - \frac{1}{N}\right)\right]^t + 2mN[\lambda_1^t - \lambda_2^t]$$

$$d_t = \left[1 - 2m\left(1 - \frac{1}{2N}\right)\right]^t + 2mN[\lambda_1^t - \lambda_2^t]$$

12.2. Solution for a cline in an infinite environment.

(a) Consider the curve for $x > 0$. Define $u \equiv dp/dx$. Then verify that $u\, du/dp = d^2p/dx^2$.

(b) Using this substitution, show that for $x > 0$ we must integrate $u\, du = (-2s/l^2)p(1-p)\, dp$.

(c) Show that the integral to this, satisfying $u \to 0$ and $p \to 1$ as $x \to \infty$, is $u^2 = (\frac{2}{3})(s/l^2)[1 - 3p^2 + 2p^3]$.

(d) Similarly, by substituting $u = dp/dx$ into the equation for $x < 0$ and integrating, show that the solution for u, satisfying $u \to 0$ and $p \to 0$ as $x \to -\infty$, is

$$u^2 = \left(\frac{2}{3}\right)\frac{s}{l^2}[3p^2 - 2p^3]$$

(e) Verify that if $p(0) = \frac{1}{2}$, then $u(0)$ from both solutions are equal, that is, that the two solutions join up correctly in terms of dp/dx, without the need for further requirements, provided that $p(0) = \frac{1}{2}$ in both solutions.

(f) Verify that $u(0) = dp/dx|_{x=0} = \sqrt{s}/(l\sqrt{3})$ from the above solutions.

(g) Now proceed to the equations for $p(x)$. Consider $x < 0$. The following integral formula is the key

$$\int \frac{dx}{x\sqrt{a+bx}} = \frac{-2}{\sqrt{a}} \text{arctanh}\left(\sqrt{\frac{a+bx}{a}}\right)$$

Show that the equation from part (d) can be written as

$$\frac{dp}{p\sqrt{3-2p}} = \sqrt{\frac{2}{3}} \frac{\sqrt{s}}{l} dx$$

Then, using the integral formula, show that

$$\text{arctanh}\left(\sqrt{\frac{2}{3}}\right) - \text{arctanh}\left(\sqrt{1-\frac{2}{3}p}\right) = \frac{\sqrt{s}}{l\sqrt{2}}x$$

is the solution satisfying $p(0) = \frac{1}{2}$.

(h) Rearrange the solution above to obtain Equation (12.52) for $x < 0$.

(i) Substitute $q = 1 - p$ into the equation for u from (c) above and obtain

$$\frac{-dq}{q\sqrt{3-2q}} = \sqrt{\frac{2}{3}} \frac{\sqrt{s}}{l} dx$$

(j) Show that the solution to this satisfying $p(0) = \frac{1}{2}$ is

$$\text{arctanh}\left(\sqrt{\frac{1+2p}{3}}\right) - \text{arctanh}\left(\sqrt{\frac{2}{3}}\right) = \frac{\sqrt{s}}{l\sqrt{2}}x$$

(k) Rearrange the solution above to obtain Equation (12.52) for $x > 0$.

12.3. Condition for cline in a finite environment.

(a) Consider

$$\frac{d^2p(z)}{dz^2} + \lambda p(z) = 0 \qquad -1 \leqslant z < 0$$

$$\frac{d^2p(z)}{dz^2} - \lambda p(z) = 0 \qquad 0 < z \leqslant \frac{c}{b}$$

(b) Show that the solution for $-1 \leqslant z \leqslant 0$ satisfying $dp/dz = 0$ at $z = -1$ is

$$p(z) = A \cos[\sqrt{\lambda}(1+z)]$$

(c) Show that the solution for $0 < z \leqslant b/c$ satisfying $dp/dz = 0$ at $z = b/c$ is

$$p(z) = B \cosh\left[\sqrt{\lambda}\left(\frac{b}{c} - z\right)\right]$$

(d) Show that the continuity of $p(z)$ and dp/dz at $z = 0$ requires, respectively,

$$A \cos(\sqrt{\lambda}) = B \cosh\left[\sqrt{\lambda}\left(\frac{b}{c}\right)\right]$$

and

$$-A\sqrt{\lambda}\sin(\sqrt{\lambda}) = -B\sqrt{\lambda}\sinh\left[\sqrt{\lambda}\left(\frac{c}{b}\right)\right]$$

(e) Show that both these conditions are satisfied by

$$\tan(\sqrt{\lambda}) = \tanh\left(\sqrt{\lambda}\frac{c}{b}\right)$$

12.4. The minimum velocity of a traveling wave of genes.

(a) Show that any solution of the form $p(x - vt)$ satisfies the relation $\partial p/\partial x = -1/v\ \partial p/\partial t$.

(b) Show that substituting $p(x - vt)$ into the diffusion equation with $h = 0$ and $g(x) = 1$ and using the above relation gives

$$\frac{l^2}{2}\frac{d^2}{dx^2}p(x) + v\frac{d}{dx}p(x) + sp(x)[1 - p(x)] = 0$$

(c) Consider locations where $p(x)$ is small so that $1 - p(x) \approx 1$. At these places $p(x)$ must satisfy

$$\frac{l^2}{2}p'' + vp' + sp = 0$$

where the primes indicate derivatives. This is a linear differential equation with constant coefficients. The solution must then be of the form $p(x) = e^{\lambda x}$. Show that λ must satisfy

$$\frac{l^2}{2}\lambda^2 + v\lambda + s = 0$$

(d) If the λ's are a pair of complex numbers, then the solution $p(x)$ must be a sum of sine and cosine and would oscillate into negative values. But $p(x)$ must be positive (or zero). Therefore, the allowable λ's must be real and not complex. Show that the condition for λ to be real is that

$$v^2 - 4\left(\frac{l^2}{2}\right)s \geq 0$$

(e) Rearrange this condition as

$$v \geq \sqrt{2sl}$$

Chapter 13
NATURAL SELECTION IN TEMPORALLY VARYING ENVIRONMENTS

IT is easy to observe that the environment of virtually every population changes from time to time. Presumably, if the environment changes, then the selective values of the various genotypes also change. The presence of variation in the environment becomes particularly interesting if the selective values actually change *direction* from generation to generation. For example, most places have an unusual type of weather every now and then. A late frost may strike, or severe rains and flooding, and so forth. Presumably, there are genes that are favored during these unusual conditions that are not favored during regular conditions. For example, a gene that confers resistance to a late frost may do so by postponing the initiation of growth at the beginning of the growing season. Then during a regular season these late starting genotypes are at a disadvantage relative to the early starters. So the question that arises is whether an environment with a selection pressure of fluctuating *direction* can maintain a polymorphism between the types favored in the different conditions. Put another way, how often must an unusual type of weather occur in order to maintain genes in the population which confer adaptation to this unusual weather type? If the unusual weather type is rare enough, then the genes that are favored in the unusual conditions will be lost.

Another question to be investigated is that of the "optimum strategy" in a variable environment. We may imagine that a population's response to a varying environment can take three forms. The population may be monomorphic for a phenotype which is adapted to the average environmental condition, or it may be monomorphic for a phenotype which is specialized to the most common type of environmental condition. Furthermore the population may evolve a polymorphic condition, with each morph adapted to one type of environment. These are all possibilities, and the question is, which is best? Our second topic in this chapter concerns some answers to this question.

Polymorphism from Temporal Variation in the Direction of Selection

The basic work on this topic is from Haldane and Jayakar (1963). Suppose that we are given a sequence of selective values for T generations. For example, $w_{AA,t}$, $w_{Aa,t}$ and $w_{aa,t}$ are the selective values in generation t. $(t = 1, 2 \cdots T)$. Then we can ask, for this given sequence of w's, could either gene be eliminated from the population during this time? If the answer is no, then obviously a polymorphic condition has been retained during the time in question. If the answer is yes, then one of the genes may have been eliminated; whether it has or not depends on the initial gene frequency at generation 0. So, we shall seek the conditions under which neither gene can be eliminated during the T generations, and these conditions are *sufficient conditions* for a polymorphism.†

There are two interesting cases. First, we assume that A is fully dominant to a. Recall that in a constant environment, no polymorphism occurs with full dominance. But as we shall soon see, polymorphism can occur in a

† Note the similarity between this problem and that of a "protected polymorphism" in the multiple-niche polymorphism of Levene considered in the last chapter.

fluctuating environment even though there is a complete dominance. Second, we consider the analogue for a fluctuating environment of polymorphism due to heterozygote superiority.

Full Dominance   The easiest way to derive the condition for a polymorphism in a fluctuating environment is to investigate the formula for the *ratio* of the frequencies of the two alleles. As always, p is the frequency of A and q of a. The ratio of the frequencies is p/q and will be denoted as u,

$$u \equiv \frac{p}{q} \tag{13.1}$$

Equivalently, p in terms of u is given by

$$p = \frac{u}{1+u} \tag{13.2}$$

Now since we know an equation that predicts p_{t+1} from p_t and the w's, we can rearrange it to predict u_{t+1} as well. The details are in the box, and the result is

$$u_{t+1} = \frac{u_t(u_t+1)}{u_t + w_t} \tag{13.3}$$

Equation for u_{t+1} where $u = p/q$:

1. Recall that

$$p_{t+1} = \frac{(p_t w_{AA} + q_t w_{Aa})}{\bar{w}} p_t$$

$$q_{t+1} = \frac{(q_t w_{aa} + p_t w_{Aa})}{\bar{w}} q_t$$

2. Consider selective values of the form
$$w_{AA} = 1, \qquad w_{Aa} = 1, \qquad w_{aa} = w_t$$

3. With these selective values p_{t+1} and q_{t+1} become

$$p_{t+1} = \frac{1}{\bar{w}} p_t$$

$$q_{t+1} = \frac{q_t w_t + p_t}{\bar{w}} q_t$$

4. Dividing the equation for p_{t+1} by that for q_{t+1} gives

$$u_{t+1} = \frac{u_t}{q_t w_t + p_t} = \frac{u_t}{w_t + p_t(1 - w_t)}$$

5. Substituting $p_t = u_t/(1 + u_t)$ and simplifying yields

$$u_{t+1} = \frac{u_t(u_t+1)}{u_t + w_t}$$

In the formula we have assumed selective values of the form

$$w_{AA,t} = 1, \qquad w_{Aa,t} = 1, \qquad w_{aa,t} = w_t \tag{13.4}$$

Part 3   **Special Topics in Evolution**

Thus we have, by convention, scaled the selective value of the dominant phenotype to equal one, and w_t is the relative selective value of the *recessive* phenotype at generation t. Also, w_t is greater than one when the recessive phenotype is favored, and less than one when it is not favored.

Protection of the a Allele First, we shall use this formula to determine the condition under which the a allele cannot be eliminated. We want the condition for the frequency of a to increase provided a is sufficiently rare. Restated in terms of the ratio, $u = p/q$, an increase of the a allele means a decrease in u, and as a becomes rare, u tends to infinity. So, we want the condition for u to decrease, provided u is sufficiently large. This condition is easy to obtain, although some of the algebra is messy. First, we arrange the formula for u_{t+1} into one for Δu. (The details are in the box.)

---

1. Develop the equation for Δu_t,

$$\Delta u_t = u_{t+1} - u_t = \frac{u_t(u_t + 1)}{u_t + w_t} - u_t$$

$$= \frac{u_t(u_t + 1) - u_t(u_t + w_t)}{u_t + w_t}$$

$$= \frac{u_t - u_t w_t}{u_t + w_t} = \frac{(1 - w_t)u_t}{u_t + w_t}$$

2. This should be rearranged still further: Expand the numerator by adding and subtracting $w_t(1 - w_t)$:

$$\Delta u_t = \frac{(1 - w_t)u_t + w_t(1 - w_t) - w_t(1 - w_t)}{u_t + w_t}$$

$$= \frac{(1 - w_t)(u_t + w_t)}{u_t + w_t} - \frac{w_t(1 - w_t)}{u_t + w_t}$$

$$= 1 - w_t - \frac{w_t(1 - w_t)}{u_t + w_t}$$

$$\Delta u_t = 1 - w_t - \frac{w_t(1 - w_t)}{u_t + w_t} \tag{13.5}$$

Now if a is rare, then u approaches infinity and therefore the term with u_t in the denominator tends to zero, giving

$$\Delta u_t \approx 1 - w_t \qquad (a \text{ rare}) \tag{13.6}$$

Next we cumulate the results over the T generations by adding all the Δu_t, therefore obtaining

$$u_T - u_0 = \sum_{t=1}^{T} \Delta u_t = \sum_{t=1}^{T} [1 - w_t] = T - \sum_{t=1}^{T} w_t \tag{13.7}$$

Then we observe that u decreases over the T generations if the right-hand side of (13.7) is negative, that is, if

$$T - \sum_{t=1}^{T} w_t < 0 \tag{13.8}$$

This condition can be simply rearranged as

$$\frac{1}{T} \sum_{t=1}^{T} w_t > 1 \qquad (13.9)$$

This formula simply states that the a allele cannot be eliminated during the T generations if the (arithmetic) average fitness over the T generations is greater than 1. This result is, of course, quite reasonable. If, on the average, the aa phenotype is more fit than the A phenotype, then the a allele is not eliminated.

Protection of the A Allele   But the next result is somewhat more surprising. Let us now determine the condition under which the A allele cannot be eliminated. We now want the condition for the frequency of A to increase, provided A is rare. Restated in terms of the ratio, $u = p/q$, an increase of A means an increase in u and if A is rare, then u is near 0. So, we seek the condition for u to increase, provided u is sufficiently small. Again the condition is easy to obtain, apart from some messy algebra. We first develop a formula for the ratio of u at consecutive times, u_{t+1}/u_t. Divide both sides of (13.3) by u_t and then rearrange as (see the box) to obtain

$$\frac{u_{t+1}}{u_t} = \frac{1}{w_t\{1 + [u_t(1+w_t)]/[(1+u_t)w_t]\}} \qquad (13.10)$$

---

1. We develop an equation for u_{t+1}/u_t,

$$\frac{u_{t+1}}{u_t} = \frac{u_t(u_t+1)}{u_t + w_t} \frac{1}{u_t}$$

$$= \frac{u_t + 1}{u_t + w_t}$$

2. Expand the denominator by adding $u_t w_t - u_t w_t$,

$$\frac{u_{t+1}}{u_t} = \frac{u_t + 1}{u_t + w_t + u_t w_t - u_t w_t}$$

$$= \frac{u_t + 1}{w_t(u_t + 1) + u_t(1 - w_t)}$$

$$= \frac{1}{w_t + [u_t(1 - w_t)]/(u_t + 1)}$$

$$= \frac{1}{w_t\{1 + [u_t(1 - w_t)]/[w_t(u_t + 1)]\}}$$

Now if A is rare then u_t is near zero and we have

$$\frac{u_{t+1}}{u_t} \simeq \frac{1}{w_t} \qquad (A \text{ rare}) \qquad (13.11)$$

Next we cumulate over the T generations by multiplying. For example, $u_{t+2}/u_t = (u_{t+2}/u_{t+1})(u_{t+1}/u_t)$, so we obtain

$$\frac{u_T}{u_0} \simeq \frac{1}{w_1 w_2 w_3 \cdots w_T} \qquad (13.12)$$

Then the condition for u to increase over the T generation is that the right-hand side must be larger than one,

$$\frac{1}{w_1 w_2 w_3 \cdots w_T} > 1 \qquad (13.13)$$

This condition can be arranged as

$$w_1 w_2 w_3 \cdots w_T < 1 \qquad (13.14)$$

And finally, if we take the Tth root of both sides of (13.14), we obtain

$$\sqrt[T]{w_1 w_2 w_3 \cdots w_T} < 1 \qquad (13.15)$$

This condition says that the A allele will not be eliminated over the T generations if the geometric mean of the fitness of the aa phenotype is less than one.†

When we compare the two conditions (13.9) and (13.15), we observe that a polymorphism can exist during the T generations, even though there is full dominance. For example, suppose that the recessive phenotype represents an early germinating annual plant and that the dominant type is late germinating. Consider three years and suppose a late frost occurs in the second year which kills most of the early germinating plants. Then the relative fitness of the recessive type for these three years might be 1.5, .3, and 1.5. The average fitness of the early germinating phenotypes then is $(1.5 + .3 + 1.5)/3 = 1.1$. Since the average fitness of the early germinating type is greater than one, it was not eliminated during the three generations (provided genetic drift was negligible). However, the geometric mean fitness of the early germinating type is $[(1.5)(.3)(1.5)]^{1/3} = .88$. Since the geometric mean is less than one, the dominant allele that produces later germination was not eliminated either during the three generations. Thus this environmental regime preserved a polymorphic condition, even though there is full dominance.

In general terms, polymorphism is promoted if the recessive type is most fit, on the average but is subject to occasional catastrophes. If it is most fit on the average, then the arithmetic mean is greater than one. If the recessive type is subject to rare catastrophes, then the geometric mean becomes less than one.

Heterozygote Superiority

The conditions for polymorphism due to heterozygote superiority in a fluctuating environment are also easy to determine. In generation t, the selective values are of the form $w_{A,t}$, 1, $w_{a,t}$. The quantities are scaled to make the heterozygote fitness equal to one. Again we are given the w's for T generations and ask whether either allele could be eliminated during these T generations. As before, we work with u instead of p and q. In this case, the standard formula for p_{t+1}, when rearranged into the formula for u_{t+1}, becomes

$$u_{t+1} = \frac{u_t w_{A,t} + 1}{w_{a,t} + u_t} u_t \qquad (13.16)$$

---

† We took the Tth root of (13.13) just to follow custom; it is not necessary for any mathematical reason. It is customary to state the polymorphism condition as a contrast between arithmetic and geometric means.

First, we determine whether the a allele can be lost. If a is rare, then u approaches infinity. We want the condition for u to decrease provided u is sufficiently large. It can be shown that u_{t+1}/u_t can be arranged in the form

$$\frac{u_{t+1}}{u_t} = w_{A,t} + \frac{1 - w_{A,t}w_{a,t}}{u_t + w_{a,t}} \tag{13.17}$$

So if a is sufficiently rare, this becomes

$$\frac{u_{t+1}}{u_t} \simeq w_{A,t} \qquad (a \text{ rare}) \tag{13.18}$$

Then multiplying to cumulate over the T generations, we obtain

$$\frac{u_T}{u_0} = w_{A,1}w_{A,2}w_{A,3} \cdots w_{A,T} \tag{13.19}$$

So for u to decrease, we require the right-hand side to be less than one. As before, we take the Tth root of the right-hand side to obtain a geometric mean,

$$\sqrt[T]{w_{A,1}w_{A,2}w_{A,3} \cdots w_{A,T}} < 1 \tag{13.20}$$

This condition asserts that the a allele cannot be lost during the T generations if the geometric mean of the fitness of the AA homozygote is less than one.

Similarly, we can determine whether the A allele can be lost. If A is rare, then u is near zero, and so we want the condition for u to increase, provided u is sufficiently small. Again it can be shown that u_{t+1}/u_t can be arranged as

$$\frac{u_{t+1}}{u_t} = \frac{1}{w_{a,t}} + \frac{u_t(w_{a,t}w_{A,t} - 1)}{w_{a,t}(u_t + w_{a,t})} \tag{13.21}$$

Now if A is rare, this becomes

$$\frac{u_{t+1}}{u_t} \simeq \frac{1}{w_{a,t}} \qquad (A \text{ rare}) \tag{13.22}$$

Then multiplying to cumulate over the T generations, we obtain

$$\frac{u_T}{u_0} = \frac{1}{w_{a,1}w_{a,2}w_{a,3} \cdots w_{a,T}} \tag{13.23}$$

So for u to increase, we require that the right-hand side exceed one. Then, as before, taking the Tth root of the right-hand side we obtain

$$\sqrt[T]{w_{a,1}w_{a,2}w_{a,3} \cdots w_{a,T}} < 1 \tag{13.24}$$

This condition asserts that the A allele cannot be lost during the T generations if the geometric mean fitness of the aa homozygotes is less than one.

Combining these conditions, we observe that a polymorphic state occurs if the geometric mean fitness of each homozygote is less than that of the hetrozygote. Once again, this requirement is easily met. Even if the (arithmetic) average fitness of one or both homozygotes exceeds that of the heterozygote, there may still be polymorphism, provided rare catastrophes are severe enough. For example, if only one out of a thousand generations is such that the homozygotes have a sufficiently low viability, then polymorphism will be retained even though the heterozygote is not highest in fitness during the other 999 generations. Because the conditions for polymorphism involve the geometric mean, the occasional rare catastrophe must be viewed as a potentially important evolutionary force.

Some Extensions

Very recently, these basic results of Haldane and Jayakar have been extended in various ways by Gillespie (1972, 1973, 1977a, and 1977b), Hartl and Cook (1973), and by Karlin and Liberman (1974 and 1975).

One extension overcomes the need to refer to a particular sequence of selective values during T generations. Instead, the three selective values in any generation, w_{AA}, w_{Aa}, w_{aa}, are regarded as random variables with a description in terms of probabilities. For example w may be assumed to have a constant mean and variance, and a fixed regime of serial correlation among the values at consecutive times. (That is, the w's may be a second-order stationary stochastic process.) With these assumptions, conditions for polymorphism which are extensions of the Haldane and Jayakar conditions have been derived. See Karlin and Liberman (1974 and 1975) for more detail.

Another extension concerns genotypes that have the same (arithmetic) *average* selective value but that may differ in variance. If w_{AA}, w_{Aa}, w_{aa} are random variables with the same mean but possibly different variances, then the alleles, A and a, are said to be *quasi-neutral*. Hartl and Cook (1973) have shown that, in spite of having the same average fitness, there is a systematic and strong selection pressure for the type with the *lowest* variance. Their result may be intuited, although not rigorously, from the Haldane and Jayakar criterion. Consider selective values in the form $1 + \sigma_{AA}x$, $1 + \sigma_{Aa}x$, $1 + \sigma_{aa}x$, where x is a random variable with zero mean and a distribution such that w is never negative. Then the arithmetic average fitness of all three is the same. But the geometric mean fitness is lowest for the type with the highest σ because this type will incur the most severe catastrophes. Thus we might expect that the type with the highest σ is selected against and that with the lowest is favored. This result may have important implications for the evolution of the stability characteristics of populations.

The Fitness Set and Strategy Reasoning

A useful and simple graphical technique for the study of variable environments was introduced by Levins (1968). The problem is to predict the outcome of evolution in a variable environment in terms of "strategy reasoning." Consider an environment that occurs in two states, for example, patches of dark and light soil, night versus day, and so forth. Also, consider a trait which is relevant to the variable environment, that is, dorsal coloration with respect to soil types, and so forth. The evolutionary response of the population could take two forms. The population might consist entirely of individuals with an intermediate phenotype, or it might consist of a mixture of two phenotypes each specialized in one of the environmental states. Other possibilities could also be imagined. Let us agree to call a conceivable evolutionary response by the population as a *strategy*. Now we also have discovered that natural selection maximizes certain quantities during evolution in a variable environment. Recall, for example, that the geometric mean fitness is maximized in the Levene multiple-niche polymorphism model. The problem, then, is to find the *strategy* which maximizes some quantity, like the geometric mean fitness, which is believed to be maximized during evolution by natural selection. Thus strategy reasoning is an attempt to predict the outcome of evolution and not the dynamics of evolution. The assumption underlying strategy reasoning is that the result of evolution is to realize an optimum strategy.

Clearly, there must be three stages in strategy reasoning. First, we must describe all the conceivable strategies, second, we must introduce the optimization principle, that is, that quantity which we believe to be maximized during the course of evolution, and third, we must find the particular strategy which is best according to the optimization principle. Levins (1968) introduced a convenient graphical scheme that helps in carrying out these stages.

Graphical Representation of the Strategies First, we consider the graphical representation of all the possible strategies. Suppose the state of the environment is measured by E. For example, let E stand for the amount of black color in the soil. Let X be a measure of the trait. For example, let X stand for the amount of black pigment in the dorsal pattern of an animal. Now, the fitness of an individual depends on both its dorsal coloration and the substrate coloration. Let $w(X, E)$ describe an individual's fitness as a function of both X and E. Typically, for any E, fitness is a more or less bell-shaped curve as a function of the phenotype, X. Figure 13.1 illustrates w as a function of X for fixed E. Presumably, the optimum amount of dorsal pigment for any fixed E is that which allows the animal to be least conspicuous to predators and so forth. Now let us consider two environments, E_1 and E_2, where E_2 is a darker substrate than E_1. Figure 13.2 illustrates w as a function of X in each environment. Obviously, the best color in E_1 is lighter than the optimum color in E_2. If E_1 and E_2 are very similar, then the optimum phenotypes in each of these environments must be similar to one another (Figure 13.3, top) while if E_1 and E_2 are very dissimilar, then the optimal phenotype for each environment will also be very dissimilar (Figure 13.3, bottom).

FIGURE 13.1 *Sketch of the fitness of different values of the character X in environment E.*

Levins has suggested that the basic information in Figures 13.1–13.3 should be graphed as follows. On the horizontal axis we plot the fitness of phenotype X in E_1 and on the vertical axis we plot the fitness of phenotype X in E_2. Then for each possible value of X we can plot a point representing the

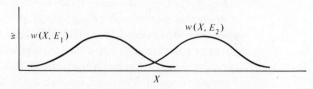

FIGURE 13.2. *Sketch of fitness as a function of X in two environments, E_1 and E_2.*

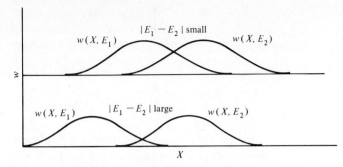

$|E_1 - E_2|$ small

$w(X, E_1)$

$w(X, E_2)$

w

$w(X, E_1)$

$|E_1 - E_2|$ large

$w(X, E_2)$

X

FIGURE 13.3. *Sketch illustrating how the curves describing fitness in each environment are placed relative to one another as a function of the difference between the environments.*

fitness of this phenotype in each of the two environments. The set of points for all possible values of X forms a curve. Indeed we can represent this curve as a vectored-valued function $\alpha_{E_1,E_2}(X)$,

$$\alpha_{E_1,E_2}(X) = (w(X, E_1), w(X, E_2)) \tag{13.25}$$

Figure 13.4 illustrates the curve which results if E_1 and E_2 are very similar. The arrow indicates the end of the curve that corresponds to the high values of X. Thus as X goes from a low to a high value, the curve starts at the origin, moves to the right, then heads northwest, and finally drops back to the origin. Levins has termed the curve, $\alpha_{E_1,E_2}(X)$, as the *fitness set*. It represents the fitness in both environments for every phenotype in the set of all possible phenotypes.

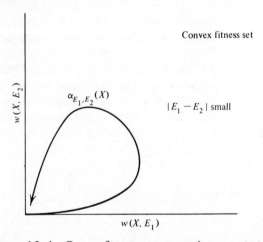

Convex fitness set

$\alpha_{E_1,E_2}(X)$

$|E_1 - E_2|$ small

$w(X, E_2)$

$w(X, E_1)$

FIGURE 13.4. *Convex fitness set representing pure strategies.*

Typically, there are two shapes to the curve $\alpha_{E_1,E_2}(X)$. If E_1 is very close to E_2, then the shape is that of Figure 13.4. This shape is called a *convex fitness set*. If E_1 and E_2 are very dissimilar, then Figure 13.5 results. This is called a *concave fitness set*. Thus similar environments produce convex curves and dissimilar environments produce concave curves.

The fitness set, $\alpha_{E_1,E_2}(X)$, represents the fitness in both environments of a single phenotype. What remains is to represent the fitness, in both environments, of *mixtures* in the population of two phenotypes. Now

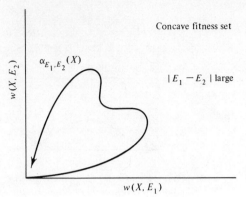

FIGURE 13.5. *Concave fitness set representing pure strategies.*

suppose phenotype X_1 is present in frequency p and X_2 is found in frequency $(1-p)$. Then the fitness of this mixture in environment E_1 is assumed to be

$$w_{E_1, X_1, X_2}(p) = pw(X_1, E_1) + (1-p)w(X_2, E_1) \qquad (13.26)$$

And similarly, the fitness of a mixture of X_1 and X_2 in environment E_2 is

$$w_{E_2, X_1, X_2}(p) = pw(X_1, E_2) + (1-p)w(X_2, E_2) \qquad (13.27)$$

We can add this information to our graph. For any p, which describes the mixture ratio, we can plot the fitness in E_1 on the horizontal axis and the fitness in E_2 on the vertical axis. Thus, for each p, we obtain a point in the graph of w_{E_1} versus w_{E_2}. We let

$$\beta_{E_1, E_2, X_1, X_2}(p) = [w_{E_1, X_1, X_2}(p), w_{E_2, X_1, X_2}(p)] \qquad (13.28)$$

$\beta(p)$ is a vectored-valued function of $p\dagger$. In fact $\beta(p)$ has a very simple shape; it is simply a straight line connecting the points $\alpha(X_1)$ and $\alpha(X_2)$. In Figure 13.6 $\beta(p)$ is sketched as a straight line between $\alpha(X_1)$ and $\alpha(X_2)$. As p varies from 0 to 1, $\beta(p)$ moves from $\alpha(X_2)$ to $\alpha(X_1)$ as indicated by the arrow. One case of special interest arises when the phenotypes in the mixture are actually the optimum phenotypes in each separate environment. If E_1 and E_2 are different enough to produce a concave fitness set, then the line, $\beta(p)$, representing mixtures of the specialists in E_1 and E_2 lies outside

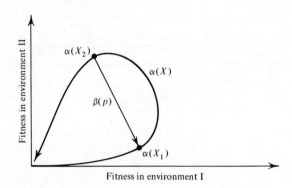

FIGURE 13.6. *Convex fitness set, together with the curve $\beta(p)$, which refers to mixed strategies.*

† Note that we omit the subscripts for ease of writing.

the $\alpha(X)$ curve as illustrated in Figure 13.7. We have now finished the first stage in this strategy reasoning approach. We have been able to represent all the strategies under consideration as points on a graph. $\alpha(X)$ contains the points for a population monomorphic for phenotype X, and $\beta(p)$ contains the points representing a polymorphism of two phenotypes in frequencies p and $1-p$.

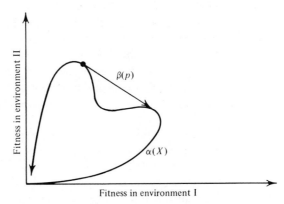

FIGURE 13.7. *Concave fitness set together with the curve $\beta(p)$, which represents mixed strategies.*

Some Optimization Principles

The second stage is to introduce some optimization principles. From our experience with selection in varying environments both in the last chapter and in this chapter, we can suggest two different optimization principles. Our first suggestion is the geometric mean fitness in the different environments. Let c_1 be the proportion of environment E_1, and c_2 the proportion of environment E_2. For example, c_1 could be the fraction of soil with color E_1, and c_2 the fraction of soil color E_2. ($c_1 + c_2 = 1$.) Also, c_1 could be the fraction of time that a place is in state E_1 and similarly for E_2. Thus c_1 and c_2 may refer to either spatial or temporal variation in the environment. In the last chapter we learned in the Levene multiple niche polymorphism model that selection maximizes the geometric mean fitness

$$A = w_1^{c_1} w_2^{c_2} \tag{13.29}$$

Also, in this chapter, we learned from the Haldane and Jayakar criteria that the relations between the geometric means of the fitness control the existence of polymorphism. Thus it is reasonable to suspect that natural selection in a variable environment occasionally maximizes the geometric mean fitness.

Another candidate is the arithmetic mean fitness,

$$A = c_1 w_1 + c_2 w_2 \tag{13.30}$$

We saw in the last chapter that the arithmetic mean fitness is maximized by natural selection under the "constant-zygote-number" formulation of the multiple-niche polymorphism model. We might also expect that the arithmetic mean is maximized if the environmental "patches" are small relative to the organisms. Suppose, for example, that the fertility of an individual is proportional to the amount of food the organism collects. Also, suppose the foraging range of an individual includes two kinds of microhabitats differing in resource level. Then the total amount harvested may be an average of the amount of each microhabitat contained in an individual foraging range. In this sense, small environmental patches may lead to effective fitnesses which are the arithmetic average of the fitnesses within each patch.

The problem of choosing the appropriate optimization principle is very tricky. There are at least two and probably many more reasonable choices of optimization principles. And even the applicability of the two principles above depends on the details of the situation; namely, the geometric principle with the constant-fertile-adult formulation and the arithmetic principle with the constant-zygote formulation of the multiple-niche polymorphism model. Nonetheless, a rough guide is to use the geometric principle if the scale of the temporal and spatial variation is large relative to the generation time and the foraging range of the organisms and to use the arithmetic principle for comparatively short-scale temporal and spatial variation. This guide is often expressed by saying that the geometric principle applies to *coarse-grained* environmental variation and the arithmetic principle to *fine-grained* environmental variation.

Finding the Optimum Strategy

The third and last stage in our program of strategy reasoning is to use the optimization principles to determine which particular strategy is best. We shall do this by plotting the optimization function, A, on the same graph together with $\alpha(X)$ and $\beta(p)$. The value of X in $\alpha(X)$ or of p in $\beta(p)$ that leads to the highest A is the optimum strategy. To see how this idea works, let us first plot the optimization function—they are plotted as contours analogous to a topographic map. For example, the arithmetic A is defined as

$$A = c_1 w_1 + c_2 w_2 \qquad (13.31)$$

Now consider a particular value of A, say A_0. Then there are many pairs, (w_1, w_2), which combine to give the same A_0. In fact, the set of pairs of (w_1, w_2) that produce a given value of A can be plotted on the graph of w_1 versus w_2. To do this, we solve (13.31) for w_2 in terms of w_1 for a given A_0,

$$w_2 = \frac{-c_1 w_1}{c_2} + \frac{A_0}{c_2} \qquad (13.32)$$

According to this formula, the pairs, (w_1, w_2), that lead to a given value of A form a straight line with slope $-c_1/c_2$ and intercept A_0/c_2. Figure 13.8 illustrates the straight lines that correspond to different values of A_0. High

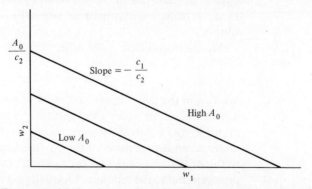

FIGURE 13.8. *Contour lines for the arithmetic mean fitness.*

values of A_0 lead to lines far away from the origin, while low A_0 leads to a line near the origin. Note how the lines can be viewed as contour lines for the optimization function.

The contour lines for the geometric optimization principle are obtained in the same way. The geometric A is

$$A = w_1^{c_1} w_2^{c_2} \tag{13.33}$$

If we solve for w_2 in terms of w_1 for a given A_0, we obtain

$$w_2 = (A_0^{1/c_2}) \frac{1}{w_1^{c_1/c_2}} \tag{13.34}$$

In this case the pairs, (w_1, w_2), which lead to a given A_0, form a hyperbola. Figure 13.9 illustrates the contour lines for this optimization function. Note again that high values of A_0 lead to curves far away from the origin while low values of A_0 produce curves near the origin.

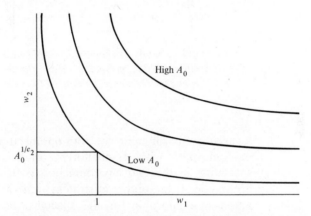

FIGURE 13.9. *Contour lines for the geometric mean fitness.*

Now all that remains is to superimpose the graph of $\alpha(X)$ and $\beta(p)$ onto a graph containing the contour lines for the optimization function. Then we can find the optimum strategy simply by inspection. Four combinations result; they are illustrated in Figure 13.10.

The optimum strategies for each of the four cases in Figure 13.10 are summarized in Table 13.1. With convex fitness sets, the optimum strategy in

Table 13.1.

	Arithmetic A Fine-Grained Variation	Geometric A Coarse-Grained Variation
Convex $\alpha(X)$ Similar environments	Monomorphic for intermediate phenotype	Monomorphic for intermediate phenotype
Concave $\alpha(X)$ Dissimilar environments	Monomorphic for specialist to most common type of environment	Polymorphic with specialists to each type of environment

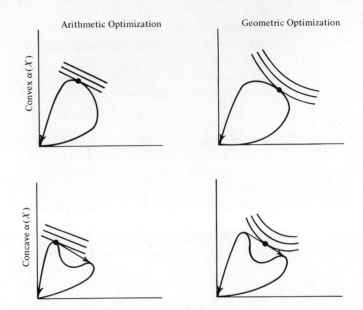

FIGURE 13.10. *All combinations of convex and concave fitness sets with the arithmetic and geometric mean fitness optimization principles. The conclusions for each case are entered in the corresponding position in Table 13.1.*

a variable environment, for both optimizations principles, is a population monomorphic with a phenotype that is intermediate between the specialists in the two environments. With a concave fitness set and an arithmetic A, the optimum strategy in a variable environment is a population monomorphic for the specialist to the most common type of environment. And with concave fitness set and a geometric A, the optimum strategy is given by a point on the $\beta(p)$ curve, thus indicating a polymorphism.

These predictions are very interesting for several reasons. First, it is easy to visualize how they relate to natural populations. Consider again an environment with patches of different substrate color. If the colors of the two patch types are not very different (convex fitness set), then the animals should have a dorsal coloration intermediate between the two substrate colors. But if the colors of the two patch types are very different (concave fitness set), then something else should happen. If the patches are small (arithmetic A), then the animals should have the same dorsal coloration as the most common substrate color, and if the patches are large (geometric A), then the population should be polymorphic with animals dorsally colored according to the two substrate colors. Second, the logic underlying these predictions is very simple. The graphical technique itself involves no biological assumptions. The basic biological assumption is that natural selection obeys the optimization principles which have been introduced. Given this, everything else follows directly, and as we have seen, the optimization principles are quite reasonable. They may not apply all of the time, but we know they must apply in at least some situations because we have specific examples where they do (e.g., the multiple niche models in a spatially varying environment and the Haldane and Jayakar conditions in a temporally varying environment). For these reasons the fitness set approach of Levins has become very popular, especially for ecological problems.

There is one proviso which Levins stressed originally when introducing the fitness set technique. Any prediction of an optimum strategy may not be achieved in nature because of constraints imposed by the genetic system. For example, if the optimum dorsal color is produced by a heterozygous genotype, then a polymorphism must result even though it is not the optimum strategy. Instead, the outcome is a compromise between the optimum strategy and genetic restraints. But the practical importance of this proviso is not well understood. We have little data in how often and for how long the outcome of evolution is constrained by details of the genetic system. Nonetheless, the proviso should be kept in mind.

Chapter 14
THE EVOLUTION OF ALTRUISM:
KIN SELECTION AND GROUP SELECTION

A POPULAR misconception about evolution is that it is a vicious struggle for survival. It would seem that cooperation and kindness are traits destined for elimination by natural selection. But this view is incorrect. First, there are a great many species where cooperation is in fact observed;† and second, we can exhibit many ways in which cooperation can evolve by more or less typical evolutionary processes. This chapter examines the evolution of cooperative relationships between individuals.

An *altruistic* act is an act that is done by an individual for the benefit of some other individual. Moreover, the act *must* involve a cost, in terms of fitness, to the individual who does the act. Thus if an individual gives food that it would otherwise eat to another, then it is an altruistic act, while if an individual gives food or materials that it would not use, then the act is not altruistic, although it could still be considered cooperative. Thus an altruistic act involves two requirements, it must be of benefit to the recipient and of cost to the donor. Our problem is to determine how altruistic behavior patterns can evolve. This chapter is basic to the study of sociobiology, the study of social behavior, as extensively reviewed by Wilson (1975).

Three Kinds of Possible Explanations

There are three kinds of explanations for the evolution of altruistic behavior. First, it may be possible to show that the altruistic act actually leads to a higher expected fitness for the particular individual who does the act. That is, although the act involved some cost when it was done, the individual who behaved altruistically will receive, on the average, a return some time later that exceeds his cost. Thus the individual who acts altruistically is in fact simply increasing his own fitness as cumulated over his own life span. This approach has been explored by Trivers (1971), who termed this idea *reciprocal altruism*. Trivers has discussed the cost-benefit relationships and the population structures that promote this kind of altruism.

The second possible explanation for the evolution of altruism involves showing that the individual benefiting from the altruistic act is a close relative of the donor. Intuitively, natural selection should favor the genotypes of individuals who leave the largest number of *genes* in the next generation. Whether these genes are left in an individual's own immediate offspring is really irrelevant. For example, suppose an individual is in a position to have only one offspring itself while, if it helps its brother or sister raise a family, then a family of 10 can be raised. It is also assumed that the brother or sister would only raise one offspring if it were not given the assistance. We may then intuit that this individual will leave more of its own *genes* in the next generation if it helps its brother or sister raise a family of 10 than if it tries to raise one offspring on its own. The altruism in this case is between close relatives. The altruistic trait evolves because it confers a higher fitness in terms of the number of genes an individual leaves in the next

† All examples, including those to be mentioned in this chapter are with animals. But I suggest that the commonly observed root connections between individual plants (i.e., plants originating from separate seeds) can be viewed in the same terms as social behavior in animals.

generation rather than in terms of the number of immediate offspring. Natural selection in this context is called *kin selection.*

The third possible explanation concerns the survival of the group of individuals who are altruistic to one another as contrasted with groups lacking altruism. We may conceive of altruism as a sacrifice by each individual for the common good. If so, the group as a whole may be expected to survive longer than groups lacking the altruism. If so, then perhaps altruism may evolve in a species because of the differential survival of groups of individuals within the species. The differential survival of groups is called *group selection.* The hypothesis that group selection leads to the evolution of altruism has had an extremely controversial history. Only recently has it become possible to discuss the process of group selection in rigorous terms and to specify the conditions under which it occurs.

The Hypotheses Are Mutually Compatible

To illustrate the interplay among these hypotheses, consider the example of warning calls. Many species of birds show flocking behavior outside of breeding season. Typically, if a bird sees a sign of danger, it sounds a warning call and then the flock escapes from danger. It is presumed that giving a warning call involves danger, and if so the warning call is an altruistic act. How has this trait evolved? First, it is conceivable that giving warning calls is a behavior that is reciprocated among members of the flock. Individuals who sound a warning today may benefit from the warning calls of others tomorrow. Second, it is also conceivable that the birds in any given flock are closely related to one another. If so, a bird who happens to find itself in danger may leave more of its genes to the next generation by warning its close relatives than by trying to escape before giving a warning. Third, if the membership of a flock is sufficiently permanent, then the flocks themselves may be the units of selection. It is conceivable (though not likely) that entire flocks whose members do not sound warning calls perish due to predation more often than flocks whose members do sound calls. Although it seems to most naturalists that entire flocks rarely if ever go extinct, this possibility cannot be dismissed out of hand. We have seen, for example, that selection much weaker than that observable by naturalists in the field nonetheless changes gene frequencies quite rapidly. Perhaps then, even very weak group selection can have an effect if given sufficient time. Thus there are three conceivable hypotheses for the evolution of warning calls, and they may all be operating together. The hypotheses are all perfectly compatible with one another. Separating their contributions to the evolution of any given trait is a difficult problem and requires a more precise theoretical understanding of the hypotheses. We now examine these hypotheses in more detail.

Reciprocal Altruism

Trivers (1971) has examined the conditions that promote the evolution of reciprocal altruism. Consider two strategists. One, the *altruist,* helps others at his own cost, but in expectation of future reciprocation. The other strategist, called the *cheater,* takes whatever benefits are given by the altruists but does *not* reciprocate. The problem is to establish the conditions under which the altruist has a higher fitness, on the average, than the cheater. Intuitively, if the altruists dispense their benefits randomly throughout the population, then the fitness of the cheaters will exceed that of the altruists. In this case all individuals, whether altruistic or not, will receive on the average the same amount of benefit from the altruists, but the cheaters will not have incurred any cost because they do not reciprocate. Therefore, the cheaters will have a higher fitness.

Conditions for Reciprocal Altruism

For altruism to be the better strategy, the following three basic requirements must be met:

1. There must be discrimination by the altruists against the cheaters. The altruists must not randomly dispense their benefits; instead they should dispense only to those who act altruistically when given the opportunity.

2. The ratio of benefit to cost for the altruistic activities must be high. Trivers uses the example of a drowning man. If someone on the shore throws the drowning man a raft, then the benefit to the recipient is huge while the cost to the donor is low. This kind of situation places the least demands on speedy reciprocation because the cost is so low.

3. The opportunity for an altruistic act must be common and occur in both directions equally often for any two parties. That is, if *A* acts altruistically to *B*, the opportunity for *B*'s reciprocation must arise.† Trivers points out that social systems with strong linear dominance hierarchies do not lend themselves to reciprocal altruism (e.g., food sharing) because of the inherent asymmetry in the relationships between individuals produced by the hierarchy.

Trivers mentions two situations that tend to satisfy these requirements. First, animals that are long lived and have low dispersal are likely to show reciprocal altruism because the cheaters can be identified. Moreover, the opportunity for reciprocation is good. Second, animals with strong family units also should show reciprocal altruism (in addition to whatever altruism evolves by kin selection). Again, the strong family units ensure the identification of cheaters and provide opportunities for reciprocation.

Warning Calls Revisited

To stress the importance of the discrimination against cheaters in the theory of reciprocal altruism, let us return to the example of warning calls in birds. Trivers argues that warning calls are *not* examples of reciprocal altruism precisely because he feels that the opportunity for discrimination against cheaters is not available. He writes "It is difficult to visualize how one would discover and discriminate against the cheater, and there is certainly no evidence that birds refrain from giving calls because neighbors are not reciprocating." Thus it is important to understand that the application of the reciprocal altruism hypothesis requires a specification of the cheater discrimination system.

In addition, bird flocks are believed to be very "open," with individuals freely entering, leaving, and interchanging flocks. If so, then as mentioned later, the hypotheses of kin selection and group selection are *also* in difficulty for explaining the evolution of warning calls. Hence Trivers feels that warning calls are *not* altruistic, but instead represent a direct vested interest of animals in preventing their conspecifics from being eaten by predators. It is known that vertebrate predators acquire experience and become more efficient and effective on the prey species they catch most often. For this reason, and perhaps others too, any potential prey individual

† Suppose that *A* acts altruistically to *B*. It is not important that *A* is reciprocated specifically by *B*, as long as someone reciprocates. But the possibility of reciprocation is enhanced if the two parties are in fact the same and that some sense of "obligation" is involved.

has a vested interest on preventing predators from taking other individuals of its own species. Trivers feels that warning calls are an expression of this vested interest.

The cases which Trivers claims do represent reciprocal altruism are the marine cleaning symbiosis and certain human behavior patterns. Trivers' original paper cited at the beginning of this section should be consulted for more details. For the most part, there is no quantitative formulation of the reciprocal altruism hypothesis—for example, what benefit/cost ratio is enough, how strong must the discrimination be against the cheaters, how often must the situations arise in which the altruistic activity occurs, and so forth. For some applications a mathematical formulation of these ideas will be needed. See, for example, Schaffer (1978).

Trivers' (1971) discussion has centered on an individual to individual reciprocation. Eshel (1972) has pointed out that the reciprocation may be spread among many individuals by the social system. The attainment of rank within a social system and the use of the privileges of rank may represent a reciprocation, provided the attainment of rank requires incurring hazard. Thus an individual who risks himself in defense of the colony is viewed as being reciprocated by promotion in rank. Obviously, for this system to make sense in terms of strategy reasoning, there must be a good correlation between rank and reproductive success.

Kin Selection and the Social Insects

Our study of kin selection will be aided by a concrete example of altruism taken from the social insects. We shall then show how kin selection considerations seem to explain surprising features of the behavioral regime in some social insects.

Who Are the Social Insects?

The social insects refer to two very different orders of insects, the Hymenoptera (bees, wasps, and ants) and the Isoptera (termites). Many species of Hymenoptera and all termites exist in large colonies and have individuals that perform different tasks. For example, certain individuals feed the young, others defend the nest and so forth, thus leading to an insect *society*. But the organization of the society is radically different in the Hymenoptera as compared to the Isoptera.

The Hymenoptera are a huge order. Except for bees, most are predatory. The most primitive are the so-called "stingless" wasps. These are typically parasitoids, that is, they deposit eggs in the victim, the larvae mature while gradually consuming the victim, and eventually the larvae kill the victim, complete their metamorphosis into adults and then leave as free-living organisms. A parasitoid is, in effect, intermediate between a parasite and a predator. Female stingless wasps possess an ovipositor, which is used to insert the eggs into the victim. In the rest of the Hymenoptera the ovipositor is modified into a sting; thus only female wasps, bees, and ants have a sting—the males do not. The stingless wasps and many of the stinging wasps are *not* social. The solitary stinging wasps typically make burrows in the ground (the digger wasps), or out of mud (the potter wasps), or in other ways. Each nest or burrow is provisioned with dead prey; then an egg is laid in the burrow and it is sealed off. There is no direct contact between the female and its offspring. The wasps with the simplest social organization are the so-called "paper-wasps," whose nests are simple structures from a paperlike material. The wasps with the most complicated social organization are the hornets and yellow jackets. The bees and ants are also in the Hymenoptera and generally have a more complex social organization than the wasps. Indeed, the biology of some ant species is unbelievably complex, with very large colonies, nests with intricate architecture, and an extensive division of labor in the society. From phylogenetic considerations it is believed that social behavior has evolved 11 independent times within the Hymenoptera. For an extensive and lucid survey of the social insects, see Wilson (1971).

A *caste* is a set of insects in the society who perform the same task. In complex societies the castes are morphologically differentiated. For example, the "soldier" individuals may have large pincers while those who tend the young do not. In simpler societies the castes are only behaviorally differentiated. The first key feature to *all* hymenopteran societies is that one caste is reproductive and the others are sterile. The reproductives are called queens and the other individuals are either workers, soldiers, and so forth. The simplest society consists of two castes, the reproductive queen and the sterile workers. The second key feature to *all* hymenopteran societies is that males are almost absent from the picture. The reproductive in the nest is a female, and the workers are females. Males are occasionally produced but play little or no role in the insect society.

In contrast, the termite society includes both sexes. The reproductive caste includes kings as well as queens, and the sterile castes also contain males. Another important fact about termites is that their diet is principally cellulose. Some species feed directly on the wood in which they nest while others nest in the soil and forage for dead wood, leaves, and so forth. To decompose this cellulose, all termites have evolved a symbiotic relationship with protozoa and bacteria. These microorganisms live in the gut and are lost whenever the exoskeleton is shed. As a result, a termite must be periodically reinfected and each termite relies upon its neighbors as the source of its new supply of symbionts.

The Paper Wasps

The paper wasps, as mentioned above, have the simplest hymenopteran social organization. There are two castes with little morphological differentiation. West Eberhard (1969) has studied the behavior of the common paper wasp of the northeastern United States, *Polistes fuscatus*, and this species provide us with a good concrete example of hymenopteran social behavior. Females survive the winter in a dessicated state. They have been fertilized prior to the winter and carry sperm packets from the males of the preceding season. In the spring one or more wasps begins the nest construction. A stalk is made and the first cell is placed at the end of the stalk. If a nest is founded by two or more wasps, a struggle for dominance soon begins, at the end of which one of the foundresses emerges as the "dominant" wasp. This dominant wasp is the queen and the subordinate wasps become workers and add cells to the nest. According to West Eberhard, the cofoundresses are sisters. The queen deposits eggs in the cells. As the larvae develop, they are fed by the workers. The first offspring to emerge are all females and they, too, function as workers. These offspring do not lay eggs of their own, but instead continue adding to the nest and feeding the larvae. Thus the offspring, who are all sisters to one another, assist their mother in the raising of more sisters rather than raise offspring of their own. But as the season nears its close, three new facts emerge. First, some of the larvae prove to be males. Second, these adult males are not fed by the workers—and the males themselves do not function as workers. Third, the workers are seen occasionally attempting to raise offspring of their own and indeed may cannibalize the nest to obtain food for the raising of their own offspring. Thus the end of the season is marked by the appearance of males and the breakdown of the social organization that had persisted for most of the season.

Three Questions to Answer Clearly, there are many puzzling features to the social behavior of paper wasps and other social Hymenoptera. Among the questions we might ask are (1) Why are there sterile castes? Why do the workers help other insects raise offspring rather than raise offspring of their own? (2) Why is the hymenopteran society female-dominated? Why are the offspring only female until the very end of the season? Why do only females function as sterile workers and not males as well? (3) Why does the society break down at the end of the season? Is the appearance of the males in some way connected with the breakdown at the end of the season? We shall find that kin selection considerations are very relevant to answering these questions.

To explain these features, Hamilton (1964) attached fundamental importance to the unusual genetics of sex determination in Hymenoptera. In the Hymenoptera, males are haploid and females are diploid. Males are formed from unfertilized eggs and females from fertilized eggs. One important consequence of this system is that sperm are formed by mitosis while eggs are formed as usual by meiosis. Therefore, all the sperm from any male are genetically identical. This fact leads to unusual schemes of genetic relationship between relatives as discussed below. In contrast, the termites do *not* have haplo-diploid sex determination; instead they are diploid in the usual fashion. Let us see if the difference in social organization between the Hymenoptera and termites can be traced to this fact.

Hamilton's Principle Hamilton (1964, 1972) introduced a principle intended for use in strategy reasoning about altruism between kin. The discussion below represents my interpretation of Hamilton's principle. Suppose I and J are related through a known pedigree, and that I may perform some altruistic act that benefits J. Then should I carry out the altruistic act? Hamilton's principle suggests a way to answer this question. To state Hamilton's principle, we need one of the quantitative measures of relationship that we developed in Chapter 10. Recall that the kinship coefficient, $K_{X,Y}$, is defined as the average number of alleles (at one locus) in Y which are identical by descent with alleles in X given the pedigree connecting Y with X. Now it is assumed that I's strategy is to leave the largest number of copies of its genes in the next generation. Let A and B represent the mates for I and J, respectively, and let I' and J' represent the offspring of I and J, respectively, as sketched in Figure 14.1. Then, given all this pedigree information, we can easily calculate $K_{I,I'}$ and $K_{I,J'}$ which are the average numbers of I's alleles that are carried by I's own offspring and by J's offspring. Suppose that the altruistic act involves a certain loss of I's fitness and gain in J's fitness. Then Hamilton's principle is that the act is profitable from I's point of view if

$$K_{I,J'}\left[\begin{matrix}\text{gain in fitness}\\ \text{to } J\end{matrix}\right] > K_{I,I'}\left[\begin{matrix}\text{loss in fitness}\\ \text{to } I\end{matrix}\right] \tag{14.1}$$

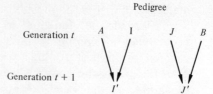

Pedigree

Generation t

Generation $t + 1$

FIGURE 14.1. *Basic setup for posing the question of whether I should do an altruistic act that benefits J.*

This condition simply says that the altruistic act is profitable to I, if the gain in the number of I's genes at $t+1$ via J's offspring exceeds the loss in the number of I's genes at $t+1$ via I's own offspring.

Before proceeding with the discussion of Hymenoptera, let us reflect somewhat on what this principle means. Consider the case where everyone is diploid, none of the parties are inbred, and that A and B are not related to either I and J. Then $K_{I,I'} = 1$ and $K_{I,J'} = \frac{1}{2}K_{I,J}$. In this important case, the condition for the altruistic act to be profitable reduces to

$$\frac{1}{2}K_{I,J}\left[\begin{array}{c} \text{gain in fitness} \\ \text{to } J \end{array}\right] > \left[\begin{array}{c} \text{loss of fitness} \\ \text{to } I \end{array}\right] \tag{14.2}$$

For example, we computed in Chapter 10 that $K_{I,J}$ between sibs is 1. Suppose I can risk himself in a way that costs, on the average, the loss of one offspring. Then, as a result of this altruistic act, J must gain more than two offspring in order for the act to be profitable from I's point of view. The quantity $K_{I,J}/2$ can be viewed as the fraction of I's genome carried by J. The relation above is often stated as, "I should 'value' J in accordance with the fraction of I's genome carried by J." But in more complicated situations where the parties are inbred, or A and B are related to I and J, and so forth, this shortcut expression becomes misleading.

A Female Wasp's
Strategy

Let us now turn to the Hymenoptera and consider the options available to a female paper wasp, I, which is born early in the growing season. (See Figure 14.2.) The first option, of course, is to have an offspring of her own, I'. If I is unfertilized, then I' is male and if I is fertilized by A, then I' is female. In either event, I always contributes exactly one allele to I' and if A is not related to I, then $K_{I,I'} = 1$. Another option is to assist the queen, Q, in the production of more of her offspring.† The queen, in effect, lives many

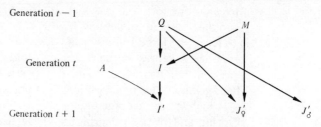

Generation $t - 1$

Generation t

Generation $t + 1$

FIGURE 14.2. *Pedigree for calculating the optimum strategy at generation t for a female member of a simple hymenopteran society.*

† It is customary to say that the female worker I is helping the queen, whereas, in effect, she is helping both the queen and her father, M. Although M is not present in the nest, his genes are—the queen carries M's sperm packet.

generations and any offspring she produces with the help of I must belong to the same generation as I's own potential offspring. To see whether I should be altruistic to Q, we need to know $K_{I,J'}$. But now the kinship coefficient depends on whether J' is male or female. If J' is female, then the probabilities of the possible configurations between I and J' are

$$\text{prob}\{I::J_\female^l\}=0$$

$$\text{prob}\{I \-- J_\female^l\}=\tfrac{1}{2} \qquad (14.3)$$

$$\text{prob}\{I = J_\female^l\}=\tfrac{1}{2}$$

The :: configuration is impossible because I and J' must carry at least one allele in common because of M. Remember, the sperm are produced by mitosis and are identical. Next, the probability is $\tfrac{1}{2}$ that I and J' carry still another allele in common because of Q. Therefore, the = configuration occurs with probability $\tfrac{1}{2}$, and last, the $\-\-$ also occurs with probability $\tfrac{1}{2}$. So the average number of alleles between I and a female J' is

$$K_{I,J_\female'} = 0 \cdot \text{prob}\{I::J_\female^l\} + 1 \cdot \text{prob}\{I \-- J_\female^l\} + 2\,\text{prob}\{I = J_\female^l\}$$

$$= 1\tfrac{1}{2} \qquad (14.4)$$

But if J' is male, then the only common ancestor between I and J' is Q. Since a male J' is haploid, the possible configurations and probabilities are

$$\text{prob}\{I : \cdot J_\male'\}=\tfrac{1}{2}$$

$$\text{prob}\{I \-- J_\male'\}=\tfrac{1}{2} \qquad (14.5)$$

so that

$$K_{I,J_\male'} = 0 \cdot \text{prob}\{I : \cdot J_\male'\} + 1 \cdot \text{prob}\{I \-- J_\male'\}$$

$$= \tfrac{1}{2} \qquad (14.6)$$

The set of kinship coefficients is summarized in Table 14.1.

Table 14.1. Average Number of Female Wasp's Alleles Carried by Various Relatives.

Coefficient	Name of Relative	Number of Alleles
$K_{I,I'}$	Offspring	1
$K_{I,J'}$	Sister	$1\tfrac{1}{2}$
$K_{I,J'}$	Brother	$\tfrac{1}{2}$

We see immediately that I should help Q, provided Q is raising I's sisters using M's sperm but not if Q is raising I's brothers. If I has offspring of her own, then each offspring net only 1 gene at time $t+1$ while if I helps Q raise sisters, then each of these sisters nets I $1\tfrac{1}{2}$ genes at time $t+1$.

A Male Wasp's Strategy But the analysis above is only part of the story. We also want to know what the best strategy is for any males who might be produced during the growing season. Consider a *male I* who is a member of generation t. Male I should act to maximize the number of his genes contributed to generation $t+1$; I has the same three options as considered before. (See Figure 14.3.) First, I can mate with A and leave an offspring of his own, I'. In fact, I' will be

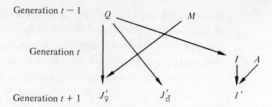

Generation $t-1$ Q M

Generation t I A

Generation $t+1$ $J'_\female$ $J'_\male$ I'

FIGURE 14.3. *Pedigree for calculating the optimum strategy at generation t for a male member of a simple hymenopteran society.*

female, but what is important is that I' carries exactly one of I's alleles. Therefore, if A is not related to I, then $K_{I,I'} = 1$. Second, I may also help the queen Q to raise a brother, $J'_\male$, or a sister $J'_\female$. The probabilities of the various configurations between I and $J'_\male$ are

$$\text{prob}\{I \cdots J'_\male\} = \tfrac{1}{2}$$
$$\text{prob}\{I - J'_\male\} = \tfrac{1}{2} \tag{14.7}$$

Therefore,

$$K_{I,J'_\male} = 0 \cdot \text{prob}\{I \cdots J'_\male\} + 1 \cdot \text{prob}\{I - J'_\male\} \tag{14.8}$$

Similarly, the probabilities of the various configurations between I and $J'_\female$ are

$$\text{prob}\{I \cdot : J'_\female\} = \tfrac{1}{2}$$
$$\text{prob}\{I \div J'_\female\} = \tfrac{1}{2} \tag{14.9}$$

Therefore,

$$K_{I,J'_\female} = 0 \cdot \text{prob}\{I \quad : J'_\female\} + 1 \cdot \text{prob}\{I \div J'_\female\} \tag{14.10}$$

This set of kinship coefficients is summarized in Table 14.2. From these coefficients, we immediately observe that the male, I, should produce offspring of its own because each offspring of its own nets 1 gene at time $t+1$ while the other options net only $\tfrac{1}{2}$ gene at time $t+1$.

Table 14.2. Average Number of Male Wasp's Alleles Carried by Various Relatives.

Coefficient	Name of Relative	Number of Alleles
$K_{I,I'}$	Daughter	1
$K_{I,J'}$	Sister	$\tfrac{1}{2}$
$K_{I,J'}$	Brother	$\tfrac{1}{2}$

The Answers to the Three Questions

Now that we have worked out all the possible strategies in terms of number of genes, each provides at time $t+1$, we can return to the puzzling features of hymenoptera social organization. The first question concerned the existence of sterile castes. We see that the sterile workers are maximizing the number of their own genes contributed to the gene pool at $t+1$ by helping the queen raise more of their sisters. Second, the society is female dominated for two reasons. It is only profitable for a worker to help a queen if the queen is raising sisters. Also, it is *not* profitable for a male to function as a worker in helping the queen raise either brothers or sisters, because its own offspring carry more of its genes. Third, the breakdown of altruism at the end of the season is connected with the appearance of males among the queen's offspring. It is then no longer profitable for the female workers to assist the queen, and instead the best strategy is to raise their own offspring.

The attractiveness of strategy reasoning with kin selection has led to its wide acceptance, especially among students of animal social behavior. But it is well to keep in mind what the approach does not explain, at least at this time—see Alexander (1974). It does not offer an explanation for the evolution of sterile castes in termites. It does not explain why many species of wasps are solitary while others are social. It does not explain why certain other insect taxa which also have haplo-diploid sex determination (e.g., certain mites and beetles) have not evolved any social organization. And of course it does not explain why Hymenoptera happen to have the haplo-diploid system to begin with.

Recent work on the topic of kin selection is of two types. The first type is the application of this strategy reasoning approach to the most complex hymenopteran societies and to more complex behavioral interactions. See the papers by Hamilton (1964 and 1972), Alexander (1974), West Eberhard (1975), and Trivers (1972 and 1974). The second type is the exploration of the actual evolution of the optimum strategy as predicted by Hamilton's principle. As always, a fundamental difficulty with any strategy reasoning approach is establishing whether natural selection actually causes the optimum strategy to evolve. Suppose, for example, that acting according to the optimum strategy as defined by Hamilton's principle is itself a trait determined by one locus with two alleles. Then will the gene that leads to acting in the optimum way always evolve? The answer to this question is not yet clear. Key papers include Maynard Smith (1964 and 1965), Orlove (1975), and Matessi and Jayakar (1976b).

Group Selection Colonies of social insects provide perhaps the best candidates for group selection, as often stressed by Wilson (1973). A nest of social insects is, as a whole, vulnerable to predators and to events like a tree or branch falling, and so forth. There is no doubt that entire nests can be eliminated if predators succeed in breaking into the nest and eating the helpless larvae inside. The task of predator defense might conceivably be regarded in terms of group selection. Colonies with good predator defense survive more often than colonies with bad defense. But this superficially simple hypothesis needs to be developed with care.

Early theory on group selection has been hampered by incorrect formulations and by polemic concerning badly chosen examples of traits that supposedly evolved by group selection. In a lucid critical review of the topic of group selection in 1966, G. C. Williams was able to write that the process of group selection had yet to receive a satisfactory theoretical formulation and that there were no known traits whose evolution could not be explained in terms of regular natural selection on individuals. Since that time the process of group selection has been examined in a series of theoretical papers. In one of the earliest papers, Williams and Williams (1957) deal with selection between groups whose members are sibs to one another. Then Levins (1970) and Eshel (1972) independently provided the major formulations of group selection, and subsequent papers include Boorman and Levitt (1973), Levin and Kilmer (1974), Wilson (1975), and Gilpin (1975). This work makes clear that group selection can occur, but the magnitude of migration between groups is critical. Group selection tends to require low migration. However, this requirement is also conducive to reciprocal altruism and to kin selection. Any trait that is a candidate for group selection is very likely to involve these more traditional modes of selection as well. Therefore, an ability to explain how a trait might have evolved without referring to group selection should not be taken to mean the group selection did not, in fact, contribute to the evolution of the trait.

The mathematics of group selection can become quite complicated, depending on the generality that is sought. It seems best to present here a simple example of group selection that reveals the basic features of the process. The example below is a special case of the model treated in generality by Eshel (1972).

A Simple Model Consider a population of haploid individuals that is divided into many discrete colonies. Assume the size of each colony is N individuals. The assumption of haploidy is not important; it merely simplifies matters. If we were to assume diploidy, then we would also have to introduce assumptions about dominance with respect both to the individual and group selection processes. We shall summarize the assumptions as we go along in Figure 14.4.

We shall suppose that there are two types, A and B, and that A is the altruist and B is the nonaltruist. Specifically, we assume the relative fitness for A is w for B is one.

Fitness

A	B
w	1

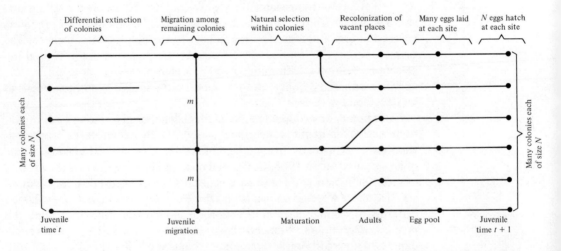

FIGURE 14.4. *The sequence of events in the Eshel model, combining group selection with natural selection.*

Since A is an altruist, w is less than one; and thus natural selection favors B. However, group selection favors A. We introduce this assumption in terms of the group survival probability, $L(p)$. The variable, p, is the frequency of A and $L(p)$ is an increasing function of p, as sketched in Figure 14.5. By convention, we can set $L(1) = 1$, since it is only the relative survival of a group that is important.

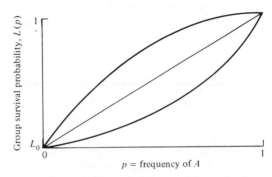

FIGURE 14.5. *Sketch of possible functional forms for the group survival probability, $L(p)$, as a function of the frequency of altruists within the group.*

To begin let us describe, in words, the sequence of events during a generation. (See Figure 14.4.) First, some of the colonies become extinct on the basis of the survival probabilities, $L(p)$. Second, the remaining colonies exchange immigrants. Third, ordinary natural selection occurs within each colony. Fourth, the vacant regions in the environment are recolonized from the existing colonies. This recolonization is assumed to occur without bias so that the probability that any given colony recolonizes a vacant place does not depend on the frequency of altruists in that colony. Fifth, at each place a large number of eggs are laid. Sixth, at each place N eggs hatch, thereby regenerating a set of colonies, each of size N. The N eggs that do hatch are chosen according to the usual binomial sampling scheme.

Figure 14.4 might correspond roughly to an insect life history as follows: At time t each colony consists of N larvae. Some of these colonies may become extinct, perhaps because of overexploiting and defoliating the host plant. Then, there is some small amount of migration between the surviving colonies on different plants. Next, there is individual selection within each colony as the larvae mature and pupate. Then, the adults from any colony oviposit on the plant in which they were raised. The adults may also oviposit in nearby plants, provided that there is negligible interchange between colonies at this point. Finally, of all the eggs deposited on any plant, only N hatch. From this point the cycle repeats itself.

The particular sequence of events during a generation is not critical to the qualitative conclusions. The above sequence is convenient mathematically because sums rather than integrals can be used. But for any quantitative application of this theory, a different sequence might be necessary, particularly one allowing for migration after selection, instead of before selection, as illustrated in Figure 14.4. Also, it might often be more realistic to have a sequence where natural selection is followed by colony extinction, and then migration. In any event only the quantitative and not qualitative results might be influenced by these alterations and the model for any sequence is developed in the same way as below.

We are concerned with predicting the proportion of colonies with each possible configuration. Let k denote the number of altruists in a colony $(k = 0, 1, 2 \cdots N)$. Then $\rho_t(k)$ denotes the proportion of colonies in the whole population that contain k altruists and $N - k$ nonaltruists. We shall predict $\rho_{t+1}(k)$, given $\rho_t(k)$ and the parameters, w, $L(p)$, N, and the migration parameter, m.

To develop the equation for $\rho_{t+1}(k)$, we proceed step-by-step through Figure 14.4. The average colony survival is

$$\bar{L} = \sum_{k=0}^{N} L\left(\frac{k}{N}\right)\rho_t(k) \tag{14.11}$$

Then the distribution of colony types after extinction, $\rho_{\text{ext}}(k)$, is

$$\rho_{\text{ext}}(k) = \frac{L(k/N)}{\bar{L}}\rho_t(k) \tag{14.12}$$

Note that the extinction process does not change gene frequencies within any colony, but the gene frequencies averaged over the whole population are changed because of the differential survival of the colonies. The gene frequency averaged over the whole population, *after* extinction, $\bar{p}_{\text{ext}}$, is

$$\bar{p}_{\text{ext}} = \frac{\bar{k}_{\text{ext}}}{N} = \frac{\sum_{k=0}^{N} k\rho_{\text{ext}}(k)}{N} \tag{14.13}$$

Next the migration takes place and the gene frequencies within colonies are changed. The gene frequency within colonies after migration, $p_m(k)$, given that the frequency before migration was k/N is

$$p_m(k) = (1 - m)\frac{k}{N} + m\bar{p}_{\text{ext}} \tag{14.14}$$

Then natural selection occurs within each colony. With haploids, the gene frequency after selection, $p_s(k)$, is related to that before selection, $p_m(k)$, as

$$p_s(k) = \frac{wp_m(k)}{wp_m(k) + q_m(k)} \tag{14.15}$$

where $q_m(k)$ is the frequency of the nonaltruist before selection. After selection, a large egg pool is formed within each colony. This egg pool contains the A and B genes in frequencies $p_s(k)$ and $q_s(k)$. Then N eggs are drawn from this pool. The probability that $n A$ eggs and $(N - n)B$ eggs are drawn is

$$\begin{bmatrix} \text{colony} \\ \text{prob} \\ k \rightarrow n \end{bmatrix} = \begin{bmatrix} \text{prob that} \\ n\ A \text{ genes} \\ \text{and } (N - n)B \\ \text{genes are} \\ \text{drawn} \end{bmatrix} = \binom{N}{n} p_s(k)^n q_s(k)^{N-n} \tag{14.16}$$

Thus any colony that began with $k A$ genes at time t, has the above probability of having $n A$ genes at time $t + 1$. Equation (14.16) is the transition probability for the transition of any colony from k to n. To conclude, we simply sum up over all the colonies. Thus

$$\rho_{t+1}(n) = \sum_{k=0}^{N} \begin{bmatrix} \text{colony} \\ \text{prob} \\ k \rightarrow n \end{bmatrix} \rho_{\text{ext}}(k) \tag{14.17}$$

This is the equation for $\rho_{t+1}(n)$, given $\rho_t(n)$. The only parameters are w, $L(p)$, and m.

Results Eshel's (1972) analysis of this model of group selection established three results:

1. For any given magnitude of w, $L(p)$ and N, if the migration is small enough, then the altruist becomes fixed throughout the entire population. (Group selection prevails.)
2. For any given magnitude of w, $L(p)$ and N, if the migration is high enough, then the nonaltruist becomes fixed throughout the entire population. (Individual selection prevails.)
3. It is possible to attain a polymorphism between the altruist and nonaltruist maintained by a balance of group selection versus individual selection.

We can give some substance to these results by some numerical examples. Figures 14.7 through 14.10 illustrate iterations of the equations for $\rho_{t+1}(k)$. In all the figures, the group survival function is assumed to be a linear function of the frequency of altruists in the group.

$$L(p) = L_0 + (1 - L_0)p \qquad (14.18)$$

The function $L(p)$ is sketched in Figure 14.6. For all the figures, the fitness of the altruist, w, equals $\frac{1}{2}$ and the survival probability of a group of nonaltruists, L_0, also equals $\frac{1}{2}$. Thus the "strength" of both modes of selection is the same. Figures 14.7 and 14.8 compare different migration rates with a colony size of $N = 2$. In these and all other figures the initial frequency of the altruists is .5 in all colonies unless otherwise indicated. With the smaller migration rate, $m = .2$, the altruist is eventually fixed and group selection prevails. In contrast, with $m = .4$ the nonaltruist is eventually fixed and individual selection prevails.

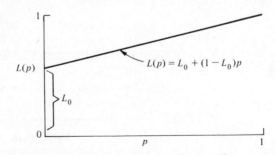

FIGURE 14.6. *Functional form of $L(p)$ used in preparing the numerical illustrations of group selection in the following figures.*

There are some surprising differences in the course of evolution presented in Figures 14.7 and 14.8. In Figure 14.7 the frequency of altruists averaged over the entire population initially decreases and only after some time has elapsed does it begin to rise. In contrast, the course of evolution in Figure 14.8 is a monotonic decrease in the frequency of altruists. The reason for this difference is that initially there is no variation between colonies for the group selection to work upon. Initially, all colonies contain an identical composition of one altruist and one nonaltruist. As a result, there is initially no differential survival of colonies and hence no progress with group selection. But as time passes, genetic drift within the colonies leads to variation among the colonies. As this "between-colony" variation builds up, group selection picks up speed and eventually leads to the elimination of the nonaltruist. In a sense, genetic drift is to group selection what mutation is to natural selection; genetic drift is a source of between-colony variation which group selection can act upon. In contrast in Figure 14.8 where individual selection prevails, there is a monotonic progression toward elimination of the altruist because genetic variation exists within each colony to begin with. The individual selection relies only on the "within-colony" variation and thus immediately progresses to the elimination of the altruists.

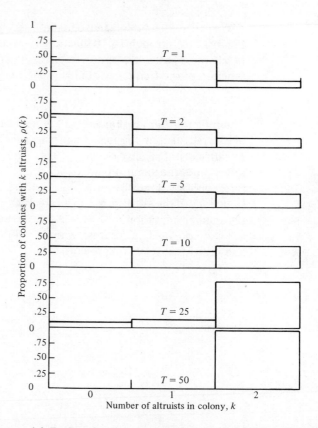

FIGURE 14.7. *Results of a computer iteration of the Eshel model. The altruist is eventually fixed and group selection prevails over individual selection in this example. Initial frequency of the altruist is .5 in all colonies for this figure and in all following figures unless otherwise indicated. Here $N = 2$, $m = .2$, $w = .5$, and $L_0 = .5$.*

Figures 14.9 and 14.10 illustrate the same points but with a larger population size, $N = 8$. There are three new features: (1) The time required for group selection to lead to fixation of the altruists is longer in Figure 14.9. With a larger N, genetic drift produces less between-colony variation for group selection to act upon, and the group selection process is slowed down.

(2) The time required for individual selection to fix the nonaltruists is slightly shorter in Figure 14.10 because the effective counteraction by the group selection is weaker. (3) The critical level of migration marking the change-over from group selection to individual selection is lowered. Again, with a higher N, the force of group selection is weakened because of the smaller variation between colonies produced by drift. Hence a smaller migration is required to allow group selection to prevail over individual selection.

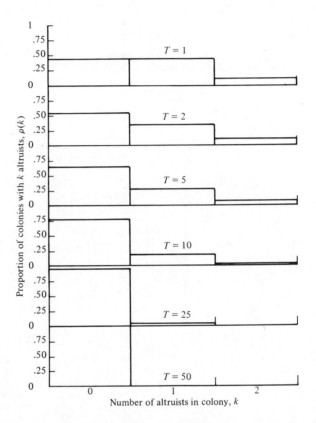

FIGURE 14.8. *Results of a computer iteration of the Eshel model. The nonaltruist is fixed and individual selection prevails over group selection. Note the higher migration rate in this example as compared with Figure 14.7. Here $N = 2$, $m = .4$, $w = .5$, and $L_0 = .5$.*

In Figures 14.9 and 14.10, the critical migration level allowing fixation of the altruist is fairly low, suggesting that for group selection to predominate, there *must* be small migration between the colonies. But this conjecture is *false*. A very high migration may still allow group selection to prevail provided the appropriate fitnesses are used. As an exercise, suppose the selection coefficients are $w = .9$ and $L_0 = .5$. It is thus assumed that a small sacrifice in an individual's fitness leads to a large improvement in the group survival probability. It can be shown that the critical level of migration is above $m = \frac{1}{2}$ if $N = 4$. Thus even if half the population consists of migrants, then group selection still prevails over individual selection with these particular fitnesses and $N = 4$. Similar examples can always be proposed for any value of N. Even when N is very large, if a tiny sacrifice in individual

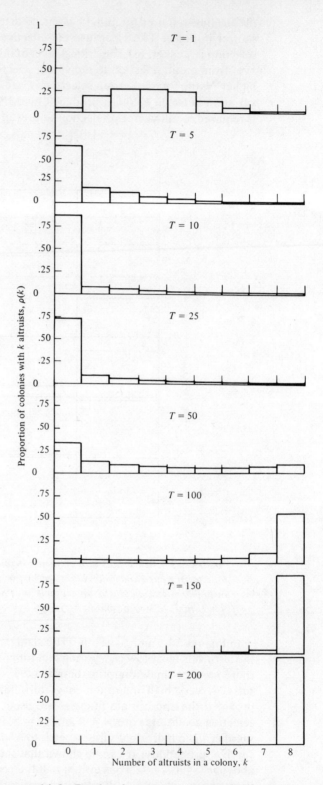

FIGURE 14.9. *Results of a computer iteration of the Eshel model. The altruist is eventually fixed and group selection prevails over individual selection. Here $N = 8$, $m = .05$, $w = .5$, and $L_0 = .5$.*

fitness leads to a substantial increase in group survival, then the trait can evolve by group selection even in the face of moderate migration between colonies.

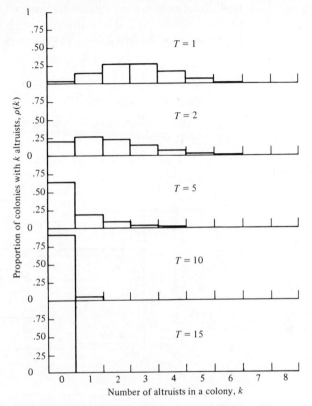

FIGURE 14.10. *Computer iteration of the Eshel model. Again with higher migration as compared with Figure 14.9 the nonaltruist is fixed. Here $N = 8$, $m = .1$, $w = .5$, and $L_0 = .5$.*

Eshel's results also establish that a polymorphism can be maintained by the balance of individual and group selection. Figure 14.11 illustrates an example of this. In general terms a polymorphism obtains if (1) the migration is so high that the altruist cannot be fixed by group selection, and (2) the group survival of nonaltruists, L_0, is so low that the nonaltruists cannot be fixed by natural selection at the given migration rate. As we have seen, if m is high *enough*, then the nonaltruists are always fixed by natural selection. But for an intermediate value of m, it can sometimes occur that neither group selection nor individual selection is powerful enough to prevail over the other. The distribution of colony types in Figure 14.11 is approached from any choice of $\rho_0(k)$, the initial distribution, at least so far as could be determined using computer iteration of the equation for $\rho_{t+1}(k)$.

There is need for substantial additional research in this area. Explicit formulas for the critical migration level and the conditions for polymorphism have not been obtained. Indeed, as yet undiscovered qualitative features of the model are likely to emerge with further analysis. Also, genetic drift *between* colonies resulting from a finite number of colonies in the population has not been studied.

It has long been a basic tenet of evolutionary theory that group selection does not occur and that any explanation for the evolution of a trait must be sought in terms of natural selection on individuals. But it is now clear that the traditional position on this matter must be reconsidered. Also, a critical problem will be to find situations in which the contributions of the various hypotheses for the evolution of altruism can be separated.

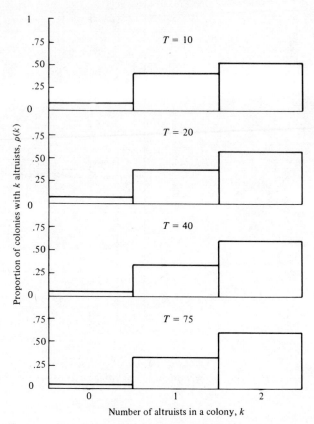

FIGURE 14.11. *Computer iteration of the Eshel model. A polymorphism is established involving both altruists and nonaltruists. The polymorphism is maintained by a balance between group and individual selection. Initial distribution in this run was $\rho(k) = (.3, .4, .3)$. Here $N = 2$, $m = .95$, $w = .5$, and $L_0 = .1$.*

PART FOUR

EVOLUTIONARY ECOLOGY OF
SINGLE POPULATIONS

Chapter 15
AN OVERVIEW OF
EVOLUTIONARY ECOLOGY

EVOLUTIONARY ecology is the fusion of population ecology with the evolutionary theory of the preceding chapters. It is a very recent field, which has developed principally over the last 10 years. The remaining chapters in this book provide an introduction to this new field. The purpose of some chapters is to present the basics of population ecology and of others is to fuse those basics with evolutionary theory. Before saying more, it will be helpful to point out where evolutionary ecology fits into the overall field of ecology.

Main Areas of Ecology — Traditionally, the field of ecology includes study at three levels of organization. The ecology *individuals*, including a large area called physiological ecology, is the study of the techniques by which individuals survive, obtain resources, and behave in their natural environment. Often the interest is focused on very harsh or extreme environments where especially fascinating behavioral and physiological mechanisms are observed. The ecology of *populations* is primarily concerned with the growth of populations and with how populations interact with one another through competition or predation. Population interactions may lead to surprising connections between the abundance and stability of all the populations living together in some place. Another area of ecology is concerned with what are called *communities* and *ecosystems*. A community is the collection of all the populations that live together in some region. (Sometimes the term *community* is used in a more restricted context. The phrase "bird community," for example, refers to the collection of all bird species in some region). An ecosystem is the community together with the surrounding environment. For instance, components of the forest ecosystem include the atmosphere of the forest and moisture in the soil of the forest floor, as well as the populations comprising the forest itself. An important question in this area of ecology is to determine whether communities and ecosystems have a common pattern of organization and function. Evolutionary ecology is essentially the fusion of evolutionary theory with the area of ecology concerned with populations. Natural selection and evolution are processes that occur at the population level, and therefore the population level of ecology is the principal entrance point for evolutionary theory into ecology.

Theory in Ecology — All the areas of ecology have developed a body of mathematical theory, although theory in the population area has the longest history. At the organism level there are mostly two kinds of models. In one kind the purpose is to characterize the dynamics of some physiological or behavioral process. For example, there are models that predict the daily activity cycle of a lizard based on the environmental temperatures and on parameters for the thermoregulatory abilities of the lizard. There are models that predict the rate of evapotranspiration and photosynthesis for a plant as a function of the water in the soil; the temperature and humidity of the air; and parameters for the leaf shape, stomatal size, and other traits of the plant. The other major kind of model in the ecology of organisms is the "optimum phenotype model." An optimum phenotype model predicts the phenotype of an organism on the basis of an optimization principle. For example, an insectivorous bird encounters many potential prey items during a day's foraging. However, a bird generally does not attempt to catch *every* potential prey item it sees; instead, a predator exhibits preferences for various types of prey. The task of an optimum phenotype model here is to predict the scheme of preferences for a predator on the basis of maximizing the expected yield per day. Such a model includes parameters for the abundance of the various prey types, the probability of a successful capture if attempted, and the caloric content of the various potential prey. Other examples of optimum phenotype models include models that predict the geometry and shapes of plants and leaves on the basis of maximizing the photosynthesis or minimizing the water loss.

At the population level, most models are intended to predict the population size of one or more populations through time. These models are called population dynamic models. The earliest and simplest population dynamic models were proposed during the 1920s by Kostitzin, Lotka, and Volterra. This early work occurred at roughly the same time as the pioneering work of Fisher, Haldane, and Wright in population genetics. We shall discuss many population dynamic models in the following chapters. It is these models which can be fused with evolutionary theory.

Models at the community and ecosystem level are typically models for the flow of material and energy between compartments. The ecosystem under study is visualized as consisting of several important compartments. For example, compartments might be the set of all plants, the set of all herbivores, the set of all carnivores, and the set of all species that decompose fallen material. Still more compartments might be identified if finer resolution is required. The model would predict the flow of material and energy between these compartments.

When all the theory in ecology is taken into account, it is clear that evolutionary ecology should be called evolutionary population ecology. To date the other levels of ecology have made little explicit use of evolutionary theory.

Organization of the Rest of the Book   The organization of the remaining chapters in the book involves a pattern of alternating between the introduction of models from population ecology, followed by the fusion of these models with evolutionary theory. For example, Chapter 16 is an introduction to the simplest models for the growth of a *single* population. Then Chapter 17 fuses this basic ecological theory with the one-locus two-allele model from basic population genetic theory. Similarly, Chapters 21 and 22 present models for the population dynamics of several populations interacting through competition and predation. Then in Chapter 23 some general theory for evolution among interacting populations is presented. Other chapters also exhibit this pattern of alternation between population dynamic models, followed by the fusion of these models with evolutionary theory.

Relevance of Evolutionary Ecology to General Ecology   Although evolutionary ecology is mainly concerned with the population level of ecology, it is also relevant to the other levels in several ways. At the organism level, the choice of an optimization principle for optimum phenotype models must be defensible in evolutionary terms. Chapters 17, 19, and

23 are intimately involved in the discovery of optimization principles that correctly predict the outcome of evolution under various ecological situations. These optimization principles can then be used in optimum phenotype models. At the community and ecosystem level, the amount of material in each compartment ultimately reduces to the sum of the population sizes comprising the various compartments. For example, the amount of material in the plant compartment is the sum of the population sizes of each plant species times the average weight of a plant for each species. Similarly, the flow between compartments is ultimately determined by predator-prey, plant-herbivore, and host-parasite interactions. The results of Chapter 23 on the theory for the evolution of interacting populations can be translated into a statement about the evolution of the flows between compartments and of the steady-state amounts in each compartment. Thus although evolutionary ecology is presently mostly relevant to population ecology, it is proving increasingly important to the study of ecology at other levels as well.

The Fusion of Genetics and Ecology

The fusion of population genetics with population ecology can be compared to a prearranged marriage between partners who speak different languages. Although both families agree that the marriage is advantageous, it is somewhat difficult to achieve because of cultural differences between geneticists and ecologists. Theory in ecology differs from that in genetics in one important way. Population genetics is blessed with a *simple and universal* mechanism of inheritance. As a result, models in genetics are usually formulated in terms of a clear mechanistic basis. There is no dispute among geneticists that the model from Chapter 3 for natural selection at one locus with two alleles in a random mating population is a good model for this particular situation. In contrast (1) ecological processes never share a *universal* mechanism, *and* (2) the mechanisms are never *simple*. Virtually every species has a unique method of obtaining resources, avoiding predation, and so forth, and these processes are expressions of complex behavior and physiology. The original basic models of population dynamics are always being subjected to challenge, and there are usually several alternative models that describe roughly the same thing but for different species and in differing degrees of detail. As a result, the fusion of population genetics with population ecology must be done in a fairly general way which is not restricted to a special choice of population dynamic models.

Simplicity and Complexity in Ecological Models

There is usually a graded series of models with increasing detail for any given population dynamic process. At one extreme are the simplest models, which were proposed in the 1920s. At the other extreme are very complex models in the tradition of "systems analysis." These models incorporate the most detailed description of an ecological process that is currently available, and the predictions from these models can be extremely accurate and reliable. Generally, the simpler models can be understood analytically while systems models must be studied with numerical techniques on a computer. In fact, some large systems models cannot be fully written out except in the form of a long computer program containing many subroutine calls. The spectrum between simple analytical models and systems models occurs at all levels of ecology. There are simple analytical as well as large computer systems models of the foraging behavior of an individual insect, and there are analytical as well as systems models of ecosystems. The degree of detail used in an ecological model depends on the purposes of the inquiry and not on the level of organization.

This book focuses on simple models of population dynamics. There are two reasons for this. First, simple models are prototypes for more detailed models. One's ultimate intention may be to develop a detailed model for some ecological process. But developing a detailed model is not only a matter of obtaining data about the process and integrating the data into a model. The mathematical structure of the model often shapes the predictions regardless of the actual data used in a numerical example. To understand how the predictions relate to the input data, it is essential to understand the implications of the mathematical structure itself. For example, in the chapter on the predator-prey interactions, we shall see several predictions that emerge from many models. These predictions arise from a common mathematical structure in spite of the fact that these models differ in many biologically interesting details. The study of simple models is about the only practical way one can discover the implications of the mathematical structure of a model because it is only comparatively simple models that offer any hope of being analytically tractable.

Second, simple models are extremely useful by themselves; they sharpen intuition and sensitize imagination. Simple models often reveal surprising results which would not have been anticipated without the model but which are seen to be very intuitive with the model. For example, we learned from population genetics that even a weak selection pressure causes rapid evolution. In a similar vein we shall see that even a small degree of competition and predation among populations can produce dramatic evolutionary results. Moreover, the combination of many populations interacting with one another often leads to surprising results because of the "feedback" among the populations. Increasing the resources of one population may lead to its net decline and to the net increase of another population because of ecological feedback. These ideas and others can best be explored with simple models where the addition of more detail would clutter up the picture without improving understanding. Simple models are tools for the discovery of possibilities.

Chapter 16
EXPONENTIAL AND LOGISTIC POPULATION GROWTH

$\mathbf{T}$HE study of population ecology begins by sharing with Malthus the realization that population growth is an explosive process. We wish to obtain equations that predict a population's size through time. Most of the equations in population ecology are fabricated in the following simple format:

$$\frac{dN(t)}{dt} = \left\{ \begin{array}{c} \text{an individual's} \\ \text{contribution to} \\ \text{population growth} \end{array} \right\} N(t) \qquad (16.1)$$

Here $N(t)$ is the population size at time t; $dN(t)/dt$ is the rate at which the population size is changing. The expression in braces will be different in each equation but whatever it is, it always represents an individual's contribution to population growth. Equation (16.1) merely says that the growth rate of the whole population equals an individual's contribution times the number of individuals.

Exponential Population Growth

For the simplest model of population growth, we shall assume an individual's contribution to population growth is a constant, r. Hence

$$\frac{dN(t)}{dt} = rN(t) \qquad (16.2)$$

The constant, r, is called the intrinsic rate of increase. Let us examine what (16.2) predicts about population size through time. Equation (16.2) is called a differential equation. To solve it means that we must find a specific function that satisfies the stated condition on its derivative. For example, the function that satisfies (16.2) *must* have the property that its slope at each point equals r times the value at that point. Any function that does not satisfy this condition is not a solution. The process of finding a function that satisfies the conditions stipulated in the differential equation is called *integrating* a differential equation. Integrating a differential equation is similar in principle to iterating a recursion equation analytically. Some differential equations can be easily integrated; others, as we shall see, cannot. Equation (16.2), however, is uncommonly easy to integrate and the details are in the box. The result is

$$N(t) = e^{rt}N(0) \qquad (16.3)$$

---

1. Arrange the equation for exponential growth as follows:

$$\frac{dN(t)}{dt} = rN(t)$$

$$dN(t) = rN(t)\,dt$$

$$\frac{dN(t)}{N(t)} = r\,dt$$

$$d\ln N(t) = r\,dt$$

---

2. The variables have now been separated. Integrate both sides yielding

$$\ln N(t) - \ln N(t_0) = rt - rt_0$$

3. Arrange as follows:

$$\ln \frac{N(t)}{N(t_0)} = r(t - t_0)$$

$$\frac{N(t)}{N(t_0)} = \exp\left[r(t - t_0)\right]$$

$$N(t) = N(t_0)\exp\left[r(t - t_0)\right]$$

4. If $t_0 = 0$ by convention, we obtain

$$N(t) = N(0)\exp(rt)$$

Again, this particular function is the solution to (16.2) *because* it has the property that its slope at every t equals r times its value at t. (You should verify this for yourself.) Equation (16.3) asserts that the population size at time t is found by raising e to the rtth power and then by multiplying by the initial population size. Figure 16.1 presents some examples for different values of r. If the population size through time obeys (16.3), then it is said to be showing *exponential growth*.

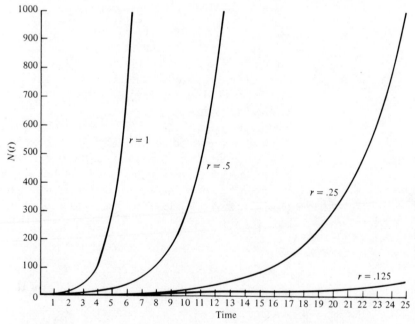

FIGURE 16.1. *Curves of population size as a function of time for exponential growth. Here $N(t) = N(0)\,e^{rt}$; $N(0) = 2$. Note that the slopes of the curves become steep. It is usually more useful to graph the logarithm of the population size on the vertical axis and time on the horizontal axis. The formula for the logarithm of the population size as a function of time is $\ln[N(t)] = rt + \ln[N(0)]$. The logarithm of the population size increases as a linear function of t; the slope of the line is r and the y-intercept is $\ln[N(0)]$.*

Because time appears as an exponent in (16.3), the population size is predicted to increase very rapidly through time. As the population size increases, the *speed* with which it increases also increases. Recently, P. Ehrlich [Ehrlich et al. (1977)] has stressed the analogy between population growth and an atomic bomb. An atomic bomb is based on a chain reaction. An atom splits into two pieces, which in turn cause more atoms to split, and so on. A chain reaction is similar to a room filled with mouse traps each loaded with a ping-pong ball. Then, if someone tosses one ball into the room, a few traps are released shooting more balls into the air. These then set off more traps and the process gathers speed eventually producing an explosion. Exponential population growth is also a chain reaction, and the analogy with the atomic bomb is exact. An *E. coli* in standard laboratory conditions divides about every 20 minutes. Starting with 100 bacteria, this rate of division will cover the earth 1 yard deep in only 32 hours. No one who has experienced locust swarms, tent caterpillar outbreaks in the eastern United States, the rabbit infestation in Australia, or any other case of a population outbreak, doubts the awesome power of exponential growth.

**Factors
That Check
Population
Growth**

All populations have the potential for exponential growth. Because of this fact, much of ecology is directed toward understanding the causal factors that keep populations in check and prevent exponential growth from being realized except for a short time in certain unusual circumstances. Two categories of factors should be considered as candidates.

*Density-
Independent
Factors*

The first category consists of factors that influence an average individual's reproductive output to an extent that is independent of the current population size. This category may include catastrophic events in the weather which levy a toll of mortality on organisms that are over wintering or that are near their time of emergence or germination in the spring. For some populations the direct effects of weather on the average individual's reproductive output are the principal causal factors that determine the temporal pattern of population size during the time the population has been under observation. Causal factors in this category are called *density-independent factors*. Populations that are probably very strongly influenced by density-independent factors include insect populations and also populations in deserts, high mountains, or other habitats with severe environmental fluctuations.

*Density-
Dependent
Factors*

The second category of factors that may check population explosion includes various mechanisms which reduce an individual's reproductive output as density increases. These mechanisms are referred to collectively as *density-dependent factors*. The basic assumption in exponential growth is that an individual's contribution to population growth is a constant. However, the presence of density dependence destroys this assumption by insuring that an individual's reproductive output decreases as the population size builds up. When the reproductive output is decreased enough, then direct replacement occurs. (Each individual produces only one offspring on the average.) Hence the population size stops growing and remains at a constant value. There are many kinds of density-dependent mechanisms, and for some populations they do seem to be the principal factors checking exponential growth. But as you might expect, many populations seem influenced both by density-independent factors and by density-dependent factors. Assessing their relative contribution to the temporal pattern of population size poses a tricky problem. See Bulmer (1976a) for an entrée to the literature on this problem.

There is an endless variety of mechanisms that *may* produce density dependence. To get a feeling for this variety, let us mention just a few.

1. Direct effects of resource depletion. As the number of organisms increases, the amount of resources available to each individual decreases. An individual's reproductive output may then drop because eggs may be resorbed if the female is in a sufficiently poor nutritional state, or juveniles may die of direct starvation if the parents are unable to find enough food. But such overt and direct effects of resource depletion are comparatively rare, and density dependence often arises in more subtle ways.

2. Increased foraging effort. As the resources become depleted, animals must spend more time looking for food. Increased foraging effort has three consequences. First, animals are particularly susceptible to predation while foraging and increased foraging times leads directly to increased predation hazard. Second, juveniles, if under parental care, are left exposed and unguarded for a longer time, thus increasing the risk of nest predation. Third, more energy is expended per unit of food captured. Hence foraging efficiency decreases.

3. Increased time devoted to social interaction. As the population size increases, animals often must allocate more time and energy to behavioral interactions. For example, animals with territories must exert energy in territorial defense roughly in proportion to the average rate of intrusion. The rate of intrusion increases with population size.

4. Use of marginal habitats. As the abundance of animals increases, the choice sites for nests and territories become used up and animals must occupy more exposed and hazardous sites. The use of marginal habitats reduces the average per capita reproductive output even though those with the best nest sites may continue to enjoy a high reproductive output.

5. Density-dependent predation. There is evidence that predators often develop a temporary behavioral preference for the most abundant species among their prey. Predators develop a so-called "search image" for certain kinds of individuals to improve their searching efficiency. This predator preference for the most abundant prey results in the predation hazard being higher on individuals whose species is abundant as compared with those whose species is rare. The result is a decrease in the average reproductive output of an individual as the population size increases.

These examples show that many kinds of density-dependent mechanisms are possible. Moreover, many more examples could have been chosen; the five given here tend to relate more to vertebrates than to invertebrates and plants. The general conclusions to be drawn are that every population *may* be influenced by density-dependent factors, and many undoubtedly are; and the particular complex of density-dependent mechanisms that impinges on any given population is unique to it. Although the phenomenon of density dependence may be very general, the detailed mechanisms in each case are probably species specific.

It should be emphasized that it is often very difficult to determine if a given cause of mortality or reduced fertility is a density-dependent or a density-independent factor. For example, suppose that mortality and lowered fertility are caused by cold weather. One might be tempted to classify this factor as density-independent. But the exposure to the bad weather may actually be the result of a high population size and the exhaustion of suitable living places caused by the high population size. If the average exposure to bad weather for members of a population varies with the population size, then even the weather would be classified as a density-dependent factor.

Density-Dependent Population Growth

Because the phenomenon of density dependence may be very general and widespread, it would be desirable to develop a theory of population growth that applies with some generality. However, the fact that each population may have its own special mechanisms of density dependence poses a fundamental dilemma. It would appear that a general theory *must* be inaccurate when applied too literally to any particular population while a theory tailored to one particular population would fail to apply to other populations with different mechanisms of density dependence. Therefore, the way to proceed is not entirely clear. What we must do is to somehow arrive at a compromise between the need for generality and the need to be faithful to the actual mechanisms of density dependence. The compromise reached depends on one's purpose. The standard theory of population dynamics represents one extreme of all possible compromises—the extreme where generality is always weighted over faithfulness to the actual mechanisms of density dependence. We shall see, however, that this theory is extremely useful and informative; in fact, surprisingly so in view of its naked simplicity.

Logistic Equation in Continuous Time

The simplest model for density-dependent population growth is the logistic equation. The existence of density dependence means that an individual's contribution to the population's growth is *not* constant but decreases as some function of the population size. But what would this function be? The function depends, of course, on the actual mechanisms of density dependence. But, as a guess, let us suppose that the function is linear, that an individual's contribution to population growth decreases as a linear function of population size, N.

$$\left\{ \begin{array}{c} \text{individual's contribution} \\ \text{to population growth} \end{array} \right\} = r - \frac{r}{K}N \qquad (16.4)$$

In this equation the y-intercept is our old friend, r, the intrinsic rate of increase. The slope of the line is $-r/K$, where K is a new parameter called the *carrying capacity* (the reason for this name will be clear shortly). Figure 16.2 illustrates this function. Now the equation for population growth with this assumption becomes

$$\frac{dN(t)}{dt} = \left[r - \frac{r}{K}N(t) \right] N(t)$$

$$= r \left[1 - \frac{N(t)}{K} \right] N(t) \qquad (16.5)$$

This equation for population growth is called the *logistic equation*. It is the simplest model of population growth with density dependence. The key assumption in the logistic equation is that an individual's reproductive output decreases as a linear function of population size.

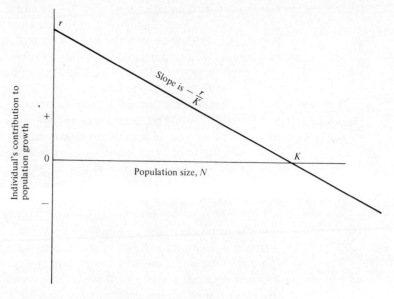

FIGURE 16.2. *An individual's contribution to population growth as a function of the population size N. In the logistic equation this function is assumed to be linear as illustrated in the figure.*

Once again we have a differential equation. Its solution predicts the population size through time. The solution is harder to obtain than that for exponential growth, but after some labor (see the box) the solution is found to be

$$N(t) = \frac{K}{1 + b e^{-rt}} \tag{16.6}$$

where b is determined from the initial condition, $N(0)$, as $b = K/N(0) - 1$.

---

1. The logistic equation is a special case of the Bernoulli equation

$$\frac{dy}{dx} + a(x)y = b(x)y^n \qquad n \neq 0, 1$$

   The solution for the logistic equation follows a standard receipe for the Bernoulli equation.

2. Arrange the logistic equation in the form

$$\frac{(dN/dt)}{N^2} + \frac{(-r)}{N} + \frac{r}{K} = 0$$

3. Define the variable

$$z(t) = \frac{1}{N(t)} - \frac{1}{K}$$

4. The substitution of variables yields

$$\frac{dz}{dt} + rz = 0$$

---

5. Integrating yields

$$z(t) = z(0) \exp(-rt)$$

6. Substituting for $z(t)$ yields

$$\frac{1}{N(t)} - \frac{1}{K} = \left(\frac{1}{N(0)} - \frac{1}{K}\right) \exp(-rt)$$

7. Solving for $N(t)$ gives

$$N(t) = \frac{K}{1 + \{[K/N(0) - 1]\} \exp(-rt)}$$

Figure 16.3 illustrates the solution, (16.6), for several values of r and K. The solution is an S-shaped curve. The initial rate of population growth is controlled by r, the intrinsic rate of increase. The population size eventually comes to equilibrium at K, the carrying capacity. *The symbol K is called the carrying capacity because it is a measure of the amount of renewable resources in the environment in units of the number of organisms these resources can support.* We can infer these roles of r and K from direct inspection of the logistic equation (16.5). In the limit as N tends to zero, the term in brackets is approximately equal to one; hence the logistic equation is approximately the same as that for exponential growth. Therefore, the population growth rate should be mostly governed by r, provided N is small. But as N approaches K, the term in brackets tends to zero, so the population growth rate tends to zero too. And when N equals K, the population growth rate is zero and the population size remains constant. Thus we see that the prediction from the logistic equation is very different from that of

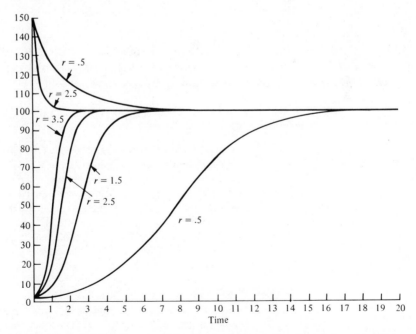

FIGURE 16.3. *Curves of population size as a function of time from the logistic differential equation. Here K = 100.*

exponential growth. Under logistic growth the population size does not grow to infinity but instead levels off at the carrying capacity of the environment.

The logistic equation is based on a very simple assumption—that an individual's reproductive output decreases as a linear function of population size. Because this assumption has been introduced without justification in terms of the actual mechanisms of density dependence, it would be surprising if the logistic equation were ever accurate. But surprisingly enough, the logistic often works very well for the growth of populations of microorganisms in laboratory conditions. Figure 16.4 presents some examples.

There is a possible explanation for the usefulness of the logistic equation in situations where we have no a priori reason to expect it to work, as in Figure 16.4. Let the effect of the actual density mechanism on reproductive output be represented as

$$\left\{ \begin{array}{c} \text{individual's contribution} \\ \text{to population's growth} \end{array} \right\} = f(N) \tag{16.7}$$

Then if we expand $f(N)$ in a Taylor series about the origin, we obtain

$$f(N) = f(0) + Nf'(0) + N^2 \frac{f''(0)}{2} + \frac{N^3 f'''(0)}{6} \cdots \tag{16.8}$$

Now suppose that the sum of all terms of second order or greater is small enough to be ignored, allowing Equation (16.8) to be truncated after $Nf'(0)$. If so, we may identify $f(0)$ with r and $-r/K$ with $f'(0)$ and produce a logistic equation. Thus so long as the mechanisms of density dependence lead to an $f(N)$ that can be safely approximated as $f(0) + Nf'(0)$, then a logistic equation will likely work. Equivalently, the logistic equation will work whenever $f(N)$ can be safely approximated with a straight line. It appears in practice that the actual form of $f(N)$ is often such that it can be approximated by a straight line for many biological purposes.

Logistic Equation in Discrete Time The conventional logistic equation is written as a differential equation in continuous time. However, it is often more realistic and useful to consider the logistic equation in discrete time,

$$\Delta N = r \left[1 - \frac{N}{K} \right] N \tag{16.9}$$

FIGURE 16.4a. [OPPOSITE] *Examples of logistic growth with two species of* Paramecium. *Three experiments using* P. aurelia *are illustrated on the top. The census data are used for the regression of $\Delta N/N$ versus N. The intercept of the regression line is the intrinsic rate of increase, r. The slope of the line is the quantity, $-r/K$. The rate of increase, r, was determined to be .99 and the slope was determined to be $-.001793$. Hence K was found to be $(-.99/-.001793) = 552$. The population size through time was projected using these values of r and K. The result is graphed above. Note that the projected growth curve roughly agrees with the data on population size through time. However, note the large spread in the data. The spread is caused both by error in estimating the population size using a sampling technique and by actual fluctuations in population size. Similar experiments using* P. caudatum *are illustrated below.* [Data from G. F. Gause (1934), *The Struggle for Existence*, reprinted in 1964 by Hafner Publishing Co., New York.]

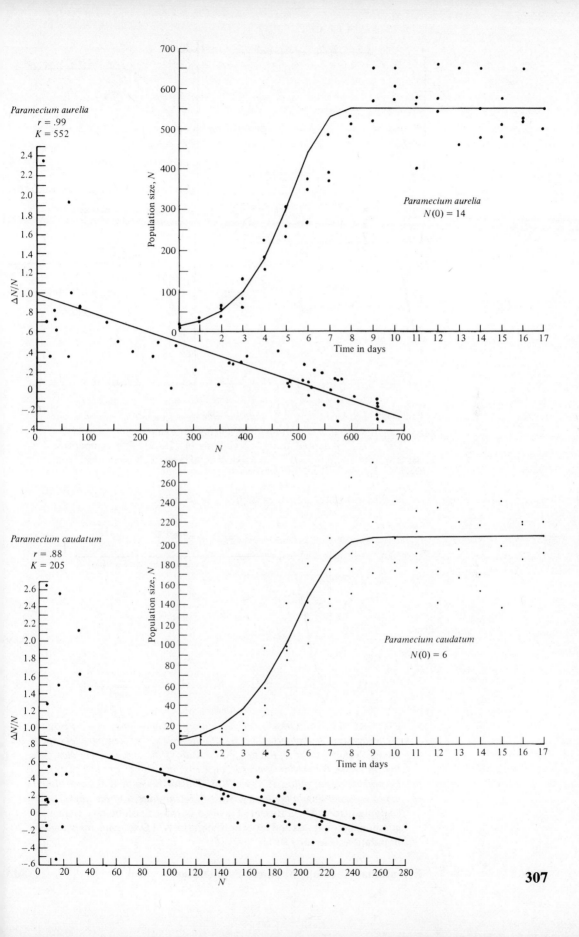

Paramecium aurelia
r = .99
K = 552

Paramecium aurelia
N(0) = 14

Paramecium caudatum
r = .88
K = 205

Paramecium caudatum
N(0) = 6

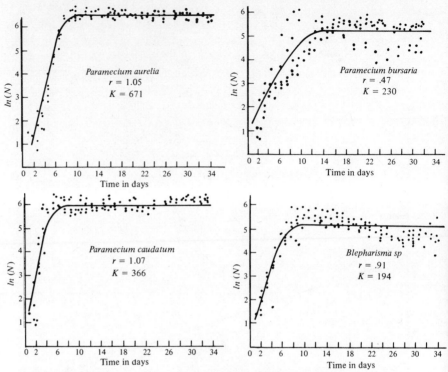

FIGURE 16.4b. *Curve of population size as a function of time for four species of protozoa. The parameters of the logistic equation for each species are also listed.* [Adapted from J. Vandermeer (1969), The competitive structure of communities: An experimental approach with protozoa. *Ecology* **50:** 302–371.]

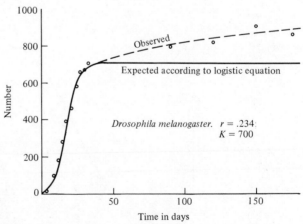

FIGURE 16.4c. *Growth of laboratory populations of* Drosophila melanogaster. *The observed values are the average of four parallel experiments. Note that during the first 30 to 40 days the curve is very close to a logistic curve. But as time proceeds, the carrying capacity appears to increase beyond its initial value of about 700. This increase in K through time is expected according to the theory that will be introduced in the next chapter.* [Adapted from A. A. Buzzati-Traverso (1955), Evolutionary changes in components of fitness and other polygenic traits in *Drosophila melanogaster* populations. *Heredity* **9:** 153–186.]

As before, ΔN is shorthand for $N_{t+1} - N_t$. Equation (16.9) is a difference equation, not a differential equation. The difference equation is more realistic in many cases because it does not presume that the population is growing continuously. In fact, most populations show seasonal reproductive activity and (16.9) could refer to the change in population size from season to season. Indeed, (16.5) requires an instantaneous feedback between an organism's reproductive output and the current population size. But in

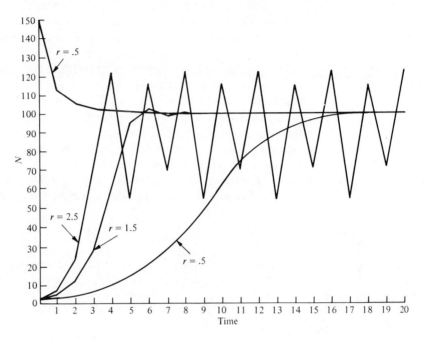

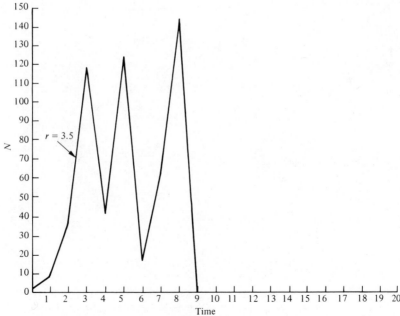

FIGURE 16.5. *Curves of population size as a function of time from the logistic difference equation in discrete time. Here $K = 100$.*

discrete time, the feedback need not be instantaneous. It is only required that the time lag in the feedback be shorter than the census interval. The price for this added realism, however, is complication in the stability of the equilibrium. Figure 16.5 illustrates some computer iterations of Equation (16.9). If r is between zero and one, the trajectories are the familiar S-shaped curves. If r is between one and two, there is overshooting, but eventually the population size converges to K. If r is between 2 and 3, then the population does not converge to K and instead oscillates between finite bounds. Finally, if r is greater than 3, the oscillations cross the $N = 0$ line thus indicating extinction. In contrast, K was a stable equilibrium for all positive r in the logistic differential equation. But in the difference equation, K is an unstable equilibrium if r is greater than two. The reason the stability depends on r in the difference equation is related to the time lag in the feedback between population size and reproductive output. Because the feedback is not instantaneous, overshoots can occur, and if r is large enough the overshoots carry the population far enough past K to prevent convergence. The detailed behavior of the oscillations in the discrete logistic equation has received much study recently. See, for example, May (1974, 1976) and Guckenheimer et al. (1977). These studies and others are discussed at length among the advanced topics presented at the conclusion of Chapter 18.

Chapter 17
DENSITY-DEPENDENT NATURAL SELECTION

THE traditional theory of population dynamics assumes that all individuals in the population are identical genetically. More specifically, the theory assumes the parameters characterizing population growth do not change during the course of population growth caused by natural selection within the population. Traditionally, the theory of population dynamics has been a separate pursuit from the theory of population genetics. This separation appears to reflect an explicit decision by the early theoreticians. It was felt that evolution by natural selection proceeds on a much longer time scale than changes in population size. For example, Lotka (1945) writes:

"The evolution of the system as a whole thus presents two aspects: interspecies evolution, the changes in the distribution of the mass among the heterogeneous gross components (i.e., among the various populations);† and intra-species evolution, the changes in the distribution of mass among the finer subdivisions of the relatively homogeneous species (i.e., among the phenotypes within the populations).

"It is characteristic of inter-group evolution that it can, and at times does, proceed at a rapid rate, ...

"Intra-group evolution, on the other hand, usually proceeds at a rate which, measured in our habitual standards, seems very slow, ..."

Thus Lotka claimed that the time scales for evolution and population growth are different. Indeed, it is still customary to refer to events in "evolutionary time" as distinct from those in "ecological time." Now if it were actually true that the time scales of evolution and ecology were really different, then we would be justified in maintaining population dynamics as a pursuit separate from population genetics and evolution. But as we saw in Chapter 0 natural selection has proved often to be far stronger than originally expected so that gene substitution can definitely occur in the same length of time needed by a population to attain equilibrium abundance. Thus the time scales of evolution and ecology need not be different and undoubtedly are often the same. In light of this realization, continuation of the separation between evolution and ecology is simply untenable. This section explains how these two fields may be combined for their mutual enrichment.

There is a practical advantage to combining population genetics with population dynamics. We no longer must regard the parameters of population growth as arbitrary. Recall from the last section that r, the intrinsic rate of increase, and K, the carry capacity were allowed to have arbitrary values. If we want to apply the logistic equation to some population, we must first measure the appropriate r and K. Indeed some early ecologists are reputed to have believed that a task for ecologists was to assemble a handbook of the values of r, K, and other parameters for various populations just as chemists do for many chemicals in their handbooks. But as we shall see, an evolutionary approach to population dynamics will allow us to infer certain properties of a population's r and K from knowledge of the environment in which the population has evolved.

† The words in parentheses have been added by the author.

The key principle in combining population genetics with population dynamics is the association of r and K with a phenotype. Usually, r and K refer to the growth properties of a *population*. But as we noted in Chapter 16, r and K also determine the function that relates an individual's reproductive output to the population size. In this sense, depending on one's point of view, r and K refer to the growth of a *population* and/or to the way an *individual's* reproductive output is influenced by the population size.

Let us now develop some machinery to handle both population growth and population genetics simultaneously. We have already cast most of the machinery back in Chapter 3. There we found, when using *absolute selective values*, a pair of simultaneous equations for allele frequency and population size,

$$p_{t+1} = \frac{p_t W_{AA} + q_t W_{Aa}}{\bar{W}} p_t \tag{17.1a}$$

$$N_{t+1} = \bar{W} N_t \tag{17.1b}$$

Thus the absolute selective values can be used to predict simultaneously the changes in both gene frequency and population size. So the remaining task is to relate the absolute selective values to the parameters of population dynamics.

A clue on how to proceed is supplied by the notion of "an individual's contribution to population growth." This notion seems suspiciously close to that of fitness—the expected number of offspring produced by an individual. Indeed, the growth of the population is caused by the production of offspring in excess of direct replacement. Thus perhaps an organism's absolute selective value is simply one plus its contribution to the population's growth;

$$W = 1 + \left\{ \begin{array}{c} \text{individual's contribution} \\ \text{to population growth} \end{array} \right\} \tag{17.2}$$

If this relationship is true, we have a recipe for connecting the W's of population genetics with the expressions concerning an individual's contribution to population growth as used in population dynamics. In particular, recall that in the logistic equation an individual's contribution to population growth was defined as

$$\left\{ \begin{array}{c} \text{individual's contribution} \\ \text{to population growth} \end{array} \right\} = r - \frac{r}{K} N \tag{17.3}$$

Therefore, if our method is correct to study how evolution influences r and K, we should consider selective values of the form

$$W = 1 + r - \frac{r}{K} N \tag{17.4}$$

According to this expression, the fitness of an individual is assumed to decrease as a linear function of population size. Now we can easily show that W's given by (17.4) will indeed be useful in studying how evolution influences r and K. At one locus with two alleles we have three W's; let us label their r's and K's as follows

$$W_{AA} = 1 + r_A - \frac{r_A}{K_A} N$$

$$W_{Aa} = 1 + r_H - \frac{r_H}{K_H} N \tag{17.5}$$

$$W_{aa} = 1 + r_a - \frac{r_a}{K_a} N$$

Suppose first that A is fixed, $(p = 1)$, then $\bar{W} = W_{AA}$ and Equation (17.1b) for N_{t+1} reduces to

$$N_{t+1} = \left[1 + r_A - \frac{r_A}{K_A} N_t \right] N_t \tag{17.6}$$

Now subtracting N_t from both sides gives

$$\Delta N = r_A \left[1 - \frac{N}{K_A} \right] N \tag{17.7}$$

This *is* a logistic equation in discrete time with r_A and K_A as parameters. Therefore, if A is fixed, the population grows logistically with the r and K of the AA phenotype. On the other hand, if a is fixed; $(p = 0)$ and $\bar{W} = W_{aa}$. Repetition of the argument shows

$$\Delta N = r_a \left[1 - \frac{N}{K_a} \right] N \tag{17.8}$$

Thus if a is fixed, the population again grows logistically with the r and K of the aa phenotype. However, if both alleles are present, growth is not exactly logistic but is similar to it. Substituting the W's from Equation (12.5) into $\bar{W}$ yields

$$\bar{W} = 1 + \bar{r} - \overline{\left(\frac{r}{K} \right)} N \tag{17.9}$$

where $\bar{r}$ is shorthand for $(p_t^2 r_A + 2 p_t q_t r_H + q_t^2 r_a)$ and $\overline{(r/K)}$ is shorthand for $(p_t^2 r_A / K_A + 2 p_t q_t r_H / K_H + q_t^2 r_a / K_a)$. Substituting $\bar{W}$ from Equation (17.9) into (17.1b) and rearranging gives

$$\Delta N = \left[\bar{r} - \overline{\left(\frac{r}{K} \right)} N \right] N \tag{17.10}$$

Thus if both alleles are present, the population growth at each time is logistic but based on the current genotypic *average* values of r and r/K in the population. We now have all the machinery we need to study the effects of evolution on r and K. We shall want to assign different values of r and K to the different phenotypes and see what natural selection then does.

Results

A Tradeoff Between r and K Is Assumed

We now introduce a fundamental assumption. We assume that there is a tradeoff between an individual's r and K. We assume that a phenotype may have a large K or a large r but not both, since a large value of r or K is purchased at the expense of the other. This assumption is justified by considering in more detail which phenotypic traits produce a large r and which a large K. To obtain a large r, an organism must rapidly allocate the energy it has acquired to the production of offspring. Recall that r controls the initial rate of population growth. If an organism initially allocates energy to other purposes and only later to seed or egg production, then the initial growth of the population will not be as rapid as if organisms immediately allocate their energy to seed or egg production. However, to obtain a large K, an organism must defer the allocation of energy to seeds and eggs and instead develop the facilities to survive under crowded conditions. For example, a plant that allocates energy to leaves will be able to survive under low light conditions. Hence a population of plants each of whom defers allocating energy into seeds until many leaves are produced will ultimately attain a larger population size even though the initial growth of the population may be slower. Similar arguments can be developed for animals. The point is that an organism cannot do both. An organism has a finite amount of energy and time during a growing season, and it can be allocated in favor of producing a high r or a high K but not both.

Undisturbed Environments Let us now examine how natural selection influences the outcome of this tradeoff. Which phenotype evolves, one with a high r or one with a high K? Let the AA phenotype have the highest r and the lowest K, and the aa-phenotype the lowest r and the highest K. Moreover, let us suppose that the heterozygote is the same as the AA homozygote. Suppose, for example, that the numerical values are

$$r_A = .8 \qquad K_A = 8,000$$
$$r_H = .8 \qquad K_H = 8,000 \qquad (17.11)$$
$$r_a = .6 \qquad K_a = 12,000$$

The W's for these values are plotted in Figure 17.1. Figure 17.2 shows how the population would grow if either allele were fixed. To see what happens if both alleles are present, we can iterate Equation (17.1) from different initial conditions using a computer. Figure 17.3 presents the results. In the figure the vertical axis represents the population size and the horizontal axis

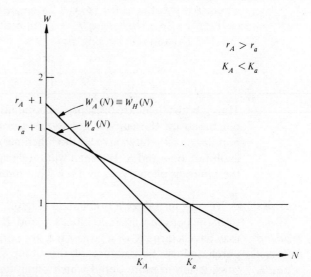

FIGURE 17.1. *Selective value as a function of population size.*

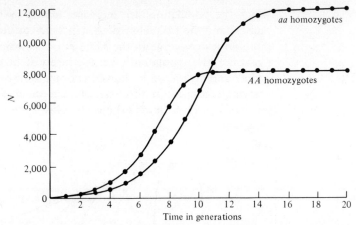

FIGURE 17.2. *Curves of population size through time for populations fixed for the A allele and the a allele. Here* $r_A = .8$, $r_a = .6$, $K_A = 8000$, *and* $K_a = 12,000$. [From J. Roughgarden (1971), Density-dependent natural selection. *Ecology* **52**: 453–468. Copyright 1971 by the Ecological Society of America.]

represents the gene frequency, *p*. The curves represent the course of the population size and gene frequency through time based on various initial conditions. The curves show several phases. First, all the curves lean to the right, indicating that the populations are growing and that the frequency of *A* is increasing. Then the curves lean to the left, indicating that the frequency of *A* is decreasing. Third, the curves seem to converge to a plateau, along

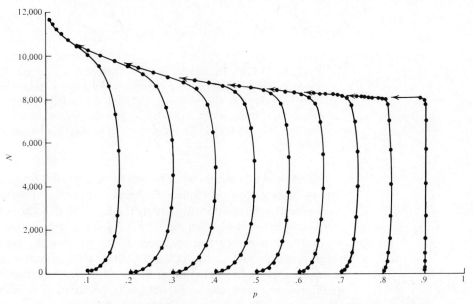

FIGURE 17.3. *Trajectories illustrating the simultaneous change in population size and allele frequency under density-dependent selection. The aa genotype has the highest K and the AA genotype has the highest r. Density-dependent selection leads to fixation of the a allele. Here* $r_H = .8$, $r_A = .8$, $r_a = .6$, $K_H = 8000$, $K_A = 8000$, *and* $K_a = 12,000$. [From J. Roughgarden (1971), Density-dependent natural selection. *Ecology* **52**: 453–468. Copyright 1971 by the Ecological Society of America.]

which the population size increases slowly as compared to the previous phases, and the frequency of A continues to decrease. Eventually, all the populations converge to the same point at which $\hat{p} = 0$ and $\hat{N} = K_a$. Thus eventually the population consists entirely of the aa phenotype, the one with a low r and high K, and achieves a population size equal to the largest of the carrying capacities. In this example, the phenotype that allocates energy to produce a large K at the expense of a low r is eventually favored by natural selection.

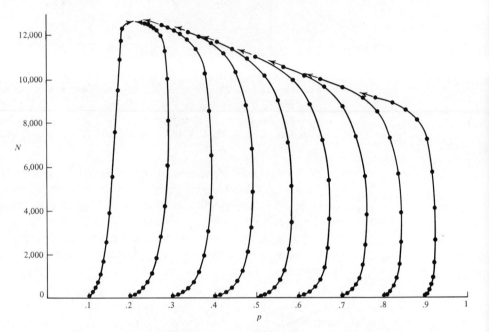

FIGURE 17.4. *Trajectories of density-dependent selection. The hetero-zygote has the highest carrying capacity and a stable polymorphism results. Here* $r_H = .7$, $r_A = .8$, $r_a = .6$, $K_H = 15{,}000$, $K_A = 8000$, *and* $K_a = 12{,}000$. [From J. Roughgarden (1971), Density-dependent natural selection. *Ecology* **52**: 453–468. Copyright 1971 by the Ecological Society of America.]

This example illustrates what proves to be a general property of density-dependent selection. As Figures 17.4 and 17.5 show, the eventual outcome is governed by the relationship among the K's. The r's do not matter. (The r's are discussed further in Chapter 23.) If K_A is the largest, then A is fixed. If K_a is the largest, then a is fixed. If K_H is largest, there is a stable polymorphism; and if K_H is the smallest, then either A or a is fixed, depending on the initial condition. Thus the K's essentially control the qualitative outcome of natural selection, provided that the population is allowed to proceed undisturbed along its full trajectory.

Disturbed Environments However, suppose that the population is not allowed to proceed along its full trajectory, and that the population exists in an environment where recurrent catastrophes cause the population size to be depressed each time to some low value. Suppose, moreover, that the average time between catastrophes is short enough so that the population never has a chance to

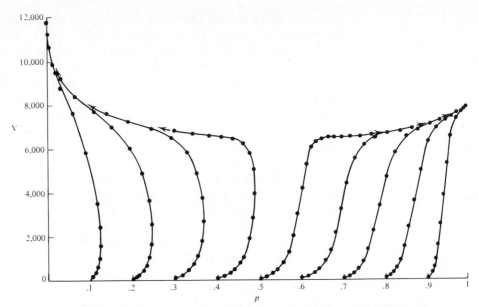

FIGURE 17.5. *Trajectories of density-dependent selection. The hetero-zygote has the lowest K. Trajectories fix on either allele depending on the initial position. Here* $r_H = .7$, $r_A = .8$, $r_a = .6$, $K_{H,} = 5000$, $K_A = 8000$, *and* $K_a = 12,000$. [From J. Roughgarden (1971), Density-dependent natural selection. *Ecology* **52**: 453–468. Copyright 1971 by the Ecological Society of America.]

approach its potential equilibrium abundance. Thus we are to consider an environment that denies a population the opportunity to approach equilibrium densities. The outcome of selection in this sort of extremely disturbed environment is also illustrated in Figure 17.3. If the population size is restricted to low values, the trajectories lean to the right, indicating that the A allele is increasing. If the population size never exceeds the region where selection favors the A allele, then eventually selection must fix the A allele and leave a population with only AA phenotypes. Hence, in this situation, the phenotype that allocates energy to produce a large r at the expense of a low K eventually evolves by natural selection. Thus the eventual outcome in this disturbed environment is opposite to that in an undisturbed environment.

The fact that the high r phenotype wins in the disturbed environment could be predicted simply by inspecting the W's in the limit as N is small. In Equation (17.5) the W's become dominated by the r's as N tends to zero. So the outcome of evolution for small N should be determined by the relationship among the r's, the K's should be essentially irrelevant. And indeed if r_A is the highest, then A is fixed. If r_a is the highest, then a is fixed; if r_H is the highest, there is a polymorphism; and if r_H is the lowest, either A or a is fixed, depending on the initial condition.

K-selection and r-selection The term *K-selection* is used to denote the situation where the K's determine the outcome of selection, irrespective of the r's. As we have seen, K selection occurs as a result of density-dependent selection in environments where a population can attain a size near the equilibrium level. The phase *r-selection* is used where the *r*'s determine the outcome of selection, irrespective of the *K*'s. The phenomenon of *r*-selection occurs in environments where density-independent mortality constrains the population size to levels far below the potential equilibrium level. The distinction between *K* selection and *r* selection originates with MacArthur (1962). The explication of these ideas in terms of density-dependent selection occurred in a series of papers by Anderson (1971), Charlesworth (1971), Roughgarden (1971), and Clarke (1972). A more recent paper which includes consideration of frequency dependence is by Smouse (1976).

If the theory above is true, we ought to be able to inspect the level of density-independent mortality in an environment and infer the magnitude of the *r*'s and *K*'s of populations in that environment. For this reason the theory would provide a powerful tool in ecology. It would remove the need to view the parameters of population growth as purely arbitrary and as requiring individual measurement in each case. Thus it is of crucial interest to test this theory in natural populations.

Test with Dandelions

Solbrig (1971) has provided a test of this theory using populations of dandelions. Solbrig established study sites in three locations. The first was heavily disturbed by regular weekly lawn mowing and was in open sunlight with dry soil. The second was moderately disturbed with monthly mowing and had moderate shade and soil moisture. The third was the least disturbed with mowing occurring only once a season and had the most shade and soil moisture. Solbrig detected several genotypes in the population, labeled *A* through *D*. In natural populations, type *A* has the largest allocation to rapid seed production, while type *D* defers seed production until after substantial leaf formation. Thus type *A* would appear to be a high *r* and low *K* phenotype while type *D* is a low *r* and high *K* phenotype. Also, types *B* and *C* are intermediate. Table 17.1 shows that these genotypes vary in frequency among the different sites.

Table 17.1. Percentage of Each of Four Genotypes in Three Populations of Dandelions (*Taraxacum officinalis*). [Data from Solbrig (1971).]

Population	A	B	C	D
1. Dry, full sun, highly disturbed	73	13	14	0
2. Dry, shade, medium disturbance	53	32	14	1
3. Wet, semishade, undisturbed	17	8	11	64

There are two hypotheses for the data so far. First, the conventional hypothesis is that type *A* is adapted to open sun and dry soil conditions while type *D* is adapted to shaded moist conditions. If so, *regardless of the level of disturbance*, type *A* should always predominate over type *D* in open sun and dry soil, and vice versa in shaded moist conditions. By the second hypothesis, involving *r*- and *K*- selection, the level of density-independent mortality is the crucial factor. Type *A* is adapted to high levels of density-independent mortality by having a high *r* and low *K*, and vice versa, for type *D*. If so, when type *A* and type *D* are grown in undisturbed greenhouse conditions, in all sorts of soil and light, then type *A* should always produce seeds first and type *D* should always defer seed production, produce many leaves, and eventually predominate over type A in all conditions. And if so, then type D should increase in frequency under undisturbed conditions regardless of the soil and water characteristics. Solbrig collected seeds from these types, and grew them in many combinations of soil and other conditions in the greenhouse. He found that, indeed, type A always was the first to seed, but that type D eventually predominated because of its greater investment in leaves. These data show that dandelion populations evolve a high r and low K under conditions of high density-independent mortality and a low r and high K in undisturbed conditions.

We shall continue to use the theory of density-dependent selection throughout our study of evolutionary ecology. After introducing models for the dynamics of several interacting populations in Chapters 21 and 22, we shall investigate the corresponding density-dependent selective values in Chapter 23.

Chapter 18
POPULATION GROWTH WITH AGE STRUCTURE

AGE is a property we are all very aware of. Friendships tend to occur among similarly aged people. Consumer goods are aimed at specific ages, and although all of us do age, few of us want to. Obviously, age is an important and general property of organisms. But so far in discussing population growth, we have considered only total population size. However, the total size is a coarse description of a population. A breakdown of the population into age classes would provide a finer and more informative description of a population. Our task in this section is to understand the relationship between age and population growth.

Four Basic Concepts

Age Distribution

There are four concepts we seek to interrelate. First, every population has an *age distribution* that indicates the proportion of the population in various age classes. Figures 18.1 and 18.2 illustrate the age distribution for two countries, one a so-called "developed country" (DC) and the other a "less-developed country" (LDC). Both populations have distributions that decrease more or less monotonically with age. There are always fewer old people than young. One of the questions we shall address is why the age distribution decreases monotonically with age. Need this be so? Are there conditions where the age distribution could show a peak at some old age rather than among babies as indicated in Figures 18.1 and 18.2. Also note

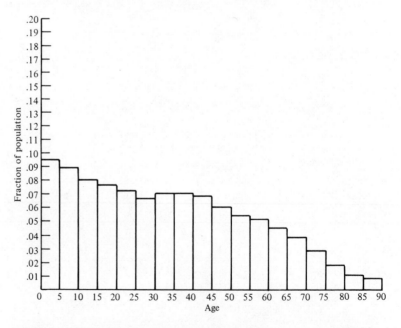

FIGURE 18.1. *Age distribution for females in Spain in 1967. The vertical axis represents the fraction of the population and the horizontal axis represents age.* [Data from N. Keyfitz, and W. Flieger (1971), *Population, Facts and Methods of Demography*, W. H. Freeman and Company, Publishers, San Francisco.]

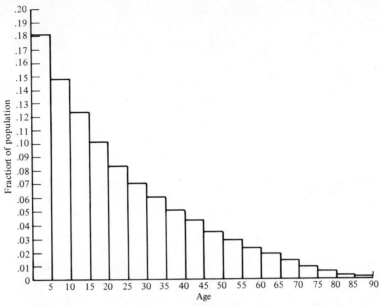

FIGURE 18.2. *Age distribution for females in Mexico in* 1966. *Note the comparatively higher proportion of young people in Mexico as compared with Spain.* [Data from N. Keyfitz, and W. Flieger (1971), *Population, Facts and Methods of Demography*, W. H. Freeman and Company, Publishers, San Francisco.]

that the LDC has a much higher fraction of young people than does the DC. Why?

Population Growth Rate

Second, every population has a *growth rate*. Figure 18.3 shows how the DC and LDC have grown in population size over 11 years. The LDC is growing much faster than the DC. Is there any connection between the age structure of the LDC and its growth rate?

Age-Specific Mortality

Third, in every population there is a regime of *age specific mortality*. Figures 18.4 and 18.5 illustrate two equivalent ways of describing the mortality incurred by individuals of different ages. The first, in Figure 18.4, is the *survivorship curve*. Suppose that 1000 individuals of age zero are placed in the environment. (A group of 1000 newborn is called a *cohort*.) The survivorship curve describes the fraction of these 1000 individuals alive through time. In humans hardly anybody lives over 100 years and the survivorship curve is nearly zero by the time 100 years are reached. Note that the mortality in the LDC occurs sooner than in the DC†. An equivalent way of presenting the age-specific mortality data is with the *age-specific conditional probability of survival*. This quantity reports the fraction of organisms of a given age who survive from one census to the next. Thus the survivorship curve directly expresses the total life spans of organisms while the conditional probability of survival directly expresses the probability of surviving over small steps of time.‡ One can always calculate the

---

† The survivorship curve thus describes the probability of surviving from age 0 to age x or more.

‡ Suppose a census occurs each year; then the conditional probability of survival is the probability of surviving from age x to $x + 1$, conditional on being alive at age x.

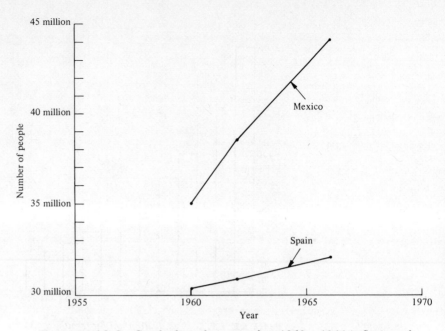

FIGURE 18.3. *Graph of population size from* 1960 *to* 1966 *in Spain and Mexico. Note that the population of Mexico is growing faster than that of Spain.* [Data from N. Keyfitz, and W. Flieger (1971), *Population, Facts and Methods of Demography*, W. H. Freeman and Company, Publishers, San Francisco.]

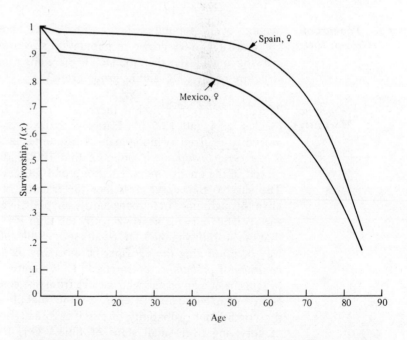

FIGURE 18.4. *Survivorship curves, $l(x)$, for females in Spain and Mexico in* 1966. [Data from N. Keyfitz, and W. Flieger (1971), *Population, Facts and Methods of Demography*, W. H. Freeman and Company, Publishers, San Francisco.]

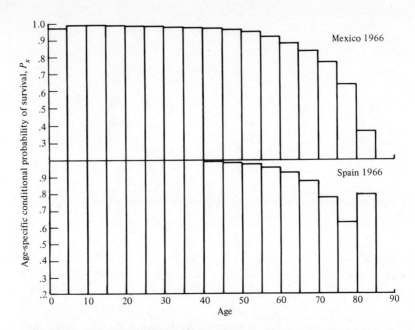

FIGURE 18.5. *Graph of the age-specific conditional probabilities of survival, P_x, for females in Spain and Mexico in* 1966. *These numbers are entered in the subdiagonal of the Leslie Matrix.* [Data from N. Keyfitz, and W. Flieger (1971), *Population, Facts and Methods of Demography*, W. H. Freeman and Company, Publishers, San Francisco.]

survivorship curve from the age specific conditional probability of survival, and vice versa, but there are times when one is more convenient than the other. Figure 18.5 illustrates the conditional probabilities of survival for the DC and LDC. Again, exactly the same data are used to prepare Figures 18.4 and 18.5; they are just different methods of presenting the data. You should closely inspect these two figures to get an intuitive feel for how to translate one curve into the other.

Age-Specific
Fertility
The fourth concept is that of *age-specific fertility*. Figure 18.6 presents data from the DC and LDC on the number of offspring from females as a function of their age. The function representing these data is called the *maternity function*. Note that females in the LDC have more offspring *and* tend to have them slightly earlier than in the DC. Our task now is to interrelate these four concepts of age structure, population growth, age-specific mortality, and age-specific fertility. We shall see that we can use the mortality and fertility information to predict the population growth and age structure in some conditions. To interrelate these four concepts, we shall develop a mathematical model that will predict the age structure through time. There are two formulations to be considered. The first, due to P. H. Leslie, is based on discrete time. It is the easiest to iterate on a computer and has found extensive practical use in projecting the population growth for various countries. The second, attributable to A. J. Lotka, is based on continuous time. It is often the easiest to solve analytically and will allow us to learn more than would be possible with the Leslie approach alone. The study of population growth with age structure is called *demography*.

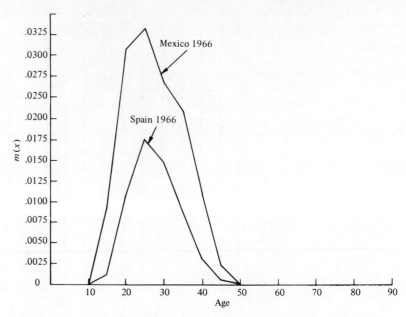

FIGURE 18.6. *Maternity function,* $m(x)$, *for female offspring in Spain and Mexico during* 1966. [Data from N. Keyfitz, and W. Flieger (1971), *Population, Facts and Methods of Demography,* W. H. Freeman and Company, Publishers, San Francisco.]

The Leslie Formulation for Discrete Time

The Leslie Model

In using the Leslie formulation, we agree to census the population at regular intervals, say 10 years, *and* we agree that the "width" of an age class is the same as the census interval. Thus, if we census every 10 years, then an age class has to be 10 years wide. Hence individuals of the youngest age class are those with ages 0 through 9; the next age class comprises those aged 10 through 19; and so forth. Next we use the age-specific mortality data in the form of the age-specific conditional probabilities of survival. Let x stand for the age class label. The youngest age class is labeled 0 by convention and the last is labeled ω. Here P_x is the fraction of those alive in age class x who survive through the interval between censuses. After they have survived the interval, they then become members of the next age class. Let $t+1$ denote time one interval later. Thus if the interval is 10 years, and if t is 1970 then $t+1$ is 1980 and $t+2$ is 1990, and so forth. Let $n_{x,t}$ denote the number of individuals in age class x at time t. Then we can use the P_x as follows

$$n_{x+1,t+1} = P_x n_{x,t} \qquad (x = 0, 1, \ldots \omega - 1) \qquad (18.1)$$

This simply states in symbols what we have already said: that a fraction P_x of those in class x at time t survive to enter class $x+1$ at time $t+1$. This expression allows us to predict the number in every age class, except the youngest at $t+1$, given that we know the mortality data in the form of P_x. To get the youngest age class at time $t+1$, we must use the fertility data. Let F_x be the average number of offspring born to individuals of age x who survive to the upcoming census. Here F_x is not simply the number of offspring born to a parent of age x but is somewhat less because of juvenile mortality during the time between birth and the next census. If we know the fertility data in this form, then the number in the youngest age class at $t+1$ is simply the sum of the offspring produced by parents of all ages,

$$n_{0,t+1} = F_0 n_{0,t} + F_1 n_{1,t} + F_2 n_{2,t} + \cdots + F_\omega n_{\omega,t} \qquad (18.2)$$

When we combine these equations, we have a complete model to predict the age structure through time. Given the mortality and fertility data, we can iterate these equations from any initial condition and see how the age structure develops through time. For example, if there are just three age classes, the equations are

$$n_{0,t+1} = F_0 n_{0,t} + F_1 n_{1,t} + F_2 n_{2,t}$$

$$n_{1,t+1} = P_0 n_{0,t} \qquad (18.3)$$

$$n_{2,t+1} = P_1 n_{1,t}$$

As a problem, you should iterate this system once or twice starting with, say, a uniform age distribution and see what happens. The iteration of these equations is very easy to implement on a computer. Moreover, Keyfitz and Flieger (1971) have tabulated the mortality and fertility data of many countries for direct use in these equations.

The Stable Age Distribution and Exponential Growth
If these equations are iterated for a long time, a remarkable result emerges as illustrated in Figure 18.7. It turns out that there is a constant λ which can be computed from the P_x and F_x data such that

$$n_{x,t+1} \approx \lambda n_{x,t} \qquad \text{(for all } x, \text{ provided } t \text{ is large)} \qquad (18.4)$$

Thus it emerges that the number in class x at $t+1$ is simply λ times the number in that class at t. Note that λ does not depend on x; the same λ applies to all classes—thus every class increases by the factor λ. There are two important consequences of this result. First, if every class increases by the same factor with each iteration, the *relationship* between the classes is unaltered; that is, the *proportion* of the total population in each class remains fixed. But recall that the proportion of the population in each age class is called the age distribution. Therefore, as t becomes large, the age distribution stops changing, that is, comes to equilibrium. This equilibrium age distribution eventually attained by the population is called the *stable age distribution*. It is determined solely by the P_x and F_x data. Thus, by iterating equations (18.1) and (18.2) together on a computer, we discover that the population age distribution, regardless of its initial configuration, eventually approaches a stable age distribution which is determined solely by the mortality and fertility data. Second, if each class increases by the same factor with every iteration, then the total population size, N_t, also increases by this factor

$$N_{t+1} \approx \lambda N_t \qquad (t \text{ large}) \qquad (18.5)$$

If (18.5) is true, then the expression for the population size through time is

$$N_t = \lambda^t N_0 \qquad (18.6)$$

Now compare (18.6) with the usual expression for exponential growth

$$N_t = e^{rt} N_0 \qquad (18.7)$$

Hence we can identify λ as

$$\lambda \equiv e^r \qquad (18.8)$$

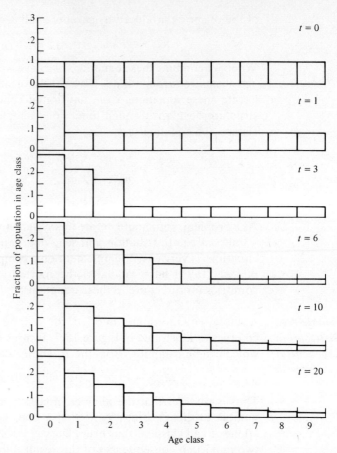

FIGURE 18.7a. *Approach to the stable age distribution. A population with* 10 *age classes was simulated using Equations* (18.1) *and* (18.2). *The population was started out with two individuals in every age class. This initial condition represents a uniform age distribution as illustrated at the top of the figure. As the equations are iterated, the age distribution rapidly approaches the stable age distribution. In the simulation* $F_x = .3$ ($x = 0, \ldots \omega$) *and* $P_x = .8$ ($x = 0, \ldots \omega - 1$).

These results show that when Equation (18.4) is true, the entire population is growing exponentially and the factor λ is related to the intrinsic rate of increase according to (18.8). The value of λ does not depend on the initial condition. It is determined solely by the P_x and F_x data. We stress that the population grows exponentially *only* when the stable age distribution is reached. During the approach to the stable age distribution, it is not growing exponentially. During the approach all sorts of things can happen to the population size depending on the initial age distribution. Thus when the stable age distribution has been attained, both the age distribution and population growth rate can be predicted from the age-specific fertility and mortality data.

For those of you who have had a course in linear algebra or who have read the appendix on matrix algebra, we wish to mention that there is an elegant way of writing the Leslie theory above. There is no new biology involved but simply a more concise formulation of the mathematics. Using matrix

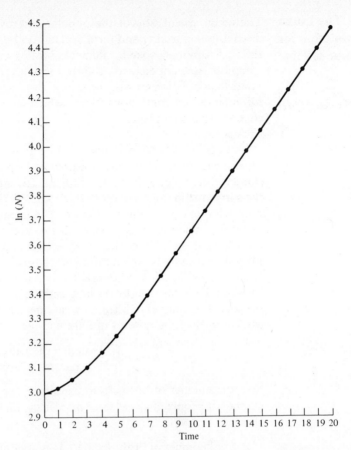

FIGURE 18.7b. *Population size as the stable age distribution is approached. The vertical axis is the natural log of the population size. The horizontal axis is time. A straight-line plot of $\ln(N_t)$ versus t indicates exponential growth. Note that the population is initially not growing exponentially. But after $t = 10$ the population has essentially achieved exponential growth. In Figure* 18.7a *we see that the population also has essentially achieved the stable age distribution at $t = 10$.*

multiplication, we can write the three equations in (18.3) as

$$
\begin{pmatrix} n_{t+1,0} \\ n_{t+1,1} \\ n_{t+1,2} \end{pmatrix} = \begin{pmatrix} F_0 & F_1 & F_2 \\ P_0 & 0 & 0 \\ 0 & P_1 & 0 \end{pmatrix} \begin{pmatrix} n_{t,0} \\ n_{t,1} \\ n_{t,2} \end{pmatrix} \tag{18.9}
$$

We are viewing the numbers in each age class as a column vector. The matrix with the F_x and P_x data is called the *Leslie matrix*. The F_x always enter in the top row and the P_x enter in the subdiagonal. If we label the column vector of rank $\omega + 1$ as $\vec{n}_t$ and the $(\omega + 1) \times (\omega + 1)$ square Leslie matrix as $\mathcal{P}$, then we can rewrite the equation in general as

$$
\vec{n}_{t+1} = \mathcal{P}\vec{n}_t \tag{18.10}
$$

Then it can be shown that the λ above is the biggest eigenvalue of $\mathcal{P}$ and that the stable age distribution is the eigenvector corresponding to λ.

The actual calculation of the population growth rate and the stable age distribution mortality and fertility data can be complicated with the Leslie theory. Although specific numerical cases can easily be worked on a computer, it is not easy to produce formulas that show how the P_x and F_x relate to population growth and the age distribution. We now turn to Lotka's formulation in continuous time, which allows derivation of formulas connecting the mortality and fertility data with population growth and the age distribution.

In Lotka's theory, we shall assume that the number of newly born offspring is being *continually* censused in the population. The term $B(t)$ will stand for the number of births at time t. The mortality data are specified as the survivorship function, $l(x)$, which you will recall, is the fraction of those born who survive to age x or more. The fertility data are specified by the maternity function, $m(x)$, which is the average number of births to an individual of age x. Since the census is continuous, no allowance is made for juvenile mortality between birth and census as was done with the F_x in the Leslie theory. The key to Lotka's theory is to recognize a certain relationship between the number of births now and the number of births at any time in the past. The number of births now, at time t, is obtained by summing up the current births from parents of different ages

$$B(t)=\int_0^{\infty} \left[\begin{array}{c}\text{number of births from} \\ \text{parents of age } x\end{array}\right] dx \qquad (18.11)$$

Now the number of births from parents of age x is simply $m(x)$ times the number of people of age x, since $m(x)$ is the number of births per individual of age x

$$\left[\begin{array}{c}\text{number of births from} \\ \text{parents of age } x\end{array}\right]=\left[\begin{array}{c}\text{number of individuals} \\ \text{of age } x\end{array}\right]m(x) \quad (18.12)$$

Finally, the number of people of age x is simply the number of babies born x years ago times the fraction that survive to age x.

$$\left[\begin{array}{c}\text{number of individuals} \\ \text{of age } x\end{array}\right]=\left[\begin{array}{c}\text{number of births} \\ x \text{ years ago}\end{array}\right]\left[\begin{array}{c}\text{fraction of newborn who} \\ \text{survive to age } x\end{array}\right]$$

$$(18.13)$$

Using our symbols, the number of births x years ago must be denoted as $B(t-x)$ and the fraction who survive to age x is $l(x)$. Therefore, in symbols (18.13) is

$$\left[\begin{array}{c}\text{number of individuals} \\ \text{of age } x\end{array}\right]=B(t-x)l(x) \qquad (18.14)$$

Now we can put it all together as

$$B(t)=\int_0^{\infty} B(t-x)l(x)m(x)\, dx \qquad (18.15)$$

This relationship is the key to Lotka's theory. It is an equation called a *renewal* equation because it also turns up in the study of a kind of stochastic process called a renewal process. Equation (18.15) is an integral equation and to solve it means to find a function, $B(t)$, which satisfies the condition on its integral as stated in (18.15).

There are direct ways of solving (18.15) but for our purposes the best technique will be to try to guess a solution and then see if it works. We learned above that when the stable age distribution is attained, the population then shows exponential growth. So a good guess for $B(t)$ is that the number of babies increases exponentially through time. Therefore, as a *trial solution*, let us take

$$B(t) = e^{rt}B(0) \tag{18.16}$$

If $B(t)$ is to be given by (18.16), then $B(t-x)$ is $e^{r(t-x)}B(0)$ and substituting into (18.15) gives

$$e^{rt}B(0) = \int_0^\infty e^{r(t-x)}B(0)l(x)m(x)\,dx \tag{18.17}$$

Upon factoring out $e^{rt}B(0)$ from within the integral, we obtain

$$e^{rt}B(0) = e^{rt}B(0)\int_0^\infty e^{-rx}l(x)m(x)\,dx \tag{18.18}$$

Therefore, if the trial solution is to satisfy (18.15), it is necessary and sufficient that the integral on the right had side equal one

$$\int_0^\infty e^{-rx}l(x)m(x)\,dx = 1 \tag{18.19}$$

Again, a solution of the form $e^{rt}B(0)$ will indeed satisfy (18.15) provided r is such that (18.19) is true. This condition involving r that allows $e^{rt}B(0)$ to be a solution is in fact the formula with which we can now relate the mortality and fertility data to the population growth rate. Once we have measured $l(x)$ and $m(x)$, we perform the integration in (18.19) and solve for r. Equation (18.19), which relates $l(x)$ and $m(x)$ to r, is called *Euler's equation*.

We can use what we have so far of Lotka's theory to discuss the conditions for zero population growth, ZPG. A population that has attained ZPG is not growing or declining. We want to know what conditions must be met for ZPG to occur. First, you should know that there is a quantity called the net reproductive rate, R_0, which is the expected total number of offspring left by an average individual throughout its life. The quantity, R_0, is defined as

$$R_0 = \int_0^\infty l(x)m(x)\,dx \tag{18.20}$$

We would expect one of the conditions for ZPG to be that direct replacement occurs; that is, R_0 equals one. This condition arises because one requirement of ZPG is that the intrinsic rate of increase must equal zero. By Euler's equation (18.19), if r is zero, then $\int l(x)m(x)\,dx$ equals one. Thus, indeed, direct replacement is equivalent to an r of zero. *But* Euler's Equation (18.19) only tells us the population growth rate, r, if there is a stable age distribution, because only then is there exponential growth (including $r = 0$). Hence the second condition of ZPG is that the population must have attained the corresponding stable age distribution. It is often possible for the $l(x)$ and $m(x)$ regime to entail that r is zero, but ZPG cannot be attained until the population attains the corresponding stable age distribution.

The stable age distribution is easily calculated once we have determined r from the $l(x)$ and $m(x)$ data. Let $c(x)$ stand for the fraction of the population with age x.

$$c(x) = \frac{[\text{number of individuals of age } x]}{[\text{total number of individuals}]} \qquad (18.21)$$

Now the total number of individuals is simply the sum of the individuals of different ages

$$\begin{bmatrix} \text{total number of} \\ \text{individuals} \end{bmatrix} = \int_0^\infty \begin{bmatrix} \text{number of individuals} \\ \text{of age } x \end{bmatrix} dx \qquad (18.22)$$

Recall from (18.14) that the number of individuals of age x is $B(t-x)l(x)$, so $c(x)$ can be expressed in symbols as

$$c(x) = \frac{B(t-x)l(x)}{\int_0^\infty B(t-x)l(x)\,dx} \qquad (18.23)$$

Finally, at the stable age distribution, we know that the population is growing exponentially with rate r so that $B(t-x) = e^{r(t-x)}B(0)$. Substituting into (18.23) gives

$$c(x) = \frac{e^{r(t-x)}B(0)l(x)}{\int_0^\infty e^{r(t-x)}B(0)l(x)\,dx} \qquad (18.24)$$

and then $e^{rt}B(0)$ cancels out leaving

$$c(x) = \frac{e^{-rx}l(x)}{\int_0^\infty e^{-rx}l(x)\,dx} \qquad (18.25)$$

This formula gives the stable age distribution. With the $l(x)$ and $m(x)$ data we first calculate r, using Euler's equation (18.19). Then we use this value of r and the $l(x)$ data to compute $c(x)$ from (18.25).

The formula for $c(x)$ tells us about the shape of the age distribution. First, if r is zero, then $c(x)$ is just a constant times $l(x)$. (The constant is $1/\int l(x)\,dx$.) Thus, if r is zero, the stable age distribution has exactly the same shape as the survivorship curve. This fact is of some practical interest. The measurement of $l(x)$ requires tagging organisms and keeping track of them for long periods of time. But if the population size is constant ($r = 0$) and if the stable age distribution obtains, then the age distribution is the same as $l(x)$ (apart from the constant factor). The age distribution can often be measured at one time by direct census. Hence under these assumptions, direct census of the age structure at one time provides an easier way to obtain information about age-specific mortality than tagging and long-term observation. Second, $c(x)$ must decrease monotonically with x for $r \geqslant 0$ because $l(x)$ decreases monotonically with x. Moreover, as r increases, then $c(x)$ decreases faster with x. This indicates that as r increases, the younger ages become an increasing proportion of the population. Recall from Figures 18.1 and 18.2 the difference in shape between the age distribution for the DC and LDC. Our formula for $c(x)$ accounts for the higher proportion of young people in the LDC than in the DC.

The remainder of this chapter treats several advanced topics concerning population dynamics and age structure.

Fisher's Reproductive Value

Fisher has introduced a concept that proves useful in several contexts. It is the concept of *the value of an organism's future offspring relative to its own current value*. The value of an offspring is the fraction of the population which that offspring represents. Thus the value of one offspring in a total population of 10 is $\frac{1}{10}$. The value of an individual's future offspring will depend on the individual's current age, as we shall see. To calculate the value of an organism's future reproduction, we must take account of the possibility that the organism may not, in fact, survive to produce all the offspring which it could produce, and the possibility that the population as a whole has been growing during the time the organism is alive. If the population grows prior to producing an offspring, then the value of that offspring drops because it is then a smaller fraction of the population.

Suppose the individual being studied is now of age a. When it was originally born the population size is N_0. If we assume exponential growth, the population size is now $N_0 e^{ra}$. The value of the future offspring is

$$\begin{bmatrix} \text{value of future} \\ \text{offspring} \end{bmatrix} = \int_a^\infty \begin{bmatrix} \text{prob of living to age } x \text{ or more} \\ \text{given alive at age } a \end{bmatrix}$$

$$\times \begin{bmatrix} \text{value of offspring} \\ \text{produced at age } x \end{bmatrix} dx \qquad (18.26)$$

The value of offspring produced at age x is simply

$$\begin{bmatrix} \text{value of offspring} \\ \text{produced at age } x \end{bmatrix} = \frac{m(x)}{N_0 e^{rx}} \qquad (18.27)$$

The probability of living to age x or more, given that the individual is alive at age a, is found from Bayes' rule

$$P(A|B) = \frac{P(A \cup B)}{P(B)} \qquad (18.28)$$

For $x > a$, $l(x)$ is the probability of being alive both at age x or more *and* at age a, and $l(a)$ is the probability of being alive at age a (or more). Therefore,

$$\begin{bmatrix} \text{prob of living to age } x \text{ or more,} \\ \text{given that it is alive at age } a \end{bmatrix} = \frac{l(x)}{l(a)} \qquad (18.29)$$

Hence

$$\begin{bmatrix} \text{value of future} \\ \text{offspring} \end{bmatrix} = \int_a^\infty \left(\frac{l(x)}{l(a)}\right)\left(\frac{m(x)}{N_0 e^{rx}}\right) dx \qquad (18.30)$$

This expression gives the value of an organism's future offspring, given that it is alive at age a. That organism's own value is

$$\begin{bmatrix} \text{value of organism} \\ \text{of age } a \end{bmatrix} = \frac{1}{N_0 e^{ra}} \qquad (18.31)$$

Therefore, the value of an organism's future offspring relative to its own value, denoted as v_a, is

$$v_a = \frac{\int_a^\infty [l(x)/l(a)][m(x)/(N_0 e^{rx})] \, dx}{1/(N_0 e^{ra})} \qquad (18.32)$$

This expression readily simplifies to

$$v_a = \frac{e^{ra}}{l(a)} \int_a^\infty e^{-rx} l(x) m(x)\, dx \qquad (18.33)$$

The value v_a always equals one when $a = 0$; that is, the reproductive value of a newborn is always exactly one. Typically, v_a rises above one and peaks near the age of maximum reproductivity and then declines steadily to zero. The typical shape of v_a as a function of a is sketched in Figure 18.8.

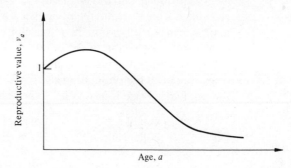

FIGURE 18.8. *Sketch of the typical shape of the curve of reproductive value, v_a, as a function of age.*

A More General Formulation of the Lotka Model

An implicit assumption in the formulation of the Lotka model for population growth with age structure is that the population must have been present in its current location for a long time. This assumption was made in the following context. We needed to know the number of parents of age x. We calculated the number of parents of age x as $l(x)B(t-x)$. In this calculation we assume that the parents who are of age x now were born into the population x years ago and were censused among the new born at that time. However, as we shall see, it is not hard to take account of the possibility that parents of a certain age and older were among the original colonists of the region and were not born into the population.

Suppose that the region containing the population being studied was originally colonized at time $t = 0$. Let the number of colonists whose ages at the time of colonization were between x and $x + dx$ be denoted as $n_0(x)\, dx$. Also, there is negligible immigration after the original colonization. (Emigration can be included in the $l(x)$ curve.) Then the basic format of the equation for $B(t)$ is

$$B(t) = \begin{bmatrix} \text{number of newborn from} \\ \text{parents who were born} \\ \text{into the population} \end{bmatrix} + \begin{bmatrix} \text{number of newborn from} \\ \text{parents who were among} \\ \text{the original colonists} \end{bmatrix}$$

$$(18.34)$$

The first term in the expression is familiar:

$$\begin{bmatrix} \text{number of newborn from parents} \\ \text{who were born into the population} \end{bmatrix} = \int_0^t B(t-x) l(x) m(x)\, dx$$

$$(18.35)$$

Note that no parent who was born into the population can be more than t

years old, since the population was founded t years ago. The second term in (18.34) is constructed as follows:

$$
\begin{bmatrix} \text{number of newborn from} \\ \text{parents who were among} \\ \text{the original colonists} \end{bmatrix} = \int_0^\infty \begin{bmatrix} \text{number of offspring from} \\ \text{parents who were of age} \\ x \text{ when they colonized} \\ \text{and are now of age } x+t \end{bmatrix} dx
$$

(18.36)

The integrand in (18.36) is seen to be

$$
\begin{bmatrix} \text{number of offspring from} \\ \text{parents who were of age} \\ x \text{ at colonization} \end{bmatrix} = m(x+t)\frac{l(x+t)}{l(x)}n_0(x) \qquad (18.37)
$$

where $m(x+t)$ is simply the fertility of individuals of age $x+t$; $l(x+t)/l(x)$ is the probability of living to age $x+t$ or more conditional on being alive at age x, as explained in (18.28) and (18.29). Finally, $n_0(x)$ gives the number of individuals of age x among the original colonists. Upon combining the above equations, we obtain

$$
B(t) = \int_0^t B(t-x)l(x)m(x)\,dx + \int_0^\infty n_0(x)\frac{l(x+t)}{l(x)}m(x+t)\,dx
$$

(18.38)

This form of the Lotka model converges to the simpler form given earlier in (18.15) in the limit as $t \to \infty$, provided that $l(x+t)$ tends to zero with t sufficiently rapidly. Equation (18.38) is called a linear Volterra integral equation, and the existence and uniqueness of solutions to (18.38) follows directly from the theory of integral equations of this form.

Density Dependence and Age Structure in Continuous Time

An extremely important topic of research in the theory of population dynamics has been to incorporate density dependence into the equations that describe population growth with age structure. The logistic equation discussed in earlier chapters omits any reference to age structure, and this omission limits its application in many situations. The Leslie and Lotka models for population growth with age structure presented above incorporate no density dependence and lead to exponential growth as the population approaches the stable age distribution. Obviously, exponential growth cannot occur indefinitely and the various mechanisms of density dependence discussed in Chapter 16 undoubtedly enter the picture. There has been very active research in recent years aimed at producing a theory of population dynamics that includes both density dependence and age structure. In this section we briefly summarize results for a continuous time model developed by Gurtin and MacCamy (1974). An additional important reference is Rorres (1976).

The basic simplifying assumption in the theory of this section is that the age-specific fertility and survivorship is supposed to depend on the total population size, N, and not on the detailed age distribution. That is, let $n(x, t)\,dx$ denote the number of organisms in the population at time t whose ages lie between x and $x+dx$. The total population size is defined as

$$
N(t) = \int n(x, t)\,dx \qquad (18.39)
$$

It is assumed that the fertility and survivorship depend only on $N(t)$ and not on $n(x, t)$.

The Density
Dependence Model

The model consists of a pair of integral equations, one for the number of newborn at time t, $B(t)$ and the other for $N(t)$.

The integral equation for $B(t)$ is essentially the same as (18.38). The only new features are allowances for the dependencies of the fertility and survivorship on $N(t)$. Specifically, the maternity function is now also a function of the population size, $m(x, N(t))$. The value of m at any time depends on the current population size $N(t)$. The survivorship function depends on the population size in a more complicated way. $l(x, 0, N(\cdot))$ is the probability of surviving from birth to age x or more given that the population size has had a certain history during the time from the individual's birth at $t - x$ up until time t. The symbol $N(\cdot)$ indicates that the survivorship curve depends not only on the the current population size, but also on its recent history.

The explicit way in which the survivorship curve depends on the $N(t)$ curves can be determined by examining the effect of the population size on the instantaneous death rate. Let $\mu(x, N(t))$ be the instantaneous death rate for an organism of age x in a population whose size is $N(t)$. Let us begin with a hypothetical cohort of newborn in the population at t_0. Let the size of the cohort be $M(t)$ where $t \geq t_0$. As the cohort gradually dies out, its rate of loss is given by the following simple differential equation:

$$\frac{dM(t)}{dt} = -\mu(t - t_0, N(t))M(t) \qquad (t \geq t_0) \qquad (18.40)$$

Because we are considering a cohort, the time and age variables vary in step with one another. Hence $\mu(t - t_0, N(t))$ is the instantaneous death rate for members of the cohort at any time t, since they have all now aged $t - t_0$ units of time. In the meantime the population size is also varying, as is described by the function $N(t)$. Hence the size of the cohort after t years is

$$M(t) = M(t_0) \exp \left\{ -\int_{t_0}^{t} \mu[\tau - t_0, N(\tau)] \, d\tau \right\} \qquad (18.41)$$

The survivorship curve is simply the ratio of the members in the cohort at time t to the members initially present. Since the cohort of those who are of age x at time t were born at time $t - x$, we can substitute $t_0 = t - x$ in (18.41) and obtain

$$l(x, 0, N(\cdot)) = \exp \left\{ -\int_{t-x}^{t} \mu[\tau - (t - x), N(\tau)] \, d\tau \right\} \qquad 0 \leq x \leq t$$
$$(18.42)$$

We shall use this survivorship curve in the integral equation for $B(t)$ in connection with determining the number of newborn from parents who were born into the population.

To determine the survivorship curve that applies to members of the original colonizing population, we use the same approach. We consider a cohort whose members are of age x_0 at time $t_0 = 0$. Then the same procedure as above leads to

$$l(t + x_0, x_0, N(\cdot)) = \exp \left\{ -\int_{0}^{t} \mu[\tau + x_0, N(\tau)] \, d\tau \right\} \qquad (18.43)$$

where $l(t+x_0, x_0, N(\cdot))$ denotes the probability of a colonist, originally of age x_0, living to age $t+x_0$ or more given the history of the population size between 0 and time t. Incidentally, the instantaneous death rate, μ, is called the "force of mortality" and is used again in the next chapter.

With the modifications to the maternity function and survivorship curve discussed above, the integral equation for $B(t)$ can be determined from (18.38) as

$$B(t) = \int_0^t B(t-x)l(x, 0, N(\cdot))m(x, N(t))\, dx$$

$$+ \int_0^\infty n_0(x_0)l(t+x_0, x_0, N(\cdot))m(t+x_0, N(t))\, dx_0 \quad (18.44)$$

The integral equation for $N(t)$ is easy to formulate. Here $N(t)$ is the sum of all individuals who have been born into the population and are still alive plus all the individuals who were among the original colonists and are still alive. Hence

$$N(t) = \int_0^t B(t-x)l(x, 0, N(\cdot))\, dx + \int_0^\infty n_0(x)l(t+x_0, x_0, N(\cdot))\, dx_0 \quad (18.45)$$

This pair of coupled integral equations defines the model. The term $n_0(x)$ is the given initial condition; $m(x, N)$ is the given density-dependent maternity function; and $\mu(x, N)$ is the given density-dependent instantaneous mortality rate. The survivorship curves are computed from $\mu(x, N)$ according to (18.42) and (18.43). The simultaneous solution of this pair of integral equations in principle yields the pair of functions, $B(t)$ and $N(t)$. Finally, if one has obtained this pair of functions, then the full picture of the age structure through time can be calculated from $B(t)$ and $N(t)$ using the survivorship functions.

Local Stability of the Equilibrium Gurtin and MacCamy studied the model above and provided results on the existence and uniqueness of solutions. They also provided an analysis of the local stability of an equilibrium solution, and the remainder of this section is an examination of their criteria for local stability of a population with density dependence and age structure in continuous time.

An equilibrium population distribution is denoted as $\hat{n}(x)$. The equilibrium number of newborn is $\hat{n}(0) = \hat{B}$. The equilibrium population size is $\hat{N} = \int \hat{n}(x)\, dx$. If the population is constant through time, then the survivorship curve (18.42) assumes a simple form

$$\hat{l}(x, \hat{N}) = \exp\left[-\int_0^x \mu(\tau, \hat{N})\, d\tau\right] \quad (18.46)$$

where $\hat{N}$ is a number, the equilibrium population size. The equilibrium population size is obtained as the root of the following equation:

$$R_0(\hat{N}) \equiv \int_0^\infty \hat{l}(x, \hat{N})m(x, \hat{N})\, dx = 1 \quad (18.47)$$

Then the unique equilibrium population distribution corresponding to $\hat{N}$ is given by

$$\hat{n}(x) = \hat{B}\hat{l}(x, \hat{N}) \quad (18.48)$$

where

$$\hat{B} = \frac{\hat{N}}{\int_0^\infty \hat{l}(x, \hat{N})\, dx} \qquad (18.49)$$

Our question now is whether this equilibrium population distribution is locally stable. Gurtin and MacCamy introduced small perturbations to this equilibrium and examined whether the perturbed distribution returned to the equilibrium distribution. The resulting condition for local stability is, in general, quite complicated. The condition is that if all the roots, λ_i, to the following transcendental equation have negative real parts, then the equilibrium is locally stable. The equation for λ is

$$1 = \int_0^\infty \bar{e}^{\lambda x} \hat{l}(x, \hat{N}) m(x, \hat{N})\, dx$$
$$+ g(\lambda)\left[\frac{\kappa}{\hat{B}} - \int_0^\infty \hat{l}(x, \hat{N}) m(x, \hat{N}) f(\lambda, x)\, dx \right] \qquad (18.50a)$$

where

$$\kappa = \hat{B} \int_0^\infty \left[\frac{\partial m(x, \hat{N})}{\partial N} \right] \hat{l}(x, \hat{N})\, dx \qquad (18.50b)$$

$$f(\lambda, x) = \int_0^x e^{-\lambda(x-\alpha)} \left[\frac{\partial \mu}{\partial N}(\alpha, \hat{N}) \right] d\alpha \qquad (18.50c)$$

$$g(\lambda) = \frac{\hat{B} \int_0^\infty e^{-\lambda x} \hat{l}(x, \hat{N})\, dx}{1 + \hat{B} \int_0^\infty f(\lambda, x) \hat{l}(x, \hat{N})\, dx} \qquad (18.50d)$$

The number κ represents the density dependence in fertility. Note that the density dependence at each age is weighted by the probability of surviving to that age. The function $f(\lambda, x)$ represents the density dependence in mortality. The number κ is zero if there is no density dependence in the fertility, and $f(\lambda, x)$ is identically zero if there is no density dependence in mortality. The equilibrium distribution given by (18.48) to (18.49) is locally stable if all roots to (18.50) have negative real parts.

The biological significance of the λ_i is as follows. We consider a perturbation, $\xi(x, t)$, which can be expressed as an expansion of the form

$$\xi(x, t) = \sum_{i=0}^\infty \xi_i(x)\, e^{\lambda_i t} \qquad (18.51)$$

where λ_i and $\xi_i(x)$ may be complex. Then each λ_i must satisfy (18.50) and the corresponding age functions $\xi_i(x)$ are

$$\xi_i(x) = \xi_i(0) \hat{l}(x, \hat{N})[e^{-\lambda_i x} - g(\lambda) f(\lambda, x)] \qquad (18.52)$$

Thus the λ_i represent the geometric rates of return for various components of a perturbation of the form of (18.51). Gurtin and MacCamy's proof does not actually rely on the expansion (18.51) and considers more general perturbations. But the significance of the λ_i is best visualized through a perturbation of the form (18.51).

A Special Case: The condition for local stability (18.50) may be greatly simplified in
Density-Dependent special cases. For example, suppose that the density dependence is only
Fertility Only through fertility and not mortality. Then $f(\lambda, x) \equiv 0$ and the denominator in

the expression for $g(\lambda)$ equals 1. Then (18.50a) becomes

$$1 = \int_0^\infty e^{-\lambda x}\hat{l}(x, \hat{N})m(x, \hat{N})\, dx + \kappa \int_0^\infty e^{-\lambda x}\hat{l}(x, \hat{N})\, dx \qquad (18.53)$$

This further simplifies to

$$1 = \int_0^\infty e^{-\lambda x}[m(x, \hat{N})+\kappa]\hat{l}(x, \hat{N})\, dx \qquad (18.54)$$

This equation is formally like the familiar Euler equation. The effective maternity function in brackets is simply $m(x, \hat{N})$ plus a constant, κ. We know the spectrum of roots to (18.54); there is one real root and it is greater than the real part of all other roots. If this root is itself negative, then all other roots have negative real parts. Hence in this special case we can focus on the dominant root, λ_0. By inspection of (18.54) we see that the dominant root increases monotonically with κ. Furthermore, if $\kappa = 0$, then $\lambda_0 = 0$ because $\hat{N}$ is chosen according to (18.47). Hence, if $\kappa < 0$, then the dominant root to (18.54) is negative. Thus we may conclude that if the density dependence affects fertility and not mortality, then the equilibrium population size and age distribution is locally stable if

$$\int_0^\infty \left[\frac{\partial m(x, \hat{N})}{\partial N}\right]\hat{l}(x, \hat{N})\, dx < 0 \qquad (18.55)$$

In particular, there may be positive dependence in some age classes so long as the total density dependence weighted as in (18.55) is negative. Note that the inequality is one-sided; that is, the density dependence may be as negative as we please consistent with stability. There is no possibility of strong negative density dependence producing diverging overshoots about the equilibrium and thereby destroying the stability. This fact that even strong negative density dependence is consistent with stability contrasts with the results based on a discrete time formulation, as discussed in the next two sections.

The Logistic Equation in Discrete Time
As a preliminary to understanding density dependence and age structure in discrete time, we need to pursue the analysis of density dependence in discrete time without age structure much more deeply than we did in Chapter 16. Recall that the logistic equation in discrete time is given by

$$\Delta N = \frac{rN(K - N)}{K} \qquad (18.56)$$

This equation exhibits a variety of solutions depending on the magnitude of r. Recall that if r lies between 0 and 1, then the solutions approach K in a sigmoid manner, which is qualitatively the same as the solution in continuous time. However, if r lies between 1 and 2, then the solutions approach K with a damped oscillation. If r lies between 2 and 3, then solutions exhibit bounded oscillation. Finally, if r is greater than 3, then extinction results. This section is an examination of the bounded oscillation in the discrete time logistic equation.

r Is Between 2 and 3
The manner in which the oscillatory pattern depends on r has received an excellent review by May (1976). When r is between 2 and approximately 2.45, solutions approach an oscillation between 2 distinct points, one above

K and the other below K. When r is between approximately 2.45 and 2.55, the solutions approach an oscillation among 4 distinct points. (See Figure 16.5 for an illustration with $r = 2.5$.) If r is slightly higher than approximately 2.55, then the solutions converge to an oscillation among 8 distinct points. As r is increased still more, the solutions approach an oscillation among 16 distinct points. This pattern whereby solutions approach an oscillation among 2^n points (n is an integer $\geqslant 1$) ends when r reaches a value of approximately 2.68, whereupon the first cycle with an odd period appears. For r slightly above 2.68, the solutions approach an oscillation among an odd number of points that has a very long period, but as r increases, the solutions approach cycles with smaller and smaller odd periods until at last a three-point cycle appears at r slightly less than 2.83. Beyond this value of r there exist cyclic solutions with every integer period as well as an uncountable number of asymptotically aperiodic solutions. These asymptotically aperiodic solutions were originally thought to imply a kind of "chaotic" dynamical behavior in the system such that there would be no regularity to the solutions whatsoever. But Smale and Williams (1976) pointed out that for almost any specified value of r, there is one unique cycle that is stable in the sense that it attracts almost all initial points. That is, the remaining (countably) infinite number of periodic solutions that exist together with the uncountably infinite number of asymptotically aperiodic solutions involve a set of points that has zero measure. So, to summarize, for almost all r in the interval $(2, 3)$ almost all solutions approach a specific oscillation that is characteristic to each value of r.

r Is Exactly 3   If r is exactly 3, then a particularly interesting form of dynamic occurs. Almost all solutions do not approach any particular cycle but instead wander throughout a certain interval. Specifically, if $r = 3$, almost all solutions remain within the interval $[0, \frac{4}{3}K]$ and do not settle down into any characteristic oscillatory pattern. Nonetheless, the solutions are not chaotic because there is a distinct kind of regularity to them. But this regularity is of an unusual type, which is treated by an area of mathematics called ergodic theory. The basic idea is to apply methods of description from the study of random variables to the study of deterministic variables. For example, with a stochastic process one is concerned with predicting, for every point in time, the probability distribution over all possible states for a random variable. In a similar vein we can seek to predict, for every point in time, the probability distribution over all values of the population size between 0 and $\frac{4}{3}K$ based on the assumption that a population is changing according to the discrete logistic equation with $r = 3$. Of course, the sequence of population sizes for any given population with a known initial condition can be predicted exactly, simply by iterating the logistic equation. But consider a very large group of logistically growing populations, that is, an ensemble. Perhaps the group as a whole exhibits certain regularities that will provide insight into regularities which are present in the dynamics of a single population.

The Logistic Equation Is Measure-Preserving if r = 3   The first key point to understand about the logistic equation in discrete time with $r = 3$ is that an ensemble of populations, each of which is growing according to the deterministic logistic equation in discrete time, settles down to a certain distribution over the states within the interval $[0, \frac{4}{3}K]$. In Figure 18.9 an ensemble of 1000 logistically growing populations is initially given a

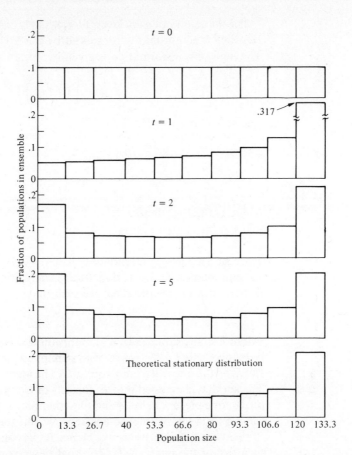

FIGURE 18.9. *Ensemble of* 1000 *populations growing according to the discrete logistic equation with* $r = 3$ *and* $K = 100$. *The initially uniform ensemble distribution rapidly approaches the theoretical stationary ensemble distribution.*

uniform distribution over the states between 0 and $\frac{4}{3}K$. After several iterations the ensemble as a whole approaches a specific distribution, which thereafter remains stationary. In ergodic theory the logistic transformation is said to be *measure-preserving*. This means that there is a specific (smooth) distribution over the states which is preserved by the transformation, that is, remains stationary through time. Figure 18.9 actually shows more than the existence of a stationary distribution; it also shows that an arbitrary initial ensemble distribution approaches the stationary distribution. The property that there exists such a smooth stationary distribution is the property described by the phrase "measure-preserving." The property that an initial ensemble distribution approaches this stationary distribution is related to the concept of "mixing," to be discussed later. In particular, the theoretical stationary density illustrated in Figure 18.9 is given by

$$\rho(x) = \frac{1}{\pi\sqrt{x(1-x)}} \qquad \text{where } x = \left(\frac{3}{4}\right)\frac{N}{K} \qquad (18.57)$$

$\rho(x)$ can be used to attach a weighting to any subinterval within the interval $[0, \frac{4}{3}K]$ and this weighting function is called a measure.

The stationary density given by (18.57) is easily verified to be exactly the measure function that is preserved by the logistic equation through time. The logistic transformation is given by

$$N_{t+1} = \left[1 + r - \frac{r}{K} N_t \right] N_t \tag{18.58}$$

The substitution

$$a = r + 1$$
$$x = \left(\frac{r}{r+1} \right) \frac{N}{K} \tag{18.59}$$

converts the logistic to

$$x_{t+1} = ax_t(1 - x_t) \equiv T(x) \tag{18.60}$$

The case $r = 3$ now corresponds to $a = 4$. $T(x)$ is a transformation defined on the unit interval $[0, 1]$. By definition a transformation is measure-preserving if there exists a smooth measure such that

$$\text{measure } (T^{-1}(E)) = \text{measure } (E) \tag{18.61}$$

where E is any subinterval within the interval over which the transformation is defined, and $T^{-1}(E)$ is a set that is mapped *into* E by the transformation T. Let E be the interval from x to $x + dx$. The measure of this interval is $\rho(x)\, dx$. The set that is mapped into E consists of two pieces as illustrated in Figure 18.10; one piece is below $\frac{1}{2}$ and the other piece is above $\frac{1}{2}$. Since the measure function $\rho(x)$ is symmetric about $\frac{1}{2}$, we can see that each of the pieces of $T^{-1}(E)$ has an equal measure. Hence let us confine ourselves to finding the measure of the piece of $T^{-1}(E)$ that lies below $x = \frac{1}{2}$. What must then be

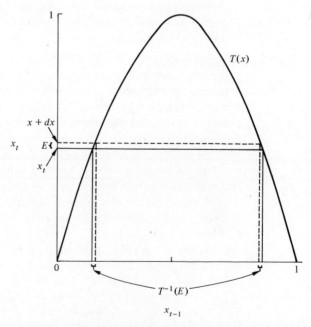

FIGURE 18.10. *Sketch showing the set E and the set, $T^{-1}(E)$, which is mapped into the set E by the transformation $T(x)$.*

shown is that

$$\rho(x)\,dx = 2\rho(T^{-1}(x))d(T^{-1}(x)) \qquad (18.62)$$

where $T^{-1}(x)$ refers to the point below $\frac{1}{2}$ at time $t-1$, which is mapped into the point x at time t. Solving (18.60) for x_{t-1} in terms of x_t yields

$$T^{-1}(x) = \tfrac{1}{2} - \tfrac{1}{2}\sqrt{(1-x)} \qquad (18.63)$$

Then evaluating $\rho(x)$ at this point yields

$$\rho(T^{-1}(x)) = \frac{2}{\pi\sqrt{x}} \qquad (18.64)$$

The differential of $T^{-1}(x)$ in terms of dx is

$$dT^{-1}(x) = \frac{1}{4}\frac{1}{\sqrt{(1-x)}}\,dx \qquad (18.65)$$

Then substituting into the right-hand side of (18.62) yields an expression that is identical to (18.57). The measure function (18.57) that is preserved is the density function of the arc-sine distribution and is discussed in a probabilistic context by Feller (1968, p. 80). The applicability of this distribution to the transformation (18.60) was pointed out by Ulam and von Neumann (1947).

The Logistic
Equation Is
Ergodic if r = 3
The next key point to understand about the logistic transformation (18.60) with $a = 4$ is that there is no subset of $[0, 1]$ which has positive measure and is invariant under the transformation. There are indeed many invariant subsets, but each has a zero measure. Specifically, the point $x = \frac{3}{4}$ is an equilibrium point. Hence $T(\frac{3}{4}) = \frac{3}{4}$ and this point is invariant under T. But a single point has a zero measure. Another invariant subset is the point $x = 0$, which is also an equilibrium point. Still another invariant subset is a two-point cycle between $(5+\sqrt{5})/8$ and $(5-\sqrt{5})/8$. There are also other invariant subsets representing cycles. Each contains a finite number of points and thus also has a zero measure. All these various subsets are attained exactly from a countably infinite number of initial conditions. For example, the initial condition $x = \frac{1}{4}$ leads, with one iteration, to $x = \frac{3}{4}$. Furthermore, there are two initial conditions that lead to $x = \frac{1}{4}$ in one iteration and thus to $x = \frac{3}{4}$ in two iterations. By working backward in this way, we can see that there is a set of points dense on $[0, 1]$ which comprises all the possible initial conditions leading to the point $x = \frac{3}{4}$. Similarly, there is a specific set corresponding to each invariant set that contains all the possible initial conditions leading into that invariant set. There is a countably infinite number of points in each set of initial conditions. But each set is a closed set and has a zero measure. Thus there are invariant subsets with a zero measure, and furthermore, although there are infinitely many initial conditions that produce a trajectory terminating in one of these subsets, these initial conditions are still a "negligible" fraction of all the possible initial conditions.

If a measure-preserving transformation possesses no invariant subsets with a positive measure, then it is said to be an *ergodic* measure-preserving transformation. As discussed in detail in Halmos (1956) and Sinai (1977), an ergodic measure-preserving transformation on a space of finite measure inherently has many interesting features. We now discuss four of these features.

Consequences of
Ergodicity 1. For almost every initial condition in $[0, 1]$, the trajectory resulting from that initial condition covers the interval $[0, 1]$ densely. That is, almost all trajectories come arbitrarily close to any given point in $[0, 1]$. The exceptions are, of course, those comparatively few trajectories which terminate at the invariant subsets.

2. The most important feature is that the average of a function as computed at any one point in time across all populations in the ensemble is numerically equal to the average of a function as computed through time along the trajectory of almost any single population. The distribution of the populations in the ensemble is assumed to be the stationary ensemble distribution. The trajectory of any single population can be used for the time averages except, again, those comparatively few which terminate at an invariant subset. In symbols, this result is stated as follows. Let $f(x)$ be the function whose average value is to be computed. For the logistic transformation (18.60) with $a = 4$

$$\lim_{n \to \infty} \left[\frac{1}{n} \sum_{j=0}^{n-1} f(T^j(x_0)) \right] = \int_0^1 f(x) \rho(x) \, dx \qquad (18.66)$$

for almost all x_0. Typical choices for $f(x)$ are

$$f(x) = x \qquad (18.67a)$$

$$f(x) = (x - \bar{x})^2 \qquad (18.67b)$$

$$f(x) = 1 \qquad \text{if } x \in E$$

$$= 0 \qquad \text{if } x \notin E \qquad (18.67c)$$

If $f(x) = x$, then the quantity being computed is the average population size. If $f(x) = (x - \bar{x})^2$, then the variance of the population size is being determined. If $f(x)$ is one for x in some set E (with positive measure) and zero for x outside E, then one is calculating the fraction of time spent in the set E. Clearly, by (18.66) the fraction of time almost any single population spends in set E equals the fraction of the whole ensemble that is in set E at any one time. Figure 18.11 illustrates the fraction of time spent at different population sizes by a single logistic population through 1000 time units. Note that the single population through time approaches the same stationary distribution illustrated earlier in Figure 18.9 for the ensemble.

3. The average length of time it takes for the populations originating in a set E (with positive measure) to return to the set E is equal to the reciprocal of the measure of E. That is,

$$\text{average time to return to } E = \frac{1}{\int_E \rho(x) \, dx} \qquad (18.68)$$

4. Trajectories tend, when averaged over time, to become statistically independent of their history. More specifically, consider the populations now in set F. We ask how many of them were also in some other set, say G, some time ago. The set $T^{-j}F$ denotes the set where populations who are now in F were at j time units ago. The set $T^{-j}F \cap G$ is the set of populations that are now in F and overlapped with set G at j time units ago. The result, for

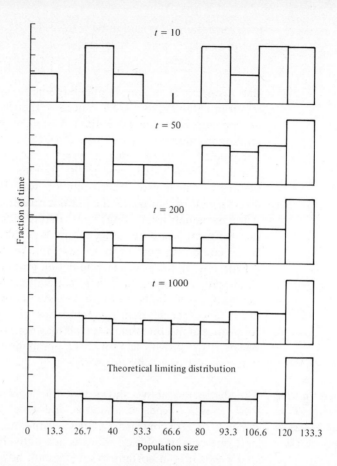

FIGURE 18.11. *One population growing according to the discrete logistic equation with r = 3 and K = 100. The initial population size was N = 10. The fraction of time the single population has spent in various intervals of population size is illustrated in the figure. Note that the distribution of time spent in various size intervals approaches a limiting distribution that is identical to the stationary ensemble distribution.*

the logistic transformation (18.60) with $a = 4$, is that

$$\lim_{n \to \infty} \left[\frac{1}{n} \sum_{j=0}^{n-1} \int_{T^{-j}F \cap G} \rho(x)\, dx \right] = \left[\int_F \rho(x)\, dx \right] \left[\int_G \rho(x)\, dx \right] \quad (18.69)$$

That is, when averaged over time, the measure of populations now in set F that overlapped with set G at times in their past equals the product of the measure of F with the measure of G. If we view the measure function as analogous to a probability density function, then we see that the joint probability of being in F now and in G in the past is simply the product of the probability of being in F with the probability of being in G.

The Logistic Equation Is Mixing if r = 3 The third key property of the logistic transformation for $r = 3$ is that trajectories are statistically independent of their history in a stronger sense than that expressed in (18.69). If a transformation is such that trajectories become independent of their history even *without averaging over time*, then the transformation is said to be mixing. Specifically, a measure-preserving transformation, T is said to be *mixing* if

$$\lim_{n \to \infty} \left[\int_{T^{-n}F \cap G} \rho(x)\, dx \right] = \left[\int_F \rho(x)\, dx \right] \left[\int_G \rho(x)\, dx \right] \qquad (18.70)$$

The property of mixing implies ergodicity; indeed (18.70) is a stronger requirement than (18.69) wherein convergence of the limit is "aided" by the mathematical averaging over time. A transformation that is mixing takes the initial conditions and stirs them throughout the space. Perhaps the most important feature of a mixing transformation is that any initial ensemble distribution tends over time to the stationary ensemble distribution. This feature of the logistic transformation was illustrated in Figure 18.9. A transformation that is ergodic but not mixing does not have this property.

It is extremely interesting to note that the features above which pertain to the statistical properties of an ensemble of deterministically growing population could also pertain to an ensemble of populations with a stochastic growth law. If, for example, population growth were modeled as a Markov chain, then an ensemble of these populations would typically exhibit statistical properties similar to those discussed above for a deterministic growth law. Hence census data on population size per se for populations that reproduce at discrete time intervals may not provide suitable information to decide if the growth law for those populations should be formulated as a deterministic or stochastic process.

An Alternative Logistic Type of Model in Discrete Time

The discussion above shows that the logistic equation in discrete time leads to many types of solutions, and which type applies in any instance depends on r. In contrast, the logistic equation in continuous time possesses only one type of solution which is qualitatively independent of r, provided that $r > 0$. It is important to understand, however, that although discrete time models *typically* have a more varied array of possible solutions than the corresponding continuous time models, they do *not necessarily* have a more varied array of solutions. For example, we can construct a logistic-like discrete time model with exactly the same qualitative solutions as the continuous time logistic model as follows. Recall that the solution to the continuous time logistic equation is

$$N(t) = \frac{K}{1 + c\, e^{-rt}} \qquad t \geq 0 \qquad (18.71)$$

where c is determined from the initial condition. Then if we examine (18.71) at $t = 0, 1, 2$, we have a solution to some growth model defined over a discrete time interval,

$$N_t = \frac{K}{1 + c\lambda^{-t}} \qquad t = 0, 1, 2 \cdots \qquad (18.72)$$

where $\lambda = e^r$. The discrete time growth model for which (18.72) is a solution is found by setting $\Delta N = N_{t+1} - N_t$ and rearranging, yielding

$$\Delta N = \left[\frac{(\lambda - 1)(1 - N/K)}{1 + (\lambda - 1)N/K} \right] N \qquad (18.73)$$

The solution to this equation approaches the equilibrium at $N = K$ in a smooth sigmoid manner for all $\lambda > 1$, that is, $r > 0$. Note that the expression in brackets in (18.73) is convex-decreasing.

Are the
Fluctuations from
These Models
Relevant to
Natural
Fluctuations?

Although it is well known that many populations, especially insect populations, exhibit an extremely erratic pattern of population size through time, often such erratic patterns are caused by fluctuations in the resource levels and/or in the severity of density-independent factors. Well-documented recent examples of butterfly population fluctuations—driven by the fluctuations of extrinsic factors—appear in Ehrlich et al. (1975). Further examples appear in Andrewartha and Birch (1954, 1960) and Clarke et al. (1967). There do not seem to be, at present, any examples in nature where erratic fluctuations in population size are known to be caused by a deterministic mechanism of population growth in discrete time that does not include environmental fluctuation. This issue receives important discussion in Hassell et al. (1976). Nonetheless the analysis of the logistic equation in discrete time demonstrates that, even in a constant environment, it is theoretically possible for a deterministic growth law to produce a pattern of population size through time which would be virtually indistinguishable from the pattern that would result from a stochastic growth law.

Density
Dependence and
Age Structure in
Discrete Time

There has not been a comprehensive study of density-dependent population dynamics with age structure in discrete time similar to that provided by Gurtin and MacCamy (1974) for continuous time. One reason is simply that many more kinds of solutions emerge with age structure and density dependence in discrete time as compared with continuous time. The second is that the basic mathematical theory which would, even in principle, allow a complete classification of the solutions does not yet exist. Even rather simple models for density dependence and age structure in discrete time rapidly lead into mathematical *terra incognita*.

An Example

The most thorough examination of a density-dependent model with age structure in discrete time appears in Guckenheimer et al. (1977). These authors analyzed a model with two age classes and density-dependent fertility

$$\begin{pmatrix} n_{0,t+1} \\ n_{1,t+1} \end{pmatrix} = \begin{pmatrix} b\,e^{-.1(n_{0,t}+n_{1,t})} & b\,e^{-.1(n_{0,t}+n_{1,t})} \\ 1 & 0 \end{pmatrix} \begin{pmatrix} n_{0,t} \\ n_{1,t} \end{pmatrix} \qquad (18.74)$$

There is one parameter, b, which controls the total reproductive output. Note that the density dependence in the fertility is based on the total population size, $n_{0,t}+n_{1,t}$. Guckenheimer et al. demonstrated that if b is sufficiently small, then the population converges to a stable equilibrium point. As b is increased, solutions are attracted to oscillations analogous to the oscillation in the discrete-time logistic equation for r between 2 and 3. However, for some values of b the stable oscillations coexist with a stable equilibrium point; for these values of b the population either will settle into a specific oscillatory pattern or will approach an equilibrium point depending on the initial condition. Furthermore, since there are two variables in the system, the possibility of "almost periodic" oscillation arises. This possibility occurs if the solutions are attracted to the points on a closed loop in state space but in a way such that the population never exactly repeats any previous value as it goes around and around the loop.

The principal finding for (18.74) is that solutions approach a so-called "strange attractor" for a significant *range* of the parameter b. A strange attractor is the name for a region of state space such that (a) solutions are attracted to the region and (b) within the region the transformation is measure-preserving and mixing. With the logistic equation if $r = 3$, then requirement (b) is satisfied in that the logistic equation is measure-preserving and mixing in the interval $[0, \frac{4}{3}K]$. This interval is not a strange attractor, however, because points below 0 and above $\frac{4}{3}K$ lead to solutions that go quickly to $-\infty$, thereby violating requirement (a). Furthermore, even requirement (b) is satisfied only if r is precisely 3. In contrast for (18.74) there appears to be a region of state space which attracts trajectories that are initially outside this region, and within this region the transformation is measure-preserving and mixing. Furthermore, the strange attractor appears to exist if b is anywhere within certain intervals, therby ensuring that the strange attractor is a structurally stable feature of the dynamics. Guckenheimer et al. illustrate how the dynamics within the strange attractor can be studied by numerically approximating the transformation with a Markov chain.

Guckenheimer et al. conjecture that complicated oscillatory solutions and strange attractors will also be commonly found in models with more age classes and in continuous time models. There is no reason to suspect that density-dependent models in discrete time with more than two age classes will fail to exhibit the kinds of solutions already found in models with two and one age class. But the possibility of generalization to continuous time is more problematic. The scheme of density dependence in (18.74) involves the total population size affecting only the fertility. But as we saw from the Gurtin-MacCamy condition (18.55), a net negative density dependence, regardless of the degree of negativity, is sufficient for local stability if the total population size is affecting only the fertility. In the physical sciences, simple continuous time models with more than one variable are known which have strange attractors, but it appears that fairly complicated assumptions will be needed to obtain strange attractors in continuous time models for single populations.

Chapter 19

AGE-SPECIFIC SELECTION AND LIFE HISTORY STRATEGIES

ONE of the most important issues in evolutionary ecology is to explain the evolution of the typical life history of individuals in a species. A life history is defined as the *schedule* that dictates the length of time an individual spends growing into an adult, the age and body size where maturity is attained, the amount and timing of reproductive activity, and the timing of the eventual death. There is a fascinating diversity of life histories in nature. At one extreme are annual plants whose schedule stipulates fast maturation, high seed output during a specific short period of life, followed by rapid senescence. At the other extreme are perennial plants whose schedule prescribes slow maturation, an initially low rate of seed output, and usually no discernible senescence. The problem is to explain why this diversity of life histories has evolved.

Evolutionary Patterns Among the Life Histories of Related Species

If we examine the life histories of related species, we often see suggestive patterns. For example, ecologists working with birds have pointed out that there is a latitudinal gradient in what is called the "clutch size" in birds. A clutch is the collection of eggs laid at approximately the same time during the breeding season. Breeding birds lay one clutch at the beginning of the reproductive season, and some species may also lay a second clutch near the middle of the season. The clutch size is the *number* of eggs in each clutch. Some birds lay only one egg in each clutch, whereas others lay up to six and seven in the clutch. Figure 19.1 adapted from Cody (1971) illustrates that in many families of birds, the clutch size increases as one moves from the tropics into the temperate zones. This pattern is called the latitudinal gradient in clutch size. Furthermore, there is a similar pattern involving island-continent comparisons. In temperate, but not in tropical, latitudes there are lower clutch sizes on islands as compared with neighboring continental locations. Figure 19.2 adapted from Cody (1971) illustrates the dependence of the island-continent comparison on latitude. Temperate islands generally are considered to have more moderate climate than neighboring continental regions, but with tropical islands this effect is not as pronounced. Both these patterns may be roughly summarized by saying that clutch size increases with environmental harshness.

The clutch size patterns above can be viewed as a pattern involving the $m(x)$ curves introduced in the last chapter. The temperate bird species often have $m(x)$ curves that indicate earlier maturity and higher reproductive output than corresponding closely related species from tropical locations. In view of these patterns involving the $m(x)$ curves we can ask, are there analogous patterns in the $l(x)$ curves? The answer is yes. Table 19.1 adapted from Lack (1968) indicates that birds with large clutch sizes and early maturation typically incur higher mortality than comparable species with lower clutch sizes and later maturation. These data may be viewed as suggesting that the $m(x)$ and $l(x)$ curves of a species evolve in relation to the environment and that the outcome of this evolution leads to a correlation of early maturation, high reproductive output, and high mortality with environmental harshness. The main task of this chapter is to understand how

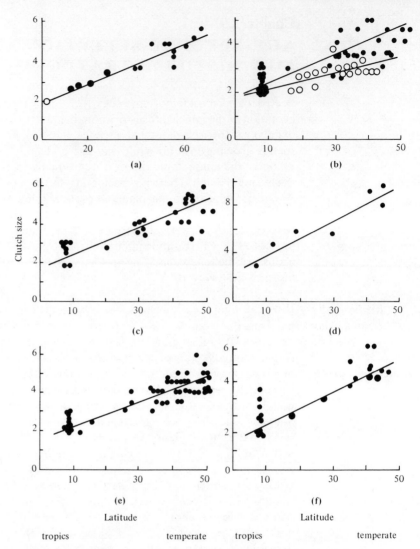

FIGURE 19.1. *The vertical axis is clutch size and the horizontal axis is latitude.* (a) *The genus* Emberiza. (b) *The family Tyrannidae. Open circles,* ○, *are South American; closed circles,* ●, *are North American. Data exclude hole nesters of the genus* Myiarchus. (c) *The family Icteridae.* (d) *The genus* Oxyura. (e) Myioborus miniatus. (f) *The family Troglodytidae.* [Adapted from M. L. Cody (1971), Ecological aspects of reproduction. *Avian Biology* **1:** 461–512.]

natural selection shapes the $m(x)$ and $l(x)$ curves and to see whether the evolutionary interpretation of the data above stands up under careful analysis.

The trends revealed in the life histories of birds do not appear appropriate for certain other groups. Specifically, it appears that there is not a simple latitudinal gradient in clutch size in lizards analogous to that in birds because clutch size is confounded with body size and the number of broods per season, as discussed by Tinkle, Wilbur, and Tilley (1970). Presumably, there exist other groups that also fail to follow the empirical example of birds. This fact must be kept in mind. At this time the theory of the evolution of life

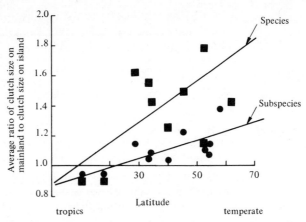

FIGURE 19.2. *The average ratio of mainland clutch size to island clutch size, in relation to latitude, for various island groups. Comparison is between nearest relatives. The ratio between nearest relatives is averaged separately for distinct species and for distinct subspecies on a particular island. The average ratio is then plotted for various islands against their latitude. Key:* ■ *comparison between species;* ● *comparison between subspecies.*

The data are based on the following island–mainland comparisons: St. Lawrence–Alaska, St. Kilda–Scotland, England-continental Europe, Ireland–England, Ireland–continental Europe, Channel Island–California, Madeira–North Africa, Canary Islands–North Africa, West Indies–Guatemala, Trinidad–Venezuela/Costa Rica, New Zealand–Australia, New Zealand Islands–New Zealand, Falkland Islands–Southern Chile. [Adapted from M. L. Cody (1971), Ecological aspects of reproduction. *Avian Biology* **1**: 461–512.]

histories is most useful in explaining phenomena in birds; its utility for other groups has yet to be established. An important review of the evolution of life histories appears in Stearns (1976).

This chapter consists of three parts: (1) A new model for population growth with age structure, called the *transport model* is introduced; (2) this

Table 19.1. Breeding Biology of Birds. [Adapted from Lack (1968).]

	Boobies and Gannets		Gulls		Terns	
	Phalacrocorax aristotelis	*Sula bassana*	*Larus ridibundus*	*Rissa tridactyla*	*Sterna hirundo* and *S. paradisaea*	*Sterna fuscata*
Feeding habit	Close inshore	Offshore	Shore	Offshore	Inshore	Offshore
Type of colony	Small	Large				
Usual clutch	3 eggs	1 egg	3 eggs	2 eggs	2 eggs	1 egg
Incubation period	31 days	44 days	23 days	27 days	21 days	29.5 days
Fledgling period	55 days	90 days	30 days	33 days	30 days	60 days
Postfledgling period of dependency	21 days	0 days				
Incubation spells	3.5 hrs	33 hrs	1–2.5 hr	1–2 per day	1/3–2 hr	5.5 days
Feeding visits	11 per day	2–3 per day	1 per hr	1 per 4 hrs	3–4 per hr	1 per day
Age of first breeding	3 years	4–7 years	2 years	3–4 years	2–4 years	6 years
Annual adult mortality	15%	5–6%	18%	12%	25%	16–20%

model is then extended to include genetic variation; and (3) finally, the evolutionary principles established from the population genetic analysis are applied to the evolution of life histories.

Transport Model for Population Growth with Age Structure

In the last chapter we studied two models for population growth with age structure, one devised by Leslie and the other by Lotka. Why are we introducing still another model? The answer is that this new model, the transport model, is the one that most easily generalizes to traits of an individual which change through time. Age can be viewed as a trait of an individual that always changes at a rate of exactly one unit of age per one unit of time. Body size is also a trait that changes through time, but the rate at which an individual grows is not a constant, as with age, but is rather a function of the individual's current size. The transport model to be presented can easily be generalized to include not only the age structure for a population, but also the size structure, that is, the distribution of body sizes of a population. Moreover, it also easily allows the incorporation of genetics. Thus, although the traditional models of Leslie and Lotka have become the foundations of demography, for ecological and evolutionary purposes, it now appears that the transport model will prove the most useful.

A transport equation in the physical sciences has the following form (in one dimension)

$$\frac{\partial n(x, t)}{\partial t} = -\frac{\partial J(x, t)}{\partial x} + f(x, t) \tag{19.1}$$

where $n(x, t)\,dx$ is the number of particles between x and $x + dx$ at time t. $J(x, t)$ is the rate of flow of particles across position x at time t, and $f(x, t)$ is a "source" term representing the rate of production or disintegration of particles at position x. This equation should already be familiar to you from the discussion of genetic drift and the study of clines. In most applications, an equation of this sort is solved subject to boundary conditions.

A version of this equation also provides an exact description of population growth with age structure. We let x represent age and view the members of a population as flowing from youth into old age. The rate of flow per individual is exactly one unit of age per one unit of time, as mentioned above. So the total population flow rate across age x is simply

$$J(x, t) = n(x, t) \tag{19.2}$$

The source term is negative in this application because it represents the death of individuals of age x. Let $\mu(x)$ denote the instantaneous death rate *per organism* of age x. This quantity, $\mu(x)$, is called the "force of mortality" in demography. It is related to the familiar survivorship curve, $l(x)$, as

$$\mu(x) = \frac{-1}{l(x)} \frac{dl(x)}{dx} \tag{19.3}$$

With the force of mortality, we can write the source term in the transport equation as

$$f(x, t) = -\mu(x)n(x, t) \tag{19.4}$$

So, with these flow and source terms the transport equation becomes

$$\frac{\partial n(x, t)}{\partial t} = -\frac{\partial}{\partial x} n(x, t) - \mu(x)n(x, t) \tag{19.5}$$

The transport equation above describes only the aging and death of individuals. To complete the picture, we must add birth. This is done in terms of the following boundary condition.

$$n(0, t) = \int_0^\infty m(x)n(x, t)\, dx \tag{19.6}$$

That is, any solution to the differential equation (19.5) is subject to the boundary condition (19.6). Equations (19.5) and (19.6) define what is called the "transport model" of population growth with age structure. This model has been introduced into biology by Von Foerster (1959) and many of its implications for ecology have been discussed by Sinko and Streifer (1967), Oster and Takahashi (1974), and Levin and Paine (1974).

Demography Revisited   The transport model above can be transformed into the Lotka's integral equation for $B(t)$ (the number of newborn at time t) customarily used in demographic theory. For more on this, see Gurtin and MacCamy (1974) and Oster and Takahashi (1974). The transport model is thus identical to the standard demographic formulation originated by A. J. Lotka. However, it will be useful to show how the classical formulas for the stable age distribution and for r follow *directly* from the differential equation above.

We seek a separable solution to the partial differential equation

$$n(x, t) = N(t)c(x) \tag{19.7}$$

where $c(x)$ will represent the stable age distribution $(\int c(x) = 1)$. Substituting (19.7) into (19.5), we obtain

$$\frac{d \ln[N(t)]}{dt} + \frac{d \ln[c(x)]}{dx} + \mu(x) = 0 \tag{19.8}$$

Then the separable solution, in terms of some as yet unknown constant, r, satisfies

$$\frac{d \ln[N(t)]}{dt} = r$$

$$\frac{d \ln[c(x)]}{dx} + \mu(x) = -r \tag{19.9}$$

The solutions for these two differential equations are

$$N(t) = N(0)e^{rt}$$

$$c(x) = \frac{e^{-rx}l(x)}{\int e^{-rx}l(x)\, dx} \tag{19.10}$$

Next, r is determined from the boundary condition. Substitution from (19.10) into (19.7) and then into (19.6) yields the familiar equation for r

$$1 = \int e^{-rx}l(x)m(x)\, dx \tag{19.11}$$

Thus the classical demographic formulas (19.10) and (19.11) follow directly upon seeking a separable solution for the partial differential equation (19.5), subject to the boundary condition (19.6).

Size, Growth, and Population Dynamics

One of the extremely useful features of the transport model is that it is easily generalized to cover the growth of individuals. In traditional demographic theory, fertility and mortality are viewed as functions of age. But for many species an organism's size is more important than its age per se. In fish, for example, fecundity often depends only on gonad size and this in turn is a function of body size. Similarly, the ability of an organism to escape predation often depends principally on size. Organisms that are very small may be ignored by certain predators, and those that are very large may also escape predation. The fact that fecundity and mortality are often directly connected with size and only indirectly with age suggests that the discussion of life histories should often be carried out in terms of an animal's size and not in terms of its age. The transport model referred to in Van Sickle (1977) allows this to be done as follows. Let y stand for some measure of body size; let $m(y)$ denote the fertility of an organism of size, y; and $\mu(y)$ the per individual death rate of an organism of size, y. Also, let $f(y)$ denote the growth rate of an organism of size, y. Then the flow term in the transport equation becomes

$$J(y, t) = f(y)n(y, t) \tag{19.12}$$

where $f(y)$ can be viewed as the speed by which each individual flows through size y and, therefore, population flow is just $n(y, t)$ times $f(y)$, where $n(y, t)$ is the number of individuals of size y in the population at time t. The transport model now becomes

$$\frac{\partial n(y, t)}{\partial t} = \frac{-\partial}{\partial y}[f(y)n(y, t)] - \mu(y)n(y, t) \tag{19.13}$$

subject to a boundary condition that expresses birth. Let y_0 denote the size of the newborn organisms and y_ω, the maximum size (y_ω can be finite or infinite). Then $n(y_0, t)$ is subject to the condition

$$n(y_0, t) = \int_{y_0}^{y_\omega} m(y)n(y, t)\, dy \tag{19.14}$$

Equations (19.13) and (19.14) directly describe the growth of a population with size structure. It also is possible to translate size into age and then to convert (19.13) and (19.14) into a model for age-specific population growth. But as discussed above, discussing life histories directly in terms of size rather than age may be more natural for many species.

The transport model for population growth with size structure may be solved for a *stable size distribution* analogous to the stable age distribution. Upon seeking a separable solution of the form

$$n(y, t) = N(t)c(y) \tag{19.15}$$

we obtain

$$N(t) = N(0)\, e^{rt} \tag{19.16a}$$

$$c(y) = (\text{const}) \exp\left[-\int_{y_0}^{y} \frac{r + f'(t) + \mu(t)}{f(t)}\, dt \right] \tag{19.16b}$$

where (const) is the normalization constant and $f'(y)$ is the derivative of the growth curve, $f(y)$. The exponential growth rate, r, is the dominant root of

$$1 = \int_{y_0}^{y_\omega} m(y) \exp\left[-\int_{y_0}^{y} \frac{r + f'(t) + \mu(t)}{f(t)}\, dt\right] dy \qquad (19.17)$$

Thus with the transport model, we have the choice of discussing life histories either in terms of age or size. If age is chosen, then a life history is represented by the pair of curves, $m(x)$ and $l(x)$, which describe the age-dependent fertility and mortality. If size is chosen, then a life history is represented by the three curves, $f(y)$, $m(y)$, and $\mu(y)$, which describe the size-dependent growth rate, fertility, and mortality. We now turn to the study of evolution in age-structured populations.

A Transport Model for Population Genetics with Age Structure

In this section we introduce and analyze a transport model for natural selection at one locus with two alleles in a population with age structure and overlapping generations. What we would like to know is whether there are conditions for polymorphism and for fixation of an allele in a population with overlapping generations that are analogous to the classical conditions for populations with discrete generations. Specifically, we know from classical population genetics that a polymorphism occurs with heterozygote superiority ($w_{12} > w_{11}, w_{22}$), that an unstable equilibrium occurs with heterozygote inferiority ($w_{12} < w_{11}, w_{22}$), and that an allele is fixed if the fitness of the three genotypes occur in an ordered sequence ($w_{11} < w_{12} < w_{22}$). Furthermore, we also know that the actual polymorphism frequency with heterozygote superiority occurs at the value of p that maximizes the mean fitness, $\bar{w}$. These classical results underlie our understanding of how natural selection works. The question we now face is whether this simple picture of natural selection is also correct when there is overlapping generations, when the fertility and survival of each genotype depend on age, and when there is a random union of gametes from parents of different ages. Clearly, the population-genetic model for a population with overlapping generations will be much more complicated than the classical equation for Δp, and it is not obvious that the classical picture of the outcomes of natural selection will be correct in this complicated model. The surprising results derived in this section are that the classical picture *is* correct, as Norton (1928), and Charlesworth (1972) noted. It will turn out that the intrinsic rates of increase of each genotype, r_{ij}, will play a role analogous to the selective values, w_{ij}, used in discrete generations. Also, there is a quantity that is maximized at a stable polymorphic equilibrium analogous to the way $\bar{w}$ is maximized in classical population genetics.

Let $n_{ij}(x, t)$ denote the number of individuals with genotype $A_i A_j$ of age x at time t. Then there are three partial differential equations, one for each genotype,

$$\frac{\partial}{\partial t} n_{ij}(x, t) = \frac{-\partial}{\partial x} n_{ij}(x, t) - \mu_{ij}(x) n_{ij}(x, t) \qquad (ij = 11, 12, 22)$$

$$(19.18)$$

These equations describe only the aging and death of individuals of each genotype. The birth of new individuals *and the genetics* are all incorporated into the boundary conditions. We assume there is random union of the gametes produced by the parents of different ages. The boundary conditions are then specified in terms of the following quantities. The total number of newborn of all genotypes is defined as

$$n(0, t) = n_{11}(0, t) + n_{12}(0, t) + n_{22}(0, t) \tag{19.19a}$$

This quantity must satisfy

$$n(0, t) = \int [m_{11}(x)n_{11}(x, t) + m_{12}(x)n_{12}(x, t) + m_{22}(x)n_{22}(x, t)] \, dx \tag{19.19b}$$

The gene frequency of A_1 among the newborn is defined as

$$p(0, t) = \frac{[n_{11}(0, t) + (\frac{1}{2})n_{12}(0, t)]}{n(0, t)} \tag{19.20a}$$

This quantity must satisfy

$$p(0, t) = \frac{\int [m_{11}(x)n_{11}(x, t) + (\frac{1}{2})m_{12}(x)n_{12}(x, t)] \, dx}{n(0, t)} \tag{19.20b}$$

and similarly for the frequency of A_2 among the newborn. Finally, by the random union of gametes assumption, the number of newborn of each genotype must satisfy

$$n_{11}(0, t) = p^2(0, t)n(0, t)$$

$$n_{12}(0, t) = 2p(0, t)q(0, t)n(0, t) \tag{19.21}$$

$$n_{22}(0, t) = q^2(0, t)n(0, t)$$

The transport model for population genetics with age structure and over-lapping generations is given by the three partial differential equations (19.18) subject to the boundary conditions (19.19), (19.20), and (19.21).

Conditions for Polymorphism in a Population with Age Structure and Overlapping Generations — Our approach is to seek a solution to the genetic model that represents a stationary polymorphic condition. That is, neither genotype should be increasing relative to the other. The total population size should be growing exponentially, and the age distribution for each genotype should be constant. Such a solution, in terms of an as yet unknown λ, must be of the form

$$n_{11}(x, t) = D(\lambda) \, e^{-\lambda x} l_{11}(x)N_0 \, e^{\lambda t}$$

$$n_{12}(x, t) = H(\lambda) \, e^{-\lambda x} l_{12}(x)N_0 \, e^{\lambda t} \tag{19.22}$$

$$n_{22}(x, t) = R(\lambda) \, e^{-\lambda x} l_{22}(x)N_0 \, e^{\lambda t}$$

where $D(\lambda)$, $H(\lambda)$, and $R(\lambda)$ are the genotype proportions among the newborn and λ is the population growth rate at the equilibrium. These forms do satisfy the partial differential equations. Virtually all our attention will now focus on the boundary conditions. By appealing to the assumption of random union of gametes, we observe that D, H, and R are given by the Hardy–Weinberg proportions based on p and q among the newborn. Hence (19.22) becomes

$$n_{11}(x, t) = p^2(\lambda) \, e^{-\lambda x} l_{11}(x)N_0 \, e^{\lambda t}$$

$$n_{12}(x, t) = 2p(\lambda)q(\lambda) \, e^{-\lambda x} l_{12}(x)N_0 \, e^{\lambda x} \tag{19.23}$$

$$n_{22}(x, t) = q^2(\lambda) \, e^{-\lambda x} l_{22}(x)N_0 \, e^{\lambda x}$$

where $p(\lambda)$ is the frequency of the A_1 allele among the newborn at the equilibrium. In (19.23) there are three unknowns to be determined from the boundary conditions, $p(\lambda)$, $q(\lambda)$, and λ where p and q are later normalized so that their sum equals one.

By substituting (19.23) into the boundary conditions (19.20a and b) for $p(0, t)$, we obtain

$$N_0 e^{\lambda t} p(\lambda)[p(\lambda)+q(\lambda)] = N_0 e^{\lambda t} p(\lambda)\left[p(\lambda) \int e^{-\lambda x} m_{11}(x) l_{11}(x)\, dx \right.$$
$$\left. + q(\lambda) \int e^{-\lambda x} m_{12}(x) l_{12}\, dx \right] \qquad (19.24)$$

There is a similar expression for the boundary condition on $q(0, t)$. To condense the notation, let

$$F_{ij}(\lambda) \equiv \int e^{-\lambda x} m_{ij}(x) l_{ij}(x)\, dx \qquad (19.25)$$

For any $l_{ij}(x)$ and $m_{ij}(x)$, $F_{ij}(\lambda)$ satisfies

(a) $F_{ij}(\lambda) > 0$

(b) $F'_{ij}(\lambda) < 0$

(c) $F_{ij}(\lambda) \rightarrow 0$ as $\lambda \rightarrow \infty$

(d) $F_{ij}(r_{ij}) = 1$ (19.26)

(e) $\lambda < r_{ij} \Rightarrow F_{ij}(\lambda) > 1$

(f) $\lambda > r_{ij} \Rightarrow F_{ij}(\lambda) < 1$

Note especially the definition of the intrinsic rate of increase for each genotype in (19.26d). The boundary conditions on $p(0, t)$ and $q(0, t)$ then become

$$p(\lambda)[F_{11}(\lambda)-1] + q(\lambda)[F_{12}(\lambda)-1] = 0 \qquad (19.27a)$$
$$p(\lambda)[F_{12}(\lambda)-1] + q(\lambda)[F_{22}(\lambda)-1] = 0 \qquad (19.27b)$$

This system has a nontrivial solution for p and q under two conditions. First, if there is some λ^* at which all $F_{ij}(\lambda^*) = 1$, then any p and q are solutions. This is the situation where there are three different life histories such that $r_{11} = r_{12} = r_{22}$. Since *any* p and q provide a stationary polymorphic solution, this case corresponds to selective neutrality. Second, the determinant of (19.27) is zero. In this situation a unique solution, p^* and q^*, is obtained upon applying the normalization condition. We stress that a polymorphism requires not only a zero determinant in (19.27), but also a positive p^* and q^*. Now consider this case in more detail.

Suppose that there is a real root, λ^*, of the equation

$$\det(\lambda) \equiv [F_{11}(\lambda)-1][F_{22}(\lambda)-1] - [F_{12}(\lambda)-1]^2 = 0 \qquad (19.28)$$

If a λ^* exists, then the solution for the gene frequencies among the newborn (p^*, q^*) is given by

$$p^* = \frac{-[F_{12}(\lambda^*)-1]}{[F_{11}(\lambda^*)-1]-[F_{12}(\lambda^*)-1]} = \frac{-[F_{22}(\lambda^*)-1]}{[F_{12}(\lambda^*)-1]-[F_{22}(\lambda^*)-1]} \qquad (19.29a)$$

$$q^* = \frac{[F_{11}(\lambda^*)-1]}{[F_{11}(\lambda^*)-1]-[F_{12}(\lambda^*)-1]} = \frac{[F_{12}(\lambda^*)-1]}{[F_{12}(\lambda^*)-1]-[F_{22}(\lambda^*)-1]} \qquad (19.29b)$$

The first expression for p^* and q^* is from (19.27a) and the second from (19.27b). Since $\det(\lambda^*) = 0$, the expressions are numerically equal. By inspection of these expressions, we observe that the positivity requirement is equivalent to the requirement that

$$\text{sign} \{F_{12}(\lambda^*) - 1\} = -\text{sign} \{F_{11}(\lambda^*) - 1\}$$
$$\text{sign} \{F_{12}(\lambda^*) - 1\} = -\text{sign} \{F_{22}(\lambda^*) - 1\}$$
(19.30)

Provided that det $(\lambda^*) = 0$, then a positive solution (p^*, q^*) exists if and only if both conditions in (19.30) are satisfied.

The positivity condition (19.30) above has many implications.

1. Consider the interval

$$I = [\min (r_{ij}), \max (r_{ij})]$$

If λ^* lies outside this interval, then the signs of $[F_{12}(\lambda^*) - 1]$ are necessarily all the same, violating (19.30). Therefore, the population growth rate at any polymorphic equilibrium is bounded within the highest and lowest of the r's of the three genotypes.

2. Suppose that $r_{11} > r_{12} > r_{22}$. Then for every λ^* between r_{22} and r_{11}, the positivity condition is violated. Therefore, there cannot be a polymorphic solution with the r's ordered as above.

3. Suppose that $r_{12} > r_{11}, r_{22}$. Then the positivity condition is satisfied if λ^* is between $\max (r_{11}, r_{22})$ and r_{12}. Thus a polymorphism exists with heterozygote superiority, provided λ^* is in a suitable interval.

4. Suppose $r_{12} < r_{11}, r_{22}$. Then the positivity condition is satisfied if λ^* lies between r_{12} and $\min (r_{11}, r_{22})$. Thus a polymorphic solution exists with heterozygote inferiority, provided λ^* is in a suitable interval.

Next, consider the equation for λ^* to determine whether roots do occur in the appropriate intervals specified above. There are two cases of interest:

1. $r_{12} > r_{11}, r_{22}$. We note that det $(r_{12}) > 0$ and that both det $(r_{11}) < 0$ and det $(r_{22}) < 0$. Therefore, there is a root, λ^*, in the interval $(\max (r_{11}, r_{22}), r_{12})$, as required. Moreover, it can easily be verified that det' $(\lambda) > 0$ in this interval; that is, det (λ) is monotone, increasing in this interval. Then, the root λ^* in this interval is unique. Hence, with heterozygote superiority, there is a unique polymorphic solution.

2. $r_{12} < r_{11}, r_{22}$, again det $(r_{12}) > 0$ and both det $(r_{11}) < 0$ and det $(r_{22}) < 0$. So again, there does exist a root, λ^*, in the interval $[r_{12}, \min (r_{11}, r_{22})]$, as required. Also it is again easy to verify that det (λ) is monotone decreasing in this interval so that λ^* is again unique.

We may summarize these results as a theorem. It was originally demonstrated by Norton (1928) and Charlesworth (1972).

THEOREM *Consider the model for natural selection at one locus with two alleles in a population with age structure and overlapping generations defined by Equations (19.18), (19.19), (19.20), and (19.21).*

1. Suppose that $r = r_{11} = r_{12} = r_{22}$. There exists a stationary polymorphic solution given by (19.23), where p and q are arbitrary and $\lambda = r$. In this case there is selective neutrality.

2. Suppose that $r_{11} \neq r_{12} \neq r_{22}$. There exists a stationary polymorphic solution, which is unique if and only if either $r_{12} > r_{11}, r_{22}$ or $r_{12} < r_{11}, r_{22}$. This solution is given by (19.23) where λ is the unique real root of (19.28) in the interval specified below and $p(\lambda)$ and $q(\lambda)$ are given by (19.29). If $r_{12} > r_{11}, r_{22}$, then λ lies within $(\max (r_{11}, r_{22}), r_{12})$ and if $r_{12} < r_{11}, r_{22}$ then λ lies within

$(r_{12}, \min(r_{11}, r_{22}))$. *Thus a polymorphic solution exists, and it is unique, if and only if there is either heterozygote superiority or inferiority as measured by the r_{ij}'s of the genotypes.*

This theorem asserts nothing about the stability of the polymorphic solutions. It is reasonable to conjecture that the polymorphic solution is stable if $r_{12} > r_{11}, r_{22}$ and unstable if $r_{12} < r_{11}, r_{22}$. Moreover, it is reasonable to conjecture that the "boundary" solution representing the fixation of an allele, say A_1, is unstable if $r_{12} > r_{11}$. However, the stability theory for equilibrium *functions* (not equilibrium points) is generally a difficult problem warranting separate treatment. It is beyond the scope of this chapter. Also the case of full dominance is not covered in this chapter.

An important paper on mathematical aspects of the convergence to the stable age distribution has been written by Golubitsky et al. (1975).

An Optimization Criterion for Computing the Polymorphic Solution

As discussed in the next section, the optimum life history strategy is regarded by most authors as that strategy with the highest intrinsic rate of increase, r. It is assumed that natural selection causes the strategy with the highest r to evolve. In this spirit we might wonder whether the exponential growth rate of a population is in some sense a maximum at the evolutionary equilibrium. If so, perhaps such a principle would also be useful in the calculation of the equilibrium solutions derived above. We now develop this idea in more detail: The equilibrium gene frequency among the newborn is that which leads to the highest exponential growth rate for the population.

Several functions will be introduced which are used in the optimization principle, and to motivate the introduction of these, consider the following situation. Imagine a polymorphic population whose genotype ratios are held constant. Moreover, suppose the genotype ratios are in Hardy–Weinberg ratios based on some p that is also a constant. This population is described by the following model

$$\frac{\partial n_{ij}(x, t)}{\partial t} = \frac{-\partial n_{ij}(x, t)}{\partial x} - \mu_{ij}(x) n_{ij}(x, t) \qquad (19.31)$$

where, as boundary conditions, we have

$$n_{11}(0, t) = p^2 n(0, t)$$
$$n_{12}(0, t) = 2pq n(0, t)$$
$$n_{22}(0, t) = q^2 n(0, t) \qquad (19.32)$$

$$n(0, t) = \sum \int m_{ij}(x) n_{ij}(x, t)\, dx$$

In these boundary conditions, p is a constant. This polymorphic population with fixed genotype ratios tends to an equilibrium age distribution at which the population grows exponentially. The solutions for the three genotypes are of the form

$$n_{11}(x, t) = p^2 e^{-\lambda x} l_{11}(x) N_0 e^{\lambda t}$$
$$n_{12}(x, t) = 2pq e^{-\lambda x} l_{12}(x) N_0 e^{\lambda t} \qquad (19.33)$$
$$n_{22}(x, t) = q^2 e^{-\lambda x} l_{22}(x) N_0 e^{\lambda t}$$

where p is a constant, and λ is to be determined by substituting (19.33) into the boundary conditions (19.32). Doing so shows that λ is the unique real root of

$$p^2F_{11}(\lambda)+2pqF_{12}(\lambda)+q^2F_{22}(\lambda)=1 \qquad (19.34)$$

Of course, since p is a constant in this equation, λ depends on p. In this way we have defined an exponential growth rate, λ, for a polymorphic population whose genotype ratios among the newborn are fixed at p^2, $2pq$, and q^2, where p is a constant.

We now introduce two functions. First, we have

$$\bar{F}(p,\lambda)\equiv p^2F_{11}(\lambda)+2p(1-p)F_{12}(\lambda)+(1-p)^2F_{22}(\lambda) \qquad (19.35)$$

Second, we introduce the function $\tilde{\lambda}(p)$ defined implicitly by

$$\bar{F}(p,\lambda)-1=0 \qquad (19.36)$$

The function, $\tilde{\lambda}(p)$, may be interpreted as the exponential growth of a polymorphic population whose genotype ratios among the newborn are fixed at p^2, $2p(1-p)$, and $(1-p)^2$, as discussed above.

Let us return to the model found in Equations (19.18) to (19.21) in which the gene frequencies are *variables*, not constants. First, we show that there is a very concise way of writing the conditions for equilibrium in terms of the function $\bar{F}(p,\lambda)$, defined above. Recall that the equilibrium solution is of the form

$$\begin{aligned}
n_{11}(x,t)&=p^2\,e^{-\lambda x}l_{11}(x)N_0\,e^{\lambda t}\\
n_{12}(x,t)&=2p(1-p)\,e^{-\lambda x}l_{12}(x)N_0\,e^{\lambda t},\\
n_{22}(x,t)&=(1-p)^2\,e^{-\lambda x}l_{22}(x)N_0\,e^{\lambda t}
\end{aligned} \qquad (19.37)$$

where p and λ were determined from the system (19.27). The system (19.27) is equivalent to the system

$$\frac{\partial \bar{F}(p,\lambda)}{\partial p}=0 \qquad (19.38a)$$

$$\bar{F}(p,\lambda)=1 \qquad (19.38b)$$

To prove this equivalency, observe that

$$\begin{aligned}
\frac{\partial \bar{F}}{\partial p}&=2(\phi-\Psi)=0\\
\bar{F}&=p\phi+(1-p)\Psi=1
\end{aligned} \qquad (19.39)$$

where

$$\begin{aligned}
\phi&\equiv pF_{11}(\lambda)+(1-p)F_{12}(\lambda)\\
\Psi&\equiv pF_{12}(\lambda)+(1-p)F_{22}(\lambda)
\end{aligned} \qquad (19.40)$$

Equation (19.39) is then equivalent to

$$\phi=1,\qquad \Psi=1 \qquad (19.41)$$

which is identical with (19.27). Therefore, (19.38) provides a concise statement of the equations from which p and λ are obtained.

Next we show that the system (19.38) is equivalent to a criterion involving the extrema of the function $\tilde{\lambda}(p)$. The function $\tilde{\lambda}(p)$ is defined implicitly by

(19.36). Differentiating implicitly, we obtain

$$\tilde{\lambda}'(p) = \frac{(\partial \bar{F}/\partial p)}{(-\partial \bar{F}/\partial \lambda)} \tag{19.42}$$

$\partial \bar{F}/\partial \lambda < 0$. Therefore, any point $(\tilde{p}, \tilde{\lambda})$, is a root of Equation (19.38) if and only if $\tilde{\lambda}'(\tilde{p}) = 0$ and $\tilde{\lambda} = \tilde{\lambda}(\tilde{p})$. Moreover, it is clear that $\tilde{\lambda}''(p) < 0$ if $r_{12} > r_{11}, r_{22}$, and $\tilde{\lambda}''(p) > 0$ if $r_{12} < r_{11}, r_{22}$; that is, $\tilde{\lambda}(p)$ is maximized at equilibrium with heterozygote superiority and minimized with heterozygote inferiority.

We can summarize these results as the following theorem originally demonstrated by Charlesworth (1972).

THEOREM *Consider the model for natural selection at one locus with two alleles in a population with age structure and overlapping generations defined by Equations (19.18), (19.19), (19.20), and (19.21). Let $\tilde{\lambda}(p)$ denote λ as a function of p defined implicitly by the equation*

$$\bar{F}(\lambda, p) - 1 = 0$$

The unique polymorphic solutions are given by

$$n_{11}(x, t) = \tilde{p}^2 \, e^{-\tilde{\lambda}x} l_{11}(x) N_0 \, e^{\tilde{\lambda}t}$$
$$n_{12}(x, t) = 2\tilde{p}(1 - \tilde{p}) \, e^{-\tilde{\lambda}x} l_{12}(x) N_0 \, e^{\tilde{\lambda}t}$$
$$n_{12}(x, t) = (1 - \tilde{p})^2 e^{-\tilde{\lambda}x} l_{22}(x) N_0 e^{\tilde{\lambda}t}$$

where $\tilde{p}$ and $\tilde{\lambda}$ satisfy

$$\tilde{\lambda}'(\tilde{p}) = 0$$
$$\tilde{\lambda} = \tilde{\lambda}(\tilde{p})$$

Moreover, with heterozygote superiority $\tilde{p}$ maximizes $\tilde{\lambda}(p)$ and with heterozygote inferiority $\tilde{p}$ minimizes $\tilde{\lambda}(p)$.

This theorem provides a convenient method for calculating a polymorphic solution. Moreover, this method lends support to the idea that natural selection, at the evolutionary equilibrium, adjusts the gene frequency to produce the highest possible population growth rate. This theorem supports the contention of many authors that the intrinsic rate of increase, r, is the appropriate optimization criterion to use in determining optimum life history strategies.

Some Examples Convenient examples of age-dependent selection are found by using the following forms for $l_{ij}(x)$ and $m_{ij}(x)$.

$$l_{ij}(x) = e^{-x/a_{ij}}$$

$$m_{ij}(x) = \frac{c_{ij}}{b_{ij}^2} x \, e^{-x/b_{ij}} \tag{19.43}$$

where a_{ij} is the time for a cohort of type, $A_i A_j$, to decay to $(1/e)$ times its original number, b_{ij} is the age of peak reproductive activity of type $A_i A_j$, and c_{ij} is the number of offspring that would be produced by someone of type $A_i A_j$ who lived forever, that is, the maximum possible number of offspring from type $A_i A_j$.

Using (19.43), we have

$$r_{ij} = \frac{\sqrt{c_{ij}} - 1}{b_{ij}} - \frac{1}{a_{ij}}$$

$$F_{ij}(\lambda) = \frac{c_{ij}}{[1 + b_{ij}/a_{ij} + b_{ij}\lambda]^2} \qquad (19.44)$$

Table 19.2 presents a numerical example of heterozygote superiority and inferiority. Figure 19.3 illustrates the function $\tilde{\lambda}(p)$ for these two examples. Observe that the equilibrium polymorphism frequency among the newborn and the exponential growth rate are

$$\text{heterozygote superiority: } \tilde{p} = .68,\ \tilde{\lambda} = .076$$

$$\text{heterozygote inferiority: } \tilde{p} = .21,\ \tilde{\lambda} = .039$$

Table 19.2. Examples of Heterozygote Superiority and Inferiority.

		A_1A_1	A_1A_2	A_2A_2
Heterozygote superiority	a_{ij}	7	10	5
	b_{ij}	5	6	3
	c_{ij}	4	5	2
	r_{ij}	.057	.106	−.062
Heterozygote inferiority	a_{ij}	7	6	7
	b_{ij}	5	5	4
	c_{ij}	4	4	3
	r_{ij}	.057	.033	.040

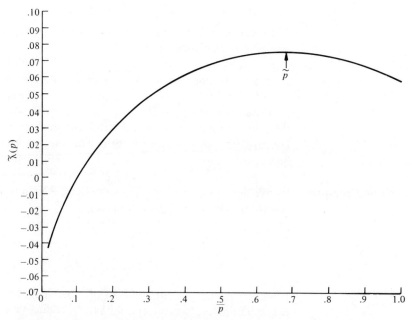

FIGURE 19.3a. *Numerical example of $\lambda(p)$ as a function of p with heterozygote superiority. The frequency of A_1 among newborn at the polymorphism is $\tilde{p}$.*

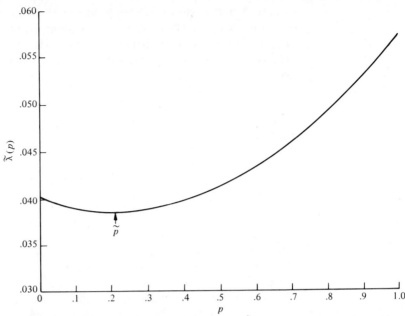

FIGURE 19.3b. *Numerical example of* $\tilde{\lambda}(p)$ *as a function of* p *with heterozygote inferiority.*

The $\tilde{\lambda}(p)$ for Figure 19.3 was obtained by solving numerically for λ in the equation $\bar{F}(\lambda) = 1$, using several values of p between 0 and 1. These examples illustrate the fact that each r_{ij} may depend in a complicated way on the features of a life history, but it is nonetheless the relation among the r_{ij}'s alone that determines the existence of a polymorphic solution.

The Evolution of Life Histories   The population genetic analysis has shown that the decisive quantities controlling the outcome of natural selection in an age-structured population are the intrinsic rates of increase of the genotypes. Individuals with a life history having the highest r will evolve by natural selection. Hence the problem of predicting the life history that will evolve in any particular situation reduces to finding the life history with the highest r in that situation. The particular life history with the highest r is called the optimum life history strategy. In this section we illustrate how optimum history strategies are relevant to explaining some of the evolutionary trends that are observed in the data on life histories of closely related species, as discussed in the introduction to the chapter.

There are two somewhat similar approaches to determining the optimum life history strategy that occur in the literature on this topic. The first approach is the most direct and is based on picking out the best life history from a specified set of possible life histories. The second approach is to assume the population has a certain life history to begin with and to inquire how that life history will be changed by modifier genes.

Direct Approach   Perhaps the simplest illustration of the direct approach is with the following example. Assume that the $m(x)$ curve is constrained to be of the form

$$m(x) = \frac{c}{b^2} x \, e^{-x/b} \tag{19.45}$$

where b is the age of peak reproductivity and c is the total number of offspring that could be produced by someone who lived forever based on this $m(x)$ curve. The area under the $m(x)$ curve, c, is a measure of an individual's potential reproductive output. Suppose that the $l(x)$ curve is constrained to be of the form

$$l(x) = e^{-x/a} \qquad (19.46)$$

where a is a measure of an individual's longevity. As mentioned in Equation (19.44), we know that r is then given explicitly by

$$r = \frac{\sqrt{c}-1}{b} - \frac{1}{a} \qquad (19.47)$$

By assuming that $m(x)$ and $l(x)$ are given by these expressions, we have partially defined the set of life histories that are assumed to be evolutionarily available to the species. But this set must be further restricted to take account of the biological "tradeoffs" involved in reproduction. We may assume that increasing reproductive activity also increases the exposure to hazard. This assumption implies that c and a are somehow inversely related. That is, an individual cannot have both a high potential reproductive output and a high longevity. Let this relation between c and a be denoted by

$$a = f(c), \qquad f'(c) < 0 \qquad (19.48)$$

Another natural tradeoff is to suppose that the potential reproductive output, c, increases with the age of peak activity, b. This assumption is typically justified by noting that egg output in many organisms directly depends on body size; for example, larger fish can carry more eggs. Hence, by postponing the age of peak reproductive activity, an organism can grow to a larger size and achieve a larger possible reproductive output. Another way to justify the assumption, which is more appropriate to birds and perhaps also to mammals, is to suppose that the ability to rear young successfully increases with experience. Organisms that postpone reproductive activity will achieve a higher potential reproductive output because of their increased experience. Let us denote the relation between the age of peak reproductive activity, b, and the total possible output, c, as

$$b = g(c) \qquad g'(c) > 0 \qquad (19.49)$$

With these assumptions, we have completely specified the set of possible life histories. Our next problem is to find that particular life history, $(m(x)^*, l(x)^*)$, from this set of possible life histories which has the highest r.

The task of finding the optimum life history in this example is easy in principle, though not always in practice. We substitute (19.48) and (19.49) into (19.47) and obtain an equation for r as a function of c

$$r(c) = \frac{(\sqrt{c}-1)}{g(c)} - \frac{1}{f(c)} \qquad (19.50)$$

Then we solve for the value of c that maximizes r. Let this value be denoted c^*. It is found by differentiating $r(c)$ and setting the derivative equal to zero. Doing this and rearranging, shows that c^* must be the root of

$$\frac{1}{2\sqrt{c}g(c)}\left[1 - \frac{2(c-\sqrt{c})g'(c)}{g(c)}\right] = -\frac{f'(c)}{f^2(c)} \qquad (19.51)$$

The root to this equation might be determined graphically (see below) or with a root-finder program on a computer. Provided that c^*, in fact, maximizes r, as established by checking the sign of $r''(c^*)$, we then determine the optimum life history strategy. We calculate $a^* = f(c^*)$ and $b^* = g(c^*)$ and the optimum life history is $[m^*(x), l^*(x)]$, based on the parameter values of a^*, b^*, and c^*.

As an explicit illustration of determining an optimum life history strategy, suppose that the longevity is linearly related to the potential reproductive output, c, as

$$a = f(c) = a_0 - \alpha c \qquad \text{for } 0 \leq c \leq \frac{a_0}{\alpha} \tag{19.52}$$

where a_0 is a measure of the basic longevity in the environment apart from any loss of longevity incurred because of reproductive activity. Suppose also that there is a linear relation between the age of peak reproductive activity, b, and c as

$$b = g(c) = \beta c \tag{19.53}$$

Next, we graphically determine the value of c^*. Let the right-hand side of Equation (19.51) be denoted as $\psi(c)$. Substituting (19.52) and (19.53), we have

$$\psi(c) \equiv -\frac{f'(c)}{f^2(c)} = \frac{\alpha}{[a_0 - \alpha c]^2} \tag{19.54}$$

Let the left-hand side (19.51) be denoted as $\phi(c)$. Substituting (19.52) and (19.53), we have

$$\phi(c) \equiv \frac{1}{2\sqrt{c}g(c)}\left[\left[1 - \frac{2(c - \sqrt{c})g'(c)}{g(c)} \right] \right]$$

$$= \frac{1}{2\beta c^{3/2}}\left(\frac{2}{\sqrt{c}} - 1 \right) \tag{19.55}$$

The value c^* is found as the intersection of $\psi(c)$ with $\phi(c)$. Figure 19.4 provides a sketch of these curves. They might be used as follows. We might assume that a mild environment is one with a high a_0 and that a harsh environment has a low a_0. That is, the base-line longevity in a harsh environment is less than in a mild environment. If the environments differ in a_0 and not very much in other respects, then the optimum life histories for the two environments will differ in the following ways. As depicted in Figure 19.4, c^* will be lower in the harsh environment than in the mild environment. Hence the age of peak reproductivity, b^*, will be earlier (i.e., the time to maturity shorter) in the harsh environment than in the mild environment. Thus, subject to our assumptions, the optimum strategy in the harsh environment is to sacrifice some of the long-term potential reproductive output, c, and to concentrate the reproductive activity into earlier ages, b.

This example illustrates the direct approach to the determination of optimum life history strategies, and also illustrates the type of qualitative biological predictions one seeks to obtain from the analysis. The three steps in the approach are (1) to incorporate assumptions that specify the set of life histories which are evolutionarily possible for a species, (2) to determine which of these possible strategies has the highest r, and (3) to relate the

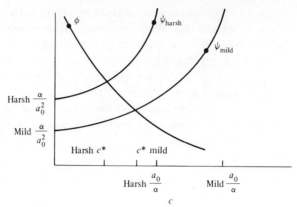

FIGURE 19.4. *Sketch of curves used in the example of how to determine an optimum life history strategy.*

analysis to the environmental conditions of species with different life histories. To use more detailed assumptions about the tradeoffs involved among the various possible strategies, and to determine which strategy has the highest r when r cannot be explicitly calculated (unlike in the above example) require techniques of optimization theory. For more information on how this is approached, consult Taylor et al. (1974).

Modifier Approach   The other approach to the evolution of life histories is to consider how modifier genes will influence a given life history. That is, suppose that we are given a particular life history $[m(x), l(x)]$. We imagine that there are alleles "chipping" away at parts of this life history and adding to other parts with "putty"; for example, there may be an allele that slightly lowers the ability to survive at age x and slightly improves the ability to survive at age y, or there may be an allele that slightly lowers survival at age x and increases the fertility at age y, and so forth. The question is to determine how such modifier alleles will mold the shape of the original $m(x)$ and $l(x)$ curves. This approach was pioneered by W. D. Hamilton (1966) in developing an evolutionary theory of senescence. An important theoretical question which we shall also consider is whether the reproductive value, v_a, introduced in the last chapter is relevant to the evolution of modifier genes. An important experimental study of the evolution of senescence appears in Mertz (1975).

In order for a modifier allele to evolve, its overall effect must be to increase r. We shall determine whether the overall effect is positive or negative by summing over each separate effect made on the fitness by the modifier. That is, the overall effect of a modifier which improves survival at age x and diminishes survival at age y is the sum of some positive effect on fitness and a negative effect. In the next several paragraphs we shall calculate the separate effects on r of a small change in $l(x)$ and in $m(x)$. Then we shall combine these to determine the net effect of modifiers on a life history.

First, we calculate the effect on fitness of a small change in survival at age a. To keep matters simple, we shall work with a discrete time and discrete age class approximation to the continuous time formulation. We approximate the integral as a sum

$$\int e^{-rx} m(x) l(x)\, dx \approx \sum_{x=1}^{\infty} e^{-rx} m_x l_x \qquad (19.56)$$

where the age class width and time interval are taken to be of unit length. Furthermore, we approximate the survivorship curve as

$$l_x = p_1 p_2 \cdots p_x \qquad (x = 1, 2, \ldots) \qquad (19.57)$$

where p_x is the probability of surviving *through* the xth age class given that one has survived to the beginning of the age class. The discrete rate of increase, λ, is defined as

$$\lambda \equiv e^r, \qquad \ln(\lambda) \equiv r \qquad (19.58)$$

Clearly, r and λ are equivalent measures of fitness. (This λ is not to be confused with the λ of the last section on population genetics.) Here λ describes the fitness of the given life history, (m_x, l_x). It is calculated as the root of the familiar equation

$$\sum_{x=1}^{\infty} \lambda^{-x} m_x l_x = 1 \qquad (19.59)$$

or equivalently,

$$\sum_{x=1}^{\infty} \lambda^{-x} m_x \left(\prod_{j=1}^{x} p_j \right) = 1 \qquad (19.60)$$

Here λ can be regarded as a function of the life history parameters, m_x and p_x. That is,

$$\lambda = \lambda(m_1, m_2 \cdots m_x, p_1, p_2 \cdots p_x) \qquad (x = 1 \cdots \infty) \qquad (19.61)$$

where λ is defined implicitly as a function of these parameters by Equation (19.60). Now consider a small change in the ability to survive at a particular age, say age a. To understand the effect of this change, we must calculate $\partial \lambda / \partial p_a$. (Actually, we shall calculate $\partial \ln(\lambda) / \partial \ln(p_a)$, which amounts to the same thing.) Differentiating implicitly in (19.60), we obtain

$$\sum_{x=a}^{\infty} \lambda^{-x} m_x \left(\prod_{j \neq a}^{x} p_j \right) + \sum_{x=1}^{\infty} \left(\prod_{j=1}^{x} p_j \right) m_x (-x) \lambda^{-x-1} \frac{\partial \lambda}{\partial p_a} = 0 \qquad (19.62)$$

Upon inserting p_a into the product in the left summation and factoring out λ^{-1} from the second summation, we obtain

$$\frac{1}{p_a} \sum_{x=a}^{\infty} \lambda^{-x} m_x l_x - \frac{1}{\lambda} \sum_{x=1}^{\infty} x \lambda^{-x} m_x l_x \frac{\partial \lambda}{\partial p_a} = 0 \qquad (19.63)$$

Rearranging yields

$$\frac{\partial \ln(\lambda)}{\partial \ln(p_a)} = \frac{\displaystyle\sum_{x=a}^{\infty} \lambda^{-x} m_x l_x}{\displaystyle\sum_{x=1}^{\infty} x \lambda^{-x} m_x l_x} \qquad (19.64)$$

Formula (19.64) indicates the effect on the log of λ (i.e., on r) for a given life history of an infinitesimal change in the log of the probability of surviving through age class a. To understand what this expression says, it is helpful to consider the quantity

$$F(x) = \lambda^{-x} m_x l_x \qquad (19.65)$$

as a probability density function. It might be called the "weighted net maternity function." Note that $\sum F(x) = 1$, as would be required of a probability density function. The denominator is then seen to be the average x, averaged with respect to this weighted net maternity function. This average age is sometimes called the "generation time." The numerator is seen to be the area in the tail of $F(x)$ to the right of a. Note that the numerator of (19.64) must be a monotonic decreasing function of the age, a, at which the effect is acting. That is, a given level of change in the ability to survive at age a has an effect on fitness which decreases monotonically with a. We shall return to this point shortly.

Next, let us calculate the effect on fitness of a small change in the reproduction of age a. Again we regard (19.59) as implicitly defining λ as a function of m_x and p_x. Then differentiating (19.59) implicitly with respect to m_a yields

$$\lambda^{-a}l_a + \sum_{x=1}^{\infty} m_x l_x (-x) \lambda^{-x-1} \frac{\partial \lambda}{\partial m_a} = 0 \tag{19.66}$$

Rearranging gives

$$\frac{\partial \ln(\lambda)}{\partial m_a} = \frac{\lambda^{-a}l_a}{\sum_{x=1}^{\infty} x\lambda^{-x} m_x l_x} \tag{19.67}$$

This formula indicates the effect on the log of λ for a given life history of an infinitesimal change in the reproductive output of age a. Note that (19.67) has the same denominator as (19.64). Note also that the numerator is again a monotonic decreasing function of a because both λ^{-a} and l_a are each monotonic decreasing functions of a.

Let us now use these formulas to study how modifier genes may influence a life history. There are several cases to consider.

GENES THAT CHIP AWAY AT SURVIVAL AT ONE AGE AND ADD TO SURVIVAL AT ANOTHER AGE. Let a_1 denote the age at which survival is increased and let a_2 denote the age at which survival is reduced. Let $\Delta \ln(p_{a_1})$ denote the amount by which the log of p_{a_1} is increased and let $\Delta \ln(p_{a_2})$ denote the amount by which survival at age a_2 is decreased. Hence

$$\Delta \ln(p_{a_1}) > 0$$
$$\Delta \ln(p_{a_2}) < 0 \tag{19.68}$$

What is the overall effect of this modifier gene on fitness? For changes sufficiently small, we have

$$\Delta r = \frac{\partial r}{\partial \ln(p_{a_1})} \Delta \ln(p_{a_1}) + \frac{\partial r}{\partial \ln(p_{a_2})} \Delta \ln(p_{a_2}) \tag{19.69}$$

Remembering that $\ln(\lambda) \equiv r$, we substitute from (19.64) yielding

$$\Delta r = \frac{\left(\sum_{x=a_1}^{\infty} \lambda^{-x} m_x l_x\right) \Delta \ln(p_{a_1}) + \left(\sum_{x=a_2}^{\infty} \lambda^{-x} m_x l_x\right) \Delta \ln(p_{a_2})}{\sum_{x=1}^{\infty} x\lambda^{-x} m_x l_x} \tag{19.70}$$

Therefore, the modifier will lead to a net increase in fitness whenever

$$\frac{\sum\limits_{x=a_1}^{\infty} \lambda^{-x} m_x l_x}{\sum\limits_{x=a_2}^{\infty} \lambda^{-x} m_x l_x} > \frac{-\Delta \ln (p_{a_2})}{\Delta \ln (p_{a_1})} \tag{19.71}$$

Hamilton (1966) has focused on the important case of equal gain and loss at the two ages. That is, the amount by which the log of the survival is reduced at age a_2 equals that gained at age a_1. In this case, the right-hand side of (19.71) equals one. In this case, provided that $\lambda > 0$, the condition is automatically satisfied whenever the age that gains survival is younger than the age that loses survival.

Hamilton (1966) discusses many interesting consequences of the result just stated. Some of these are: (1) There is an inherent selection pressure favoring genes that are pleiotropic in the sense of improving survival at early ages at the expense of survival in old age. The evolution of these genes should lead to many physiological causes of aging. There should not be any single physiological mechanism responsible for aging. This is a critical prediction for medical research in geriatrics and, if correct, destroys the hope of a medical breakthrough leading to a fountain of youth. Genetically determined senescence is an unfortunate but inevitable consequence of natural selection in every population. (2) The evolution of senescence is facilitated by a high r. Rapidly growing populations permit condition (19.71) to be satisfied for a larger class of modifier genes than slowly growing or stationary populations. (3) The *"reproductive value," v_a, introduced by Fisher is irrelevant to the evolution of the survivorship curve in this case.* Hamilton constructs examples where reference to v_a and the correct relation based on (19.71) above lead to contradictory results. (4) The fact that humans and many organisms show high infant mortality poses a problem for a theory based solely on the modification of the l_x curve. Under a theory based solely on the modification of l_x, a gene that reduces survival by one unit during the age of peak reproductive activity and increases survival by one unit during any prereproductive age will satisfy condition (19.71) and is predicted to evolve by natural selection. Thus, by such a theory, there is no reason to expect higher mortality during prereproductive ages than during reproductive ages. Because of this difficulty, Hamilton suggests that the evolution of high infant mortality represents selection on the family level. He refers to the phenomenon of sibling replacement whereby parents replace offspring who die with a new child. He conjectures that selection on the family level will select for the early expression of very deleterious or lethal genes so as to facilitate sibling replacement. However, if modification occurs simultaneously on both m_x and l_x, then v_a is relevant, as discussed below.

GENES THAT CHIP AWAY AT REPRODUCTION AT ONE AGE AND ADD TO REPRODUCTION AT ANOTHER AGE. Again let a_1 denote the age at which reproduction is increased and a_2 the age when it is decreased. Let Δm_{a_1} denote the amount by which reproduction at a_1 is increased and Δm_{a_2} the amount by which it is decreased at a_2, then

$$\Delta m_{a_1} > 0$$
$$\Delta m_{a_2} < 0 \tag{19.72}$$

For changes sufficiently small we have

$$\Delta r = \frac{\partial r}{\partial m_{a_1}} \Delta m_{a_1} + \frac{\partial r}{\partial m_{a_2}} \Delta m_{a_2} \qquad (19.73)$$

Substituting from (19.67) leads to the following condition for the gene to increase fitness

$$\frac{\lambda^{-a_1} l_{a_1}}{\lambda^{-a_2} l_{a_2}} > \frac{-\Delta m_{a_2}}{\Delta m_{a_1}} \qquad (19.74)$$

Again the case of equal gain and loss at the two ages is instructive. In this case, provided $\lambda > 0$, the condition is automatically fulfilled whenever $a_1 < a_2$.

Relation (19.74) leads to several conclusions: (1) There is an inherent selection pressure in favor of genes that sacrifice late reproduction in order to improve early reproduction. This result may explain why the $m(x)$ curves of so many species drop in older ages. (2) This tendency is enhanced in rapidly growing populations relative to slowly growing or stationary populations. Thus there is a stronger pressure for the evolution of early reproduction in rapidly growing populations. (3) The fact that $m(x)$ curves do not pile up at $x = 0$ may be explained by reference to the increased fertility that is possible at later ages due to increased size and experience, as discussed earlier in connection with optimum life history strategies. If sacrificing a small amount of fertility at a very young age leads to a large increase in fertility shortly thereafter, then relation (19.74) could be satisfied even if $a_1 > a_2$. It must be remembered that the general tendency of selection as revealed in the equal-gain-equal-loss case can always be reversed, provided the gain versus loss situation is sufficiently unequal.

GENES THAT CHIP AWAY AT SURVIVAL AND ADD TO REPRODUCTION. Let a_1 be the age at which reproduction is increased and a_2 the age at which survival is reduced. Let the increments be

$$\Delta m_{a_1} > 0$$
$$\Delta \ln (p_{a_2}) < 0 \qquad (19.75)$$

The condition for this gene to increase is

$$\frac{\lambda^{-a_1} l_{a_1}}{\sum\limits_{x=a_2}^{\infty} \lambda^{-x} m_x l_x} > \frac{-\Delta \ln (p_{a_2})}{\Delta m_{a_1}} \qquad (19.76)$$

This condition can be used to address the following important issue. Generally speaking, increasing reproductive output involves increased exposure to hazard. The question is, how much hazard can be incurred per unit of increased reproductive output and still show a net profit in fitness? However, this question must be sharpened to include the ages at which the hazard is incurred and at which the increased reproductive output is realized, although the case of most interest is where the hazard and the increased reproduction occur at the same age. Let Δm_{a_1} equal 1 so that we refer to a unit increase in the reproduction at age a_1. The hazard incurred at age a_2 is measured by $-\Delta \ln (p_{a_2})$. Therefore, from (19.76) we find that r

shows a net increase whenever

$$\begin{bmatrix} \text{hazard at age } a_2 \text{ per unit} \\ \text{increase in reproduction} \\ \text{at age } a_1 \end{bmatrix} < \frac{\lambda^{-a_1} l_{a_1}}{\sum\limits_{x=a_2}^{\infty} \lambda^{-x} m_x l_x} \qquad (19.77)$$

Conversely, we may ask how much gain in reproduction at age a_1 resulting from a unit loss of survival at age a_2 leads to a net increase in fitness. We set the hazard $-\Delta \ln (p_{a_2}) = 1$ and measure the gain in reproduction in terms of Δm_{a_1}. From (19.76) we then find that r shows a net increase whenever

$$\begin{bmatrix} \text{gain in reproduction} \\ \text{at age } a_1 \text{ per unit} \\ \text{hazard at age } a_2 \end{bmatrix} > \frac{\sum\limits_{x=a_2}^{\infty} \lambda^{-x} m_x l_x}{\lambda^{-a_1} l_{a_1}} \qquad (19.78)$$

These relations are fundamental to understanding how tradeoffs between survival and reproduction are related to natural selection.

Special interest centers on the case where the increased hazard and increased fertility occur at the same age. Imagine, for example, that the hazard is actually encountered during the act of obtaining food to raise more offspring. In situations like this, $a_1 = a_2 = a$ in the formulas above. Condition (19.77) for a net increase in fitness therefore becomes

$$\begin{bmatrix} \text{hazard at age } a \text{ per unit} \\ \text{increase in reproduction} \\ \text{at age } a \end{bmatrix} < \frac{\lambda^{-a} l_a}{\sum\limits_{x=a}^{\infty} \lambda^{-x} m_x l_x} \qquad (19.79)$$

Now recall that the reproductive value of Fisher for someone of age a is defined as

$$v_a = \frac{\sum\limits_{x=a}^{\infty} \lambda^{-x} m_x l_x}{\lambda^{-a} l_a} \qquad (19.80)$$

Therefore, (19.79) becomes

$$\begin{bmatrix} \text{hazard at age } a \text{ per unit} \\ \text{increase in reproduction} \\ \text{at age } a \end{bmatrix} < \frac{1}{v_a} \qquad (19.81)$$

This remarkable result shows that the reproductive value v_a controls the extent to which it is profitable to incur hazard in order to raise reproductive output at any given age. It shows that the maximum hazard per unit increase in reproduction which yields a net increase in fitness varies inversely with the reproductive value, v_a.

In discussing the first case, we noted that the occurrence of high infant mortality posed a problem for a theory of senescence which operates through genes which only modify survival. To overcome this difficulty, Hamilton made recourse to the idea of sib replacement and selection at the family level. However, Fisher (1958) had long ago noticed the general inverse relationship between the force of mortality and the reproductive value but did not offer any rationale for this relationship. Condition (19.81) predicts that this relationship is caused in part by the evolution of traits that sacrifice survival ability at some age in order to increase reproductive output at that age. This explanation could account in part for the evolution of high infant mortality because v_a is low for infants and rises to its peak near the beginning of the age of major reproductive activity.

Concluding Remarks on the Evolution of Life Histories Recall from the introduction to this chapter that data on the life histories of related bird species suggest that evolution of the $m(x)$ and $l(x)$ curves leads to a correlation of early maturation, high reproductive output, and high mortality with environmental harshness. Is this predicted from the evolutionary theory above? Yes, provided we interpret a harsh environment as one where the populations are often cut back to a low abundance so that the populations are more often in a state of growth (high r) than are populations in mild environments. If this assumption is granted, then the general features of the bird data follow from the modification theory above. In a harsh environment the higher r causes increased evolution of senescence and an earlier and larger reproductive output than in mild environments. However, a more detailed evolutionary prediction of the life histories in an environment would require a correspondingly more detailed specification of the tradeoffs involved.

Chapter 20

STOCHASTIC ENVIRONMENTS: EXTINCTION, RESOURCE TRACKING, AND PATCHINESS

As any backpacker knows, the physical environment is always fluctuating. Sometimes the fluctuations are predictable and sometimes they are not. Obviously, the coming of the seasons is predictable; indeed, the shortening of day length during the summer very reliably predicts the coming of fall. However, weather changes of shorter duration are often unpredictable. Most models of population dynamics are based on the assumption that the *parameters* in the models are constants. Yet these parameters typically describe features of the environment that, in fact, are continually in fluctuation. In particular, parameters like the carrying capacity describe the level of resources in the environment, and this quantity certainly fluctuates in time. Therefore, it is important to extend population dynamic theory to address issues that arise because of fluctuations in the parameters of the traditional ecological models.

The Issues   Three biological issues raised by environmental fluctuation have recently received much attention. The first is the possibility of *population extinction*

Extinction   induced by environmental fluctuation. In a constant environment, the logistic equation in continuous time indicates that a population will exist at a population size equal to K, provided only that r is positive. But suppose the carrying capacity fluctuates. Is this conclusion still true? That is, in spite of having a high average carrying capacity and a positive r, is it likely that a population will go extinct and, if so, what is the average "waiting time" to extinction?

Tracking   The second issue that has received attention involves "resource tracking" In a constant environment a logistic population attains a size that exactly accords with the amount of available resources. But in a fluctuating environment, provided extinction does not occur, the population size will occasionally be above the current carrying capacity, and occasionally below. A fluctuating environment entails that there will typically be a discrepancy between the population size and the amount of available resources. A population is said to "track" its resources well if this discrepancy is usually small and to "track" poorly if this discrepancy is large. We shall discover some interesting relationships between a population's ability to track fluctuating resources and the predictability of the environment.

Patchiness   The third biological issue to be discussed concerns the spatial pattern of a population in a fluctuating environment. Most populations occupy a large area relative to the size of an individual organism in the population. However, virtually no population is equally abundant throughout its range. Typically, the members of a population are more concentrated in certain places within its range than others. The surprising part about this observation is that the regions where individuals happen to be concentrated often do not correspond with places where there is a permanent high abundance of resources. Perhaps the regions where organisms are concentrated represent places with abundant resources in the past, but it is rarely possible to find out by direct reference to historical records. A *patch* is an area within a species range where the organisms are more abundant than average. A patch may be large or small depending on whether the region containing more than average abundance is itself large or small in area. The problem is to predict the steady-state distribution of patch sizes *in space*, based on assumptions about how the resource level at each place within the species range varies *in time*. Thus the problem is to predict the spatial distribution based on assumptions of stochastic fluctuations in time throughout the species range.

In addition to discussion of the issues above, this chapter will present the basic vocabulary used to describe a fluctuating environment. Concepts like "predictability" and "pattern" can be made very precise and intuitive, and also are defined in ways which are straightforward to measure. Recently, Botkin and Sobel (1975) have also suggested a concept of stability for time-varying ecosystems.

Before continuing I would like to mention that research on population dynamics in variable environments has increased tremendously during the last two to three years. Although this chapter is quite long, it only introduces this literature. It focuses on the biological topics to which the literature is relevant, rather than on the mathematical techniques of analysis (which are unfortunately quite complex).

Some Basic Vocabulary

There is a basic distinction between two types of causes of random population fluctuation. One possible cause of population fluctuation is internal to the population and has nothing to do with environmental variation. It arises from the finiteness of the population, together with the simple fact that offspring come in integers. Suppose that the *average* number of offspring per family is two so that, on the average, direct replacement occurs. But, in fact, some families are having 0 offspring, some 1 offspring, some 2 offspring, and others 3, and so forth. Now suppose, also, that the size of this population is currently 20 individuals. Then one generation later it is very unlikely that the population will again be exactly 20 individuals. For this to happen would require that the number of families with more than two offspring exactly balance the number of families with less than two offspring so that the total number of offspring *exactly* replace the parental generation. More typically, the population size fluctuates because the exact number of offspring is either slightly more or less than that needed for replacement of the adults, although on the *average*, exact replacement does occur. This cause of population fluctuation is called *demographic stochasticity* (May 1973). It is a traditional topic in the theory of probability. A common model for this phenomenon is a *branching process*, but this topic is *not* covered in this chapter. For more on this, see any textbook on stochastic processes such as Cox and Miller (1968). For biological applications, see Pielou (1969) and Ludwig (1974).

The other major cause of population fluctuation is variation in parameters that describe the environment, for example, the carrying capacity as already mentioned. This cause of population fluctuation is called *environmental stochasticity* by May (1973). This chapter is exclusively devoted to studying this particular cause of population fluctuation. For a treatment in which both sources of stochasticity are present (based on the theory of branching processes in random environments), see Keiding (1975).

Variation   The vocabulary for describing population fluctuations is drawn from statistics, especially an area called time series analysis. A sequence of data points collected at regularly spaced intervals is called a *time series*. Figure 20.1 illustrates four time series—each curve represents the history of some quantity, X, which varies through time. There are three important measures used to describe a time series. The first is a measure of the overall amount of variation in the series. The overall variation is measured simply by the *variance* of all the numbers in the time series. Clearly, a high variance indicates a high overall amount of variation.

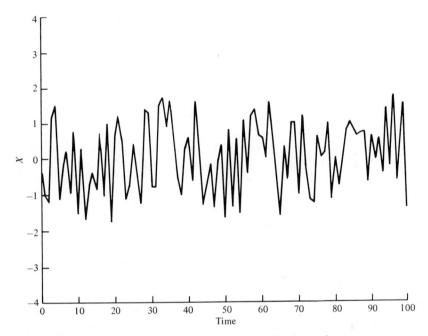

FIGURE 20.1a. *An unpredictable time series. The vertical axis represents a quantity, X, whose value changes through time and the horizontal axis is time. In this figure the correlation coefficient between consecutive values of X equals zero.*

Predictability   The other two measures that describe a time series are closely related to one another: One describes the *predictability* in a time series; the other describes the *pattern* in a time series. The predictability is described in terms of the *autocorrelation function*. A synonym for this is the *serial correlation function*. (The pattern is described by the variance spectrum, as will be discussed later in the section on patchiness.) To determine if there is any predictability inherent in a time series, we must determine whether there is any correlation between consecutive data points. If there is no correlation between consecutive data points, then the sequence is completely unpredictable. But if there is correlation, then this correlation could be used to make a prediction about the future based on past observations. The autocorrelation function, denoted $\rho(h)$, is the correlation coefficient between

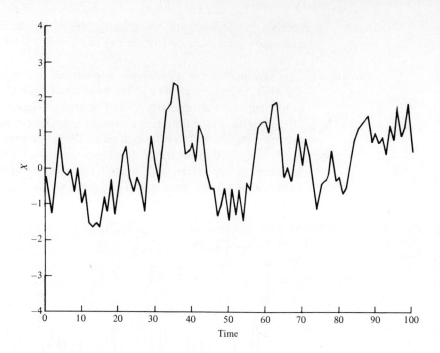

FIGURE 20.1b. *A predictable time series. The correlation coefficient between consecutive values of X equals .75. Generated by a first-order autoregressive process.*

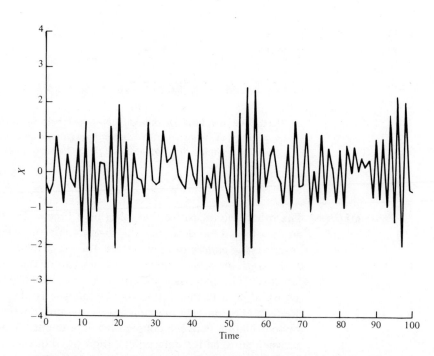

FIGURE 20.1c. *A predictable time series. The correlation coefficient between consecutive values of X equals −.75. Generated by a first-order autoregressive process.*

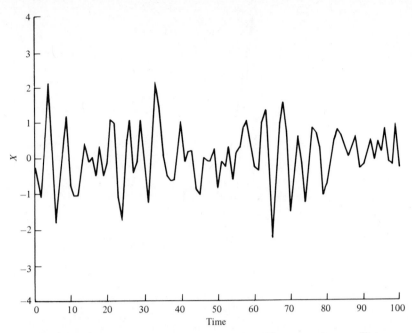

FIGURE 20.1d. *A predictable time series. The regression coefficient between values of X separated by one time interval is .5 and that between values separated by two time intervals is −.75. Generated by a second-order autoregressive process.*

observations separated by h time intervals. Hence $\rho(1)$ denotes the correlation coefficient between adjacent data points, $\rho(2)$ denotes the correlation between points separated by two time intervals, and so forth.

The autocorrelation function $\rho(h)$ is entirely a descriptive quantity. Its use does not commit one to any underlying model or assumptions. Figure 20.2 (top) illustrates the autocorrelation function for an unpredictable sequence. Here $\rho(0)$ equals one because a number is always perfectly correlated with itself; $\rho(h)$ equals zero for every other h.† Figure 20.2 (top) also illustrates a typical $\rho(h)$ for a predictable sequence where the level of correlation decreases geometrically with h. Thus data points close to one another are more correlated than points far apart. Figure 20.2 (top) illustrates still another typical $\rho(h)$ for a predictable sequence—this time there is negative correlation between adjacent points. A "bust-boom" environment could be represented in this way. If there is negative correlation between adjacent points, then there will be positive correlation between points separated by an even number of time intervals and negative correlation between points separated by an odd number of time intervals. Note the absolute level of correlation again decreases as h increases. Finally, Figure 20.2 (bottom) illustrates another typical $\rho(h)$ for a predictable sequence—but here there is a long-term oscillation inherent in the pattern. There is positive correlation between times separated by even multiples of the half wavelength and negative correlation between times separated by odd multiples of the half

† An unpredictable sequence thus has an autocorrelation given by $\rho(h) = \delta_{0,h}$, where δ is the Kronecker delta. For this reason an unpredictable sequence is sometimes said to be "delta correlated."

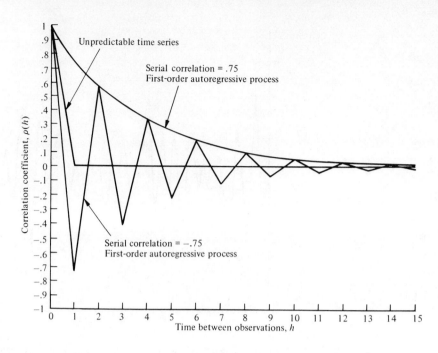

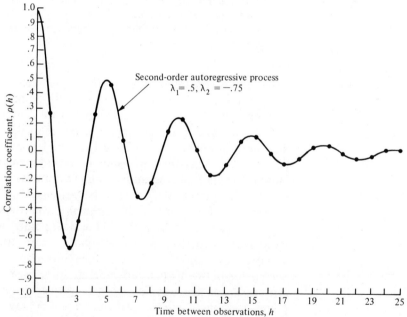

FIGURE 20.2. *Autocorrelation functions for the time series illustrated in Figure 20.1.*

wavelength. This kind of $\rho(h)$ would apply especially to populations that show a long-term cyclic character in their abundance through time.

Pattern The *pattern* of a time series is very closely tied to its predictability. It is helpful to think of the relation as follows. If there were a discernible pattern in a time series, then this pattern itself could be used to predict future data points on the basis of the past. For example, if a cyclic character were inherent in the data, then a prediction of any future data could be made by extending into the future the cycle inherent in the data obtained so far. Conversely, if there is no predictability, then there should be no consistent pattern in the data, for if there were, it could be used to make a prediction. Data that show no consistent pattern nonetheless do represent fluctuations—the data move up and down as seen in Figure 20.1. The point is, *in an unpredictable sequence, there is no particular cycle or period that is present more than any other cycle or period*. This is a basic point. An unpredictable sequence has a uniform distribution of all cyclic components so that no particular cyclic pattern stands out over any other cyclic pattern. In contrast, a predictable sequence necessarily does have some cyclic components present in higher amounts than other components. It is this nonuniform distribution of cyclic components which makes prediction possible. We shall discuss how to quantify the distribution of cyclic components later when we discuss the topic of patchiness and spatial pattern. For the moment all you need to know is that there is a fundamental connection between predictability and pattern.

A final item of jargon is that an unpredictable sequence is often called a "white noise" sequence because, as mentioned previously, there is an equal mixture of all cyclic components in an unpredictable sequence. It is known that white light is an even mixture of light of all colors. By analogy, an unpredictable sequence is called "white noise." Thus a white noise sequence is one with zero autocorrelation and a uniform distribution of all possible cyclic components.

Extinction

One of the most important implications of environmental fluctuation is that population extinction can result. Even if population extinction does not occur, the population size will typically be displaced from the value it would obtain in a constant environment. As a result the population size may occupy many possible values and we shall want to know the probability distribution for the various possible values of population size. This section briefly introduces the use of diffusion theory in studying extinction and in determining the probability distribution of population sizes. To understand this section, you should have read the chapter on genetic drift. Key references for this section are Turelli (1977) and Feldman and Roughgarden (1975).

Let $p(n, t)$ denote the probability density function for the population's size at time t.† This quantity will be found from a diffusion equation

$$\frac{\partial p(n, t)}{\partial t} = -\frac{\partial}{\partial n}[m(n)p(n, t)] + \frac{1}{2}\frac{\partial^2}{\partial n^2}[v(n)p(n, t)] \qquad (20.1)$$

The equilibrium probability distribution of population sizes, if it exists, is the equilibrium solution to the equation and is given by

$$\hat{p}(n) = \frac{c}{v(n)}\exp\left[2\int\frac{m(n)}{v(n)}dn\right] \qquad (20.2)$$

where c is a normalization constant. These formulas should be familiar from the chapter on genetic drift.

---

† The notation convention is that n denotes the population size when the size is a random variable, and N denotes the population size when it is a constant or a deterministic variable.

Our problem here is to relate the quantities $m(n)$ and $v(n)$ to the biological model under study. Recall from the analysis of genetic drift that $m(n)$ represented the deterministic part of the process; that is, $m(n)$ is the *average of the change in n* per unit time based on starting at n. For genetic drift this term included the standard deterministic formulas for mutation and selection. Recall also that $v(n)$ represented the stochastic part of the process; that is, $v(n)$ is the *variance of the change in n* per unit time based on starting at n. For genetic drift the stochasticity arose from independent binomial sampling of the gene pool each generation. However, our problem here is to relate $m(n)$ and $v(n)$ to the deterministic and stochastic parts of population growth.

The problem of connecting a population dynamic model with $m(n)$ and $v(n)$ in the diffusion equation involves several subtleties. A diffusion equation inherently applies only to completely unpredictable stochastic fluctuations in continuous time. In genetic drift, for example, the fluctuations (in discrete time) are completely unpredictable because the binomial samples at consecutive generations are completely independent; that is, the sampling error at time t is completely uncorrelated with the error at time $t + 1$. Thus if we inspect the sequence of sampling errors realized by a population undergoing genetic drift through time, we obtain an unpredictable sequence, provided fixation has not occurred. But in population dynamic models the environmental fluctuations being studied may not be unpredictable— whether they are or not is an empirical matter. The requirement of diffusion theory for unpredictable fluctuations in continuous time raises two difficulties. First, the analysis simply may not apply to empirical situations where there is strong autocorrelation in the fluctuations. However, a stochastic process with predictable fluctuations can *sometimes* be *approximated* by a diffusion process. Second, the concept of completely unpredictable fluctuations in continuous time is not itself a natural concept and must be understood in terms of the limit of some other process. For example, there is no difficulty at all in understanding the concept of an unpredictable sequence of discrete events. Tossing a single coin every minute yields a familiar and obvious example of a sequence of separate events with no serial correlation. But in continuous time, it is another matter. Can you imagine tossing a coin fast enough so that there is no correlation in its state over some very tiny time interval? Clearly, the idea that the state of a coin in continuous time has zero correlation over *any* finite time interval, however small, is an abstraction. Since diffusion theory does use this assumption, we must understand that diffusion theory at best yields an approximate solution to the problem being studied. We must use it with care even when the empirical assumption of unpredictable environmental fluctuation seems plausible on biological grounds.

There are currently two principal methods of connecting a population dynamic model with a diffusion equation. Both refer only to stochastic variation in parameters that enter in a linear fashion into the original population dynamic model. The remainder of this section consists of four parts. The first two discuss the two principal methods of connecting a population dynamic model with a diffusion equation. The third presents the method used to determine from a diffusion equation whether extinction occurs. The fourth part presents six examples of stochastic population models.

Ito Method of
Connecting a
Population
Dynamic
Model with a
Diffusion Equation

THE DETERMINISTIC MODEL. The first method applies to a population dynamic model which refers to a process originally defined over a discrete time interval Δt. We write the original deterministic model as

$$\Delta N = N_{t+\Delta t} - N_t = f(N_t)\,\Delta t \tag{20.3}$$

Of course, a familiar example is the logistic model where

$$f_1(N) = rN - \frac{rN^2}{K} \tag{20.4}$$

The function, $f_1(N)$, is a linear function of the parameter r. Therefore, we can use the method to study stochastic fluctuations in r. However, $f_1(N)$ is *not* a linear function of K and we cannot use the method to study stochastic fluctuations in K. However, we can do what may amount to the same thing by studying fluctuations in a new parameter, c, in the formula

$$f_2(N) = rN - \frac{crN^2}{K} \tag{20.5}$$

Since $f_2(N)$ is a linear function of c, we can legitimately discuss fluctuations in c. In a constant environment c equals one, but in a fluctuating environment c varies around one, which, to some extent, is the same thing as allowing K to fluctuate.

ADDING THE STOCHASTICITY. Once we have a population dynamic model defined in discrete time, we allow one of the parameters to fluctuate randomly. Let m denote a parameter. Then we let

$$m_t = m_0 + \sigma z_t$$

where m_0 denotes the average value of the parameter and z_t is a white noise fluctuation defined in discrete time. The variable, z_t, has zero mean and its variance is Δt. That is, over a unit time interval, z_t has unit variance; but as the time interval is lengthened, there is more time for a large change in the parameter. Conversely over a small time interval, the variance is low. Furthermore, there is no correlation between consecutive values of z_t by the assumption of white noise. The quantity, σ, controls the typical range of fluctuations around m_0. For example, if r fluctuates, we take

$$r_t = r_0 + \sigma z_t \tag{20.6}$$

If the parameter c in $f_2(N)$ fluctuates, we take

$$c_t = 1 + \sigma z_t \tag{20.7}$$

Next we substitute the parameter and its fluctuation into $f(N)$ and arrange the result as follows:

$$\Delta n \equiv n_{t+\Delta t} - n_t = f(n_t)\,\Delta t + \sigma g(n_t) z_t \tag{20.8}$$

We now use a lowercase n because the population size is now a random variable because of the presence of z_t. For example, if r fluctuates in $f_1(N)$, we have

$$\Delta n = \left(r_0 n - \frac{r_0 n^2}{K} \right) \Delta t + \sigma \left(n - \frac{n^2}{K} \right) z_t \tag{20.9}$$

and if c fluctuates we have

$$\Delta n = \left(rn - \frac{rn^2}{K} \right) \Delta t - \sigma \left(\frac{rn^2}{K} \right) z_t \qquad (20.10)$$

Thus, after the fluctuating parameter is introduced, we can always arrange the equation for Δn in the form of (20.8). The first term, $f(n)$, reflects the deterministic influence on n and the second term, $\sigma g(n)z_t$ reflects the stochastic influence on n. We shall analyze the examples (20.9) and (20.10) in more detail later.

THE DIFFUSION CONNECTION. With the stochastic population dynamic model in the form of (20.8), a diffusion equation can be developed to predict the probability distribution of population size through time. The population dynamic model here is defined in discrete time, and the stochastic fluctuation is white noise. However, a diffusion equation applies in continuous time. The way around this problem is essentially the same as that used in the study of genetic drift. We shift to a new time scale so that many iterations of the original equation for Δn corresponds to the unit time interval in the new time frame. Hence, in a suitable limit, as the time frame expands, we can equate $m(n)$ and $v(n)$ from a diffusion equation with the $f(n)$ and $g(n)$ from a population dynamic model as follows.

$$\frac{\text{average of } \Delta n \text{ conditioned on starting at } n}{\text{unit time}} \equiv m(n) = f(n) \quad (20.11a)$$

$$\frac{\text{variance of } \Delta n \text{ conditioned on starting at } n}{\text{unit time}} \equiv v(n) = \sigma^2 g^2(n) \quad (20.11b)$$

These relations can be obtained formally from (20.8) simply by setting $m(n)$ equal to the expectation of Δn conditional on n_t and $v(n)$ equal to the variance of Δn conditional on n_t. These formulas for $m(n)$ and $v(n)$ are actually the same as those used in our analysis of genetic drift. They apply to a process which is originally defined in discrete time and for which the concept of white noise fluctuation is natural. This method of relating a population dynamic model to a diffusion equation, based in the formulas in (20.11), is called the Ito method. Later in this section we shall work explicit examples of the Ito method together with examples using the second method, which follows.

Stratonovich Method of Connecting a Population Dynamic Model with a Diffusion Equation

THE DETERMINISTIC MODEL. A second method applies to a population dynamic model, which refers to a process that is originally defined in continuous time. We write the original deterministic model as

$$\frac{dN}{dt} = f(N) \qquad (20.12)$$

The examples, (20.4) and (20.5), for $f(N)$ introduced for the Ito method are appropriate here too.

ADDING THE STOCHASTICITY. Again we introduce stochasticity into a parameter. Let m denote a parameter that enters linearly into $f(N)$. Then we consider

$$m(t) = m_0 + \sigma z(t) \qquad (20.13)$$

where $z(t)$ is a random variable in continuous time. The variable, $z(t)$, should be as close to white noise as is practically possible in continuous time. That is, the autocorrelation between $z(t)$ and $z(t+Y)$ should fall off very rapidly as Y increases. Upon substituting the stochastic parameter (20.13) into the deterministic model (20.12), we can arrange the result as

$$\frac{dn}{dt} = f(n) + \sigma g(n)z(t) \qquad (20.14)$$

This equation is called a stochastic differential equation because it is a differential equation that is satisfied by a random variable, $n(t)$.

THE DIFFUSION CONNECTION. With a stochastic population dynamic model in the form of (20.14) we can again associate a diffusion equation to predict $p(n, t)$. Again, the diffusion equation applies in a certain limit. If we expand the unit time interval—then the interval over which there is autocorrelation in $z(t)$ will become relatively small. In a suitable limit the process $z(t)$ will approach a white noise process in continuous time.† Then the functions $m(n)$ and $v(n)$ in the diffusion equation can be identified with the functions, $f(n)$ and $g(n)$, in the biological model as follows:

$$m(n) = f(n) + \frac{\sigma^2}{2} g(n)g'(n) \qquad (20.15a)$$

$$v(n) = \sigma^2 g^2(n) \qquad (20.15b)$$

where $g'(n)$ is the first derivative of $g(n)$. This method of relating the stochastic population model to a diffusion equation, based on the formulas in (20.15), is called the *Stratonovich method*. The two methods are always the same for $v(n)$. They differ in the formulas for $m(n)$. Furthermore, if $g(n)$ is itself a constant and not a function of n, then the methods become the same because $g'(n) \equiv 0$ in this case. The Stratonovich formula for $m(n)$ asserts that the stochasticity may bias the *average* dn/dt; that is, push $m(n)$ away from the value caused solely by the deterministic part of the model. In the Ito method the stochasticity influences only the variance of Δn whereas the average of Δn is given only the deterministic part of the model. But in the Stratonovich formula, the stochasticity also influences the average of dn/dt.

The intuitive rationale for the Stratonovich method appears to be that the noise, $z(t)$, in the stochastic model defined in continuous time, (20.14), is not completely white noise to begin with. Figure 20.3a (top and middle) is a sketch of possible realizations of $z(t)$. If $z(t)$ has autocorrelation, then it will not tend to cross the x axis quite as often per unit time as in a true white noise process. The average total time above and below the axis will be the same with or without any autocorrelation because the average of $z(t)$ is zero. However, the presence of autocorrelation prolongs the visit of $z(t)$ to any place relative to a true white noise process. An autocorrelated $z(t)$ shows a smoother path, whereas a truly white noise $z(t)$ fluctuates with an inconceivable abruptness.‡ We can illustrate the effect of this autocorrelation with an analogy. We can mimic a small piece of the path of an

† That is, in the limit the actual autocorrelation function approaches a Dirac delta function.

‡ In fact, it can be shown that realizations of a so-called Brownian motion process in continuous time are *continuous* functions and yet are *not* differentiable in the usual sense at virtually *any* point. However, Brownian motion is *even smoother* than white noise in continuous time, for white noise *is* considered to be the "derivative" of Brownian motion where special concepts of the derivative must be introduced in order to make this proposition meaningful. The concept of a truly white noise process in continuous time is very enigmatic.

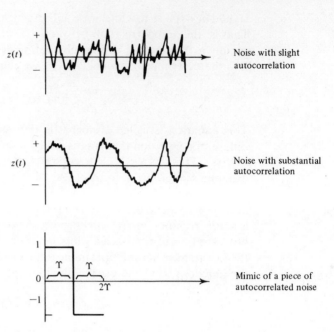

FIGURE 20.3a. *Sketch of possible forms of $z(t)$ as a function of time.*

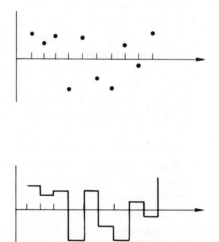

FIGURE 20.3b. *Sketch illustrating different methods of obtaining the limit of white noise.*

autocorrelated $z(t)$ with the curve in Figure 20.3a (bottom). For a small time interval, Y, $z(t)$ equals 1 and then it drops to −1 for another interval. We shall calculate the average value of (dn/dt) during the total interval of $2Y$. This quantity represents the bias to the $m(n)$ term introduced by the autocorrelated stochasticity, $z(t)$. The noise $z(t)$ drives dn/dt and hence $n(t)$ itself through the stochastic part of the Equation (20.14), namely

$$\frac{dn}{dt} = \sigma g(n)z(t) \qquad (20.16)$$

For Y sufficiently small we may expand $g(n)$ to first order in n as

$$g(n) = an \qquad (20.17)$$

where a is a constant.

We denote the time average of dn/dt during the first interval as $\langle \ \rangle_1$. The population grows exponentially during the first interval, so

$$\left\langle \frac{dn}{dt} \right\rangle_1 = \left(\frac{1}{Y} \right) \int_0^Y \sigma a n(t) \, dt$$

$$= \left(\frac{1}{Y} \right) \int_0^Y \sigma a \, e^{\sigma a t} n(0) \, dt$$

$$= \left(\frac{1}{Y} \right) (e^{\sigma a Y} - 1) n(0) \qquad (20.18)$$

During the next interval the population decays exponentially from the value it had attained at time Y. The average growth rate during this interval is

$$\left\langle \frac{dn}{dt} \right\rangle_2 = \left(\frac{1}{Y} \right) \int_Y^{2Y} \sigma a n(t)(-1) \, dt$$

$$= \left(\frac{1}{Y} \right) \int_Y^{2Y} (-\sigma a) \, e^{-\sigma a t} \, e^{\sigma a Y} n(0) \, dt$$

$$= \left(\frac{1}{Y} \right) (e^{-\sigma a Y} - 1) n(0) \qquad (20.19)$$

Then the average growth rate over both intervals is

$$\left\langle \frac{dn}{dt} \right\rangle = \frac{1}{2} \left\langle \frac{dn}{dt} \right\rangle_1 + \frac{1}{2} \left\langle \frac{dn}{dt} \right\rangle_2 \qquad (20.20)$$

Substituting from (20.18) and (20.19) and expanding e to second order in Y yields

$$\left\langle \frac{dn}{dt} \right\rangle = \left(\frac{1}{Y} \right) \left(\frac{1}{2} \right) \sigma^2 a^2 n(0) Y^2 \qquad (20.21)$$

But remembering that $an(0)$ is $g(n_0)$ and a is $g'(n_0)$, we have

$$\left\langle \frac{dn}{dt} \right\rangle = \frac{\sigma^2}{2} g(n_0) g'(n_0) Y \qquad (20.22)$$

This is the bias to the average growth rate caused by the fluctuation of $z(t)$ from 1 to -1 during the total time interval of 2Y. Note that (20.22) is, apart from the factor Y, the same as the bias term in the Stratonovich formula for $m(n)$ given by (20.15a). The formula shows that if $Y \neq 0$ then there certainly is a bias introduced in the average of (dn/dt) by the stochastic fluctuation. Therefore, it is reasonable for the average of (dn/dt) to include a term for the stochastic contribution in addition to the deterministic contribution.

The analogy above is useful in visualizing the differences in how the Ito and Stratonovich limits lead to white noise in continuous time. Figure 20.3b (top) illustrates a series of discrete white noise values of a random variable. By expanding the time frame, the dots are compressed together making a very jagged process in continuous time. This represents how the Ito limit leads to white noise. Figure 20.3b (bottom) illustrates a random variable in continuous time which has autocorrelation. By expanding the time frame, this curve is also compressed, again making a very jagged process in continuous time. This represents how the Stratonovich limit leads to white noise. In the Ito method no term representing the stochasticity is introduced into $m(n)$. This case is analogous to having Y equal to zero in the preceding discussion. But in the Stratonovich method there is a term for the stochasticity in $m(n)$ analogous to having Y equal one in the preceding discussion.

Using the Diffusion Equation to Determine Whether Extinction Occurs

Once we have the diffusion equation corresponding to a population dynamic model, either using the Ito or the Stratonovich method, then we can analyze the diffusion equation. We can determine the stationary distribution of population sizes if it exists. Also we can determine whether populations are tending to extinction as a result of the environmental fluctuations, and if so, whether the extinction is occurring rapidly (in finite time) or slowly (in infinite time). The basic technique for this analysis is to classify the boundaries of the diffusion process into one of five categories. For all population processes the lower boundary is at $n = 0$. The upper boundary is either at $n = K$ or $n = \infty$. Figure 20.4 presents a flow chart for boundary classification.

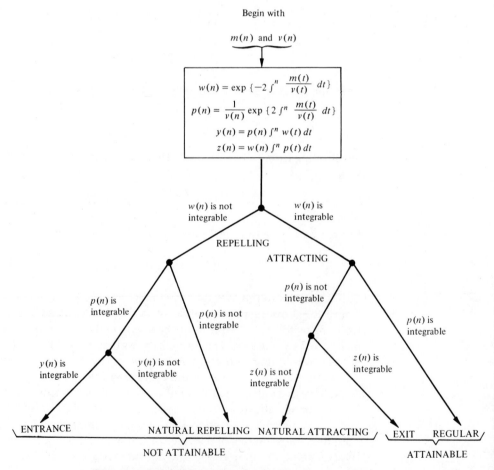

FIGURE 20.4. *Flow chart for the classification of boundaries in a one-dimensional diffusion process.*

The vocabulary for boundary classification is due to Feller (1952) and Prohorov and Rozanov (1969). A boundary, say at $n = a$, is said to be *attracting* if

$$\text{prob}\,\{\lim_{t \to \infty} n(t) = a\} > 0$$

That is, a boundary is attracting if there is a positive probability that the limit of a sample path, as time tends to infinity, is at the boundary. A boundary is said to be *repelling* if

$$\text{prob}\,\{\lim_{t \to \infty} n(t) = a\} = 0$$

A boundary is said to be *attainable* or accessible if there are sample paths that actually hit the boundary in *finite* time. A boundary is said to be *unattainable* or inaccessible if sample paths do not hit the boundary in *finite* time. The five types of boundaries are

1. Entrance boundary. The boundary *cannot* be reached from the interior in finite time; the interior *can* be reached from the boundary in finite time. Boundary is repelling and unattainable.

2. Natural repelling boundary. Boundary *cannot* be reached from the interior in finite time; the interior *cannot* be reached from the boundary in finite time. Boundary is repelling and unattainable.

3. Natural attracting boundary. Boundary *cannot* be reached from interior in finite time; the interior *cannot* be reached from the boundary in finite time. Boundary is attracting the unattainable.

4. Exit boundary. Boundary *can* be reached from interior in finite time; interior *cannot* be reached from boundary in finite time. Boundary is attracting and attainable.

5. Regular boundary. Boundary *can* be reached from interior in finite time; interior *can* be reached from boundary in finite time. Boundary is attracting and attainable.

Boundary classification is important in determining the biological implications of a diffusion equation. For example, the boundary at $n = 0$ represents population extinction. If $n = 0$ is attracting, then populations are approaching extinction and are, at least in the limit as $t \to \infty$, becoming extinct. Furthermore, if $n = 0$ is actually attainable, then populations are becoming extinct in a finite time due to the stochastic fluctuation. Thus, by carrying out the boundary classification, we determine the implications for population extinction of the environmental fluctuation.

In practice the way to classify a boundary is to determine whether several functions, which are introduced below, are integrable in the neighborhood of the boundary. For a function to be integrable in the neighborhood of a boundary, say $n = 0$, means that the integral is finite in that neighborhood. For example, the function $y(n) = n^{-2}$ is not integrable in the neighborhood of $n = 0$ whereas the function $y(n) = n^{-1/2}$ is. Indeed, for the function $y(n) = n^{\alpha}$ to be integrable at $n = 0$, we require $\alpha > -1$. There are a total of four functions whose integrability at a boundary determines the classification of that boundary.

The first function is called the scale density and is given by

$$w(n) = \exp\left[-2 \int^{n} \frac{m(t)}{v(t)}\,dt\right] \tag{20.23}$$

The second function is called the speed density,

$$p(n) = \frac{1}{v(n)w(n)} = \frac{1}{v(n)} \exp\left[2\int^n \frac{m(t)}{v(t)}\,dt\right] \qquad (20.24)$$

We have labeled the speed density as $p(n)$ because the speed density differs only by the normalization constant from the stationary distribution equation (20.2). If the speed density, $p(n)$ is integrable at both boundaries, then the stationary distribution exists in the sense that it can be normalized to unit area. Thus, in the course of classifying both boundaries for diffusion process, we automatically determine whether a stationary distribution exists. The next two functions are built up from the first two,

$$y(n) = p(n)\int^n w(t)\,dt \qquad (20.25)$$

$$z(n) = w(n)\int^n p(t)\,dt \qquad (20.26)$$

Figure 20.4 illustrates the flow chart for the classification. Note that the integrability of $w(x)$ is sufficient to determine whether a boundary is repelling or attracting. But completing the classification requires checking on the integrability of one or two more functions.

Examples and Discussion
We now use the diffusion machinery presented above to answer some biological questions about the fate of a population in a stochastic environment. There are three interesting examples to study, first where only the r in the logistic equation fluctuates, next where only the K fluctuates, and last where both fluctuate together. In addition, we shall consider both discrete and continuous time versions of the problem so that the biological results of the Ito and Stratonovich methods can be contrasted. All the cases are recorded in Table 20.1.

ONLY r FLUCTUATES—ITO METHOD. The stochastic population model, where only r fluctuates, was given earlier in (20.9) as

$$\Delta n = r_0 n\left(1 - \frac{n}{K}\right)\Delta t + \sigma n\left(1 - \frac{n}{K}\right)z_t \qquad (20.27)$$

Using the Ito method, we obtain

$$m(n) = r_0 n\left(1 - \frac{n}{K}\right)$$
$$v(n) = \sigma^2 n^2\left(1 - \frac{n}{K}\right)^2 \qquad (20.28)$$

The diffusion process is restricted to the interval $[0, K]$; 0 and K are absorbing boundaries. A population at K will remain there regardless of any fluctuation in r. Similarly, a population at 0 is extinct. We shall regard the initial condition, $n(0)$, as somewhere between 0 and K and ask about the fate of such a population. The deterministic part of the model is causing populations to flow away from 0 and up to K. However, near the 0 boundary this flow is opposed by the stochastic fluctuation. The problem is to determine whether all of the populations eventually approach 0, or K, or whether some approach one boundary and the rest approach the other boundary. This question is answered by classifying the boundaries, $n = 0$ and $n = K$.

In the neighborhood of $n = 0$ the four functions become

$$w(n) \approx n^{-2r_0/\sigma^2}$$

$$p(n) \approx n^{2r_0/\sigma^2 - 2}$$

$$y(n) \approx n^{-1} \qquad (\sigma^2 \neq 2r_0)$$

$$z(n) \approx n^{-1} \qquad (\sigma^2 \neq 2r_0)$$

(20.29)

We now observe that $w(n)$ is not integrable if $\sigma^2/2 < r_0$ and is integrable if $\sigma^2/2 > r_0$. The function $p(n)$ is integrable if $\sigma^2/2 < r_0$ and is not integrable if $\sigma^2/2 > r_0$. Furthermore, the functions $y(n)$ and $z(n)$ are not integrable in the neighborhood of $n = 0$. Therefore, according to the classification in Figure 20.4, we conclude that $n = 0$ is a natural repelling boundary for $\sigma^2/2 < r_0$ and is a natural attracting boundary for $\sigma^2/2 > r_0$.

The corresponding analysis at $n = K$ demonstates that this boundary is always natural attracting. The function $p(n)$ is not integrable at $n = K$, and thus a stationary distribution does not exist for this diffusion process.

Thus, if only r fluctuates in the logistic equation and not K, and if the Ito method is used, then there are two possibilities. With slight stochasticity $(\sigma^2/2 < r_0)$ all populations converge to a close neighborhood of K. With strong stochasticity $(\sigma^2/2 > r_0)$ some populations converge to a close neighborhood of K, the others to a close neighborhood of 0, that is, these populations approach extinction. There is no stationary distribution in either case.

ONLY r FLUCTUATES—STRATONOVICH METHOD. The stochastic population model is now given by

$$\frac{dn}{dt} = r_0 n \left(1 - \frac{n}{K}\right) + \sigma n \left(1 - \frac{n}{K}\right) z(t)$$

(20.30)

Using the Stratonovich method, we obtain

$$m(n) = r_0 n \left(1 - \frac{n}{K}\right) + \left(\frac{\sigma^2}{2}\right) n \left(1 - \frac{n}{K}\right)\left(1 - \frac{2n}{K}\right)$$

$$v(n) = \sigma^2 n^2 \left(1 - \frac{n}{K}\right)^2$$

(20.31)

In this model it is easy to verify that at $n = 0$, $w(n)$ is not integrable, $p(n)$ is integrable, and $y(n)$ is integrable. Hence the $n = 0$ boundary is always natural repelling. In a corresponding manner it can be shown that $n = K$ is always natural attracting and that $p(n)$ is never integrable at $n = K$.

Thus, if only r fluctuates in the logistic equation and not K, and if the Stratonovich method is used, then there is only one possibility. All populations converge to a close neighborhood of K. No populations approach extinction.

The reason that extinction does not occur in the Stratonovich formulation is that a favorable bias is introduced by the stochasticity into the formula for the average growth rate. As $n \to 0$, $m(n)$ becomes

$$m(n) \approx r_0 n + \left(\frac{\sigma^2}{2}\right) n$$

$$\approx \left(r_0 + \frac{\sigma^2}{2}\right) n$$

(20.32)

Thus the average growth rate near $n = 0$ is equivalent to exponential growth at a rate equal to $r_0 + \sigma^2/2$. Hence, for a given r_0, there is a stronger force pushing the populations away from extinction in the Stratonovich formulation than in the Ito formulation.

ONLY K FLUCTUATES—ITO METHOD. We can study the effects of fluctuation in K with the model introduced in (20.10) as

$$\Delta n = rn\left(1 - \frac{n}{K_0}\right)\Delta t - \sigma\left(\frac{rn^2}{K_0}\right)z_t \tag{20.33}$$

By the Ito method we have

$$m(n) = rn\left(1 - \frac{n}{K_0}\right)$$

$$\tag{20.34}$$

$$v(n) = \frac{\sigma^2 r^2 n^4}{K_0^2}$$

If the carrying capacity fluctuates, there is no specific nonzero equilibrium point defined as there was if only r fluctuates. There is only one equilibrium point, $n = 0$, representing extinction. Therefore, $m(n)$ and $v(n)$ above apply to a diffusion equation that operates over the entire positive line $(0, \infty)$.

The boundary analysis for this model reveals that $n = 0$ is natural repelling and $n = \infty$ is an entrance boundary. Also, a stationary distribution exists, which is given by

$$p(n) = \frac{\text{const}}{n^4}\exp\left\{\frac{2K_0}{\sigma^2 r}\frac{1}{n}\left(1 - \frac{K_0}{2}\frac{1}{n}\right)\right\} \tag{20.35}$$

The mean and variance of the population size according to this distribution are finite.

The reason that $n = 0$ is a natural repelling boundary if only K fluctuates under the Ito formulation is that fluctuations in K produce a much weaker stochasticity near $n = 0$ than do fluctuations in r. Notice in Equation (20.33) that z_t is preceded by the factor n^2. As n tends to zero, this term vanishes relative to the term representing deterministic exponential growth. Hence the deterministic force pushing away from $n = 0$ overpowers the stochastic force resulting in a repelling boundary.

ONLY K FLUCTUATES—STRATONOVICH METHOD. The stochastic population model is

$$\frac{dn}{dt} = rn\left(1 - \frac{n}{K_0}\right) - \sigma\left(\frac{rn^2}{K_0}\right)z(t) \tag{20.36}$$

and $m(n)$ and $v(n)$ are

$$m(n) = rn\left(1 - \frac{n}{K_0}\right) + \frac{\sigma^2 r^2 n^3}{K_0^2}$$

$$\tag{20.37}$$

$$v(n) = \frac{\sigma^2 r^2 n^4}{K_0^2}$$

The boundary analysis for this model reveals that $n = 0$ is again natural repelling, but $n = \infty$ is now a regular boundary. A stationary distribution always exists, which is given by

$$p(n) = \frac{\text{const}}{n^2} \exp\left\{\frac{2K_0}{\sigma^2 r}\frac{1}{n}\left(1 - \frac{K_0}{2}\frac{1}{n}\right)\right\} \qquad (20.38)$$

The mean and variance of the population size according to this distribution are infinite.

In this model the bias introduced into $m(n)$ by the stochasticity greatly increases the flow away from $n = 0$ and toward $n = \infty$. This model cannot be meaningful in that a regular boundary at ∞ implies that some populations explode to infinity in finite time.

BOTH r AND K FLUCTUATE—ITO METHOD. Environmental fluctuation may influence mostly r or mostly K, or it may significantly influence both r and K together. For example, consider fluctuations in the amount of food available to a bird population. This quite probably leads to fluctuation in K. But it may or may not lead to fluctuation in r, depending on the reproductive biology of the species involved. Some birds have a very rigid clutch size. For these species fluctuation in food could influence the degree of density-dependent mortality and hence K but not influence the intrinsic rate of increase, that is, the rate of increase apart from density-dependent effects. Other bird species have a vary plastic clutch size and hence food fluctuations could influence both r and K in a parallel fashion. As a different example, consider the impact of temperature fluctuations on cold-blooded animals. These fluctuations may influence the rate at which eggs are laid and yet not influence the carrying capacity. For example, lower temperatures may lead to a lower rate of egg laying, but food requirements also drop and more time is spent hidden in burrows or crevices. Thus temperature fluctuations may not influence the steady-state number of animals which can be supported in an area but could influence the rate at which the population could grow to that steady-state level. Thus all of the cases we have discussed so far are potentially empirically meaningful and what remains is to consider joint fluctuations in r and K.

We can successfully analyze the following scheme of joint fluctuation. Suppose that both r and K vary together in such a way that their ratio remains constant. This is achieved by assuming

$$r_t = r_0 + \sigma z_t$$
$$K_t = K_0 + \sigma\left(\frac{K_0}{r_0}\right)z_t \qquad (20.39)$$

With this assumption the ratio of r_t/K_t is a constant, r_0/K_0. r_0 is the average intrinsic rate of increase; and K_0 is the average carrying capacity. The stochastic population model then becomes

$$\Delta n = r_0 n\left(1 - \frac{n}{K_0}\right)\Delta t + \sigma n z_t \qquad (20.40)$$

With the Ito method we have

$$m(n) = r_0 n\left(1 - \frac{n}{K_0}\right)$$
$$v(n) = \sigma^2 n^2 \qquad (20.41)$$

The boundary classification reveals that $n = 0$ is natural attracting for $\sigma^2/2 > r_0$ and is natural repelling for $\sigma^2/2 < r_0$; $n = \infty$ is always an entrance boundary. If $n = 0$ is natural repelling, then a stationary distribution exists given by

$$p(n) = \text{const } n^{2r_0/\sigma^2 - 2} \exp\{-2r_0 n/(\sigma^2 K_0)\} \qquad (20.42)$$

This distribution has a finite mean and variance. Thus if the stochasticity is strong $(\sigma^2/2 > r_0)$, then all populations eventually tend to extinction whereas if the stochasticity is weak $(\sigma^2/2 < r_0)$, then the populations are spread out in the equilibrium distribution given above.

BOTH r AND K FLUCTUATE—STRATONOVICH METHOD. The stochastic population model for this formulation is

$$\frac{dn}{dt} = r_0 n\left(1 - \frac{n}{K_0}\right) + \sigma n z(t) \qquad (20.43a)$$

Hence, with the Stratonovich method,

$$m(n) = r_0 n\left(1 - \frac{n}{K_0}\right) + \left(\frac{\sigma^2}{2}\right)n$$
$$v(n) = \sigma^2 n^2 \qquad (20.43b)$$

In this model $n = 0$ is always a natural repelling boundary and $n = \infty$ is always an entrance boundary. A stationary distribution always exists given by

$$p(n) = \text{const } n^{2r_0/\sigma^2 - 1} \exp\{-2r_0 n/(\sigma^2 K_0)\} \qquad (20.44)$$

Table 20.1.

	Formulation	
Kind of Fluctuation	Ito	Stratonovich
r fluctuates	Strong stochasticity: some populations become extinct while others converge to K Weak stochasticity: all populations converge to K	All populations converge to K
K fluctuates	Populations attain an equilibrium distribution of population sizes; $E(n)$ and Var (n) finite	Populations attain an equilibrium distribution of population size; $E(n)$ and Var (n) infinite
Both r and K fluctuate with r/K remaining constant	Strong stochasticity: all populations eventually become extinct Weak stochasticity: populations attain an equilibrium distribution of population sizes	Populations attain an equilibrium distribution of population sizes

The mean and variance of this distribution are finite. The reason $n = 0$ is always repelling is the same as discussed in the case where only r fluctuates. The bias introduced by the stochasticity in the Stratonovich formulation increases the effective intrinsic rate of increase, thus increasing the speed with which populations tend to rebound after falling, by chance, to a low value.

All the cases have been summarized in Table 20.1. It is interesting to see some qualitative trends on the table. From the standpoint of the persistence and long-term constancy of population size, the best state of affairs occurs when only r fluctuates and the Stratonovich method applies. The worst states occur when both r and K fluctuate together and the Ito method applies and when only K fluctuates and the Stratonovich method applies. The other cases are more or less in between these extremes.

Resource Tracking in an Autocorrelated Environment

There are two biological issues addressed in this section. First, we want to know the effect of an autocorrelated stochastic environment on a population's dynamics. The diffusion methods of the last section have applied to unpredictable fluctuations. But, obviously, a basic feature of population biology is that many environmental changes that influence organisms are predictable. Indeed, organisms have many traits that enable them to perceive cues that are good predictors of future change. The use of day length as a cue to end growth and to begin reproduction is one very common example. The diffusion methods of the preceding section predict, in some cases, the equilibrium distribution of population sizes, $\hat{p}(n)$. We can always calculate the variance of $\hat{p}(n)$ and thereby determine the variability of population size as a function of the degree of environmental variability. Such a calculation would, however, mainly apply to white noise environmental fluctuation. The first issue to be discussed in this section is how the population's variability is affected by autocorrelation in the environment. In addition, we shall determine how the autocorrelation of the population itself is related to the autocorrelation in the environment.

The second issue is that of resource tracking. Fluctuations in resource level cause, on the average, a discrepancy between the population size and the carrying capacity. The average size of this discrepancy depends on the autocorrelation in the environment. We shall also see that there is an optimum value of the intrinsic rate of increase, r, which minimizes the average discrepancy between N and K.

The mathematics in this section (and the next) is much easier and more straightforward than the diffusion methods of the last section. This occurs because we shall condition upon nonextinction of the populations under study. That is, we assume in the discussion that we are referring only to those populations which are not at or near extinction. For this set of realizations, the populations are fluctuating near the mean population size, because those which have departed far from the mean are likely to be almost or already extinct. With this assumption we can replace the nonlinear population dynamic models with linear approximations that are easy to solve. Actually, the linear model used in this section and the next may be as good a biological model as the logistic equation itself. It must be remembered that the logistic equation is not derived from basic first principles but is an approximation to begin with. A priori, there is no reason prohibiting a still simpler equation than the logistic one from being as good or better biologically in some situations. The key reference to this section is Rough-garden (1975a). The mathematical techniques of this section are standard methods in an area of statistics called time series analysis. The importance of these techniques in ecology is illustrated and discussed in Poole (1976, 1978) and Anderson (1978).

The Linear Model for Population Dynamics

The model analyzed is one possible approximation to the logistic equation in a fluctuating environment. The logistic equation in discrete time is

$$N_{t+1} = \left(r + 1 - \frac{r}{K_t}N_t\right)N_t, \tag{20.45}$$

where K_t is the carrying capacity at time t. If $K_t = K$ for all t, then K is an equilibrium point. The stability of this equilibrium is controlled by r. If r is in $(0, 1)$, K is stable and the approach to K is a sigmoid curve. If r is in $(1, 2)$, K is still stable but the approach to K successively overshoots and undershoots resulting in a damped oscillation. If r is in $(2, 3)$, K is unstable and bounded nonlinear oscillation results, given a suitable initial condition. We assume that r is less than 2.

In this section, K_t varies in time. It is assumed, however, that there is regularity of a statistical sort in the environment. Specifically, it is assumed that K_t is drawn from a "second-order stationary stochastic process" with mean $\bar{K}$. Qualitatively, this assumption requires that the mean and variance of the possible K_t are constant in time and that any ability to predict a future carrying capacity, say K_{t+h}, from information about the present, say K_t, depends only on the length of the interval over which the prediction is desired. That is, the ability to predict K_{t+h} given K_t depends only on h and not on t.

Given this assumption, it is convenient to speak of the population size and the carrying capacity in terms of their deviations from $\bar{K}$, since $\bar{K}$ itself is constant. So, introducing

$$k_t = K_t - \bar{K}$$
$$n_t = N_t - \bar{K} \tag{20.46}$$
$$n_{t+1} = N_{t+1} - \bar{K},$$

substituting into (20.45) and noting that $[1 - \bar{K}/(\bar{K} + k_t)] = [k_t/(\bar{K} + k_t)]$ gives

$$n_{t+1} = \left[1 - r\left(\frac{\bar{K}}{\bar{K} + k_t}\right)\right]n_t + r\left(\frac{\bar{K}}{\bar{K} + k_t}\right)k_t + r\left(\frac{k_t - n_t}{\bar{K} + k_t}\right)n_t \tag{20.47}$$

Equation (20.47) is, apart from the change of variable, identical with (20.1), which is the full nonlinear logistic model. Consider now another model termed the "linear model":

$$n_{t+1} = (1 - r)n_t + rk_t \qquad (0 < r < 2) \tag{20.48}$$

This linear model should approximate the full logistic model for realizations during which both $\bar{K}/(\bar{K} + k_t)$ is nearly equal to one and $(k_t - n_t)n_t/(\bar{K} + k_t)$ is negligibly small. Thus the linear model should usually approximate the full logistic model in an environment consisting of small fluctuations in a large carrying capacity.

The linear model is easily iterated. Suppose, for example, that $n_0 = 0$; then

$$n_1 = rk_0$$

$$n_2 = (1-r)rk_0 + rk_1$$

$$n_3 = (1-r)^2 rk_0 + (1-r)rk_1 + rk_2 \qquad (20.49)$$

$$n_4 = \cdots$$

If the process began at some very distant time in the past, then the pattern apparent in (20.49) can be extended back into the past, yielding

$$n_t = \sum_{i=1}^{\infty} r(1-r)^{i-1}k_{t-i} \qquad (20.50)$$

Equation (20.50) shows that n_t is an exponentially weighted running average of all the previous values of the carrying capacity. Intuitively, in a fluctuating environment, the population size at any time should reflect the history of previous carrying capacities. Equation (20.50) indicates that past values of k are weighted exponentially in their contribution to the present value of n. Moreover, the weights extend farther and more uniformly into the past as r tends to zero. This too is intuitive, for a low r represents a "sluggish" population, which would be expected to retain the influence of past carrying capacities longer than a more "responsive" population.

Figure 20.5 presents the weight function for several illustrative values of r. Note that the sum of the weights is one:

$$\sum_{i=0}^{\infty} r(1-r)^i = r\left[\frac{1}{1-(1-r)}\right] = 1 \qquad (20.51)$$

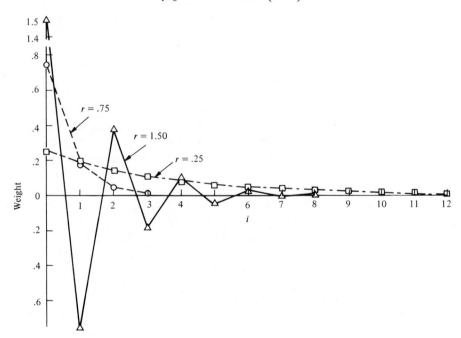

FIGURE 20.5. *The weight function $r(1-r)^i$ used in forming N_t from an exponentially weighted running average of the previous values of the carrying capacity. From Equation (20.51).* [From J. Roughgarden (1975), A simple model for population dynamics in stochastic environments. *Amer. Natur.* **109**: 713–736. Copyright 1975 by the University of Chicago.]

It is important to understand that r describes the population's responsiveness to carrying capacity fluctuations; it does not refer to an individual's responsiveness. The model refers to fluctuations in resource level which influence mostly K and not r. For example, birds with fixed clutch sizes might be considered to have little responsiveness as *individuals* to fluctuations. Yet a population with large fixed clutch sizes would respond more rapidly to changes in K than a population with small fixed clutch sizes.

An Example of a Predictable Environment

The next step is to apply the linear model (20.48) to a predictable environment, that is, to a sequence k_t where there is autocorrelation among consecutive values of the carrying capacity. To make the discussion easier to visualize, it will help to consider population dynamics in a particular kind of autocorrelated environment. [General formulas applicable to any autocorrelated environment are derived in Roughgarden (1975). The following discussion concerns this particular example of an autocorrelated environment.]

Let us assume the stochastic variation in the carrying capacity can be represented as a "first-order autoregressive process,"

$$k_t = \lambda k_{t-1} + Z_t \qquad (-1 < \lambda < 1) \tag{20.52}$$

Each Z_t is an independent, identically distributed random variable with zero mean and variance σ_z^2. The distribution for Z_t is arbitrary, and it need not be Gaussian. Equation (20.52) asserts that the carrying capacity at t is λ times the value at $t-1$ plus a random component; λ controls the predictability. If λ is zero, then each k_t is simply an independent draw of the random variable Z. In this case, k_t is completely unpredictable, since at k_{t-1} or at any other time it is useless to predict what k_t will be. But if λ is positive, then some of the value of k_{t-1} persists into k_t and, therefore, serial correlation will occur. If λ is negative, there is also predictability but with negative serial correlation. A negative λ indicates a "bust-boom" or oscillatory environment. Biologically, λ also can be viewed as a measure of the speed (relative to the population) with which the level of renewable resources recovers from a perturbation. If the weather causes K_t to drop below $\bar{K}$, then a high λ indicates slow restoration of the resource base. Similarly, if favorable weather causes K_t to exceed $\bar{K}$, a high λ indicates long persistence of additional resources. Thus, in addition to controlling the predictability of k_t, λ also reflects the difference in life span between say, predator and prey, and herbivore and host plant.

The process defined by (20.52) is called a first-order autoregressive process because λ is the regression coefficient of k_t on k_{t-1}. Thus the process is being regressed against itself, that is, autoregressive. It is a first-order process because k_t is only being regressed against k_{t-1}. A second-order autoregressive process is defined as

$$k_t = \lambda_1 k_{t-1} + \lambda_2 k_{t-2} + Z_t \tag{20.53}$$

The first three examples in Figure 20.1 represent a first-order autoregressive process. The fourth example is a second-order autoregressive process. In this section we shall study the population dynamics that result from a first-order carrying capacity process in detail.

The autocorrelation function for a first-order autoregressive process is

$$\rho_k(h) = \lambda^{|h|} \quad \begin{matrix} (-1 < \lambda < 1) \\ (h = \cdots -1, 0, 1, \cdots) \end{matrix} \tag{20.54}$$

This is a geometric sequence in λ. If $\lambda = 0$, the environment is fully unpredictable and $\rho_k(h)$ equals zero unless h is zero. If λ is a positive integer, $\rho_k(h)$ assumes the profile of an exponential decay. If λ is negative, $\rho_k(h)$ oscillates in sign, being negative for odd and positive for even h. Three examples of $\rho(h)$ from (20.54) are included in Figure 20.2. The variance of the carrying capacities is related to λ and the variance of the random component as

$$\sigma_k^2 = \frac{\sigma_z^2}{1 - \lambda^2} \tag{20.55}$$

Thus, if we assume a first-order autoregressive process for k_t, we obtain a particularly simple description of the environment: σ_k^2 is a measure of the variability; λ becomes a measure of the predictability; and λ can be used in this way because the function, $\rho_k(h)$, which describes the predictability, is a simple one-parameter curve.

Population Dynamics for a First-Order, Autoregressive, Carrying Capacity Process If the carrying capacity process is a first-order autoregressive process, then the variance of the population size through time, σ_n^2, is given by

$$\sigma_n^2 = \left(\frac{r}{2-r}\right)\left(\frac{1 + (1-r)\lambda}{1 - (1-r)\lambda}\right)\sigma_k^2 \tag{20.56}$$

The autocorrelation function of the population size is given by the following recursive formulas

$$\rho_n(0) = 1$$
$$\rho_n(h) = (1-r)\rho_n(h-1) + \frac{r(2-r)\lambda^h}{1 + (1-r)\lambda} \tag{20.57}$$

These results are derived in Roughgarden (1975a) using standard methods of time series analysis. These results have many biological implications.

1. If r equals one, then the population exactly follows or "tracks" the changes in carrying capacity. By Equation (20.48), if $r = 1$ then $n_{t+1} = k_t$ so that n always equals the preceding k. As a result, the variability and predictability of the population are identical with those of the carrying capacities. Hence, when $r = 1$,

$$\sigma_n^2 = \sigma_k^2$$
$$\rho_n(h) = \rho_k(h)$$

2. If λ equals zero, then the environment is a completely unpredictable (white noise) environment. In this case

$$\sigma_n^2 = \left(\frac{r}{2-r}\right)\sigma_k^2$$
$$\rho_n(h) = (1-r)^{|h|}$$

The point to notice here is that apart from the constant, σ_k^2, the variability and predictability of the population are determined by r. As r tends to zero, the population becomes more sluggish in its response to changes in k. This sluggishness lowers the variance, σ_n^2, and increases the serial correlation, $\rho_n(h)$, for any h. So, in an unpredictable environment, whatever predictability occurs in a population is an expression of its own dynamics, particularly of its responsiveness to changes in k. In general terms we see that the variability and predictability of a population are potentially attributable both to its own dynamics and to the environment. The two special cases above are unusual in that in each case only one of these factors is important.

3. Figure 20.6 illustrates the dependence of σ_n^2 on both r and λ. Here σ_n^2 is always an increasing function of r because the responsiveness of the population to fluctuations in k is controlled by r. Augmenting r increases the responsiveness of the population and therefore also σ_n^2. Dependence on λ is particularly curious. High values of λ cause σ_n^2 to approximately equal σ_k^2 over a wide range of r's. A high λ indicates strong positive serial correlation in the k's. Hence, with a high λ, the changes in k between *consecutive* times are rarely large. Instead the total variation in k, as measured by σ_k^2, is realized by a rather slow undulation in the value of k. This fact keeps σ_n^2 close to σ_k^2. The slow course of the overall variation with a high λ provides time for even a sluggish population $(r<1)$ to "catch up" to the current state of k and also does not induce "jitter" in an over-responsive population $(r>1)$.

4. Figure 20.7 illustrates the dependence of $\rho_n(h)$ on both r and λ. Consider first the situation where λ is zero (i.e., the environment is completely unpredictable). As discussed in (1) above, in this situation any autocorrelation is controlled only by the degree of population responsiveness, r. Sizable autocorrelation extends far into the past when r is near zero. If r is one, $\rho_n(h)$ equals one for $h=0$ and equals zero for $h>0$; $\rho_n(h)$ is identical with $\rho_k(h)$ in this case. As r exceeds one, the autocorrelation function oscillates, reflecting the overshoots and undershoots by the population.

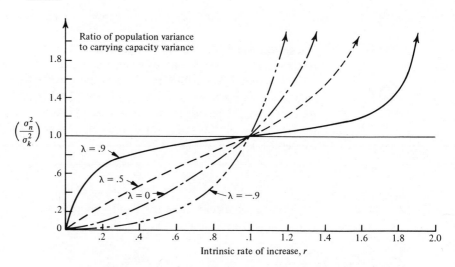

FIGURE 20.6. *The variance of population size relative to the variance of carrying capacity as a function of population responsiveness, r, and environmental predictability, λ. From Equation* (20.56). [From J. Roughgarden (1975), A simple model for population dynamics in stochastic environments. *Amer. Natur.* **109:** 713–736. Copyright 1975 by the University of Chicago.]

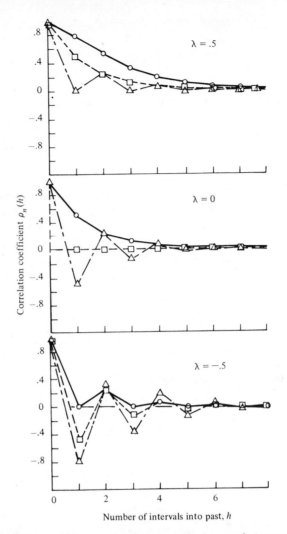

FIGURE 20.7. *The autocorrelation function for population census data where the carrying capacity of the population is fluctuating according to a first-order autoregressive process. From Equation (20.57). Circles refer to $r = .5$, squares to $r = 1$, and triangles to $r = 1.5$.* [Adapted from J. Roughgarden (1975), A simple model for population dynamics in stochastic environments. *Amer. Natur.* **109:** 713–736. Copyright 1975 by the University of Chicago.]

The situations where λ is not zero are natural extensions of those where λ is zero. With positive environmental autocorrelation ($\lambda > 0$), some positive autocorrelation is "added" to the population's autocorrelation. Similarly, when λ is negative, the population's autocorrelation function acquires more of an oscillatory character because of the "bust-boom" nature of the environment.

Note that throughout Figure 20.7 many combinations of r and λ produce the same population pattern: $\lambda = .5$ and $r = 1.5$ yield the same $\rho_n(h)$ as $\lambda = -.5$ and $r = .5$; $\lambda = .5$ and $r = 1.0$ are, together, the same as $\lambda = 0$ and $r = .5$; also, $\lambda = 0$ and $r = 1.5$ are the same as $\lambda = -.5$ and $r = 1.0$. Thus any given degree of population variability and predictability can be produced by many combinations of population responsiveness and environmental predictability. To separate the causes requires observations on *both* the population and the environment through time.

Resource Tracking with a First-Order, Autoregressive, Carrying Capacity Process

The issue of resource tracking concerns the average discrepancy between the population size and the carrying capacity. A natural measure of the tracking "error" is the average value of $(k_t - n_t)^2$. Furthermore, we may define the relative tracking error, e, as

$$e = \frac{\overline{(k_t - n_t)^2}}{\sigma_k^2} \tag{20.58}$$

This quantity is the average tracking error relative to the environmental variability. If we assume that k_t is given by a first-order autoregressive process, then we can determine how e depends on both the population responsiveness, r, and the environmental predictability, λ. The mathematical details are in the boxed material. The result is

$$e(r, \lambda) = \frac{2(1-\lambda)}{2(1-\lambda) + r[(3\lambda - 1) - r\lambda]} \tag{20.59}$$

$$e = \frac{E(k_t - n_t)^2}{\sigma_k^2} = \frac{E[k_t^2 - 2k_t n_t + n_t^2]}{\sigma_k^2} \tag{1}$$

Then substituting

$$n_t = \sum_{i=1}^{\infty} r(1-r)^{i-1} k_{t-i} \tag{2}$$

we obtain

$$e = \frac{E\left\{ k_t^2 - 2 \sum_{i=1}^{\infty} r(1-r)^{i-1} k_t k_{t-i} + \sum_{i=1}^{\infty} \sum_{j=1}^{\infty} r^2 (1-r)^{i-1}(1-r)^{j-1} k_{t-i} k_{t-j} \right\}}{\sigma_k^2}$$

$$= \frac{\sigma_k^2 - 2 \sum_{i=1}^{\infty} r(1-r)^{i-1} \gamma_k(i) + \sum_{i=1}^{\infty} \sum_{j=1}^{\infty} r^2 (1-r)^{i-1}(1-r)^{j-1} \gamma_k(i-j)}{\sigma_k^2} \tag{3}$$

where $\gamma_k(h)$ is the autocovariance function for k_t. Equation (3) gives the relative tracking error for any carrying capacity process. In the special case of a first-order autoregressive process for the carrying capacities, we have

$$\gamma_k(h) = \sigma_k^2 \lambda^{|h|} \tag{4}$$

Substituting (4) into (3) and doing the sums yields

$$e = 1 - \frac{2r\lambda}{[1-(1-r)\lambda]} + \left[\frac{r}{(2-r)} \right] \frac{[1+(1-r)\lambda]}{[1-(1-r)\lambda]}$$

$$= \frac{2(1-\lambda)}{2(1-\lambda) + r[(3\lambda - 1) - r\lambda]} \tag{5}$$

Equation (5) is the relative tracking error in the case of a first-order, autoregressive, carrying capacity process.

Figure 20.8 illustrates the relative tracking error as a function of r for several values of λ. There is a curious threshold at $\lambda = \frac{1}{3}$. If $\lambda < \frac{1}{3}$, the error increases monotonically with r. Hence the error is minimized at $r = 0$, provided $\lambda < \frac{1}{3}$. But when $\lambda > \frac{1}{3}$ the error is no longer monotonic in r but exhibits a minimum at a value of r between 0 and 1. Thus the average discrepancy between the population size and the carrying capacity is a curious function of the population responsiveness and the environmental predictability.

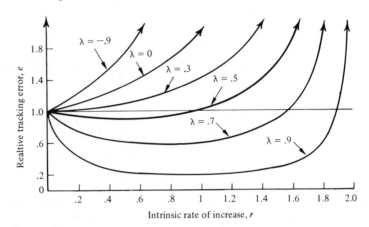

FIGURE 20.8. *The relative tracking error as a function of the intrinsic rate of increase. From Equation (20.60).*

The r that minimizes the tracking error is defined as the optimum r. It is found by differentiating e with respect to r, setting equal to zero and solving, yielding

$$r_{opt} = 0 \qquad \lambda < \frac{1}{3}$$

$$= \frac{3\lambda - 1}{2\lambda} \qquad \lambda \geq \frac{1}{3} \qquad (20.60)$$

The corresponding minimum tracking error is

$$e_{min} = 1 \qquad \lambda < \frac{1}{3}$$

$$= \frac{8\lambda(1 - \lambda)}{(1 + \lambda)^2} \qquad \lambda \geq \frac{1}{3} \qquad (20.61)$$

Figure 20.9 illustrates both r_{opt} and e_{min} as functions of λ. The intuitive rationale for these results is as follows: In an unpredictable environment, responding to a change in k is not likely to improve the tracking because at the next interval k is as likely to be some very different value as it is to remain near its previous value. In this situation, the least tracking error arises from an unresponsive population whose population size simply remains fixed at the average value of the carrying capacity. However, if the environment is predictable in the sense that when changes occur, they are likely to persist, then responding to a change leads to a reduction in the tracking error.

The linear model of population dynamics in an autocorrelated stochastic environment is easily generalized to sets of interacting species. See Rough-garden (1975b) for the extension to N competing species. The topic of resource tracking has also been explored in models applying to spatially

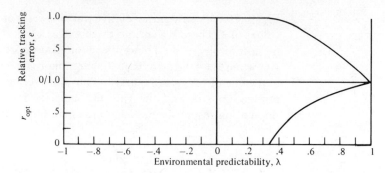

FIGURE 20.9. (Top) *The relative tracking error as a function of λ under the assumption that* $r = r_{opt}(\lambda)$. *(Bottom)* The optimum intrinsic rate of increase $r_{opt}(\lambda)$ as a function of λ.

varying environments that are constant in time. See Roughgarden (1974c). In these models the issue of how the dispersal of organism leads to discrepancies between the spatial pattern of abundance and the spatial pattern of resources is analyzed.

Patchiness in a Random Environment

Populations are not distributed uniformly over their range; instead they occur in higher concentrations at some places within the range relative to other places. What makes this fact interesting is that it is true even when the environment is apparently uniform. Organisms are not uniformly distributed even in a uniform environment. One of the clearest examples involves the small-scale spatial distribution of plankton. Figure 20.10, from Wiebe (1970), presents the abundance of several species of plankton along a transect 500 meters long. Within this transect the water mass is essentially homogeneous. Yet the organisms are distributed in a very nonhomogeneous way. A patch is a region of the environment where the abundance of organisms is above average. From Figure 20.10 we see that a typical patch length is about 30 m long. The distinctiveness of the patches depends on the difference between the typical abundance in the peaks and that in the troughs. We shall discuss how to measure the patch length and distinctiveness in ways that directly relate to the theoretical predictions.

Another empirical example of patchiness that is particularly clear is found in the distribution of New Guinea birds by Diamond (1975), as illustrated in Figure 20.11. In the case of bird census data, a patch is a region where the species is present; it is absent in regions between patches. It is possible that a patch is generally a more discrete entity with birds than with plankton, but it is difficult to tell because of the differences in census techniques used in measuring bird and plankton abundance. Notice that the patch size varies substantially among species. The starling, *Mino anais*, is present throughout the lowlands that comprise its suitable habitat, except for a comparatively small region in the Northeast. The tree creeper, *Climacteris leucophaea*, is absent from a large region of its suitable habitat in the central mountains. Near the other extreme, the logrunner, *Cinclosoma ajax*, is absent from almost all of the suitable lowland habitat and occurs only in four rather small patches.

The term *patchiness* is currently applied to the distributions of both plankton and New Guinea birds, although the scale of the phenomenon is enormously different. It is not yet clear whether the same causal processes

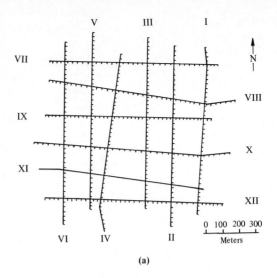

(a)

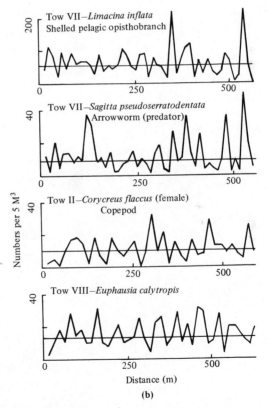

Tow VII—*Limacina inflata*
Shelled pelagic opisthobranch

Tow VII—*Sagitta pseudoserratodentata*
Arrowworm (predator)

Tow II—*Corycreus flaccus* (female)
Copepod

Tow VIII—*Euphausia calytropis*

Numbers per 5 M³

Distance (m)

(b)

FIGURE 20.10. *Patchiness in oceanic zooplankton.* (a) *Twelve tows were carried out and they together form a grid as illustrated. The location of each sample is indicated by short lines perpendicular to the tow path.* (b) *Examples of the plots of abundance versus distance obtained from the tows. The plots show large changes in abundance over short distances. The median value is also plotted.* [Adapted from P. H. Wiebe (1970), Small scale spatial distribution in oceanic zooplankton. *Limnol. Oceanogr.* **15:** 205–217.]

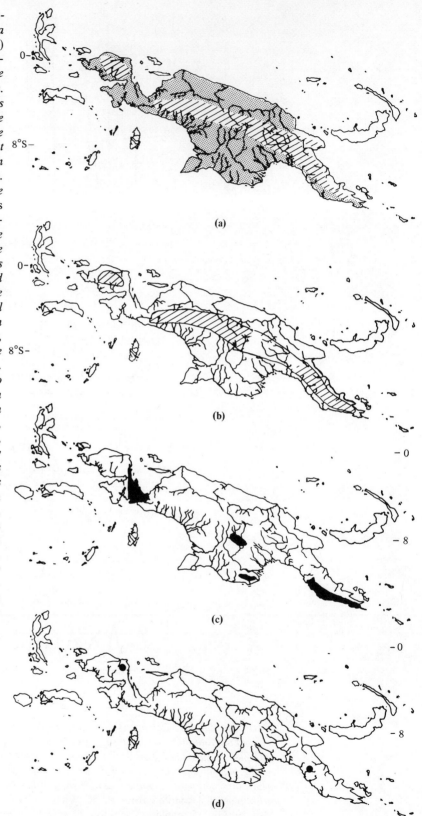

FIGURE 20.11. *Patchiness in New Guinea bird distributions.* (a) *Distribution of the starling* Mino anais *in the lowlands of New Guinea. Unsuitable mountainous areas are hatched. The species is present in the stippled area and absent from the blank area in northeast New Guinea.* (b) *Distribution of the tree creeper,* Climacteris leucophaea, *in mountains of New Guinea. The Central Dividing Range is outlined. This species is present in the hatched area and absent from the blank of the Central Range.* (c) *Distribution of the log runner,* Cinclosoma ajax, *in the lowlands of New Guinea. The species is confined to the four blocks shown in black.* (d) *Distribution of the berry pecker,* Melanocharis arfakiana, *in the mountains of New Guinea. The species is known from two localities 1000 miles apart.* [From J. Diamond (1975), Assembly of species communities, in M. L. Cody and J. M. Diamond, eds., *Ecology and Evolution of Communities.* Belknap Press of Harvard University Press.]

are involved in both the plankton and bird distributions. As we shall see, the scale of the patchiness is set largely by the dispersal distances of the organisms involved. It may be that 30 m is to a planktonic crustacean what 200 km is to a bird. If so, the phenomena may be the same in spite of the difference in scale, but if not, then separate mechanisms must be sought.

In this section we introduce a technique for measuring patchiness and some of the theory that predicts the occurrence of patchiness even in a uniform environment. This section is largely taken from Roughgarden (1977b).

Measuring Patchiness with the Variance Spectrum In the next several paragraphs we develop a natural and practical method for quantifying the concept of pattern and of patchiness in particular. This approach has been taken by Platt and Denman (1975) and Powell et al. (1975). Let us begin with a sample hypothetical spatial pattern of abundance,

$$\ldots 100, 150, 150, 100, 50, 50, 100, 150, 150, 100, 50, 50, 100 \ldots$$

These are the numbers of animals in each quadrat from a sequence of contiguous quadrats. The variance of these 13 numbers is 1667. We can naturally define the *patch length* in this case as 3, where the unit length is the quadrat length. The patch distinctiveness is related to the difference in abundance between peak and trough. The overall variance of the abundance is a good measure of patch distinctiveness because the variance tells us the size of the difference between peaks and troughs.

The basic concept in quantifying spatial pattern is that any spatial pattern (empirical or theoretical) can be decomposed into the sum of regular patterns of patches. Consider another simple pattern,

$$\ldots 100, 125, 125, 125, 125, 125, 100, 75, 75, 75, 75, 75, 100 \ldots$$

Here the patch length is 6 and the variance is 521. This pattern can be added to the previous pattern yielding

$$\ldots 200, 275, 275, 225, 175, 175, 200, 225, 225, 175, 125, 125, 200 \ldots$$

The variance is now $1667 + 521 = 2188$. It is natural to view this pattern as representing the effects of two different patch lengths, with a larger contribution from patches of length 3 than from those of length 6. In this spirit we shall say that the *amount* of variance "explained by" patches of length 3 is 1667 and the amount explained by patches of length 6 is 521. The total overall variance, which is the measure of patch distinctiveness, is the sum of the variances explained by all contribution patch sizes. These are the essential ideas in the terminology discussed below.

The pattern of variation in the examples can be viewed as a wave, and the definition of the patch length is simply half the wavelength of the wave pattern. (The wavelength is the distance between adjacent peaks.) There is a continuum of patch lengths that can contribute to any spatial pattern. The smallest patch length that can be resolved is simply equal to the width of 1 quadrat. In practice, the largest patch length that can be resolved equals one half the length of the entire study area. For theoretical models, however, we assume that the study site is infinite and therefore there is no maximum patch length. Let L denote a patch length, and let the unit of length be the quadrat length. Thus we assume L can take any value between one and infinity. We introduce the quantity

$$\omega = \frac{\pi}{L} \tag{20.62}$$

where ω has the property of varying between 0 and π as L varies from ∞ to 1. The symbol ω is called the frequency of the wave pattern and has the dimensions of radians per length. In time series analysis it is standard practice to use ω in the derivation of formulas, and we may always convert back to L using Equation (20.62), that is, $L = \pi/\omega$. With this terminology we may describe the nature of the patchiness with the function, $g(\omega)$. By definition $g(\omega)$ is the amount of variance explained by a wave pattern of frequency, ω, that is the amount of variance explained by patches of length $L = \pi/\omega$. The function $g(\omega)$ is named the *variance spectrum* of the pattern. For example, patches of length 3 correspond to $\omega = \pi/3 = 1.05$ and patches of length 6 correspond to $\omega = \pi/6 = .349$. So $g(\omega)$ for the composite pattern above is

$$g(1.05) = 1667$$
$$g(.349) = 521 \tag{20.63}$$

There happens to be only two patch lengths in the example given, but in general there is a continuum of patch lengths.

We may introduce two indices, one for the average patch length in the pattern and one for the patch distinctiveness. First, we need to define the average frequency, $\bar{\omega}$ by

$$\bar{\omega} = \frac{\int_0^\pi \omega g(\omega)\, d\omega}{\int_0^\pi g(\omega)\, d\omega} \tag{20.64}$$

Hence the index of the average patch size is

$$\bar{L} = \frac{\pi}{\bar{\omega}} \tag{20.65}$$

where $\bar{\omega}$ is obtained from (20.64). In our example, $\bar{\omega}$ is

$$\bar{\omega} = \frac{[1.05(1667) + .349(521)]}{[1667 + 521]} = .883 \tag{20.66}$$

Therefore, the index of average patch length is $\bar{L} = \pi/.883 = 3.56$. The index of patch distinctiveness is simply the variance of the abundance in all of the quadrats. The variance is given by

$$\sigma^2 = \int_0^\pi g(\omega)\, d\omega \tag{20.67}$$

The total variance is simply the sum of the variances explained by all possible frequencies. Thus, if we have the variance spectrum $g(\omega)$ we can compute indices of the average patch length and the patch distinctiveness using these formulas. The theory presented later is based on these quantities. The model assumes that $g(\omega)$ for the spatial pattern of resource abundance is given, and predicts $g(\omega)$ for the spatial pattern of population abundance.

Before proceeding with the model, it is worth noting the patch structure associated with a random distribution. A random distribution *does* lead to patches, but there is no preferred patch length. Recall, for example, that any realization of a Poisson distribution leads, by chance, to some neighboring quadrats with high abundances, and to some neighboring quadrats with low abundance. Thus, by chance, there will be regions of the environment with high abundance and regions with low abundance and therefore, by definition, a patch structure occurs. But it can be shown that there is no preferential patch length in a random distribution. Specifically, for a random spatial distribution,

$$g(\omega) = c \qquad (0 \leq \omega \leq \pi) \tag{20.68}$$

The function $g(\omega)$ is constant for all ω indicating that all patch lengths contribute equally to the pattern. The patch structure indices for a flat $g(\omega)$ are readily computed, and they are

$$\bar{L} = 2$$
$$\sigma^2 = c\pi \tag{20.69}$$

So the average patch length with a random distribution is 2. *We shall say that a pattern is more patchy than a random distribution if $\bar{L} > 2$, and less patchy if $\bar{L} < 2$.* For example, the pattern composed of two patch lengths introduced above is more patchy than random because $\bar{L} = 3.56$. With this terminology we can proceed to see how patchiness is influenced by migration in a spatially varying stochastic environment.

Patchiness Caused by Stochastic Fluctuation of Resources

We now analyze how random (white noise) fluctuation in the carrying capacity from place to place leads to patchiness. That is, even if the distribution of resource abundance is randomly patchy (i.e., $\bar{L} = 2$ for the resources), nonetheless the population *does* become more patchy than random (i.e., $\bar{L} > 2$ for the population). Patchiness is an inevitable consequence of dispersal in a fluctuating environment. To show these results, we must extend the linear stochastic population to include spatial variation.

The model based at one location is

$$n_{t+1} = (1 - r)n_t + rk_t \tag{20.70}$$

Now consider a model where there are many locations and migration among the locations. Let d_{ij} be the probability that an individual at i originated from j. Because any individual must have originated from some place, the d_{ij} satisfy

$$\sum_{j=-\infty}^{\infty} d_{ij} = 1 \tag{20.71}$$

Moreover, we assume that d_{ij} depends only on $u \equiv |i--j|$, that is, the distance separating location i from location j. The model then is given by

$$n_{i,t+1} = \sum_{u=-\infty}^{\infty} d_u[(1-r)n_{i+u,t} + rk_{i+u,t}] \tag{20.72}$$

where $n_{i,t+1}$ is the number at location i at time $t+1$; $n_{i+u,t}$ is the number at the location u units of distance away at time t; and $k_{i+u,t}$ is the resource level at the location u units away at time t. As always, n and k refer to deviations from their long-term averages values. This model can be solved to predict the spatial pattern of the population caused by stochastic fluctuation in the resource levels at various places throughout the population's range. This model couples the spatial and temporal dimensions.

Our intention is to predict the spatial pattern of the population in terms of the function, $g(\omega)$, given a similar description of the spatial distribution of resources. Let $g_n(\omega)$ denote the variance spectrum of the population's pattern, and $g_k(\omega)$ that of the resource pattern. We assume that the statistical pattern of patchiness in the resource distribution is constant; that is, although the locations of patches of resources may move around, it is assumed that $g_k(\omega)$ is constant in time. In addition, we assume that the resource patterns at different times are independent. With these assumptions the population's spatial pattern approaches an equilibrium, $g_n(\omega)$, given by

$$g_n(\omega) = \left[\frac{r^2 d^2(\omega)}{1 - (1-r)^2 d^2(\omega)} \right] g_k(\omega) \tag{20.73}$$

where $d(\omega)$ describes the smoothing which occurs because of dispersal. The detailed shape of the function depends on the dispersal characteristics of the organisms. The derivation of this formula appears in Roughgarden (1977b).

Two familiar examples of dispersal models are the "stepping stone" and "geometric" dispersal models. Specifically, the stepping stone model and its properties are

$$
\begin{aligned}
d_u &= m/2 \quad && \text{for } u = -1, 1 \\
d_u &= (1-m) \quad && \text{for } u = 0 \\
d_u &= 0 \quad && \text{for } u \neq -1, 0, 1 \\
E[|u|] &= m \\
E[u^2] &= m \\
d(\omega) &= (1-m) + m \cos(\omega)
\end{aligned}
\tag{20.74}
$$

where $0 < m < 1$. The geometric dispersal model is given by

$$
\begin{aligned}
d_u &= (\tfrac{1}{2})(1-m)m^{|u|} \\
d_0 &= 1 - m \\
E[|u|] &= m/(1-m) \\
E[u^2] &= m(1+m)/(1-m)^2 \\
d(\omega) &= \frac{(1-m)[1 - m \cos(\omega)]}{1 - 2m \cos(\omega) + m^2}
\end{aligned}
\tag{20.75}
$$

where again $0 < m < 1$. The expressions $E[|u|]$ and $E[u^2]$ mean the expected value of the dispersal distance and of the square of the dispersal distance, respectively. The quantity in brackets in Equation (20.73) is often called the *transfer function* in the literature. It indicates how well any wavelength present in the resource pattern is passed into the pattern of the population abundance. In the important special case where the resource pattern is random, then $g_k(\omega)$ equals a constant and therefore $g_n(\omega)$ is simply equal to a constant times the transfer function itself.

The qualitative results of this model can be inferred from inspecting the graph of the transfer function. Figure 20.12 illustrates the transfer function for several values of m and r, using a geometric dispersal function. First, suppose the intrinsic rate of increase, r, is fixed and examine the effect

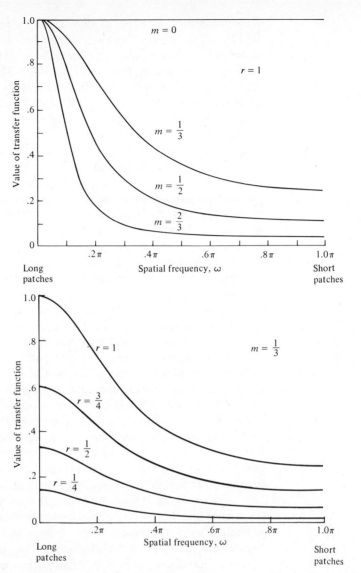

FIGURE 20.12. *Transfer function relating the spatial pattern of the population abundance to the spatial pattern of the resource abundance. Geometric dispersal is assumed. Increasing dispersal, m, produces longer but less distinct patches. Lowering the intrinsic rate of increase, r, produces shorter and less distinct patches.* [From J. Roughgarden (1977), Patchiness in the spatial distribution of a population caused by stochastic fluctuation in resources. *Oikos* **29:** 52–59.]

of varying the dispersal parameter, *m*. Clearly, as *m* increases (i.e., dispersal increases) the area under the curve decreases. Hence as dispersal increases, the variance in the abundance from place to place decreases. This result means that the "patch distinctiveness" decreases as dispersal increases. However, as *m* increases, the index of average patch length also increases. As *m* increases, the curve becomes concentrated near $\omega = 0$. Therefore, the average spatial frequency, $\bar{\omega}$, becomes smaller and hence $L \sim 1/\bar{\omega}$ becomes larger. Thus we see that the *qualitative effect of increasing the dispersal distance is to produce longer but less distinct patches of population abundance.*

Next consider the effect of varying r for a fixed m. By inspecting Figure 20.12 (bottom) we see that decreasing r with fixed m reduces the total area under the curve *and* produces a flatter curve. Hence lowering r reduces the variance in abundance from place to place. This result means that "patch distinctiveness" decreases as r decreases. Also, as r decreases the spectrum becomes flatter. Hence the typical patch length decreases as r decreases. Thus *the qualitative effect of decreasing r is to produce shorter and less distinct patches of population abundance.*

The overall picture that emerges is that patchiness produced by this mechanism should be prominent in organisms with both a high r and moderate dispersal. A low r wipes out patch distinctiveness and shortens the patch length to levels expected with a random spatial distribution, even if the dispersal is high. Even with a high r very short dispersal distances lead to short patch lengths which would be close to that expected by chance, whereas very long dispersal would produce patches which lack distinctiveness and which would probably not be detected. The best combination of parameters to produce patchiness is therefore a high r combined with moderate dispersal.

As mentioned above, if the spatial distribution of the resources is random then, by definition, $g_k(\omega)$ is flat. In this case, Figure 20.12 actually illustrates $g_n(\omega)$, the representation of the population's spatial pattern. All the curves indicate a distribution that is more patchy than a random spatial distribution, provided there is any dispersal at all ($m > 0$). Thus the distribution of the population is inevitably more patchy than the resource distribution, even if the resource distribution itself is random.

The model shows that patchiness in a population's spatial distribution is an inevitable consequence simply of dispersal in a stochastic environment. It is not necessary to assume that the environment itself exhibits underlying patchiness or that there is some special aggregation behavior in the organisms, although, if present, these will also be contributing factors. In light of this analysis, it is not surprising that the observation of patchiness in the spatial distribution of populations is so ubiquitous.

The theory of patchiness based on stochastic fluctuation of sources has been extended to include the interaction of two competitors in Roughgarden (1978b). The topic of patchiness has also been treated in models of interacting populations where there is dispersal in space but the environment is constant in time. Okubo (1974) and Levin and Segel (1976) have shown that an "interaction-diffusion instability" may lead to patchiness in an environment which is constant in space and time, provided that special conditions on the dispersal and population interactions are satisfied. Very recently, Whittaker and Levin (1977) have considered patchiness in land plant communities and have investigated the role of succession and other community processes in the formation of patchiness.

PART FIVE

EVOLUTIONARY ECOLOGY OF
INTERACTING POPULATIONS

Chapter 21
COMPETITION

No population is alone. Populations interact with one another in many ways. These interactions are important. We will see that we cannot understand many population phenomena without considering the interactions. Even if we want to ignore population interactions, we cannot. But above all, the interactions are intrinsically interesting because they produce perhaps the most intricate and fascinating patterns in biology. We begin our exploration of these patterns with this chapter on competition.

A Classic Example   Let us begin with a classic example of a pattern caused by interspecific competition. A ubiquitous organism of rocky coasts is the barnacle, a sessile, filter-feeding crustacean with planktonic larvae. Even a casual glance at the rocks in the intertidal region reveals that barnacles occur in a zone at the top of the intertidal. In a classic study of competition, Connell (1961a, b) examined in detail the causes of barnacle zonation of the Scottish coast. Two species are involved. *Chthamalus* occupies the highest zone and *Balanus* occupies a zone immediately below. Connell established that *Chthamalus* was limited from above by dessication and from below by competition with *Balanus*. In the absense of *Balanus*, *Chthamalus* readily lives throughout the zone used by *Balanus*. But in the presence of *Balanus*, *Chthamalus* is restricted to the top of the intertidal. Connell discovered that *Balanus* individuals are larger and grow faster, and are able literally to grow over and/or pry loose *Chthamalus* individuals in the region where *Balanus* can live. However, they are more sensitive to dessication than *Chthamalus* and, therefore, are not able to live as high in the intertidal as *Chthamalus* is. Thus *Chthamalus* occurs in the top zone as a result of a competitive interaction with *Balanus*. In effect, it is "pushed" into a marginal region of the environment by its competitor. Thus a population interaction is essential to explaining the conspicuous zonation pattern of barnacles. If we had simply determined the population growth characteristics of *Chthamalus* in the laboratory under various environment conditions and then used this information to predict its distribution in nature, we would have failed miserably.

The example of competition provided by the barnacles involves "aggressive" contact between *Balanus* and *Chthamalus*. But to an ecologist, competition encompasses more than overt aggression between individuals of different species. It may involve no "conflict" at all. Indeed, perhaps the most important form of competition is the joint use of common limiting resources by members of two species. The criterion for competition between two species is that the fitness of an individual of each species must be a decreasing function of the abundance of the other species. An operational test is whether the abundance of each species increases upon removal of the other species. By this general criterion, competition is simply a negative interaction betweeen the individuals of two species, regardless of how the interaction is actually accomplished.

More recent experimental studies of competition in nature include Hall, Cooper, and Werner (1970), Dayton (1971), Menge (1972), Wilbur (1972), and Lubchenco (1978).

Many ecologists like Miller (1967) classify the mechanisms of competition as exploitative or as interference. Exploitative competition is the joint exploitation of limiting resources. Interference competition refers to behavior involving physical contact or conflict. Which mechanisms of competition actually occur in any case depend on the identities of the species involved. Some populations competing for limited resources seem to exhibit no interference mechanisms at all, or at least no mechanisms have been detected. Others employ both kinds of mechanisms. However, as discussed in the chapter on coevolution, there is reason to suspect that interference competition can only evolve if there is prior exploitative competition. If so, interspecific aggression or conflict should only arise with regard to a limiting commodity; there should not be conflict over nothing.

The Lotka–Volterra Competition Model

Because the mechanisms of competition are special to the particular species involved, we are faced with a familiar dilemma when proposing a model for competition. If we are very faithful to the mechanisms, then we ensure that the model cannot apply to many cases and may be intractable besides, whereas if we propose a simple model, we risk irrelevance to the real world. Again, a compromise is needed and the result will depend, in part, on one's purpose. As with the logistic equation, we shall choose a model that is very simple and, admittedly, does not do justice to our knowledge of the mechanisms of competition. Nonetheless, we shall explore this model and discover that it is surprisingly rich with biological predictions. At the very least, it is a prototype for more realistic models of competition and at best, it is a useful tool for suggesting new field experiments and for interpreting the results already in hand.

The Basic Assumption

In Chapter 16, we introduced the logistic equation by assuming that an individual's reproductive output decreased as a linear function of population size. The traditional model for competition is a natural extension of this assumption. We assume that an individual's reproductive output decreases as a linear function *both* of the size of its population *and* the size of its competitor's population. Specifically, we assume that for an individual of species-1

$$\left\{ \begin{array}{l} \text{individual's contribution} \\ \text{to population growth rate} \end{array} \right\} = r_1 - \frac{r_1}{K_1} N_1 - \frac{r_1}{K_1} \alpha_{12} N_2 \qquad (21.1)$$

The first two terms here are identical to the logistic equation, where r_1 and K_1 are the intrinsic rate of increase and carrying capacity of species-1. Only the last term is new. The term, α_{12}, is called the competition coefficient for the effect of an individual of species-2 against an individual of species-1. The factor, r_1/K_1, is involved in the last term for the following reason. By definition, we want α_{12} to indicate the strength of interspecific competition from species-2 *relative* to the strength of intraspecific competition within species-1. Specifically, if α_{12} equals one, then the coefficient in front of N_2 is the same as that in front of N_1. If α_{12} equals one, the effect of competition from an individual of species-2 on the reproductive output of an individual of species-1 is exactly the same as that from another individual of species-1. A shorthand way of saying this is that interspecific competition is exactly as strong as intraspecific competition. However, if α_{12} is less than one, then interspecific competition has less between-individual effect than intraspecific competition, while if α_{12} is greater than one, then interspecific competition has more effect than intraspecific competition. We want to be able to use α_{12} in this way as a measure of the *strength* of interspecific competition *relative* to intraspecific competition. However, to do so, we need the factor, $-r_1/K_1$, in the third term in order to ensure that when α_{12} equals 1, then the overall coefficient in front of N_2 is the same as that in front of N_1.

To summarize, we assume that the reproductive output of an individual decreases as a linear function of the size of each population. We introduce the competition coefficient, α_{12}, as a measure of the strength of interspecific competition, relative to intraspecific competition.

The Equations   Because the growth rate of the whole population is simply the individual's contribution to population growth times the number of individuals, we have, for species-1

$$\frac{dN_1}{dt} = \left(r_1 - \frac{r_1}{K_1}N_1 - \frac{r_1}{K_1}\alpha_{12}N_2 \right)N_1 = r_1\frac{(K_1 - N_1 - \alpha_{12}N_2)}{K_1}N_1 \qquad (21.2a)$$

and similarly for species-2,

$$\frac{dN_2}{dt} = \left(r_2 - \frac{r_2}{K_2}N_2 - \frac{r_2}{K_2}\alpha_{21}N_1 \right)N_2 = r_2\frac{(K_2 - N_2 - \alpha_{21}N_1)}{K_2}N_2 \qquad (21.2b)$$

These equations are usually written in the form on the right, but in this form, the assumptions concerning the effect of competition on an individual's reproductive output are not as obvious as when written in the form on the left. These equations are called the *Lotka–Volterra competition equations*. We shall refer to them as the LV equations. They are a pair of differential equations that relate the growth rate of each population to the abundance of both. Of course, when α_{12} and α_{21} are zero, both populations grow logistically, but if they are not zero, then all kinds of things can happen.

From your experience with the model for natural selection at one locus with two alleles, you will recall that four different outcomes are possible. We might suppose, then, that four different outcomes may be possible here too: Species-1 could always win, species-2 could always win, species-1 and species-2 may permanently coexist, either species-1 or species-2 could win, depending on which is initially more abundant. Now, let us analyze the LV equations to see whether any or all of these possibilities can occur.

A Graphical   The traditional way of analyzing the LV competition equations is with a
Analysis   graphical approach. However, this approach does not generalize to competition among three or more competing species. It does work, however, for two competing species and is very simple. We shall, therefore, follow tradition here. More general results are presented in Chapter 24. Consider a graph, in which N_1 is plotted on the horizontal axis and N_2 on the vertical axis. Each point on this graph indicates a specific population size for each species. The technique has three parts. First, we separate the graph into several regions. Next, we determine what happens in each region. Last, we combine our analyses for each region to obtain the overall predictions.

To separate the N_1-N_2 graph into regions, we set the expressions for dN_1/dt and dN_2/dt equal to zero. If dN_1/dt equals zero (and assuming r_1

and N_1 do not equal zero), then the expression in parentheses in (21.2a) must equal zero

$$K_1 - N_1 - \alpha_{12}N_2 = 0 \qquad (21.3a)$$

This equation is a line in the N_1-N_2 plane. Since two points determine a line, we can graph this equation simply by finding the intercepts on the N_1 and N_2 axes and connecting the intercepts with a straight line. The intercept on the N_1 axis is found by letting N_2 equal zero and solving for N_1. Thus the N_1 intercept is simply K_1. Similarly, the N_2 intercept is found by setting N_1 equal to zero and solving for N_2. The N_2 intercept is then K_1/α_{12}. Figure 21.1 graphs the equation (21.3a) on the basis of the intercepts. The line separates the positive part of the N_1-N_2 plane into two regions labeled A and B. *In region A, the growth rate of species-1 is positive, in region B, it is negative.* Of course, any point exactly on the line is associated with no change in species-1. Similarly, if we set dN_2/dt equal to zero we obtain

$$K_2 - N_2 - \alpha_{21}N_1 = 0 \qquad (21.3b)$$

The N_2 and N_1 intercepts are found to be K_2 and K_2/α_{21}, respectively. This equation is graphed as a dotted line in Figure 21.2. Again, the plane is separated into two regions, labeled C and D. In C, dN_2/dt is positive and in D, dN_2/dt is negative. The lines separating the regions are called *isoclines*.

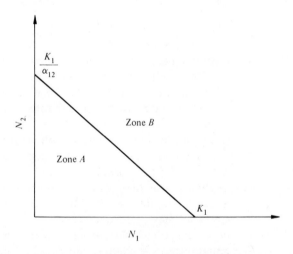

FIGURE 21.1. *The line along which dN_1/dt equals zero.*

Now, let us draw lines on the same graph. We have a problem, however. We can orient the lines in four different ways, as illustrated in Figure 21.3. The dotted line of species-2 may or may not intersect the solid line of species-1 in the positive region of the N_1-N_2 plane. Moreover, if the lines do not intersect, the dotted lines can be above the solid line, or vice versa. If they do intersect, there are also two ways this can occur. Which of the four illustrations is appropriate in any particular case depends on the position of the intercepts. For example, an illustration like Figure 21.3a results whenever $K_1 > K_2/\alpha_{21}$ and $K_2 < K_1/\alpha_{12}$. Similarly, an illustration like Figure 21.3c requires $K_1 < K_2/\alpha_{21}$ and $K_2 < K_1/\alpha_{12}$, and so on. Because we have four cases, we shall have to consider each, one by one.

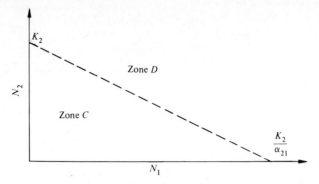

FIGURE 21.2. *The line along which dN_2/dt equals zero.*

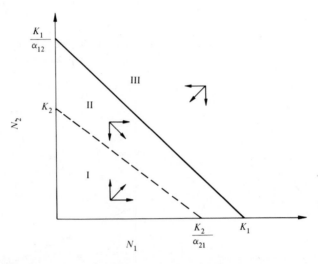

FIGURE 21.3a. *Case I. $\hat{N}_1 = K_1$ and $\hat{N}_2 = 0$ is the only stable equilibrium point.*

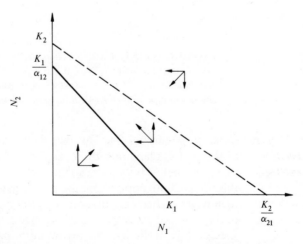

FIGURE 21.3b. *Case II. $\hat{N}_1 = 0$ and $\hat{N}_2 = K_2$ is the only stable equilibrium point.*

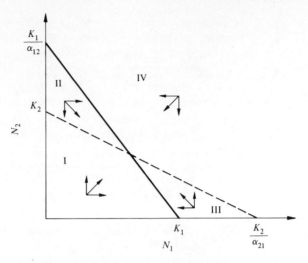

FIGURE 21.3c. *Case III. The only stable equilibrium is at the intersection point of the two lines. Both species coexist at this point.*

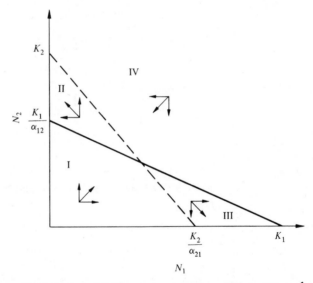

FIGURE 21.3d. *Case IV. There are two stable equilibria. One is $\hat{N}_1 = K_1$ $\hat{N}_2 = 0$ and the other is $\hat{N}_1 = 0$, $\hat{N}_2 = K_2$. Which of these points is actually reached by a two-species system depends on the initial condition. The intersection point of the two lines is an unstable equilibrium.*

Cases Where the Isoclines Do Not Intersect Consider Figure 21.3a. The two lines, together, divide the positive part of the N_1-N_2 plane into three regions, labeled I, II, and III. Any point in region I is below both lines, so that both dN_1/dt and dN_2/dt are positive; in this region, both populations are growing. Let us place arrows in each region to indicate whether populations are growing or declining. An arrow pointing to the right indicates species-1 is growing; an arrow to the left indicates that species-1 is declining. An arrow pointing up indicates that species-2 is growing and an arrow pointing down indicates that species-2 is declining. With these conventions, we can place one arrow pointing up and one pointing to the right in region I of Figure 21.3a. Now, we can combine these arrows, in the sense of vector addition, to give a net arrow that points both up and to the right. This arrow indicates how the two-species system changes, given that it starts from the point that is at the foot of the vector. All the points within region I are associated, in this sense, with a vector that has components pointing up and to the right. Therefore, from any point within region I, the system moves toward and, finally, into region-II. Next, any point in region III is above both lines, so that both dN_1/dt and dN_2/dt are negative; that is, both populations are declining. Hence each point in region III is associated with a vector pointing down and to the left. Therefore, from any point within region III the system moves down toward and finally into region II. Thus trajectories, originating either in I or III, lead into II. What, then, happens in region II? Points in this region are below the solid line of species-1 but above the dotted line of species-2. Therefore, dN_1/dt is positive and dN_2/dt is negative in this region. Hence the vectors from points in this region aim down and to the right. Once the system is in region II, it never leaves, so that the trend of the vectors within region II is to take the system to the lower right-hand corner of the region. As a result, all the trajectories converge to the point where N_1 equals K_1, and N_2 equals zero. Thus whenever the illustration in Figure 21.3a is appropriate, the outcome of competition is that species-1 wins and species-2 is eliminated, regardless of the initial abundances.

The situation in Figure 21.3b is essentially the same as above, only in this case species-2 always wins and species-1 is eliminated. Thus in both Figures 21.3a and b, the species with the most exterior line is the eventual winner.

Cases Where the Isoclines Do Intersect Let us discuss Figures 21.3c and d together. In both figures, the regions I and IV are the same. In I, the arrows point up and to the right. In IV, they point down and to the left. Therefore, trajectories from I or IV will take the system into either II or III. In Figure 21.3c region II is below the solid line of species-1 and above the dotted line of species-2. Thus in this region, dN_1/dt is positive, while dN_2/dt is negative. Therefore, the vectors in this region will point down and to the right, leading to the point of intersection between the lines. Moreover, in region III of Figure 21.3c, dN_1/dt is negative, while dN_2/dt is positive. So the vectors of region III point up and to the left, leading once again to the intersection point. Thus whenever the illustration in Figure 21.3c is appropriate, the trajectories lead to the intersection point. This result means that the two species coexist with one another at equilibrium. Neither excludes the other and the equilibrium abundance of each species is given by its coordinate at the intersection point. Turning to Figure 21.3d, we find that region II in this illustration is above the solid line of species-1 and below the dotted line of species-2. Therefore, at this point dN_1/dt is negative and dN_2/dt is positive. Thus vectors in region II of Figure 21.3d lead up and to the left; they lead, in fact, to the upper left-hand corner of region II and, thus, away from the intersection point. Similarly, in region III of Figure 21.3d, the vectors are found to lead to the lower right corner of region III, again, away from the intersection point. In this case, either species-2 wins or species-1, depending on whether the system arrives first in region II or region III. Thus which species wins depends on the initial abundances.

Examples of the trajectories for the cases discussed here appear in Figure 21.4. The trajectories were obtained by integrating the Lotka–Volterra equations numerically with a computer.

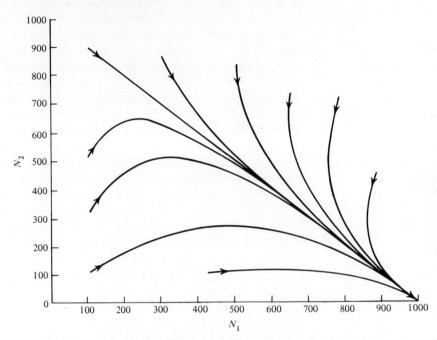

FIGURE 21.4a. *Example of the trajectories that occur in Case I where species-1 wins. Here $K_1 = K_2 = 1000$, $r_1 = r_2 = .5$, $\alpha_{12} = .5$, and $\alpha_{21} = 1.5$.*

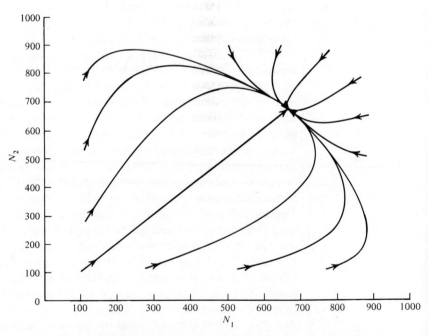

FIGURE 21.4b. *Examples of the trajectories that occur in Case III where both species coexist. Here $K_1 = K_2 = 1000$, $r_1 = r_2 = .5$, and $\alpha_{12} = \alpha_{21} = .5$.*

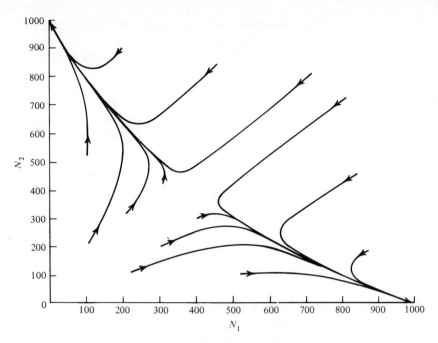

FIGURE 21.4c. *Examples of the trajectories that occur in Case IV where the final outcome depends on the initial position. Here $K_1 = K_2 = 1000$, $r_1 = r_2 = .5$, and $\alpha_{12} = \alpha_{21} = 1.5$.*

Interpretation of the Conditions for Each of the Cases

These paragraphs show that the outcome of competition between two species depends on which of the graphs in Figure 21.3 is appropriate. Figure 21.3a predicts that species-1 should always exclude species-2; Figure 21.3b predicts that species-2 should always exclude species-1; Figure 21.3c predicts that there should be coexistence between the species with abundances given by the coordinates of the intersection point; and Figure 21.3d predicts that either species can exclude the other, depending on the initial abundances. This result is surprising, so let us try to understand what underlying biology produces each of the graphs in Figure 21.3. Each graph is determined by the location of the intercepts. To obtain Figure 21.3a, for example, we need $K_1 > K_2/\alpha_{21}$, and $K_2 < K_1/\alpha_{12}$. Similarly, each graph corresponds to a specific positioning of the intercepts. All four cases are

$$\text{Figure 21.3a} \leftrightarrow K_1 > \frac{K_2}{\alpha_{21}},\ K_2 < \frac{K_1}{\alpha_{12}}$$

$$\text{Figure 21.3b} \leftrightarrow K_1 < \frac{K_2}{\alpha_{21}},\ K_2 > \frac{K_1}{\alpha_{12}}$$

$$\text{Figure 21.3c} \leftrightarrow K_1 < \frac{K_2}{\alpha_{21}},\ K_2 < \frac{K_1}{\alpha_{12}} \qquad (21.4)$$

$$\text{Figure 21.3d} \leftrightarrow K_1 > \frac{K_2}{\alpha_{21}},\ K_2 > \frac{K_1}{\alpha_{12}}$$

These are the intercepts that generate each of the four graphs in Figure 21.3.

Let us now rearrange them so that the α's are separated from the K's:

$$\text{Figure } 21.3a \leftrightarrow \alpha_{21} > \frac{K_2}{K_1}, \alpha_{12} < \frac{K_1}{K_2}$$

$$\text{Figure } 21.3b \leftrightarrow \alpha_{21} < \frac{K_2}{K_1}, \alpha_{12} > \frac{K_1}{K_2}$$

$$\text{Figure } 21.3c \leftrightarrow \alpha_{21} < \frac{K_2}{K_1}, \alpha_{12} < \frac{K_1}{K_2}$$

$$\text{Figure } 21.3d \leftrightarrow \alpha_{21} > \frac{K_2}{K_1}, \alpha_{12} > \frac{K_1}{K_2}$$

(21.5)

Suppose That The conditions, as written in (21.5), can be interpreted biologically. For
$K_1 = K_2$ the sake of illustration, let us first assume that $K_1 = K_2$. Then Figure 21.3a is
produced whenever $\alpha_{21} > 1$ and $\alpha_{12} < 1$. This means that species-1 excludes
species-2, if the effect of species-1 against species-2 is stronger than the
intraspecific competition within species-2 *and* if the effect of species-2
against species-1 is weaker than the intraspecific competition within species-
1. Thus for species-1 always to win, it is not enough simply to be a better
competitor, that is, $\alpha_{21} > \alpha_{12}$. Instead, individuals of species-1 must be so
effective that an individual of species-2 incurs stronger interspecific
competition than intraspecific competition. And in addition, individuals of
species-2 must be so ineffective that an individual of species-1 incurs weaker
interspecific than intraspecific competition. Thus the conditions for Figure
21.3a prove to be very meaningful biologically; they lay explicit requirements on the balance between inter- and intraspecific competition for *both*
species.

Figure 21.3b is, in principle, just the same as Figure 21.3a, except that
species-2 always excludes species-1. Therefore, let us move on to Figure
21.3c. Here again, illustrating the point with equal K's, the condition is that
$\alpha_{21} < 1$ and $\alpha_{12} < 1$. This means that the species coexist whenever the
between-individual effect of intraspecific competition is stronger than interspecific competition in both species. If, say, species-1 is a better competitor
than species-2, they may still both coexist, provided both α_{12} and α_{21} are less
than one. If species-1 is a better competitor, it will, of course, be more
abundant at equilibrium, but it cannot *exclude* species-2 until its effect is
strong enough to exceed the intraspecific competition within species-2.

Figure 21.3d is produced, assuming equal K's, when both α_{21} and α_{12} are
greater than one. This means that either of the species can exclude the other,
depending on the initial condition, when the interspecific competition is
stronger than the intraspecific competition in both species. The attention of
ecologists has been mostly directed to the first three cases, whereas Figure
21.3d has been comparatively neglected. But the situation in Figure 21.3d
may be much more important than previously believed. First, many plants
release chemicals into the soil that inhibit the growth of their competitors.
These chemicals are called alleleopathic substances, as reviewed by
Whittaker and Feeny (1971). These chemicals may result in interspecific
competitive effects that are stronger than intraspecific effects. Second, this
case may be a prototype for what are called alternative stable communities.
In the intertidal, for example, a patch of substrate is often eventually
dominated by the species of organism, such as mussel or barnacle, which

Evolutionary Ecology of Interacting Populations

colonizes first in sufficient abundance, as described by Sutherland (1974). It is a matter of chance whether one or another species happens to arrive at a newly opened patch of substrate first. But whichever species does arrive first is likely to exclude the second arrival. In this sense, alternative stable communities are possible, and Figure 21.3d is the simplest population dynamic model that produces this result. We shall return to this idea several paragraphs later, when discussing zonation.

Suppose That $K_1 \neq K_2$ The discussion above shows that the outcome of competition depends on the balance of intraspecific to interspecific competition in both species. However, for purposes of illustration, we assumed equal carrying capacities. When the carrying capacities are not equal, the outcome is still determined by the balance of intraspecific with interspecific competition. The importance of any numerical value of this balance, however, is determined by the ratio of the carrying capacities. When the K's are equal, then Figure 21.3a requires that $\alpha_{21} > 1$ and $\alpha_{12} < 1$. But if the K's are not equal, then for example, both α_{21} and α_{12} might be less than one. If K_2 is sufficiently lower than K_1, then species-1 will exclude species-2. Thus a species cannot tolerate a nearly equal balance of inter- and intraspecific competition if its carrying capacity is sufficiently low relative to the other species. Thus, in general, the outcome of competition between two species depends jointly on the inter–intra specific competition balance in both species and on the ratio of the carrying capacities.

With Two Species the r's Do Not Influence the Final Outcome One of the most immediate predictions of the LV equations (for two species) is that the qualitative outcome of competition is determined only by the α's and K's; and r's do not enter into the conditions for any of the cases in Figure 21.3. Moreover, if the α's and K's are such that coexistence should occur, then the equilibrium abundances also depend only on the α's and K's and do not depend on the r's. Hence the LV equations predict that observations during the initial phase of population growth involving two competitors are irrelevant to anticipating the ultimate outcome.

Now that we have developed a model for interspecific competition, we can begin to use it to predict the results of competition between populations in the laboratory and in nature. We shall discuss three kinds of applications in the following paragraphs. First, are some results concerning laboratory experiments with competing populations, second, results concerning what is called "resource partitioning" between competitors, and third, we shall return to the barnacle example and obtain some results on zonation.

Laboratory Experiments The application of the LV equations to experiments in the laboratory with competing populations is illustrated in Figure 21.5 (see pages 422–426). The figure captions summarize some of the classic work of Gause (1934) and the more recent work of Vandermeer (1969).

The Concept of Limiting Similarity The next kind of application involves the conditions for coexistence. We would like to know to what extent two species can use the same resources and still coexist. Suppose the mechanism of competition is purely exploitative. If so, then the more that two species overlap in the resources which they exploit, then the higher the interspecific competition. We may phrase this

[*Text continued on page 426.*]

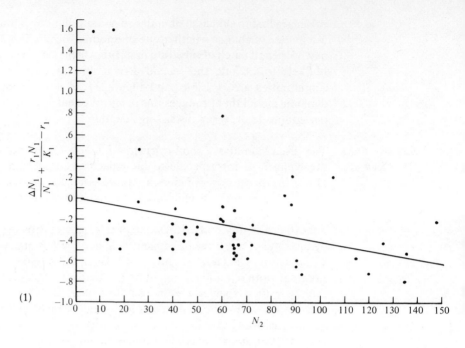

(1)

FIGURE 21.5a. *Example of three competition experiments involving two species of* Paramecium. *The values of r_i and K_i were estimated from experiments in which each species was growing alone, as previously illustrated in Figure 16.4a. The competition experiments illustrated here provided the additional information needed to estimate the competition coefficients. For species-1* (P. aurelia), *the quantity*

$$\frac{\Delta N_1}{N_1} + \frac{r_1 N_1}{K_1} - r_1$$

was regressed against N_2. The intercept of the regression is zero by convention. The slope of the regression line is the quantity $-\alpha_{12} r_1 / K_1$. Once the slope has been determined, α_{12} can be computed, using the known values of r_1 and K_1. The corresponding procedure was also used with species-2 (P. caudatum) *to determine α_{21}. Based on the estimates of r_i, K_i, and α_{ij} obtained in this way, the growth of the two populations in competition was projected using the Lotka–Volterra competition equations. The result appears at the bottom of the figure. The dots and crosses are the experimental points for species-1 and species-2, respectively. Note that the projected curves agree quite well with the data, although there is a lot of variability in the data.* (1) $r_1 = .99$, $K_1 = 552$, *and* $\alpha_{12} = 2.31$; (2) $r = .88$, $K_2 = 205$, $\alpha_{21} = .40$. [*Data from G. F. Gause* (1934), The Struggle for Existence, *reprinted by Hafner Pub. Co., 1964.*]

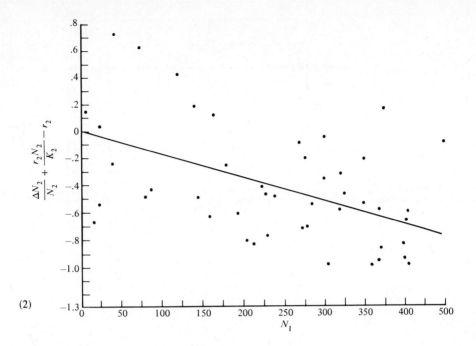

$$(2)$$

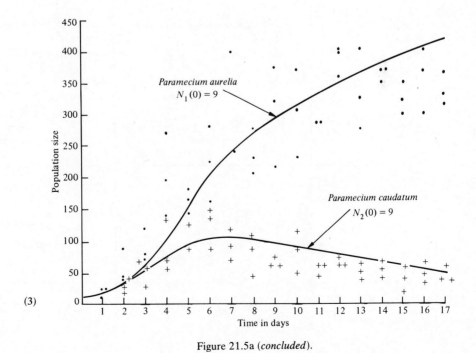

$$(3)$$

Figure 21.5a (*concluded*).

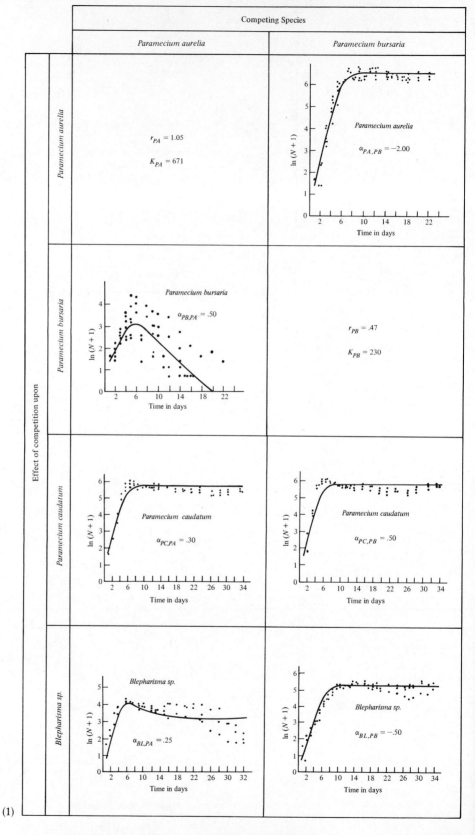

(1)

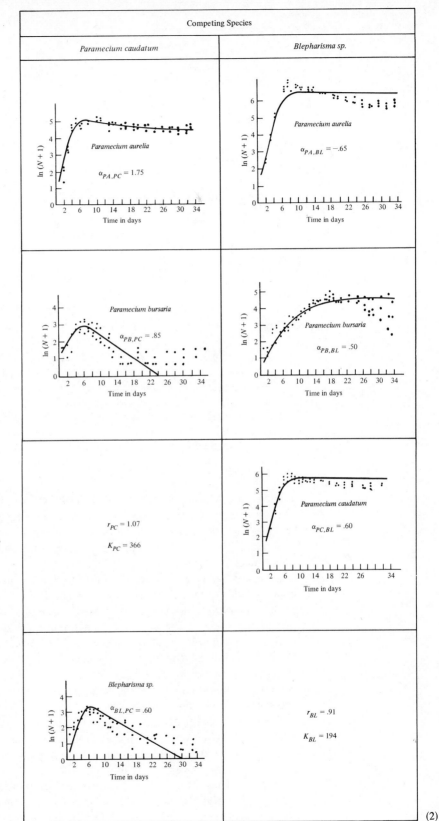

FIGURE 21.5b.
Results of experiments on the growth of mixed cultures of protozoa. Each mixed culture consists of two species; there are four species altogether. Mixed cultures of all possible pairs of these four species appear here. The data for their growth in pure culture were presented in Figure 16.4b. Data from these mixed culture pairs, together with the data from growth in pure culture, were used to calculate all the elements of the matrix of competition coefficients. [Data from J. Vandermeer (1969), The competitive structure of communities: An experimental approach with Protozoa. *Ecology* **50**: 362–371.]

(2)

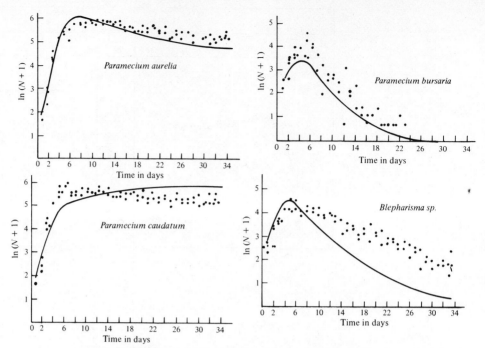

FIGURE 21.5c. *As a test of the Lotka–Volterra equations, J. Vandermeer prepared a mixed culture involving all four species that were growing together. The coefficients, r_i, K_i, and α_{ij}, estimated from the pure cultures, and all the pair-wise mixed cultures were used to predict the outcome of the mixed culture involving all four species. In this figure the solid line is the predicted growth curve using the Lotka–Volterra model for four species. The points are the data.* [Adapted from J. Vandermeer (1979), The competitive structure of communities: An experimental approach with Protozoa. *Ecology* **50**: 362–371.]

idea more precisely by saying that the competition coefficient, α, increases monotonically with the amount of resource overlap. In addition, let us assume that the amount of resource overlap between two species is a function of the phenotypic similarity between the species. Common examples concern the size of a bird's bill and the size and shape of a lizard's jaw. Animals with a large bill or jaw often take larger prey than do those with smaller bills or jaws. Then the degree of similarity between the bills or jaws of the members of two species indicate the extent of overlap in the resource use of the species. Let d denote the phenotypic difference between the species, for example, d can be simply the difference, in millimeters, between the average bill length in the two species. We shall assume, then, that α is a decreasing function of d. We can now use the LV equations to predict the minimum value d can have consistent with coexistence, that is, how similar the species can be and still coexist.

For the sake of illustration, suppose that α_{12} and α_{21} are equal, that is, that each species is influenced to an equal extent by competition from the other. Since α decreases monotonically as a function of the difference between the species, it will look something like that in Figure 21.6. The maximum value of α is one. This obtains when d is zero because if the species are exactly the same in phenotype, then the effect of interspecific competition is the same as that of intraspecific competition. The function $\alpha(d)$ decreases monotoni-

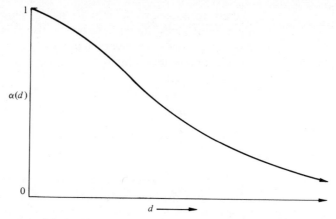

FIGURE 21.6. *Sketch of the competition function, $\alpha(d)$. The vertical axis is the competition coefficient, and the horizontal axis is a measure of the average phenotypic difference between members of different species.*

cally with d, indicating that as overlap decreases, so does competition. If the species coexist, then their abundances are given by the coordinates of the intersection point in Figure 21.3c. These coordinates are (see boxed material)

$$\hat{N}_1 = \frac{K_1 - \alpha(d)K_2}{1 - \alpha(d)^2}$$

$$\hat{N}_2 = \frac{K_2 - \alpha(d)K_1}{1 - \alpha(d)^2}$$

(21.6)

---

1. The intersection point $(\hat{N}_1, \hat{N}_2)$ for the lines in Figure 21.3c must, by definition, satisfy the equations for both lines at the same time. Hence $\hat{N}_1$ and $\hat{N}_2$ satisfy

$$\hat{N}_1 + \alpha_{12}\hat{N}_2 = K_1$$

$$\alpha_{21}\hat{N}_1 + \hat{N}_2 = K_2$$

2. If we solve the bottom equation for $\hat{N}_2$ in terms of $\hat{N}_1$, α_{21} and K_2, we obtain

$$\hat{N}_2 = K_2 - \alpha_{21}\hat{N}_1$$

3. Now we substitute this expression for $\hat{N}_2$ into the top equation, yielding

$$\hat{N}_1 + \alpha_{12}(K_2 - \alpha_{21}\hat{N}_1) = K_1$$

4. Hence

$$\hat{N}_1 + \alpha_{12}K_2 - \alpha_{12}\alpha_{21}\hat{N}_1 = K_1$$

5. Solving for $\hat{N}_1$, we obtain

$$\hat{N}_1 = \frac{K_1 - \alpha_{12}K_2}{1 - \alpha_{12}\alpha_{21}}$$

6. In a similar way, we can solve the top equation for $\hat{N}_1$ and substitute into the bottom equation, yielding

$$\hat{N}_2 = \frac{K_2 - \alpha_{21}K_1}{1 - \alpha_{12}\alpha_{21}}$$

In these formulas, we have written the competition coefficient as $\alpha(d)$ to remind ourselves that α, and hence also $\hat{N}_1$ and $\hat{N}_2$, are functions of the *difference* between the average phenotypes of the species, d. Our answer to the question of how similar the species can be and still coexist is found by finding the smallest d which allows both $\hat{N}_1$, and $\hat{N}_2$ to remain positive.

By convention, let us agree to label the species with the smaller carrying capacity as species-2. Hence, without loss of generality we can assume

$$\frac{K_2}{K_1} \leqslant 1 \tag{2.17}$$

Therefore, as d becomes smaller, the first species to have its abundance hit zero will be species-2. Indeed, the value of d that will make $\hat{N}_2$ equal zero is found from

$$K_2 - \alpha(d)K_1 = 0 \tag{21.8}$$

or

$$\alpha(d) = \frac{K_2}{K_1} \tag{21.9}$$

If we know the functional form of $\alpha(d)$, we can solve explicitly for the value of d, which makes $\hat{N}_2$ equal zero. Otherwise, we can determine the value graphically, as illustrated in Figure 21.7. We shall call the value of d, which makes N_2 equal zero, the *limiting similarity*, and denote it as $\tilde{d}$. We see in Figure 21.7 that $\tilde{d}$ is found where $\alpha(d)$ intersects the K_2/K_1 line.

From Figure 21.7 we can see how the limiting similarity, $\tilde{d}$, between two species is a function of the ratio of their carrying capacities. Clearly, the limiting similarity decreases as K_2/K_1 tends to one. This result means that if two species have nearly the same carrying capacities, they may be very similar and still coexist, while if they have very different carrying capacities, they *cannot* be very similar and still coexist. How similar species *should* be to one another, is an evolutionary question that we will consider in Chapter 23. The theory above, however, predicts a threshold or limit to the possible

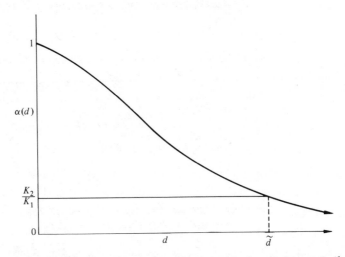

FIGURE 21.7. *Graphical determination of the limiting similarity between two competing species.*

similarity between competitors that is consistent with coexistence. It predicts a lower bound to the similarity that depends on the ratio of the carrying capacities. Remember, however, that we explicitly assumed only exploitative competition. If interference mechanisms are also allowed, then the analysis above must be further refined. We shall return to the idea of limiting similarity in detail in Chapter 24.

Zonation For a third application of the LV equations, let us return to the topic of zonation, with which we began our study of competition. The idea behind zonation is that the carrying capacities and perhaps also the competition coefficients of each species change from place to place. For example, among barnacles, the carrying capacity of *Balanus* drops to zero at the height in the intertidal where that species cannot withstand dessication. Meanwhile, the *Chthamalus* carrying capacity does not reach zero until the very top of the intertidal, because of its dessication tolerance. Thus each position in space is associated with particular values for the carrying capacities. Similarly, the competitive abilities may also differ from place to place.

To understand what should happen at each place due to competition, we should compare α_{12} with K_1/K_2 and α_{21} with K_2/K_1 at each place. Let us denote $K_1(x)$, $K_2(x)$, $\alpha_{12}(x)$ and $\alpha_{21}(x)$ as the values for these parameters at *location x*. In our example, x refers to height in the intertidal. We suppose one population is adapted to locations at the left, the other to the right as diagramed in Figure 21.8. Let us now plot the ratio $K_1(x)/K_2(x)$ and $\alpha_{12}(x)$ with solid lines and the ratio $K_2(x)/K_1(x)$ and $\alpha_{21}(x)$ with dotted lines, as illustrated in Figure 21.9. For the purposes of illustration, we assume that α_{12} and α_{21} do not vary much from place to place and are drawn as more or less straight lines. By comparing the solid curves with each other and dotted curve with each other in Figure 21.9, we conclude that species-1 should exclude species-2 at all places to the left of x_1. Both species coexist between x_1 and x_2, and species-2 should exclude species-1 at all locations to the right of x_2. Thus in this example, the predicted zonation pattern would include a stable zone of overlap between x_1 and x_2.

Figure 21.9 presents only one of the many imaginable examples. Let us consider another. In the barnacle study, *Balanus* individuals exhibited an interference mechanism whereby it pushed aside or grew over *Chthamalus* individuals. If we let species-1 be *Chthamalus* and species-2 be *Balanus*, then α_{12} is greater than one and α_{21} is less than one. If this situation leads to a graph like Figure 21.10, then there is again a zone of stable coexistence

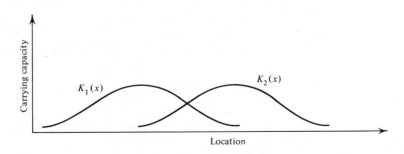

FIGURE 21.8. *Sketch of the carrying capacity as a function of location for two species.*

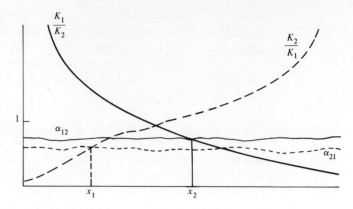

FIGURE 21.9. *Graph of the conditions that determine the outcome of competition as a function of location. According to this figure, there is a stable zone of species overlap between locations x_1 and x_2.*

between x_1 and x_2. However, if α_{12} is large enough, then x_2 can actually lie to the left of x_1 as in Fig. 21.11. In this case, the zone between x_2 and x_1 is a zone where either species can win, depending on initial abundances. For example, in the intertidal, if the interference of *Balanus* against *Chthamalus* is large enough (i.e., α_{12} is large enough), then there should be a zone where either *Balanus* or *Chthamalus* can exclude each other. Within this zone, the species composition should be quite patchy, reflecting chance events in the colonization of empty regions of substrate that become available from time to time. It would be predicted that the patchiness within this zone would contrast with a smooth intergradation between species, as produced in Figures 21.9 and 21.10.

In general terms, many possible schemes of zonation patterns could be produced by allowing multimodal forms for $K(x)$ and also allowing the α's to differ substantially from place to place. More important, the process of dispersal has not been explicitly included and its significance in zonation patterns is unknown. The graphical approach above is exact in the limit of

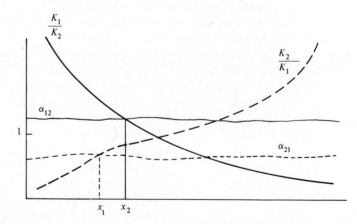

FIGURE 21.10. *Graph of the conditions that determine the outcome of competition as a function of location. According to this figure, there is a stable zone of species overlap between locations x_1 and x_2.*

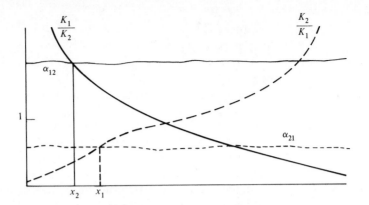

FIGURE 21.11. *Graph of the conditions that determine the outcome of competition as a function of location. According to this figure, there is an unstable zone of species overlap between locations x_2 and x_1.*

zero dispersal between zones, and should be qualitatively correct, provided the dispersal is not too high. The use of the LV equations, especially when extended to include dispersal, in explaining the spatial pattern of populations is essentially unexplored and is a fruitful area for further research.

To conclude this section, I wish to stress that the LV equations apply to competition as mediated through density-dependent mechanisms. But some populations appear to be in environmental situations where density-dependent competitive effects seem small in contrast to known density-independent effects. Also, many populations appear to receive heavy predation pressure so that they rarely, if ever, attain abundances where density-dependent competitive interactions are important. A characteristic of such populations is a high susceptibility to extinction—a population's existence at any place is a precarious affair. In this situation, the dynamics of the entire species can perhaps be viewed as a continuing process of extinction and recolonization of locations within the species range. The LV equations are not really helpful to species with this sort of population dynamics and recently a colonization-extinction model for competition has been developed by Levins and Culver (1971). This model has been further analyzed by Slatkin (1974) and Horn and MacArthur (1972). There has also been great interest in developing competition models that may be more faithful to the details of the density-dependent mechanisms that do occur. For alternatives to the Lotka–Volterra equations, consult Gilpin and Ayala (1973 and 1976), Schoener (1973 and 1976), and Hirsch and Smale (1974, Chapter 12). A competition model in discrete time has been analyzed by Hassell and Comins (1976). Experimental studies which reveal population dynamics that differ quantitatively from the Lotka–Volterra model include Neill (1974) and Smith–Gill and Gill (1978).

Chapter 22
PREDATION

PERHAPS the most dramatic of the population interactions is the predator-prey interaction. Predation conjures up images of lions hunting wildebeest on the plains of Africa and eagles swooping down upon their prey. In addition to being occasionally dramatic, predation is a very common population interaction, for, obviously, all carnivores must prey on something in order to live. Indeed, most predation is carried out by comparatively less dramatic actors—insectivorous birds, lizards and frogs, numerous fishes, myriads of arthropods from mites and spiders to dragonflies, robber flies and wasps, and many marine invertebrates—just to name a few.

There are many questions concerning predation that we would like to answer. To what extent can predation reduce the abundance of prey? Or conversely, if the predators are removed, will the prey population explode or only increase slightly? What happens to the predator-prey balance, if a general insecticide like DDT is added, or at the opposite extreme, if excess fertilizer like phosphate is added? Can the predator-prey interaction lead to population oscillations, and if so, what determines the frequency and amplitude of the oscillations? In this chapter, we shall focus on these and other questions.

The Basic Format of Predator-Prey Models

The theory of predator-prey interactions cannot rely on one model to nearly the extent possible with the theory of competition. Instead, we shall consider a series of models based on a common format. Also, for most of the section, we shall discuss a system with one predator and one prey; the extension to multiple predator and prey species is largely unknown.

Let V (for victim) denote the size of the prey population and P the size of the predator population. Then, the common format for a two-component predator-prey model is

$$\frac{dV}{dt} = \begin{bmatrix} \text{growth rate of prey} \\ \text{population in absence} \\ \text{of predator} \end{bmatrix} - \begin{bmatrix} \text{capture rate} \\ \text{of prey per} \\ \text{predator} \end{bmatrix} P$$

$$\frac{dP}{dt} = \begin{bmatrix} \text{rate at which each predator} \\ \text{converts captured prey} \\ \text{into predator births} \end{bmatrix} P - \begin{bmatrix} \text{rate that predators} \\ \text{die of in absence} \\ \text{of prey} \end{bmatrix} \qquad (22.1)$$

The first term in the equation for dV/dt describes the growth of the prey population in the absence of the predator and usually represents either exponential or logistic growth.

The Functional Response

The second term in dV/dt couples the prey population to the predators. The expression in brackets representing the capture rate of prey per predator is called the *functional response* of a predator. It describes how a predator functions in its capturing of prey. The functional response of a predator summarizes a great deal of biological information about the predator.

First, the functional response of a predator depends on the prey abundance, V, as illustrated in Figure 22.1. The simplest assumption is that the rate of prey capture by a predator increases linearly with the abundance of prey. This assumption is unrealistic, however, in the limit where the prey are very abundant relative to a predator's needs. It is more realistic to assume that the rate asymptotes at some maximum value. This maximum is determined by the minimum time needed to catch and handle a prey item and by the satiation level of a predator. Another form of functional response is a sigmoid curve, illustrated in Figure 22.1. This curve refers to a special class of predators whose hunting efficiency for a given type of prey depends on how often the prey are encountered. For example, capturing efficiency may improve with experience.

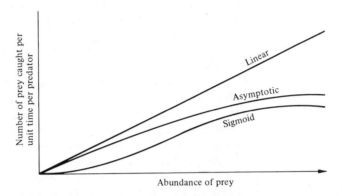

FIGURE 22.1. *Sketch of types of functional response by predators to prey abundance. The vertical axis is the number of prey caught per unit time by an individual predator. The horizontal axis is the prey abundance.*

Second, the functional response of a predator may depend on the predator abundance, P. If there is aggression between predators, then the rate of capturing prey per predator decreases with P. If predators hunt in packs, then the per capita rate of capture increases with P when P is small. However, if there are no behavioral interactions of these sorts between predators, then the functional response is independent of P.† Also, the functional response may depend on the history of the system in a special way. Some predators have what is called *indeterminate growth*, such that an animal's size is determined by the amount of prey it has captured during its life. Thus two animals of the same age could have very different sizes, depending on the prey abundance experienced since birth. Moreover, the rate of prey capture per predator is often directly related to size. As a result, the functional response of the predators can depend on the regime of prey abundances they have experienced since birth, thus making the functional response at any time dependent on the recent history of the system. This complication may be quite important in marine systems where so many organisms, including especially fish, mollusks and echinoderms, show indeterminate growth. Nonetheless, history-dependent functional responses are essentially unexplored, theoretically. In the rest of this chapter, we shall assume that the functional response depends only on V.

† In particular, if the predators only have intraspecific exploitative competition, then the functional response is independent of P. Exploitative competition is mediated by lowering the abundance of prey and is accounted for by the dependence of the functional response on V.

The first term in dP/dt couples the predators to the prey. The expression in brackets, representing the rate at which an individual predator converts captured prey into reproductive output, is called the *numerical response* of a predator. It describes how the numbers of new predators relate to the captured prey.† Less is known about the numerical response than the functional response, because determining the numerical response requires observing the reproductive patterns of predators under different regimes of prey availability. A common and often reasonable assumption is that a constant fraction of the energy in the captured prey is allocated to the production of eggs. If so, the rate of production of new predators by an individual predator is simply a constant times the rate at which it is capturing prey; that is, the numerical response is a constant times the functional response. We shall use this assumption throughout this chapter.

The second term in dP/dt describes the growth of the predator population in the absence of prey. If the predators were not solely dependent on one species of prey, then a logistic term might be used here. But if the predators are dependent on solely one prey species, then this term must indicate a death rate. The second term in dP/dt is usually assumed to indicate an exponential decay of the predator population in the absence of prey. This assumption means that there is a constant death rate among the predators. But if the predators fight with one another during or after the capture of prey, then the death rate increases with P; and if the death rate depends on the nutritional state of a predator, then it may also depend on the prey abundance, V. Thus again, many assumptions are possible and the simplest is that of a constant death rate among predators.

As the paragraphs above demonstrate, many models can be proposed for the predator-prey interaction. We shall consider several of the possible models. We shall see that certain predictions follow from all of them and other predictions follow from several of them. When we are done, we shall see a common pattern that we could not have discerned from any one model alone. Note also that we shall consider models defined in continuous time. Predator-prey models defined in discrete time are also important, particularly where both predator and prey are seasonal arthropods. A key reference to models for insect parasite-prey populations dynamics is Hassell and May (1973).

The Volterra Model

The simplest imaginable model for the predator-prey system, according to the above format, is to assume (1) exponential growth of the prey in the absence of the predator, (2) a linear functional response for the predators, (3) a numerical response for the predators that is a constant times the linear functional response, and (4) a constant death rate among the predators. With these assumptions we have

$$\frac{dV}{dt} = rV - (aV)P$$

$$\frac{dP}{dt} = b(aV)P - dP$$

(22.2)

† The numerical response is thus a function of the functional response. Thus if $F(V, P)$ is the function response, and B is the numerical response, then $B = B[F(V, P)]$.

These equations are called the *Volterra predator-prey equations*. The variable, r, is the intrinsic rate of increase of the prey; (aV) is the functional response and a is the slope of the predator's functional response curve; b is the constant that relates the numerical response to the functional response; and d is the death rate of the predators. Even though this model is the simplest one imaginable, we see that it has certain features that are present in all the more realistic models.

The Oscillatory Tendency To conduct preliminary analysis of the Volterra equations, we shall consider a graphical technique similar to that used with the competition equations. First, set dV/dt equal to zero. Then the V's divide out, leaving

$$\hat{P} = \frac{r}{a} \tag{22.3}$$

This value of P is plotted as a horizontal line in Figure 22.2a. If there are more predators than $\hat{P} = r/a$, then the prey population declines. If there are fewer predators than $\hat{P} = r/a$, then the prey population increases. These trends are illustrated with arrows in Figure 22.2a. The important point here is that *whether the prey increase or decrease is determined only by the number of predators*. Now let dP/dt equal zero. Then the P's divide out leaving

$$\hat{V} = \frac{d}{ab} \tag{22.4}$$

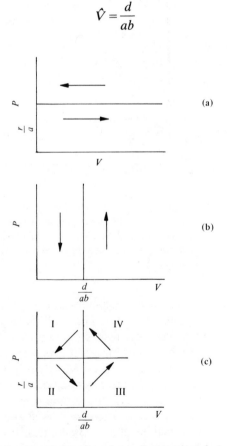

FIGURE 22.2. *Graphs used in the preliminary analysis of the Volterra predator-prey equations.*

This value of V is plotted as a vertical line in Figure 22.2b. If there are fewer prey than $\hat{V} = d/(ab)$ then the predators decrease and if there are more prey than $\hat{V} = d/(ab)$, then the predators increase. These trends are illustrated with arrows in Figure 22.2b. The important point now is that *whether the predators increase or decrease is determined only by the number of prey*. Since the predators, alone, control the prey and the prey, alone, control the predators, we might expect an oscillation to result. And indeed, if we combine both Figures 22.2a and b into Figure 22.2c, and combine the arrows as though they were vectors, then a circular pattern is suggested. The arrows lead from region I into II, from II into III, from III into IV, and finally from IV back into I. This pattern suggests that the predator and prey population sizes oscillate with the prey "leading" the predators.

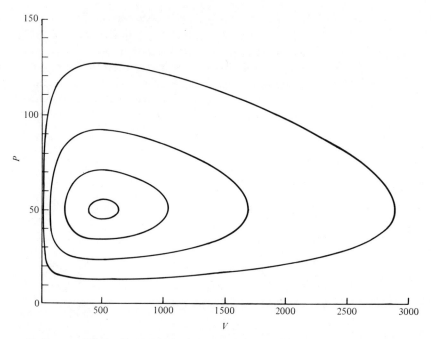

FIGURE 22.3. *Trajectories of the Volterra equations. The trajectories are closed orbits indicating sustained predator-prey cycles. The amplitude of the cycles depend on the initial conditions. The system moves counterclockwise around the orbits. Here $r = .5$, $a = .01$, $b = .02$, and $d = .1$.*

One can imagine, however, that the oscillations are not sustained indefinitely but "damp out" through time. Sustained oscillations require a sequence of arrows from regions I, II, III, IV, and back to I, which *exactly* rejoin the starting point. But if the arrows leading back into region I do not rejoin the starting point, then a spiral results. And if the arrows eventually lead into the intersection point on the graph, then a damped oscillation is indicated. When the system is at the intersection point, the oscillation has ceased because then *both dP/dt and dV/dt* are zero. However, although we can imagine these possibilities, it can be shown that the oscillations *are* sustained in the Volterra equations above and that the arrows *do exactly* rejoin any starting point. Trajectories are illustrated in Figure 22.3.

The Classic Lynx Oscillation
The Volterra equations were for a long time of interest as a possible model for some natural populations that show oscillations. One of the most intensely studied examples of population oscillation involve the lynx of Canada. Elton and Nicholson (1942) examined the fur trapping records of the Hudson Bay Fur Trading Company and compiled the data illustrated in Figure 22.4. The oscillations are more regular in some regions of Canada than in others, but generally speaking, there are seven years separating the peaks. The possible utility of the Volterra equations in explaining this oscillation seemed enhanced by the fact that lynx are known to prey almost exclusively on a hare, *Lepis americanus*, which oscillates in synchrony with the lynx. Indeed the lynx' dependence on this prey is so complete that according to Seton (1911), as quoted in Elton and Nicholson (1942); "It lives on Rabbits, follows the Rabbits, thinks Rabbits, tastes like Rabbit, increases with them, and in their failure dies of starvation in the unrabbited woods." Moreover, the lynx oscillations cause its own predators also to oscillate and, all in all, provide a complex system of interacting oscillating populations. See Butler (1953) and Keith (1963) for information on the coupled oscillations.

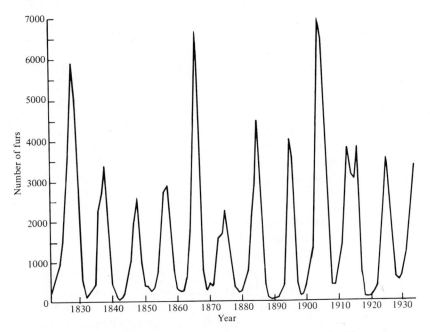

FIGURE 22.4. *Oscillations of a lynx population in Canada. The vertical axis represents the number of furs that were collected; the horizontal axis depicts time.* [Data from C. Elton and M. Nicholson (1942), The ten-year cycle in numbers of the lynx in Canada. *J. Anim. Ecol.* **11**: 215–244.]

Criticisms of the Volterra Model
The Volterra equations, alone, are quite inadequate, however, as a model for natural predator-prey oscillations, although they are the first step in what appears to be the right direction. First, the amplitudes of the oscillations in the Volterra model are completely arbitrary—they are determined only by the initial conditions and not by any of the parameters in the model. (The frequency, however, *is* set by the parameters in the model.) Since the

amplitude is determined only by the initial conditions, the oscillations are said to be *neutrally stable*. This means that any perturbation of the amplitude results in a new cycle beginning at the point to which the perturbation moved the system. As illustrated in Figure 22.3, if the predator and prey abundances are changed to new values, then the cycling continues from this point, just as if it has started there. Hence in a Volterra model, accidents of history alone explain the amplitude of the oscillations, although biological considerations are responsible for the frequency of the oscillations. Moreover, the practical result of neutral stability is the eventual extinction of the populations involved. Given sufficient time, it is certain that a sequence of perturbations to the system will produce a cycle that goes so close to the axes that population extinction results. Thus the Volterra model must be a very improbable explanation of any sustained predator-prey oscillation.

A second inadequacy of the Volterra model is that it is *structurally unstable*. This means that slight alterations of the model, itself, change the predictions in important respects. Specifically, as we shall see, the inclusion of density dependence and more realistic predator-prey functional response curves does alter the model in important ways. Third, the Volterra model, alone, must be inadequate because it predicts only one result—that predator-prey systems oscillate. But, obviously, many predator-prey systems are not in continual oscillation, as illustrated by the lynx, and we would hope that a more realistic theory would predict other kinds of outcomes, depending on the values of the biological parameters.

Factors That Influence the Equilibrium Abundance of Predator and Prey

Before proceeding with more advanced predator-prey models, however, let us examine the Volterra model further to obtain some results that will also emerge in other models. Recall that the population sizes of the predator and prey at the intersection of the lines in Figure 22.2c are

$$\hat{P} = \frac{r}{a}$$

$$\hat{V} = \frac{d}{ab}$$

(22.5)

If both predator and prey have these values, then they remain constant through time. Also, if the oscillations happen to have a small amplitude, then these values roughly equal the population sizes averaged over a cycle. So, for the sake of argument, let us assume (22.5) predicts, approximately, the average predator and prey population sizes. We notice immediately that the predator abundance does not depend on its own death rate, d, and that the prey abundance does not depend on its own growth rate, r. Instead the number of predators is controlled by the prey's growth rate, and the number of prey by the predator's death rate. This property reappears in later models too. It often turns out that the predator abundance is controlled more by the growth rate of the prey than by anything else, and that the prey abundance is controlled more by the predator death rate than by anything else. We notice, next, that the parameter measuring the feeding rate, a, in the functional response is in the denominator of both $\hat{P}$ and $\hat{V}$. Hence the lower the feeding rate of the predators, the higher is the prey abundance *and* the predator abundance too. It seems intuitive that prey abundance should increase as the predator feeding rate becomes lower, but it is surprising, at first, that the predator population size should also increase as each predator's feeding rate decreases. But at second thought, of course, a lower feeding rate per predator means more predators are necessary to effect a given predation rate on the prey population.† Finally, the constant b, which indicates the relationship between the rate of food captured by a predator (the functional response) and the rate of birth of new predators (the numerical response), appears in the denominator for $\hat{V}$. A large b can be caused by several factors. It may indicate a large allocation of captured food into reproductive output by the predator, that predators are small in size relative to prey so that few prey are required to synthesize a predator, or that the prey are calorically or nutritionally well suited to biochemical conversion, and so forth. Whatever the reason, the rapid production of new predators from captured prey (high b) reduces the prey abundance but leaves the predator abundance unaffected. We shall want to compare these equilibrium solutions for $\hat{P}$ and $\hat{V}$ with those later on.

The Volterra
Principle
and General
Pesticides

The equilibrium solutions for $\hat{P}$ and $\hat{V}$ above can be used to suggest a surprising prediction about the effect of general insecticides. Usually, the organisms regarded as pests are plant feeders like aphids, caterpillars, weevils, beetles, and so forth. These pests are usually the prey of insect predators like wasps and predatory beetles as well as of avian predators. A general insecticide is a chemical that is deleterious to all insects, regardless of whether they are herbivores or predators. The application of a general insecticide will increase the death rate of predators and reduce the intrinsic rate of increase of their prey. Now, if the predator and prey are in balance, then by (22.5), the *prey* will *increase* because of increasing d, the predator death rate, while the *predators* will *decrease* because of decreasing r, the intrinsic rate of increase of the prey. Thus the result of the general insecticide will be an increase in the abundance of the pests, themselves, whereas the number of predators on the pests will decrease. This effect is called the *Volterra principle*. It shows just one of the several reasons why the result of the insecticide may often be exactly the opposite of that desired by those who applied the chemical. The Volterra principle arises in any model in which the abundance of predators is controlled mostly by the growth rate of the prey and the abundance of prey by the death rate of the predators. Although first derived in the Volterra model, it is found in many predator-prey models.

The Volterra principle understates the actual harm that insecticides do. In practice, chemicals like DDT accumulate in the higher trophic levels and cause more direct damage to the predators than to the prey. Hence the effect is not that of a *general* insecticide but the effect of one that is *specific* for the predators on the pests. In addition, the generation time of the prey is usually faster than that of the predator and so the pests evolve insecticide resistance faster than their predators. It is not surprising, then, that truly massive doses of insecticide are usually required for pest "control." Thus, in the use of insecticides, it is crucial to assess whether the pests are under insect predator control to begin with. If so, the insecticide will bring results opposite to those desired.

---

† We observe from (22.3) that the prey can support any number of predators up until $(r - aP)$ equals zero. If a is low, then a large P is required to make this expression equal zero.

To summarize our results with the Volterra model, we have seen that there is an inherent tendency in the predator-prey interaction for oscillations. The Volterra model always leads to oscillations the amplitudes of which are arbitrary but the frequency of which depends on the parameters of the model. The approximate average abundance of the predator is controlled by the intrinsic rate of increase of the prey; and the abundance of the prey, by the death rate of the predator. These formulas for the approximate average abundances lead to the Volterra principle concerning general insecticides.

A Model with Prey Density Dependence   Now let us consider some more realistic models of the predator-prey system. Assume that the prey grow logistically, instead of exponentially. Then, upon keeping all the other assumptions of the Volterra equation, we have

$$\frac{dV}{dt} = r\left(1 - \frac{V}{K}\right)V - aVP$$

$$\frac{dP}{dt} = abVP - dP$$

(22.6)

Even though all that has been added is prey density dependence, this model has many new features. The equilibrium predator-prey abundances are

$$\hat{V} = \frac{d}{ab}$$

$$\hat{P} = \frac{r}{a}\left(1 - \frac{\hat{V}}{K}\right) = \frac{r}{a}\left(1 - \frac{d}{abK}\right) \qquad (\hat{V} < K)$$

(22.7)

The equilibrium prey abundance is identical to that in the Volterra model. The equilibrium predator abundance includes a new factor $(1 - \hat{V}/K)$. Thus the carrying capacity of the prey influences the predator's abundance, not the prey's. On the whole, the prey abundance is still set by properties of the predator, and the predator's abundance by properties of the prey. The prey's carrying capacity influences the predator abundance because it affects the renewability of the prey as a resource to the predator. If $\hat{V}$ is near K, then the prey population is undergoing slow turnover because of predation and captured prey are only slowly replaced. Accordingly, when $\hat{V}$ is near K, only a small predator population size is sustained. The factor of $(1 - \hat{V}/K)$ in the expression for $\hat{P}$ implies that the predator cannot coexist with the prey unless $abK > d$. In contrast, in the Volterra model, the predator and prey always coexist.

Recall that the Volterra principle concerns an insecticide that increases the death rate of the predator and lowers the intrinsic rate of increase of the prey. In this model, as before, such a chemical leads to an increase in $\hat{V}$ and a drop in $\hat{P}$. Thus the Volterra principle is also found when prey density-dependence is incorporated.

A Prudent Predator   The formula for $\hat{P}$ shows a complex dependence on the feeding rate a. As a is lowered, $\hat{P}$ can increase, as before, because more individuals can be sustained on a given level of predation pressure. But also, as a is lowered, $\hat{V}$ approaches K, thereby diminishing the renewability of the prey as a resource. Because of these counteracting effects, there is an optimum feeding rate, a_0, which maximizes the predator population size. Differentiating $\hat{P}$ with respect to a and setting equal to zero gives

$$a_0 = \frac{2d}{bK} \tag{22.8}$$

With the logistic equation the growth rate of the entire population is highest when $N = K/2$. (See the boxed material.) We might expect, then, that the optimum feeding rate, a_0, is that which causes $\hat{V}$ to equal $K/2$. And, indeed, upon substituting a_0 into $\hat{V}$, we obtain

$$\hat{V}_0 = \frac{K}{2} \tag{22.9}$$

To find the population size that yields the highest population growth rate based on the logistic equation, we first differentiate (dN/dt) with respect to N,

$$\frac{\partial}{\partial N}(dN/dt) = \frac{\partial}{\partial N}\left[\frac{rN(K-N)}{K}\right] = r - \frac{2rN}{K}$$

Then we set the derivitive equal to zero and solve for N, yielding $N = K/2$.

Slobodkin (1961) termed a predator with the optimum feeding rate, a_0, a *prudent predator*. The prudent predator does not deplete its resources by overexploitation, but instead, preys at the rate that provides the maximum sustainable yield. We shall consider in the next chapter whether natural selection can cause the evolution of prudent predators.

Coexistence Is the Only Outcome Provided That $\bar{V} < K$

As just mentioned, one new feature which arises with prey density dependence is the existence of an optimum feeding rate, a_0. Another new feature concerns the stability of the equilibrium. The equilibrium, $\hat{V}$ and $\hat{P}$ given by (22.7), is always stable. In this model, we never have sustained oscillations. It is possible, however, to have damped oscillations during the approach to equilibrium. Whether there are such damped oscillations or not depends on the relationship between K and the other parameters of the model. If K is lower than a certain value, then there are no oscillations whatever in the approach to equilibrium, while if K is larger than this certain value, then a damped oscillation occurs. Examples are illustrated in Figure 22.5. Increasing K is said to have a destabilizing effect on the system. Increasing K reduces the effective amount of density dependence in the system and "allows" the essential oscillatory character of the predator-prey interaction to appear.

The fact that even the slightest density dependence, that is, any finite value of K, prevents sustained oscillations, illustrates the idea that the simple Volterra model is not structurally stable. Even the slightest density dependence destroys the sustained oscillations and causes a stable equilibrium to be attained.

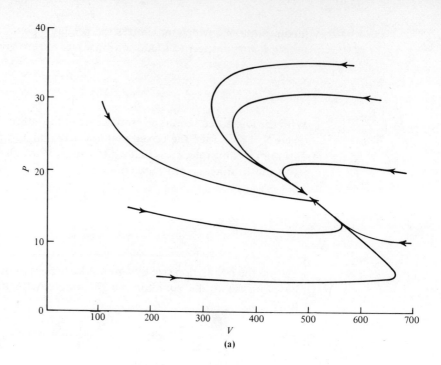

(a)

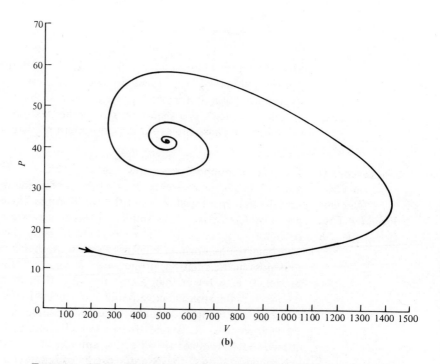

(b)

FIGURE 22.5. *Trajectories of the predator-prey model with logistic prey growth and a linear functional response.* (a) *If the carrying capacity of the prey is sufficiently low, an equilibrium is approached without a spiral, whereas* (b) *if the carrying capacity is high enough, the equilibrium is approached by trajectories that spiral around the equilibrium point. In any event an equilibrium is always approached. In* (a), *r* = .5, *a* = .01, *b* = .02, *d* = .1, *K* = 750; *in* (b), *r* = .5, 1 = .01, *b* = .02, *d* = .1, *K* = 3000.

Although introducing prey density dependence has led to some new insight, it appears that we still lack an adequate predator-prey model. Using density dependence, we could conceivably account for the predator-prey systems which do not oscillate. But what about those systems that do oscillate? Could a still more realistic model conceivably account both for oscillations and stable coexistence?

Suppose now that we add another element of realism, namely, the phenomenon of predator satiation when the prey are very abundant. Predator satiation is described in Figure 22.6 as a curve that asymptotes at a maximum feeding rate per predator. A typical functional response curve that incorporates satiation is

$$c(1-e^{-aV/c}) \tag{22.10}$$

where c is the maximum rate of prey capture per predator. The parameter, a, describes how easily a predator satiates with prey, as illustrated in Figure 22.6. Using this functional response curve and still assuming logistic prey growth, we have

$$\frac{dV}{dt} = r\left(1-\frac{V}{K}\right)V - c(1-e^{-aV/c})P$$

$$\frac{dP}{dt} = bc(1-e^{-aV/c})P - dP \tag{22.11}$$

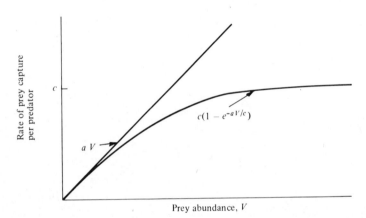

FIGURE 22.6. *Sketch of a functional response curve that incorporates predator satiation.*

A Limit Cycle The analysis of this model leads to several results. All the predictions of the last model (22.6) are among the possible results of this model. The equilibrium predator population size is maximized when the prey abundance equals $K/2$. Also the Volterra principle obtains. If K is low enough, the equilibrium is stable and there is no damped oscillation in the approach to equilibrium. If K is somewhat larger, there is a damped oscillation during the approach to equilibrium. There is, in addition, a new feature in this model. If K is made larger than a certain value, then a special kind of sustained oscillation called a *limit cycle* occurs as discussed in May (1972) and Kolmogorov (1936). When the limit cycle exists, the equilibrium *point* itself is unstable and the system enters a permanent oscillatory mode. Figure 22.7 illustrates all these possibilities.

A limit cycle is an oscillation whose frequency *and* amplitude are determined by the parameters of the model. The limit cycle that arises from (22.11) is not dependent on the initial condition of the system. Provided the parameters have values that cause a limit cycle to exist, then the system

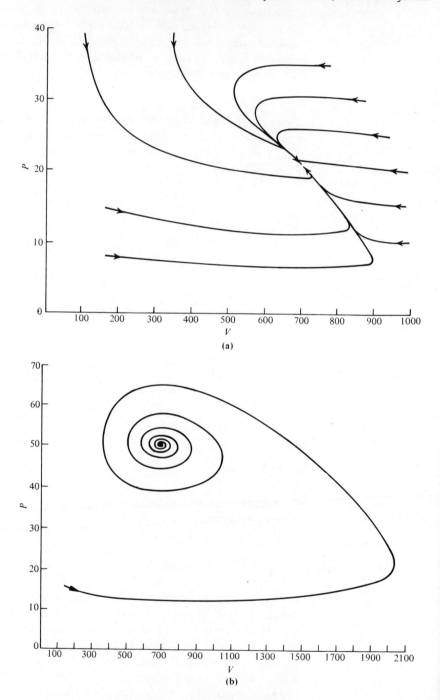

(a)

(b)

Figure 22.7. (*Continued on the next page.*)

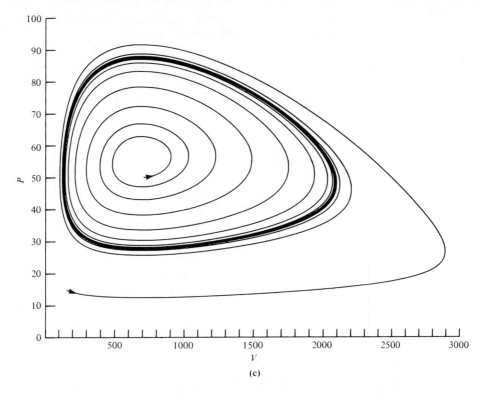

FIGURE 22.7. *Trajectories of the predator-prey model with logistic prey growth and a functional response that incorporates predator satiation. (a) Trajectories with low carrying capacity for the prey (K = 1000). (b) Trajectories for moderate prey carrying capacity (K = 2500). (c) With a high prey carrying capacity, the trajectories approach a stable limit cycle (K = 3500). A stable limit cycle is a closed orbit that other nearby trajectories approach. The amplitude and frequency of the oscillation are independent of the initial condition and are determined by the values of the biological parameters in the model. In all three figures, r = .5, a = .01, b = .02, d = .1, and c = 10.*

converges to the limit cycle from any initial condition. Also, if the system is perturbed away from the limit cycle, then the new trajectory returns to the limit cycle. In this sense, the oscillation itself is *stable* to perturbation. You should contrast the idea of a limit cycle with the neutrally stable oscillations in the Volterra model, because the difference between a limit cycle and neutrally stable oscillations is fundamental.

The Biology of the Limit Cycle   From a biological point of view, the limit cycle oscillation occurs because predator satiation is a destabilizing factor. If the prey abundance is increasing fast enough, predation cannot check the increase, because the predators are satiated. So the prey escape from predator control and begin to approach their carrying capacity. Meanwhile, the predator population grows because some of the prey that have been captured are being steadily converted into new predators. Then, as the predators increase, the prey become predator-regulated again and their abundance drops. This, in turn, causes the predator abundance to drop and the cycle repeats itself. However, if K is low, then the stabilizing effect of prey density dependence prevents the prey from ever increasing fast enough to escape predator control. Hence the oscillation can only occur if K is sufficiently large.

The Mathematics of the Limit Cycle   A limit cycle is a result unique to nonlinear differential equations; it does not arise in linear systems.† One of the principal theorems in the theory of nonlinear differential equations is called the Poincaré–Bendixon theorem. It concerns the fate of a trajectory in the graph of P versus V. It says that *if a trajectory enters, and thereafter remains, within some finite region of the P-V plane, and if the trajectory does NOT move to an equilibrium POINT in this region, then the trajectory converges to a limit cycle.* This theorem is important because it means that if a trajectory permanently enters a finite region of the P-V plane, then it does not simply wander around "aimlessly." In order to use this theorem to show that there is a limit cycle in the model with predator satiation, we must *first* know that there is a finite region of the P-V plane which, once entered, is never left by any trajectory. This fact should seem intuitive, biologically, because (1) neither population reduces the other to zero abundance, so that no trajectory crosses either the P or V axis, ‡ and (2) neither population can explode to infinite abundance because the prey become resource limited if their abundance is high enough. An infinite number of prey is a precondition for an infinite number of predators. Therefore, we can delineate a finite region of the P-V plane from which no trajectory leaves. Furthermore, we know there are trajectories that begin outside the region, which do enter it. *Second*, we must show that there are no stable equilibria within the region. There is one equilibrium point for Equations (22.11); *but*, for K large enough, it is unstable. Therefore, when the equilibrium *point* is unstable, the trajectories must tend to the limit cycle. Such analysis of predator-prey models with limit cycles traces back to Kolmogorov (1936). See also a summary in Minorsky (1962), p. 69.

The limit cycle never exists at the same time as the stable equilibrium. The precise value of K that is the condition for the point to be unstable is also the condition for the limit cycle to exist—a stable equilibrium and stable limit cycle never "coexist" in this model. A diagrammatic summary is shown in Figure 22.8. Thus when K equals K_0, both the equilibrium point becomes unstable and a stable limit cycle appears. In mathematics, this property of a model is called a *Hopf bifurcation* and K_0 is the bifurcation point. This

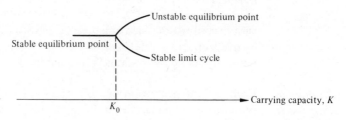

FIGURE 22.8. *The predator-prey system bifurcates at $K = K_0$.*

† Technically, we refer to models where the parameters do not vary in time. Of course, a cycle could occur with a periodic forcing function in a linear system.

‡ In the limit of small V, $c(1 - e^{-aV/c}) \to aV$, so that the model reduces to the Volterra model. As you know, the trajectories in the Volterra model do not cross the axes. The property that neither species drives the other to extinction does not require special assumptions about prey refuges, and so forth.

model, with prey density dependence and predator satiation, is an entrée to the theory of nonlinear differential equations. Unfortunately, however, the mathematics of nonlinear differential equations is comparatively recent and rather limited from a practical point of view. However, it is known that there is no generalization of the Poincaré–Bendixson theorem to more than two equations. Biological models involving more than one predator and prey may easily pass into mathematical *terra incognita*. Good introductions to the contemporary theory of systems of nonlinear differential equations are Arnold (1973) and Hirsch and Smale (1974). The Hopf bifurcation is discussed very thoroughly in Marsden and McCracken (1976).

The Paradox of Enrichment The importance of the prey carrying capacity, K, in controlling the stability properties of predator-prey models has been stressed by Rosenzweig (1972) in what he calls the *paradox of enrichment*. If the environment is enriched, for example, by adding fertilizer, then the carrying capacity of the prey is increased. If the prey are a population of fresh water algae, for example, then adding phosphates to the pond will, likely, increase the algae's K. The result of such an enrichment will not, however, be an improvement to the inhabitants of the pond. Increasing the K *destabilizes* the *point* of coexistence and, if K is increased enough, produce sustained oscillations. Moreover, in practice, these mathematically sustained oscillations often lead to extinction of a population while it is at its low point in the cycle. The paradox is that enrichment of this system is often deleterious to the members of the system. The paradox of enrichment is still another example of the nonobvious consequences of altering a predator-prey balance.

The Models Compared The properties of the predator-prey models considered so far are summarized in Table 22.1. It is clear that several conclusions appear fairly general. First, in all the equilibrium solutions, the number of predators increases with the intrinsic rate of increase of the prey and the number of prey increases with the death rate of the predators. Hence a factor reducing the prey's r and increasing the predator's d will increase the absolute number of prey; that is, the Volterra principle obtains throughout. Second, in all models except the Volterra model, there is a predation rate which maximizes the predator population size. Any other predation rate either over- or underexploits the prey. Third, there is an inherent tendency for oscillation in the predator-prey system. Fourth, this tendency is opposed by the presence

Table 22.1.

Name	Equation Number	Prey Growth	Functional Response	Volterra Principle	Existence of Optimum Predation Rate	Neutrally Stable Cycles	Stable Equil. Point	Limit Cycle
Volterra model	(22.2)	Exponential	Linear	Yes	No	Yes	No	No
Logistic prey	(22.6)	Logistic	Linear	Yes	Yes	No	Yes	No
Logistic prey and predator satiation	(22.11)	Logistic	Asymptotic	Yes	Yes	No	Yes	Yes

Note: All have linear numerical responses and constant predator death rates.

of density dependence, with the result that increasing the prey's K destabilizes the system by reducing the effective amount of density dependence. Fifth, predator satiation is a factor that tends to destabilize the system and allow the inherent oscillatory character to emerge. Sixth, in all but the structurally unstable Volterra model, the nature of predator-prey coexistence is either a specific stable equilibrium point or a specific stable oscillation called a limit cycle.

Lynx Revisited   There are not many tests of these models with data from natural populations. We can return to the lynx-hare system, however, to ask if their oscillations appear to be limit cycles. Because of the large dependence by lynx on hare, it is, perhaps, the closest natural example we shall ever find to a one-predator one-prey system, even though it is not a perfect example.

The alternative hypothesis to a limit cycle for the lynx is to claim that the oscillations really are not true oscillations at all but are caused by random events in the environment that occur in a special way. The lynx pattern is obviously too regular to be generated by, for example, the toss of a coin each year. It would be very improbable to obtain the pattern of lynx abundance through time solely by a random number generator. A more subtle hypothesis is plausible. To draw an analogy, consider a damped spring. In a constant environment, the spring simply settles into its resting position. But, suppose that the spring is put in a chamber with little pellets or buckshot raining down on it. Then, the spring will be agitated or energized slightly whenever a pellet strikes. As a result, the spring will bounce around and its motion will have an oscillatory character to it, as well as a random character. Returning to the lynx situation, if the predator-prey system has sufficient density dependence, then, in a constant environment, a stable equilibrium will be attained. But if the population sizes are altered by an environmental disturbance, then the system is displaced away from equilibrium and it will tend to spiral back into equilibrium until another disturbance occurs. The predator-prey system with sufficient density dependence might be analogous to the damped spring and, by "energizing" the system, the environmental disturbances might be the basic cause of the oscillatory pattern. According to this hypothesis, the oscillatory pattern represents the response to extrinsic factors, whereas in the limit cycle hypothesis, the oscillations are intrinsic to the predator-prey system.

The Fluctuating   This alternative hypothesis involving a fluctuating environment has been
Environment   investigated by Moran (1953), using the lynx data. The model incorporating
Model   the random environmental disturbances is called a second-order auto-regressive process† and is written as follows. Let P_t be the deviation of the predator population size from its average value. Then, the model is

$$P_t = \lambda_1 P_{t-1} + \lambda_2 P_{t-2} + Z_t \qquad (22.12)$$

According to this model, the predator abundance at time t is λ_1 times its value at $t-1$ plus λ_2 times the value at $t-2$ plus a random component, Z_t. This quantity, Z_t, represents the random environmental disturbance introduced at time t. We have discussed stochastic models of this sort in Chapter 20. For our present purposes, it suffices to state that this model can yield

† The name, autoregressive process, arises because we can view (22.12) as a multiple regression of P_t against P_{t-1} and P_{t-2} with λ_1 and λ_2 as the regression coefficients. The term Z_t, then stands for the variation that is not explained by the regression.

oscillatory patterns and has been used with a wide variety of quasi-oscillatory phenomena, including sunspot "cycles," and so forth. It is possible to derive a relationship between λ_1 and λ_2 and the parameters in the predator-prey model with logistic prey growth (22.6), provided the fluctuations are small. The model is tested by seeing whether the observed oscillatory pattern agrees with the one predicted from the best fit provided by this model.

To test whether the autoregressive model fits the data, we compute a function called the *serial correlation function* (also called the autocorrelation function). This function presents the correlation between the predator abundance at time t and the abundance at earlier times. More specifically, the serial correlation is the correlation between P_t and P_{t-h}; h is called the *lag*. The serial correlation tends to zero as h becomes sufficiently large because in a stochastic environment the correlation between the present abundance and the abundance in the sufficiently distant past is zero. If the population abundance has an oscillatory character, it will be reflected in the serial correlation function. The oscillatory character in the population causes the serial correlation function also to oscillate because there is positive correlation between population sizes at peaks in the cycle and negative correlation between sizes at peaks and troughs.

Moran (1953) found the best fit of Equation (22.12) with the lynx data and then generated the predicted serial correlation function. He also calculated the serial correlation function directly from the lynx data. Moran considered the data from the Mackenzie River district of Canada, illustrated in Figure 22.4. His results appear in Figure 22.9. The actual serial correlation function is substantially different from that expected from the autoregressive model (22.12). The oscillations in the data are too regular and sustained to be explained by the autoregressive model. Using the physical analogy again, particles randomly energizing a spring would indeed produce an oscillatory pattern, but *not* one with oscillations as sustained and regular as those found in the lynx pattern.

The comparison above of the lynx data with an autoregressive model is fairly strong evidence that the lynx oscillations really represent a limit cycle. As we have seen, a functional response with predator satiation can lead to limit cycle behavior. Undoubtedly, other plausible models do, also. So the issue of exactly what mix of mechanisms causes the lynx oscillations is still open, although the work surveyed above has made a definite contribution to understanding this problem. An important recent discussion of this issue appears in Bulmer (1976b).

Other Issues   In conclusion, we mention three other issues. First, the extension of predator-prey theory to systems with multiple prey and multiple predators is a very important challenge because one qualitatively new feature must be incorporated that cannot be understood in the context of one predator and one prey. This feature is called *switching*. There is evidence that predators alter the predation intensity on different prey populations in accordance with their comparative abundance. It is often believed that predators specifically utilize the most abundant prey, and switch among species as the relative abundance among different prey changes. For detailed discussion, see Tinbergen (1960), Royama (1970) and Tullock (1971) for birds, and Ivlev (1961), Murdoch (1969), and Murdoch et al. (1975). As a result, the

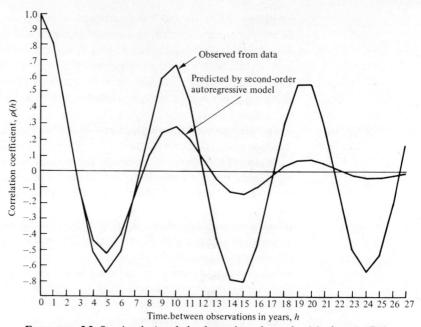

FIGURE 22.9. *Analysis of the lynx data from the Mackenzie River District. A log transformation was applied to the lynx data illustrated in Figure 22.4. The autocorrelation function calculated from these transformed data is plotted in the figure. Then, a second-order autoregressive model was fitted to the transformed data. The Yule–Walker estimate for the parameters leads to the following model:* $P_t = 1.41\,P_{t-1} - .77\,P_{t-2} + Z_t$ *where Z_t has a variance of .46. The autocorrelation function predicted by this model is* $\rho(h) = .88^h\,\sin(.64h + 1.40)/\sin(1.40)$. *This formula is also graphed in the figure. Note that the data show a much more regular cycle than expected by this model.* [Original analysis by P. A. P. Moran (1953), The statistical analysis of the Canadian lynx cycle I. Structure and prediction. *Aust. J. Zool.* **1**: 163–173.]

functional response of a predator with respect to a given prey species will depend on the abundance of that species relative to other prey species. Depending on the functional form of the functional response, many interesting phenomena may emerge which are impossible even in a complicated one predator-one prey system. See Murdoch and Oaten (1975) and Oaten and Murdoch (1975) for elaboration of this point. Second, an explicit assumption in all the predator-prey theory is that the predators are capturing prey that are integral reproducing members of the prey population. However, Errington (1946) has adduced evidence that some vertebrate predators capture mostly those prey individuals who would "die anyway"— that is, those who are aged and diseased and who would die shortly even if the predators were absent. Obviously, predators of this sort are similar to scavengers and could not be responsible for keeping a prey population under predator ccontrol. Removal of such predators or interfering with the predator-prey balance in this case presumably would not lead to the many surprising results we have derived in this chapter. Finally, there has been much interest recently in the connection between patchiness and predation, especially with regard to the effect of patchiness on the stability of predator-prey coexistence. See Hassell (1974), Steele (1974a, b), and Oaten (1977) for information on this topic.

Chapter 23

COEVOLUTION IN ECOLOGICAL SYSTEMS

POPULATIONS that live together evolve together. Populations evolutionarily shape one another. "Coevolution" is the term applied to the simultaneous evolution of interacting populations. As a result of coevolution, the structure of an entire ecological community may be molded by natural selection. Some populations may evolve extremely intricate interdependencies with one another. Although coevolution is one of the most important processes in population biology, it has received little theoretical attention until very recently. This chapter begins with a brief review of the empirical issues that motivate the study of coevolution and then introduces some of the mathematical theory in this area. This chapter is based mostly on my own papers and presents my view on how a comprehensive quantitative theory of coevolution may be developed. However, other and perhaps better approaches may be developed in the future.

Some Empirical Topics in Coevolution

Coevolution is potentially a huge topic and, as a result, it has been approached from several viewpoints. However, most issues concerned with coevolution can be classified into the following scheme:

Coupled Speciation of Plants and Plant-Eating Insects

COEVOLUTION AND SPECIES DIVERSITY. One of the earliest papers on coevolution, by Ehrlich and Raven (1965), was principally concerned with the synchronized adaptive radiation of plants and butterflies. A major, perhaps the most important, defense by plants against herbivores, is to produce toxic chemicals. Generally speaking, the reason why plants have so many unusual chemicals in them, including chemicals used as drugs, is that they are antiherbivore defenses. To digest plant material a plant-eating insect must use rather specialized biochemical techniques. As a result, the relationship between plants and their insect herbivores tends to be very close and biochemically specific. People working with vertebrates are accustomed to birds, mammals, and lizards eating many prey species and to grazing mammals eating many plant species. In contrast, the food of a butterfly larva is often restricted to one or two plant species because it can detoxify the toxins in only these species. This close biochemical connection between plants and plant-eating insects has led to a remarkably synchronous radiation between butterflies and plants. One of the principal issues concerned with coevolution is to understand how the speciation and diversification of plants is coupled to those of plant-eating insects. This issue is not yet addressed by any of the mathematical theories. A nontechnical review of these issues appears in Ehrlich and Raven (1967).

The Search for Generalizations About the Entire Community

COEVOLUTION AND COMMUNITY PROPERTIES. Another issue concerned with coevolution is the search for general principles about the effect of coevolution on the entire community. It has been suggested that coevolution maximizes a community's ability to withstand external disturbances and perturbations and that coevolution maximizes the efficiency by which solar energy is converted into the aggregate mass of organisms in the community. It has also been said that coevolution leads eventually to a complete usage of the available resources and that coevolution produces a community with maximum resistance to invasion by a species from somewhere else. In this chapter, we present some general principles about coevolution that are derived from a firm population genetic basis. These principles do not usually imply the suggestions above and, indeed, the suggestions above will appear rather naïve in retrospect.

Symbiosis   COMPLEX COADAPTATION. One result of coevolution is the occasional formation of intricate and fascinating relationships between species. In fact, the majority of the interest in coevolution to date is in connection with these relationships. A physically close and relatively permanent relationship between two or more individuals of different species is called a *symbiotic relationship*. Three kinds of symbiotic relationships are usually distinguished: In a *parasitic relationship*, one party benefits at the clear expense of the other. In a *commensal relationship*, one party, the guest, benefits from the other, the host, in a way that brings negligible harm to the host. (Tiny hydroids, which attach to the shell of a hermit crab, might be an example.) In a *mutualistic relationship*, both parties clearly benefit each other. Moreover, there are examples, as mentioned later, where the individuals in a mutualistic relationship clearly incur some cost in doing activities that benefit one another.

Parasitism   A point of special interest with parasitism is that parasites may evolve to reduce the harm they inflict upon their hosts. An example has been documented with the myxoma virus which infects rabbits in Australia (Fenner and Ratcliffe, 1965, Fenner, 1971). Rabbits are not native to Australia; they were introduced from Europe in 1859 and eventually became a major pest. To control the rabbits, a South American myxoma virus carried by mosquitoes was introduced in 1950. This virus was lethal to European rabbits, although its effect on South American rabbits was not severe. Over the next 20 years, *both* the rabbits and virus evolved; the rabbits evolved more resistance to the virus and the virus evolved lower virulence. This case is one of the best studies of coevolution presently available.

Mutualism   Mutualistic associations have always captured the fascination of biologists and it is worth citing several examples here. Most examples are tropical. Janzen (1966 and 1967) discovered an obligate mutualistic association between ants and acacia trees. The acacia species harbors ant colonies in modified thorns and actually provides food for the ants with nectaries at the bases of their leaves. The ants in turn provide 24-hour protection against plant-eating insects. Janzen demonstrated increased mortality in plants from which ants were excluded. Another terrestrial example of a complex mutualistic association occurs between bird species. Smith (1968) discovered that a species of nest parasites (a cow-bird) actually increased the fitness of the host (a colonial species—the chestnut headed oropendola) in certain locations. The eggs of the cowbird generally hatch earlier than those of the oropendola. The cowbird chicks protected the oropendola chicks from insect damage. The marine biological literature is especially replete with examples of mutualistic symbiosis. One of the earliest studies (Verwey 1930) concerns a very common association between damsel fish and sea anemones in the Pacific. Sea anemones are, themselves, predators and often eat fish who are stung and killed by contact with their tentacles. Nevertheless, many species of damsel fish have developed the ability to live within the tentacles of large sea anemones. These fish secrete a mucuouslike substance that leads to their protection, perhaps, by raising the physical threshold required to stimulate the discharge of the nematocysts in the anemone tentacles. As a result, the damsel fish can hide among the tentacles and use them as a shelter from predators. It has also been established that damsel fish occasionally catch prey, which they feed to their anemones. Another common example of a mutualistic symbiosis from marine biology is the "cleaning symbiosis." Small fish and shrimp are known to glean the ectoparasites on large fish. These large fish present themselves at fixed locations on coral reefs called cleaning stations, where they are cleaned by the cleaner fish or shrimp. After the cleaning, the large fish leaves. For more on this fascinating example, see Feder (1966) and Limbaugh (1961).

Other Examples of Coadaptation   These examples of complex coadaptations between species refer to symbiotic associations. There may be intricate coadaptation between species, even though no symbiotic association is involved. Consider, again, the plant-insect herbivore interaction. We may visualize a plant as being faced with a choice of allocating energy into rapid growth and seed production or into the manufacture of toxic chemicals. The optimum choice will depend, in part, on the herbivore pressure and on the effectiveness per unit of the available toxins in reducing the pressure. However, the insect is faced with a choice of many potential plant species. The insect must allocate energy to detoxifying the toxins in each plant species it might use and, then, must allocate its own grazing pressure accordingly. Hence the realized regime of insect predation on plants may represent an evolutionary equilibrium which is the resultant of all parties pursuing what appears, to each, as its own optimum strategy. In this sense there may be complex coadaptations between species, even though there is no symbiosis involved. An important review of issues in the coevolution of animals and plants appears in Gilbert and Raven (1975).

The Design of Coadaptation   In each of these cases of complex coadaptations, we may ask whether natural selection is, so to speak, a good engineer. For example, suppose the damsel fish, mentioned previously, provided even more food to their anemone hosts. Could this increase the population size of the sea anemones and, in turn, lead to an increase in the damsel fish population size? And if so, will natural selection under these circumstances cause the evolution of more feeding of the anemones by the fish? Although there are many examples known of complex coadaptations between species, there is no case where we know how well they are engineered. It will be predicted from the theory that natural selection is at best a very imperfect engineer and, thus, data on this point would be extremely important.

The preceding paragraphs illustrate the range of empirical issues to which the theory of coevolution is potentially relevant. In this chapter, we will not concentrate on models for specific coadaptations. [However, for a simple cost-benefit model for the evolution of mutualism, see Roughgarden (1975c).] Instead, we shall be concerned with general properties of population-genetic models of interacting population. Nonetheless, several models for specific coadaptations will be introduced along the way for illustrative purposes.

The Models The models we shall study are models that fuse typical population dynamic models with equations for gene frequency change within each of the interacting populations. Specifically, for a set of S interacting populations, a coevolutionary model is of the form

$$\Delta N_i = (\bar{W}_i - 1)N_i$$

$$\Delta p_i = p_i(1 - p_i)\frac{\partial \bar{W}_i/\partial p_i}{2\bar{W}_i} \qquad i = 1 \cdots S \qquad (23.1)$$

where N_i and p_i are the population size and gene frequency in species-i. For $S = 1$ this model reduces to the model for density-dependent selection, discussed in Chapter 17. Recent studies of this type of model for two species appear in Leon (1974) and Levin and Udovic (1977).

With these coevolutionary models, we can address questions about the equilibrium result of coevolution, and also, to a more limited extent, questions about the dynamics of coevolution. The equilibrium analysis is relevant to understanding how natural selection shapes the final equilibrium configuration of an ecological community. The dynamic analysis concerns questions like the time lag inherent in the evolutionary response by predators to the evolution of new defenses in the prey. The evolutionary arms race between predator and prey is essentially about dynamics of coevolution, whereas questions about how natural selection shapes a community are essentially about the equilibrium result of coevolution. An extensive summary appears at the end of the chapter.

Density-Dependent Evolution in a Single Species Some of the important ideas in the theory of coevolution are first met in the study of density-dependent evolution in a single species. In the following paragraphs we generalize the idea of K-selection to general single population models. This section is based on Roughgarden (1976). In the next section we shall further generalize this work to interacting populations. The model to be analyzed is

$$\Delta p = \frac{p(1-p)}{2\bar{W}}\frac{\partial \bar{W}}{\partial p}$$

$$\Delta N = (\bar{W} - 1)N \qquad (23.2)$$

where

$$\bar{W} = \bar{W}(p, N) = p^2 W_{11}(N) + 2p(1-p)W_{12}(N) + (1-p)^2 W_{22}(N) \qquad (23.3)$$

It is assumed that W_{ij} are functions only of N and not also of p. A familiar choice for $W_{ij}(N)$ is based on the logistic equation.

$$W_{ij}(N) = 1 + r_{ij} - \frac{r_{ij}N}{K_{ij}} \qquad (23.4)$$

We are interested in the equilibria of this system. There are two types of equilibria. *Boundary equilibria* represent the fixation of one of the alleles.

Boundary equilibria are points $(\hat{p}, \hat{N})$, satisfying

$$A_1 \text{ fixed:} \quad \hat{p} = 1, \quad W_{11}(\hat{N}) = 1$$

$$A_2 \text{ fixed:} \quad \hat{p} = 0, \quad W_{22}(\hat{N}) = 1 \tag{23.5}$$

An *interior equilibrium* represents a polymorphism. An interior equilibrium $(\hat{p}, \hat{N})$ satisfies

$$\bar{W}(\hat{p}, \hat{N}) = 1$$

$$\frac{\partial}{\partial p} \bar{W}(\hat{p}, \hat{N}) = 0 \tag{23.6}$$

The interesting result is that the stability of these equilibria can be expressed in terms of how natural selection influences the equilibrium population size.

Consider the equilibrium population size that is attained if the gene frequency is somehow held constant. In this situation, the population dynamics are given by

$$\Delta N = [\bar{W}(p_0, N) - 1]N \tag{23.7}$$

In this equation, the gene frequency is held constant at the value p_0, whereas N is a variable which may change through time. For any constant value of p, there is an equilibrium population size $\hat{N}$, which is found by solving for N in

$$\bar{W}(p, N) = 1 \tag{23.8}$$

In fact $\hat{N}$ is a function of p and Equation (23.8) can be regarded as an implicit definition of the function, $\hat{N}(p)$. Sometimes, we can also write down $\hat{N}(p)$ explicitly as with logistic fitnesses, but this is generally not possible. In any event $\hat{N}(p)$ is defined implicitly by (23.8). If p equals 0, then $\hat{N}$ is the equilibrium population size that results if A_2 is fixed. If p equals 1, then $\hat{N}$ is the equilibrium population size that results if A_1 is fixed; and if p is some constant between 0 and 1, then $\hat{N}$ is the equilibrium population size of a polymorphic population with gene frequency, p. Since we are interested in how natural selection influences the equilibrium population size in an ecological model, it is necessary to assume that there exists a unique stable equilibrium population size to begin with. Otherwise, we may not have anything to talk about. Therefore, we assume that, for any fixed p the equilibrium, $\hat{N}(p)$, defined implicitly by (23.8), is the unique, stable equilibrium for the population dynamic model, (23.7). Under these conditions we can establish the following results.

RESULT 1 *Suppose that $\hat{N}(p)$ is the unique stable equilibrium population size for the population dynamic model (23.7), in which the gene frequency is constant. Then*

1. *A point (p^*, N^*) is a locally stable interior equilibrium point in the model for density-dependent selection (23.2), in which both p and N are variable if and only if p^* locally maximizes $\hat{N}(p)$ and $N^* = \hat{N}(p^*)$.*
2. *A point (p^*, N^*) with $p^* = 0$ or 1 is a locally stable boundary equilibrium point in the model for density-dependent selection if and only if $\hat{N}(p)$ is higher at the boundary $p = p^*$ than at any nearby interior p. (That is, for stability at $p^* = 0$, the slope of $\hat{N}(p^*)$ must be negative and at $p^* = 1$ the slope must be positive.)*

Points (1) and (2), above, really say the same thing, namely, that natural selection causes the gene frequency, eventually, to assume a value which maximizes the equilibrium population size. It must be emphasized that it is the *equilibrium* population size that is maximized, relative to other possible *equilibrium* population sizes. During the approach to equilibrium, the population may frequently exceed the eventual equilibrium value because of successive overshoots. This theorem provides a generalization to the idea of K-selection, which was originally developed in the context of logistic fitnesses. The proof of the result, above, is presented in Appendix I of Roughgarden (1976). The result has been extended to one locus with n-alleles by Ginzburg (1977a). This model has also been solved in the context of parthenogenetic populations by Templeton (1974). The way in which this result is altered when frequency dependence is present is discussed in an important recent paper by Slatkin (1978).

Example To illustrate this result, consider the example of the logistic fitness (23.4), which was extensively discussed in Chapter 17. To apply Result 1, we must confirm that $\hat{N}(p)$ is a locally stable equilibrium population size for any fixed value of p. With logistic fitnesses, $\hat{N}(p)$ can be determined explicitly as

$$\hat{N}(p) = \frac{p^2 r_{AA} + 2p(1-p)r_{Aa} + (1-p)^2 r_{aa}}{p^2(r_{AA}/K_{AA}) + 2p(1-p)(r_{Aa}/K_{Aa}) + (1-p)^2(r_{aa}/K_{aa})}$$

(23.9)

Furthermore, for any fixed, p, this equilibrium population size is stable if and only if

$$0 < \bar{r} < 2$$

(23.10)

Therefore, subject to condition (23.10), we can use Result 1. Part (1) tells us how to determine a stable polymorphism. It is easy to verify that $\hat{N}(p)$ has a maximum between 0 and 1 if and only if the heterozygote has the highest K. To find the polymorphism frequency, we differentiate $\hat{N}(p)$ with respect to p and set the derivative equal to zero, yielding

$$Ap^2 + Bp + C = 0$$

(23.11)

where

$$A = (r_{Aa} - r_{aa})\left(\frac{r_{AA}}{K_{AA}}\right) + (r_{AA} - r_{Aa})\left(\frac{r_{aa}}{K_{aa}}\right) + (r_{aa} - r_{AA})\left(\frac{r_{Aa}}{K_{Aa}}\right)$$

(23.12a)

$$B = (2r_{Aa} - r_{AA})\left(\frac{r_{aa}}{K_{aa}}\right) + r_{aa}\left(\frac{r_{AA}}{K_{AA}} - \frac{2r_{Aa}}{K_{Aa}}\right)$$

(23.12b)

$$C = -r_{aa}r_{Aa}\left(\frac{1}{K_{aa}} - \frac{1}{K_{Aa}}\right)$$

(23.12c)

p^* is the root of (23.11), which is between 0 and 1. Part (2) of Result 1 tells us the condition for which the fixation of an allele is stable. With logistic fitnesses, Part (2) translates as

$$K_{AA} > K_{Aa} \Rightarrow A \text{ fixation is stable}$$
$$K_{aa} > K_{Aa} \Rightarrow a \text{ fixation is stable}$$

(23.13)

Thus if a homozygote for an allele has a higher K than the heterozygote, then fixation of that allele is locally stable. Of course, if the heterozygote has a lower K than both homozygotes, then both fixation states are stable. Which state is actually attained in an experiment depends on the initial condition. These results are all familiar from Chapter 17. The interesting feature, here, is that they are obtained from a result that applies to any single species population dynamic model. Figure 23.1 presents a numerical illustration of a polymorphism that maximizes $\hat{N}(p)$.

The major qualitative importance of Result 1 is that natural selection has been shown to mold the parameters of an ecological model, so as to yield the highest equilibrium population size. Those parameter values that provide the highest equilibrium population size produce what we may call an *evolutionarily stable population*.

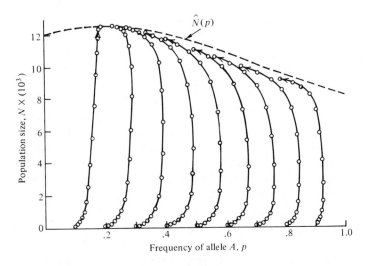

FIGURE 23.1. *Trajectories illustrating density-dependent selection in a single population. The heterozygote has the largest carrying capacity, and a polymorphism results. The equilibrium population size as a function of allele frequency, $\hat{N}(p)$, is plotted as a dashed line. Note that the equilibrium point coincides with the peak of the $\hat{N}(p)$ function.* [From J. Roughgarden (1976), Resource partitioning among competing species—A coevolutionary approach. *Theor. Pop. Biol.* **9**: 388–424.] *Here* $r_H = .7$; $r_A = .8$; $r_a = .6$; $K_H = 15,000$; $K_A = 8000$; *and* $K_a = 12,000$.

Purely Density- Dependent Coevolution

In this section, we generalize the preceding result on a single population to a collection of interacting populations. Our aim will be to identify an evolutionarily stable configuration for a set of interacting populations. This section is taken from Roughgarden (1977a).

The theory of population ecology is mostly concerned with predicting equilibrium population sizes on the basis of assumptions about the interactions between populations. Familiar population interactions are interspecific competition, predation, symbiosis including parasitism and mutualism, and others. Most population dynamic models predict that the interacting populations will attain stable equilibrium abundances, provided the parameters in the model satisfy certain requirements that are special to each model. Many models also allow for other possibilities, including cycling of various forms. Nonetheless, almost all models contain stable coexistence at an equilibrium point as *one* of the possibilities.

Whenever species are coexisting at a stable equilibrium point, it is clear that their equilibrium abundances are functions of the parameters in the model. For example, the equilibrium abundance of each competititor in a Lotka–Volterra competition system is a function of all the competition coefficients, α_{ij}, and all the carrying capacities, K_i. However, the parameters in a population dynamic model are, themselves, subject to evolutionary modification by natural selection. Therefore, natural selection, also indirectly, controls the equilibrium population sizes because natural selection directly controls the parameters. In this context, the basic question an ecologist may ask of coevolutionary theory is: *To what value does natural selection set each parameter in an ecological model and what are the equilibrium population sizes attained as a result?* To answer this question is to understand how the combined effects of natural selection in each of the interacting populations shapes the *final* configuration of the whole ecological community.

There are two separate clauses in the question posed above. The first clause concerns the parameter values, themselves. In some circumstances, this clause is the focus of interest. For example, the competition coefficients, α_{ij}, are often interpreted in terms of the similarity of resource use by two competitors. A coevolutionary theory might predict the value of α_{ij} and thereby predict the degree of similarity of two competitors. In this case, the degree of similarity itself is the primary feature of interest. The population sizes of the competitors are of secondary interest. In other circumstances, the parameters in the model are of secondary interest to the equilibrium population sizes that they cause. This is especially true, if it can be shown that the effect of evolution on the parameters of a model leads to a lowering of the equilibrium population size of a species. Showing this raises the possibility of natural selection causing the extinction of a population. Another circumstance where the population sizes are of primary interest, is where the amount of standing crop at different trophic levels (the energy pyramid) remains to be explained. In this case, data on the summed abundance of several interacting populations are available but the details of the interactions are largely unknown. Thus the interest focuses on the population abundances.

For the question above to be meaningful, it must be true that there exists a stable equilibrium point to talk about. That is, we view natural selection in each of the populations as influencing the parameters in the model, provided that the parameters are restricted to values where a stable equilibrium point exists. We shall be able to detect when selection is *tending* to destroy an initially stable equilibrium (see Example 2), but the evolutionary analysis itself is restricted to the region of parameter space in which there exists a stable equilibrium point.

In any theory for the evolution of the parameters in ecological models, one must specify that species exert *evolutionary control* over the parameters. Consider a Lotka–Volterra system again. We, typically, expect evolution within species-i to control its own carrying capacity, K_i. It is also possible, however, that some other species pollutes the environment, thereby causing K_i to be under the control of another species. Less far fetched, the parameter, α_{ij} may be under the joint evolutionary control of both species-i

and species-*j*. The assignment of the evolutionary control of the parameters is the most critical assumption in the theory of coevolution.

The simplest assignment of evolutionary control is to *assume that each species controls only the parameters in its own equation for population growth.* This assumption rules out the possibility of species-*i* controlling the carrying capacity of species-*j* and, also, implies that no parameter is under the joint control of two or more species. For example, α_{ij} is assumed to be under the sole control of species-*i*. This assumption is *equivalent to assuming that the fitness of genotype A_iA_j in species-s, $W_{s,ij}$, is a function only of the various population sizes and not of the gene frequencies in the various species.* This assumption means that the *coevolution is purely density-dependent; there is no interspecific frequency dependence.*

The Graph of a Community
An ecological system consists of the populations and their interactions. The following paragraphs present a useful graphical representation of an ecological system.

To develop a coevolutionary model, one begins with a population dynamic model defined in discrete time with the time interval set at one generation. The original population dynamic model is of the form

$$\Delta N_i = [W_i(N_1 \cdots N_S) - 1]N_i \qquad (i = 1 \cdots S) \qquad (23.14)$$

where N_i is the abundance of species-*i* at time *t* and W_i is the average number of offspring produced per individual in species-*i*. If $W_i = 1$ there is direct replacment by each individual and $\Delta N_i = 0$. By definition $W_i \geq 0$. By assumption, this population dynamic model has a positive, locally stable equilibrium. This means that the matrix whose elements are

$$a_{ij} = \frac{\partial(\Delta N_i)}{\partial N_j}\bigg|_{N_i = \hat{N}_i} \qquad (23.15)$$

has eigenvalues that lie in a unit circle in the left half of the complex plane, as shown in Figure 23.2.

The stability criteria from continuous time models are necessary, but not sufficient, for the discrete time models discussed here. The elements may also be computed in terms of the *W*'s as

$$a_{ij} = \hat{N}_i \frac{\partial W_i}{\partial N_j} \qquad (23.16)$$

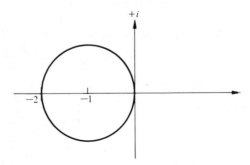

FIGURE 23.2. *The eigenvalues must lie in the circle for ecological stability.*

where $\partial W_i/\partial N_j$ is evaluated at $N_i = \hat{N}_i$ for all i. With these quantities, Levins (1974) has posed a graphical representation of an ecological system at equilibrium, which proves very helpful. Each species abundance at equilibrium is placed at a node and lines are drawn between nodes to indicate the population interactions. For example, with three species, we have Figure 23.3. A line terminating with an arrow indicates a positive effect. For example, a_{21} indicates that N_1 has a positive effect on N_2, and a_{22} indicates that N_2 grows autocatalytically. A line terminating with a dot indicates a negative effect. This graph is also called a Coates graph in the electrical engineering literature. See Chen (1971).

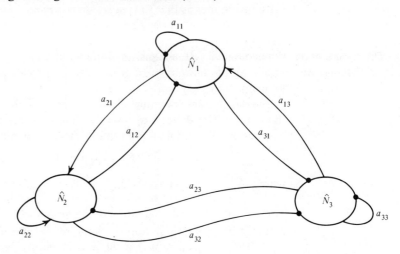

FIGURE 23.3. *The graph G representing the ecological interactions among the components of a community.*

There are several important measures defined on this graph and on certain subgraphs formed from it.

1. The *feedback* in the community is defined in Levins (1974) as

$$F \equiv (-1)^{S+1} \det (a_{ij}) \qquad (23.17)$$

where S is the number of species in the community. To see this more explicitly, note that

$$
\begin{pmatrix} a_{11} & \cdots & a_{1S} \\ \vdots & & \vdots \\ a_{S1} & \cdots & a_{SS} \end{pmatrix} = \begin{pmatrix} \hat{N}_1 & & 0 \\ & \ddots & \\ 0 & & \hat{N}_S \end{pmatrix} \begin{pmatrix} \dfrac{\partial W_1}{\partial N_1} & \cdots & \dfrac{\partial W_1}{\partial N_S} \\ \vdots & & \vdots \\ \dfrac{\partial W_S}{\partial N_1} & \cdots & \dfrac{\partial W_S}{\partial N_S} \end{pmatrix} \qquad (23.18)
$$

Therefore,

$$\det (a_{ij}) = \left(\prod_{j=1}^{S} \hat{N}_j \right) \det \left(\frac{\partial W_i}{\partial N_j} \right) \qquad (23.19)$$

Hence

$$F = (-1)^{S+1} \left(\prod_{j=1}^{S} \hat{N}_j \right) \det \left(\frac{\partial W_i}{\partial N_j} \right) \qquad (23.20)$$

If $S = 1$, then F reduces to

$$F = \hat{N} \frac{\partial W}{\partial N} \qquad (23.21)$$

Thus when $S = 1$, the sign of F is identical to the sign of the density dependence in the population.

2. A *subcommunity* (of order $S - 1$) is a subsystem obtained by deleting one species from the system. The variable, G_i, denotes the graph of the subcommunity obtained by deleting species-*i*. For example, G_1 from the preceding example is in Figure 23.4.

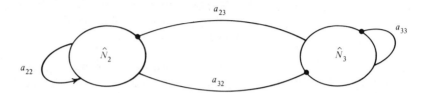

FIGURE 23.4. *The graph G_1 representing the subcommunity obtained by deleting species-1.*

The *feedback of the subcommunity* is defined, just as above, by

$$F_i \equiv (-1)^S \det{}_{ii} (a_{ij}) \qquad (23.22)$$

where the symbol, $\det_{ii} (a_{ij})$, means the determinant of the matrix of rank $S - 1$, obtained by deleting the *i*th column and *i*th row from the matrix (a_{ij}). In terms of the W's, we have

$$F_i = (-1)^S \left(\prod_{j \neq i} \hat{N}_j \right) \det{}_{ii} \left(\frac{\partial W_i}{\partial N_j} \right) \qquad (23.23)$$

If $S = 2$, then the sign of F_i is identical to the sign of the density dependence in the undeleted species.

3. The *connecting web from species-j to species-i* is represented by the graph, G_{ij}, obtain by deleting all the lines leading into j and all the lines leading out of i. The variable, G_{12} in the example, is found in Figure 23.5.

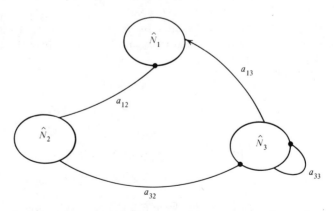

FIGURE 23.5. *The graph G_{12} representing the connecting web running from species-2 to species-1.*

Species-2 is seen to influence species-1 directly through a_{12} and also indirectly via species-3. The *feedback of a connecting web* is defined by Roughgarden (1977a) as

$$F_{ij} \equiv (-1)^{S+1}(-1)^{i+j} \det_{ji} (a_{ij}) \qquad (23.24)$$

where $\det_{ji} (a_{ij})$ means the determinant of the matrix of rank S-1, obtained by deleting the jth row and the ith column from (a_{ij}). In terms of the W's, we have

$$F_{ij} = (-1)^{S+1}(-1)^{i+j}\left(\prod_{k \neq j} \hat{N}_k\right) \det_{ji} \left(\frac{\partial W_i}{\partial N_j}\right) \qquad (23.25)$$

If $S = 2$, the sign of F_{ij} is identical to the sign of $\partial W_i/\partial N_j$; that is, it is negative if j competes with or preys upon i, and positive if i preys upon j, and so forth. By saying that j exerts a negative (positive) effect upon i, we mean that $F_{ij} < (>) 0$.

These feedback measures completely specify whether the evolution within some species increases or decreases its own equilibrium population size, and increases or decreases the equilibrium population sizes of other species under pure density-dependent coevolution.

A key lemma follows.

LEMMA 1   *For any community at a stable equilibrium point*

$$
\begin{aligned}
F &< 0 \\
\sum_i F_i &< 0
\end{aligned}
\qquad (23.26)
$$

That is, a necessary condition for local stability is that the feedback of the entire community should be negative and that the sum of the feedback from all subcommunities should be negative. This result is an immediate consequence of the conditions on the coefficients of the characteristic equation that are necessary for the eigenvalues to have negative real parts.

Recently, Holt (1977) has suggested that the community graph and its appropriate subgraphs may be useful in understanding the conditions under which a population may appear to be competing with another as a result of its being embedded in a community in a special way.

The Coevolutionary Model and the Associated Pure Ecological Model   A *coevolutionary model* is obtained by fusing the standard formulas for natural selection at one locus with two alleles with a population dynamic model, yielding

$$\Delta N_i = (\bar{W}_i - 1)N_i$$

$$\Delta p_i = \frac{p_i(1-p_i)}{2\bar{W}_i} \frac{\partial \bar{W}_i}{\partial p_i} \qquad (i = 1 \cdots S) \qquad (23.27)$$

where $\bar{W}_i$ is the mean fitness in species-i,

$$\bar{W}_i = p_i^2 W_{i,11} + 2p_i(1-p_i)W_{i,12} + (1-p_i)^2 W_{i,22} \qquad (23.28)$$

and p_i is the allele frequency in species-i. By assumption,

$$W_{i,jk} = W_{i,jk}(N_1 \cdots N_S) \qquad (23.29)$$

that is, the fitnesses are purely density-dependent. (Of course, by definition, $W_{i,jk} \geqslant 0$.) The assumption of pure density dependence implies

$$\frac{\partial \bar{W}_i}{\partial p_j} \equiv 0 \qquad j \neq i \qquad\qquad (23.30)$$

A coevolutionary equilibrium satisfies, for every i, one of the following conditions

$$\bar{W}_i = 1, \qquad \frac{\partial \bar{W}_i}{\partial p_i} = 0 \qquad\qquad (23.31a)$$

$$W_{i,11} = 1, \qquad \hat{p}_i = 1 \qquad\qquad (23.31b)$$

$$W_{i,22} = 1, \qquad \hat{p}_i = 0 \qquad\qquad (23.31c)$$

In the first condition, species-i is at a polymorphic or "interior" equilibrium; in two other conditions, species-i is at a fixation or "boundary" equilibrium.

As a guide to analyzing the coevolutionary model above, we refer continually to another model, called the *pure ecological model*. The pure ecological model is defined by taking the gene frequency in every species as a constant, while allowing the population size to vary. Thus the pure ecological model is a population dynamic model of the form

$$\Delta N_i = (\bar{W}_i - 1)N_i \qquad\qquad (23.32)$$

Since the p's are constants in this model, the equilibrium population sizes can be regarded as functions of the p's. Generally, different equilibrium population sizes will result for different fixed values of the p's. Whenever this pure ecological model leads to a stable equilibrium, the resulting equilibrium community can be represented by the community graph, as previously discussed. For example, with two populations we might have Figure 23.6.

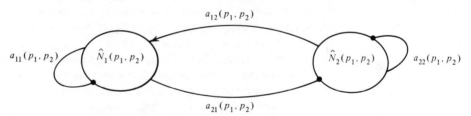

FIGURE 23.6. *A community graph between two populations, illustrating the dependencies on the gene frequencies.*

Clearly, both the equilibrium population sizes and the interactions are functions of the p's. The $\hat{N}$'s are defined as functions of the p's implicitly by the system of equations

$$\bar{W}_i(p_i, N_1 \cdots N_S) = 1 \qquad (i = 1 \cdots S) \qquad\qquad (23.33)$$

The a_{ij} are defined by

$$a_{ij}(p_1 \cdots p_S) = \hat{N}_i(p_1 \cdots p_S) \frac{\partial \bar{W}_i(p_i, \hat{N}_1 \cdots \hat{N}_S)}{\partial N_j} \qquad\qquad (23.34)$$

Furthermore, for every set of p's, we can evaluate the feedback in the various subgraphs of the equilibrium community. Thus for every set of p's, we can obtain a complete picture of the equilibrium ecological system that results. For this discussion to be well posed, we *assume that the pure ecological model has a unique, positive, locally stable equilibrium point for any set of fixed p's.* This assumption, together with that stipulating purely density-dependent fitnesses, are the two principal assumptions in this section.

The set of equilibrium communities generated by the pure ecological model for all possible p's, comprise the set of communities that are *evolutionarily possible*. The question to be solved is: Which of the evolutionarily possible communities will result from coevolution? That is, if the gene frequencies are not, in fact, held constant, but instead evolve to some values, say $(\hat{p}_1 \cdots \hat{p}_S)$, then the coevolutionary equilibrium community that results is identical to the community that would have resulted from the pure ecological model, if the p's were held constant at $(\hat{p}_1 \cdots \hat{p}_S)$ to begin with. Thus coevolution ultimately leads to one of the evolutionarily possible communities and the question is, which?

Results   We first state the central population genetic result for density-dependent coevolution.

LEMMA 2   *Assume that the fitnesses are purely density-dependent and that the associated pure ecological model has a unique, locally stable equilibrium for any set of fixed p's. Then, an equilibrium point in the coevolutionary model $(p_1^* \cdots p_S^*, N_1^* \cdots N_S^*)$ is locally stable, if and only if, the mean fitness in each species is a local maximum with respect to the gene frequency in that species; that is, for each i, $\bar{W}_i$, is a local maximum with respect to p_i.*

This lemma applies, regardless of whether all species are polymorphic, all fixed, or a mixture with some polymorphic and the rest fixed. This result is derived in Roughgarden (1976 Appendix II and 1977a). It is derived by examining the eigenvalues of the matrix, (a_{ij}), at the various equilibria. The absence of interspecific frequency dependence guarantees that the matrix (a_{ij}) has many zeros. The characteristic equation always factors in a way such that stability of the coevolution system is separable into the stability criteria for the ecological part and for the genetic part of the model.

Armed with the preceding lemma, we can identify which of the evolutionarily possible communities actually result from coevolution. We term an equilibrium community that represents a stable equilibrium in the co-evolution model a *coevolutionarily stable community*. We derive two different, but equivalent, ways of identifying a coevolutionarily stable community. The first is based on examining, for each species, whether the equilibrium gene frequency within the species maximizes or minimizes the population size of that species.

RESULT 2   *Condition for a coevolutionarily stable community from the effect of p_i on $\hat{N}_i$ for each i: Assume that the fitnesses are purely density-dependent and that the associated pure ecological model has a unique, locally stable equilibrium $(\hat{N}_1 \cdots \hat{N}_S)$ for any set of fixed gene frequencies $(p_1 \cdots p_S)$.*

A. An equilibrium community, $(\hat{p}_1 \cdots \hat{p}_S, \hat{N}_1 \cdots \hat{N}_S)$, is coevolutionarily stable if and only if, for each species-i, $\hat{N}_i$ is maximized or minimized at $\hat{p}_i$ according to the following criterion:

1. *If the sign of the feedback in the subcommunity obtained by deleting species-i is negative [i.e., $F_i(\hat{p}_1 \cdots \hat{p}_s) < 0$] then $\hat{N}_i$ is a local maximum with respect to p_i.*
2. *If the sign of the feedback in the subcommunity obtained by deleting species-i is positive [i.e., $F_i(\hat{p}_1 \cdots \hat{p}_s) > 0$], then $\hat{N}_i$ is a local minimum with respect to p_i.*

B. *If the feedback in the subcommunity obtained by deleting species-i is identically zero [i.e., $F_i(p_1 \cdots p_s) \equiv 0$], then evolution within species-i has no effect on the equilibrium abundance realized by species-i.*

C. *At least one species in a community must satisfy condition A1 above. That is, at least one species is such that pure density-dependent coevolution leads to the maximization of its population size.*

Part C restates the fact that $\sum F_i < 0$ in a stable community. Therefore, at least one F_i must be negative and hence at least one population satisfies condition A1 above. This result is proved in Roughgarden (1977a; see also Appendix II of 1976).

Condition A2 is especially important, because when it is met, evolution within a species reduces its own equilibrium population size. This is true, even though the mean fitness is being maximized. Also, this result is not to be confused with phenomena like the evolution of predation rates that overexploit the prey, and so forth. Such phenomena are a reflection of the effects of *interspecific frequency dependence*. As we shall see in the next sections, interspecific frequency dependence, generally, causes p_i to assume a value that leads to an $\hat{N}_i$, which is neither a local maximum nor minimum, but somewhere in between. In contrast, the result above shows that, even without interspecific frequency dependence, the result of evolution within a species may be to *minimize* its abundance. This possibility arises solely from the ecological network in which the species is embedded and does not involve schemes of frequency-dependent selection.

The next result presents a criterion for identifying a coevolutionarily stable community, based on examining the effect of every species upon a given species.

RESULT 3 *Condition for a coevolutionarily stable community from the effect of every gene frequency, p_j, on the population size of a given population N_i:*

Assume that the fitnesses are purely density-dependent and that the associated pure ecological model has a unique locally stable equilibrium $(\hat{N}_1 \cdots \hat{N}_S)$ for any set of fixed gene frequencies $(p_1 \cdots p_S)$.

A. *An equilibrium community $(\hat{p}_1 \cdots \hat{p}_S, \hat{N}_1 \cdots \hat{N}_S)$ is coevolutionarily stable, if and only if, for any chosen species, $\hat{N}_i$, that every p_j influences $\hat{N}_i$ as follows:*

1. *For $j = i$, then $\hat{N}_i$ is maximized or minimized with respect to p_i, accordingly as $F_i < 0$ or $F_i > 0$, as discussed above in Result 1; and*
2. *For all $j \neq i$, then*
   a. *$\hat{N}_i$ is minimized with respect to p_j if the feedback of the connecting web from j to i is negative (i.e., $F_{ij} < 0$) or*
   b. *$\hat{N}_i$ is maximized with respect to p_j if the feedback of the connecting web from j to i is positive (i.e., $F_{ij} > 0$).*

B. *If the feedback in the connecting web from j to i is identically zero [i.e., $F_{ij}(p_1 \cdots p_S) \equiv 0$], then evolution within species-j has no effect on the equilibrium abundance realized by species-i.*

This result is also proved in Roughgarden (1977a). Both results may be used together to give a complete picture of how evolution within all of the populations shapes the abundance of one another and their interactions. These results have also been obtained by Ginzburg (1977b), who has identified the feedback measures with minors of the inverse of the fitness matrix. We now illustrate these results with three examples.

Examples Three examples are presented to illustrate the results above. In each example, there are three variables, N_1, p_1, and N_2. Only species-1 is evolving, but its evolution may affect the equilibrium abundances realized by both species.

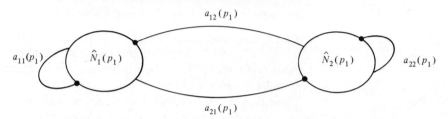

FIGURE 23.7. *The community graph G for Example 1. Both populations are competitors.*

EXAMPLE 1. TWO COMPETITORS. Consider a community with the graph illustrated in Figure 23.7. From this graph we can infer the configuration of a coevolutionarily stable community. To apply Result 2, we inspect the sign of the feedback in the subcommunity that is obtained by deleting species-1. The graph, G_1, is illustrated in Figure 23.8. The feedback in this subcommunity is negative ($a_{22} < 0$). Therefore, by Result 2, the coevolutionarily stable community is the configuration associated with a value of p_1 that *maximizes* $\hat{N}_1$. Alternatively, we may focus attention on species-2. To apply Result 3, we inspect the sign of the feedback of the connecting web from 1 to 2. The graph, G_{21}, appears in Figure 23.9. The feedback from this connecting web is negative. Therefore, by Result 3, the coevolutionarily stable community is the one associated with a value of p_1 that minimizes $\hat{N}_2$. The graphs in Figure 23.10 illustrate a coevolutionary process leading to the

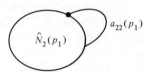

FIGURE 23.8. *The subgraph G_1 in Example 1.*

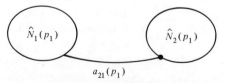

FIGURE 23.9. *The subgraph G_{21} in Example 1.*

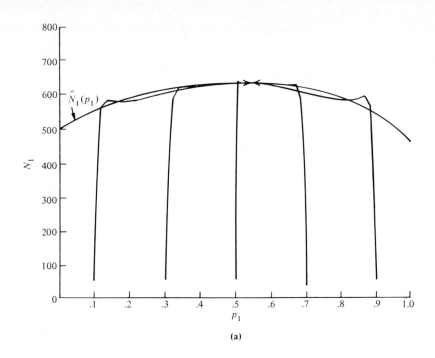

(a)

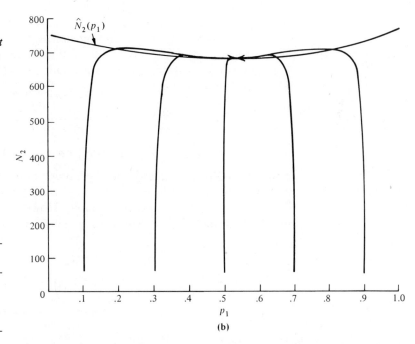

FIGURE 23.10.

Trajectories found in Example 1 in text of purely density-dependent coevolution between two competing populations. (a) *Note that the trajectories converge to the peak of the* $\hat{N}_1(p_1)$ *function, and* (b) *note that the trajectories converge to the low point of the* $\hat{N}_2(p_1)$ *function. Here* $r_1 = 1$, $r_2 = 1$, $\alpha_{21} = .5$, *and* $K_2 = 1000$.

α_{12}	K_1
A_1A_1: .7	1000
A_1A_2: .05	750
A_2A_2: .4	800

(b)

[From J. Roughgarden (1977), Coevolution in ecological systems: Results from "loop analysis" for purely density-dependent coevolution, pp. 499–518 in F. B. Christiansen and T. Fenchel, eds., *Measuring Selection in Natural Populations*, Vol. 19 in *Lecture Notes in Biomathematics*, Springer-Verlag, New York, Inc.]

community characterized above. The W's are assumed to be

$$W_{1,ij} = 1 + r_1 - \frac{r_1}{K_{1,ij}} N_1 - r_1 \frac{\alpha_{1ij,2}}{K_{1,ij}} N_2$$

$$W_2 = 1 + r_2 - \frac{r_2}{K_2} N_2 - r_2 \frac{\alpha_{2,1}}{K_2} N_1 \tag{23.35}$$

Notice that p_1 evolves to a value that simultaneously maximizes $\hat{N}_1$ and minimizes $\hat{N}_2$.

EXAMPLE 2. PREDATOR-PREY SYSTEM WITH PREY EVOLVING. Consider the standard Volterra predator-prey model with the additional assumption of negative density dependence in the prey. This additional assumption is needed for stability in a discrete-time model. There is no density dependence in the predator, however. The community graph is illustrated in Figure 23.11.

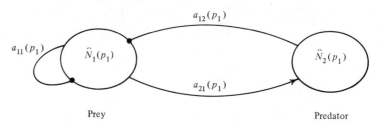

FIGURE 23.11. *The community graph in Example* 2.

The appropriate subgraphs appear in Figure 23.12.

By Result 2, evolution within species-1 does *not* affect the equilibrium abundance of species-1, because the feedback in G_1 is zero for all p. By Result 3, evolution within species-1 maximizes $\hat{N}_2$ because the feedback of the connecting web, G_{21}, is positive. Figure 23.13 illustrates the co-evolutionary process leading to the coevolutionarily stable community. The W's in the illustration are

$$W_{1,ij} = 1 + r_1 - \frac{r_1}{K_{1,ij}} N_1 - aN_2$$

$$W_2 = 1 + baN_1 - d \tag{23.36}$$

In the illustration the heterozygous genotype in the prey was assigned the highest K, thereby producing a polymorphism. But note that, as p_1 evolves to its final value, there is no effect on $\hat{N}_1$. However, p_1 does maximize the predator's abundance. As Rosenzweig (1972) has pointed out, increasing

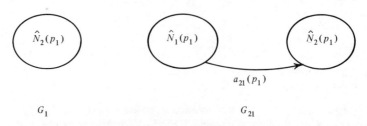

FIGURE 23.12. *The subgraphs* G_1 *and* G_{21} *for Example* 2.

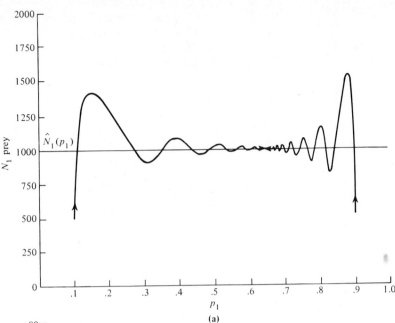

(a)

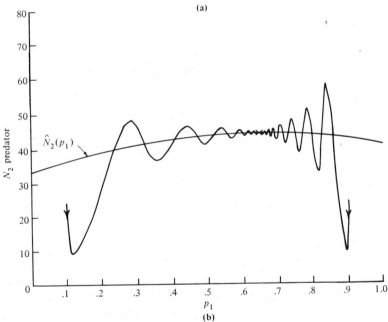

FIGURE 23.13.

Trajectories found in Example 2 in text of purely density-dependent coevolution between predator and prey populations. (a) Note that evolution within the prey does not influence the equilibrium prey abundance. The graph of the function, $\hat{N}_1(p_1)$, is flat and therefore evolutionary change in the prey does not affect $\hat{N}_1$. (b) In contrast, evolution within the prey leads to an increase in the equilibrium predator population size. Initial abundances: prey, 500; predator, 25.

		Prey Carrying Capacity, $K_{1,ij}$	
$r_1 = 1$	prey r		
$a = .01$	predator coefficient	A_1A_1	1700
$b = .1$	predator conversion coefficient	A_1A_2	2000
$d = 1$	predator death rate	A_2A_2	1500

[From J. Roughgarden (1977), Coevolution in ecological systems: Results from "loop analysis" for purely density-dependent coevolution, pp. 499–518 in F. B. Christiansen and T. Fenchel, eds., *Measuring Selection in Natural Populations*, Vol. 19 in *Lecture Notes in Biomathematics*, Springer-Verlag, New York, Inc.]

the K of the prey has a destabilizing effect on the system. Indeed, if K evolved to a sufficiently high level, the system would no longer have a stable equilibrium point.

It is well known that evolution of high predation rates may destabilize a predator-prey system by causing overexploitation of the prey. But here, a new effect is involved: Evolution of increased productivity by the prey is destabilizing the system.

EXAMPLE 3. A PIONEER AND A COMPETITOR. Consider the community graph in Figure 23.14. The pioneer species exerts a positive effect on itself and on the other species, while the competitor exerts a negative effect on itself and on the pioneer. The pioneer species might be viewed as a species that stabilizes the soil and causes the buildup of organic content in the soil, whereas the competitor is the second to arrive. The appropriate subgraphs appear in Figure 23.15.

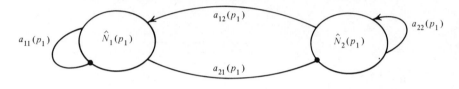

Competitor   Pioneer

FIGURE 23.14. *Community graph in Example 3.*

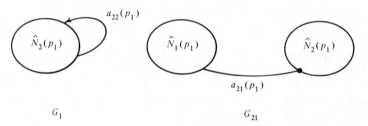

G_1 G_{21}

FIGURE 23.15. *The subgraphs G_1 and G_2 in Example 3.*

By Result 2 evolution within species-1 *minimizes* its own abundance because $F_1 > 0$. By Result 3 evolution within species-1 also minimizes the abundance of species-2, because $G_{21} < 0$. Figure 23.16 illustrates the co-evolutionary process. The W's are

$$W_{1,ij} = 1 + r - \frac{r}{K}N_1 + \alpha_{1ij,2}\frac{r}{K}N_2$$

$$W_2 = 1 + r + \frac{r}{K}N_2 - \alpha_{21}\frac{r}{K}N_1 \tag{23.37}$$

In the illustration, the heterozygote was assigned the highest α_{12}, thereby producing a polymorphism. Note that coevolution leads to the lowest population size for both species.

This example highlights the possibility of coevolution causing the extinction of a species. Evolution within a species may reduce *its own* equilibrium

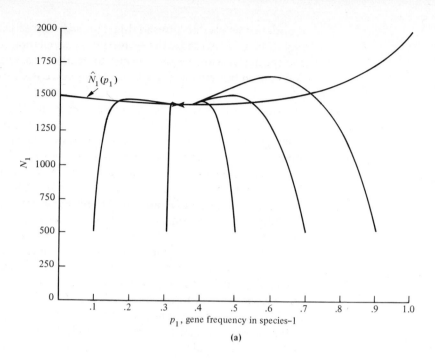

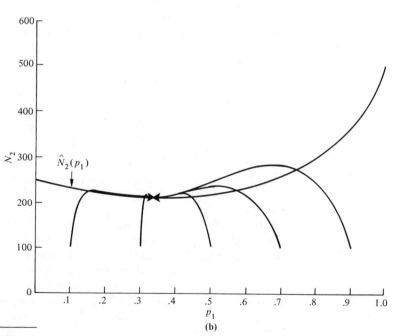

FIGURE 23.16.
Trajectories for Example 3 in text of purely density-dependent coevolution between a pioneer species and a competing species. (a) *Note that evolution within species-1 leads to a minimum of the $\hat{N}_1(p_1)$ function.* (b) *Note that evolution within species-1 also leads to a minimum of the $\hat{N}_2(p_1)$ function. Initial abundances: $N_1 = 500$; $N_2 = 100$. In both figures, $r = .25$, $K = 500$, $\alpha_{21} = .5$ in species-2.*

α_{12} for genotypes in species-1	
A_1A_2	3
A_1A_2	5
A_2A_2	4

[From J. Roughgarden (1977), Coevolution in ecological systems: Results from 'loop analysis' for purely density-dependent coevolution, pp. 499–518 in F. B. Christiansen and T. Fenchel, eds., *Measuring Selection in Natural Populations*, Vol. 19 in *Lecture Notes in Biomathematics*, Springer-Verlag, New York, Inc.]

abundance below the threshold level which causes extinction. Of course, evolution may also cause the extinction of populations through the effect of one species causing another species to decline to a level that brings extinction. But in this example, a species may evolve itself to extinction.

Evolutionary Reciprocation

Evolution in a species that is interacting with other species causes two effects. First, evolution within a species causes the population sizes of the other species to change. Second, it also causes the gene frequencies in the other species to change. The two results presented so far deal only with the first effect, which is called the "direct effect." Note that the results presented formulas only for the partial derivatives, $\partial \hat{N}_i / \partial p_j$, representing the effect of a small change in p_j on $\hat{N}_i$, assuming the other p's are held constant. In the preceding examples, only one species was evolving (although both underwent changes in N). Thus the results involving the partial derivatives happen to describe the process completely in these cases. But when two or more species are evolving, then the second effect of evolution within a species also enters the picture and we must take this into account: The term *evolutionary reciprocation* refers to the effect of evolution on the population size of a species, mediated by changes in the gene frequencies of other species.

It is easy to distinguish evolutionary reciprocation from the direct effect of evolution on population size as follows. Consider S interacting species, where all but one of them are evolving. Let the gene frequency in species-i be fixed at p_i, while all the other gene frequencies and all the population sizes are variables. In this situation, every equilibrium gene frequency becomes a function of p_i. Let $\hat{p}_k(p_i)$ denote the equilibrium gene frequency in species-k as a function of the gene frequency in species-i. Then, the *total effect* of evolution within species-i is the sum of the direct effect, plus the effect attributable to evolutionary reciprocation as follows:

$$\frac{d\hat{N}_i}{dp_i} = \frac{\partial \hat{N}_i}{\partial p_i} + \sum_{k \neq i} \frac{\partial \hat{N}_i}{\partial p_k} \frac{d\hat{p}_k}{dp_i} \quad (23.38)$$

total effect   direct   evolutionary
of p_i on   effect   reciprocation
$\hat{N}_i$

This formula shows that the evolutionary reciprocation arises from the summed effect of a change in p_i on $\hat{p}_k$ times the effect that changing p_k, in turn, has upon $\hat{N}_i$. A similar formula exists for $d\hat{N}_i/dp_j$.

The presence of evolutionary reciprocation raises a very important question. Can evolutionary reciprocation counteract or even overpower the direct effect of evolution within a species? For example, suppose that the direct effect of evolution within a species maximizes its population size, $(F_i < 0)$. Can the evolutionary reciprocation lead to a net decline in the population size? Similarly, suppose the direct effect of evolution within a species minimizes its population size, $(F_i > 0)$. Can the evolutionary reciprocation save it and produce a net increase in population size?

To answer this question, we shall inspect, for each i, the properties of $\hat{N}_i(p_i) = \hat{N}_i[\hat{p}_1(p_i) \cdots p_i \cdots \hat{p}_S(p_i)]$ at a coevolutionarily stable community. This function describes the equilibrium population size of species-i as a function of p_i, assuming that the rest of the community has come to both ecological *and evolutionary* equilibrium, relative to p_i. For simplicity, we assume that there is polymorphism in each species. Since $\partial \hat{N}_i / \partial p_j = 0$ for all i and j in a coevolutionarily stable community, we can infer from inspection of (23.38) that $d\hat{N}_i / dp_i = 0$. This means that in a coevolutionarily stable community, evolution within each species has either totally minimized or totally maximized its population size, when *both* the direct effect of evolution and of evolutionary reciprocation are taken into account.

The question moves, therefore, to the second derivatives, $d^2 N_i / dp_i^2$, for each i. We know the direct effect of p_i on $\hat{N}_i$ is to maximize $\hat{N}_i$, if $F_i < 0$ and to minimize it, if $F_i > 0$. The question, now, is whether the sign of the second derivatives can be changed by including the reciprocation. The second derivatives are given by

$$\frac{d^2 \hat{N}_i}{dp_i^2} = \frac{\partial^2 \hat{N}_i}{\partial p_i^2} + \sum_{k \neq i} \frac{\partial^2 \hat{N}_i}{\partial p_i \partial p_k} \frac{d\hat{p}_k}{dp_i} \qquad (23.39)$$

The effect of the reciprocation on the sign of the second derivative is given by the second term. To evaluate this term, we must determine $d\hat{p}_k / dp_i$. There are two steps to this. First, we define the equilibrium gene frequencies as a function of the N's, implicitly, by

$$\frac{\partial \bar{W}_k}{\partial p_k}(p_i, N_1 \cdots N_S) = 0 \qquad i = 1 \cdots S \qquad (23.40)$$

This equation implicitly defines $\hat{p}_i(N_1 \cdots N_S)$. The partial derivative with respect to the N_j is

$$\frac{\partial \hat{p}_k}{\partial N_j} = \frac{\partial^2 \bar{W}_k / \partial N_j \partial p_k}{-\partial^2 \bar{W}_k / \partial p_k^2} \qquad (23.41)$$

Next, we determine $d\hat{p}_k / dp_i$ by combining the known derivatives for $\partial \hat{N}_j / \partial p_i$ with the derivatives for $\partial \hat{p}_k / \partial N_j$, as just defined. Writing

$$\hat{p}_k = \hat{p}_k \{ \hat{N}_j [\hat{p}_1(p_i) \cdots : p_i \cdots \hat{p}_S(p_i)] \} \qquad j, k = 1 \cdots S \qquad (23.42)$$

and then differentiating the entire system with respect to p_i, leads to a system of $(S-1)$ linear equations for $d\hat{p}_k / dp_i$

$$\left[1 - \sum_j \frac{\partial \hat{p}_k}{\partial N_j} \frac{\partial \hat{N}_j}{\partial p_k} \right] \frac{d\hat{p}_k}{dp_i} - \sum_{h \neq k, i} \left[\sum_j \frac{\partial \hat{p}_k}{\partial N_j} \frac{\partial \hat{N}_j}{\partial p_h} \right] \frac{d\hat{p}_h}{dp_i} = \sum_j \frac{\partial \hat{p}_k}{\partial N_j} \frac{\partial \hat{N}_j}{\partial p_i} \qquad h = 1 \cdots S$$

$$(23.43)$$

This system can be solved explicitly by $d\hat{p}_k / dp_i$. However, it is sufficient to note here that, at a coevolutionary stable community, $\partial \hat{N}_j / \partial p_i = 0$ for all i and j so that (23.43) reduces to

$$\frac{d\hat{p}_k}{dp_i} = 0 \qquad \text{for all } i \qquad (23.44)$$

Equation (23.44) means that the effect of evolutionary reciprocation on the total second derivative is zero. That is, (23.39) becomes

$$d^2 \hat{N}_i = \left(\frac{\partial^2 \hat{N}_i}{\partial p_i^2} \right) dp_i^2 \qquad (23.45)$$

Equivalently, we may say that evolutionary reciprocation is always a lower order effect than the direct effect of evolution within a species on its own population size.

We may summarize the above paragraphs in the following statement.

RESULT 4 *Consider purely density-dependent coevolution. In a coevolutionarily stable community, evolution within each species totally maximizes its own population size if $F_i < 0$ and totally minimizes its own population size if $F_i > 0$.*

This statement is a strong and interesting generalization of Result 2. It means that if the direct effect of evolution within a species is to maximize its size, then evolutionary reciprocation cannot alter the fact. If the direct effect is to minimize population size, then evolutionary reciprocation cannot save the population from evolving itself to extinction. A similar generalization of Result 3 is possible.

A convenient spin-off of the statment above is that it is possible to present a simple graphical description of a community of S species under density-dependent coevolution. Instead of a single graph of dimension $2S$, we may write S separate graphs in two dimensions, one graph for each species. For species-i the vertical axis is N_i and the horizontal axis is p_i. The function of one variable, $\hat{N}_i(p_i) = \hat{N}_i[\hat{p}_i(p_i) \cdots p_i \cdots \hat{p}_S(p_i)]$, can be plotted in the graph for species-i. Even as the rest of the community is changing in both N and p, the trajectory for species-i on this graph will converge to a maximum or minimum of $\hat{N}_i(p_i)$, according to the sign of F_i.

Discussion   We conclude this section with some general remarks about how coevolution influences community structure. The traditional view of coevolution is that it leads to mutually accommodated and coadapted species. It is presumed that a community of coevolved species satisfies one or more of the following descriptions: It is maximally resistant to invasion by a new species, maximally resistant to destruction from environmental perturbations, and maximally likely to persist through time. These presumptions have resurfaced frequently in the discussion of complexity and stability in ecosystems. R. May (1973) has shown that complex ecosystems are less likely to be stable than simple ones, in contrast to the previous supposition. See also the experimental results of Hairston et al. (1968) and the cautionary note of Lawlor (1978). However, the complexity-stability issue has generally been discussed in terms of purely ecological models with no possibility of coevolution among the interacting populations. Critics of these models have often dismissed the conclusions on the basis that they apply only to "random assemblages" and not to a coevolved biological system. Hence the question remains, when coevolution *is* taken explicitly into account, does coevolution generally improve community stability in any of the senses mentioned above?

On the basis of this section, the answer is no. Coevolution certainly restructures and molds a community, but its end result is not generally more stable than the initial configuration. This message is especially clear in this section, because only the best-behaved coevolution has been considered. It is, a priori, evident that coevolution with complicated interspecific frequency dependence can greatly destabilize a community. Indeed, examples are easily made in which the mean fitness is minimized at a stable polymorphic equilibrium and the mean fitness maximized at an unstable polymorphic equilibrium, all because of interspecific frequency dependence. In purely density-dependent coevolution, however, the mean fitness is always maximized at a stable equilibrium. In this case, our intuitions about the optimum strategies of evolution, *within* each species, are correct. Nonetheless, the optimum strategy within each species for maximizing fitness, even when natural selection will in fact maximize fitness, does not necessarily improve the abundance of that species. Whether it does or not, depends on the ecological network in which it is embedded.

In two of the examples coevolution tends to destabilize the community. In the predator-prey example the evolution of increased productivity by the prey had no effect on the realized prey abundance, but maximized the predator's abundance. Increasing the K of the prey, destabilizes the system, as termed the "paradox of enrichment" by Rosenzweig (1972).

But in the third example coevolution destabilizes the system in a much stronger sense. In that example species-1 is the species that stabilizes the system to begin with. If species-1 were removed, the subcommunity that remains could not exist because of its inherent positive feedback. However, evolution within species-1 lowers its own abundance. It may evolve itself to extinction and, if so, the remaining community is inherently unstable. *Whenever* a species is connected to an unstable subcommunity (i.e., $F_i > 0$), then evolution within that species will lower its own population size. Thus the coevolutionary fate of a community composed of unstable subcommunities connected to an occasional stabilizing species is bleak indeed.

Of course, we may imagine that some combined process of population turnover, together with coevolution, will eventually produce communities composed only of interconnected stable subcommunities. Such a community would presumably be very stable. But it requires a great leap of optimism and faith to believe that this really happens.

Interspecific Frequency-Dependent Coevolution	In this section, we address the evolution of parameters under the joint control of two species. This section is taken from Roughgarden (1978a). Most of the traditional topics of coevolution, in fact, do involve the evolution of parameters under the joint control of two interacting species. For example, the parameters characterizing the rate at which prey are captured by the predator population are under the control of both predator and prey. The prey control the extent to which they exhibit toxicity, defensive armor, crypticness, and the extent to which they hide in order to avoid predation while seeking food in comparatively open areas. The predators control their own preferences for prey of different types and the amount of time spent looking for it. The realized predation rate thus reflects the traits which have evolved in both predator and prey and hence is under the evolutionary control of both species. Similarly, the evolution of character displacement between two competitors involves parameters, the competition coefficients, which are under the joint control of the two species involved. This section extends the ideas and techniques developed earlier in this chapter to coevolution of parameters that are jointly controlled by the interacting populations.

Assuming that a parameter is under joint control is equivalent to assuming that there is interspecific frequency dependence. In general, *interspecific frequency dependence means that the fitness of genotype, A_iA_j, in species-s, $W_{s,ij}$, is potentially a function of all the population sizes and of the gene frequencies of all the other species.* (That is, with S species, $W_{s,ij}$ is a function of N_k, $k = 1 \cdots S$, and p_k, $k = 1 \cdots S$ and $k \neq s$.) We shall see that interspecific frequency dependence introduces two very important considerations that were not present in purely density-dependent coevolution. First, our belief that natural selection maximizes fitness must be strongly qualified. Second, even when natural selection does maximize fitness, the result does not maximize or minimize the population sizes of the species involved. What does happen can be understood in terms of extensions to the graphical terminology introduced in the last section, but the overall result is often surprising.

The Maximization of Fitness Under Interspecific Frequency Dependence

The basic coevolutionary model remains unchanged,

$$\Delta N_i = (\bar{W}_i - 1)N_i$$

$$\Delta p_i = \frac{p_i(1-p_i)}{2\bar{W}_i} \frac{\partial \bar{W}_i}{\partial p_i} \qquad (i = 1 \cdots S) \qquad (23.46)$$

where

$$\bar{W}_i = p_i^2 W_{i,11} + 2p_i(1-p_i)W_{i,12} + (1-p_i)^2 W_{i,22} \qquad (23.47)$$

What is new is that

$$W_{s,ij} = W_{s,ij}(N_1 \cdots N_S, p_1 \cdots p_{i \neq s} \cdots p_S) \qquad (23.48)$$

Again, a coevolutionary equilibrium (stable or unstable) satisfies, for every i, one of the following

$$\bar{W}_i = 1, \qquad \frac{\partial \bar{W}_i}{\partial p_i} = 0 \qquad (23.49\text{a})$$

$$W_{i,11} = 1, \qquad \hat{p}_i = 1 \qquad (23.49\text{b})$$

$$W_{i,22} = 1, \qquad \hat{p}_i = 0 \qquad (23.49\text{c})$$

In (23.49a), there is polymorphism, and in (23.49b) and (23.49c), there is fixation of an allele.

As before, we associate, with the coevolutionary model, another model that is the pure ecological model. This model is obtained, formally, from (23.46) by taking the p's as constants, thus leaving us with the equations for the variable population sizes

$$\Delta N_i = (\bar{W}_i - 1)N_i \qquad (23.50)$$

Again, the set of equilibrium communities resulting from this model for all fixed gene frequencies comprises the set of communities that are evolutionarily possible. The question remains as to which of the evolutionarily possible communities will occur at the coevolutionary equilibrium.

We can provide a partial answer to this question on the basis of the following lemma.

LEMMA 3 *Assume that the fitnesses have density dependence and interspecific frequency dependence, and that the associated pure ecological model has a unique locally stable equilibrium for any set of fixed p's. Then,*

(a) An equilibrium point of the coevolutionary model, $(\hat{p}_1 \cdots \hat{p}_S, \hat{N}_1 \cdots \hat{N}_S)$, where every species is fixed (i.e., $\hat{p}_i = 0, 1$ for all i), is

locally stable if and only if the mean fitness in each species is maximized with respect to the gene frequency in that species.

(b) *An equilibrium point of the coevolutionary model, $(\hat{p}_1 \cdots \hat{p}_S, \hat{N}_1 \cdots \hat{N}_S)$, in which one or more of the species is polymorphic, is not necessarily stable if the mean fitness is maximized in every species. Nor is it necessarily unstable if the mean fitness is minimized in some species.*

This lemma is much weaker than its counterpart for purely density-dependent coevolution. It is proved by methods identical to those used in Roughgarden (1977a). If an equilibrium involves a polymorphism in one or more species, there simply is not any general intuitive criterion to determine its stability. All the polymorphic equilibria can always be located, using formula (23.49a). This stability must be settled, however, in each example by detailed reference to the eigenvalues of the stability matrix at each equilibrium point. At present, there is no general optimization principle that determines stability, when polymorphism is involved.

It is of fundamental interest to know whether fitness is maximized at equilibrium. If it is, then we are entitled to use strategy reasoning to determine what kinds of traits will result from evolution. Part (a) of the above result tells us that we are entitled to use strategy reasoning at least some of the time. It says that there exist situations in which coevolution with interspecific frequency dependence does lead to fitness maximization in each species. Furthermore, part (a) asserts that a community configuration, in which fitness is maximized in each species, is certain to be coevolutionarily stable, provided that this configuration is realized without invoking polymorphism in the species involved.

Part (a) of the above result provides a connection with the concept of an "evolutionarily stable strategy" or ESS introduced by Maynard Smith (1974). An ESS is defined as a strategy such that if the population consists entirely of individuals with the ESS, then no mutant allele producing some slightly different strategy can increase. The criterion for an ESS is thus a criterion for the stability of a boundary equilibrium. Part (a) shows that a community configuration, in which fitness is maximized within each of the species, is the community analogue of an ESS. As we saw in the last section, a much stronger concept exists for purely density-dependent coevolution. With interspecific frequency dependence, however, we must be satisfied with a justification for strategy reasoning in terms of the stability of boundary equilibria. The importance of the criteria for the initial increase of alleles in a coevolutionary model is further discussed in Allen (1976) and Schaffer (1977).

Part (b) in the above result is also interesting. It indicates a generally destabilizing role for polymorphism in ecological contexts in which traits that govern species interactions are involved. It is possible for polymorphism to destabilize a strategy in which fitness is maximized in every species and that would certainly be stable if attained without recourse to polymorphism. It is also possible for polymorphism to stabilize strategies that minimize fitness.

In the next section, we restrict attention to communities where fitness is maximized in each species. We shall see that interspecific frequency dependence influences the equilibrium population sizes of the species in a community in surprising ways, even when fitness is maximized in each species.

The Theory of Coadaptation

Natural selection in interacting species causes the evolution of traits in each species that are adaptations to the presence of the other species. This section explains the evolutionary design of coadaptation. It explains how the evolution of coadaptation influences the abundance of the species that are coadapted to one another. Here, we abandon explicit reference to gene frequencies and take a strategy reasoning approach because we can most easily visualize the theory in this context.

We shall assume that there is a *trait* in each species whose value is X_s for species-s. For example, X_s could represent the amount of energy that a prey allocates into defensive armor, or X_s could be the amount of time, per 24 hours, that a predator actually searches for prey, and so forth. The interaction coefficient between the species are functions of these traits. For example, the standard predator-prey coefficient, a, in the predator-prey model is a function of both X_1 and X_2; that is, $a = a(X_1, X_2)$. Thus the interaction coefficients between the species, and hence ultimately, their equilibrium population sizes, are influenced by the coevolution of these traits. The interaction coefficients are jointly controlled by the two interacting species.

We shall determine the optimum value of X_s in each species by maximizing the fitness in each species. We are justified in doing this because, as discussed in the last section, the community configuration associated with fitness maximization is coevolutionarily stable, provided the configuration occurs at a genetic boundary equilibrium. That is, if, for every s, species-s is fixed for an allele that produces a value of the trait, X_s, which maximizes the fitness in species-s, then, a mutant allele that produces a slightly different value of X_s cannot increase. Our probem is to discover what the community looks like, which does satisfy the criterion that the value of the trait, X_s, in each species maximizes the fitness, W_s, in that species.

The Coevolutionary Strategy Model

A strategy model is not, itself, a dynamic model, although it can be used to predict the equilibrium configuration of a dynamic model. The strategy model is defined to be the set of S expressions for the fitness in each species.

$$W_s(X_1 \cdots X_S, N_1 \cdots N_S) \qquad s = 1 \cdots S \qquad (23.51)$$

The set of traits representing the coevolutionary equilibrium, $(\hat{X}_1 \cdots \hat{X}_S)$, is defined as the set of traits, such that W_s is maximized with respect to X_s at $X_s = \hat{X}_s$, subject to the constraint that $W_s = 1$, for all s. Furthermore, the equilibrium population sizes, $\hat{N}_i(X_1 \cdots X_S)$, are defined implicitly by the equation

$$W_s = 1 \qquad s = 1 \cdots S \qquad (23.52)$$

These equilibrium population sizes must be positive and unique for all values of the traits under consideration. The coevolutionarily stable community is defined as the set of coevolutionary equilibrium traits, $(\hat{X}_1 \cdots \hat{X}_S)$, together with the set of equilibrium population sizes corresponding to those traits, $(\hat{N}_1(\hat{X}_1 \cdots \hat{X}_S) \cdots \hat{N}_S(\hat{X}_1 \cdots \hat{X}_S))$.

In addition to the definitions above, there are additional constraints that the coevolutionary strategy model must satisfy. These constraints are that the equilibrium community, $\hat{N}_1(X_1 \cdots X_S)$, represent a stable ecological community for all values of the traits under consideration. This constraint translates into requiring that the matrix, $\partial W_i/\partial N_j$, evaluated at $N_i = \hat{N}_i(X_1 \cdots X_s)$, have eigenvalues with negative real parts for all values of the traits under consideration. This is a technical requirement, which is automatically satisfied, if the ecological model within which the coevolution is embedded is well defined to begin with.

The Coevolutionary Strategy Graph The coevolutionary strategy model can be represented by a graph. There are nodes for the equilibrium population sizes, $\hat{N}_i$, and also for the equilibrium values of the traits $\hat{X}_i$. A coevolutionary strategy graph for two coevolving species is illustrated in Figure 23.17.

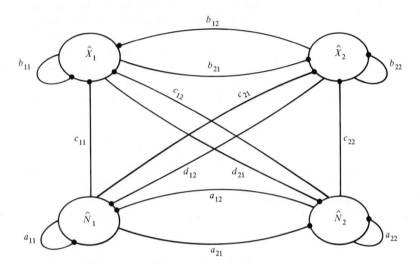

FIGURE 23.17. *The graph H used to determine the effect of evolving strategies that maximize fitness on the population sizes of the members of a community.*

The connections between the $\hat{N}_i$ are familiar from our earlier study of pure density-dependent coevolution. We define the weights attached to each connection between the $\hat{N}_i$ as

$$a_{ij} = \frac{\partial W_i}{\partial N_j} \tag{23.53}$$

The element a_{ij} refers to the direct effect on N_j upon the fitness in species-i. (These definitions of a_{ij} in the strategy graph are the same, in principle, as those in the community graph. The only differences are in the absence of $\hat{N}_i$ as a factor.) The variables, a_{ij}, must satisfy the conditions that ensure that the eigenvalues of the matrix $\{a_{ij}\}$ have negative real parts, mentioned in the paragraph above. Such conditions will involve the sign of the feedback, defined on this part of the graph, in a manner completely analogous to that in density-dependent coevolution.

The connections from the X's to the N's are defined as

$$d_{ij} = \frac{\partial W_i}{\partial X_j} \tag{23.54}$$

The element d_{ij}, refers to the direct effect of X_j upon the fitness in species-i. An important feature of the graph is that d_{ii} always equals zero in a coevolutionary stable community. This fact is true because the $\hat{X}_i$ are defined as those values of X_i that maximize $\hat{W}_i$, and, therefore, $\partial W_i / \partial X_i = 0$ at $X_i = \hat{X}_i$.

The connections among the X's are defined as

$$b_{ij} = \frac{\partial}{\partial X_j}\left(\frac{\partial W_i}{\partial X_i}\right) \tag{23.55}$$

The element, b_{ij}, refers to the direct effect of X_j upon the slope of W_i, with respect to X_i. This quantity is very important, as illustrated in Figure 23.18. The fact that W_i is maximized at $\hat{X}_i$ means that $\partial W_i / \partial X_i = 0$ at $X_i = \hat{X}_i$. If this slope becomes positive as X_j is increased to $X_j + dX_j$, then it means that the value of X_i that maximizes W_i is moved to the right. This is illustrated above. Thus, if b_{ij} is positive, it means that the direct effect of increasing X_j is to increase $\hat{X}_i$. For this reason, the term b_{ij} describes the connections between the $\hat{X}$'s. Note that b_{ii} is necessarily negative because $\hat{X}_i$ maximizes W_i.

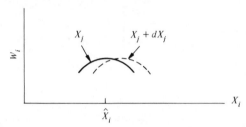

FIGURE 23.18. *Sketch illustrating the interpretation of b_{ij} as a mixed second derivative.*

The connections from the N's to the X's are, similarly, defined as

$$c_{ij} = \frac{\partial}{\partial N_j}\left(\frac{\partial W_i}{\partial X_i}\right) \tag{23.56}$$

The element, c_{ij}, refers to the direct effect of N_j on the slope of W_i, with respect to X_i.

The Measures of Feedback
There are measures of the feedback in this graph and in various subgraphs. We begin with the subgraphs that are familiar for the pure density-dependent section, as summarized in Table 23.1.

In Table 23.1, $\{a_{ij}\}$ means the matrix of dimension, $S \times S$, formed by the elements $\partial W_i / \partial N_j$. The symbol, $\det_{ji}\{a_{ij}\}$, means the determinant of the

Table 23.1.

Graph		Feedback Measure
G	Graph obtained by deleting all the X's	$F = (-1)^{S+1} \det\{a_{ij}\}$
G_i	Subgraph to G obtained by deleting N_i	$F_i = (-1)^{S} \det_{ii}\{a_{ij}\}$
G_{ij}	Subgraph of G representing the connecting web from species-j to species-i	$F_{ij} = (-1)^{S+1}(-1)^{i+j} \det_{ji}\{a_{ij}\}$

matrix of dimension, $(S-1) \times (S-1)$, obtained by deleting the ith column and the jth row from $\{a_{ij}\}$. By the assumption of ecological stability, $F < 0$ and $\sum F_i < 0$.

The full strategy graph corresponds to the matrix of dimension $2S \times 2S$, given by

$$Z = \begin{pmatrix} B & C \\ D & A \end{pmatrix} \tag{23.57}$$

The feedback of various subgraphs of the full strategy graph are defined in terms of the Z matrix above. The matrix Z^{ii} is defined as the matrix of dimension $(2S-1) \times (2S-1)$, obtained by striking out the ith row and column from Z. Table 23.2 presents the new subgraphs and feedback measures.

Table 23.2

	Graph	Feedback Measure
H	The full coevolutionary strategy graph	$I = -\det(Z)$
$H^{(X_i)}$	The subgraph of H obtained by deleting X_i	$I^{(X_i)} = \det(Z^{ii})$
$H_{N_j}^{(X_i)}$	The subgraph of $H^{(X_i)}$ obtained by deleting N_j	$I_{N_j}^{(X_i)} = -\det_{N_j, N_j}(Z^{ii})$
$H_{X_j}^{(X_i)}$	The subgraph of $H^{(X_i)}$ obtained by deleting X_j $(j \neq i)$	$I_{X_j}^{(X_i)} = -\det_{X_j, X_j}(Z^{ii})$
$H_{N_k, X_j}^{(X_i)}$	The subgraph of $H^{(X_i)}$ representing the connecting web from X_j to N_k $(j \neq i)$	$I_{N_k, X_j}^{(X_i)} = (-1)^{1+k+j} \det_{X_j, N_k}(Z^{ii})$
$H_{N_k, N_j}^{(X_i)}$	The subgraph of $H^{(X_i)}$ representing the connecting web from N_j to N_k $(j \neq k)$	$I_{N_k, N_j}^{(X_i)} = (-1)^{k+j} \det_{N_j, N_k}(Z^{ii})$
$H_{X_k, X_j}^{(X_i)}$	The subgraph of $H^{(X_i)}$ representing the connecting web from X_j to X_k $(j \neq k \neq i)$	$I_{X_k, X_j}^{(X_i)} = (-1)^{k+j} \det_{X_j, X_k}(Z^{ii})$
$H_{X_k, N_j}^{(X_i)}$	The subgraph of $H^{(X_i)}$ representing the connecting web for N_j to X_k $(k \neq i)$	$I_{X_k, N_j}^{(X_i)} = (-1)^{1+k+j} \det_{N_j, X_k}(Z^{ii})$

In Table 23.2 the symbol $\det_{X_j, N_k}(Z^{ii})$ means that the determinant of the matrix of dimension $(2S-2) \times (2S-2)$ is obtained by deleting the column, corresponding to N_k, and the row, corresponding to X_j from the matrix, Z^{ii}. In calculating the feedback of a connecting web, the practical points to remember are to delete the column of the receiving entity and the row of the sending entity. If the entities are of the same type (i.e., if both are N's or both X's), then the coefficient is $(-1)^{k+j}$. If the entities are of different types, however, the coefficient is $(-1)^{1+k+j}$. These definitions of feedback follow the same idea as those introduced previously, and are set out in the table above for completeness.

Results Our task is to discover what a coevolutionarily stable community looks like when interspecific frequency dependence is involved. In one sense, this task is accomplished by determining the values of the traits that maximize the fitness in each species and then by evaluating the corresponding equilibrium population sizes. A more interesting question is to examine the nature of the mismatch between coadapted species. The graph and feedback measures introduced above allow us to study this issue with simplicity and generality. We shall derive expressions for the direct effects of changing a trait in a species, upon the abundances of all the species. For example, suppose that $\hat{X}_i$ is the value of the trait in species-i that maximizes W_i. Then, we want to know the quantity $\partial \hat{N}_i / \partial \hat{X}_i$ evaluated at this point. If this derivative is positive, it means that if X_i were increased beyond $\hat{X}_i$, then the direct effect is to increase the abundance of the species involved. Furthermore, we want to know the quantity $d\hat{N}_i / dX_i$, which represents the total effect on $\hat{N}_i$ of changing X_i. If this quantity is positive, then raising X_i above $\hat{X}_i$ will raise the abundance of species-i after all the evolutionary reciprocation is taken into account. Recall that in a coevolutionarily stable community under pure density-dependent coevolution, the quantities $\partial \hat{N}_i / \partial X_i$ and $d\hat{N}_i / dX_i$ are zero. With pure density-dependent coevolution, the final result is either the best or the worst for the species involved, depending on F_i. But with interspecific frequency dependence, the final result is somewhere in between the best and worst. The values of $\partial \hat{N}_i / \partial X_i$ and of $d\hat{N}_i / dX_i$ at the equilibrium help to understand exactly where that somewhere is. These derivatives help to specify the character of the mismatch inherent in coadaptation.

THE DIRECT EFFECTS. The direct effect on the $\hat{N}$'s of varying X_i around $\hat{X}_i$ is given by the following result:

RESULT 5   *In a coevolutionarily stable community, the direct effects of varying X_i are*

$$\frac{\partial \hat{N}_i}{\partial X_i} = -\sum_{k \neq i} \left(\frac{F_{ik}}{F} \right) d_{ki} \tag{23.58a}$$

$$\frac{\partial \hat{N}_i}{\partial X_j} = -\sum_{k \neq i,j} \left(\frac{F_{ik}}{F} \right) d_{kj} + \left(\frac{F_i}{F} \right) d_{ij} \tag{23.58b}$$

Obviously, the quantities $\partial \hat{N}_i / \partial X_j$, are generally not equal to zero, and, therefore, even when selection is assumed to maximize fitness in each species, the result does not maximize (or minimize) the abundance of the species involved. This result depends solely on the interspecific frequency-dependence. Without frequency dependence, the d_{ij} equal zero and the results of the section on pure density-dependent coevolution are regained. Equations (23.58) are derived by differentiating the fitness functions implicitly and then, employing the appropriate feedback definitions.

Although the equilibrium population size is usually not maximized (or minimized) in the presence of interspecific frequency dependence, there is, nonetheless, a quantity similar to the equilibrium population size that is maximized and that helps to explain what interspecific frequency dependence is doing in the system. Consider a new quantity that we will call the *conditional equilibrium population size*, $\tilde{N}_i$, discussed in Roughgarden (1976). It is defined implicitly for each species, say species-i, by the single equation

$$W_i(X_1 \cdots X_S, N_1 \cdots N_S) = 1 \tag{23.59}$$

$\tilde{N}_i$ is a function of all the traits and of all the population sizes, except that of species-i. That is,

$$\tilde{N}_i = \tilde{N}_i(X_1 \cdots X_S, N_1 \cdots N_{j \neq i} \cdots N_S) \tag{23.60}$$

The derivatives of W_i and $\tilde{N}_i$ are related to each other, as found by differentiating implicitly, which yields

$$\frac{\partial \tilde{N}_i}{\partial X_i} = \frac{(\partial W_i / \partial X_i)}{(-\partial W_i / \partial N_i)}, \quad \frac{\partial^2 \tilde{N}_i}{\partial X_i^2} = \frac{(\partial^2 W_i / \partial X_i^2)}{(-\partial W_i / \partial N_i)} \tag{23.61}$$

These equations show that, *in a species with negative density dependence, maximizing fitness is equivalent to maximizing the conditional equilibrium population size.*

The result that a population (with negative density dependence) evolves so as to maximize its conditional equilibrium population size, $\tilde{N}_i$, is very interesting, biologically. *The quantity, $\tilde{N}_i$, is the equilibrium population size that species-i would attain, if the population sizes and traits of the other species remained constant.* The fact that evolution within a species maximizes $\tilde{N}_i$ means that natural selection within a species is, so to speak, operating as though the properties of other species in the community were constants. It means that natural selection within a species does not distinguish between the abiotic and biotic parts of its environment. Natural selection within a species causes traits to evolve that promote adaptation to the presence of other species, exactly as it causes traits to evolve that promote adaptation to permanent physical features of the environment. Yet, herein, lies the basic dilemma of coadaptation. The other species are not constants—they change in abundance and in phenotype. As a result, coadaptation generally leads to a slight mismatch. Although natural selection does maximize the conditional equilibrium population size, it does not maximize the actual equilibrium population size. What it does to the actual population size is generally not the worst of all possibilities, nor the best. Instead, natural selection achieves an imperfect state of coadaptation between interacting species.

EVOLUTIONARY RECIPROCATION. Interspecific frequency dependence makes evolutionary reciprocation into an important and surprising phenomenon. Recall that with pure density-dependent coevolution the total effect of a species on its own population size is dominated by the direct effect. In a coevolutionarily stable community, the contribution from the evolutionary reciprocation was zero. Reciprocation could not counteract or overpower the direct effect of evolution within a species on its own population size. But, with interspecific frequency dependence, the situation changes dramatically. In general terms, two new facts emerge. First, the evolutionary reciprocation is of the same order of magnitude as the direct effect and so the total outcome of evolution upon a species is never determined by the direct effect, alone. Second, the direction in which the reciprocation acts is often surprising. As we learned above, the direct effect of evolution does not, generally, lead to the optimal value of the traits involved. This problem is compounded, when reciprocation is taken into account. The direct effect of evolution within a species will lead to a nonoptimal value of a trait, but it will also cause an interacting species to produce a nonoptimal value of *its* trait. The net effect of both species evolving nonoptimal values of their traits can be very surprising and can differ for each model and situation. We now explore this issue in more detail.

The total effect of a trait upon the abundance of a species is defined as the sum of the direct effect and the evolutionary reciprocation

$$\frac{d\hat{N}_i}{dX_i} = \frac{\partial \hat{N}_i}{\partial X_i} + \sum_{k \neq i} \frac{\partial \hat{N}_i}{\partial X_k} \frac{d\hat{X}_k}{dX_i} \qquad (23.62)$$

$$\underset{\text{total}}{} \underset{\text{direct}}{} \underset{\text{evolutionary}}{}$$
$$\underset{\text{effect}}{} \underset{\text{effect}}{} \underset{\text{reciprocation}}{}$$

In rare circumstances, this formula is actually the most efficient way to calculate the total effect. Usually, however, the most practical way to calculate the total effect is with the formulas below, using the feedback measures introduced earlier.

RESULT 6 *In a coevolutionarily stable community, the total effects of varying X_i are*

$$\frac{d\hat{N}_i}{dX_i} = -\sum_{j \neq i} \left(\frac{I^{(X_i)}_{N_i, X_j}}{I^{(X_i)}} \right) b_{ji} - \sum_{j \neq i} \left(\frac{I^{(X_i)}_{N_i, N_j}}{I^{(X_i)}} \right) d_{ji} \qquad (23.63a)$$

$$\frac{d\hat{N}_k}{dX_i} = -\sum_{j \neq i} \left(\frac{I^{(X_i)}_{N_k, X_j}}{I^{(X_i)}} \right) b_{ji} - \sum_{j \neq i, k} \left(\frac{I^{(X_i)}_{N_k, N_j}}{I^{(X_i)}} \right) d_{ji} + \left(\frac{I^{(X_i)}_{N_k}}{I^{(X_i)}} \right) d_{ki} \qquad (23.63b)$$

$$\frac{d\hat{X}_k}{dX_i} = -\sum_{j \neq i, k} \left(\frac{I^{(X_i)}_{X_k, X_j}}{I^{(X_i)}} \right) b_{ji} + \left(\frac{I^{(X_i)}_{X_k}}{I^{(X_i)}} \right) b_{ki} - \sum_{j \neq i} \left(\frac{I^{(X_i)}_{X_k, N_j}}{I^{(X_i)}} \right) d_{ji} \qquad (23.63c)$$

The formulas for the important case of two interacting species are summarized in Table 23.3.

Table 23.3. Coevolutionary Mismatch for Two Interacting Populations.

Direct effects:

$$\frac{\partial \hat{N}_1}{\partial X_1} = -\left(\frac{F_{12}}{F} \right) d_{21} \qquad \frac{\partial \hat{N}_2}{\partial X_1} = \left(\frac{F_2}{F} \right) d_{21}$$

Total effects:

$$\frac{d\hat{N}_1}{dX_1} = \frac{(I^{(X_1)}_{N_1, X_2}) b_{21} + (I^{(X_1)}_{N_1, N_2}) d_{21}}{-I^{(X_1)}}$$

$$\frac{d\hat{N}_2}{dX_1} = \frac{(I^{(X_1)}_{N_2, X_2}) b_{21} - (I^{(X_1)}_{N_2}) d_{21}}{-I^{(X_1)}}$$

$$\frac{d\hat{X}_2}{dX_1} = \frac{-(I^{(X_1)}_{X_2}) b_{21} + (I^{(X_1)}_{X_2, N_2}) d_{21}}{-I^{(X_1)}}$$

The effects of X_2 on N_1 and N_2 are given by symmetrical formulas.

Examples We illustrate the use of the formulas introduced in Table 23.3 with two examples. The first example is of the coevolution between predator and prey. The second example is of the coevolution of character displacement between two competing species.

EXAMPLE 1. PREDATOR-PREY COEVOLUTION. Perhaps the most classic topic in the biology of coevolution is the coevolution of predator and prey populations in conjunction with one another. Papers on this topic include Pimental et al. (1968, 1975, 1978) and Rosenzweig (1969, 1973). Here we visualize that X_1 denotes the effort allocated by a prey individual to predator defense and avoidance. The variable X_2 denotes the effort allocated by a predator to the search for prey and to the breakdown of the prey defenses. We may want to know what value X_1 and X_2 will attain at the coevolutionary equilibrium, what the net predation rate is at the coevolutionary equilibrium and what the equilibrium population sizes of predator and prey will be. But more importantly, we have discovered in general terms from the preceding theory, that the prey defense effort, X_1, and the predator search effort, X_2, will not evolve to the optimum values that maximize the population sizes of the respective species. What we now want are more specifics about the mismatch of coevolution. Do the prey evolve a level of defense that is too much, in the sense that lowering the defense effort would increase their population size? Or is the reverse true? Do the predators search too hard for their prey, or the reverse? These questions can all be answered by using the theoretical approach developed above.

For concreteness, consider the familiar predator-prey model with logistic growth in the prey,

$$W_1 = 1 + r - \frac{r}{K(X_1)} N_1 - a(X_1, X_2)N_2 \qquad (23.64a)$$

$$W_2 = b(X_2)\, a(X_1, X_2)\, N_1 - d + 1 \qquad (23.64b)$$

We assume that the predation coefficient, $a(X_1, X_2)$, is both a monotonic decreasing function of X_1, the prey-defense effort; and a monotonic increasing function of X_2, the predator-search effort. Furthermore, we assume the prey-defense effort, X_1, involves some cost. Therefore, let the prey carrying capacity, $K(X_1)$, be a monotonic decreasing function of X_1. Similarly, predator-search effect, X_2, involves some cost. Thus we let the parameter for the efficiency of conversion of captured prey into predator reproduction, $b(X_2)$, be a monotonic decreasing function of X_2. These assumptions are summarized as,

$$\frac{\partial a(X_1, X_2)}{\partial X_1} < 0$$

$$\frac{dK(X_1)}{dX_1} < 0$$

$$\frac{\partial a(X_1, X_2)}{\partial X_2} > 0 \qquad (23.65)$$

$$\frac{db(X_2)}{dX_2} < 0$$

These assumptions specify a strategy model for predator-prey coevolution in sufficient detail so that the coevolutionary strategy graph can be written out by inspection and all the lines in the graph have definite signs. The graph is given in Figure 23.19.

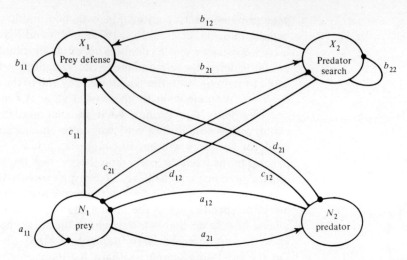

FIGURE 23.19. *The graph H for predator-prey coevolution.*

Within this model, we can answer the questions posed above. The equilibrium population sizes, as a function of X_1 and X_2, are given by

$$\hat{N}_1(X_1, X_2) = \frac{d}{b(X_2)a(X_1, X_2)}$$

$$\hat{N}_2(X_1, X_2) = \frac{r}{a(X_1, X_2)}\left[1 - \frac{\hat{N}_1(X_1, X_2)}{K(X_1)}\right]$$

$$= \frac{r}{a(X_1, X_2)}\left[1 - \frac{d}{K(X_1)b(X_2)a(X_1, X_2)}\right] \quad (23.66)$$

The traits at the coevolutionary equilibrium are the roots, $\hat{X}_1, \hat{X}_2$, of the following system of equations

$$\frac{r\hat{N}_1(X_1, X_2)}{K^2(X_1)} \frac{dK(X_1)}{dX_1} - \hat{N}_2(X_1, X_2) \frac{\partial a(X_1, X_2)}{\partial X_1} = 0$$

$$a(X_1, X_2)\frac{db(X_2)}{dX_2} + b(X_2) \frac{\partial a(X_1, X_2)}{\partial X_2} = 0 \quad (23.67)$$

Generally, this system cannot be solved explicitly and will require a computer. In any event, when one has $\hat{X}_1$ and $\hat{X}_2$ from this system, then the coevolutionary equilibrium is given by $\hat{X}_1$, $\hat{X}_2$, $\hat{N}_1(\hat{X}_1, \hat{X}_2)$ and $\hat{N}_2(\hat{X}_1, \hat{X}_2)$. Furthermore, the total predation rate on the prey population is $a(\hat{X}_1, \hat{X}_2)\hat{N}_1(\hat{X}_1, \hat{X}_2)\hat{N}_2(\hat{X}_1, \hat{X}_2)$. Further interpretation of these results will depend on the detailed functional form of $a(X_1, X_2)$, $K(X_1)$, and $b(X_2)$ functions.

The nature of the coevolutionary mismatch is, perhaps, the most interesting aspect of predator-prey coevolution. We can learn much about this simply by inspecting the coevolutionary strategy graph. The nature of the mismatch is largely determined by the sign pattern of the coevolutionary situation. A qualitative understanding of the problem does not require our solving for $\hat{X}_1$ and $\hat{X}_2$ explicitly in equations above. This fact brings out the power of the graph and its measures. Often, all we want is a qualitative sense of how natural selection shapes the coadaptation between populations. The numerical values, $\hat{X}_1$ and $\hat{X}_2$, are unmeasurable, in themselves and are difficult to estimate, even on a computer. Therefore, the graphical approach, which does not rely on calculating $\hat{X}_1$ and $\hat{X}_2$, becomes very useful.

At the coevolutionary equilibrium, the direct effects on the population sizes of varying the traits away from their equilibrium values are given by

$$\frac{\partial \hat{N}_1}{\partial X_1} = -\left(\frac{F_{12}}{F}\right)d_{21} = -\frac{d_{21}}{a_{21}} > 0 \tag{23.68a}$$

$$\frac{\partial \hat{N}_2}{\partial X_1} = \left(\frac{F_2}{F}\right)d_{21} = \frac{a_{11}}{a_{12}a_{21}}d_{21} < 0 \tag{23.68b}$$

$$\frac{\partial \hat{N}_2}{\partial X_2} = -\left(\frac{F_{21}}{F}\right)d_{12} = -\frac{d_{12}}{a_{12}} < 0 \tag{23.68c}$$

$$\frac{\partial \hat{N}_1}{\partial X_2} = \left(\frac{F_1}{F}\right)d_{12} \equiv 0 \tag{23.68d}$$

The result for $\partial \hat{N}_1/\partial X_1$ means that increasing the level of prey defense beyond the point that evolves by natural selection will actually increase the prey population size. This result does not, of course, include the possibility of evolutionary reciprocation by the predator, as discussed below. Thus, provided X_2 remains at $\hat{X}_2$, the equilibrium level of prey defense, $\hat{X}_1$, is suboptimal. The allocation of more effort into defense, at the cost of lowering K, would consequently yield net increase in the prey's abundance.

The result for $\partial \hat{N}_2/\partial X_1$ is the other side of the story. If the prey were to allocate more effort into defense, then the equilibrium predator abundance would show a net decline. The result for $\partial \hat{N}_2/\partial X_2$ means that if the predators were to reduce their searching effort, X_2, below that which evolves by natural selection, then their abundance would actually increase (provided X_1 remains at $\hat{X}_1$). The result for $\partial \hat{N}_1/\partial X_2$ means that there is no net effect on the prey equilibrium abundance by varying the predator searching effect. This phenomenon arises because of the absence of density dependence in the predator.

These direct effects of varying predator search effort versus defense on size of the predator population may be summarized by saying that at a coevolutionary equilibrium, *the prey underdefend* and *the predators over-search*.

The total effects of varying the traits X_1 and X_2 away from their equilibrium values includes the effect of evolutionary reciprocation, in addition to the direct effects calculated above. First, we calculate the total effect of varying X_1. The basic quantity in these calculations is $I^{(X_1)}$, the feedback in the subsystem obtained by deleting X_1. The graph of this quantity appears in Figure 23.20.

The feedback of the subsystem in Figure 23.20 is

$$I^{(X_1)} = -a_{12}b_{22}a_{21} < 0 \tag{23.69}$$

The feedback in this subsystem is always negative. Next, we inspect various subgraphs within this subsystem. For $d\hat{N}_1/dX_1$, we require the graphs in Figure 23.21.

The corresponding feedback measures are

$$I^{(X_1)}_{N_1,X_2} \equiv 0 \qquad I^{(X_1)}_{N_1,N_2} = -b_{22}a_{12} < 0 \tag{23.70}$$

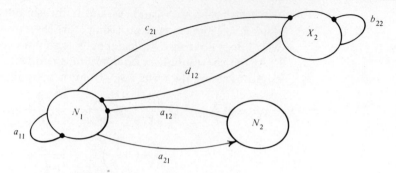

FIGURE 23.20. *The subgraph $H^{(X_1)}$ obtained by deleting the variable, X_1, from the system.*

therefore

$$\frac{d\hat{N}_1}{dX_1} = \frac{(0)b_{21} + (-b_{22}a_{12})d_{21}}{-(-a_{12}b_{22}a_{21})} = \frac{-d_{21}}{a_{21}} > 0 \qquad (23.71)$$

These results mean that the total effect on prey abundance of varying the level of prey defense is identical to the direct effect, $\partial \hat{N}_1/\partial X_1$, as calculated earlier. Hence the effect of evolutionary reciprocation is zero in this case. This result could also be obtained by writing

$$\frac{d\hat{N}_1}{dX_1} = \frac{\partial \hat{N}_1}{\partial X_1} + \frac{\partial \hat{N}_1}{\partial X_2}\frac{d\hat{X}_2}{dX_1} \qquad (23.72)$$

However the quantity, $\partial \hat{N}_1/\partial X_2$, is identical to zero, as calculated above, and, therefore, the term describing the effect of evolutionary reciprocation is zero. Thus we see that even when evolutionary reciprocation is taken into account in this model, it is predicted that the prey are underdefended at the coevolutionary equilibrium.

The quantity, $d\hat{N}_2/dX_1$, is obtained from the graphs in Figure 23.22. The feedback quantities are

$$I_{N_2,X_2}^{(X_1)} = -d_{12}a_{21} > 0 \qquad I_{N_2}^{(X_1)} = -(a_{11}b_{22} - c_{21}d_{12}) \qquad (23.73)$$

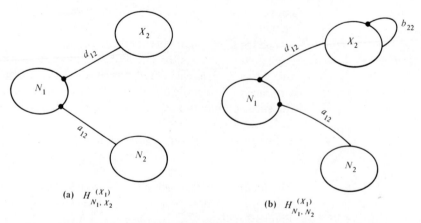

(a) $H_{N_1,X_2}^{(X_1)}$ (b) $H_{N_1,N_2}^{(X_1)}$

FIGURE 23.21. *The subgraphs of $H^{(X_1)}$ needed in calculating $d\hat{N}_1/dX_1$. Here $H_{N_1,X_2}^{(X_1)}$ represents the web within $H^{(X_1)}$ that runs from X_2 to N_1; $H_{N_1,N_2}^{(X_1)}$ represents the web within $H^{(X_1)}$ that runs from N_2 to N_1.*

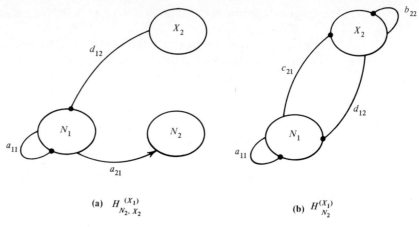

(a) $H^{(X_1)}_{N_2, X_2}$

(b) $H^{(X_1)}_{N_2}$

FIGURE 23.22. *The subgraphs $H^{(X_1)}_{N_2, X_2}$ and $H^{(X_1)}_{N_2}$.*

Therefore, $d\hat{N}_2/dX_1$ is found to be

$$\frac{d\hat{N}_2}{dX_1} = \frac{(-b_{21}d_{12})}{a_{12}b_{22}} + \frac{(-c_{21}d_{12}d_{21})}{a_{12}a_{21}b_{22}} + \frac{a_{11}d_{21}}{a_{12}a_{21}} \tag{23.74}$$

The first two terms are always positive and the second is always negative. Therefore, the sign of $d\hat{N}_2/dX_1$ is not definitely known without more detail on the functional forms for $K(X_1)$, $a(X_1, X_2)$ and $b(X_2)$. Thus the total effect of increasing prey defense effort, X_1, on the equilibrium predator abundance involves a direct effect, which is aways negative. Also involved is the evolutionary reciprocation, which may or may not counteract the direct effect, depending on the detailed assumptions.

The total effect of increasing prey defense on the equilibrium level of predator search effect, is obtained from the graphs in Figure 23.23.

The required feedback quantities are

$$I^{(X_1)}_{X_2} = F = a_{21}a_{12} < 0 \qquad I^{(X_1)}_{X_2, N_2} = -a_{12}c_{21} < 0 \tag{23.75}$$

Therefore,

$$\frac{d\hat{X}_2}{dX_1} = \frac{-a_{21}a_{12}b_{21} - a_{12}c_{21}d_{21}}{a_{12}b_{22}a_{21}} > 0 \tag{23.76}$$

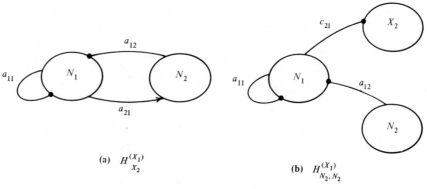

(a) $H^{(X_1)}_{X_2}$

(b) $H^{(X_1)}_{N_2, N_2}$

FIGURE 23.23. *The subgraphs $H^{(X_1)}_{X_2}$ and $H^{(X_1)}_{X_2, N_2}$.*

Hence the total effect of increasing prey defense effort on predator search effort is positive.

Next, consider the total effect of changing the predator search effort X_2 on the other variables in the system. The subsystem obtained by deleting X_2 is illustrated in Figure 23.24.

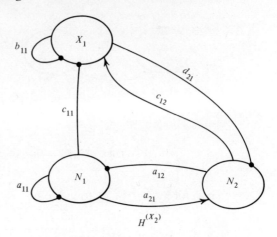

FIGURE 23.24. *The subgraph $H^{(X_2)}$.*

The feedback of this subsystem is

$$I^{(X_2)} = (-b_{11}a_{12}a_{21}) + (d_{21}c_{11}a_{12}) + (-d_{21}a_{11}c_{12}) < 0 \qquad (23.77)$$

All three terms are negative. Hence $I^{(X_2)}$ is negative.

To find the total effect of varying the search effect on the predator's equilibrium population size, we require the feedback of the connecting webs in Figure 23.25.

The necessary quantities are

$$I^{(X_2)}_{N_2,X_1} = -a_{11}d_{21} < 0 \qquad I^{(X_2)}_{N_2,N_1} = (c_{11}d_{21} - a_{21}b_{11}) > 0 \qquad (23.78)$$

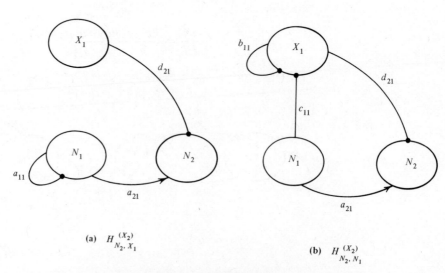

(a) $H^{(X_2)}_{N_2, X_1}$

(b) $H^{(X_2)}_{N_2, N_1}$

FIGURE 23.25. *The subgraphs $H^{(X_2)}_{N_2,X_1}$ and $H^{(X_2)}_{N_2,N_1}$.*

Hence we have

$$\frac{d\hat{N}_2}{dX_2} = \frac{-a_{11}d_{21}b_{12} + (c_{11}d_{21} - a_{21}b_{11})d_{12}}{(-I^{(X_2)})} \tag{23.79}$$

The numerator is always negative. Therefore, $d\hat{N}_2/dX_2$ is negative. Recall that the direct effect, $\partial\hat{N}_2/\partial X_2$, is also always negative. This result shows that the total effect, which includes the evolutionary reciprocation from the prey, has the same sign as the direct effect. Hence we again predict that $\hat{X}_2$ is too high. That is, if the predators were to allocate less effort to the search for prey, then their population size would show a net increase, even when evolutionary reciprocation by the prey is taken into account.

To determine the total effect of X_2 on the equilibrium prey abundance, we inspect Figure 23.26.

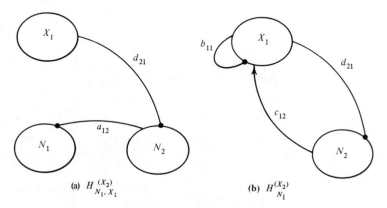

(a) $H^{(X_2)}_{N_1, X_1}$ (b) $H^{(X_2)}_{N_1}$

FIGURE 23.26. *The subgraphs $H^{(X_2)}_{N_1, X_1}$ and $H^{(X_2)}_{N_1}$.*

The feedback quantities are

$$I^{(X_2)}_{N_1, X_1} = d_{21}a_{12} > 0 \qquad I^{(X_2)}_{N_1} = d_{21}c_{12} < 0 \tag{23.80}$$

Hence

$$\frac{d\hat{N}_1}{dX_2} = \frac{(d_{21}a_{12}b_{12}) + (-d_{21}c_{12}d_{12})}{(-I^{(X_2)})} \tag{23.81}$$

The first term in the numerator is positive and the second is negative. Therefore, the total effect of X_2 on $\hat{N}_1$ depends on the detailed shapes of functions $K(X_1)$, $a(X_1, X_2)$, $b(X_2)$. The interesting aspect of this result is that the direct effect of X_2 on $\hat{N}_1$, $\partial\hat{N}_1/\partial X_2$, is identical to zero and, therefore, the total effect above is caused solely by the evolutionary reciprocation.

Similarly, the total effect of varying the predator-search effort on the level of defense evolved by the prey is found to be

$$\frac{d\hat{X}_1}{dX_2} = \frac{(-a_{12}a_{21}b_{12}) + (a_{21}c_{12}d_{12})}{(-I^{(X_2)})} \tag{23.82}$$

The first term in the numerator is positive and the second is negative. The total effect of varying X_2 on $\hat{X}_1$ depends on more detail than is contained in the signs of the connections in the strategy graph.

To summarize, at a coevolutionary equilibrium *first, the prey underdefend themselves* $(d\hat{N}_1/dX_1 > 0)$. *Second, increasing prey defense will cause increased predator searching effort to evolve* $(d\hat{X}_2/dX_1 > 0)$; *and third, the predators search too hard* $(d\hat{N}_2/dX_2 < 0)$. To determine the total effect of varying X_1 and X_2 on the equilibrium values of the other variables in the system requires information not only on the signs, but also the magnitudes of the connections in the strategy graph.

Additional theoretical studies that pertain to the coevolution of predator and prey include papers by Mode (1958), Jayakar (1970), and Stewart (1971).

EXAMPLE 2. COEVOLUTION OF CHARACTER DISPLACE-MENT. This example concerns the evolution of the competition coefficients between two competitors, according to the scheme diagramed in Figure 23.27. In each species, there is a trait that indicates the kind of resources used by that species. The carrying capacity of a species depends on the value of its trait, according to the function $K(X)$. The competition coefficient between the competitors is determined by the value of the traits in both species, according to the function $\alpha(X_1 + X_2)$. Thus the strategy model is given by the following formulas:

$$W_1 = 1 + r - \frac{r}{K(X_1)}N_1 - \frac{r\alpha(X_1 + X_2)}{K(X_1)}N_2$$

$$W_2 = 1 + r - \frac{r}{K(X_2)}N_2 - \frac{r\alpha(X_1 + X_2)}{K(X_2)}N_1$$

(23.83)

There is full symmetry in the model. This model leads to the strategy graph in Figure 23.28. With this example, we illustrate two interesting biological points. First, the concept of evolutionary reciprocation depends upon the assumption of a coevolutionary *path*. Second, we shall determine a very general fact about the mismatch in the coevolution of character displacement.

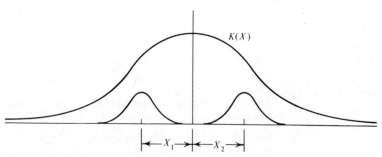

FIGURE 23.27. *Sketch of the setup of a model for the coevolution of character displacement.*

The total effect of varying a trait, say X_1, on the population size of a species, say $\hat{N}_1$, is formally determined by the chain rule in calculus

$$d\hat{N}_1 = \frac{\partial \hat{N}_i}{\partial X_1} dX_1 + \frac{\partial \hat{N}_1}{\partial X_2}\left(\frac{dX_2}{dX_1}\right) dX_1$$

(23.84)

The quantity dX_2/dX_1 describes the path that X_2 takes as X_1 varies. In the entire preceding discussion, we have assumed that X_2 moves along the

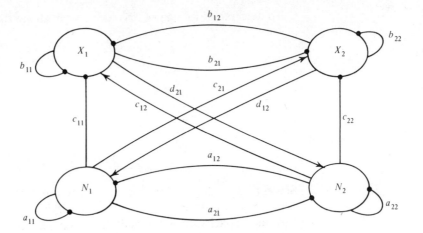

FIGURE 23.28. *The graph H for the coevolution of character displacement between two competitors.*

function $\hat{X}_2(X_1)$ as X_1 is varied. This is one meaningful choice of a path, and probably the most important choice, for most purposes. Movement of X_2 along the path just described is not the only possible and useful choice of a path, however, as we shall see.

Consider one of the possible dynamic paths taken by the competitors during their coevolution of character displacement. Suppose that both species start out with exactly symmetrical initial positions very near the center of the $K(X)$ function, that is, $X_1(0) = X_2(0) = \varepsilon$. Then, as time proceeds, both species will move away from one another until they stop at the coevolutionary equilibrium point. Along this path each step moved by species-1 is matched by an equal step taken by species-2 in the opposite direction, so that

$$\frac{dX_1}{dX_2} \equiv \frac{dX_2}{dX_1} \equiv 1 \tag{23.85}$$

The total change in the equilibrium population size of species-1 along this path then reduces to

$$d\hat{N}_1 = \left(\frac{\partial \hat{N}_1}{\partial X_1} + \frac{\partial \hat{N}_1}{\partial X_2} \right) dX_1 \tag{23.86}$$

At the coevolutionary equilibrium, the direct effects are known in terms of the feedbacks, so that

$$\frac{d\hat{N}_1}{dX_1} = \frac{-F_{12}}{F} d_{21} + \frac{F_1}{F} d_{12} \tag{23.87}$$

This expression gives the total effect of varying X_1 on $\hat{N}_1$ based on the assumption that X_2 shifts in a symmetric way. If $d\hat{N}_1/dX_1$ is positive at the coevolutionary equilibrium, it means that both species have not moved far enough apart. If they both moved farther apart, then both species would realize a net increase in population size. Note d_{12} equals d_{21} here.

To determine the sign of $d\hat{N}_1/dX_1$, we evaluate the feedbacks as

$$F_{12} = a_{12} = \frac{\partial W_1}{\partial N_2} = \frac{-r\alpha(X_1+X_2)}{K(X_1)}$$

$$F_1 = a_{22} = \frac{\partial W_2}{\partial N_2} = \frac{-r}{K(X_2)} = \frac{-r}{K(X_1)} \qquad (23.88)$$

So we obtain

$$\frac{d\hat{N}_1}{dX_1} = \frac{-r}{K(X_1)}[1-\alpha(X_1+X_2)]\frac{d_{12}}{F} > 0 \qquad (23.89)$$

Since F is negative, $d\hat{N}_1/dX_1$ is positive, provided that $1 > \alpha(X_1+X_2)$, which is always true. Thus, without specifying the functional form of the carrying capacity and competition functions, apart from symmetry, we conclude, generally, that *the process of character displacement between two symmetrical competing species does not proceed far enough.* A very special case of this result was derived in Roughgarden (1976), based on a Gaussian $K(X)$ and $\alpha(X)$.

The fact that two symmetrical coevolving species never displace far enough apart from one another is all the more surprising when contrasted with the evolutionary result for a single species, evolving with respect to an evolutionarily rigid competitor. Suppose that X_2 has some fixed value and that species-1 is evolving. At the evolutionary equilibrium for species-1, we have

$$\frac{d\hat{N}_1}{dX_1} \equiv \frac{\partial \hat{N}_1}{\partial X_1} = \frac{-F_{12}}{F}d_{21} < 0 \qquad (23.90)$$

Thus, *a single species, displacing with respect to an evolutionarily rigid competitor, displaces too far.*

Other studies that treat the evolution of competing species include Bulmer (1974), Gill (1974), Leon (1974), and Lawlor and Maynard Smith (1976).

Summary   This chapter presents models that combine population genetics with population dynamics. The key feature of these models is that both population sizes and gene frequencies are variables. The typical form of these models is, for species-s,

$$\Delta p_s = p_s(1-p_s)\frac{\partial \bar{W}_s/\partial p_s}{2\bar{W}_s}$$

$$\Delta N_s = (\bar{W}_s - 1)N_s \qquad (23.91)$$

In this type of model, p_s is an allele frequency in species-s. The variable N_s represents the population size of species-s. There is a pair of these equations for each species in the system. For example, with two species, there are four equations. The equation for Δp_s is the standard population-genetic formula (provided there is no intraspecific frequency dependence). The second equation is the ecological part of the model and expresses the population dynamics. The parameters in the model and the species interactions are introduced by specifying the functional form of the selective values. $W_{s,ij}$ is the fitness of genotype A_iA_j in species-s. By specifying how this fitness varies with the population sizes and gene frequencies of the various species in the system, models for the coevolution between many types of interacting populations can be formulated and studied.

Many results about the way coevolution influences community structure can be derived directly from the equations above, that is, prior to adding detailed additional assumptions. The results emerge from examining whether and how the familiar fundamental theorem of natural selection from classical population genetics can be extended into models in which the population sizes of one or more species are also variables. The general issues are

1. Is fitness maximized in each coevolving species?
2. If so, what are the implications of this for the equilibrium population size that is attained by each of the interacting species?

Density-
Dependent
Selection in a
Single
Population

Studies on density-dependent selection in single population models show one key result. The result consists of (1) under density-dependent selection, at any locally stable equilibrium the mean fitness is a local maximum with respect to variation of the gene frequency, p, about its equilibrium value, $\hat{p}$. (2) Furthermore, property (1) above, is valid both for polymorphic equilibria, $\hat{p} \neq 0, 1$; and for boundary equilibria, $\hat{p} = 0, 1$. (3) The ecological implication of this fitness maximization is that the equilibrium population abundance, $\hat{N}$, is a local maximum. That is, fitness maximization is realized on the ecological side as abundance maximization. There are three key assumptions used to derive this result: (a) The genetic system must be, to begin with, one that permits fitness maximization under classical density-independent selection. Thus certain two-locus schemes are ruled out. (b) In the same vein, there can be no intraspecific frequency dependence. It is known in classical population genetics that frequency dependence generally destroys the possibility of fitness maximization, and (c) on the ecological side, the population dynamics must permit a stable equilibrium to begin with. More technically, if we hold the gene frequency, p, constant, while leaving the population size, N, as the variable, then N must approach an equilibrium. Of course, most single species population dynamic models usually show oscillation for some parameter values. Thus this assumption means that the genotypes must have growth parameters restricted to the region of parameter space that permits an equilibrium population size to exist. These three assumptions are sufficient to entail the key result on density-dependent selection in a single population, as stated above.

Purely Density-
Dependent
Coevolution

The simplest type of coevolution that can occur among interacting populations is pure density-dependent coevolution. In pure density-dependent coevolution, the fitness of the carrier of a genotype, say A_iA_j in species-s, is a function of all the population sizes in the community

$$W_{s,ij} = W_{s,ij}(N_1 \cdots N_S)$$

By definition, the fitnesses are not functions of any allele frequencies in any of the species. This situation arises naturally, when traits that do not directly govern the species interactions are considered. For example, consider the coevolution between a predator and a prey population. There are potentially a great many traits in the phenotype of the predator, which evolve in response to environmental pressures, but which do not, per se, improve a predator's ability to catch prey. Evolution of such traits, nonetheless, may affect the prey population. In particular, evolution of predator traits that are unrelated to the act of predation may lead to an increased abundance of predators and, thereby, lead to an increased overall predation pressure on the prey population. Traditionally, the term coevolution has been applied to the evolution of the traits that govern species interactions. However, the study of the traits that do not govern the interactions is just as important.

The main finding for purely density-dependent coevolution is that (1) a boundary equilibrium in any species is locally stable if and only if the mean fitness of that species is higher at the boundary than at nearby interior points; (2) a polymorphic equilibrium in any species is locally stable if the mean fitness of that species is at a local maximum at $\hat{p}$; and (3) maximizing fitness does *not* necessarily entail maximizing abundance. Maximizing fitness in any particular species either maximizes or *minimizes* the equilibrium abundance of that species, depending on the details of how the species is interacting with the other species in the system. There is a simple criterion, involving a subdeterminant computed from the matrix of interaction coefficients, whose sign determines whether fitness maximization leads to abundance maximization or minimization. The assumptions used in deriving this result are identical to those stated earlier for the single population result. In particular, there are no assumptions of Lotka–Volterra relationships or any other specific functional form.

Coevolution and Community Stability This result is relevant to the continuing discussion of the relation between complexity and stability in ecosystems originated by May (1973). May pointed out that several lines of analysis of mathematical models of communities all pointed to the conclusion that complexity diminished the chance for a community becoming stable. Roughly speaking, the more species there are in a system, the more tricky it is to adjust the connections among them so as to enable the species to coexist. One objection to all the arguments pointing to this conclusion is that coevolution among the species was not explicitly taken into account. Perhaps, one might hope, a general or typical trend of coevolution is to adjust the connections among species in ways which enhance the opportunity for coexistence. But the result above, for purely density-dependent coevolution, refutes this conjecture for the following reason.

Purely density-dependent coevolution is the mathematically "best-behaved" form of coevolution. The introduction of interspecific frequency dependence has a destabilizing effect in coevolutionary models. Hence, if purely density-dependent coevolution does not improve chances for the coexistence of many species, when frequency dependence is added the situation becomes worse. The result above establishes that there is a condition under which a species undergoing maximization of fitness as a result of evolution attains the lowest population size available to it. Obviously, the fate of this species is not promising. Indeed, as its population progressively declines, its extinction is increasingly likely. And if the species does so, it is likely to bring others with it. The general conclusion is that some interacting populations that succeed in passing the already strict criteria for stable coexistence according to a purely population dynamic analysis are destined to extinction as a result of purely density-dependent coevolution. Furthermore, it can be shown that the likelihood that a species will evolve itself into extinction increases as the system becomes more complex and interdependent. Thus an explicit analysis of the effect of coevolution on the complexity-stability issue fails not only to reverse or nullify May's original argument, but strongly supports it.

Interspecific Frequency Dependence

The next section on coevolutionary models is an examination of the effect of interspecific frequency dependence. Interspecific frequency dependence occurs whenever the fitness of carriers of a genotype in any particular species depends on the gene frequencies in other species. This situation arises naturally when the evolution of traits that govern species interactions are considered. Typical exampes of coevolutionary problems with inherent interspecific frequency dependence are the evolution of the amount of predator searching "effort," of the evolution of prey defenses, and the evolution of character displacement between two competitors.

There are three main findings in this section. First, under the same assumptions used in the preceding work, it is not true that a polymorphic equilibrium in a species is stable if and only if the mean fitness of that species is at a local maximum. The stability of a polymorphic equilibrium with interspecific frequency dependence is not tied to any general principle, but must be settled in each case by referring to the model. Second, again, under the same assumptions, it *is* true that a boundary equilibrium for all species (i.e., $\hat{p}_s = 0, 1$ for *all s*) is locally stable if and only if the mean fitness in every particular species becomes lower as the gene frequency for that species is moved away from the boundary. Thus, with interspecific frequency dependence, a stable equilibrium exists where one of the alleles is fixed in every species if the mean fitness is maximized in every species. This result supplies a *limited* justification of strategy reasoning, based on maximizing fitness in coevolutionary problems that involve interspecific frequency dependence. Although the justification for strategy reasoning is limited with interspecific frequency dependence, with intraspecific frequency dependence strategy reasoning cannot be justified. Interspecific frequency dependence proves to be more tractable, in general terms, than intraspecific frequency dependence but is much less tractable than purely density-dependent coevolution.

The Imperfect Design of Coadaptation

The third finding is the following. With interspecific frequency dependence, even when fitness is maximized, for example, because of evolution producing an equilibrium where one allele is fixed in every species involved in the system, the processes of fitness maximization and maximizing (or minimizing) the abundance of a species show no correlations. Recall that with pure density-dependent coevolution, the result of fitness maximization was either to maximize or to minimize the population size. With interspecific frequency dependence, fitness maximization, even when it does occur, generally does not translate into either maximizing or minimizing the population size. Instead, fitness maximization translates into some intermediate condition.

There is great interest in determining where the evolutionary equilibrium abundance is, relative to the maximum abundance. For example, suppose X refers to the amount of energy allocated by a prey to producing toxins. X is, thus, one of the traits that govern the predator-prey interaction and, hence, will evolve in a coevolutionary context with interspecific frequency dependence. Let $\hat{X}$ be the level at the coevolutionary equilibrium. We want to know the sign of $d\hat{N}/dX|_{\hat{X}}$. If it is positive, it means that $\hat{X}$ is below the value that maximizes $\hat{N}$. If $d\hat{N}/dX$ is positive, it would mean, qualitatively, that coevolution leads to prey that underdefend themselves; that is, to a situation such that if all prey did allocate more energy to prey defense than they will have evolved to do, then their abundance would rise. Such a proposition is the other side of the coin of the familiar issue of the evolution of a prudent predator. The section on coadaptation and equilibrium population size develops the machinery to determine the signs of quantities, like $d\hat{N}/dX$, as a result of interspecific frequency dependence and illustrates the application of the machinery to a predator-prey model and to a model for displacement between two competitors.

Chapter 24

NICHE THEORY AND ISLAND BIOGEOGRAPHY

NICHE theory and the theory of island biogeography have come to occupy a curiously pivotal position in population biology. This theory provides models for the phenomenon of resource partitioning among competing species, and for the pattern of species diversity on oceanic islands. Although the phenomena addressed by this theory are fairly common and empirically conspicuous, a major reason for the pivotal position of this topic is that it serves as a melting pot for much of the theory presented in the preceding chapters of this book. In this chapter we meet, again, the population dynamic theory for competing species (Chapter 21), the theory for population dynamics in a stochastic environment (Chapter 20), the theory for the evolution of a quantitative character (Chapter 9), and the theory of coevolution in ecological systems (Chapter 23). Furthermore, we might hope, in the future, that theory using the transport model for population growth and individual growth in size (Chapter 19) will also be included in niche theory. Before presenting this theory, it is important to dwell in detail on some empirical examples that illustrate what niche theory and island biogeography are trying to explain. I will again focus on the ecology of *Anolis* lizards from the islands of the Caribbean, because I am most familiar with this system from my own field work. Other systems could serve equally well as illustrations, however. Some other systems which are particularly illustrative of resource partitioning include grassland birds discussed by Cody (1968), the birds of New Guinea studied by Diamond (1975), sparrows in Pulliam (1975), continental lizards analyzed by Pianka (1973), mud snails by Fenchel (1975), and *Conus* snails by Kohn (1959 and 1968). For a wide ranging review of the literature of resource partitioning, see Schoener (1974) and Pianka (1976). For a discussion of whether the theory to be presented in this chapter is appropriate for coral reef fish communities, see Sale (1977) and Anderson et al. (1979). Also, the plethodontid salamanders of tropical America share many similarities with the *Anolis* lizard radiation of tropical America; see Wake and Lynch (1976) for studies of these salamanders.

Examination of the Empirical Issues

An Overview of the Anolis *Populations of the Caribbean*

Lizards or the genus *Anolis* are small, mainly insect-eating lizards, which forage during the day. Most species are arboreal, in the sense that they perch on tree trunks or bushes. Their food is often caught on the ground, although they also eat insects and small berries found in the trees. *Anolis* lizards may often be observed perching on tree trunks and scanning the ground waiting for an insect to land within catching distance. They are called "sit and wait" predators to distinguish them from "searchers" who are in constant motion to find prey. *Anolis* lizards are usually territorial. *Anolis* lizards are in the iguanid family. The genus is defined by (1) the presence of toe pads on each toe (a sign of arboreality), and (2) the presence, in males, of an often colorful flap of skin under the chin and neck which is extended and retracted in a complex sequence during territorial disputes and during courtship. This flap of skin is called a dewlap or throat fan. These animals are a dominant and conspicuous component of the ecosystem in the islands in the

Caribbean. They are found throughout central and tropical South America, although they are not nearly as abundant in continental situations. People from continental places in the temperate zone are accustomed to thinking of lizards as rare curiosities and are surprised to find them occupying such a prominent position in the total vertebrate community on the islands of the Caribbean.

Figure 24.1 presents a map of the Caribbean region. The islands can be divided into four categories:

1. The first region consists of the islands beginning with St. Croix in the north, then running south through Grenada, then curving west and ending at Aruba. These islands have long been isolated from one another. Each island contains either one or two native anole species that are highly differentiated from one another. Most of these islands are known as the Lesser Antilles.

2. The region marked as two consists of the largest islands in the Caribbean, Cuba, Jamaica, Hispaniola, and Puerto Rico. Each of these islands is the site of an internal radiation of many species. These islands are known as the Greater Antilles.

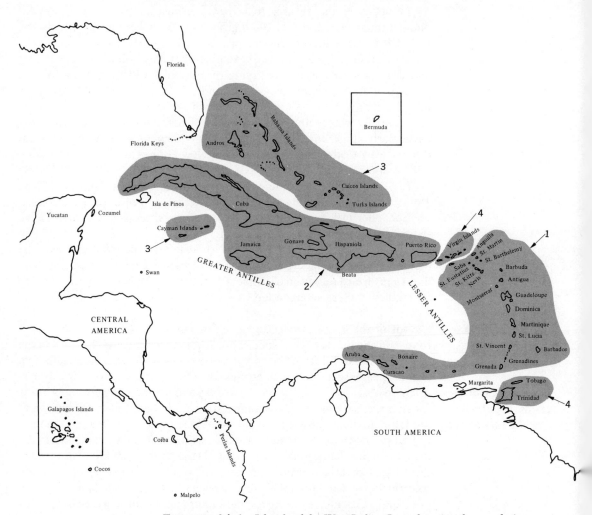

FIGURE 24.1. *Islands of the West Indies. In each region the populations of* Anolis *lizards exhibit a distinctive ecology.*

3. Category three consists of low-lying oceanic islands, including the Bahamas and the Caymen Islands, whose anole fauna appears recent. The species on these islands are relatively slightly differentiated from their parent populations.

4. Category four consists of islands that are satellites or members of a larger bank. This category includes the British and U.S. Virgin Islands (except St. Croix), which belong to the Puerto Rico Bank, and various other islands near the Greater Antilles and the continent. This category includes Trinidad and Tobago. These islands differ from those in the other categories by having had a land bridge connection to a source region, a fact which has strongly influenced their eçology. We now examine the ecology in the first *three* of these categories in more detail.

Category 1 Islands: Resource Partitioning Between Species

The Anoles within category 1 comprise three lineage groups. The northern islands contain lizards from the *bimaculatus* group; the islands from Martinique south and west through Bonaire contain lizards from the *roquet* group; and the islands of Curaçao and Aruba contain lizards from the *lineatus* group. This fact is important because the ecological patterns discussed below have evolved independently on each of the islands and also independently in each lineage group.

The major generalization within the group 1 islands is the evolution of characteristic body sizes which depend on whether a species is solitary on an island or coexists with a competitor. Figure 24.2 illustrates the sizes of adult animals from the native species on islands where there is only one anole species. Observe the similarity in sizes among these species. This similarity in size contrasts with their extreme differentiation in color and other traits. Figure 24.3 illustrates the sizes of adult animals from species on islands where two competing species co-occur. Note that one species is smaller than or equal to the solitary size, and the other species is far larger than the solitary size. This displacement of size on the two-species islands, relative to the solitary islands, presumably reflects the competitive interaction between the two species. Again, this generalization transcends the lineage groups that occur in these islands.†

The generalization above presents a biogeographic pattern in the morphology of the animals. A connection between the morphological pattern and actual resource utilization was provided by Schoener and Gorman (1968) who showed, from analysis of stomach contents, that the larger species on Grenada consumed significantly larger sizes of prey than the smaller species. Furthermore, they showed that any taxonomic differences in the food consumed by the two species was largely a reflection of the prey size differences. This work and others led to the interpretation that the morphological pattern expressed in Figures 24.2 and 24.3 is caused, in some way, by competition for prey of different sizes. By this interpretation, the competition between two species is indirectly indicated by the difference in size between the species. Two species with similar size are presumed to have a higher competition coefficient than two species with

---

† The generalization depicted in Figures 24.2 and 24.3 was originally documented by Schoener (1969), based on measurements of preserved museum specimens. Figures 24.2 and 24.3 are based on measurements of live specimens. The generalizations in Figures 24.2 and 24.3 apply to all the islands in category 1 except Marie Galante and St. Maarten.

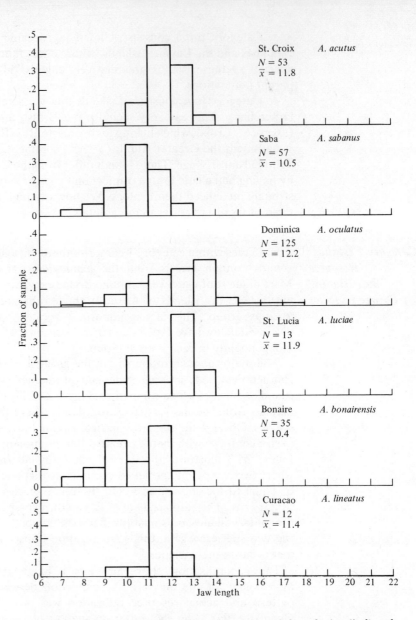

FIGURE 24.2. *Distribution of jaw lengths in adult male* Anolis *lizards on single-species islands in category* 1. *Note the convergent evolution to a mean jaw length of approximately* 11 mm *in all these populations. For the only exception to this pattern see Figure* 24.7.

dissimilar size, because the degree of similarity in size is assumed to indicate the amount of overlap in the use of limiting resources. It should always be remembered, however, that this assumption has never been experimentally tested, either with this group of animals or any other. The general issue of the relationship between body size and resources has been discussed by Wilson (1975b).

There are several additional facts that complement the generalizations above concerning size. First, on islands with two species, the competitors

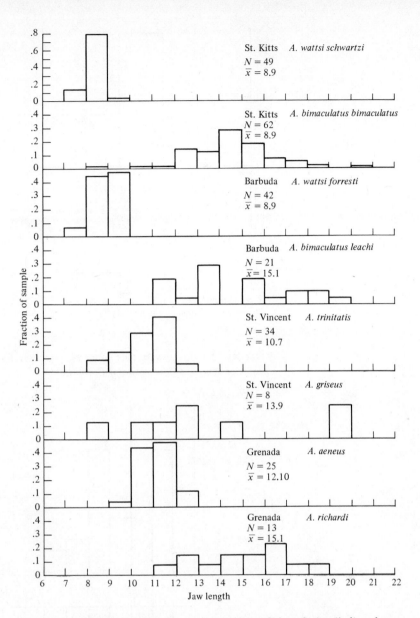

FIGURE 24.3. *Distribution of jaw length in adult male* Anolis *lizards on two-species islands in category 1. Note that each island has a species whose average size is much larger than the solitary size, and another species whose average size is equal to or less than the solitary size. For the only exception to this pattern see Figure 24.6.*

always show differential characteristics with respect to at least one other dimension. On Barbuda the species show an obvious difference in perch height. Figure 24.4, from Roughgarden et al. (1979), illustrates the fact that the large species on Barbuda perches much higher than the smaller species. On Grenada, the species show a consistent difference in sun-shade preference. Figure 24.5 from Schoener and Gorman (1968) shows that the smaller species, consistently, has higher body temperatures than the larger

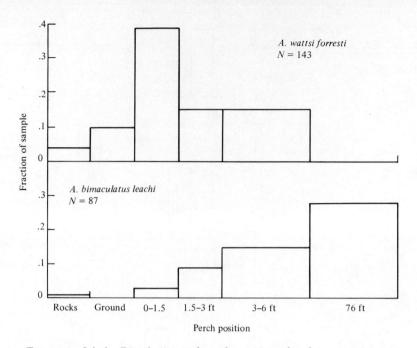

FIGURE 24.4. *Distributions of perch positions for the two sympatric species of* Anolis *on the island of Barbuda.*

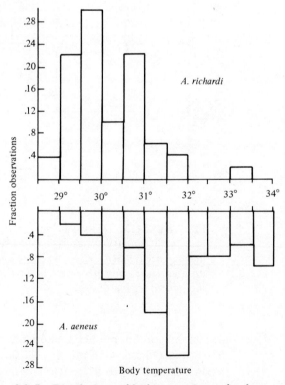

FIGURE 24.5. *Distributions of body temperatures for the two sympatric species of* Anolis *on the island of Grenada.* [Adapted from T. Schoener and G. Gorman (1968), Some niche differences in three Lesser-Antillean lizards of the genus *Anolis. Ecology* **49**: 819–830.]

species, indicating that the smaller species perches more often in the sun than the larger species. There does not seem to be any way to predict which dimension, other than size, will differentiate the species. We shall see that niche theory is preoccupied with models that use only one dimension, but eventually as many as, perhaps, four dimensions will be needed.

A second set of facts concerning size and competition stresses that there are several exceptions to the pattern involving body size presented above. Williams (1972) summarized the locales from which specimens had been collected that violated the overall size pattern. Roughgarden and Fuentes (1977) examined, in the field, some of the places where the violations occurred in the size of *solitary* populations and showed that the violations could be explained on the basis of an unusually high local productivity. That is, they found that lizards grew to larger sizes in regions of high productivity and, therefore, in spots of extremely high productivity, lizards from a solitary population could approach the size of the larger species on a two-species island. The largest specimens for Dominica in Figure 24.2 illustrate this fact. One exception, which could not be explained in this way, remained. The solitary anole on the small island of Marie Galante is extremely large and no explanation currently exists. There is also a very interesting exception to the pattern with the *two species* from the island of St. Maarten. As Figure 24.6 shows, the larger species is of solitary size. It is, then, a very significant additional fact that the smaller species is very restricted in its distribution on the island. The smaller species is found in the hills in the center of the island. In contrast, on all the other two species islands, both species co-occur over almost all of the island's area and, as mentioned before, do show the size differentiation depicted in Figure 24.3. St. Maarten may be the exception that proves the rule. Size differentiation of the degree depicted in Figure 24.3 may be a necessary condition for widespread coexistence. Table 24.1 provides a summary of these points. For further study of differences in the ecology of the various two-species islands, see Roughgarden et al. (1979).

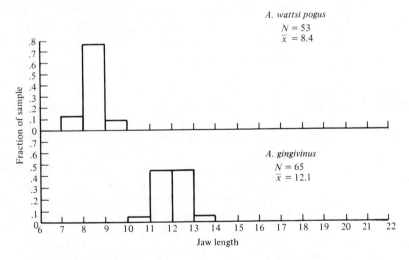

FIGURE 24.6. *Distributions of jaw lengths of adult males for the two species of* Anolis *on the island of St. Maarten. This island is the exception to the pattern illustrated in Figure 24.3.*

Table 24.1. Summary of Size Pattern in Category 1.

	Islands Obeying Size Rules	Islands Violating Size Rules
Islands with one species	1. St. Croix 2. Sombrero 3. Anguilla 4. St. Bartholomew 5. Saba 6. Redonda 7. Montserrat 8. Guadeloupe 9. Dominica 10. Martinique 11. St. Lucia 12. Barbados 13. La Blanquilla 14. Bonaire 15. Curacao	1. Marie Galante
Islands with two species	1. Barbuda 2. Antigua 3. St. Eustatius 4. St. Kitts 5. Nevis 6. St. Vincent 7. Grenada	1. St. Maarten

To summarize, among the category 1 islands: (1) 15 of 16 species that are solitary on an island have evolved a characteristic body size that is called the "solitary size"; (2) on 7 of 8 islands where two species coexist, one species is much larger than the solitary size and it co-occurs with another species that is equal to or smaller than the solitary size; (3) there is no island with two native populations that are both of solitary size; (4) the differences in body size indicate resource partitioning of food with respect to prey size; and (5) there may also be partitioning of space with respect to tree height and/or degree of exposure to the sun.

Category 2 Islands The second category includes the four large islands of the Caribbean, which are collectively termed the Greater Antilles. On the basis of osteological features, Etheridge (1960) has determined that there are two major lineage groups in the Greater Antilles (and in the genus as a whole). One lineage group (the α group) occupies the islands of Puerto Rico and Hispaniola. The other lineage group (the β group) occupies Jamaica. Cuba contains representatives of both groups. Except for recent introductions, there are no species in common among any of the Greater Antilles. The general picture is one of each island containing an internal radiation of native species. The number of species on each island is summarized in Table 24.2.

The ecology in the Greater Antilles has not yet been studied in as much comparative detail as have the islands in category 1. Nonetheless, several

Table 24.2. Summary of Maximum Snout-Vent Length from Williams (1972) and Lazell (1972).

Island	Faunal Size	Largest ♂ Specimen in Largest Species	Largest ♂ Specimen in Smallest Species
Cuba	28 species	191 mm	38 mm
Hispaniola	23 species	175 mm	38 mm
Puerto Rico	11 species	137 mm	40 mm
Jamaica	7 species	124 mm	57 mm
Barbuda	2 species	113 mm	52 mm
Antigua	2 species	111 mm	58 mm
St. Eustatius	2 species	90 mm	49 mm
St. Kitts	2 species	114 mm	~45 mm
Nevis	2 species	112 mm	~45 mm
St. Vincent	2 species	136 mm	74 mm
Grenada	2 species	115 mm	77 mm
St. Croix	1 species	65 mm	
Saba	1 species	69 mm	
Redonda	1 species	81 mm	
Montserrat	1 species	70 mm	
Guadeloupe	1 species	82 mm	
Dominica	1 species	96 mm	
Martinique	1 species	86 mm	
St. Lucia	1 species	91 mm	
Barbados	1 species	83 mm	

new phenomena are evident in the data currently available. The following is based on observations in Hispaniola, Puerto Rico, and Jamaica and is attributable, originally, to Rand (1964 and 1967).

1. There is a rather standard architecture to an anole community in the lowlands (as opposed to the mountains) of the Greater Antilles. The co-occurring species typically consists of (a) a "trunk-crown anole," which is green but with the ability to turn brown, is moderately large, and tends to occupy the top regions of trees; (b) a "trunk anole," which is quite small and which perches on thick tree trunks at moderate heights; (c) a "trunk-ground anole," which is moderately large, gray to brown, and which perches from the ground to one meter high; and (d) a "grass anole" which is quite small, is camouflaged with white longitudinal stripes, and perches on small bushes and twigs near the ground. Each class, above, is called a "structural niche" (Rand, 1962). Thus the standard architecture of a lowland anole community consists of four structural niches.

2. On each of the islands, except Jamaica, there is at least one structural niche for which there are several species. On Puerto Rico, there are three species of grass anoles and three species of trunk-ground anoles. On Hispaniola, there are at least two species of grass anoles, three of the lowland trunk-ground anoles, two of trunk anoles, and two of the lowland trunk-crown anoles. The important fact is that *species from the same structural niches never coexist*, except in narrow zones of overlap at the species borders. Indeed, the border between two species belonging to the same structural niche is often extremely sharp and may occur within a span of 100 meters of apparently homogenous habitat. This fact suggests that

species that are sufficiently similar to one another *cannot* coexist, for it is repeatedly observed that they do not.

The fact that species in the same structural niche do not coexist is the basis of what Diamond (1975) has termed "rules of community assembly." It is worth quoting the observations of Diamond (1973) on New Guinea birds, which parallel the observations above:

"Differences in body size provide the commonest means by which closely related species can take the same type of food in the same space at the same time . . . Among congeners sorting by size in New Guinea, the ratio between the weights of the larger bird and the smaller bird is on the average 1.90; it is never less than 1.33 and never more than 2.73. Species with similar habits and with a weight ratio less than 1.33 are too similar to coexist locally (that is, to share territories) and must segregate spatially. For instance, the cuckoo—shrikes, *Coracina tenuirostris* and *C. papuensis*, segregate by habitat on New Guinea where their average weights are 73 grams and 74 grams, respectively, but they often occur together in the same tree on New Britain, where their respective weights are 61 g and 101 g. New Guinea has no locally coexisting pairs of species with similar habits and with a weight ratio exceeding 2.73, presumably because a medium sized bird of relative weight $\sqrt{2.73} = 1.65$ can coexist successfully with both the large species and with the small species. Thus one finds a sequence of three or more species rather than just two species of such different sizes."

It should be stressed that the rules of community assembly that emerge from the *Anolis* data and from the data on the birds of New Guinea are empirical rules. As we shall see, a major aim of niche theory is to provide an explanation for such rules.

3. Another phenomenon that occurs in the Greater Antilles is the occurrence of huge-sized and of tiny-sized species. Table 24.2 summarizes the sizes of the largest male specimen in the smallest and largest species from each of the Greater Antilles. These species on the extremes of the size range are not common and in some cases are very rare and in danger of extinction. They are not classed with the four typical structural niches discussed above. Their presence establishes the genetic potential of the genus to evolve sizes of lizards other than those which commonly occur.

Category 3 Islands The islands in category 3 comprise mainly the low-lying islands of the Bahamas and the small Cayman islands. No existing Bahaman island exceeds 220 feet in height and most are lower. These islands were wholly under water before and during the early Pleistocene and suffered periodic flooding or submergences during the Pleistocene, as Williams (1969) mentioned. These geological points accord with the biological observation that the species on these islands have only slightly differentiated from populations that are native to the Greater Antilles. In contrast, the populations on the permanent and mountainous islands in category 1 are all very differentiated from the species of the Greater Antilles. These facts suggest that the ecological patterns that occur in the old category 1 islands are the result of both population dynamics *and* coevolution on each island, whereas the patterns on the newer category 3 islands represent only the population-dynamic outcome of interactions between species whose basic traits were evolved somewhere else.

Another feature of great interest in the Bahamas is that the current area above water in the Bahamas is a small fraction of the formerly exposed area. As a result, many small islands today have many more species than would be expected on the basis of their current area. For example, the island of Bimini which is only 5 square miles in area, has four anole species, three of which are Cuban and one is Hispaniolan. Such an island might test theory about the maximum number of competing species which can coexist. The key paper on resource partitioning in the Bahama islands is Schoener (1968).

Resource Partitioning Within Species

Populations of *Anolis* lizards have also been used to illustrate the phenomenon of resource partitioning within a species. A population of lizards consists of individuals of many different sizes. Adult females are often substantially smaller than adult males, and of course juveniles and hatchlings are smaller than adults of either sex. Schoener (1967) has shown, from analysis of stomach contents that females use smaller prey than males. He has conjectured, along with Selander (1966), that one function of sexual dimorphism is to promote a wider use of the available prey sizes in the environment than would occur if both sexes were the same size. See also the review by Selander (1972).

Resource partitioning between the size classes within a species may occur to a much finer degree than that revealed by comparing the resource use of different sexes. Roughgarden (1974a) studied the resource partitioning between adult males of different sizes. An adult male was defined as one that was reproductively mature and was defending a territory. Adult males presumably differ in size, in part, because of age (young adult males may grow still further); in part, because of developmental plasticity in size; and, in part, because they are genetically different. In any event, it is observed from analysis of stomach contents that there is a correlation between prey size and jaw size among adult males.

Figure 24.7 presents the distribution of jaw lengths among adult males for the solitary population on Marie Galante. Figure 24.8 presents the distribution of prey sizes for different size classes of adult male lizards. Figure 24.9 presents more information on the prey used by the entire population of adult males. Clearly, the cumulative data for the entire population, Figure 24.10, includes the fact that individuals of different jaw sizes within the population have very different prey utilization curves, as we have seen in Figure 24.8.

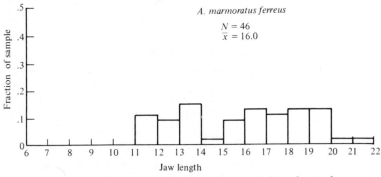

FIGURE 24.7. *Distribution of jaw lengths of adult males in the population of* Anolis *on the island of Marie Galante. Also, this island is the exception to the pattern illustrated in Figure 24.2.*

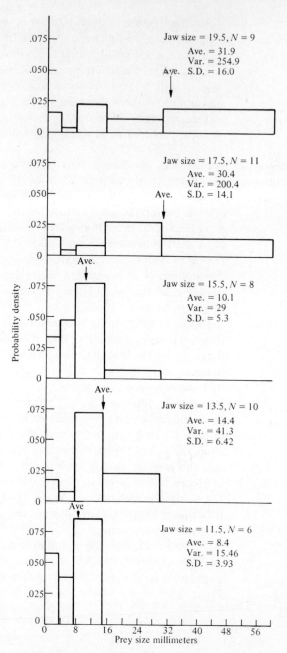

FIGURE 24.8. *Distribution of prey sizes collected from stomachs of adult male* Anolis *lizards. The lizards span several jaw-length classes. The population is from Marie Galante. The distribution is with respect to the weight of stomach content from the various prey-length classes. The total area of the bars in each graph equals one.*

The variance of the entire population's prey-size utilization curve measures what is called the "niche width" of the population. The niche width of a population describes whether that population is specialized or generalized, with respect to its resource use on the dimension under study. A narrow niche width is equivalent to a ecologically specialized population. A

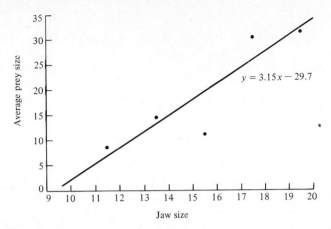

FIGURE 24.9a. *Plot of average prey size as a function of jaw length for adult males in the population of* Anolis *lizards on Marie Galante.*

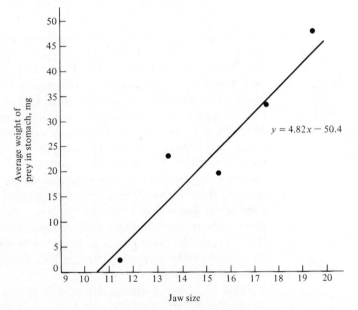

FIGURE 24.9b. *The average total prey weight in stomachs of adult male lizards, as a function of jaw length in the population of* Anolis *lizards on Marie Galante.*

classic question in ecology has been to understand to what extent a generalist population (i.e., a population with a wide niche width) is really made up from cumulating the activities of many specialized individuals. One may imagine two extremes—that an overall generalist population consists of only one type of individual, which is, itself, very generalized in its resource use. Such a population is said to be "monomorphic with generalists." At the other extreme, an overall generalist population could consist of many kinds of individuals, each one of which is very specialized. Such a population is said to be "polymorphic with specialists." Clearly, there are also intermediate conditions between these extremes.

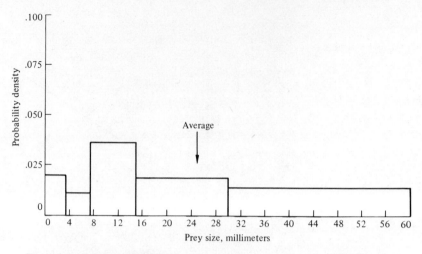

FIGURE 24.10. *Distribution of prey sizes in stomachs of adult males cumulated over all jaw-length classes. Specimens are* Anolis *lizards from Marie Galante. Here N* = 44, Ave = 24.9, Var = 243.3, *and* S.D. = 15.5.

Roughgarden (1972) introduced a simple analysis of variance technique that partitions the total population's niche width into two components. One component measures the contribution of the variety of different kinds of individuals to the population's niche width. This component is called the "between-phenotype component" or BPC. The other component measures the average variance of resource use of the individuals themselves. This component is the "within-phenotype component" or WPC. The population's total niche width is the sum of the BPC and WPC. With these components, one can quantify a population's position on the spectrum between the extremes of being monomorphic with generalists and being polymorphic with specialists. One examines the *percent* of the population's total niche width that is due to the BPC. Zero percent BPC means the population is monomorphic with one generalist phenotype; 100 percent BPC means the population is polymorphic with extremely specialized individuals; and intermediate percentages describe the intermediate conditions between these extremes.

Table 24.3 from Roughgarden (1974a) presents the total niche width and its breakdown into the BPC and WPC components for the adult males of

Table 24.3.

	Productivity	Total Number of *Anolis* Species Present	Sample Size	% BPC	% WPC	Total Niche Width
A. ferreus— Marie Galante	High	1	44	32.5	67.5	243 mm²
A. cybotes— Hispaniola lowlands	High	5	73	11.8	88.2	36 mm²
A. shrevei— Hispaniola high mountain cloud forest	Low	1	41	1.4	98.6	38 mm²

three *Anolis* populations. Both *A. ferreus* and *A. cybotes* are lowland populations, but *A. cybotes* coexists with three other competitors, whereas *A. ferreus* is a solitary population. The comparison between these two species suggests that both the total niche width and the percent BPC both increase as the number of competing species decreases, provided the comparison is between populations from mesic lowland locations. *A. shrevei* is a solitary population in the cold, cloud forest of the high mountains of Hispaniola; its environment is presumably one of low productivity. *A. shrevei* and *A. cybotes* are both trunk-ground lizards and are closely related from a taxonomic standpoint. Note that *A. shrevei* is a population nearly at the extreme of being monomorphic with one phenotype, so far as resource use is concerned. The comparison between *A. shrevei* and *A. cybotes* suggests that the percent BPC drops substantially and the total niche width is relatively unaffected if both the productivity and number of competing species are decreased simultaneously.

The data in Table 23.3 suggest that there are biogeographic patterns in the basic features of the resource partitioning within species. The extent to which populations, as a whole, are specialized or generalized in their resource use depends on the competitive milieu and on the environmental productivity. Furthermore, a population's position between the extremes of zero percent BPC and 100 percent BPC also depends on these factors. At present, the data on resource partitioning within species are very limited, as compared with those on the partitioning between species. One reason is that large sample sizes are required to document partitioning within a species, whereas the differences between species are usually quite large and hence more easily studied in a limited amount of time.

The study of the possible relationship between intraspecific morphological variation in characters like bill depth and jaw length and the phenomenon of intraspecific resource partitioning has had a contentious history. Key references include Van Valen (1965), Selander (1966), Willson (1969), Soulé and Stewart (1970), Grant (1971), Selander (1972), Rothstein (1973), and Lister (1976).

Island Biogeography

The biogeography of lizards in the Caribbean contrasts rather sharply with what has become the traditional picture of island biogeography in birds and insects. First, we review the patterns exhibited by birds and insects and then present the lizard data. The lizard data may be typical of groups that are poor over-water dispersers relative to birds and insects.

Two of the basic features to bird and insect island biogeography are the *area effect* and the *distance effect*. Figures 24.11 and 24.12 from Diamond (1973) illustrate these effects for bird species on islands in the southwest Pacific. The number of land and fresh-water bird species coexisting on these tropical islands varies from 1, for some isolated atolls, up to 513 for all of New Guinea. Note in this analysis, that sea birds are excluded. Figure 24.11 illustrates the "area effect." The figure shows the number of bird species at sea level as a function of island area for islands between 8 and 500 km from New Guinea. The data are summarized by the formula

$$S = 12.3A^{0.22} \tag{24.1}$$

Thus a tenfold increase in area increases the species diversity at sea level by somewhat less than a factor of 2, provided the island is close, but not too

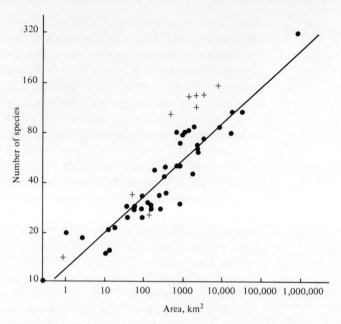

FIGURE 24.11. *Number of resident land and fresh-water bird species on New Guinea satellite islands, plotted as a function of island area on a double logarithmic scale. The symbol "+" indicates land bridge islands. The symbol "•" indicates all other islands. The straight line was fitted by least squares through points for all islands except land bridge islands.* [Adapted from J. Diamond (1973). Distributional ecology of New Guinea birds. *Science* **179**: 759–769. Copyright 1973 by the American Association for the Advancement of Science.]

close, to New Guinea. Figure 24.12 illustrates the "distance effect." The figure refers to islands that are farther than 500 km from New Guinea. The vertical axis represents the ratio of the actual species diversity at sea level to the species diversity predicted on the basis of its area according to Equation (24.1). The horizontal axis refers to the distance of the island from New Guinea. The figure shows that the ratio of actual species diversity to the species diversity predicted by (24.1) decreases exponentially with distance, by a factor of $1/e$ for each 2600 km from New Guinea. Furthermore, the mountains of higher southwest Pacific islands have additional bird species that do not occur at sea level. On the average each 1000 m of elevation, L, enriches an island's avifauna by a number of montane species equal to 8.9 percent of its avifauna at sea level. Diamond has summarized his data on bird species diversity with the formula

$$S = (12.3)[1 + (8.9 \times 10^{-5})L] \, e^{-D/2600} \, A^{0.22} \qquad (24.2)$$

In this formula, S is the total number of species on an island, including those not present at sea level. The variable L is the maximum elevation on the island; D is the distance of the island from New Guinea; and A is the island's area.

As we shall see, one explanation for these data is a model in which there is a continual flow of new species arriving on an oceanic island from some source region and also a continual process of species extinction on an island. By this hypothesis, studied by MacArthur and Wilson (1963 and 1967), the

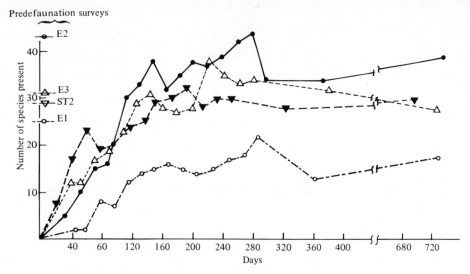

FIGURE 24.13. *The colonization curves of four small mangrove islands in the lower Florida Keys. The original faunal size of each island is indicated for reference near the vertical axis. Then the islands were defaunated with methyl bromide fumigation. The subsequent faunal buildup is recorded in the figure.* [From D. S. Simberloff and E. O. Wilson (1970), Experimental zoogeography of islands. A two-year record of colonization. *Ecology* **51**: 934–937. Copyright 1970 by the Ecological Society of America.]

The biogeography of lizards in the Caribbean does not, however, appear to parallel that of New Guinea birds, nor does it show any sign of turnover as documented with insects in mangrove islands. The category 1 islands simply do not show a significant correlation between area and numbers of species of anoles. When lizards from other families are added, the correlation does not improve. The category 1 islands do not show a distance effect and also there is no clear source region from which to measure the distances. Furthermore, there is no evidence of turnover, because the endemism of the species on these islands implies that they have been on the islands for a very long time. Similarly, the Greater Antilles (category 2) do not show a distance effect. They do shown an area effect in the sense that there is a perfect rank correlation between area and number of anole species, but there is no evidence that species turnover is involved in this fact, again, because the fauna within each island is endemic to that island. These and other points led Williams (1969) to state that "There is at present no evidence to indicate that extinction as a continuing or continual process has had the importance for *Anolis* in the West Indies that Mayr (1965a, b) finds for it in birds or that MacArthur and Wilson (1963 and 1967) have postulated it generally has for island faunas."

More recently, Schoener has examined the island biogeography within the Bahamas (category 3). He found a remarkable relationship between the number of diurnal lizard species and island area, as illustrated in Figure 24.15. The tiniest islands contain only one species. At a threshold area, a second species is added; at a still larger area, a third species is added. This pattern is described by saying that the faunal list on small islands is a "nested subset" of the faunal list of larger islands. This observation supplies a new and important fact to be accounted for by theories of island biogeography.

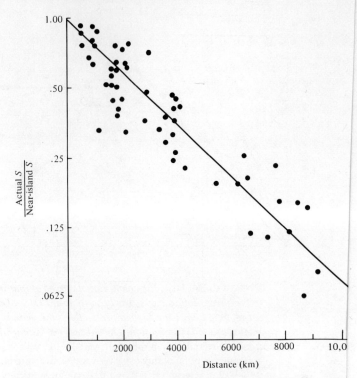

FIGURE 24.12. *Vertical axis (logarithmic scale) is the ratio of the number of resident land and fresh-water bird species on tropical s Pacific islands more than 500 km from New Guinea, divided by the of species expected on an island of equivalent area located closer than from New Guinea. The horizontal axis is the distance from New* [From J. Diamond (1973), Distributional ecology of New Guin Science **179**: 759–769. Copyright 1973 by the American Associ the Advancement of Science.]

actual level of species diversity observed on an island is an representing a balance between the input of new species and e existing ones. The key feature of this hypothesis is the exi steady-state *turnover* of species. That is, a necessary feature pothesis is that the total number of species on an island stays ap constant at a level that depends on its area, distance from region, and elevation, but the identity of those species must cha time as a result of the continuing immigration and extinction.

Simberloff and Wilson (1969 and 1970) have provided direct the existence of species turnover. They studied the insects on formed by mangrove trees in the Florida keys. Their experir eliminate all the insects from several islands and to observe th recolonization. Figure 24.13 shows the buildup of species dive islands after the defaunation. Note that the islands do attain c levels of species diversity. Figure 24.14 is a record of the p absence of various species. Note that the identities of the sp island do show substantial change through time, even thou number of species remains approximately constant, as depict 24.13.

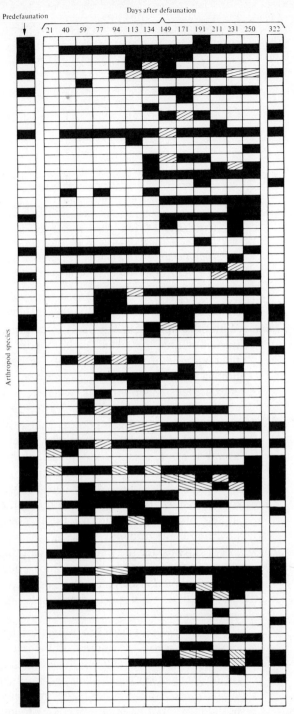

FIGURE 24.14. *The colonists of an experimentally defaunated island, ST2. Each line refers to a specific arthropod species. The column headings in the graph identify the number of days after defaunation for each census. Solid entries indicate that a species was seen; shaded entries show that the species was inferred to be present from other evidence; open entries indicate that the species was not seen, or was inferred to be absent.* [Adapted from D. S. Simberloff and E. O. Wilson (1969), Experimental zoogeography of islands: The colonization of empty islands. *Ecology* **50**: 278–296.]

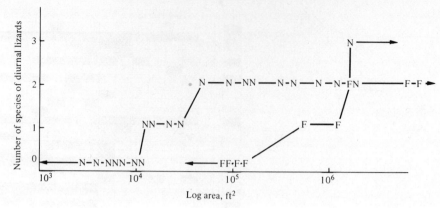

FIGURE 24.15. *Species-area relation for diurnal lizards from small islands in the Bahamas. An "N" refers to a near island and an "F" to a far island relative to the presumed source region. The faunal list on small islands is a nested subset of that on larger islands. [Unpublished data courtesy of T. Schoener.]*

Posing the Issues   With the empirical background of the preceding discussion, we can pose the questions that we want niche theory to address. First, consider the phenomenon of resource partitioning between species. The theory should predict the limit to the similarity between species that is consistent with their coexistence. This prediction would be testable against empirical "rules of community assembly." Furthermore, the theory should predict the niche separation between competitors that results from their coevolution. The coevolved niche separation must be wider than the minimum niche separation that would permit coexistence, but how much wider?

Second, consider the phenomenon of resource partitioning within a species. Here, the theory should predict the total niche width of a species and the breakdown of that niche width into the BPC and WPC components. It should predict in qualitative terms how these quantities change along environmental gradients.

Third, consider the patterns of island biogeography. The theory should explain the area effect and distance effect in groups where these effects occur. It should also explain the existence of nested faunal lists and address the issue of whether species turnover is a necessary feature of the area effect and distance effect.

As we shall see, niche theory and the theory of island biogeography, when viewed as a whole, do address all the issues posed here. Within the overall framework of these theories, however, there is substantial disagreement as to the mechanism for explaining the phenomenon under discussion. The controversy in this area has made niche theory and island biogeography one of the most exciting areas of ecology today.

Preliminaries on the Lotka–Volterra Competition Equation   Niche theory is based on the Lotka–Volterra competition equations, together with the assumption that the competition coefficient between two species is related to their overlap in the use of different types of limiting resources. The assumption that competition is a function of overlap greatly simplifies the stability analysis in all niche theory models. This section presents some of the ideas and results that are common to all models in niche theory. Most results discussed in this section originate with MacArthur, Levins, and May.

Each species is assumed to have a *resource utilization curve*. If the resources are ordered along a continuous resource axis, as with prey-size in the preceding examples, the resource utilization curve is a probability density function defined on that axis. Specifically, if $u_i(x)$ is the utilization curve for species-i, it means that $u_i(x)\,dx$ is the probability that species-i obtains a unit of resource from the interval x and $x + dx$ on the resource axis. Generally, $u_i(x)$ is a multivariate probability density function. An example of a one-axis $u_i(x)$ curve is presented in Figure 24.10 for the entire population of adult males in *Anolis ferreus*. Figure 24.8 presented the utilization curves for specific phenotypes within the population of adult males. The resource utilization curve is an empirical quantity and it describes the actual niche of the species. The first two moments of the curve have special meanings. The *niche position* of species i is defined as the average resource taken by species i,

$$\frac{\text{niche position}}{\text{of species-}i} = \int x u_i(x)\,dx \tag{24.3}$$

The *niche width* of species-i is defined as the variance in resources taken by species-i

$$\frac{\text{niche width of}}{\text{species-}i} = \int (x - \bar{x})^2 u_i(x)\,dx \tag{24.4}$$

(Equivalently, it is sometimes more convenient to use the standard deviation rather than the variance as a measure of niche width.) If the resources are classified by discrete categories, rather than ordered along a continuous axis, then the utilization curve is the probability density function for a discrete random variable. In this case the niche position is not defined and the niche width is measured by a more general measure of dispersion, for example, the Shannon–Weiner information measure H. More on this later, when the topic of niche width is discussed in detail.

Competition and Overlap   The competition between two species is assumed to be determined by the overlap between the utilization curves. Furthermore, it is usually assumed that the competition coefficients are related to the utilization curves by the formula

$$\alpha_{ij} = \frac{\int u_i(x) u_j(x)\,dx}{\int u_i^2(x)\,dx} \tag{24.5a}$$

The numerator in this formula describes the probability of an individual from species-i meeting an individual of species-j at resource position-x, and then integrated over all resource positions. Similarly, the denominator describes the probability of an individual from species-i meeting another individual from species-i at resource location-x and then integrated over all x. This formula is motivated by the interpretation of the competition coefficient as a measure of the amount of interspecific competition relative to the amount of intraspecific competition. By this formula, $\alpha_{ii} = 1$ for all i. However, the α-matrix is not necessarily symmetric, nor are the off-diagonal elements necessarily less than one. Also, the formula can apply to a multivariate resource classification. It is perfectly proper to visualize x in

Equation (24.5) as a vector and the integration as the volume integral over several dimensions. Thus the formula is not so restrictive as to be without interest. Nonetheless, the formula is an assumption with enormous implication for the stability of coexisting species.

Furthermore, it is also possible to generalize the overlap formula (24.5a) to situations where each species not only uses resources from different positions, but also uses different total amounts of resource. Let A_j be the total amount of resource used by species-j. Then, a weighted overlap formula is given by

$$\alpha_{ij} = \frac{A_j \int u_i(x)u_j(x)\, dx}{A_i \int u_i^2(x)\, dx} \tag{24.5b}$$

The weights further increase the asymmetry of the α-matrix. Nonetheless, all the results of this section apply to formula (24.5b) as well as (24.5a). The proofs that follow use formula (24.5a); the extension to (24.5b) is left as an exercise.

Properties of the α-Matrix Although the α-matrix generated by (24.5) is generally not symmetric, it is closely related to a symmetric matrix. Let the matrix ϕ be defined as

$$\phi_{ij} = \int u_i(x)u_j(x)\, dx \tag{24.6}$$

The matrix ϕ is symmetric. Let the diagonal matrix C be defined as

$$C_{ii} = \frac{1}{\phi_{ii}} = \frac{1}{\int u_i^2(x)\, dx}, \qquad C_{ij} = 0 \tag{24.7}$$

Then, the α-matrix is given by diagonal matrix times a symmetrix matrix

$$\alpha = C\phi \tag{24.8a}$$

We shall call ϕ the overlap matrix. With (24.5b), the α-matrix is given by

$$\alpha = A^{-1}C\phi A \tag{24.8b}$$

where A is a diagonal matrix, whose diagonal elements are the total amount of resource used by each species.

The overlap matrix ϕ is not only symmetric, it has real and positive eigenvalues. A symmetric matrix with real positive eigenvalues is called a *positive definite matrix*. To prove that ϕ is positive definite, we introduce a certain quadratic form and show that it is positive definite. The details are illustrated in the box.

$$\Psi(y_1 \cdots y_N) = \sum_i \sum_j y_i \phi_{ij} y_j$$

$$= \sum_i \sum_j y_i \left[\int u_i(x)u_j(x)\, dx \right] y_j$$

$$= \int \sum_i \sum_j y_i u_i(x) y_j u_j(x)\, dx$$

$$= \int \left[\sum_i y_i u_i(x) \right]\left[\sum_j y_j u_j(x) \right] dx$$

$$= \int \left[\sum_i y_i u_i(x) \right]^2 dx$$

$$> 0$$

Thus the assumption of the formula (24.5a), relating resource utilization to the competition coefficients, entails that the α-matrix is structured so that it can be written as diagonal matrix times a positive definite matrix.

Now, consider the implications of the α-matrix for the coexistence of competitors in the Lotka–Volterra competition model. The N-species Lotka–Volterra competition model is

$$\frac{dN_i}{d_t} = r_i N_i \left(K_i - \sum_j \frac{\alpha_{ij} N_j}{K_i} \right) \tag{24.9a}$$

The equilibrium point at which all species coexist satisfies

$$\alpha \hat{N} = K \tag{24.9b}$$

where α is the α-matrix, $\hat{N}$ is a column vector of equilibrium population sizes, and K is a column vector of carrying capacities. All the elements of $\hat{N}$ must be positive for the equilibrium to be considered. If any elements are zero or negative, then, without any further analysis, we can conclude that all the species cannot coexist. If all the elements are positive, then we may proceed with the stability analysis.

Because (24.9b) is a linear equation, the solution is unique (provided the determinant of α is nonzero, as it almost always is). Therefore, there can be only one equilibrium point at which all species coexist. Of course, there are many "boundary" equilibria in the model, that is, equilibria where one or more species are missing and it is perfectly possible to model the existence of alternative communities with the Lotka–Volterra equations. However, the alternative communities in a Lotka–Volterra model represent communities with different species lists. To model the occurrence of alternative communities with identical species composition but different abundances requires a more complicated model than the Lotka–Volterra system.

The stability of a positive equilibrium , $\hat{N}$, is determined by examining the eigenvalues of the matrix

$$S = -D\alpha \tag{24.10}$$

where D is a diagonal matrix whose elements are

$$d_{ii} = \frac{r_i \hat{N}_i}{K_i} \qquad d_{ij} = 0 \qquad \text{for } i \neq j \tag{24.11}$$

The eigenvalues of the matrix S must have negative real parts for local stability of the equilibrium.

You will recall from Chapter (21) that for *two* competing species we determined whether a positive equilibrium was stable or not by inspecting a graph with intersecting isoclines. There N_1 and N_2 were plotted on the horizontal and vertical axes, and the equations for the isoclines included K_1, K_2, α_{12}, and α_{21} but did *not* include r_1 and r_2. Strobeck (1973) was the first to point out that the stability analysis for a three-species equilibrium could not be carried out with any extension of the graphical technique that is popular for the two-species model. The reason is that the stability depends on the r's as well as the K's and α's and the graphical technique does not include any mention of the r's. Strobeck provided an example with three species and a very unsymmetric α matrix where the r's, in fact, control the stability of the three-species equilibrium. His example is particularly interesting in that all the boundary equilibria are always unstable. Hence, when the internal equilibrium is also unstable, there is no stable equilibrium point anywhere and the three species oscillate in a complicated way without ever converging to an equilibrium point. They may, however, converge to a *region* which is called an "attractor." The analysis of coexistence between competitors under conditions of oscillation is an important topic of active current research. For examples of complex dynamical behavior in multispecies competition systems, see Gilpin (1975a), May and Leonard (1975), Armstrong and McGehee (1976), and Smale (1976).

Positivity Implies Global Stability

An extremely important fact for niche theory is that the use of an α-matrix that is obtained as the product of a diagonal matrix times a positive definite matrix implies that any positive equilibrium *point* is stable. Thus cycling does not occur, and moreover the r's are not important in the determination of stability. Indeed, the result is even stronger: The mere fact that the equilibrium point is positive implies that it is stable. The quickest but least interesting proof of this result is to note that the stability matrix (24.10) must have negative real parts, because it is given by

$$S = -DC\phi \tag{24.12}$$

where D and C are positive diagonal matrices and ϕ is positive definite.† A more interesting proof is to exhibit a global Liapunov function that applies to positive trajectories whenever the α-matrix is of the form $C\phi$.

Consider the function for S species, defined as

$$Q(t) = \sum_i \sum_j (N_i(t) - \hat{N}_i)\phi_{ij}(N_j(t) - \hat{N}_j) \tag{24.13}$$

where $Q(t)$ is a positive definite quadratic form. It equals 0 if and only if $N_i = \hat{N}_i$ for all i and is greater than zero for any other set of N_i. This latter property is entailed by the assumption that ϕ is a positive definite matrix. We now show that $dQ(t)/dt \le 0$, that is, all trajectories minimize $Q(t)$. By the product rule of differentiation

$$\frac{dQ(t)}{dt} = \sum_i \sum_j \left[\phi_{ij}(N_i - \hat{N}_i)\frac{dN_j}{dt} + \phi_{ij}(N_j - \hat{N}_j)\frac{dN_i}{dt} \right]$$

$$= \sum_i \sum_j \left\{ \phi_{ij}(N_i - \hat{N}_i)\left[\frac{r_j}{K_j}N_j\left(K_j - \sum_k C_j\phi_{jk}N_k \right) \right] \right.$$

$$\left. + \phi_{ij}(N_j - \hat{N}_j)\left[\frac{r_i}{K_i}N_i\left(K_i - \sum_k C_i\phi_{ik}N_k \right) \right] \right\} \tag{24.14}$$

We now substitute

$$K_j = \sum_k C_j\phi_{jk}\hat{N}_k, \qquad K_i = \sum_k C_i\phi_{ik}\hat{N}_k$$

† Let A be the diagonal matrix DC. Let $A^{1/2}$ be the diagonal matrix where $A_{ii}^{1/2} = \sqrt{A_{ii}}$ and similarly $A^{-1/2}$. Then $S = -A\phi$ is similar to $-A^{-1/2}\phi A^{1/2}$ which, itself, is similar to $-\phi$. Hence, except for sign, the eigenvalues of S are the same as the eigenvalues of ϕ.

These rearrange as

$$\frac{dQ(t)}{dt} = -\sum_j \frac{r_j N_j}{K_j} C_j \sum_i \phi_{ij}(N_i - \hat{N}_i) \sum_k \phi_{jk}(N_k - \hat{N}_k)$$

$$\qquad (24.15)$$

$$-\sum_i \frac{r_i N_i}{K_i} C_i \sum_j \phi_{ji}(N_j - \hat{N}_j) \sum_k \phi_{ik}(N_k - \hat{N}_k)$$

Finally, we note that $\phi_{ij} = \phi_{ji}$, since ϕ is symmetric (even though the α-matrix may not be). If we substitute ϕ_{ji} for ϕ_{ij} in the first line of (24.15), then we see that the two lines in (24.15) are actually identical, apart from the names of the index variables of i and j. Therefore, (24.15) becomes

$$\frac{dQ(t)}{dt} = -2 \sum_i \frac{r_i C_i}{K_i} N_i(t) J_i^2(t) \qquad (24.16a)$$

where

$$J_i(t) = \sum_j \phi_{ij}(N_j(t) - \hat{N}_j) \qquad (24.16b)$$

By inspection of (24.16a), we see that $dQ(t)/dt$ is always negative unless $N_i = \hat{N}_i$ for all i. Therefore, along all positive trajectories, Q, continually decreases until equilibrium is reached, at which point Q is zero. The discovery of this function traces through May (1973) to MacArthur (1970).†

We may summarize the discussion so far in terms of the following result

RESULT 1 *Consider the Lotka–Volterra competition equations*

$$\frac{dN_i}{dt} = r_i N_i \frac{\left(K_i - \sum_j \alpha_{ij} N_j\right)}{K_i} \qquad (i = 1 \cdots S) \qquad (24.17)$$

with $\alpha_{ii} = 1$, $\alpha_{ij} \geq 0$. Suppose the competition coefficients are determined from the utilization curves $u_i(x)$, of the species by the formula

$$\alpha_{ij} = \int u_i(x) u_j(x)\, dx \Big/ \int u_i^2(x)\, dx \qquad (24.18)$$

(It is understood that x can be multidimensional and that the α-matrix need not be symmetric.)

(a) *There is global convergence to an equilibrium point at which all species coexist if and only if the unique equilibrium solution, $\hat{N}$, of*

$$\alpha \hat{N} = K \qquad (24.19)$$

is strictly positive, that is, if and only if $\hat{N}_i > 0$ for all i.

(b) *The determinant of the α-matrix is positive.*

(c) *All principal minors of the α-matrix are positive.*

Parts (b) and (c) follow directly from the fact that the α-matrix is a product of a diagonal matrix with a positive definite matrix.

The key to part (a) of the result above is the existence of the Liapunov function, $Q(t)$, given by (24.13). As defined in (24.13), this function can only be interpreted as a measure of the total deviation from equilibrium. Saying

† The proviso that trajectories must remain positive is unimportant in practice. It can be shown that any trajectory which is initially positive remains positive for all finite time in the LV model and virtually all other models of competition as well.

that the trajectories continually reduce $Q(t)$ only says that a certain measure of the total deviation from equilibrium continually decreases. It would be desirable to attach a more biological interpretation to the fact that $Q(t)$ is minimized. To do this, MacArthur (1970) and May (1973) introduce several additional assumptions: (1) There exists a normalized utilization function, $f_i(x)$, for each species, such that $\alpha_{ij} = \int f_i(x)f_j(x)\,dx$, thereby making the α-matrix itself positive definite; and (2) there exists a function, $R(x)$, which describes the amount of each resource such that the carrying capacities are determined as $K_i = \int R(x)f_i(x)\,dx$. With these assumptions, it can be shown that $Q(t)$, as defined previously, is identical to another expression which may be interpreted as a measure of the difference between the available resources in the environment and the total amount of resources actually used by all organisms of all species. In this context, the fact that trajectories continually decrease $Q(t)$ can be taken to mean that the difference between resource availability and total resource use is continually decreasing until equilibrium is attained. It must be stressed, however, that this interpretation requires justification of both additional assumptions above. Furthermore, the interpretation above applies only to the *population dynamics* of the approach to equilibrium for several competitors. It does not, in any case, apply to the *evolution* of competing species.

Result 1 above is also true of the α-matrix and is given by

$$\alpha = A^{-1}C\phi A$$

where A and C are positive diagonal matrices and ϕ is positive definite. This form arises if the overlap formula (24.5b) is used. In this case, the Q function is defined as

$$Q(N_1 \cdots N_S) = \sum_i \sum_j (N_i - \hat{N}_i)A_i\phi_{ij}A_j(N_j - \hat{N}_j)$$

To prove this generalization, simply repeat the steps in the proof above but begin with this definition of Q instead of that in (24.13).

Invasion When Rare Implies Positivity   We turn now to another general result that applies to all models in niche theory. The result assumes that the determinant of the α-matrix is positive and that all principal minors obtained by striking out one row and the corresponding column of the α-matrix are also positive. This condition is met when the α-matrix is generated by the resource overlap formula (24.18). The result obtained traces to Strobeck (1973).

RESULT 2   *Consider the Lotka–Volterra competition equations for S species. Assume the determinant and all principal minors of order S − 1 of the α-matrix are positive. Let $\hat{N}_i$ denote the equilibrium population size of species-i at the interior equilibrium point. That is, $\hat{N}_i$ is the ith component of the vector $\hat{N}$ obtained as the solution of*

$$\alpha\hat{N} = K \tag{24.20}$$

Let $(dN_i/dt)^*$ denote the rate of increase of species-i when it is very rare and when the remaining species are present in the equilibrium abundance, which they would attain in the absence of species-i.

Then, for each i,

$$\hat{N}_i > 0 \qquad \textit{if and only if} \qquad \left(\frac{dN}{dt}\right)^* > 0 \qquad (24.21)$$

This result asserts that the positivity of the equilibrium solution for species-i is mathematically equivalent to the ability of species-i to increase when it is rare, provided the α-matrix satisfies certain conditions. To prove this result, we can, without loss of generality, focus on the first species, $i=1$. By Cramer's rule

$$\hat{N}_1 = \frac{\begin{vmatrix} K_1 & \alpha_{12} & \cdots & \alpha_{1S} \\ \vdots & & & \\ K_S & \alpha_{S2} & \cdots & 1 \end{vmatrix}}{\det \alpha} > 0 \qquad (24.22)$$

where $\det \alpha$ means the determinant of the α-matrix. By assumption $\det \alpha$ is positive; therefore, to make the condition of $\hat{N}_1$ positive the determinant in the numerator of (24.22) must be positive. Now expand the numerator along the top row leaving

$$K_1 \begin{vmatrix} 1 & \cdots & \alpha_{2S} \\ \vdots & & \vdots \\ \alpha_{S2} & \cdots & 1 \end{vmatrix} - \alpha_{12} \begin{vmatrix} K_2 & \alpha_{23} & \cdots & \alpha_{2S} \\ \vdots & \vdots & & \vdots \\ K_S & \alpha_{S3} & \cdots & 1 \end{vmatrix}$$

$$+ \alpha_{13} \begin{vmatrix} K_2 & 1 & \cdots & \alpha_{2S} \\ \vdots & \vdots & & \vdots \\ K_S & \alpha_{S2} & \cdots & 1 \end{vmatrix} \cdots > 0$$

$$(24.23)$$

Then for each determinant multiplied by α_{1j} place the column of K's in the jth position

$$(24.24)$$

Thus the column of K's is extracted from its leading position, as illustrated in (24.23), and is inserted in the position shown in (24.24) and all the columns before j are moved over to the left by one to make room. This interchange of columns leaves

$$K_1 \begin{vmatrix} 1 & \cdots & \alpha_{2S} \\ \vdots & & \vdots \\ \alpha_{S2} & \cdots & 1 \end{vmatrix} - \alpha_{12} \begin{vmatrix} K_2 & \alpha_{23} & \cdots & \alpha_{2S} \\ \vdots & \vdots & & \vdots \\ K_S & \alpha_{S3} & \cdots & 1 \end{vmatrix}$$

$$- \alpha_{13} \begin{vmatrix} 1 & K_2 & \cdots & \alpha_{2S} \\ \vdots & \vdots & & \vdots \\ \alpha_{S2} & K_S & \cdots & 1 \end{vmatrix} \cdots > 0 \qquad (24.25)$$

Note the net sign change before α_{1j}, where j is odd and no sign change where j is even. Now let $\tilde{N}_j^{(1)}$ denote the equilibrium population size of species-j in the absence of species-1. This number may be negative. The determinant multiplied by K_1 is a principal minor of order $S-1$ of the α-matrix and is positive by assumption. Then, dividing through by this principal minor yields

$$K_1 - \sum_j \alpha_{1j} \tilde{N}_j^{(1)} > 0 \tag{24.26}$$

Finally, note that (24.26) is exactly the condition for species-1 to increase when rare; that is, $N_1 \approx 0$, and when all the other species have the equilibrium values they would attain in the absence of species-1. The equivalence of (24.22) with (24.26) proves the equivalence of the positivity condition with the condition for invasion when rare under the stated assumptions.

Results one and two are often combined in practice as follows. Suppose the α-matrix is generated by the resource overlap formula (24.18). Suppose also that we determine that each species can increase when rare by showing that criterion (24.26) is satisfied for each species. Then by Result 2 there is a positive equilibrium point at which all species coexist. Also, by Result 1 that equilibrium point is globally stable and all trajectories converge there. Thus, in niche theory models, the entire qualitative dynamical picture for competing species can be determined by analyzing whether each species can increase when rare.

There is another result of general interest in niche theory models. It concerns the number of discrete resource categories required to obtain a nonsingular α-matrix. Suppose that a total of H resource categories are distinguished for S species. Let u_{ih} be the fraction of resource use by species-i in resource h. Then the α's are determined as

$$\alpha_{ij} = C_i \sum_{h=1}^{H} u_{ih} u_{jh} \tag{24.27}$$

where

$$C_i = \frac{1}{\sum\limits_{h=1}^{H} u_{ih}^2} \tag{24.28}$$

Let the matrix U be defined as

$$U = \begin{pmatrix} u_{11} & \cdots & u_{1H} \\ \vdots & & \vdots \\ u_{S1} & \cdots & u_{SH} \end{pmatrix} \tag{24.29}$$

Then in matrix form the α-matrix is given by

$$\alpha = CUU^T \tag{24.30}$$

where U^T indicates the transpose of U.

If $H < S$, the rank of α is H, and if so, then α is singular (i.e., det $\alpha = 0$). See Levins (1968). Therefore, to calculate a nonsingular α-matrix, one must use more resource categories than species. For example, someone studying five competitors must obtain data on resource use from five or more resource categories or otherwise the α-matrix is necessarily singular.

Another way the above result has been interpreted is to say that no more than H species can coexist on H resources. For, it is argued, if there were more than H species on H resources, there would be a curve of possible equilibria (there is no unique equilibrium point if det $\alpha = 0$) and presumably one of the species would become extinct. For a series of papers on the relation between the minimum resource number and species number, see MacArthur (1968), Levin (1970), and Haigh and Maynard Smith (1972). However, I should add that the issue of whether a total of S species can exist on fewer than S resources is moot in many, perhaps almost all circumstances. In practice, there are a great many more independent resource types than species competing for them. For example, insectivorous birds and lizards eat literally hundreds of species of insects, and furthermore, they may obtain them from different microhabitats. However, the total insectivorous community, including vertebrates, wasps and spiders, consists of only, say, one half as many species as are available in the prey community, and the issue can rarely arise that the number of competing species is limited by the *number* of distinct resources. In contrast, as we have seen, competitors that use many resources in *too similar* a fashion, do not coexist. As we shall see, they cannot coexist according to competition theory.

Resource Partitioning Within Species

We begin coverage of the models in niche theory with models for the resource partitioning within species. We shall be specifically concerned with the forces that shape the niche width of a population. As we shall see, models that predict the niche width of a population assume that the "niche widths" of the individual members of the population are given in advance. Similarly, models for resource partitioning between species assume that the niche widths for each of the species are given. There, presently, is no theoretical treatment that considers both within- and between-species resource partitioning simultaneously.

The Components of Niche Width

A population potentially consists of many types of individuals, each with a characteristic pattern of resource use. As mentioned in the introduction to this chapter, we can partition the variance of resource use by the whole population into two components, one measuring the variety of individuals (BPC), and the other measuring the average variance in an individual's resource use (WPC). To define these quantities more precisely, let y label an individual whose average resource use is from position y on a resource axis. Let $v(x|y)$ denote the resource utilization curve of individual, y. By our labeling convention

$$y = \int x v(x|y)\, dx \qquad (24.31)$$

$v(x|y)$ is the probability distribution of resource use of an individual, conditional on the fact that its average use is from location y. Let $p(y)$ denote the phenotype distribution in the population. That is, $p(y)\, dy$ is the proportion of the population that have their average resource use between y and $y + dy$. As before, let $u(x)$ denote the utilization curve of a whole population. Then $u(x)$ is obtained by cumulating the resource use of all individuals in the population

$$u(x) = \int p(y) v(x|y)\, dy \qquad (24.32)$$

With this notation, the niche width of the whole population is defined as

$$w^2 = \int (x - \bar{x})^2 u(x)\, dx \qquad (24.33)$$

The "niche width" of an individual whose average use is from location y on the axis is given by

$$\sigma_v^2(y) = \int (x-y)^2 v(x|y)\,dx \qquad (24.34)$$

The components of the population's niche width are then defined as

$$\text{BPC} = \int (y-\bar{y})^2 p(y)\,dy \qquad (24.35a)$$

$$\text{WPC} = \int \sigma_v^2(y) p(y)\,dy \qquad (24.35b)$$

Thus the BPC is the variance of the phenotype distribution, and the WPC is the average variance of the individual utilization curves. It is an identity that

$$w^2 = \text{BPC} + \text{WPC} \qquad (24.36)$$

If the phenotypes use different total amounts of resource then the formulas above should be modified slightly. Let $A(y)$ denote the total amount of resource used by phenotype y. Then, a "weighted" phenotype distribution $\tilde{p}(y)$ is defined as

$$\tilde{p}(y) = \frac{A(y)p(y)}{\int A(y)p(y)\,dy} \qquad (24.37)$$

In this case $\tilde{p}(y)$ should be used instead of $p(y)$ in Equation (24.32) and (24.35). See Roughgarden (1974a) for an explicit example.

The partitioning of the total niche width into these two components is also possible with discrete unordered resource categories. Also, the phenotype classes may be unordered. In this situation, the mean and variance have no meaning and the measures of niche width are based on the Shannon information theory formula. Let X_h denote a resource category, $h = 1 \cdots H$, and let Y_i denote a phenotype class, $i = 1 \cdots S$. The original utilization data can be organized as a joint distribution function, $P(X_h, Y_i)$. This function is the probability that a unit of used resource is of type, X_i, and was used by phenotype, Y_i. This function must satisfy

$$\sum_h \sum_i P(X_h, Y_i) = 1 \qquad (24.38)$$

From this function, which expresses the original utilization data, we need to define two marginal distributions and two conditional distributions. One marginal distribution represents the population's phenotype distribution, irrespective of their resource use. This distribution is denoted $p(Y_i)$ and is calculated as

$$p(Y_i) = \sum_h P(X_h, Y_i) \qquad (24.39)$$

The other marginal distribution represents the whole population's resource use, irrespective of what each phenotype does. This distribution is denoted $u(X_h)$ and is calculated as

$$u(X_h) = \sum_i P(X_h, Y_i) \qquad (24.40)$$

Next, we require the conditional distributions. First, we introduce the distribution of resource use for each given phenotype, $v(X_h|Y_i)$, as

$$v(X_h|Y_i) = \frac{P(X_h, Y_i)}{p(Y_i)} \qquad (24.41)$$

Finally, we introduce the distribution of phenotypes that use any given resource, $\rho(Y_i|X_h)$, as

$$\rho(Y_i|X_h) = \frac{P(X_h, Y_i)}{u(X_h)} \tag{24.42}$$

We now take, as our measure of the niche width for the total population,

$$w^2 = -\sum_h u(X_h) \ln u(X_h) \tag{24.43}$$

This formula is the analogue of (24.33). It measures the evenness of the resource distribution of the whole population. The within-phenotype component is given by

$$\text{WPC} = \sum_i p(Y_i) \left[-\sum_h v(X_h|Y_i) \ln v(X_h|Y_i) \right] \tag{24.44}$$

The expression in brackets is simply the "niche width" for phenotype Y_i, and the WPC is the average of the niche widths of the individual phenotypes. Equation (24.44) is the same in spirit as (24.35b). The between-phenotype component is given by

$$\text{BPC} = \left[-\sum_i p(Y_i) \ln p(Y_i) \right]$$
$$- \left\{ \sum_h u(X_h) \left[-\sum_i \rho(Y_i|X_h) \ln \rho(Y_i|X_h) \right] \right\} \tag{24.45}$$

This formula looks complicated, but it really is not. The first term measures the evenness of the population's phenotype distribution among the various phenotypic classes. The second term measures the average evenness of the phenotype distribution, as "seen" by the different resources. If the average phenotype distribution that encounters a resource is narrow, compared with the phenotype distribution in the population as a whole, then the different phenotypes must be using different resources. The BPC is the measure of the extent to which the phenotypes use different resources. It is, again, an identity that the total niche width partitions into its components as

$$w^2 = \text{BPC} + \text{WPC} \tag{24.46}$$

These formulas do not require modification if the phenotypes use different total amounts of resource, because this possibility is built into the joint probability function to begin with.

For further readings on the measurement of niche width, see Colwell and Futuyma (1971), Pielou (1972), and Roughgarden (1974a). It must be stressed that measuring the niche width and its components, with any of the formulas above, is model-free. One is not committed to any assumption of niche theory models in using these formulas.

The Clone Selection Model

In this section, we present the simplest model that predicts the niche width of a population, together with the breakdown of that niche width into the two components. The model presents an *idealization* of the phenomenon of resource partitioning within a species—the idealization is framed along lines suggested by the data on resource partitioning with respect to prey size, as illustrated in Figure 24.8.

The Utilization Curves We imagine that the population potentially consists of individuals with utilization curves as depicted in Figure 24.16. The individuals whose average prey size is small, also have a small variance in prey size, as suggested by the data in Figure 24.8. A useful functional form with this property is the "log-normal" distribution given by

$$v(z|z_o) = \frac{\exp\{-\frac{1}{2}[\ln(z) - \ln(z_0)/\sigma_v]^2\}}{z\sqrt{2\pi\sigma_v^2}} \tag{24.47a}$$

$v(z)$ is a distribution with unit area between 0 and ∞. The mean of $v(z|z_0)$ is

$$\bar{z} = z_0\, e^{\sigma_v^2/2} \tag{24.47b}$$

and the variance is

$$\sigma_z^2 = z_0^2\, e^{\sigma_v^2}(e^{\sigma_v^2} - 1) \tag{24.47c}$$

For details on fitting the log-normal distribution to field data, see Pielou (1975, pp. 49 ff.). The curves in Figure 24.16 are plotted from (24.47a). It is called the log-normal distribution because under the transfirmation

$$x = \ln(z), \qquad x_0 = \ln(z_0) \tag{24.48}$$

The curve becomes a normal distribution in x

$$v(x|x_0) = \frac{\exp\{-\frac{1}{2}[(x - x_0)/\sigma_v]^2\}}{\sqrt{2\pi\sigma_v^2}} \tag{24.49}$$

Thus, if z is distributed according to (24.47), the log of z is distributed normally as sketched in Figure 24.17.

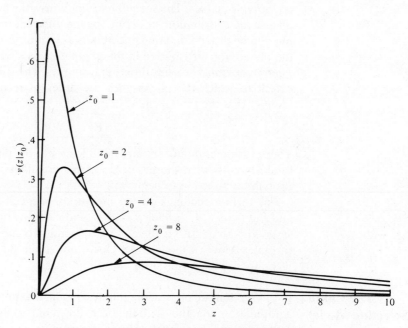

FIGURE 24.16. *The log-normal distribution: $\sigma_v^2 = 1$. Also,*

$$v(z\mid z_0) = \frac{\exp[-\frac{1}{2}(\ln z - \ln z_0)^2/\sigma_v^2]}{z\sqrt{2\pi\sigma_v^2}}$$

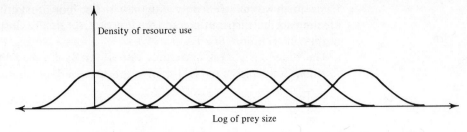

FIGURE 24.17. *Sketch of utilization curves for individuals of different sizes after the log transformation.*

Furthermore, we want to allow for the possibility that the individuals whose average prey size is large are also consuming more total food, as was illustrated in Figure 24.9. There is experimental evidence, presented by Hemmingsen (1960), and Bartholomew and Tucker (1964), which suggests that the energy consumption by a lizard is proportional to its weight, raised to approximately the $\frac{3}{4}$ power. Assuming an animal's weight varies as its length is cubed, we arrive at the conjecture that the total resource consumption of an animal of size z is

$$A(z) = A_0 z^{\kappa}$$
$$= A_0 \, e^{\kappa \ln z} \tag{24.50}$$

where κ is approximately $\frac{9}{4}$.

The Competition Function With the individual utilization curves given by (24.47) and the total resource use given by (24.50) we can calculate the competition coefficient between two individuals, using the overlap formula (24.5b). The competition coefficient, representing the effect of individual z_j on z_i, is given by

$$\alpha(z_i, z_j) = \frac{A(z_j)}{A(z_i)} \frac{\int v(z|z_i)v(z|z_j)\,dz}{\int v^2(z|z_i)\,dz} \tag{24.51}$$

And upon substituting (24.47) and (24.50) into (24.51), calculating the integral, and rearranging, eventually yields

$$\alpha(z_i, z_j) = \exp\left[-\kappa(\ln z_i - \ln z_j)\right] \exp\left[\frac{-\frac{1}{2}(\ln z_i - \ln z_j)^2}{2\sigma_v^2}\right]$$
$$= \exp(\sigma_v^2 \kappa^2) \exp\left\{-\frac{1}{2}\left[\frac{(\ln z_i - \ln z_j) + 2\sigma_v^2 \kappa]^2}{2\sigma_v^2}\right]\right\} \tag{24.52}$$

Note that the competition coefficient between z_i and z_j depends only on the difference between their logs. Indeed, because of the form of (24.47) and (24.52), we can simplify the entire discussion by applying a log transformation to the original variables. Therefore, we take (24.49) as the description of the utilization curves and rewrite (24.52) as

$$\alpha(d) = \exp(\sigma_v^2 \kappa^2) \exp\left[\frac{-\frac{1}{2}(d + 2\sigma_v^2 \kappa)^2}{2\sigma_v^2}\right] \tag{24.53}$$

where

$$d = x_i - x_j, \, x_i = \ln z_i, \, x_j = \ln z_j$$

Henceforth we shall deal only with the x's. The individual utilization curves are then normal distributions and the function $\alpha(d)$ is also closely related to a normal distribution. See Figure 24.18.

The function $\alpha(d)$ is extremely important. It is called the *competition function*. It expresses the fact that, in this idealized model, the competition

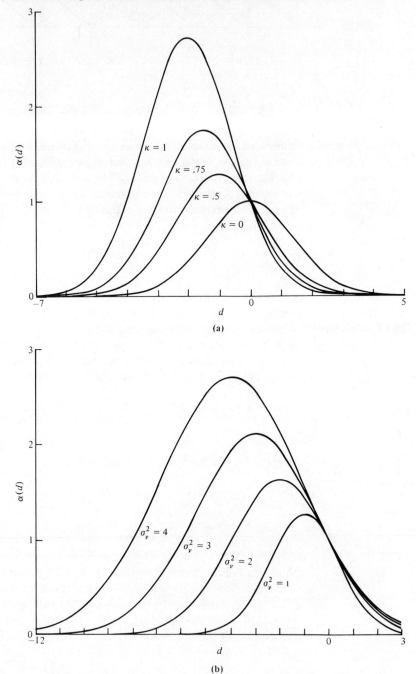

(a)

(b)

FIGURE 24.18. (a) *Competition function for several values of the parameter,* κ. (b) *Competition function for several values of the parameter,* σ_v^2. *In* (a) $\sigma_v^2 = 1$; *in* (b) $\kappa = .5$.

coefficient between two individuals, say x_i and x_j, depends only on the *difference* between x_i and x_j and not on the particular values of x_i and x_j, themselves. When it exists, the competition function provides a complete summary of all the entries in an α-matrix. If there is a competition function, then the *ij*th entry of an α-matrix is simply $\alpha_{ij} = \alpha(x_i - x_j)$.

The *shape* of the competition function expresses a lot of information. The competition function (24.53) differs from a Gaussian curve only in the normalization constant. The area under (24.53) is given by

$$A = (2\sqrt{\pi})\sigma_v \, e^{\sigma_v^2 \kappa^2} \tag{24.54}$$

$\alpha(d)$ never has unit area and so it is helpful for descriptive purposes to associate with $\alpha(d)$ another function with the same shape, but which has been normalized to unit area. This normalized function is a standard Gaussian,

$$\alpha^*(d) = \frac{\exp\left[-\frac{1}{2}(d + 2\sigma_v^2 \kappa)^2 / (2\sigma_v^2)\right]}{\sqrt{2\pi 2\sigma_v^2}} \tag{24.55}$$

The mean and variance of $\alpha^*(d)$ are

$$\begin{aligned} \bar{x}_\alpha &= -2\sigma_v^2 \kappa \\ \sigma_\alpha^2 &= 2\sigma_v^2 \end{aligned} \tag{24.56}$$

The three quantities A, $\bar{x}_\alpha$, and σ_α^2 describe the overall shape of the competition function. Here A measures the total amount of competition, and $\bar{x}_\alpha$ measures the asymmetry in the competitive relationships; a negative value to $\bar{x}_\alpha$ means that larger individuals have a competitive advantage over small individuals. The quantity, σ_α^2, measures the range of the competitive interaction. A high σ_α^2 means that even quite dissimilar phenotypes exhibit strong competition with one another. Figure 24.18 illustrates $\alpha(d)$ for various choices of κ and σ_v^2.

The Carrying Capacity Function

The other major element which must be included in the model is an assumption about the amount of resource that is available to each phenotype. This assumption is introduced in terms of the carrying capacity function, $K(z)$, defined as follows: $K(z)$ is the carrying capacity of the environment for phenotype-z, provided the population consists only of phenotype-z. Obviously, the presence of other phenotypes that compete with phenotype-z would alter abundance of phenotype-z from the value it would have if those competitors were absent. Therefore, $K(z)$ is standardized so that it refers to the equilibrium population size of a population consisting only of individuals of phenotype-z in the environment under consideration. Thus $K(z)$ measures the amount of renewable resources in the environment in units of the steady-state numbers of organisms of phenotype-z, which those resources can support. Clearly, we are not likely to possess any good data on the shape of $K(z)$ in the near future so any assumption about the shape of $K(z)$ is quite conjectural. However, there are certain general features that the curve must have. If z refers to the typical prey size used by a lizard, then $K(z)$ must equal zero for z sufficiently small and for z sufficiently large. The carrying capacity of the environment for individuals who eat insects much smaller than ants, say 10 or 100 times smaller, is certainly low, because of the low net caloric yield per prey item, rarity, and difficulty of finding. On the other side, the carrying capacity of individuals using huge insects must also be low because of the rarity of huge insects. Perhaps the carrying capacity is highest for lizards whose average prey size is about 3 to 6 mm. In any event, it seems reasonable at this time to use the log-normal distribution again, that is, to suppose that

$$K(x) = K \frac{\exp\{-\frac{1}{2}[\ln(z) - \ln(z_k)]^2/\sigma_k^2\}}{z\sqrt{2\pi\sigma_k^2}} \tag{24.57}$$

And again, if we discuss in terms of the log of prey size, we have

$$K(x) = K \frac{\exp[-\frac{1}{2}(x - \bar{x}_k)^2/\sigma_k^2]}{\sqrt{2\pi\sigma_k^2}} \tag{24.58}$$

Note that $K(x)$ is a normal distribution apart from the factor, K. The parameter K measures the *total amount* of resources in the environment; $\bar{x}_k$ denotes the *position* with the highest carrying capacity; and σ_k^2 measures the extent to which available resources are *spread* over the resource axis. These three parameters describe the major features of the carrying capacity function.

The Model The simplest model that uses the competition function, $\alpha(d)$, and the carrying capacity function, $K(x)$, to predict the population's total niche width and its breakdown into the components, is a clone selection model for a quantitative character. Let $n(x)\,dx$ denote the number of organisms whose phenotype is between x and $x + dx$. Then, if organisms of every phenotype reproduce asexually, so that each phenotype is represented by a clone, we have

$$\frac{\partial n(x, t)}{\partial t} = r(x)\left[K(x) - \int \alpha(x - x')\,n(x', t)\,dx'\right]\frac{n(x, t)}{K(x)} \tag{24.59}$$

If this model has a positive and stable equilibrium distribution, $\hat{n}(x)$, it satisfies

$$\int \alpha(x - x')\hat{n}(x')\,dx' = K(x) \tag{24.60}$$

(In this analysis we assume all functions, including the initial condition, are continuous functions of x.) We define the total population size, N, and the phenotype distribution $p(x, t)$ as

$$N(t) = \int n(x, t)\,dx$$
$$p(x, t) = \frac{n(x, t)}{N(t)} \tag{24.61}$$

If this model has a positive stable equilibrium distribution, $\hat{n}(x)$, then the niche width and its components can be found in terms of the corresponding equilibrium phenotype distribution, $\hat{p}(x)$. The way to do this follows equations (24.32) to (24.37), as discussed earlier in connection with the measurement of niche width. Specifically,

$$\tilde{p}(x) = \frac{A(x)\hat{p}(x)}{\int A(x)\hat{p}(x)\,dx} \tag{24.62}$$

Then the population's utilization curve is

$$u(x) = \int \tilde{p}(x') v(x|x') \, dx'$$

The population's niche position is the mean of $u(x)$, the total niche width, w^2, is the variance of $u(x)$, the BPC is the variance of $\tilde{p}(x)$, and the WPC is σ_v^2. [The WPC is the average variance of the $v(x)$ curves.]

Thus the equilibrium solution to (24.59) will enable us to predict all the features of the population's niche width.

Approach to   The first point to note about (24.59) is that a positive solution is stable in *Equilibrium*   analogy to Result 1 from the section on the Lotka–Volterra competition equations. To show this, we define the functional

$$\Psi(y) = \int_{x_1} \int_{x_2} y(x_1) A(x_1) \phi(x_1, x_2) A(x_2) y(x_2) \, dx_1 \, dx_2 \qquad (24.63)$$

where

$$\phi(x_1, x_2) = \int v(x|x_1) v(x|x_2) \, dx \qquad (24.64)$$

Upon substituting (24.64) into (24.63), we establish that

$$\Psi(y) \geqslant 0 \qquad (24.65)$$

with equality holding if and only if $y(x)$ is identically zero. Next, we define the continuous analogue to the Q function as

$$Q(t) = \int_{x_1} \int_{x_2} [n(x_1, t) - \hat{n}(x_1)] A(x_1) \phi(x_1, x_2) A(x_2) [n(x_2, t) - \hat{n}(x_2)] \, dx_1 \, dx_2$$
$$(24.66)$$

Then differentiating $Q(t)$ with respect to t, and paralleling the argument in Equations (24.13) to (24.16), we obtain

$$\frac{dQ(t)}{dt} = -2 \int_x \frac{r(x) \, C(x)}{K(x)} n(x, t) J^2(x, t) \, dx \leqslant 0 \qquad (24.67a)$$

where

$$J(x, t) = \int_{x_1} A(x_1) \phi(x, x_1) [n(x_1, t) - \hat{n}(x_1)] \, dx_1 \qquad (24.67b)$$

$$C(x) = \frac{1}{\phi(x, x)}$$

The equality in (24.67a) holds only when equilibrium is reached. Thus the system tends to the equilibrium distribution given by (24.60).

The Equilibrium   The next point is to examine the solution to (24.60). The integral in (24.60) is called a convolution integral and it is known that the solution to this equation can be written in terms of Fourier transforms. Let $A(\omega)$, $\hat{N}(\omega)$, and $K(\omega)$ denote the Fourier transforms of the functions in (24.60). Then, in terms of Fourier transforms, (24.60) becomes

$$A(\omega) \hat{N}(\omega) = K(\omega) \qquad (24.68)$$

Hence the transform of the equilibrium distribution is given by

$$\hat{N}(\omega) = \frac{K(\omega)}{A(\omega)} \qquad (24.69)$$

The full equilibrium distribution in terms of the original coordinate, x, is then obtained by finding the Fourier inverse of (24.69). However, even without inverting (24.69), we can obtain expressions for the total population size, $\hat{N}$, and the mean $\bar{x}_n$ and variance σ_n^2 of the phenotype distribution,

$$\hat{N} = \frac{K}{A} \qquad (24.70a)$$

$$\bar{x}_n = \bar{x}_k - \bar{x}_\alpha \qquad (24.70b)$$

$$\sigma_n^2 = \sigma_k^2 - \sigma_\alpha^2 \qquad (24.70c)$$

The first formula asserts that the total population size at equilibrium equals the total area under the carrying capacity function divided by the total area under the competition function. The second formula asserts that the mean of the phenotype distribution is displaced away from the mean of the carrying capacity function by any asymmetry in the competition function. Specifically, if $\bar{x}_\alpha$ itself is negative, thereby indicating the competitive advantage of large over small individuals, the $\bar{x}_n$ is shifted above $\bar{x}_k$, as we would expect. The third formula asserts that the variance of individuals in the population is decreased as the range of the competitive interaction, σ_α^2, increases. Formulas (24.68) to (24.70) are derived in detail in Roughgarden (1972).

Results Let us return to the specific example involving the Gaussian competition function (24.55) and carrying capacity function (24.58). In this case the solution, $\hat{n}(x)$, is itself Gaussian in shape with mean, variance, and total area given by (24.70). Furthermore, A, $\bar{x}_\alpha$, and σ_α^2 are related to σ_v^2 and κ by (24.54) and (24.56). Putting all this together, we obtain

$$\hat{N} = \frac{K}{(2\sqrt{\pi})\sigma_v e^{\sigma_v^2 \kappa^2}} \qquad (24.71a)$$

$$\bar{x}_n = \bar{x}_k + 2\sigma_v^2 \kappa \qquad (24.71b)$$

$$\sigma_n^2 = \sigma_k^2 - 2\sigma_v^2 \qquad (24.71c)$$

Thus the equilibrium phenotype distribution $\hat{p}(x)$ is Gaussian with mean and variance given by (24.71b) and (24.71c), and the total population size is given by (24.71a). To calculate the niche width and its components, we convert $\hat{p}(x)$ into the weighted phenotype distribution, $\tilde{p}(x)$, according to (24.62). The result is that $\tilde{p}(x)$ is again Gaussian. Its mean and variance are the niche position and BPC, respectively,

$$\text{niche position} = \bar{x}_n + \sigma_n^2 \kappa$$

$$= (\bar{x}_k + 2\sigma_v^2 \kappa) + (\sigma_k^2 - 2\sigma_v^2)\kappa$$

$$= \bar{x}_k + \sigma_k^2 \kappa \qquad (24.72)$$

$$\text{BPC} = \sigma_n^2$$

$$= \sigma_k^2 - 2\sigma_v^2 \qquad (24.73)$$

Recall that, in this case, the

$$\text{WPC} = \sigma_v^2 \qquad (24.74)$$

The sum of the BPC and WPC, then, equals the total niche width as

$$w^2 = (\sigma_k^2 - 2\sigma_v^2) + \sigma_v^2$$
$$= \sigma_k^2 - \sigma_v^2 \qquad (24.75)$$

The total population's utilization curve, $u(x)$, is Gaussian with the mean given by (24.72) and variance given by (24.75). Thus, in this case, we obtain a complete explicit solution that predicts the niche position, the total niche width, and the breakdown into components. These predictions are based on the mean and variance of the available resources, $\bar{x}_k$ and σ_k^2; on the variance in the individual utilization curves σ_v^2; and on the coefficient, κ, which describes how an individual's total resource requirement increases with size. Note that the effect of κ is solely on the niche position. The niche position shifts up as the carrying capacity variance increases.

It is interesting to dwell on the biological interpretation underlying the formula for the BPC. Equation (24.73) states, in words, that

$$\begin{bmatrix} \text{variety of individuals} \\ \text{using different} \\ \text{resources} \end{bmatrix} = \begin{bmatrix} \text{variety of} \\ \text{available} \\ \text{resources} \end{bmatrix} - 2 \begin{bmatrix} \text{average variety of} \\ \text{resources used by} \\ \text{an individual} \end{bmatrix}$$

The first term on the right asserts that the variety of phenotypes will increase if the variety of available resources is increased. The second term asserts that the variety of phenotypes will decrease if the average variety of resources used by an individual increases. The first term seems obvious. The second term is not as obvious but makes good sense nonetheless. The reason why a population cannot support a wide variety of phenotypes when each of those phenotypes is using a wide variety of resources (although differing in their means) is that the individuals based at marginal food categories are excluded by exploitative competition from the more numerous individuals who are based at the common food categories. The key point is that the presence of the comparatively few individuals based at marginal food categories is very sensitive to any increases in the range of resource use by the comparatively large number of individuals who are based at the common food categories. An increase in the range of resource use by every individual will cause a loss of the individuals based at the marginal food categories and thereby leave a population with a lower degree of intraspecific resource partitioning.

Possible Relevance to Niche Width in Anolis *Populations*   Earlier, we presented data on the niche width and its components for some *Anolis* lizard populations (Table 24.3). The data suggest (1) that both the total niche width and the percentage BPC increase as the number of competing species decreases, provided the comparison is between populations from environments of the same productivity, and (2) that the percentage BPC drops substantially, although the total niche width is relatively unaffected, as *both* the number of competing species decreases *and* the environmental productivity decreases.

Roughgarden (1974a) interpreted these observations in terms of the preceding model, based on two additional conjectures. First, it is conjectured that σ_k^2 increases as the number of competitors decreases. One imagines that the removal of competing species opens up more resource types to the remaining species, so σ_k^2 increases for the remaining species as competitors are removed. Second, it is conjectured that σ_v^2 increases as the productivity decreases. One imagines that each individual is forced to use a wider range of prey sizes, as prey become increasingly rare. (Notice from Table 24.3 that the absolute level of WPC *is* higher in *A. shrevei* than in *A. cybotes*, so that the conjecture is, in fact, true in this empirical case.) These two conjectures, together with the formulas for the niche width and its components, do entail the observations cited here. Specifically, the total niche width and its breakdown into the components is

$$w^2 = \sigma_k^2 - \sigma_v^2 \qquad (24.76a)$$

$$\text{percent BPC} = \left[\frac{\sigma_k^2 - 2\sigma_v^2}{\sigma_k^2 - \sigma_v^2}\right](100)$$

$$= \left[1 - \frac{\sigma_v^2}{\sigma_k^2 - \sigma_v^2}\right](100) \qquad (24.76b)$$

$$\text{percent WPC} = \left[\frac{\sigma_v^2}{\sigma_k^2 - \sigma_v^2}\right](100) \qquad (24.76c)$$

First, notice that as the number of competitors decreases and provided the productivity remains constant (σ_k^2 increasing with σ_v^2 fixed), w^2 increases, the percentage BPC increases, and the percentage WPC decreases. This result is the first of the two points suggested by the data. Second, notice that as the number of competitors decreases and the productivity also decreases (σ_k^2 increasing and σ_v^2 increasing), the changes in σ_k^2 and σ_v^2 counteract one another in the formula for w^2, whereas the percent BPC decidedly decreases and the percent WPC increases. This result is the second of the two points suggested by the data. This agreement of theory with data does not *prove* that the model above is on the right track, with respect to explaining biogeographic patterns in niche width. Obviously, the clonal selection model ignores many aspects of the biology of the organisms to which the data refer. Nonetheless, this preliminary agreement is encouraging.

Extensions There are several extensions to the model for the evolution of niche width presented above, and there is only space to mention these briefly here. The extensions deal with more complex genetic systems than that of clonal selection.

One approach found in Roughgarden (1972) has been to view the evolution of niche width in the context of selection on a quantitative character. Specifically, the selection pressure on an individual with character value x is assumed to be

$$W(x) = 1 + r - \frac{r}{K(x)} \int \alpha(x - x') n(x') \, dx' \qquad (24.77)$$

Such a selection pressure is the same as that used in the clone selection model above. The new feature is that this selection pressure operates in the model for a quantitative character, discussed in Chapter 9,

$$p_{t+1}(x) = \frac{\iint L[x - h^2(y+z)/2 - (1-h^2)\bar{x}_t] \, W(y)W(z)p_t(y)p_t(z) \, dy \, dz}{\bar{W}^2}$$

$$(24.78a)$$

$$N_{t+1} = \bar{W}N_t \tag{24.78b}$$

where

$$\bar{W} = \int p_t(x)W(x)\,dx. \tag{24.78c}$$

$L(x)$ is the segregation kernel with variance, σ_L^2, and h^2 is the heritability. As we saw in Chapter 9, the variance of the equilibrium phenotype distribution in the quantitative inheritance model is strongly influenced by σ_L^2. This result is true here also. The variance of the equilibrium phenotype distribution in the model generally depends both on σ_α^2, σ_k^2, *and* on σ_L^2. Hence the population's niche width and its breakdown into components generally depend on the variance of the segregation kernel as well as on the parameters that describe the resource use per individual and the resource availability. However, it can also be shown that if the segregation kernel has a *special* shape, then the equilibrium phenotype distribution in this model will coincide with the equilibrium distribution of the clone selection model. Roughgarden (1972) conjectured that natural selection would mold the segregation kernel, $L(x)$, to that special shape, that is, will cause the genetic system itself to evolve, so that the equilibrium phenotype distribution of the quantitative character does coincide with that from the clone selection model.

The other approach has been to view the evolution of niche width in the context of a one-locus-two-allele genetic system. Matessi and Jayakar (1976a) have studied the polymorphism that can result where the three genotypes show resource partitioning. They show that, with a symmetric intraspecific competition matrix, the equilibrium gene frequency minimizes the average degree of competition between the genotypes. This paper is unusual in that it presents a thorough analysis of density-dependent selection mixed with frequency-dependent selection. Their analysis includes polymorphic equilibria and boundary equilibria. Another important recent paper using a one-locus system in niche theory is by Fenchel and Christiansen (1977). They studied the conditions for the increase of a rare allele which may (a) widen the niche of a population or which may (b) cause the niche position to shift. One of their results is that the strength of the selection pressure that drives niche shift is stronger, by an order of magnitude, than that which drives niche widening. The qualitative implication of this result is that the evolution of niche position is faster than the evolution of niche width. Roughgarden (1972) also asserted this point but on the basis of an entirely different argument. Roughgarden's argument was that, with selection on a quantitative character in a random mating population, natural selection must also mold the segregation kernel in order to mold the niche width to any significant degree, whereas it is not necessary to mold the segregation kernel to shift niche position. Furthermore, Roughgarden (1974a) found some evidence that the niche width is a more conservative character than the niche position in *Anolis* lizards and interpreted the data as support for his argument. However, the model of Fenchel and Christiansen leads to the same conclusion, but with fewer assumptions. The model contains an argument based directly on the magnitudes of the selection pressures involved and does not invoke assumptions about constraints imposed by the genetic system. Hence it presently appears that the Fenchel–Christiansen model offers simpler and better grounds for the prediction that the evolution of niche shift should be faster than the evolution of niche width.

One final note is that a model for the evolution of niche width based on growth is sorely needed. In fish and reptiles, the size of an animal may be largely a reflection of its age, especially in temperate environments and in environments with high density-independent mortality. It is easy to formulate a model for the niche width of a population that combines the transport model of population growth with size structure introduced in Chapter 19 with the niche theory assumptions introduced in this chapter. Unfortunately, this problem has defied analysis so far.

Resource Partitioning Between Species

Now we move on to the theory of resource partitioning between species. There are two issues to be addressed by this theory. First, we want to know the *limit* to the similarity between competing species consistent with their being able to coexist. Second, we want to know the actual degree of similarity between competitors that results from their coevolution with respect to one another. The degree of niche separation that results from coevolution must be larger than the minimum niche separation that permits coexistence, but how much larger? This section is organized into three subsections. The first two subsections present two different approaches to the determination of the limiting similarity. The third presents the co-evolution of the niche separation.

The setup for between-species niche theory models is essentially the same as that from the last section on within-species resource partitioning, but the questions asked are different. We assume that each species has a utilization curve, $u(x|x_i)$, which is Gaussian. The common variance of $u(x|x_i)$ is the common population niche width, w^2, and the mean, x_i, is the niche position of species-i. Again, we can assume that the total amount of resource consumed by a species may vary with the niche position as

$$A(x_i) = A_0 \, e^{\kappa x_i} \tag{24.79}$$

When these assumptions are substituted into the overlap formula (24.15b), we obtain the familiar Gaussian-shaped competition function

$$\alpha(d) = e^{w^2 \kappa^2} \exp\left\{ \frac{-.5[d + 2w^2\kappa]^2}{(2w^2)} \right\} \tag{24.80}$$

This competition function applies between species. The function, $\alpha(d)$, is the interspecific competition coefficient for the effect of a species at $x_i + d$ against a species at x_i. In some models, there is also a carrying capacity function, $K(x_i)$, which denotes the carrying capacity of a species whose niche position is x_i; $K(x_i)$ will also be taken as Gaussian.

Limiting Similarity by the Positivity Criterion

The most fundamental criterion for coexistence of species in niche theory models is the requirement that the equilibrium solution be positive. As we found in the section on mathematical preliminaries, a positive equilibrium solution is necessarily stable. Furthermore, the condition for a positive solution is equivalent to the condition that each species be able to increase when rare. The criterion of positivity provides the most basic approach to determining a limiting similarity. In this approach we determine how close the species can be to one another consistent with each being able to increase, when rare. Let us now consider several examples that illustrate the variety of models and results that are possible in this framework.

1. TWO SPECIES. *In this model the α's vary with position and the K's are fixed.* By convention, we label as species-1, that species with the lower niche position. Let the carrying capacities be K_1 and K_2, and we assume the carrying capacities are not functions of the niche positions. The condition for species-1 and species-2 to increase when rare are, respectively, that

$$K_1 - \alpha(-d)K_2 > 0 \qquad (24.81a)$$

$$K_2 - \alpha(d)K_1 > 0 \qquad (24.81b)$$

where $d = x_2 - x_1$. Both species coexist at a stable equilibrium point if and only if both these conditions are satisfied. Let d_1 denote the closest distance to which species-1 can approach species-2. The variable, d_1, is found as the root of

$$K_1 - \alpha(-d_1)K_2 = 0 \qquad (24.82a)$$

Similarly, let d_2 denote the closest distance to which species-2 can approach species-1. The variable, d_2, is the root of

$$K_2 - \alpha(d_2)K_1 = 0 \qquad (24.82b)$$

(If either d_1 or d_2 is found to be negative or complex, then, by convention, it is set equal to zero.) Using the competition function given by (24.80), we find that the appropriate root to (24.82a) is

$$\frac{d_1}{w} = 2\left[w\kappa + \sqrt{w^2\kappa^2 + \ln\left(\frac{K_2}{K_1}\right)} \right] \qquad (24.83a)$$

The closest distance that species-1 can approach species-2 is expressed relative to the niche width, w. Similarly, the appropriate root to (24.82b) is

$$\frac{d_2}{w} = 2\left[-w\kappa + \sqrt{w^2\kappa^2 - \ln\left(\frac{K_2}{K_1}\right)} \right] \qquad (24.83b)$$

The *limiting similarity*, $\hat{d}$, is then the larger of d_1 and d_2. If the niche separation, $x_2 - x_1$, were smaller than the larger of d_1 and d_2, then one of the species would have a negative population size. Thus the limiting similarity in this example is found as

$$\hat{d} = \max(d_1, d_2) \qquad (24.84)$$

Let us examine this result in some special cases. First, suppose $K_1 = K_2$. Then the limiting similarity is

$$\frac{\hat{d}}{w} = 4w\kappa \qquad (24.85)$$

With equal K's, species-1 cannot approach species-2 any closer than this value because it is disadvantaged in competition. However, notice that if $\kappa = 0$ so that reciprocal competition coefficients are equal, then the limiting similarity is zero. In the very special case of complete symmetry, the positivity criterion is always satisfied and there is no limit to the similarity. As we shall see shortly, under the same assumption of complete symmetry, the second criterion for limiting similarity *does* yield a nonzero limit, thereby leading to different predictions.

Second, suppose $\kappa = 0$ and $K_1 \neq K_2$. By convention, we can let $K_1 < K_2$. Then, the limiting similarity is determined only by the ratio of the K's,

$$\frac{\hat{d}}{w} = 2\sqrt{\ln\left(\frac{K_2}{K_1}\right)} \qquad (24.86)$$

If species-1 is closer than this value, it incurs a negative population size. Third, suppose $\kappa > 0$ and $K_1 < K_2$, then the limiting similarity is given by (24.83a). In this case, species-1 is disadvantaged both in competition and carrying capacity and these two factors reinforce each other so that the largest niche separation is required. Fourth, suppose that $\kappa > 0$ and $K_1 > K_2$, so that species-1 is disadvantaged in competition but has a higher carrying capacity. In this case there is a curious discontinuity in the system, as sketched in Figure 24.19. As $\ln(K_2/K_1)$ is varied from zero down to $-w^2\kappa^2$, the limiting similarity continuously decreases. This is intuitive; increasing K_1 relative to K_2 allows species-1 to move closer to species-2 in spite of its competitive disadvantage. But as K_1 is increased slightly further, so that $\ln K_2/K_1$ moves through $-w^2\kappa^2$, then the limiting similarity jumps from $2w\kappa$ down to $2(-1+\sqrt{2})w\kappa$, that is, down to $.83\, w\kappa$. As K_1 is increased still more relative to K_2, then the limiting similarity continuously rises. In this region, it is species-2 that must move away from species-1 in order to coexist.

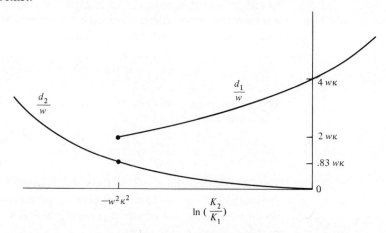

FIGURE 24.19. *Limiting similarity, d/w, between two species as a function of the log of their carrying capacity ratio, $\ln(K_2/K_1)$. The limiting similarity is the higher of the two curves. Note that as $\ln(K_2/K_1)$ is varied below $-w^2\kappa^2$, the limiting similarity drops discontinuously from $2w\kappa$ to $.83w\kappa$.*

2. THREE EVENLY SPACED SPECIES. *Here the α's are symmetric and vary with position and the K's are fixed.* The setup for this model is sketched in Figure 24.20. By convention, the position of the middle species is labeled as 0 and the outside species are located at $-d$ and d. Also, by convention, we label as species-1, the outside species with the lower carrying capacity,

$$K_1 \leqslant K_3 \qquad (24.87)$$

Futhermore, we denote the average of the two outside carrying capacities as K,

$$K = \tfrac{1}{2}(K_1 + K_3) \qquad (24.88)$$

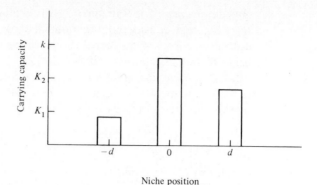

FIGURE 24.20. *Sketch of the setup of carrying capacities used in the problem of limiting similarity with three species on one axis.*

and the carrying capacity of the middle species as k. In this model, the competition function is assumed to be symmetric. If the utilization curves are Gaussian, then the competition function is given by (24.80), with $\kappa = 0$. The necessary and sufficient condition for these three species to coexist in stable equilibrium is that each must be able to increase when rare. The conditions are

$$\text{Species-1:} \quad K_1 - \alpha(d)\tilde{N}_2^{(1)} - \alpha(2d)\tilde{N}_3^{(1)} > 0 \tag{24.89a}$$

$$\text{Species-2:} \quad k - \alpha(d)\tilde{N}_1^{(2)} - \alpha(d)\tilde{N}_3^{(2)} > 0 \tag{24.89b}$$

$$\text{Species-3:} \quad K_3 - \alpha(2d)\tilde{N}_1^{(3)} - \alpha(d)\tilde{N}_2^{(3)} > 0 \tag{24.89c}$$

where $\tilde{N}_i^{(j)}$ denotes the equilibrium population size of species-i in the absence of species-j. The $\tilde{N}_i^{(j)}$ are given by

$$\tilde{N}_2^{(1)} = \frac{k - \alpha(d)K_3}{1 - \alpha^2(d)}, \qquad \tilde{N}_3^{(1)} = \frac{K_3 - \alpha(d)k}{1 - \alpha^2(d)}$$

$$\tilde{N}_1^{(2)} = \frac{K_1 - \alpha(2d)K_3}{1 - \alpha^2(2d)}, \qquad \tilde{N}_3^{(2)} = \frac{K_3 - \alpha(2d)K_1}{1 - \alpha^2(2d)} \tag{24.90}$$

$$\tilde{N}_1^{(3)} = \frac{K_1 - \alpha(d)k}{1 - \alpha^2(d)}, \qquad \tilde{N}_2^{(3)} = \frac{k - \alpha(d)K_1}{1 - \alpha^2(d)}$$

Upon substituting these expressions into (24.89) and rearranging, we obtain .

$$\text{Species-1:} \quad \frac{-(K_3 - K_1)}{1 - \alpha(2d)} + \tilde{N}_3^{(1)} > 0 \tag{24.91a}$$

$$\text{Species-2:} \quad \alpha(2d) - \frac{2}{(k/K)}\alpha(d) + 1 > 0 \tag{24.91b}$$

$$\text{Species-3:} \quad \frac{(K_3 - K_1)}{1 - \alpha(2d)} + \tilde{N}_1^{(3)} > 0 \tag{24.91c}$$

These conditions are necessary and sufficient for the three species to coexist. This model has proved to be surprisingly rich. It was introduced by MacArthur and Levins (1967), who studied condition (24.91b) for the middle species to increase on the assumption that $K_1 = K_3$. Roughgarden (1974b) extended the analysis to cover the conditions pertaining to the outside species (24.91a) and (24.91c), allowed for the possibility that $K_1 \leqslant K_3$, and considered the influence of the assumption of Gaussian utilization curves on the conclusions. These papers report five conclusions from the model, summarized as follows.

(a) *Condition for invasion by the middle species with equal carrying capacities.* In the very special case where $K_1 = k = K_3$, then it can be verified that the limiting similarity is the root, $\hat{d}$, of

$$\alpha(2d) - 2\alpha(d) + 1 = 0 \qquad (24.92)$$

If $d > \hat{d}$, then all species can increase when rare. With equal K's the conditions (24.91a) and (24.91c) are automatically satisfied for any $d > 0$, and the limiting similarity in the system is determined by (24.91b). For the case of Gaussian utilization curves

$$\frac{\hat{d}}{w} = 1.6 \qquad (24.93)$$

This root to (24.92) is found with a root-finder program on a computer and rounded to one decimal place. This case contrasts sharply with the corresponding two-species case. As we saw earlier, with symmetric $\alpha(d)$ and with equal K's, two species can be arbitrarily close to one another and still coexist. But with three evenly spaced species, the niche separation must exceed 1.6 times the niche width, w, in order to coexist.

(b) *Condition for invasion by the middle species as a function of k/K and the "similarity barrier."* Whether the middle species can invade depends on the relation of d to the root(s), $\hat{d}$, of

$$\alpha(2d) - \frac{2}{(k/K)}\alpha(d) + 1 = 0 \qquad (24.94)$$

If $d > \hat{d}$, the middle species can enter the community, but it may exclude one or more of the outside species as a result. First, let us simply consider whether the middle species can invade the community and then turn to the fate of the outside species. The roots to (24.94) as a function of k/K are graphed in Figure 24.21. If $k/K < 1$, that is, the invader has a carrying capacity lower than the average of the outside carrying capacities, then the limiting similarity exceeds 1.6. If k/K is in a small interval between 1 and a number slightly larger than 1, then (24.94) has two roots, $\hat{d}_{lower}$ and $\hat{d}_{upper}$. If the outside species are far enough apart so that $d > \hat{d}_{upper}$, then invasion can occur; *and* if the outside species are sufficiently close, so that $d < \hat{d}_{lower}$, then again the invasion can occur. But for intermediate spacing, $\hat{d}_{lower} < d < \hat{d}_{upper}$, invasion cannot occur. The zone of similarities between $\hat{d}_{lower}$ and $\hat{d}_{upper}$, that is, the interval $[\hat{d}_{lower}, \hat{d}_{upper}]$ is called the "similarity barrier." The middle species can invade above or below this barrier.

When the middle species does invade below the similarity barrier, it is almost certain that one or both of the outside species will be excluded as a result. Whether only one or both of the outside species is excluded depends on the difference between K_1 and K_3. Figure 24.22 illustrates the effect of invasion by the middle species upon the outside species. Note that whenever there is invasion below the similarity barrier, and provided at least one of the outside species can remain in the community, then the result is a community with close niche separation distances, a community with close "packing." In particular, when replacement occurs, the invader excludes species-1 and the

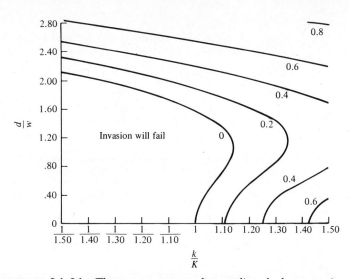

FIGURE 24.21. *These curves are used to predict whether a species can successfully invade between two resident species. To use the figure first, one must establish that the resource utilization curves of the species are approximately normal and with the same variance. The common variance is the niche width, w. Next, determine the ratio of the invader's carrying capacity to the average of the residents, k/K. Third, determine d as one-half of the distance separating the niches of the two residents. Then, locate the point corresponding d/w and k/K. If the point is in the region at the lower left, the invasion will fail. If the point is in the successful region, then the rate of increase per individual in the propagule is proportional to the value indicated on the nearest contour line.* [From J. Roughgarden (1974), Species packing and the competition function with illustrations from coral reef fish. *Theor. Pop. Biol.* **5**: 163–186.]

system, again, consists of two species, but now separated by half the separation distance that was originally present.

(c) *Dependency of conclusions on the shape of the utilization curves.* It is clear that the numerical values of the limiting similarity for any value of (k/K) depend on the functional form of the utilization curves. For example, suppose the utilization curves are back to back exponentials, as given by

$$u(x) = \frac{\sqrt{2}}{2w} \exp(-\sqrt{2}|x|/w) \tag{24.95}$$

Then the competition function is found to be

$$\alpha(d) = \exp(-\sqrt{2}|d|/w)\left(1 + \sqrt{2}\frac{|d|}{w}\right) \tag{24.96}$$

With this competition function, the analogue to Figures 24.21 to 24.22 appears in Figure 24.23. Note that the figures are qualitatively the same in the sense that there still is a limiting similarity and similarity barrier. However, with these back to back exponential $u(x)$ curves, the limiting similarity is generally lower than that required by Gaussian $u(x)$ curves. For example, with all equal K's the limiting similarity is

$$\frac{\hat{d}}{w} = .9 \tag{24.97}$$

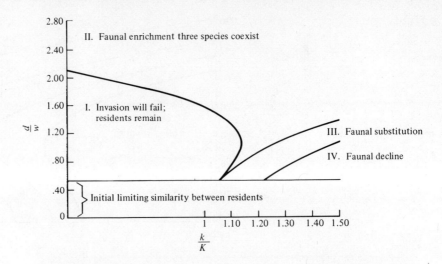

FIGURE 24.22. *This graph is used to predict the fate of the resident species, subsequent to invasion by a middle species. By convention, it is assumed that K_3 is larger than K_1. In this example, $K_1 = (.75) K_3$. If the point $(k/K, d/w)$ plots in Region* I, *then the invader cannot enter the community as discussed before in Figure 24.21. In Region* II *the successful invader coexists with the two residents and a stable 3-species community results. In Region* III *the invader eliminates species-1 resulting in faunal substitution. The community remains at the two-species level, but the niches have moved closer. That is, 2 and 3 are closer to one another than 1 and 3 were. In Region* IV *the invader eliminates both residents resulting in a single-species community. Because $K_1 \neq K_3$ the residents cannot approach one another too closely to begin with. The horizontal line near the bottom of the graph marks the initial limiting similarity between the residents.* [Adapted from J. Roughgarden (1974), Species packing and the competition function with illustrations from coral reef fish. *Theor. Pop. Biol.* **5**: 163–186.]

Thus with equal K's, the species can be almost twice as close to one another with this $u(x)$ curve, as compared with Gaussian $u(x)$ curves.

(d) *Forbidden paths of faunal buildup.* This point and the next arise when species-2 can exclude species-1 in the absence of species-3 but not in the presence of species-3. This situation arises when all of the conditions in (24.91) are satisfied and, moreover, the following condition is also met

$$\frac{-(K_3 - K_1)}{1 - \alpha(2d)} < \tilde{N}_1^{(3)} < 0 \qquad (24.98)$$

Thus $\tilde{N}_1^{(3)}$ is assumed negative but not enough so as to violate (24.90c). In this situation species-2 excludes species-1 unless species-3 is present. When this situation obtains, then certain paths of faunal buildup are forbidden. All the possible sequences by which the three species could enter the community are

$$1, 2, 3$$

$$2, 1, 3$$

$$1, 3, 2$$

Evolutionary Ecology of Interacting Populations **546**

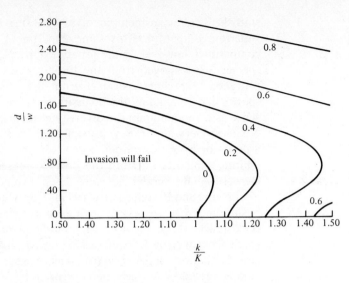

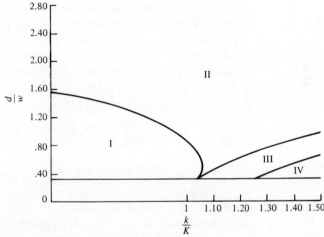

FIGURE 24.23. *Graphs analogous to Figures* 24.21 *and* 24.22 *but based on different utilization curves. The utilization curves are assumed to be back-to-back exponential functions.*

$$2, 3, 1$$
$$3, 1, 2$$
$$3, 2, 1$$

However, the first two of these sequences involve the simultaneous presence of species-1 and species-2 in the absence of species-3. These sequences are forbidden. The other paths of faunal buildup are allowed because, in them, species-1 never faces species-2 alone; species-3 is there also.

(e) *A keystone competitor.* Since the experiments of Paine (1966), it has been customary to view predation pressure as increasing species diversity. Paine removed a predatory starfish from the intertidal area and observed the species diversity of the prey drop. Mussels, the preferred prey of the starfish, caused competitive exclusion of other prey species in the absence of the starfish. This experiment involves choosing a predator that prefers the superior competitor. This same experiment could be carried out with competitors, however. In the situation above, species-3 is the key species. If species-3 is removed from the three-species community, then species-2 proceeds to exclude species-1 from the remaining community. Thus a species can "maintain" the diversity of its competitors in exactly the same sense as it can maintain the diversity of its prey.

This model with three evenly spaced species has been extended to include a predator by Roughgarden and Feldman (1975). It was shown that the limiting similarity between the prey decreases as the predation pressure increases. This model has also been extended to two dimensions by Yoshiyama and Roughgarden (1977). They studied the geometry of species positioning with respect to two resource axes that are consistent with coexistence. Also, the role of predator-mediated coexistence has been studied recently by Caswell (1978) based on a nonequilibrium condition. Another important paper on the effect of predation on resource partitioning among prey has recently been written by Vance (1978). The effect of interference competition on niche theory has been studied by Case and Gilpin (1974). For the details, the original papers should be consulted.

3. TWO SPECIES. *Here both α and K vary with niche position.* Another formulation of limiting similarity models is to assume that the carrying capacity also varies with the niche position. The setup for the simplest illustration of this type of model is sketched in Figure 24.24. Suppose that (1) the carrying capacity function is Gaussian in shape,

$$K(x) = \frac{K_0 \exp\left[-\frac{1}{2}(x^2/\sigma_k^2)\right]}{\sqrt{2\pi\sigma_k^2}} \qquad (24.99)$$

and (2) that the competition function is Gaussian and given by (24.80) with $\kappa = 0$. Suppose also that the niche position of species-1 is directly under the peak of the $K(x)$ function, that is, at $x = 0$. Then, the question is what are the niche positions at which invasion by species-2 can occur? Clearly, the farther the invader is from species-1, the less competition is incurred, but also, the farther it is from the peak of the carrying capacity function. The answer is, again, found by seeking the condition for species-2 to increase when rare

$$K(x_2) - \alpha(x_2)K(0) > 0 \qquad (24.100)$$

It is readily verified that (24.100) is satisfied for any value of x_2 if

$$\sigma_k^2 > 2w^2 \qquad (24.101)$$

Thus invasion can occur from any position if the variance of the carrying capacity function exceeds twice the niche width, whereas the community is

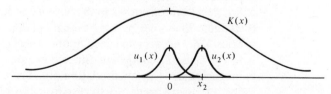

FIGURE 24.24. *Setup for a limiting similarity model in which both α and K vary with position.*

closed to invasion by a second species if $\sigma_k^2 < 2w^2$. This type of model can also be investigated on the assumption that $K(x)$ and $\alpha(d)$ have different functional forms and/or are not symmetric, and so forth. In these cases, there are intervals on the axis where the community is susceptible to invasion, and other intervals where it is not. Still another interesting extension is to consider two species, displaced by the same amount but in opposite directions from the peak of the $K(x)$ curve. Again, one determines the intervals on the axis from which invasion by a third species is possible.

Limiting Similarity by Criteria of Persistence in a Stochastic Environment   May and MacArthur (1972) introduced another criterion for determining the limiting similarity between competing species, a criterion that is more stringent than the positivity criterion. The intuitive idea is that species that are far enough apart to coexist in a constant environment must be still further apart to coexist in a randomly fluctuating environment. The case that best illustrates this issue is that of two species with equal K's and with a symmetric $\alpha(d)$. In this case, according to the positivity criterion, the limiting similarity is zero, that is, the species can be arbitrarily close to one another and still coexist in a constant environment. Yet, it seems intuitive that their coexistence is precarious if the two species are very close to one another.

To develop this idea in more detail, we require a model of population dynamics in a stochastic environment. The problem will be to determine the limit to the similarity of two species that is consistent with their continued coexistence in a *stochastic* environment. May and MacArthur focused upon the logistic equation for a single species, in which both r and K fluctuate stochastically according to

$$r_t = r_0 + \sigma z_t$$

$$K_t = K_0 + \sigma\left(\frac{K_0}{r_0}\right) z_t \tag{24.102}$$

where z_t is a white noise random variable and σ is the standard deviation of the variation. With this scheme of variation in r and K, the ratio of r_t/K_t is constant and equals r_0/K_0. Furthermore, May and MacArthur consider the Ito calculus as appropriate for the analysis of this stochastic population model. We learned in Chapter 20 that under these assumptions, the population will persist in a stochastic environment if

$$\frac{\sigma^2}{2} < r_0 \tag{24.103}$$

The researchers used this result as the basis to propose a criterion for the coexistence of two or more species in a stochastic environment.

To undertand the proposal, we first should note that r_0 is related to the eigenvalue for the logistic equation, which describes how rapidly the system returns to equilibrium after a perturbation. Let us denote the eigenvalue as λ. By definition λ is

$$\lambda = \frac{\partial(dN/dt)}{\partial N}\bigg|_{N=K} \tag{24.104}$$

As defined above, λ is the eigenvalue that controls the rate of return to equilibrium following a perturbation. With the logistic equation, the variable, λ evaluates as

$$\lambda = -r_0 \qquad (24.105)$$

With this terminology, we can rewrite the condition (24.102) for persistence of a solitary species in a stochastic environment as

$$\frac{\sigma^2}{2} < |\lambda| \qquad (24.106)$$

The proposal of May and MacArthur is to extend this condition to cover interacting species.

Consider a two-species competition model where the r and K of both species vary stochastically as

$$r_{i,t} = r_{i,0} + \sigma z_{i,t} \qquad (i = 1, 2)$$
$$K_{i,t} = K_{i,0} + \sigma(K_{i,0}/r_{i,0})z_{i,t} \qquad (24.107)$$

Again, the ratio $r_{i,t}/K_{i,t}$ is constant and equals $r_{i,0}/K_{i,0}$; $z_{i,t}$, is the white noise fluctuation in the parameters for species-i. To find a criterion for persistence of the two species, we form the matrix

$$\begin{vmatrix} \dfrac{\partial(dN_1/dt)}{\partial N_1} & \dfrac{\partial(dN_1/dt)}{\partial N_2} \\[2mm] \dfrac{\partial(dN_2/dt)}{\partial N_1} & \dfrac{\partial(dN_2/dt)}{\partial N_2} \end{vmatrix}_{N_i = \hat{N}_i} \qquad (24.108)$$

This matrix is evaluated at the deterministic equilibrium point and it is presupposed that there is a positive equilibrium point from which the species are being perturbed. Thus the positivity criterion must be satisfied as a precondition of this analysis. Assuming that there is a positive equilibrium, we determine the two eigenvalues of (24.108). In niche theory models, both eigenvalues will be real and negative because of the structure of the α-matrix. Let $\lambda_{\min}$ denote the eigenvalue that is smallest in absolute value. Then, May and MacArthur propose that both species will persist in the stochastic environment if

$$\sigma^2 \ll |\lambda_{\min}| \qquad (24.109)$$

and conversely, one or both species are bound for extinction if the inequality is reversed. The symbol $\ll$ means that $|\lambda_{\min}|$ must be an *order of magnitude* or more greater than the environmental variability in the parameters of the model.

It is important to understand that (24.109) is a proposal and that it has not been proved. The stochastic model leads to a two-dimensional diffusion equation, which at present, has received little analysis. It should also be added that (24.109) has been strongly criticized on mathematical grounds by Turelli (1978).

The criterion (24.109) can be used to find a limiting similarity as follows. For a given level of environmental variability, σ^2, we find the niche separation that causes $|\lambda_{\min}|$ to equal σ^2. Then, the species should persist if the separation is much wider than this value. One or both should become extinct if the separation is much smaller than this value.

To illustrate the use of the criterion, we consider the completely symmetric example, in which the α's depend on position and the K's are fixed.

As mentioned before, this is the case in which the positivity criterion predicts a limiting similarity of zero. The matrix (24.108) in this case is, from Equation (24.10),

$$\begin{pmatrix} \dfrac{-r_0}{1+\alpha(d)} & \dfrac{-r_0}{1+\alpha(d)}\alpha(d) \\[3mm] \dfrac{-r_0}{1+\alpha(d)}\alpha(d) & \dfrac{-r_0}{1+\alpha(d)} \end{pmatrix} \tag{24.110}$$

The two eigenvalues are

$$\lambda_{\min} = -r_0\left[\frac{1-\alpha(d)}{1+\alpha(d)}\right] \tag{24.111a}$$

$$\lambda_{\max} = -r_0 \tag{24.111b}$$

Therefore, the limiting similarity, $\hat{d}$, is found as the root of

$$|\lambda_{\min}| = r_0\left[\frac{1-\alpha(d)}{1+\alpha(d)}\right] = \sigma^2 \tag{24.112a}$$

Rearranging, we have

$$\alpha(d) = \left[\frac{1-(\sigma^2/r_0)}{1+(\sigma^2/r_0)}\right] \tag{24.112b}$$

With a Gaussian competition function, the limiting similarity is explicitly

$$\frac{\hat{d}}{w} = 2\sqrt{\ln\left[(1+\sigma^2/r_0)/(1-\sigma^2/r_0)\right]} \tag{24.113}$$

For x small, the quantity $\ln\left[(1+x)/(1-x)\right]$ is approximately $2x$, so for σ^2/r_0 sufficiently small, we have

$$\frac{\hat{d}}{w} \approx 2\sqrt{2}\,\frac{\sigma}{\sqrt{r_0}} \tag{24.114}$$

The limiting similarity is plotted as a function of $\sigma/\sqrt{r_0}$ in Figure 24.25.

An interesting feature of Figure 24.25 is the dependence of the limiting similarity by this criterion on $\sigma/\sqrt{r_0}$. According to that result, the minimum niche separation necessary for coexistence increases with the environmental variance, σ^2, and decreases with the intrinsic rate of increase, r_0. Thus, for a given r_0, the niche separation must be larger in a more variable environment than in a comparatively constant environment; and for a given σ^2, the niche separation must be larger for populations with a low r_0, as compared to populations with a high r_0. Prediction of the factors that control the limiting similarity is clearly very different from that based on the positivity criterion. On the positivity criterion, the variable, $\hat{d}$, is controlled by the ratio of the K's and by the asymmetry of $\alpha(d)$. Here the factors of σ^2 and r_0 are added to the picture.

May (1973) has also extended the analysis of limiting similarity, based on the criterion (24.109), to three, and more evenly spaced species, with equal K's. For more on this point, the original work should be consulted.

It must be stressed that the existence of the criterion (24.109) is, itself, strongly model-dependent, whereas the existence of the positivity criterion is *not* model-dependent. The *detailed predictions* of limiting similarity based on

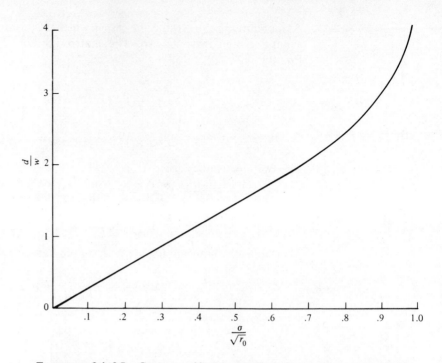

FIGURE 24.25. *Conjectured limiting similarity between two symmetrical species in an environment where* (a) *both r and K vary stochastically but retain a constant ratio* [*see Equation* (24.102)]; (b) *the Ito calculus is chosen for the analysis; and* (c) *Gaussian utilization curves are assumed.*

the positivity criterion are of course model-dependent. They depend on the symmetry and shape of $\alpha(d)$ and on any other assumptions that are brought in. But the *criterion* of positivity of the population sizes would remain a necessary condition for coexistence in *any* population model. In contrast, the *criterion* (24.109) for persistence in a stochastic environment is itself only a feature of a special stochastic model. We saw in Chapter 20 that this criterion does not arise if the Stratonovich calculus is used or if the stochastic fluctuation occurs only in K and not in r. In these stochastic models, the populations do persist, even though criterion (24.109) is violated. Furthermore, even if (24.109) does apply and, if it is violated, then the typical waiting time to extinction may be very long, thus making its empirical importance uncertain. Also, if criterion (24.109) does apply, then the detailed predictions will, of course, depend on the assumptions of the symmetry and shape of $\alpha(d)$, just as with the positivity criterion. Thus an empirical test of the limiting similarity, $\hat{d}$, given by (24.113) is a check both on the detailed assumptions used to calculate $\hat{d}$ *and* of the criterion (24.109) itself for persistence in a stochastic environment.

The dependence of $\hat{d}$ on r_0 in (24.113) is a qualitative prediction that seems especially susceptible to empirical examination. It is predicted by (24.113) for a given σ^2, that two species with a low r_0 must be more widely separated than two species with a high r_0. However, we can argue to the opposite conclusion, based on a different stochastic model. Consider the following linear stochastic model of the population dynamics of two competing species in discrete time.

$$n_{1,t+1} = (1 - r^*)n_{1,t} + r^*(k_{1,t} - \alpha(d)n_{2,t})$$
$$n_{2,t+1} = (1 - r^*)n_{2,t} + r^*(k_{2,t} - \alpha(d)n_{1,t})$$

(24.115)

where $r^* = r\hat{N}/K = r/(1+\alpha)$; $n_{i,t}$ denotes the *deviation* of the population size of species-i at time t from its long-term average value, $\hat{N}$; and $k_{i,t}$ denotes the *deviation* of the carrying capacity of species-i at time t from its long-term average value K. Here r^* is called the "effective rate of increase," The single species version of this model was discussed extensively in Chapter 20. Let σ_n^2 denote the variance of the population size. Then, in this model, we may take as a criterion for persistence that

$$Q\sigma_n < \hat{N}$$

(24.116)

where Q is a number set by convention. If Q is 3 and if $n_{i,t}$ is distributed normally (which it is if both $k_{i,t}$ are), then the probability is less than .001 that the population size is negative. A larger value of Q indicates a more stringent criterion. Now σ_n^2 is itself a function of $\alpha(d)$ and we may determine the limiting similarity as the value of d that causes $\hat{N} = Q\sigma_n$. If the $k_{i,t}$ are independent, white noise random variables with variance σ_k^2, then the population variance, σ_n^2, is found to be

$$\sigma_n^2 = \left\{ \frac{1}{2} \frac{r^{*2}}{1 - [(1-r^*) - r^*\alpha(d)]^2} + \frac{1}{2} \frac{r^{*2}}{1 - [(1-r^*) + r^*\alpha(d)]^2} \right\} \sigma_k^2$$

(24.117)

[The derivation of this result and others concerning the model defined by (24.115) occurs in Roughgarden (1975b, 1978b).] One important point to notice is that the variance blows up, $\sigma_n^2 \to \infty$, as $\alpha(d) \to 1$, that is, as the niches move closer together. Hence, for any r, the persistence criteria, (24.116) above, will be violated, if the niches are sufficiently close. Thus, by the criterion (24.116), a limiting similarity does exist for any value of r. Another important point is that for any value of $\alpha(d)$, the variance tends to zero as $r \to 0$, that is, two species can coexist arbitrarily close to one another, according to criterion (24.116), provided the r is sufficiently small. Thus, in this stochastic model, for a given σ_k^2, two species with high r must be more widely separated than two species with a low r. This result is clearly opposite to that predicted by criterion (24.109). The reason for these different results traces to the role of r in the models. In the models defined by (24.102) and (24.107), a high r prevents extinction by causing a high speed of population growth, whenever the population size is temporarily reduced to a value near zero. In those models, high r prevents a population from staying at a dangerously low level for a long time. But in the model defined by (24.115), the role of r is different. The fluctuation in k is multiplied by r and, if r is very low, then a given fluctuation in k has little effect in changing the population size. A low r refers to a population that is not very responsive to fluctuations in carrying capacity and hence those fluctuations are less likely to cause population extinction. Yet, it must be remembered that both these stochastic models represent the logistic and Lotka–Volterra competition equations in a stochastic environment. There are usually many stochastic models that correspond to any particular deterministic model.

Roughgarden (1976) introduced an evolutionary approach to the subject of resource partitioning between species. The problem facing all models of limiting similarity, regardless of the criterion used, is that it is not obvious that the actual niche separations between co-occurring species are the minimum separations that will allow coexistence. Limiting similarity models assert that species, which are sufficiently similar to one another, cannot coexist. Such models are silent, however, on the matter of the similarity among species that do coexist. Clearly, species that do coexist must be separated from one another by at least a bit more than the limiting similarity. Indeed, it is conceivable that they are separated by *much* more than the limiting similarity. One approach to determining the *realized similarity*, as distinct from the limiting similarity, is to assume that the realized similarity is the result of coevolution of the two (or more) competing species. One envisions the set up illustrated in Figure 24.26. Natural selection favors niche separation because separation lowers the competition coefficient. As the species move apart, however, it is assumed that their carrying capacities decrease, as sketched in Figure 24.26. The realized similarity will, then, be the degree of similarity attained at the coevolutionary eqilibrium. The equilibrium will reflect the counteracting forces of character displacement and of the costs of shifting to peripheral resources.

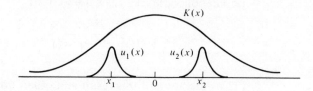

FIGURE 24.26. *Setup for a coevolutionary model for the realized similarity between two species.*

The coevolutionary model for the realized similarity is a coevolutionary strategy model of the type discussed in the section on coadaptation in Chapter 23. Specifically, the fitness in species-1 and in species-2 are

$$W_1(X_1, X_2, N_1, N_2) = 1 + r_1 - \frac{r_1}{K(X_1)} N_1 - \frac{r_1}{K(X_1)} \alpha(X_1 - X_2) N_2$$

$$(24.118a)$$

$$W_2(X_1, X_2, N_1, N_2) = 1 + r_2 - \frac{r_2}{K(X_2)} N_2 - \frac{r_2}{K(X_2)} \alpha(X_2 - X_1) N_1$$

$$(24.118b)$$

We find the coevolutionarily stable niche positions, $\tilde{X}_1$ and $\tilde{X}_2$, simultaneously by maximizing W_1 with respect to X_1 and W_2 with respect to X_2, subject to the constraints that $W_1 = 1$ and $W_2 = 1$. As an example to compare with the preceding results, we assume both the competition function and the carrying capacity functions are Gaussian. With symmetry, $-\tilde{X}_1 = \tilde{X}_2 \equiv \tilde{X}$ and the realized niche separation, $\tilde{d}$, is simply $2\tilde{X}$. In the Gaussian case, the realized separation is found to be

$$\frac{\tilde{d}}{w} = 2 \sqrt{\ln \left[\frac{-1 + \sigma_k^2}{w^2} \right]}$$

$$(24.119)$$

The realized niche separation from (24.119) is plotted in Figure 24.27. Observe that the curve starts at σ_k/w equal to $\sqrt{2}$. Recall from our third example of limiting similarity by the positivity criterion, that $\sigma_k^2 > 2w^2$ is required for a second species to enter an island where the first species is positioned under the peak of the $K(X)$ curve. If this condition is satisfied, then the second species can invade and the two species then coevolve a degree of similarity given by Equation (24.119). If $\sigma_k^2 < 2w^2$, then the island never attains two species to begin with and hence the coevolution of two species never arises. Roughgarden (1976) also treats the case of the coevolution of resource partitioning between three species and presents the analogue to (24.119) for several other shapes of competition and carrying capacity functions.

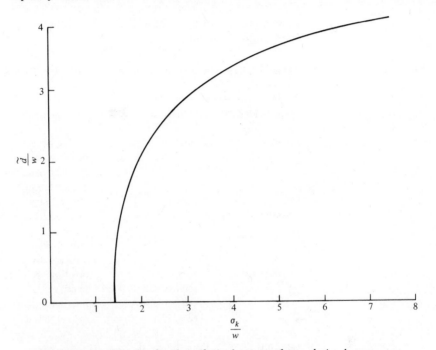

FIGURE 24.27. *Realized similarity because of coevolution between two symmetrical competing species assuming Gaussian utilization curves and carrying capacity function.*

An interesting additional feature of this model is that the species do not displace far enough from one another to maximize their population sizes. The niche separation that would maximize the population sizes, d_m, is (in the Gaussian case),

$$\frac{d_m}{w} = 2\sqrt{\ln\left[-1 + \frac{2\sigma_k^2}{w^2}\right]} \qquad (24.120)$$

Recall that we also deduced this qualitative result using the graphical approach introduced in Chapter 23. If the species were to move farther apart than they do, as a result of natural selection, then their population sizes would increase.

To conclude this section, we summarize the predictions about niche separation between competing species. Let us focus on symmetrical models for two species because the major differences between the niche theory models are already apparent at this level. See Table 24.4. The models also differ in their predictions concerning the partitioning for more than two species and with asymmetry and so forth, but all this additional structure is not needed to differentiate the models. These predictions should be testable with good comparative data on resource partitioning. For example, predictions three and four in the table should be distinguishable by comparing the niche separation between a pair of low r species and that from a pair of high r species, provided the environments have a comparable degree of variability. Similarly, prediction five could be checked by seeing how the niche separation between two species varies along an environmental gradient in the variety of resources, σ_k^2. Most empirical work on resource partitioning between species simply documents the existence of the phenomenon. What is needed are data on how that partitioning varies with the parameters implicated by the models on this topic.

Table 24.4. Prediction of Niche Separation for Two Symmetrical Competing Species.

Type of Model	Criterion	Assumptions	Predictions
Limiting similarity	Positivity	α varies with position, K's fixed, constant environment	Can always coexist, minimum niche separation is $\hat{d} = 0$
Limiting similarity	Positivity	α and K vary with position, constant environment	Coexistence requires variety of resources, σ_k^2, greater than $2w^2$
Limiting similarity	Persistence in stochastic environment: criterion (20.109)	α varies with position, K's fixed, fluctuating environment	Minimum niche separation increases with environmental variance and *decreases* with r
Limiting similarity	Persistence in stochastic environment: criterion (20.116)	α varies with position, K's fixed, fluctuating environment	Minimum niche separation increases with environmental variance and *increases* with r
Realized similarity	Coevolutionary stability	α and K vary with position, constant environment	Realized niche separation increases with variety of resource, σ_k^2

Island Biogeography

Our problem in this section is to provide models to explain the area effect, distance effect, and other observations concerning species diversity of oceanic islands. There is a useful review of the literature on island biogeography in Simberloff (1972) and interested readers should consult this paper for more detail than we have space to provide in this section. Also, the original work in this area, which pioneered what has become a revolution in biogeography, consists of the papers by Preston (1962) and MacArthur and Wilson (1963) and also the book, *The Theory of Island Biogeography* by MacArthur and Wilson in (1967). Recent theoretical extensions of this work include Richter-Dyn and Goel (1972), Diamond, Gilpin and Mayr (1976), Gilpin and Diamond (1976), and Diamond and May (1977). In general terms, there are three different hypotheses to explain the fact that the species diversity of an island is an increasing function of area and a decreasing function of distance from a source region; these are presented below.

The Nonequilibrium Hypothesis　By this hypothesis, distant islands contain fewer species than nearby islands, simply because the rate of arrival of new species to distant islands is slower than to nearby islands. Similarly, the effect of area is explained by assuming that the rate of arrival of new species to large islands is higher than to small islands. In this hypothesis the extinction of species that do arrive on an island is not a major factor, nor is an island considered closed to invasion. Of course, a propagule may go extinct if members of the propagule arrive in poor physiological condition, as a result of injury sustained during their journey. But the point to this hypothesis is that islands are currently in a state of accumulating fauna. If one could return one thousand years from now, then, according to the nonequilibrium hypothesis, the islands would have more species than they do now, and the increase in fauna will be caused by the arrival of new species and perhaps also to speciation within the island. Furthermore, one thousand years from now, most of the species currently on the islands would still be there. The islands would simply have added more species, reflecting the arrivals during the thousand years.

It is clear that if this hypothesis applies to any groups at all, it applies only to those which are extremely poor dispersers, like snakes and large mammals. In certain groups, this hypothesis is probably true in the initial phases of faunal buildup. It is sometimes observed that when the first species from a group is introduced to an island, it rapidly spreads over the whole island. There are many examples of introductions of novel species to islands and in such cases the faunal buildup on the island had been limited simply by the slowness of dispersal by natural means. However, it is often difficult to introduce a second or third species from a group into an island that already contains one member of that group. A clear example is provided by the history of the introduction of *Anolis* lizards to the island of Bermuda studied by Wingate (1965). In 1940, an anole from Jamaica was deliberately introduced to Bermuda to reduce the fruit fly population. It was the first anole species on the island and rapidly spread throughout the entire island. Some years later, the large anole from Antigua was introduced, but its distribution is restricted to the center of the island where a small forest occurs. Third, the solitary anole from Barbados was introduced and its distribution is extremely restricted to the northern periphery of the island. Clearly, Bermuda was not a faunal equilibrium, with respect to Anoline lizards, prior to the introduction. Thus it presently appears that the nonequilibrium hypothesis applies only to poor dispersers and, then, only in the very initial phases of faunal buildup.

The Turnover Hypothesis　This hypothesis was introduced to island biogeography by Preston (1962) and by MacArthur and Wilson (1963). It is an equilibrium hypothesis. One envisions that there is not only a rate of arrival of new species to the island, but also significant rate of extinction of whatever species are already on the island. Thus the species pool on an island has an input and output, and the steady-state level of the species pool is that attained when the overall immigration rate equals the overall extinction rate. Specifically, let $S(t)$ denote the number of species at time t on an island. Let $I(S)$ denote the rate at which *new* species arrive on the island, and let $E(S)$ denote the rate at which species already on the island are going extinct. Then, the rate of change of the species diversity on the island is simply:

$$\frac{dS}{dt} = I(S) - E(S) \tag{24.121}$$

The *equilibrium number of species*, $\hat{S}$, is the root of the equation

$$I(S) - E(S) = 0 \tag{24.122}$$

Furthermore, the *equilibrium turnover rate*, T, is defined as

$$T \equiv I(\hat{S}) = E(\hat{S}). \tag{24.123}$$

This hypothesis has two major points. (1) The number of species on an island approaches and attains an equilibrium level; if we return to the island in a thousand years, then, in principle, it will have the same total number of species. (2) There is a regular turnover of species on the island. If we return to the island in a thousand years, then, the identities of the species will have changed even though the total number of species will have remained constant. This hypothesis has been developed in some detail, and we now sketch some of the interesting predictions that follow from it. Recall that these two basic points have been verified experimentally with the insect fauna of small mangrove islands of the Florida keys. This work by Simberloff and Wilson was summarized in Figures 24.13 and 24.14.

The first explicit mathematical example that was solved using Equation (24.121) uses the assumption that the immigration and extinction rates are linear functions of the species number, S. Specifically, it is assumed that

$$I(S) = I_0 - \left(\frac{I_0}{P}\right)S \tag{24.124a}$$

$$E(S) = \left(\frac{E_p}{P}\right)S \tag{24.124b}$$

The constants in these expressions have the following interpretation: P is the total number of species in the source area, which is the maximum number of species that are available to colonize the island; and I_0 is the immigration rate of *new* species to the island in the limit as $S \to 0$. By definition, if $S = P$, then there are no new species left in the source pool so that $I(P) = 0$. $I(S)$ is a decreasing function of S. The extinction rate on the island is E_p when all P species are present on the island; it is the upper bound to the extinction rate. Also, $E(S)$ is an increasing function of S, and E_p/P is the per-species probability of extinction. The total extinction rate on the island is then E_p/P times S. With these assumptions, the equilibrium species diversity is given by

$$\hat{S} = \frac{I_0 P}{E_p + I_0} \tag{24.125}$$

The steady-state turnover rate that occurs at this equilibrium is

$$T = (E_p/P)\hat{S} = \frac{E_p I_0}{E_p + I_0} \tag{24.126}$$

To explain the area effect and the distance effect with this model, two additional assumptions are introduced. First, it is assumed that I_0 is a decreasing function of the distance from the source region, D,

$$I_0 = I_0(D), \quad \frac{dI_0(D)}{dD} < 0 \tag{24.127}$$

Second, it is assumed that E_p is a decreasing function of island area, A,

$$E_p = E_p(A), \quad \frac{dE_p(A)}{dA} < 0 \tag{24.128}$$

With these assumptions, both $\hat{S}$ and T are, themselves, functions of island area and distance from the source

$$\hat{S}(A, D) = \frac{I_0(D)P}{E_p(A) + I_0(D)} \tag{24.129a}$$

$$T(A, D) = \frac{E_p(A)I_0(D)}{E_p(A) + I_0(D)} \tag{24.129b}$$

With these assumptions, we can verify that there is an area effect

$$\frac{\partial \hat{S}(A, D)}{\partial A} > 0 \tag{24.130a}$$

and there is also a distance effect

$$\frac{\partial \hat{S}(A, D)}{\partial D} < 0 \tag{24.130b}$$

These results only restate the assumptions in slightly different terms, for we designed the model to predict an area and distance effect to begin with. What is interesting is that this model also entails many new predictions that are not so obvious and require additional data to check out.

The first set of new predictions concerns the derivatives of T with respect to A and D. One can verify that

$$\frac{\partial T(A, D)}{\partial A} < 0 \tag{24.131a}$$

That is, the steady-state turnover rate decreases with area. Similarly,

$$\frac{\partial T(A, D)}{\partial D} < 0 \tag{24.131b}$$

Thus the turnover rate is lower on more distant islands. The important point to notice is that, according to this model, these trends in the turnover rate as a function of A and D must also occur in addition to the area and distance effects involving $\hat{S}$. The observation of area and distance effects does not, by itself, confirm the turnover model. The predictions involving T (and/or those mentioned in the following discussion) must also be checked out.

The second set of new predictions concerns the mixed derivatives of $\hat{S}$. It can be verified that

$$\frac{\partial}{\partial D}\left(\frac{\partial \hat{S}(A, D)}{\partial A}\right) = \frac{P[I_0(D) - E_p(A)][dI_0(D)/dD][dE_0(A)/dA]}{[E_p(A) + I_0(D)]^3}$$

(24.132)

If $I_0(D) > E_p(A)$, then this quantity is positive, whereas if $I_0(D) < E_p(A)$, then the quantity is negative. Now recall that $\partial \hat{S}/\partial A$ is positive to begin with (the area effect). Recall also that the steepness of $I(S)$ line is controlled by I_0, and the steepness of the $E(S)$ line is controlled by E_p. Of course, the slope is positive for $E(S)$ and negative for $I(S)$. The absolute values of the slopes, however, that is, the steepness, are controlled by I_0 and E_p. Hence we can state (24.132) as follows: *If the slope of the immigration line is steeper than the extinction line for all the islands under study, then the number of species varies faster with area on distant islands. If the slope of the extinction line is steeper than the immigration line for all the islands under study, then the number of species varies faster with area on the near islands.*

Similarly, we can reverse the order of differentiation in (24.132), so that

$$\frac{\partial}{\partial A}\left(\frac{\partial \hat{S}}{\partial D}\right) = \frac{\partial}{\partial D}\left(\frac{\partial \hat{S}}{\partial A}\right)$$

(24.133)

Now recall that $\partial \hat{S}/\partial D$ is negative to begin with (the distance effect). Therefore, we can state: *If the slope of the immigration line is steeper than the extinction line for all the islands under study, then the number of species varies faster with distance on small islands. Conversely, if the slope of the extinction line is steeper than the immigration line for all the islands under study, then the number of species varies faster with distance on large islands.*

We can compare these predictions with the data obtained by Diamond on the number of bird species from islands near New Guinea. Recall from Equation (24.2) that the data are summarized by an equation of the form

$$S(A, D, L) = a(1 + bL)e^{-D/c}A^d$$

(24.134)

where a, b, c, and d are empirical constants. It is readily verified for this expression that

$$\frac{\partial}{\partial D}\left(\frac{\partial S}{\partial A}\right) < 0$$

(24.135)

that is, species number varies faster with area on nearby islands. For this result to be consistent with the above version of the turnover model of island biogeography, it is necessary that the slope of the extinction line be greater than the slope of the immigration line for most of the islands that were censused.

The third set of predictions concerns the time required for the equilibrium to be attained. Because the equation for $S(t)$ is a linear differential equation with constant coefficients, the full time-dependent solution is easily found. It is

$$S(t) = \hat{S}[1 - e^{-[(I_0 + E_p)/P]t}] + S_0 e^{-[(I_0 + E_p)/P]t}$$

(24.136)

where S_0 is the initial number of species on the island and $\hat{S}$ is the equilibrium number of species given by (24.125). If the island is initially empty, then S_0 is zero and the second term in (24.136) drops out. This equation may be solved for the so-called time constant of the system. The time constant, in this situation, is defined as the time needed to approach $(1 - 1/e)$, which is 63 percent, of the equilibrium value, given that the island is initially empty. The time constant is denoted as τ and by inspection of (24.136), it is given by

$$\tau(A, D) = \frac{P}{I_0(D) + E_p(A)} \qquad (24.137)$$

From this expression, we infer that

$$\frac{\partial \tau}{\partial A}(A, D) > 0 \qquad (24.138a)$$

$$\frac{\partial \tau}{\partial D}(A, D) > 0 \qquad (24.138b)$$

The first inequality means that more time is needed for large islands to approach equilibrium. The second inequality means that more time is needed for distant islands to approach equilibrium.

The turnover model of island biogeography can be enhanced in several ways. First the linear immigration and extinction rates can be replaced with other functional forms. Second, the model can be enlarged to allow for species-specific immigration and extinction rates. Third, the model can be formulated as a stochastic process, instead of as a deterministic process. For these extensions see the references cited at the beginning of this section.

The Niche Theory Hypothesis   The niche theory hypothesis is intended to apply to competing organisms that are also poor dispersers. The criterion for a poor disperser is that the time interval between the arrival of a new competing species on an island is as long or longer than the time for the species currently on the island to achieve coevolutionary equilibrium with one another. This hypothesis is also an equilibrium hypothesis, but it does not involve turnover. At present, it is only a slightly developed hypothesis, when compared to the turnover model, but it is implicit in the theory of resource partitioning between species presented earlier in the chapter.

The niche theory hypothesis is best visualized as a flow diagram or algorithm (Table 24.5). The algorithm fuses the criterion for invasion with

Table 24.5. Algorithm for Faunal Buildup Based on Niche Theory.

Step 1	Species-1 enters island
Step 2	Species-1 evolves niche position under the peak of carrying capacity function
Step 3	Test: can species-2 invade at any position?; if no then STOP at 1 species
Step 4	If yes, then introduce species-2 at an allowable position
Step 5	Both species coevolve niche position
Step 6	Can species-3 invade at any position?;
Step 7	If no, then STOP at 2 species
Step 8	If yes, then continue.
	$\vdots$
Step N	STOP at S species

the formulas for the coevolution of resource partitioning, as presented in the section on species packing.

Step 1 consists simply of the introduction of the first species. Step 2 is to allow time for this species to evolve the niche position under the peak of the $K(x)$ function. Then step 3 is to test whether a second species can invade at any niche position, given the position of the resident from the preceding step. If no species can invade at this point, then the faunal buildup has come to completion. But if invasion is possible, then the invader is introduced and the new fauna is allowed time to come to coevolutionary equilibrium. Then the fauna is again tested for invasibility, and so forth.

This algorithm has yet to be explored in detail, but it clearly holds much potential. In addition to predicting the final species diversity on the island, it can reveal the continual contraction of available invasion positions as the fauna accumulates and it also can reveal alternative pathways to a stopping point and the presence of forbidden pathways.

The algorithm outlined above uses only the invasion criterion and the formulas for the coevolution of niche positions. Both are results from the theory of species packing. It may also be possible to include the evolution of the niche width in the algorithm. This is certainly possible for the first stage, where only a solitary species is involved. The preliminary result is suggestive. With symmetrical Gaussian competition and carrying capacity functions, the niche width of a solitary population at equilibrium is $w^2 = \sigma_k^2 - \sigma_v^2$ (see Equations 24.75). The condition for a second species to invade is that $\sigma_k^2 - 2w^2 > 0$ (see Equation 24.101). When these formulas are combined, the condition for invasion by a second species, given that both the niche position and width of the first species have come to evolutionary equilibrium, is $2\sigma_v^2 - \sigma_k^2 > 0$. But recall also that the BPC at equilibrium is $\sigma_k^2 - 2\sigma_v^2$ so that the combined invasion criterion becomes $-\text{BPC} > 0$. Since the BPC is a nonnegative quantity, this condition can never be satisfied. Therefore, the very possibility of an island passing beyond the one-species state with respect to a congeneric group depends on the first species *not* having the time to evolve its equilibrium BPC. If it does have enough time, then there will not be any space left for a second species to enter. Hence the discussion following Equation (24.78) on constraints to the speed of the evolution of the niche width may be very important in the biology of faunal buildup.

As the length of this chapter testifies, niche theory and the theory of island biogeography consist of a large body of theoretical work. Yet it is clear that the development of this theory has just begun. The implications of niche theory for faunal buildup have not received much attention. Empirical tests of the various criteria for limiting similarity have yet to be made. The inclusion of predators into the models together with a more thorough exploration of asymmetrical competition models needs attention. All of these problems are tractable, and presumably there will be much further progress in this area in the near future.

PART SIX
APPENDICES

Appendix 1
THE MEAN AND VARIANCE

ALTHOUGH you have been calculating averages since those weekly spelling tests in second grade, you may not be familiar with the notation for the averaging process as used in this book.

The Mean For our purposes, the words *mean, average,* and *expected value* are synonymous in the sense that they all refer to the same computational process. The only difference is semantic and concerns the intended use and context of the calculation. It is computed as follows: Suppose we wish the average of three numbers, 12, 14, 12. Let us label the average as $\bar{X}$. Then the average is

$$\bar{X} = \frac{12 + 14 + 12}{3} = 12.67 \tag{A1.1}$$

This is the way you have been doing it all along. Instead, let us rewrite (A1.1) as

$$\bar{X} = \tfrac{2}{3}(12) + \tfrac{1}{3}(14) \tag{A1.2}$$

We have lumped the two occurrences of 12 together. Now here is the new twist. We shall call X a random variable when it serves as a *numerical label* for something which, by chance, can occur in different states. In (A1.2) our random variable has values 12 and 14. Moreover, we assume X has any given value with a specific probability. In particular, on the available evidence, we would say X has value 12 with probability $\tfrac{2}{3}$ and value 14 with probability $\tfrac{1}{3}$. In this terminology, the average value of X can be written as

$$\bar{X} = X_1 P(X_1) + X_2 P(X_2) + X_3 P(X_3) \cdots \tag{A1.3}$$

where $P(X_1)$ is the probability that X has value X_1, and so forth. $P(X)$ is called the probability density for the random variable X. Note the summation is over the different possible values of X. Let us take an example. Consider a die with two sides, each with 2 dots; 3 sides with 7 dots; and 1 side with 11 dots. Let X stand for the number of dots on a side. Then, the expected value of X (note usage) is found as

$$\bar{X} = \tfrac{2}{6}(2) + \tfrac{3}{6}(7) + \tfrac{1}{6}(11) = 6 \tag{A1.4}$$

Now, if we tossed this die 10 times, the average of X from these 10 tosses might be a little more or less than 6. But if we tossed the die many many times, the average of X would very closely approach 6. Note the slight difference in usage. If we compute the average, using the probability density, it is often called an expected value or mean. Whereas, if computed from a finite sample (say 10 tosses), it is called an average. It is very important that you learn to recognize a formula like

$$\bar{X} = \sum_{i=1}^{n} X_i P(X_i) \tag{A1.5}$$

as being just another way of expressing the averaging process you have been using since second grade. Incidentally, what does the sum of all the probabilities equal?

$$\sum_{i=1}^{n} P(X_i) = ?\qquad\qquad (A1.6)$$

The Variance If you understand averages as presented above, then variances are easy. Consider a probability density function $P(X_i)$. Then, the average of X was defined as

$$\bar{X} = \sum X_i P(X_i)\qquad\qquad (A1.7)$$

(It is understood that the summation is over the values that X is allowed to have.) We can generalize this idea. Let $f(X)$ be any function of X. Then, we define the average of $f(X)$ with respect to $P(X)$ as

$$\overline{f(X)} = \sum f(X_i)\,P(X_i)\qquad\qquad (A1.8)$$

The term $\bar{X}$ is then the special case where $f(X) = X$. Now the variance involves another choice of $f(X)$. Let $f(X)$ be the square of the *deviation* of X from its average,

$$f(X) = (X - \bar{X})^2\qquad\qquad (A1.9)$$

The variance is the average of this function; that is, the variance is the average of the squared deviation from the mean and is denoted as, σ^2.

$$\sigma^2 = \sum (X_i - \bar{X})^2 P(X_i)\qquad\qquad (A1.10)$$

The reason for squaring is that, otherwise, deviations above and below the mean would often cancel out. By squaring, we work only with positive numbers. The variance is a very interesting quantity. It tells us how close the values of X usually are to the mean. If the variance is small, the values of X are usually very close to the mean. That is, there is little spread. Conversely, the variance is large if the range of values of X about the mean is wide.

There is another quantity called the *standard deviation*, which is simply the square root of the variance. Since the variance is denoted σ^2, the standard deviation is σ. The reason for using σ, is that σ^2 has units of X^2. Sometimes a quantity with the same units as X and $\bar{X}$ is needed for calculations. We then use the standard deviation.

Sample Estimates of the Variance In biology, we rarely are able to observe an entire natural population—most often, only a small sample of it. Therefore, the problem arises of trying to infer the properties of a population from observations on the sample. Our first impulse is to calculate the average and variance of X, based on the sample and to suppose these numbers are reasonable estimates of the average and variance within the entire population. This procedure is fine for estimating the population average but not for the population variance. Roughly speaking, a small sample cannot adequately encompass the full range of variation present in the population and, consequently, the variance estimated from the sample using formula (A1.10) usually slightly underestimates the true population variance. To correct for this bias, the following formula is used. Let $N(X_i)$ be the number of times value X_i occurs in the sample. The total number of items in the sample is $N = \sum N(X_i)$.

The formula for the sample estimate of the population's variance, s^2, is

$$s^2 = \frac{\sum (X_i - \bar{X})^2 N(X_i)}{N-1}\qquad\qquad (A1.11)$$

We can show that (A1.11) is almost the same as (A1.10). Since $P(X_i)$ is the fraction of the items with value X_i, we could write $P(X_i) = N(X_i)/N$, and then (A1.10) becomes

$$\sigma^2 = \frac{\sum (X_i - \bar{X})^2 N(X_i)}{N} \qquad (A1.12)$$

Thus the difference between (A1.11) and (A1.12) is that $N-1$ is in the denominator, instead of N. This makes a value computed from (A1.11) slightly larger than that from (A1.12), thus compensating for the bias introduced by the limitations of a small sample size. It is a theorem in statistics that s^2, from (A1.11), is an "unbiased estimator" of the population variance, σ^2. Thus one calculates a variance, using (A1.10), when a theoretical density function is given or if an entire population is known. One calculates with (A1.11) if only a sample is given.

We present two final points. First, the sample estimate of the standard deviation is simply the square root of s^2 calculated from (A1.11). Second, formula (A1.11) is somewhat tedious on a hand calculator and it may be rearranged to a form that is less tedious to implement. Suppose there are N items in the sample and X_j is the value to the jth item. Then an alternative formula is

$$s^2 = \frac{1}{N-1} \left[\sum_{j=1}^{N} X_j^2 - \frac{1}{N} \left(\sum_{j=1}^{N} X_j \right)^2 \right] \qquad (A1.13)$$

Note that the summation here is over the *items* in the sample. In all the preceding formulas, the summation was over each different *value* of X and we took into account multiple occurrences of the same value with the $P(X_i)$. Equation (A1.13), however, is an item-by-item summation, without regard to whether some values are being used several times. To make sure you understand, the variance of X estimated from the sample (12, 14, 12) is 1.33. See if you can get this answer from both (A1.11) and (A1.13).

Appendix 2
HOW TO WRITE A COMPUTER PROGRAM IN BASIC

A computer is essentially just a machine that follows directions. A sequence of directions that specifies how some task is to be accomplished is called an algorithm. When the sequence of directions, the algorithm, is written in a "computer language," it is called a program.

A computer language, called BASIC, is especially designed for use when communicating with a computer via a teletype. This language is discussed below.

Most tasks a computer is ordered to carry out have essentially three parts. First, the computer must be told to receive some initial numbers; second, it must perform some calculations with them; and third, it must type out the result of the computations in an aesthetically pleasing format.

In BASIC every separate instruction must have a statement number. The computer then carries out the instructions, starting with the lowest numbered statement, and works up through the higher numbered statements.

A program to average three numbers is

$$
\begin{array}{ll}
100 & \text{INPUT } A, B, C \\
110 & \text{LET } X = (A + B + C)/3 \\
120 & \text{PRINT } X \\
130 & \text{END}
\end{array}
$$

This program has the three parts mentioned above. When the computer carries out the program, it comes first to statement 100. It types a "?" and the person at the teletype should type back 3 numbers separated by commas and then carriage return. The computer then assigns the first number as the value for A, the second for B, and the third for C. It then goes to statement 110, which tells it to add the three numbers, divide by 3, and assign the result as the value of X. Next, statement 120 tells the computer to type the value of X, which is our answer. Finally, statement 130 tells the computer that the task is completed.

In the following paragraphs we shall review the concepts of programming in BASIC.

Entering Numbers The INPUT statement was mentioned above. Another way to enter data is with a combination of two statements as illustrated below.

$$
\begin{array}{ll}
90 & \text{DATA } 3, 4.1695, 1.54\,E + 6 \\
100 & \text{READ } A, B, C
\end{array}
$$

The computer understands these two commands together; it assigns 3 to A, 4.1695 to B, and 1.54×10^6 to C.[†] The principal advantage of using the DATA and READ statements in preference to the INPUT statement is that with the DATA statements the data is actually incorporated into the program and is stored with the program whenever the program is saved on a storage device. In contrast, if the data is entered with an INPUT statement, it

---

† Note that 1.54E + 6 is the way the computer represents 1.54×10^6.

must be entered anew every time the program is executed and cannot be saved with the program on the storage device.

Output   The output of the computer is programmed with the PRINT statement. Consider some illustrations.

> 70   PRINT "PROGRAM TO AVERAGE 3 NUMBERS"
> 75   PRINT
> 120   PRINT "AVERAGE IS" X

Statement 70 makes the computer type out everything enclosed in the quotation marks. Statement 75 makes it skip a line. Statement 120 makes it type characters enclosed in the quotation marks followed by the number which was computed for X. For example, suppose that the three numbers to be averaged are 4, 5, and 6; then the average, X, is 5. In response to statement 120, the computer would type

AVERAGE IS 5

Some more points are illustrated below:

> 500   PRINT A, B, C
> 510   PRINT A; B; C
> 520   PRINT "AVERAGE";
> 530   PRINT "IS" X

Statement 500 causes the values of A, B, and C to be types in columns with about 15 spaces in between. Statement 510 leads to a more compact output where the numbers have only two spaces in between. So, by using semicolons, more numbers can be typed on one line than by using commas. Putting a semicolon at the end of a line prevents the computer from returning the carriage and starting a new line. Thus statements 520 *and* 530 together would result in

AVERAGE IS 5

Computation   Most computations are done using the LET statement. Examples are presented in the following table:

	Statement	Formula Being Computed
600	LET $X = A + B + C$	$A + B + C$
610	LET $X = A - B - (-C)$	$A - B + C$
620	LET $X = A * B * C$	ABC
630	LET $X = A * B / C$	$\dfrac{AB}{C}$
640	LET $X = A / B * C$	$\dfrac{A}{B} C = \dfrac{AC}{B}$
650	LET $X = A / (B * C)$	$\dfrac{A}{BC}$
660	LET $X = A \uparrow 2 + B \uparrow .(2/3)$	$A^2 + B^{(2/3)}$

Branching   We now come to one of the most important concepts in computer programming—a command that tells the computer to do some other command.

An unconditional branching statement is of the form 300 GOTO 400. An example of its use is illustrated in line 140 below.

```
100     INPUT A, B, C
110     LET X = A + B + C
120     PRINT "SUM IS" X
130     PRINT
140     GOTO 100
150     END
```

This program repeatedly finds the sum of three numbers.

The computer asks for the first 3 numbers in 100, finds their sum in 110, prints the result in 120, skips a line in 130, and then 140 tells it to go back to line 100 and repeat the process. The program will do this indefinitely until you manually stop it at the teletype.

Conditional
Branching

A conditional branching statement tells the computer to follow some other statement, provided that some condition is met. For an example of its use, consider a program which divides two numbers, provided that the denominator is not zero. The conditional branching statement is in line 110.

```
100     INPUT N, D
110     IF D = O THEN 150
120     LET X = N/D
130     PRINT "QUOTIENT IS" X
140     GOTO 160
150     PRINT "DENOMINATOR IS ZERO"
160     END
```

The program prints the result of N divided by D unless $D = 0$, in which case it prints that the denominator is zero. Whenever the condition is not met, the computer ignores the conditional branching statement and proceeds to the next higher statement. The kinds of conditions that one can place in a conditional branch are listed in the following table.

Statement		Relation
600	IF $X < Y$ THEN 900	X less than Y
610	IF $X < = Y$ THEN 900	X less than or equal to Y
620	IF $X > Y$ THEN 900	X greater than Y
630	IF $X > = Y$ THEN 900	X greater than or equal to Y
640	IF $X = Y$ THEN 900	X equals Y
650	IF $X \# Y$ THEN 900	X not equal to Y

Also, in a conditional branching statement, one can substitute for Y in the table any complicated expression. For example, one possible statement is

$$600 \quad \text{If } X > = B{\uparrow}2 - 4 * A * C \text{ THEN } 900$$

which transfers the computer from line 600 to line 900, provided X is greater than or equal to $B^2 - 4AC$.

Loops

A loop is a word in computer jargon for a part of a program that is carried out over and over again. We previously illustrated a never-ending loop to compute the sum of 3 numbers. Almost always, however, one wants part of the program repeated only a certain number of times. As an example, suppose we want to find the sum of 3 numbers only 5 times. The way to write a program with this built in restriction is

Explicit Loops

```
90     LET I = 1
100        INPUT A, B, C
110        LET X = A + B + C
120        PRINT "SUM IS" X
130        PRINT
140    LET I = I + 1
150    IF I < = 5 THEN 100
160    END
```

The computer starts out with $I = 1$ from line 90. Then, it computes and prints the sum of the first three numbers in lines 100 through 130. Next, in line 140 it makes the new value of I equal to the old value plus 1. Thus, *after* computing the first sum, I is equal to 2 in anticipation of computing the second sum. Line 150 then returns the computer to line 100, if I is less than or equal to 5. The computer then computes the second sum of 3 numbers. When line 140 is reached, I is made equal to 3. And so on, until at the end of the fifth summation of 3 numbers, $I = 6$. Since the condition in line 150 is not met, the computer proceeds to line 160, which makes it stop.

Implicit Loops

The commands that make a computer do certain parts of the program over and over again are used so often that shorthand commands for setting up a loop have been devised in many computer languages. In BASIC, the FOR and NEXT commands are used together for setting up a loop quickly. The program below is completely equivalent to the one just considered

```
90     FOR   I = 1 to 5
100        INPUT A, B, C
110        LET X = A + B + C
120        PRINT "SUM IS" X
130        PRINT
140    NEXT I
150    END
```

The FOR and NEXT statements bracket the section of the program to be carried out repeatedly.

Variables and Subscripts

Finally, we consider how much flexibility we have in choosing variables. Until now, our variables have been simply letters like A, B, C, X, I, J, and so forth. Clearly, if these were our only allowed variables, we would be restricted to a total of 26 variables. We can also choose as a variable any letter followed by a single number, for example, $A1$, $Z9$, $X7$, $I2$, and so forth. One could write, for example:

```
100    LET    A1 = 7
110    LET    Z5 = 6
120    LET    X9 = A1 * Z5
130    PRINT  X9
140    END
```

The program would duly compute the product of 6 and 7 and print out the answer as 42.

In addition, there exists the capability for letting any of the 26 regular letters stand for an entire list or table of numbers. For example, suppose that we have a column of 4 numbers.

$$3.1415$$
$$2.876$$
$$1.414$$
$$7.543$$

We can let the letter A stand for the column of four numbers as a whole, and if we do, we can refer to the first number in the column as $A(1)$, the second as $A(2)$, the third as $A(3)$, and the fourth as $A(4)$. The following command, for example, adds up all the numbers in the column

$$100 \quad \text{LET } S = A(1) + A(2) + A(3) + A(4)$$

Whenever you intend to let a letter stand for an entire list, you should say so with an explicit statement, the DIM statement

$$90 \quad \text{DIM } A(4)$$

DIM is short for dimension. The statement tells the computer that A henceforth will stand for a list 4 terms long.

The special advantage of using lists is that the subscript (number inside the parentheses) can, itself, be computed, and this feature allows a very easy handling of a large quantity of data. For example, the following sequence computes the sum of N numbers; in it, S accumulates the sum

```
95       LET S = 0
100      FOR I = 1 to N
110          LET S = S + A(I)
115      NEXT I
```

The computer reads statement 110 as saying "Let the new value of S equal the old value of S plus $A(I)$".

A program that would compute the average of 100 items of data follows, assuming that 10 items were entered per line in 10 lines at the beginning of the program:

```
101      DATA ...
  ⋮
110      DATA ...
115      DIM A(100)
170      FOR I = 1 to 100
130          READ A(I)
140      NEXT I
150      LET S = 0
160      FOR I = 1 to 100
170          LET S = S + A(I)
180      NEXT I
190      LET M = S/100
200      PRINT "AVERAGE IS" M
210      END
```

If you can follow the above program, you understand enough to write your own programs to accomplish a huge variety of tasks. As an exercise, you should now write a program to calculate the sample estimate of a population's variance, s^2, based on 100 data points.

Appendix 3
MATRIX ALGEBRA AND STABILITY THEORY

PROBABLY the most common mathematical problem in population biology is to determine the solutions to a *dynamical model*. A dynamical model is a model that predicts how some quantity, or quantities, change through time. For example, the logistic equation

$$\frac{dN}{dt} = rN\left(1 - \frac{N}{K}\right) \tag{A3.1}$$

predicts how the population size changes through time, and the familiar equation from population genetics

$$\Delta p = \frac{p(1-p)}{2\bar{w}} \frac{d\bar{w}}{dp} \tag{A3.2}$$

predicts how the gene frequency changes through time. The majority of the models in population biology are dynamical models. Solving a dynamical model means finding the value of the quantity at every time, based on any initial condition. For example, solving the logistic equation means finding the function, $N(t)$, that describes the population size for all t, based on any initial value of the population size. Similarly, solving the equation for Δp means finding the sequence $p_0, p_1, p_2, \ldots p_t$, which describes the gene frequency for all times, based on any initial gene frequency, p_0.

In practice, the process of solving a dynamical model usually involves combining many mathematical tools, together with computer analysis. Solving a dynamical model is an art and there is no simple way to go about it that always works. There is one tool, however, that is almost always useful. This tool is called a *local stability analysis*. The purpose of a local stability analysis is to determine whether there are special points, called stable *equilibrium points*, to which many solutions converge. When a local stability analysis is successful, it allows us to understand the asymptotic behavior of solutions, that is, what happens to solutions like $N(t)$ and p_t for t sufficiently large. The local stability analysis provides, at best, only a partial solution to a dynamical model, but nonetheless this partial solution is often, in practice, all we really need to know.

In some situations a global stability analysis is also possible. See, for example, the niche theory models of Chapter 24, the papers by Goh (1976) and Hastings (1978), and Chapter 7 on the $\bar{w}$ function.

In this appendix, we first explain how to do a local stability analysis in a one-variable system. Then, we turn to two variable systems. We shall see that the mathematical techniques of an area of mathematics called *linear algebra* are needed whenever a dynamical model involves two or more variables. After presenting the basics of linear algebra for two-variable systems, we return to solve the problem of a local stability analysis for a two-variable system. We are not going beyond two variables in this appendix. The generalization to more than two variables is not hard, but it is well covered in books on linear algebra and differential equations, such as Hirsch and Smale (1974).

We begin with a dynamical model, defined in continuous time

$$\frac{dN}{dt} = f(N) \tag{A3.3}$$

Here, N is the variable and $f(N)$ is a continuous function, which is differentiable with respect to N. [That is, $df(N)/dN$ is not infinite for any value of N.] Note that $f(N)$ is assumed to depend only on N and not on t. Furthermore, we are usually interested in models where only positive values of N are meaningful.

Our first purpose is to determine whether an equilibrium point is stable. An equilibrium point is defined as a point, $\hat{N}$, such that

$$f(\hat{N}) = 0 \tag{A3.4}$$

For example, in the logistic equation, $f(N) = rN(1 - N/K)$ and there are two equilibrium points, $\hat{N}_1 = 0$ and $\hat{N}_2 = K$. *By definition, when we say that an equilibrium point $\hat{N}$ is locally stable, we mean that all solutions which begin from an initial condition sufficiently close to $\hat{N}$, converge to $\hat{N}$ as $t \to \infty$.* Local stability, as defined in this way, is sometimes called local asymptotic stability. There are other definitions of local stability that are weaker and are sometimes used in mathematics, but the definition above is best for our purposes.

Our second purpose is to determine *how* the solutions approach an equilibrium if it is stable, and how they depart from a point if it is unstable. With the techniques of local stability analysis, we can satisfy both these purposes; that is, not only can we find whether an equilibrium is stable, but we can also determine how the solution behaves in the neighborhood of the equilibrium.

The term $f(N)$ is generally a nonlinear function. The basic idea behind local stability analysis is to work with the linear part of $f(N)$, which results from a Taylor series expansion of $f(N)$ about an equilibrium point. Specifically, we introduce the variable, $n = N - \hat{N}$, which describes the extent to which N deviates from a given equilibrium point. In terms of this variable, the dynamical model becomes

$$\frac{dN}{dt} = \frac{dn}{dt} = f(n + \hat{N}) \tag{A3.5}$$

Now, a Taylor series expansion of $f(N)$ about $\hat{N}$ yields

$$\frac{dn}{dt} = f(\hat{N}) + f'(\hat{N})n + \left[\begin{array}{c} \text{higher-order terms in } n \\ \text{e.g., } n^2, n^3, \text{ etc.} \end{array} \right] \tag{A3.6}$$

where $f'(\hat{N})$ means $df(N)/dN$ evaluated at $N = \hat{N}$. Now, since $\hat{N}$ denotes an equilibrium point, $f(\hat{N}) = 0$. Furthermore, if the initial condition is sufficiently close to $\hat{N}$, then n is very small. Hence if the initial condition is sufficiently close to $\hat{N}$, the higher-order terms in n are very small compared to the term, $f'(\hat{N})n$, provided $f'(\hat{N}) \neq 0$. Thus, provided that $f'(\hat{N}) \neq 0$, we can use the following model to obtain the solutions that begin from initial conditions very close to $\hat{N}$,

$$\frac{dn}{dt} = f'(\hat{N})n \tag{A3.7}$$

This model simply leads to exponential growth or decay, with rate $f'(\hat{N})$. The solution to (A3.7) is

$$n(t) = n(0) e^{f'(\hat{N})t} \tag{A3.8}$$

Obviously, if $f'(\hat{N})$ is negative, then $n(t) \to 0$ as $t \to \infty$. That is, if $f'(\hat{N}) < 0$, then solutions that begin close to $\hat{N}$ converge to $\hat{N}$ as $t \to \infty$. Thus if $f'(\hat{N}) < 0$, then we can conclude that $\hat{N}$ is a locally stable equilibrium. Furthermore, the deviation between the solution and $\hat{N}$ decreases exponentially through time with rate $f'(\hat{N})$. In contrast, if $f'(\hat{N})$ is positive, then all solutions sufficiently close to $\hat{N}$ leave that neighborhood of $\hat{N}$, thereby indicating that $\hat{N}$ is an unstable equilibrium point.

We may summarize our analysis above in the following theorem.

THEOREM A3.1 *Consider a model $dN/dt = f(N)$ where $f(N)$ is continuous and differentiable for $N \geq 0$. Let $\hat{N}$ denote an equilibrium point, that is, $f(\hat{N}) = 0$.*

 a. *If $f'(\hat{N}) < 0$, then $\hat{N}$ is locally stable and all solutions that begin sufficiently close to $\hat{N}$ converge to $\hat{N}$ exponentially with rate $f'(\hat{N})$.*
 b. *If $f'(N) > 0$, then $\hat{N}$ is unstable and all solutions that begin sufficiently close to $\hat{N}$ diverge from $\hat{N}$ exponentially with rate $f'(\hat{N})$.*
 c. *If $f'(\hat{N}) = 0$, then the stability of $\hat{N}$ can only be determined through further analysis.*

To solidify this result in your mind, you should do the following problem.

PROBLEM A3.1 For the logistic equation, show (a) that $\hat{N}_1 = 0$ is locally stable for $r < 0$ and unstable for $r > 0$, and (b) that $\hat{N}_2 = K$ is locally stable for $r > 0$ and unstable for $r < 0$.

Next, we move into the analysis of a dynamical model defined in discrete time. There are two equivalent forms in which a model in discrete time is expressed. The first is

$$\Delta N = f(N) \tag{A3.9}$$

and the second is

$$N_{t+1} = g(N_t) \tag{A3.10}$$

These two forms are equivalent because we can add N_t to both sides of (A3.9), yielding

$$N_{t+1} = f(N_t) + N_t = g(N_t) \tag{A3.11}$$

Thus $f(N) + N$ is identical to $g(N)$. For future reference, we also note that their derivatives are also closely related

$$f'(N) + 1 = g'(N) \tag{A3.12}$$

The procedure for determining whether an equilibrium point, $\hat{N}$, is locally stable in a discrete model proceeds in a very similar way to our derivation above. For convenience, we shall work with the model in the form of (A3.10). An equilibrium, $\hat{N}$, satisfies, by definition

$$g(\hat{N}) = \hat{N} \tag{A3.13}$$

To determine if $\hat{N}$ is locally stable, we consider the variable, $n_t = N_t - \hat{N}$, which describes the deviation of N from $\hat{N}$ at time t. Substituting into (A3.10) yields

$$n_{t+1} + \hat{N} = g(n_t + \hat{N}) \tag{A3.14}$$

Expanding $g(N)$ in a Taylor series about $\hat{N}$ yields

$$n_{t+1} + \hat{N} = g(\hat{N}) + g'(\hat{N}) \, n_t + \left[\begin{matrix} \text{higher-order terms} \\ \text{in } n_t \end{matrix} \right] \tag{A3.15}$$

Upon remembering that $g(\hat{N}) = \hat{N}$ and dropping the higher-order terms in n_t, provided that $g'(\hat{N}) \neq 0$, we obtain

$$n_{t+1} = g'(\hat{N}) \, n_t \tag{A3.16}$$

This equation can now be used to determine the behavior of solutions that begin sufficiently near $\hat{N}$. Equation (A3.16) is a simple linear recursion equation that is easily iterated for yield

$$n_t = [g'(\hat{N})]^t n_0 \tag{A3.17}$$

This solution describes geometric growth. By inspection of (A3.17), we can now determine the fate of solutions that begin near $\hat{N}$. All the possibilities are tabulated in Theorem A3.2.

THEOREM A3.2 *Consider a model $N_{t+1} = g(N_t)$ [or equivalently $\Delta N = f(N)$], where $g(N)$ is continuous and differentiable for $N \geq 0$. Let $\hat{N}$ denote an equilibrium point, that is, $g(\hat{N}) = \hat{N}$ [or equivalently $f(\hat{N}) = 0$].*

(a) *If $g'(\hat{N})$ is between zero and one [or $f'(\hat{N})$ is between -1 and 0], then $\hat{N}$ is locally stable and all solutions that begin sufficiently close to $\hat{N}$ converge to $\hat{N}$ geometrically with rate $g'(\hat{N})$.*

(b) *If $g'(\hat{N})$ is between -1 and 0 [or $f'(\hat{N})$ is between -2 and -1], then $\hat{N}$ is locally stable. All solutions that begin sufficiently close to $\hat{N}$ converge to $\hat{N}$ in an oscillatory manner. The solutions successively overshoot and undershoot $\hat{N}$, but the absolute value of the distance between the solution and $\hat{N}$ decreases geometrically through time with rate $|g'(\hat{N})|$.*

(c) *If $g'(\hat{N})$ is greater than one [or $f'(\hat{N}) > 0$], then $\hat{N}$ is unstable and all solutions initially close to $\hat{N}$ leave the neighborhood of $\hat{N}$ geometrically with rate $g'(\hat{N})$.*

(d) *If $g'(\hat{N})$ is less than -1 [or $f'(\hat{N}) < -2$], then $\hat{N}$ is unstable. All solutions initially close to $\hat{N}$ leave the neighborhood of $\hat{N}$ by successively overshooting and undershooting $\hat{N}$. The absolute value of the distance between the solution and $\hat{N}$ increases geometrically with rate $|g'(\hat{N})|$.*

(e) *If $g'(\hat{N}) = 0$ [or $f'(\hat{N}) = -1$], then $\hat{N}$ is locally stable, but further analysis is required to determine the manner in which the equilibrium is approached. If $g'(\hat{N}) = -1$ or 1 [or $f'(\hat{N}) = -2$ or 0], then further analysis is required to determine if $\hat{N}$ is locally stable.*

To become familiar with the theorem above, you should work out the following two problems.

PROBLEM A3.2 For the logistic equation in discrete time, show that the equilibrium $\hat{N} = K$ has the following stability characteristics as a function of r:

$r < 0$: $\hat{N} = K$ is unstable; nonoscillating exit from the neighborhood of equilibrium.

$0 < r < 1$: $\hat{N} = K$ is locally stable; nonoscillating approach to equilibrium.

$1 < r < 2$: $\hat{N} = K$ is locally stable; oscillatory approach to equilibrium.

$r > 2$: $\hat{N} = K$ is unstable; oscillating exit from the neighborhood of equilibrium.

PROBLEM A3.3 From the equation for Δp given by (A3.2), show that the approach to equilibrium with heterozygote superiority does not oscillate in the neighborhood of $\hat{p}$.

The Setup for a Local Stability Analysis with Two Variables It is easy to set up the analysis of the local stability of an equilibrium involving two variables. We simply develop, again, the same type of argument we used before with one variable. Specifically, in continuous time, we begin with a model with the form

$$\frac{dN_1}{dt} = f_1(N_1, N_2)$$

$$\frac{dN_2}{dt} = f_2(N_1, N_2) \tag{A3.18}$$

By definition, an equilibrium, $(\hat{N}_1, \hat{N}_2)$, satisfies the two equations simultaneously

$$f_1(\hat{N}_1, \hat{N}_2) = 0$$

$$f_2(\hat{N}_1, \hat{N}_2) = 0 \tag{A3.19}$$

Now, we consider the variables

$$n_1 = N_1 - \hat{N}_1$$

$$n_2 = N_2 - \hat{N}_2 \tag{A3.20}$$

Substituting into (A3.18) and noting that $dN_i/dt = dn_i/dt$, we obtain

$$\frac{dn_1}{dt} = f_1(n_1 + \hat{N}_1, n_2 + \hat{N}_2)$$

$$\frac{dn_2}{dt} = f_2(n_1 + \hat{N}_1, n_2 + \hat{N}_2) \tag{A3.21}$$

We now Taylor-expand both f_1 and f_2 about $(\hat{N}_1, \hat{N}_2)$. Remembering (A3.19) and keeping only the linear terms yields

$$\frac{dn_1}{dt} = \left[\frac{\partial f_1(\hat{N}_1, \hat{N}_2)}{\partial N_1}\right] n_1 + \left[\frac{\partial f_1(\hat{N}_1, \hat{N}_2)}{\partial N_2}\right] n_2$$

$$\frac{dn_2}{dt} = \left[\frac{\partial f_2(\hat{N}_1, \hat{N}_2)}{\partial N_1}\right] n_1 + \left[\frac{\partial f_2(\hat{N}_1, \hat{N}_2)}{\partial N_2}\right] n_2 \tag{A3.22}$$

We now have a system of two linear first-order differential equations with constant coefficients. This system describes how solutions behave when they are close to the equilibrium point under examination. Before, when we were at this juncture, we were able to write out the general solution for one variable and to deduce from that solution the important sufficient condition for local stability. But now, it is best to postpone the solution to (A3.22) until we have learned some linear algebra. Our need to solve (A3.22) provides us with a concrete motivation for introducing the ideas of linear algebra.

In a similar way, we can set up the equations we need to solve for the local stability analysis of the two-variable system, defined in discrete time. We begin with a model of the form

$$N_{1,t+1} = g_1(N_{1,t}, N_{2,t})$$
$$N_{2,t+1} = g_2(N_{1,t}, N_{2,t}) \tag{A3.23}$$

Again, we introduce the variables

$$n_{1,t} = N_{1,t} - \hat{N}_1$$
$$n_{2,t} = N_{2,t} - \hat{N}_2 \tag{A3.24}$$

Upon substituting these variables into (A3.23), Taylor-expanding both g_1 and g_2 about $(\hat{N}_1, \hat{N}_2)$, and keeping only the linear terms in the expansion, we obtain

$$n_{1,t+1} = \frac{\partial g_1(\hat{N}_1, \hat{N}_2)}{\partial N_1} n_{1,t} + \frac{\partial g_1(\hat{N}_1, \hat{N}_2)}{\partial N_2} n_{2,t}$$

$$n_{2,t+1} = \frac{\partial g_2(\hat{N}_1, \hat{N}_2)}{\partial N_1} n_{1,t} + \frac{\partial g_2(\hat{N}_1, \hat{N}_2)}{\partial N_2} n_{2,t} \tag{A3.25}$$

This is the system we must solve for the two-variable discrete time local stability analysis. Now, let us move into the basics of linear algebra. After developing some key ideas in linear algebra, we shall return to finish the two-variable stability analysis.

Linear Algebra

Linear algebra supplies us with the basic conceptual and arithmetic tools to solve many problems involving several simultaneous variables. It provides the machinery that allows us to visualize and to solve many multidimensional problems. It is limited in scope in that the problems that can be solved with the tools of linear algebra involve linear equations. In contrast, ordinary univariate algebra contains tools that apply to many classes of nonlinear equations and there are often no multidimensional analogues for the univariate nonlinear tools. Thus linear algebra offers us an extension to multivariate problems but is restricted to the context of linear equations.

Matrix Arithmetic

Our first task in linear algebra is simply to organize all the variables into a common format and to develop a compact notation for all the calculations. We lump together our variables and constants into *vectors* and *matrices*. An example for a row vector is

$$(X_1, X_2)$$

An example of a column vector is

$$\begin{pmatrix} X_1 \\ X_2 \end{pmatrix}$$

An example of a matrix is

$$\begin{pmatrix} a & b \\ c & d \end{pmatrix}$$

We shall define rules of arithmetic for vectors and matrices, which you should memorize. Vector and matrix *addition* is defined as

$$(a, b) + (c, d) \equiv (a + c, b + d)$$

$$\begin{pmatrix} a \\ b \end{pmatrix} + \begin{pmatrix} c \\ d \end{pmatrix} \equiv \begin{pmatrix} a + c \\ b + d \end{pmatrix}$$

$$\begin{pmatrix} a & b \\ c & d \end{pmatrix} + \begin{pmatrix} e & f \\ g & h \end{pmatrix} \equiv \begin{pmatrix} a + e & b + f \\ c + g & d + h \end{pmatrix}$$

A number is sometimes called a scalar. We can multiply a vector or matrix by a number. The rule to do this is called the rule of *scalar multiplication*. Let α be a number. Then scalar multiplication is defined as

$$\alpha(a, b) \equiv (\alpha a, \alpha b)$$

$$\alpha \begin{pmatrix} a \\ b \end{pmatrix} \equiv \begin{pmatrix} \alpha a \\ \alpha b \end{pmatrix}$$

$$\alpha \begin{pmatrix} a & b \\ c & d \end{pmatrix} \equiv \begin{pmatrix} \alpha a & \alpha b \\ \alpha c & \alpha d \end{pmatrix}$$

Finally, vectors and matrices can also be multiplied with one another by a rule of *matrix multiplication*. The rule is based on a simple idea. Consider two matrices.

$$\begin{pmatrix} a & b \\ c & d \end{pmatrix} \begin{pmatrix} e & f \\ g & h \end{pmatrix}$$

Each matrix can be viewed as being made from two row vectors placed on top of each other. The rule for matrix multiplication combines scalar multiplication with vector addition as follows: The rule is that we take "a" times the row vector (e, f), and "b" times the row vector (g, h), and add these to obtain the top row of the new matrix. That is,

$$\begin{bmatrix} \text{top row of} \\ \text{new matrix} \end{bmatrix} = a(e, f) + b(g, h)$$

Similarly, we scalar-multiply "c" times (e, f), and "d" times (g, h), and add these to obtain the bottom row of the new matrix. That is,

$$\begin{bmatrix} \text{bottom row of} \\ \text{new matrix} \end{bmatrix} = c(e, f) + d(g, h)$$

Hence the product of the two matrices is given by

$$\begin{pmatrix} ae + bg & af + bh \\ ce + dg & cf + dh \end{pmatrix}$$

Although this rule may seem complicated at first, with a little practice, it becomes automatic.

The idea of matrix multiplication can be used with matrices of many different sizes. What is required is that the number of columns of the matrix on the left be equal to the number of rows of the matrix on the right. For example, we can legitimately multiply a 2×2 square matrix on the left with a column vector on the right, and we obtain another column vector as follows,

$$\begin{pmatrix} a & b \\ c & d \end{pmatrix} \begin{pmatrix} e \\ f \end{pmatrix} = \begin{pmatrix} ae + bf \\ ce + df \end{pmatrix}$$

Perhaps now, you can begin to see the utility of matrix arithmetic for simplifying the way multivariate problems are formulated. For example, let us return briefly to the equations we will have to solve for the local stability analysis of a two-variable system. For the discrete time model, we let n_t denote a column vector

$$n_t \equiv \begin{pmatrix} n_{1,t} \\ n_{2,t} \end{pmatrix} \tag{A3.26}$$

and we let A denote a 2×2 matrix with the following entries

$$A \equiv \begin{vmatrix} \dfrac{\partial g_1(\hat{N}_1, \hat{N}_2)}{\partial N_1} & \dfrac{\partial g_1(\hat{N}_1, \hat{N}_2)}{\partial N_2} \\ \dfrac{\partial g_2(\hat{N}_1, \hat{N}_2)}{\partial N_1} & \dfrac{\partial g_2(\hat{N}_1, \hat{N}_2)}{\partial N_2} \end{vmatrix} \tag{A3.27}$$

With this notation, Equation (A3.25) becomes simply

$$n_{t+1} = An_t \tag{A3.28}$$

Obviously, the rules of matrix arithmetic have greatly simplified the notation because (A3.28) is much more concise than (A3.25), although both describe the same actual calculations.

The Basic Ideas of Linear Algebra It is possible to summarize the essence of linear algebra with four main ideas: First, Figure A3.1 illustrates what we call a *vector space* in two dimensions. Notice the presence of the usual horizontal and vertical axes and the origin. A column vector, say $\begin{pmatrix} a \\ b \end{pmatrix}$, is represented on this graph by a line from the origin to the point, (a, b), on the plane. *Each point* on this plane

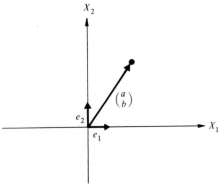

FIGURE A3.1. *Graphical representation of a vector space. A column vector,* $\begin{pmatrix} a \\ b \end{pmatrix}$ *is represented as an arrow from the origin to the point, (a, b). The "basis vectors," e_1 and e_2, are vectors that determine where the axes are. The basis vector, e_1 and e_2, correspond to the column vectors* $\begin{pmatrix} 1 \\ 0 \end{pmatrix}$ *and* $\begin{pmatrix} 0 \\ 1 \end{pmatrix}$, *respectively.*

represents a column vector because the line drawn from the origin to each point on the plane represents a unique vector. The process of vector addition is represented graphically in Figure A3.2. Suppose that we want to add the vector $\begin{pmatrix} 1 \\ 2 \end{pmatrix}$ to the vector $\begin{pmatrix} 1 \\ 1 \end{pmatrix}$. The way to do this is to take the first vector and move it to the head of the other vector, while keeping its length and direction unchanged. The head of the vector that has been moved now points to the head of the vector that represents the sum, as illustrated in Figure A3.2. The process of scalar multiplication is even simpler to visualize. Scalar multiplication lengthens a vector in all directions by the same factor. For example, if we multiply $\begin{pmatrix} 1 \\ 1 \end{pmatrix}$ by a constant, say 2, we obtain $\begin{pmatrix} 2 \\ 2 \end{pmatrix}$, which represents a vector pointing in exactly the same direction as does $\begin{pmatrix} 1 \\ 1 \end{pmatrix}$, except that it is longer. Similarly, multiplying a vector by -1 simply reflects it through the origin. Thus the first idea of linear algebra is to visualize each point on a plane as representing a column vector. Once we have done this, we can provide a graphical interpretation to vector addition and scalar multiplication.

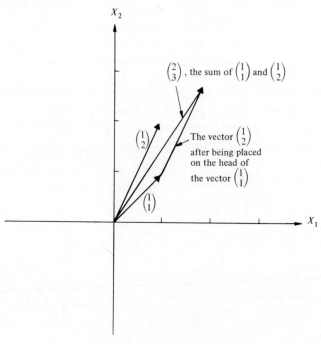

FIGURE A3.2. *Graphical representation of vector addition. To add* $\begin{pmatrix} 1 \\ 2 \end{pmatrix}$ *to* $\begin{pmatrix} 1 \\ 1 \end{pmatrix}$, *we move the arrow representing* $\begin{pmatrix} 1 \\ 2 \end{pmatrix}$ *to the head of the arrow representing* $\begin{pmatrix} 1 \\ 1 \end{pmatrix}$ *while keeping its length and direction unchanged. The head of the arrow that we moved now points to the head of the arrow representing the sum of the two column vectors* $\begin{pmatrix} 2 \\ 3 \end{pmatrix}$.

For the second main idea of linear algebra, we provide a graphical interpretation of the effect of matrix multiplication upon a vector. Matrix multiplication converts a vector into a new vector. A matrix is said to represent the *operation* whereby vectors are converted or transformed into other vectors. Consider three examples. First, let us take a matrix with zero on the off-diagonal positions. By convention, "the diagonal" refers to the entries on the diagonal running from top left to bottom right; A_1 is called a "diagonal matrix."

$$A_1 = \begin{pmatrix} 2 & 0 \\ 0 & \frac{1}{2} \end{pmatrix}$$

We may understand what this matrix does to vectors by using it with a typical vector. Let X be $\begin{pmatrix} a \\ b \end{pmatrix}$. Then A_1 converts X into Y as follows

$$\begin{pmatrix} 2 & 0 \\ 0 & \frac{1}{2} \end{pmatrix}\begin{pmatrix} a \\ b \end{pmatrix} = \begin{pmatrix} 2a \\ \frac{1}{2}b \end{pmatrix} = Y$$

Thus A_1 represents the operation of stretching the first coordinate, while shrinking the second. See Figure A3.3(a). As a second example, consider

$$B_1 = \begin{pmatrix} 2 & 0 \\ 1 & 2 \end{pmatrix}$$

Again let $X = \begin{pmatrix} a \\ b \end{pmatrix}$. Then B_1 converts X into Y as follows:

$$\begin{pmatrix} 2 & 0 \\ 1 & 2 \end{pmatrix}\begin{pmatrix} a \\ b \end{pmatrix} = \begin{pmatrix} 2a \\ a + 2b \end{pmatrix}$$

Thus B_1 stretches both coordinates by the same factor and adds the prestretched value of the first coordinate to the second. See Figure A3.3(b). As a third example consider

$$C_1 = \begin{pmatrix} .87 & -.5 \\ .5 & .87 \end{pmatrix}$$

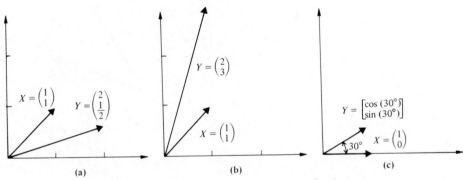

FIGURE A3.3. (a) *Action of the operator* $\begin{pmatrix} 2 & 0 \\ 0 & \frac{1}{2} \end{pmatrix}$. (b) *Action of the operator* $\begin{pmatrix} 2 & 0 \\ 1 & 2 \end{pmatrix}$. (c) *Action of the operator* $\begin{pmatrix} \cos(30°) & -\sin(30°) \\ \sin(30°) & \cos(30°) \end{pmatrix}$. *In all of the examples, Y is the result of the operator acting on the vector X.*

Now, it happens that .5 is sin (30°) and .87 is cos (30°) so that we can rewrite C_1 as

$$C_1 = \begin{pmatrix} \cos(30°) & -\sin(30°) \\ \sin(30°) & \cos(30°) \end{pmatrix}$$

The operation of C_1 on $X = \begin{pmatrix} a \\ b \end{pmatrix}$ is

$$\begin{pmatrix} \cos(30°) & -\sin(30°) \\ \sin(30°) & \cos(30°) \end{pmatrix} \begin{pmatrix} a \\ b \end{pmatrix} = \begin{pmatrix} a\cos(30°) - b\sin(30°) \\ a\sin(30°) + b\cos(30°) \end{pmatrix} = Y$$

By elementary trigonometry, one can verify that Y is simply the vector X rotated through 30° in a counterclockwise direction. See Figure A3.3(c). Thus we see that matrix multiplication upon a vector transforms it into another vector. We see, also, that there are a variety of possible operations upon vectors that matrices can represent. Hence the study of matrices becomes, in effect, the study of what matrices *do* to vectors when vectors are used in matrix multiplication. A matrix is not just a table of numbers; it signifies a specific *action* or operation that it performs when used in multiplication. This idea, that matrices represent operations upon vectors, is the second main idea in linear algebra.

The third main idea of linear algebra is that there is great freedom in the choice of the coordinate frame. In Figure A3.1, we have chosen as the coordinate axes, the usual rectilinear coordinate frame. The coordinate frame is specified by the "basis vectors." The standard basis vectors are

$$e_1 = \begin{pmatrix} 1 \\ 0 \end{pmatrix} \quad \text{and} \quad e_2 = \begin{pmatrix} 0 \\ 1 \end{pmatrix}.$$

These basis vectors are labeled in Figure A3.1. Turn now to Figure A3.4. We may use, as basis vectors, any two vectors that are not parallel to one

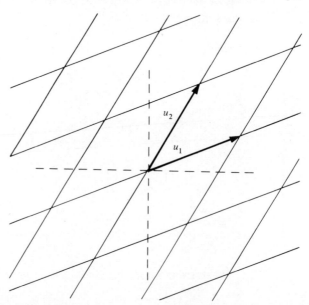

FIGURE A3.4. *The coordinate frame resulting from the choice of u_1 and u_2 as basis vectors.*

another. For example, two vectors, u_1 and u_2, are drawn in Figure A3.4. The coordinate frame that these vectors produce is also indicated. Clearly, any point in the plane can be located by using a nonstandard coordinate frame, just as well as it can by using the standard coordinate frame. Such freedom to choose the coordinate frame means that any given vector or operator has a different numerical representation, depending on the coordinate frame that is used. For example, the vector u_1 might be represented as

$$\begin{pmatrix} 1 \\ \frac{1}{2} \end{pmatrix}$$

in the standard basis. But if u_1 is itself taken as the first basis vector, then its coordinates in this coordinate frame become $\begin{pmatrix} 1 \\ 0 \end{pmatrix}$. Thus the coordinates that represent a vector, and, by extension, a matrix that operates on vectors, are *relative* to a set of basis vectors.

The fourth main idea of linear algebra is that *all* 2×2 matrices can be viewed as one of the three types of operations discussed previously as examples, provided the basis vectors are appropriately chosen. Specifically, let $A = \begin{pmatrix} \lambda_1 & 0 \\ 0 & \lambda_2 \end{pmatrix}$, where λ_1 and λ_2 are real numbers (also λ_1 may equal λ_2). The operator, A, represents a stretching (or shrinking) by a factor of λ_1 along the first basis vector, and by a factor of λ_2 along the second basis vector. The matrix, A, is called a diagonal matrix. Now, the important point to understand is that many matrices that are not in diagonal form when they are originally in the standard basis become diagonal in some other basis. That is, a great many matrices represent the operation of stretching (or shrinking) in one direction, by a factor λ_1, and in another direction, by a factor λ_2, although the directions do not coincide with the directions of the standard basis vectors. But if we use a coordinate frame with the natural directions of such a matrix as our basis, then the matrix assumes a diagonal form in that basis. Similarly, other matrices can, by a suitable choice of basis vectors, be put in the form of

$$B = \begin{pmatrix} \lambda & 0 \\ 1 & \lambda \end{pmatrix}$$

As we saw before, this type of matrix indicates a stretching (or shrinking) by the same factor, λ, in both coordinate directions, together with the addition of the prestretched value of the first coordinate to the second. And, finally, a great many matrices can, by choosing suitable basis vectors, be put in the form

$$C = |\lambda| \begin{pmatrix} \cos \theta & -\sin \theta \\ \sin \theta & \cos \theta \end{pmatrix}$$

This type of matrix rotates a vector through an angle θ in a counterclockwise direction and then stretches or shrinks it in all directions equally by the positive factor $|\lambda|$. It is a truly remarkable result that all 2×2 matrices can be uniquely classified into one of these three types, simply by choosing the natural coordinate frame. This finding expresses the fourth main idea of linear algebra. Much of the "guts" of linear algebra consists of the machinery that allows us to classify into one of these three types any matrix which we are originally given in the standard basis form.

How to Classify
2 × 2 Matrices   Linear algebra provides us with several tools which, when used together, allow us to classify any matrix into one of the three categories discussed above. Perhaps the most basic tool in linear algebra is the formula for a determinant of a matrix. *The determinant of a 2 × 2 matrix*

$$A = \begin{pmatrix} a & b \\ c & d \end{pmatrix}$$

is computed by the formula

$$\det (A) = ad - cb \tag{A3.29}$$

For example, the determinant of the matrix $\begin{pmatrix} 1 & 2 \\ 3 & 4 \end{pmatrix}$ is $4 - 6 = -2$. This formula was invented because it has the property that it is zero if one row is simply a constant times the other row. Also, it is zero if one column is a constant times the other column. For example, the determinant of $\begin{pmatrix} 5 & 15 \\ 1 & 3 \end{pmatrix}$ is zero because the top row is 5 times the bottom row. The formula for a determinant was invented because it serves as a numerical indicator of whether the rows (or columns) in a matrix are different from one another or whether they are the same to within a multiplicative constant.

The reason we want to know whether or not the rows of a matrix are the same to within a multiplicative constant arises as follows. Suppose that we have two simultaneous linear equations that both equal zero

$$aX_1 + bX_2 = 0$$
$$cX_1 + dX_2 = 0 \tag{A3.30}$$

In terms of matrix multiplication the equations can be rewritten as

$$\begin{pmatrix} a & b \\ c & d \end{pmatrix}\begin{pmatrix} X_1 \\ X_2 \end{pmatrix} = \begin{pmatrix} 0 \\ 0 \end{pmatrix} \tag{A3.31}$$

Now, if the two equations are really different from one another, then the only solution to (A.30) is that $X_1 = X_2 = 0$. We can see this result graphically in Figure A3.5. The first equation can be rewritten as

$$X_2 = \frac{-a}{b}X_1 \tag{A3.32}$$

This is a line, which goes through the origin with slope $-a/b$, and is plotted as a solid line in Figure A3.5. The second equation is also a line through the origin

$$X_2 = \frac{-c}{d}X_1 \tag{A3.33}$$

and is plotted as a dashed line in Figure A3.5. The only place the lines intersect is at the origin, provided that the lines do not coincide. Thus provided that

$$\frac{-a}{b} \neq \frac{-c}{d} \tag{A3.34}$$

then the only simultaneous solution is $X_1 = X_2 = 0$. However, if the lines coincide, then it means that the first equation is just a multiple of the second.

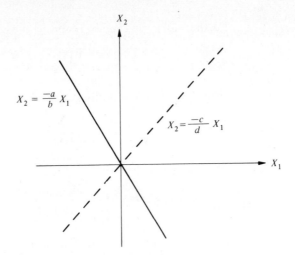

FIGURE A3.5. *The graph of two simultaneous linear equations that both pass through the origin. The origin is their only point of intersection unless they coincide.*

We then have really only one equation and the second is redundant. If the lines are parallel, we have

$$\frac{-a}{b} = \frac{-c}{d} \tag{A3.35}$$

which we may rearrange as

$$ad - bc = 0 \tag{A3.36}$$

The expression on the left is the determinant of the matrix of coefficients. Hence, if the determinant of the matrix of coefficients in (A3.31) is zero, then the two equations do have many simultaneous solutions that are nonzero. Indeed, every point along the line (A3.32) is a solution because the line (A3.33) coincides with (A3.32) when the determinant is zero.

The approach to classifying a matrix is analogous to classifying the contents of the nucleus of an atom by firing electrons at it. What we shall do with a matrix is to bombard it with a column vector and look for a very specific kind of collision. Specifically, we shall *look to see if there are any column vectors such that the only effect of the matrix on them is to stretch (or shrink) them*. Its surprisingly easy to answer this question. Let the matrix we are trying to classify be $\begin{pmatrix} a & b \\ c & d \end{pmatrix}$ and let $\begin{pmatrix} X_1 \\ X_2 \end{pmatrix}$ be a column vector. Let λ denote a real number. We want to know *if* there exists one or more column vectors such that

$$\begin{pmatrix} a & b \\ c & d \end{pmatrix}\begin{pmatrix} X_1 \\ X_2 \end{pmatrix} = \lambda \begin{pmatrix} X_1 \\ X_2 \end{pmatrix} \tag{A3.37}$$

λ represents the factor by which the matrix stretches or shrinks the column vector without changing its direction. We can rearrange (A3.37) as

$$\begin{pmatrix} a & b \\ c & d \end{pmatrix}\begin{pmatrix} X_1 \\ X_2 \end{pmatrix} - \lambda \begin{pmatrix} X_1 \\ X_2 \end{pmatrix} = 0 \tag{A3.38}$$

Furthermore, we can always insert the so-called identity matrix $\begin{pmatrix} 1 & 0 \\ 0 & 1 \end{pmatrix}$ as

$$\begin{pmatrix} a & b \\ c & d \end{pmatrix}\begin{pmatrix} X_1 \\ X_2 \end{pmatrix} - \lambda \begin{pmatrix} 1 & 0 \\ 0 & 1 \end{pmatrix}\begin{pmatrix} X_1 \\ X_2 \end{pmatrix} = 0 \tag{A3.39}$$

Now, we can scalar multiply the $-\lambda$ and the identity matrix giving

$$\begin{pmatrix} a & b \\ c & d \end{pmatrix}\begin{pmatrix} X_1 \\ X_2 \end{pmatrix} + \begin{pmatrix} -\lambda & 0 \\ 0 & -\lambda \end{pmatrix}\begin{pmatrix} X_1 \\ X_2 \end{pmatrix} = 0 \tag{A3.40}$$

Next, we can factor out $\begin{pmatrix} X_1 \\ X_2 \end{pmatrix}$ leaving

$$\left[\begin{pmatrix} a & b \\ c & d \end{pmatrix} + \begin{pmatrix} -\lambda & 0 \\ 0 & -\lambda \end{pmatrix} \right]\begin{pmatrix} X_1 \\ X_2 \end{pmatrix} = 0 \tag{A3.41}$$

Finally, by matrix addition, we have

$$\begin{pmatrix} a-\lambda & b \\ c & d-\lambda \end{pmatrix}\begin{pmatrix} X_1 \\ X_2 \end{pmatrix} = 0 \tag{A3.42}$$

Thus we have rearranged (A3.37) into (A3.42) and the question now becomes, are there column vectors such that (A3.42) is satisfied? Now, we can answer this question immediately. From the paragraph above, we learned that a system of two linear equations like (A3.42) has a nonzero solution—indeed many of them—if and only if the determinant of the matrix of coefficients in (A3.42) is zero. That is, (A3.42) has a nonzero solution if and only if

$$(a-\lambda)(d-\lambda)-bc = 0 \tag{A3.43}$$

Let me emphasize that, in this equation, λ is an unknown. We are given the matrix to begin with. If the matrix is such that there happens to exist a real λ satisfying (A3.43), then there does exist a vector with the property that the action of the matrix on it is simply to stretch it or to shrink it. Of course, if there is a vector that the matrix simply stretches or shrinks, then its length is arbitrary and all vectors of that direction would also be simply stretched or shrunk. Hence we may, if we wish, impose an artificial convention and restrict ourselves to vectors of unit length.

Our problem now moves to determining if there are solutions to (A3.43). Equation (A3.43) is simply a quadratic equation in λ, which may be rearranged as

$$\lambda^2 - (a+d)\lambda + (ad-bc) = 0 \tag{A3.44}$$

The roots to this equation are

$$\lambda_1 = \frac{(a+d)+\sqrt{(a-d)^2+4bc}}{2}$$

$$\lambda_2 = \frac{(a+d)-\sqrt{(a-d)^2+4bc}}{2} \tag{A3.45}$$

We now have the basic machinery to classify 2×2 matrices, as discussed below. For the record, Equation (A3.43) or (A3.44) is called the *charac-teristic equation* of the matrix $\begin{pmatrix} a & b \\ c & d \end{pmatrix}$, the roots to the characteristic equation are called the *eigenvalues* of the matrix, and a column vector that represents a direction along which the action of a matrix is to stretch or shrink vectors is called an *eigenvector* of the matrix.

Since (A3.44) is a quadratic equation, we know that there are three possibilities concerning the solutions (A3.45).

CASE 1. Both roots are real and distinct (i.e., $\lambda_1 \neq \lambda_2$). This possibility requires that $(a-d)^2 + 4bc > 0$. This finding means that the matrix stretches or shrinks by a factor of λ_1 in one direction, and by a factor of λ_2 in another direction. The direction corresponding to λ_1 is called the eigenvector corresponding to the eigenvalue λ_1. Similarly, there is an eigenvector pointing in a different direction that corresponds to λ_2. It is possible to solve explicitly for the eigenvectors, but that is not necessary for our purposes. All we need to know is that they exist. If we were to take the two distinct eigenvectors as the basis vectors for our coordinate frame, then the matrix would assume a diagonal form, $\begin{pmatrix} \lambda_1 & 0 \\ 0 & \lambda_2 \end{pmatrix}$ with the eigenvalues lying on the diagonal. [You should note, of course, that if the matrix is diagonal to begin with, for example, $\begin{pmatrix} 2 & 0 \\ 0 & \frac{1}{2} \end{pmatrix}$, then $\begin{pmatrix} 1 \\ 0 \end{pmatrix}$ is the eigenvector corresponding to the eigenvalue of 2, and $\begin{pmatrix} 0 \\ 1 \end{pmatrix}$ to $1/2$.]

CASE 2. Both roots are complex numbers. This case requires that $(a-d)^2 + 4bc < 0$ so that Equation (A3.45) indicates λ_1 and λ_2 are a conjugate pair of complex numbers. If a matrix satisfies this condition, it means that there are not any (real) eigenvectors, that is, there are not any directions such that action of the matrix is solely to stretch or shrink along these directions. Now, it can be shown in this case that a coordinate frame can always be found that allows the matrix to be represented in the form which expresses a counterclockwise rotation, followed by a uniform stretching or shrinking in all directions. That is, whenever $(a-d) + 4bc < 0$, then the matrix $\begin{pmatrix} a & b \\ c & d \end{pmatrix}$ can by a suitable choice of basis vectors be put in the form

$$|\lambda| \begin{pmatrix} \cos\theta & -\sin\theta \\ \sin\theta & \cos\theta \end{pmatrix} \tag{A3.46}$$

where $|\lambda|$ is a positive number and θ is an angle expressed in units of radians. The quantities $|\lambda|$ and θ are related to the elements of the original matrix as

$$|\lambda| = \sqrt{ad - bc}$$

$$\theta = \text{arc cos} \frac{a+d}{2\sqrt{ad-bc}} \tag{A3.47}$$

Note that $|\lambda|$ is simply the square root of the determinant. Again the computational details of how to find the basis vectors that lead to the form (A3.46) are irrelevant here. What is important is that they can be found.

CASE 3. Both roots are real and equal (i.e., $\lambda_1 = \lambda_2$). This case requires that $(a-d)^2 + 4bc = 0$. This is a very rare case and requires a very special symmetry in the matrix for the quantity $(a-d)^2 + 4bc$ to equal zero *exactly*. When this case does arise, it may mean one of two different situations. First, there may be more than one direction along which the matrix stretches or shrinks vectors, and it just happens that the factor by which the stretching or shrinking occurs is the same for all the directions. For example, the matrix $\begin{pmatrix} 2 & 0 \\ 0 & 2 \end{pmatrix}$ has two different eigenvectors, $\begin{pmatrix} 1 \\ 0 \end{pmatrix}$ and $\begin{pmatrix} 0 \\ 1 \end{pmatrix}$. Indeed, *all* vectors are eigenvectors of this matrix because it stretches *any* vector by a factor of two. This matrix essentially represents scalar multiplication by a factor of 2. Second, if the eigenvalues are equal, it may mean that there is only *one* eigenvector. For example, the matrix $\begin{pmatrix} 2 & 0 \\ 1 & 2 \end{pmatrix}$ has only one eigenvector, namely $\begin{pmatrix} 1 \\ 0 \end{pmatrix}$, and the eigenvalue corresponding to this direction is 2.

Whenever the eigenvalues are equal, we must decide which of these two situations is occurring. For 2×2 matrices, it can be shown that the first situation arises if and only if the matrix is in a diagonal form to begin with. That is, the first situation above arises if and only if the matrix is originally of the form $\begin{pmatrix} \lambda & 0 \\ 0 & \lambda \end{pmatrix}$. This form is especially trivial because it means the original matrix is merely a constant times the identity matrix. The second situation applies to 2×2 matrices, whenever the eigenvalues are equal *and* the matrix is not originally in a diagonal form. In this second situation, a basis can always be found that puts the matrix in the form $\begin{pmatrix} \lambda & 0 \\ 1 & \lambda \end{pmatrix}$, where λ is the eigenvalue.

Thus we see that the eigenvalues of a matrix are the main diagnostic characters for the taxonomy of matrices. In practice, one calculates the eigenvalues, using the formulas in (A3.45), and then applies the classification into the three cases discussed above. The flow chart for the classification is summarized in Figure A3.6.

Conclusion to Local Stability Analysis with Two Variables; Continuous Time

The conceptual and computational tools of linear algebra presented above allow us to conclude our analysis of local stability for equilibria in two variable dynamical models. We begin with the continuous time model. Recall from Equation (A3.22) that we must solve the system of differential equations

$$\begin{pmatrix} \dfrac{dn_1}{dt} \\[2mm] \dfrac{dn_2}{dt} \end{pmatrix} = \begin{pmatrix} \dfrac{\partial f_1}{\partial N_1} & \dfrac{\partial f_1}{\partial N_2} \\[2mm] \dfrac{\partial f_2}{\partial N_1} & \dfrac{\partial f_2}{\partial N_2} \end{pmatrix} \begin{pmatrix} n_1 \\[2mm] n_2 \end{pmatrix} \qquad (A3.48)$$

Recall that $n_1(t)$ and $n_2(t)$ are the deviation of the variables from the equilibrium point under examination. If both $n_1(t)$ and $n_2(t)$ tend to zero as $t \to \infty$, then the equilibrium point is locally stable. We want to know if the equilibrium point is locally stable and also *how* the solutions approach the equilibrium if it is stable or *how* they depart from it if it is unstable.

The matrix in (A3.48) is called the *gradient matrix* or the *Jacobian matrix*. We have learned from linear algebra that any 2×2 matrix may be represented in one of three forms, provided we choose the appropriate coordinate frame. This fact implies that we need to study (A3.48) above for only three

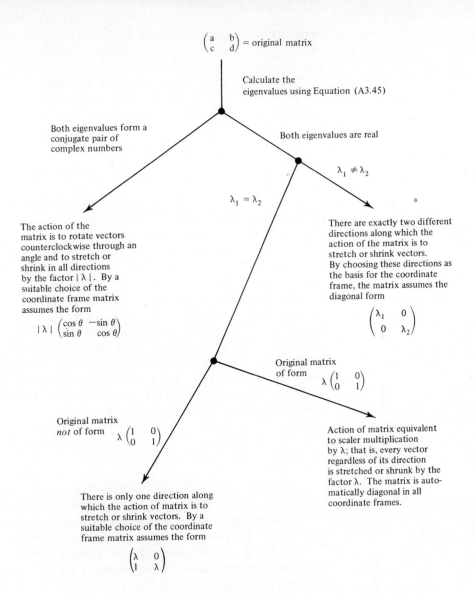

FIGURE A3.6. *Flow chart for the classification of* 2×2 *matrices. The main diagnostic characters for the taxonomy of matrices are the eigenvalues. The eigenvalues are computed using Equation* (A3.45).

representative gradient matrices. The solution to (A3.48) will always be qualitatively the same as the solutions for these three representative examples.

Example I   *The gradient matrix is diagonalizable.* As illustrated in Figure A3.6, this example arises if the eigenvalues of the gradient matrix are real and distinct; it may also arise if the eigenvalues are real and equal to one another. Let $\begin{pmatrix} X_1(t) \\ X_2(t) \end{pmatrix}$ denote the solution through time as represented in the coordinate frame that puts the gradient matrix in a diagonal form. In this coordinate frame, (A3.48) assumes the form

$$\begin{vmatrix} \dfrac{dX_1(t)}{dt} \\ \dfrac{dX_2(t)}{dt} \end{vmatrix} = \begin{pmatrix} \lambda_1 & 0 \\ 0 & \lambda_2 \end{pmatrix} \begin{pmatrix} X_1(t) \\ X_2(t) \end{pmatrix} \qquad \text{(A3.49)}$$

where λ_1 may equal λ_2. The general solution to this system is

$$X_1(t) = e^{\lambda_1 t} X_1(0)$$
$$X_2(t) = e^{\lambda_2 t} X_2(0) \qquad \text{(A3.50)}$$

By inspecting these solutions, we can determine their behavior in the neighborhood of the equilibrium point.

CASE IA. $\lambda_1 \neq \lambda_2, \lambda_1 < 0$, and $\lambda_2 < 0$. The solutions approach the equilibrium point from the directions of both basis vectors. The deviation from equilibrium along the first basis vector decays exponentially at rate λ_1 and along the second basis at rate λ_2. Most solutions appear to approach the equilibrium along the basis vector that has the slowest decay rate. For example, if $|\lambda_1| > |\lambda_2|$, then the deviation from equilibrium with respect to the first basis decays away faster than that measured with respect to the second basis. As a result, the solution appears to approach the equilibrium along the second basis vector because the deviation from equilibrium measured in this direction decays more slowly. In this case the equilibrium is termed a *stable node*. See Figure A3.7a.

CASE IB. $\lambda_1 \neq \lambda_2, \lambda > 0$, and $\lambda_2 > 0$. This case is the reverse of case IA above. The solutions leave the equilibrium along both basis directions. The equilibrium point is called an *unstable node*.

CASE IC. $\lambda_1 \neq \lambda_2$ and $\lambda_1 < 0$ and $\lambda_2 > 0$ or $\lambda_1 > 0$ and $\lambda_2 < 0$. The solutions approach the equilibrium along the basis vector corresponding to the negative eigenvalue and depart from it along the basis vector corresponding to the positive eigenvalue. The equilibrium point is called a *saddle point*. The equilibrium is, of course, unstable. See Figure A3.7b.

CASE ID. $\lambda_1 = \lambda_2 < 0$. The solutions approach the equilibrium from all directions at the same rate. The equilibrium is called a *stable star*. See Figure A3.7c.

CASE IE. $\lambda_1 = \lambda_2 > 0$. The solutions leave the equilibrium in all directions at the same rate. The equilibrium is called an *unstable star*.

Example II *The gradient matrix is of the form* $\begin{pmatrix} \lambda & 0 \\ 1 & \lambda \end{pmatrix}$. As illustrated in Figure A3.6, this example arises if the eigenvalues of the gradient matrix are equal and the matrix is not diagonalizable. In the appropriate coordinate frame, (A3.48) assumes the form

$$\begin{vmatrix} \dfrac{dX_1(t)}{dt} \\ \dfrac{dX_2(t)}{dt} \end{vmatrix} = \begin{pmatrix} \lambda & 0 \\ 1 & \lambda \end{pmatrix} \begin{pmatrix} X_1(t) \\ X_2(t) \end{pmatrix} \qquad \text{(A3.51)}$$

Note that the second basis vector is the single eigenvector of the matrix. The

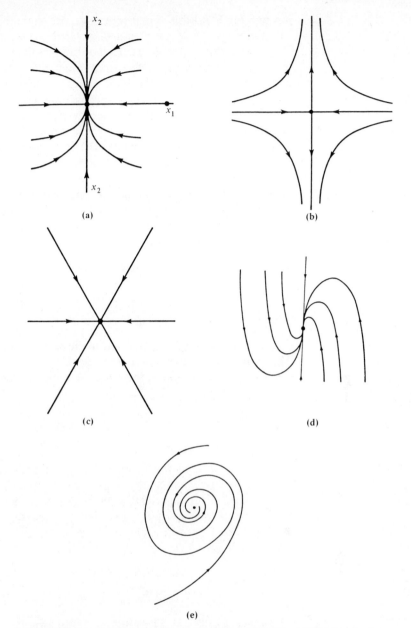

FIGURE A3.7. *Patterns of the trajectories in the neighborhood of an equilibrium.* (a) *Stable node;* (b) *saddle point;* (c) *stable star;* (d) *stable improper node; and* (e) *stable focus.* [Adapted from M. W. Hirsch and S. Smale (1974). *Differential Equations, Dynamical Systems, and Linear Algebra.* Academic Press, Inc., New York.]

general solution to this system is known to be

$$X_1(t) = e^{\lambda t} X_1(0)$$
$$X_2(t) = t\, e^{\lambda t} X_1(0) + e^{\lambda t} X_2(0)$$

(A3.52)

CASE IIA. $\lambda < 0$. The solutions approach the equilibrium along the direction of the second basis vector. Furthermore, solutions may cross through the first basis vector (but not more than once) on their way to the equilibrium point. If λ is negative, all terms in (A3.52) go to zero as $t \to \infty$. The term with $t\,e^{\lambda t}$ tends to zero more slowly than the terms with $e^{\lambda t}$. As a result, the distance from the equilibrium along the second basis vector decreases more slowly than that along the first basis vector. Hence the solution has the appearance of approaching along the second vector. The equilibrium point is called a *stable improper node*. See figure A 3.7d.

CASE IIB. $\lambda > 0$. The solutions leave the equilibrium with the reverse pattern to that discussed above in IIA. The equilibrium is called an *unstable improper node*.

Example III   *The gradient matrix is a rotation.* From Figure A3.6, we observe that this example arises if the eigenvalues of the gradient matrix form a conjugate pair of complex numbers. In the appropriate coordinate frame, the system of differential equations assumes the form

$$
\left|\begin{matrix} \dfrac{dX_1(t)}{dt} \\ \dfrac{dX_2(t)}{dt} \end{matrix}\right| = \begin{pmatrix} \alpha & -\beta \\ \beta & \alpha \end{pmatrix}\begin{pmatrix} X_1(t) \\ X_2(t) \end{pmatrix} \tag{A3.53}
$$

The numbers α and β are found from the eigenvalues of the gradient matrix. The number α is the real part of the eigenvalues and β is the imaginary part. That is, the eigenvalues of the gradient matrix are related to the α and β as

$$
\begin{aligned} \lambda_1 &= \alpha + \beta i \\ \lambda_2 &= \alpha - \beta i \end{aligned} \tag{A3.54}
$$

where α and β are real numbers. As discussed in the section on linear algebra, a matrix in the form in (A3.53) represents a counterclockwise rotation through an angle that depends on α and β, followed by a uniform stretching or shrinking in all directions by a factor that also depends on α and β. For our present purposes, we do not need to compute the angle and stretch factor from the α and β. Instead, we leave the matrix in the form expressed in (A3.53). The general solution to (A3.53) is

$$
\begin{aligned} X_1(t) &= e^{\alpha t} \cos(\beta t) X_1(0) - e^{\alpha t} \sin(\beta t) X_2(0) \\ X_2(t) &= e^{\alpha t} \sin(\beta t) X_1(0) + e^{\alpha t} \cos(\beta t) X_2(0) \end{aligned} \tag{A3.55}
$$

CASE IIIA. $\alpha < 0$. The solutions spiral around and into the equilibrium point. The angular velocity is determined by β, and the rate at which the radial distance from the equilibrium shrinks is determined by α. The equilibrium point is called a *stable focus*. See Figure A3.7e.

CASE IIIB. $\alpha > 0$. The solutions spiral around and away from the equilibrium point. The equilibrium is called an *unstable focus*.

We can discern a common theme in the stability criteria from the three examples discussed.

THEOREM A3.3   *If both eigenvalues of the gradient matrix are real, then the equilibrium is locally stable if they are both negative. If both eigenvalues are complex, then the equilibrium is locally stable if the real parts of the eigenvalues are negative.*

If both eigenvalues are zero or purely imaginary, or if one eigenvalue is negative and the other is zero, then we cannot determine whether the equilibrium is locally stable from the linearized system (A3.48) and further analysis would be needed. However, if one eigenvalue is zero and the other is positive, then the equilibrium is certainly unstable, although we could not fully describe how the solutions leave the equilibrium point without further analysis.

PROBLEM A3.3   Consider the Lotka–Volterra competition equation

$$\frac{dN_1}{dt} = r_1 N_1 \frac{(K_1 - N_1 - \alpha_{12} N_2)}{K_1}$$

$$\frac{dN_2}{dt} = r_2 N_2 \frac{(K_2 - \alpha_{21} N_1 - N_2)}{K_2}$$

(a) Verify that the interior equilibrium point is a stable node if

$$\alpha_{12} < \frac{K_1}{K_2} \qquad \text{and} \qquad \alpha_{21} < \frac{K_2}{K_1}$$

(b) Verify that the interior equilibrium point is a saddle point if

$$\alpha_{12} > \frac{K_1}{K_2} \qquad \text{and} \qquad \alpha_{21} > \frac{K_2}{K_1}$$

PROBLEM A3.4   Consider the following predator-prey model

$$\frac{dV}{dt} = rV \left(1 - \frac{V}{K}\right) - c(1 - e^{-aV/c})P$$

$$\frac{dP}{dt} = bc(1 - e^{-aV/c})P - dP$$

(a) Verify that, provided $d/(bc) < 1$, there is a K_0 such that if $K > K_0$, then a unique positive interior equilibrium exists.
(b) Verify that there is a K_1 such that if K is in the interval (K_0, K_1), then the interior equilibrium point is a stable node.
(c) Verify that if $K = K_1$, then the equilibrium is a stable improper node.
(d) Verify that there is a $K_2 > K_1$ such that if K is in the interval (K_1, K_2), then the equilibrium is a stable focus.
(e) Verify that if $K > K_2$, then the equilibrium is an unstable focus.
(f) As an additional advanced exercise, you might wish to apply the Poincaré-Bendixson theorem to show that if the equilibrium point is an unstable focus according to (e) above, then the solutions approach a stable limit cycle that encloses the unstable equilibrium point.

Local Stability in Discrete Time   To conclude this appendix, we provide a brief discussion of the local stability analysis of an equilibrium point in a two-variable dynamical model defined in discrete time. Recall that we now are to analyze the system of recursion equations

$$\begin{pmatrix} n_{1,t+1} \\ n_{2,t+1} \end{pmatrix} = \begin{pmatrix} \dfrac{\partial g_1}{\partial N_1} & \dfrac{\partial g_1}{\partial N_2} \\ \dfrac{\partial g_2}{\partial N_1} & \dfrac{\partial g_2}{\partial N_2} \end{pmatrix} \begin{pmatrix} n_{1,t} \\ n_{2,t} \end{pmatrix} \tag{A3.56}$$

Again, the matrix in this system is called a gradient or Jacobian matrix. And again, we can determine the qualitative properties of the solutions by considering the three representative examples.

Example I   *The gradient matrix is diagonalizable.* In the natural coordinate frame for the gradient matrix, the system (A3.56) becomes

$$\begin{pmatrix} X_{1,t+1} \\ X_{2,t+1} \end{pmatrix} = \begin{pmatrix} \lambda_1 & 0 \\ 0 & \lambda_2 \end{pmatrix} \begin{pmatrix} X_{1,t} \\ X_{2,t} \end{pmatrix} \tag{A3.57}$$

where λ_1 and λ_2 are the eigenvalues (which must be real) of the gradient matrix, and where $\begin{pmatrix} X_{1,t} \\ X_{2,t} \end{pmatrix}$ represents the solution at time t as expressed in that coordinate frame which diagonalizes the gradient matrix. The general solution to this system expressed in matrix terms is

$$\begin{pmatrix} X_{1,t} \\ X_{2,t} \end{pmatrix} = \begin{pmatrix} \lambda_1^t & 0 \\ 0 & \lambda_2^t \end{pmatrix} \begin{pmatrix} X_{1,0} \\ X_{2,0} \end{pmatrix} \tag{A3.58}$$

Clearly, if both λ_1 and λ_2 are between -1 and 1, then the solution approaches the equilibrium point. We may summarize this sufficient condition for local stability as

$$|\lambda_i| < 1 \qquad i = 1, 2 \tag{A3.59}$$

where $|\lambda_i|$ means the absolute value of λ_i. By further inspection of (A3.58) we could also determine *how* the solutions approach the equilibrium. We will not do so here, however, because this information has not been used elsewhere in the book.

Example II   *The gradient matrix is of the form $\begin{pmatrix} \lambda & 0 \\ 1 & \lambda \end{pmatrix}$.* In this example the system (A3.56) can be represented as

$$\begin{pmatrix} X_{1,t+1} \\ X_{2,t+1} \end{pmatrix} = \begin{pmatrix} \lambda & 0 \\ 1 & \lambda \end{pmatrix} \begin{pmatrix} X_{1,t} \\ X_{2,t} \end{pmatrix} \tag{A3.60}$$

The general solution to this system is

$$\begin{pmatrix} X_{1,t} \\ X_{2,t} \end{pmatrix} = \begin{pmatrix} \lambda^t & 0 \\ t\lambda^{t-1} & \lambda^t \end{pmatrix} \begin{pmatrix} X_{1,0} \\ X_{2,0} \end{pmatrix} \tag{A3.61}$$

Again, we observe that this solution converges to the equilibrium point if

$$|\lambda| < 1 \tag{A3.62}$$

where $|\lambda|$ means the absolute value of λ.

Example III   *The gradient matrix is a rotation.* If the gradient matrix is a rotation, then there is a coordinate frame in which the recursion system takes the form

$$\begin{pmatrix} X_{1,t+1} \\ X_{2,t+1} \end{pmatrix} = \begin{pmatrix} \alpha & -\beta \\ \beta & \alpha \end{pmatrix} \begin{pmatrix} X_{1,t} \\ X_{2,t} \end{pmatrix} \tag{A3.63}$$

The numbers, α and β, are found from the eigenvalues of the gradient matrix as

$$\begin{aligned} \lambda_1 &= \alpha + \beta i \\ \lambda_2 &= \alpha - \beta i \end{aligned} \tag{A3.64}$$

As you know, the matrix represents a counterclockwise rotation through an angle θ, followed by a stretching or shrinking by a (positive) factor $|\lambda|$. The angle θ is found in terms of α and β as

$$\theta = \text{arc cos}\left(\frac{\alpha}{\sqrt{\alpha^2 + \beta^2}}\right) \tag{A3.65}$$

The stretching factor is simply the norm of either of the eigenvalues in the complex plane. That is,

$$|\lambda| = \sqrt{\lambda_1 \bar{\lambda}_1} = \sqrt{\lambda_1 \lambda_2} = \sqrt{\alpha^2 + \beta^2} \tag{A3.66}$$

where the bar indicates the complex conjugate. Hence we can also write (A3.63) in the form

$$\begin{pmatrix} X_{1,t+1} \\ X_{2,t+1} \end{pmatrix} = |\lambda| \begin{pmatrix} \cos\theta & -\sin\theta \\ \sin\theta & \cos\theta \end{pmatrix} \begin{pmatrix} X_{1,t} \\ X_{2,t} \end{pmatrix} \tag{A3.67}$$

In this form, we see that iterating (A3.63) up to time t is simply a matter of successively rotating the initial vector through an angle θ for t times and, also, shrinking or expanding by the factor, $|\lambda|$, for t times. Hence we have that

$$\begin{pmatrix} X_{1,t} \\ X_{2,t} \end{pmatrix} = |\lambda|^t \begin{pmatrix} \cos(\theta t) & -\sin(\theta t) \\ \sin(\theta t) & \cos(\theta t) \end{pmatrix} \begin{pmatrix} X_{1,0} \\ X_{2,0} \end{pmatrix} \tag{A3.68}$$

Clearly, the solution approaches the equilibrium if

$$|\lambda| < 1 \tag{A3.69}$$

where $|\lambda|$ indicates the norm of an eigenvalue.

We observe that the stability condition is essentially the same for the three examples. We may summarize the analysis in Theorem A3.4.

THEOREM A3.4 *If both eigenvalues of the gradient matrix in (A3.56) lie within the unit circle in the complex plane centered at the origin, then the equilibrium point is locally stable. (See Figure A3.8.)*

Finally, we note that dynamical models in discrete time are sometimes formulated as difference equations rather than as recursion equations. If

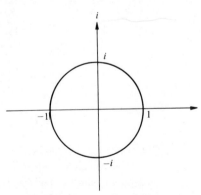

FIGURE A3.8. *An equilibrium in a dynamical model defined in discrete time is locally stable if both eigenvalues of the gradient matrix in (A3.56) lie within the unit circle centered at the origin in the complex plane.*

the model is formulated as a difference equation, then the Taylor expansion about the equilibrium leads to

$$\begin{pmatrix} \Delta n_1 \\ \Delta n_2 \end{pmatrix} = \begin{vmatrix} \dfrac{\partial f_1}{\partial N_1} & \dfrac{\partial f_1}{\partial N_2} \\ \dfrac{\partial f_2}{\partial N_1} & \dfrac{\partial f_2}{\partial N_2} \end{vmatrix} \begin{pmatrix} n_1 \\ n_2 \end{pmatrix} \tag{A3.70}$$

The gradient matrix in this system looks very similar to the one we analyzed for the continuous time model. Of course, (A3.70) can also be put in the form as (A3.56) yielding

$$\begin{pmatrix} n_{1,t+1} \\ n_{2,t+1} \end{pmatrix} = \begin{vmatrix} \dfrac{\partial f_1}{\partial N_1}+1 & \dfrac{\partial f_1}{\partial N_2} \\ \dfrac{\partial f_2}{\partial N_1} & \dfrac{\partial f_2}{\partial N_2}+1 \end{vmatrix} \begin{pmatrix} n_{1,t} \\ n_{2,t} \end{pmatrix} \tag{A3.71}$$

Hence we can translate our previous results as follows.

THEOREM A3.5 *If both eigenvalues of the gradient matrix in (A3.70) lie within a unit circle in the complex plane centered at the point -1, then the equilibrium is locally stable.*

The utility of this version of the local stability criterion for discrete time systems is that it facilitates the comparison with the stability criterion for continuous time systems. For continuous time the eigenvalues of the gradient matrix analogous to that in (A3.70) must be anywhere in the left half of the complex plane in order to guarantee local stability. But for discrete time, the eigenvalues of the gradient matrix in (A3.70) must lie

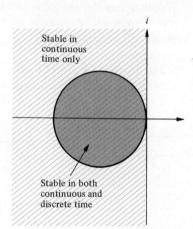

FIGURE A3.9. *Comparison of stability conditions for continuous time and discrete time dynamical models. In continuous time if both eigenvalues of the gradient matrix in (A3.48) lie anywhere within the left half of the complex plane, then the equilibrium is locally stable. In contrast, in discrete time, if both eigenvalues of the analogous gradient matrix in (A3.70) lie within a unit circle centered at -1 in the left half of the complex plane, then the equilibrium is locally stable.*

within the confines of a certain unit circle in the left half of the complex plane in order to guarantee local stability. Thus the discrete time criterion is much more stringent. See Figure A3.9. We can conclude that if an equilibrium point in a difference equation model is locally stable, then that point is also locally stable in the corresponding differential equation model. The converse, on the other hand, is clearly false.

BIBLIOGRAPHY

ALEXANDER, R. D., 1974. The evolution of social behavior. *Ann. Rev. Syst. and Ecol.* **4**: 373–383.

ALLAN, P. M., 1976. Evolution, population dynamics, and stability. *Proc. Nat. Acad. Sci. (USA)* **73**: 665–668.

ALLARD, R. W., 1975. The mating system and microevolution. *Genetics* **79**: 115–126.

ALLARD, R. W., S. K. JAIN, and P. L. WORKMAN, 1968. The genetics of inbreeding populations. *Adv. Genet.* **14**: 55–131.

ANDERSON, G., A. EHRLICH, P. EHRLICH, J. ROUGHGARDEN, B. RUSSELL, and F. TALBOT, 1979. The community structure of coral reef fishes. Unpublished ms.

ANDERSON, O. D., 1978. On sorting out Poole's paper, "Stochastic difference equation predictors of population fluctuation," about the Box-Jenkins analysis and forecasting of ecological time series. *Theor. Pop. Biol.* **13**: 179–189.

ANDERSON, W. W., 1971. Genetic equilibrium and population growth under density-regulated selection. *Amer. Natur.* **105**: 489–498.

ANDREWARTHA, H. G., and L. C. BIRCH, 1954. *The Distribution and Abundance of Animals.* University of Chicago Press, Chicago.

ANDREWARTHA, H. G., and L. C. BIRCH, 1960. Some recent contributions to the study of the distribution and abundance of insects. *Ann. Rev. Ent.* **5**: 219–242.

ARMSTRONG, R. A., and R. MCGEHEE, 1976. Coexistence of species competing for shared resources. *Theor. Pop. Biol.* **9**: 317–328.

ARNOLD, V. I., 1973. *Ordinary Differential Equations.* Trans. by R. A. Silverman. The M.I.T. Press, Cambridge, Mass.

AYALA, F., 1977. Protein evolution: Nonrandom patterns in related species. In *Measuring Selection in Natural Populations*, F. B. Christiansen and T. M. Fenchel, eds., Vol. 19 in *Lecture Notes in Biomathematics.* Springer-Verlag New York, Inc., pp. 177–205.

AYALA, F. J., M. L. TRACEY, L. G. BARR, J. F. MCDONALD, and S. PEREZ-SALAS, 1974. Genetic variation in natural populations of five *Drosophila* species and the hypothesis of the selective neutrality of protein polymorphisms. *Genetics* **77**: 343–384.

BARKER, J. S. F., 1977. Cactus breeding *Drosophila*—A system for the measurement of natural selection. In *Measuring Selection in Natural Populations*, F. B. Christiansen and T. M. Fenchel, eds., Vol. 19 in *Lecture Notes in Biomathematics.* Springer-Verlag New York, Inc., pp. 403–430.

BARTHOLOMEW, G. A., and V. A. TUCKER, 1964. Size, body temperature, thermal conductance, oxygen consumption, and heart rate in Australian Varanid lizards. *Physiol. Zool.* **37**: 341–354.

BAUM, L. E., and J. A. EAGON, 1967. An inequality with applications to statistical estimation for probabilistic functions of Markov processes and to a model for ecology. *Bull. Ann. Math. Soc.* **73**: 360–363.

BELLMAN, R., 1970. *Introduction to Matrix Analysis*, 2nd ed. McGraw-Hill Book Company, New York.

BERNSTEIN, S. C., L. H. THROCKMORTON, and J. L. HUBBY, 1973. Still more genetic variability in natural populations. *Proc. Nat. Acad. Sci. (USA)* **70**: 3928–3931.

BLAIR, W. F., 1960. *The Rusty Lizard.* University of Texas Press, Austin.

BODMER, W. F., 1960. The genetics of homostyly in populations of *Primula vulgaris. Phil. Trans. Roy. Soc.* **B242**: 517–549.

BODMER, W. F., 1965. Differential fertility in population genetic models. *Genetics* **51**: 411–424.

BODMER, W. F., and A. W. F. EDWARDS, 1960. Natural selection and the sex ratio. *Ann. Hum. Genet.* **24**: 239–244.

BODMER, W. F., and J. FELSENSTEIN, 1967. Linkage and selection: Theoretical analysis of the deterministic two locus random mating model. *Genetics* **57**: 237–265.

BOORMAN, S. A., and P. R. LEVITT, 1973. Group selection on the boundary of a stable population. *Theor. Pop. Biol.* **4**: 85–128.

BOSSERT, W. H., 1963. *Simulation of Character Displacement.* Unpublished Ph.D. Thesis. Harvard University, Cambridge, Mass.

BOTKIN, D. B., and M. J. SOBEL, 1975. Stability in time-varying ecosystems. *Amer. Natur.* **109**: 625–646.

BRADSHAW, A. D., 1965. Evolutionary significance of phenotypic plasticity in plants. *Adv. Genet.* **13**: 115–155.

BULMER, M. G., 1972. Multiple niche polymorphism. *Amer. Natur.* **106**: 254–257.

BULMER, M. G., 1974. Density-dependent selection and character displacement. *Amer. Natur.* **108**: 45–58.

BULMER, M. G., 1976a. The statistical analysis of density dependence. *Biometrics* **31**: 901–911.

BULMER, M. G., 1976b. The theory of prey-predator oscillations. *Theor. Pop. Biol.* **9**: 137–150.

BUNDGAARD, J., and F. B. CHRISTIANSEN, 1972. Dynamics of polymorphisms I. Selection components in an experimental population of *Drosophila melanogaster. Genetics* **71**: 439–460.

BUTLER, L., 1953. The nature of cycles in populations of Canadian mammals. *Can. J. Zool.* **31**: 242–262.

CAIN, A. J., and J. D. CURRAY, 1963. Area effects in *Cepaea. Phil. Trans.* **B246**: 1–81.

CANNINGS, C., 1971. Natural selection at a multiallelic autosomal locus with multiple niches. *J. Genetics* **60**: 255–259.

CASE, T., and M. GILPIN, 1974. Interference competition and niche theory. *Proc. Nat. Acad. Sci. (USA)* **71**: 3073–3077.

CASWELL, H., 1978. Predator-mediated coexistence: A non-equilibrium model. *Amer. Natur.* **112**: 127–154.

CAVALLI-SFORZA, L. L., and W. F. BODMER, 1971. *The Genetics of Human Populations.* W. H. Freeman and Company, Publishers, San Francisco.

CHARLESWORTH, B., 1971. Selection in density regulated populations. *Ecology* **52**: 469–474.

CHARLESWORTH, B., 1972. Selection in populations with overlapping generations III. Conditions for genetic equilibrium. *Theor. Pop. Biol.* **3**: 377–395.

CHARNOV, E. L., 1978. Sex-ratio selection in eusocial hymenoptera. *Amer. Natur.* **112**: 317–326.

CHEN, W., 1971. *Applied Graph Theory.* North-Holland Publishing Co., Amsterdam.

CHRISTIANSEN, F. B., 1974. Sufficient conditions for protected polymorphism in a subdivided population. *Amer. Natur.* **108**: 157–166.

CHRISTIANSEN, F. B., 1975. Hard and soft selection in a subdivided population. *Amer. Natur.* **109**: 11–16.

CHRISTIANSEN, F. B., 1977. Population Genetics of *Zoarces viviparus* (L.): A review. In *Measuring Selection in Natural Populations,* F. B. Christiansen and T. M. Fenchel, eds., Vol. 19 in *Lecture Notes in Biomathematics.* Springer-Verlag New York, Inc., pp. 21–47.

CHRISTIANSEN, F. B., and T. M. FENCHEL, eds., 1977. *Measuring Selection in Natural Populations,* Vol. 19 in *Lecture Notes in Biomathematics.* Springer-Verlag New York, Inc.

CHRISTIANSEN, F. B., and O. FRYDENBERG, 1973. Selection component analysis of natural polymorphisms using population samples including mother-offspring combinations. *Theor. Pop. Biol.* **4**: 425–445.

CLARK, L. R., P. W. GIER, R. D. HUGHES, and R. F. MORRIS, 1967. *The Ecology of Insect Populations in Theory and Practice.* Methuen & Co. Ltd, London.

CLARKE, B., 1966. The evolution of morph cline ratios. *Amer. Natur.* **100**: 389–402.

CLARKE, B., 1972. Density-dependent selection. *Amer. Natur.* **106**: 1–13.

CLAUSEN, J., D. KECK, and W. HIESEY, 1941. Regional differentiation in plant species. *Amer. Natur.* **75**: 231–250.

CLEGG, M. T., and R. W. ALLARD, 1972. Patterns of genetic differentiation in the slender wild oat species, *Avena barbata. Proc. Nat. Acad. Sci. (USA)* **69**: 1820–1824.

CLELAND, R. E., 1962. The cytogenetics of *Oenothera. Adv. Genet.* **11**: 147–237.

COCKERHAM, C. C., and B. S. WEIR, 1973. Descent measures for two loci with some applications. *Theor. Pop. Biol.* **4**: 300–330.

CODY, M. L. 1968. On the methods of resource division in grassland bird communities. *Amer. Natur.* **102**: 107–148.

CODY, M. L., 1971. Ecological aspects of reproduction. *Avian Biology.* **1**: 461–512.

COLWELL, R., and D. FUTUYMA, 1971. On the measurement of niche breadth and overlap. *Ecology* **52**: 567–576.

CONNELL, J., 1961a. Effects of competition, predation by *Thais lapillus* and other factors on natural populations of the barnacle *Balanus balanoides. Ecol. Mon.* **31**: 61–104.

CONNELL, J., 1961b. The influence of interspecific competition and other factors on the distribution of the barnacle *Chthamalus stellatus. Ecology* **42**: 710–723.

COUCH, F. G., and J. A. BEARDMORE, 1959. Assortative mating and reciprocal difference in the Blue-Snow Goose complex. *Nature* **183**: 1833–1834.

COX, D. R., and H. D. MILLER, 1968. *The Theory of Stochastic Processes.* John Wiley & Sons, Inc., New York.

COX, E. C., and T. C. GIBSON, 1974. Selection for high mutation rates in chemostats. *Genetics* **77**: 169–184.

CROW, J. F., 1954. Breeding structure of populations II. Effective population number. *Statistics and Mathematics in Biology*, O. Kempthorne, ed. Hafner, New York, pp. 543–556.

CROW, J. F., and M. KIMURA, 1964. The number of alleles that can be maintained in a finite population. *Genetics* **49**: 725–738.

CROW, J. F., and M. KIMURA, 1965. Evolution in sexual and asexual populations. *Amer. Natur.* **99**: 439–450.

CROW, J. F., and M. KIMURA, 1969. Evolution in sexual and asexual populations: A reply. *Amer. Natur.* **103**: 89–91.

CROW, J. F., and M. KIMURA, 1970. *An Introduction to Population Genetics Theory.* Harper & Row, Publishers, New York.

DARLINGTON, C. D., and K. MATHER, 1949. *The Elements of Genetics.* George Allen & Unwin Ltd., London.

DAYTON, P. K., 1971. Competition, disturbance, and community organization: The provision and subsequent utilization of space in a rocky intertidal community. *Ecol. Monogr.* **41**: 351–389.

DEAKIN, M. A. B., 1966. Sufficient conditions for genetic polymorphism. *Amer. Natur.* **100**: 690–692.

DEAKIN, M. A. B., 1968. Genetic polymorphism in a subdivided population. *Aust. J. Biol. Sci.* **21**: 165–168.

DEMPSTER, E. R., 1955. Maintenance of genetic heterogeneity. *Cold Spring Harbor Symp. Quant. Biol.* **70**: 25–32.

DESSAUER, H. C., and E. NEVO, 1969. Geographic variation of blood and liver proteins in cricket frogs. *Biochem. Genet.* **3**: 171–188.

DIAMOND, J. M., 1973. Distributional ecology of New Guinea birds. *Science* **179**: 759–769.

DIAMOND, J. M., 1975. Assembly of species communities. In *Ecology and Evolution of Communities*, M. L. Cody and J. M. Diamond, eds. Belknap Press of Harvard University Press, Cambridge, Mass.

DIAMOND, J., and R. MAY, 1977. Species turnover rates on islands: Dependence on census interval. *Science* **197**: 266–270.

DIAMOND, J., M. GILPIN, and E. MAYR, 1976. Species-distance relation for birds of the Solomon-Archipelago, and the paradox of the great speciators. *Proc. Nat. Acad. Sci. (USA)* **73**: 2160–2164.

DOBZHANSKY, TH., 1951. *Genetics and the Origin of Species*, 3rd ed. Columbia University Press, New York.

DOBZHANSKY, TH., 1954. Evolution as a creative process. Proc. 9th Int. Cong. Genet., in *Caryologia*, pp. 435–449.

DOBZHANSKY, TH., and S. WRIGHT, 1946. Genetics of Natural Populations XII, Experimental reproduction of some changes caused by natural selection in certain populations of *Drosophila pseudo-obscura*. *Genetics* **31**: 125–156.

EAST, E. M., and A. J. MANGELSDORF, 1925. A new interpretation of the heredity behavior of self sterile plants. *Proc. Nat. Acad. Sci. (USA)* **11**: 166–171.

EHRLICH, P. R., and P. H. RAVEN, 1965. Butterflies and plants: A study in co-evolution. *Evolution* **18**: 586–608.

EHRLICH, P. R., and P. H. RAVEN, 1967. Butterflies and plants. *Sci. Amer.* **216**: 104–113.

EHRLICH, P. R., A. H. EHRLICH, and J. P. HOLDREN, 1977. *Ecoscience: Population, Resources, and Environment.* W. H. Freeman and Company, Publishers, San Francisco.

EHRLICH, P. R., R. R. WHITE, M. C. SINGER, S. W. MCKECHNIE, and L. E. GILBERT, 1975. Checkerspot butterflies: A historical perspective. *Science* **188**: 221–228.

ELTON, C., and M. NICHOLSON, 1942. The ten-year cycle in numbers of the lynx in Canada. *J. Anim. Ecol.* **11**: 215–244.

ENDLER, J., 1977. *Geographic Variation, Speciation, and Clines.* Princeton University Press, Princeton, N.J.

ERRINGTON, P. L., 1946. Predation and vertebrate populations. *Quart. Rev. Biol.* **21**: 144–177, 221–245.

ESHEL, I., 1972. On the neighborhood effect and the evolution of altruistic traits. *Theor. Pop. Biol.* **3**: 258–277.

ESHEL, I., 1975. Selection on the sex-ratio and the evolution of sex-determination. *Heredity* **34**: 351–361.

ESHEL, I., and M. W. FELDMAN, 1970. On the evolutionary effect of recombination. *Theor. Pop. Biol.* **1**: 88–100.

ETHERIDGE, R., 1960. *The Relationships of the Anoles (Reptilia : Sauria : Iguanidae): An Interpretation Based on Skeletal Morphology.* University Microfilms, Inc., Ann Arbor, Mich.

EWENS, W. J., 1965. A note on Fisher's theory of the evolution of dominance. *Ann. Hum. Genet. Lond.* **79**: 85–88.

EWENS, W. J., 1968. A genetic model having complex linkage behavior. *Theor. and Appl. Genet.* **38**: 140–143.

EWENS, W. J., 1969. *Population Genetics.* Methuen & Co. Ltd., London.

EWENS, W. J., 1972. The sampling theory of selectively neutral alleles. *Theor. Pop. Biol.* **3**: 87–112.

EWENS, W. J., 1977. Selection and neutrality. In *Measuring Selection in Natural Populations,* F. B. Christiansen and T. M. Fenchel, eds., Vol. 19 in *Lecture Notes in Biomathematics.* Springer-Verlag New York, Inc., pp. 159–175.

EWENS, W. J., and M. W. FELDMAN, 1976. The theoretical assessment of selective neutrality. In *Population Genetics and Ecology.* S. Karlin and E. Nevo, eds. Academic Press, Inc., New York, pp. 303–337.

EWENS, W. J., and F. H. GILLESPIE, 1974. Some simulation results for the neutral allele model with interpretations. *Theor. Pop. Biol.* **6**: 35–57.

FALCONER, D. S., 1960. *Introduction to Quantitative Genetics.* The Ronald Press Company, New York.

FEDER, H. M., 1966. Cleaning symbiosis in the marine environment. In *Symbiosis,* Vol. 1., M. S. Henry, ed. Academic Press, Inc., New York, pp. 327–380.

FELDMAN, M. W., 1971. Equilibrium studies of two locus haploid populations with recombination. *Theor. Pop. Biol.* **2**: 299–318.

FELDMAN, M. W., 1972. Selection for linkage modification: I. Random mating populations. *Theor. Pop. Biol.* **3**: 324–346.

FELDMAN, M. W., I. R. FRANKLIN, and G. THOMSON, 1974. Selection in complex genetic systems: I. The symmetric equilibrium of the three locus symmetric viability model. *Genetics* **76**: 135–162.

FELDMAN, M. W., and S. KARLIN, 1971. The evolution of dominance: A direct approach through the theory of linage and selection. *Theor. Pop. Biol.* **4**: 482–492.

FELDMAN, M. W., and J. ROUGHGARDEN, 1975. A population's stationary distribution and chance of extinction in a stochastic environment with remarks on the theory of species packing. *Theor. Pop. Biol.* **7**: 197–207.

FELLER, W., 1952. The parabolic differential equations and the associated semigroups of transformations. *Ann. Math.* **55**: 468–519.

FELLER, W., 1968. *An Introduction to Probability Theory and Its Applications*, 3rd ed. John Wiley & Sons, Inc., New York.

FELLOWS, D. P., and W. B. HEED, 1972. Factors affecting host plant selection in desert-adapted cactiphilic *Drosophila*. *Ecology* **53**: 850–858.

FENCHEL, T., 1975. Character displacement and coexistence in mud snails (Hydrobiidae). *Oecologia* **20**: 19–32.

FENCHEL, T., and F. B. CHRISTIANSEN, 1977. Selection and interspecific competition. In *Measuring Selection in Natural Populations*, F. B. Christiansen and T. M. Fenchel, eds., Vol. 19 in *Lecture Notes in Biomathematics*. Springer-Verlag New York, Inc., pp. 477–798.

FENNER, F., 1971. Evolution in action: Myxomatosis in the Australian wild rabbit. In *Topics of the Study of Life, the Bio Source Book*, A. Krämer, ed. Harper & Row, Publishers, New York, pp. 463–471.

FENNER, J., and F. N. RATCLIFFE, 1965, *Myxomatosis*. Cambridge University Press, New York.

FISHER, R. A., 1918. The correlation between relatives on the supposition of Mendelian inheritance. *Trans. Roy. Soc. Edinburgh* **52**: 379–433.

FISHER, R. A., 1929. The evolution of dominance; reply to Professor Sewall Wright. *Amer. Natur.* **63**: 553–556.

FISHER, R. A., 1937. The wave of advance of advantageous genes. *Ann. Eugen.* **7**: 355–369.

FISHER, R. A., 1958. *The Genetical Theory of Natural Selection*, 2nd rev. ed. Dover Publications, Inc., New York.

FLEMING, W. H., 1975. A selection-migration model in population genetics. *J. Math. Biol.* **2**: 219–233.

FORD, E. B., 1964. *Ecological Genetics*. Methuen & Co. Ltd., London.

FRANKLIN, I., and R. C. LEWONTIN, 1970. Is the gene the unit of selection? *Genetics* **65**: 707–734.

GALTON, F., 1889. *Natural Inheritance*. Macmillan Publishing Co., Inc., New York.

GANTMACHER, F. R., 1959. *The Theory of Matrices*, Vols. 1 and 2. Chelsea Publishing Co., Inc., New York.

GAUSE, G. F., 1934. *The Struggle for Existence*. Reprinted in 1964 by Hafner Press, New York.

GIBSON, J., 1970. Enzyme flexibility in *D. melanogaster*. *Nature* **227**: 959–960.

GILBERT, L. E., and P. H. RAVEN, eds., 1975. *Coevolution of Animals and Plants*. University of Texas Press, Austin.

GILL, D. E., 1974. Intrinsic rates of increase, saturation density, and competitive ability II. The evolution of competitive ability. *Amer. Natur.* **108**: 103–116.

GILLESPIE, J., 1972. The effects of stochastic environments on allele frequencies in natural populations. *Theor. Pop. Biol.* **3**: 241–248.

GILLESPIE, J., 1973. Polymorphism in random environments. *Theor. Pop. Biol.* **4**: 193–195.

GILLESPIE, J., 1977a. Sampling theory for alleles in a random environment. *Nature* **266**: 443–445.

GILLESPIE, J., 1977b. Multilocus behavior in random environments II. *Genetics* **87**: 569–579.

GILPIN, M. E., 1975. *Group Selection in Predator-Prey Communities.* Princeton University Press, Princeton, N.J.

GILPIN, M. E., 1975. Limit cycles in competition communities. *Amer. Natur.* **109**: 51–60.

GILPIN, M. E., and F. J. AYALA, 1973. Global models of growth and competition. *Proc. Nat. Acad. Sci. (USA)* **70**: 3590–3593.

GILPIN, M. E., and F. J. AYALA, 1976. Schoener's model and Drosophila competition. *Theor. Pop. Biol.* **9**: 12–14.

GILPIN, M., and J. DIAMOND, 1976. Calculation of immigration and extinction curves from the species-area-distance relation. *Proc. Nat. Acad. Sci. (USA)* **11**: 4130–4134.

GINZBURG, L. R., 1977a. The equilibrium and stability for n alleles under density-dependent selection. *J. Theor. Biol.* **68**: 545–550.

GINZBURG, L. R., 1977b. Local consideration of polymorphisms for populations coexisting in stable ecosystems. *J. Math. Biol.* **5**: 33–41.

GOH, B. S., 1976. Global stability in two species interactions. *J. Math. Biol.* **3**: 313–318.

GOLUBITSKY, M., E. KEELER, and M. ROTHSCHILD. 1975. Convergence of the age structure: Applications of the projective method. *Theor. Pop. Biol.* **7**: 84–93.

GOODHARDT, C. B., 1963. "Area effects" and non-adaptive variation between populations of *Cepaea* (*Mollusca*). *Heredity* **18**: 459–465.

GRANT, P. R., 1971. Variation in the tarsus length of birds in island and mainland regions. *Evolution* **25**: 599–614.

GUCKENHEIMER, J., G. OSTER, and A. IPAKTCHI, 1977. The dynamics of density dependent population models. *J. Math. Biol.* **4**: 101–147.

GURTIN, M. E., and R. C. MACCAMY, 1974. Non-linear age-dependent population dynamics. *Arch. Rat'l. Mech. Anal.* **54**: 281–300.

HADELER, K. P., and U. LIBERMAN, 1975. Selection models with fertility differences. *J. Math. Biol.* **2**: 19–32.

HAIGH, J., and J. MAYNARD SMITH, 1972. Can there be more predators than prey? *Theor. Pop. Biol.* **3**: 290–299.

HAIRSTON, N. G., J. D. ALLAN, R. K. COLWELL, D. J. FUTUYMA, J. HOWELL, M. D. LUBIN, J. MATHIAS, and J. H. VANDERMEER, 1968. The relationship between species diversity and stability. An experimental approach with protozoa and bacteria. *Ecology* **49**: 1091–1101.

HALDANE, J. B. S., 1930. A note on Fisher's theory of the origin of dominance. *Amer. Natur.* **64**: 87–90.

HALDANE, J. B. S., 1957. The cost of natural selection. *J. Genet.* **55**: 511–524.

HALDANE, J. B. S., 1966. *The Causes of Evolution.* Cornell University Press, Ithaca, N.Y.

HALDANE, J. B. S., and S. D. JAYAKAR, 1963. Polymorphism due to selection of varying direction. *J. Genet.* **58**: 237–242.

HALL, D. J., W. E. COOPER, and E. E. WERNER, 1970. An experimental approach to the production dynamics and structure of freshwater animal communities. *Limnol. Oceanogr.* **15**: 187–928.

HALMOS, P. R., 1956. *Lectures in Ergodic Theory*. Chelsea Publishing Co., New York.

HAMILTON, W. D., 1964. The genetical theory of social behavior I, II. *J. Theor. Biol.* **7**, 1–52.

HAMILTON, W. D., 1966. The moulding of senescence by natural selection. *J. Theoret. Biol.* **12**: 12–45.

HAMILTON, W. D., 1967. Extraordinary sex ratios. *Science* **156**: 477–488.

HAMILTON, W. D., 1972. Altruism and related phenomena, mainly in the social insects. *Ann. Rev. Syst. and Ecol.* **3**: 193–232.

HAMILTON, W. D., and R. M. MAY, 1977. Dispersal in stable habitats. *Nature* **269**: 578–581.

HAMRICK, J. L., and R. W. ALLARD, 1972. Microgeographic variations in allozyme frequencies in Avena barbata. *Proc. Nat. Acad. Sci. (USA)* **69**: 2100–2104.

HANCOCK, H., 1960. *Theory of Maxima and Minima*. Dover Publications, Inc., New York.

HARRIS, H., 1966. Enzyme polymorphisms in man. *Proc. Roy. Soc. Lond.* **B164**: 298–310.

HARTL, D. L., and R. D. COOK, 1973. Balanced polymorphism of quasineutral alleles. *Theor. Pop. Biol.* **4**: 163–172.

HASSELL, M. P., 1974. Aggregation of predators and insect parasites and its effect on stability. *J. Anim. Ecol.* **43**: 567–587.

HASSELL, M. P., and H. M. COMINS, 1976. Discrete time models for two-species competition. *Theor. Pop. Biol.* **9**: 202–221.

HASSELL, M. P., and R. M. MAY, 1973. Stability in insect-parasite models. *J. Anim. Ecol.* **42**: 693–726.

HASSELL, M., J. LAWTON, and R. MAY, 1976. Pattern of dynamical behavior in single-species populations. *J. Anim. Ecol.* **45**: 471–486.

HASTINGS, A., 1978. Global stability of two species systems. *J. Math. Biol.* **5**: 399–403.

HELLER, C., and E. POULSEN. 1972. Altitudinal zonation of chipmunks (*Eutamias*): Adaptation to temperature and high humidity. *Amer. Mid Natur.* **87**: 296–313.

HEMMINGSEN, A. M., 1960. Energy metabolism as related to body size and respiratory surfaces and its evolution. *Rep. Steno. Mem. Hosp. Nord. Insulin Lab.* **9**: 3–110.

HERSHFIELD, M. S., and N. G. NOSSAL, 1973. *In vitro* characterization of mutator T4 DNA polymerase. *Genetics* **73** (Suppl.): 131–136.

HILL, W. G., 1974. Disequilibrium among several linked neutral genes in finite populations: II. Variances and covariances of disequilibria. *Theor. Pop. Biol.* **6**: 181–198.

HIRSCH, M. W., and S. SMALE, 1974. *Differential Equations, Dynamical Systems and Linear Algebra*. Academic Press, Inc., New York.

HOCHACHKA, P. W., and G. N. SOMERO, 1973. *Strategies of Biochemical Adaptation*. W. B. Saunders Company, Philadelphia.

HOLT, R. D., 1977. Predation, apparent competition, and the structure of prey communities. *Theor. Pop. Biol.* **12**: 197–229.

HORN, H. S., and R. H. MACARTHUR, 1972. Competition among fugitive species in a harlequin environment. *Ecology* **53**: 749–752.

IVLEV, V. S., 1961. *Experimental Ecology of the Feeding of Fishes.* Yale University Press, New Haven, Conn.

JAIN, S. K., 1976. The evolution of inbreeding in plants. *Ann. Rev. Ecol. and Syst.* **7**: 469–495.

JAIN, S. K., and A. D. BRADSHAW, 1966. Evolutionary divergence among adjacent plant populations I. The evidence and its theoretical analysis. *Heredity* **21**: 407–441.

JANZEN, D. H., 1966. Coevolution of mutualism between ants and acacias in Central America. *Evolution* **20**: 249–275.

JANZEN, D. H., 1967. Interaction of the Bull's-Horn Acacia (*Acacia cornigera L.*) with an ant inhabitant (*Pseudomyrmex furruginea* F. Smith) in eastern Mexico. *Univ. Kansas Sci. Bull.* **47**: 315–558.

JAYAKAR, S. D., 1970. A mathematical model for interaction between gene frequencies in a parasite and its host. *Theor. Pop. Biol.* **1**: 140–164.

JOHNSON, G., 1973. Enzyme polymorphism and biosystematics: The hypothesis of selective neutrality. *Ann. Rev. Ecol. and Syst.* **4**: 93–116.

JOHNSON, G., 1974. Enzyme polymorphism and metabolism. *Science* **184**: 28–37.

JOHNSON, G., 1976. Polymorphism and predictability at the α-glycero phosphate dehydrogenase locus in *Colias* butterflies: Gradients in allele frequency within single populations. *Biochem. Genet.* **14**: 403–426.

JOHNSON, G., 1977. Hidden heterogeneity among electrophoretic alleles. In *Measuring Selection in Natural Populations*, F. Christansen and T. M. Fenchel, eds., Vol. 19 in *Lecture Notes in Biomathematics.* Springer-Verlag, New York, Inc., pp. 223–244.

JOHNSON, M. S., 1971. Adaptive lactate dehydrogenase variation in the crested blenny *Anoplorchus. Heredity* **27**: 205–226.

JOHNSON, W. E., and R. K. SELANDER, 1971. Protein variation and systematics in kangaroo rats (genus *Dipodomys*). *Syst. Zool.* **20**: 377–405.

JOHNSTON, J. S., and W. B. HEED, 1975. Dispersal in *Drosophila*: The effect of baiting on the behavior and distribution of natural populations. *Amer. Natur.* **109**: 207–216.

KARLIN, S., 1968a. Rates of approach to homozygosity for finite stochastic models with variable population size. *Amer. Natur.* **102**: 443–455.

KARLIN, S., 1968b. Equilibrium behavior of population genetic models with non-random mating. *J. App. Prob.* **5**: 231–313, 487–566.

KARLIN, S., 1972. Some mathematical models of population genetics. *American Mathematical Monthly* **79**: 699–739.

KARLIN, S., 1973. Sex and infinity: A mathematical analysis of the advantages and disadvantages of recombination. In *The Mathematical Theory of the Dynamics of Biological Populations*, M. S. Bartlett and R. W. Hiorns, eds. Academic Press, Inc., New York, pp. 155–194.

KARLIN, S., 1975. General two-locus selection models: Some objectives, results, and interpretations. *Theor. Pop. Biol.* **7**: 364–398.

KARLIN, S., 1976. Population subdivision and selection-migration inter-action. In *Population Genetics and Ecology*, S. Karlin and E. Nevo, eds. Academic Press, Inc., New York, pp. 617–657.

KARLIN, S., 1977a. Gene frequency patterns in the Levene subdivided population model. *Theor. Pop. Biol.* **11**: 356–385.

KARLIN, S. 1977b. Protection of recessive and dominant traits in a sub-divided population with general migration structure. *Amer. Natur.* **111**: 1145–1162.

KARLIN, S., and D. CARMELLI, 1975a. Numerical studies on two-loci selection model with general viabilities. *Theor. Pop. Biol.* **7**: 397–421.

KARLIN, S. and D. CARMELLI, 1975b. Some population genetic models combining artificial and natural selection pressures: II. Two-locus theory. *Theor. Pop. Biol.* **7**: 123–148.

KARLIN, S., and M. W. FELDMAN, 1968. Further analysis of negative assortive mating. *Genetics* **59**: 117–136.

KARLIN, S., and M. W. FELDMAN, 1970. Linkage and selection: Two locus symmetric viability model. *Theor. Pop. Biol.* **1**: 37–71.

KARLIN, S., and U. LIBERMAN, 1974. Random temporal variation in selection intensities: Case of large population size. *Theor. Pop. Biol.* **6**: 355–382.

KARLIN, S., and U. LIBERMAN, 1975. Random temporal variation in selection intensities: One-locus two allele model. *J. Math. Biol.* **2**: 1–17.

KARLIN, S., and J. MCGREGOR, 1964. Direct product branching pro-cesses and related Markov chains. *Proc. Nat. Acad. Sci. (USA)* **51**: 598–602.

KARLIN, S., and J. MCGREGOR, 1968. The role of the Poisson progeny distribution in population genetics models. *Math. Biosciences* **2**: 11–17.

KARLIN, S., and J. MCGREGOR, 1972a. Polymorphisms for genetic and ecological systems with weak coupling. *Theor. Pop. Biol.* **3**: 210–238.

KARLIN, S., and J. MCGREGOR, 1972b. Application of method of small parameters to multi-niche population genetic models. *Theor. Pop. Biol.* **3**: 186–209.

KARLIN, S., and J. MCGREGOR, 1974. Towards a theory of the evolution of modifier genes. *Theor. Pop. Biol.* **5**: 59–103.

KARLIN, S., and N. RICHTER-DYN, 1976. Some theoretical analyses of migration selection interaction in a cline: A generalized two range environment. In *Population Genetics and Ecology*, S. Karlin and E. Nevo, eds. Academic Press, Inc., New York. pp. 659–706.

KARLIN, S., and F. M. SCUDO, 1969. Assortative mating based on pheno-type II. Two autosomal alleles without dominance. *Genetics* **63**: 499–510.

KEIDING, N., 1975. Extinction and exponential growth in random environments. *Theor. Pop. Biol.* **8**: 49–63.

KEITH, L. B., 1963. *Wildlife's Ten-Year Cycle*. University of Wisconsin Press, Madison, Wisc.

KEMPTHORNE, O., 1957. *An Introduction to Genetic Statistics*. John Wiley & Sons, Inc., New York.

KENDALL, D. G., 1948. A form of wave propagation associated with the equations of heat conduction. *Proc. Cambridge Phil. Soc.* **44**: 591–594.

KERR, W., and S. WRIGHT, 1954. Experimental studies of the distribution of gene frequencies in very small populations of *Drosophila melanogaster*: I. Forked. *Evolution* **8**: 172–177.

KETTLEWELL, H. B. D., 1958. Industrial melanism in the Lepidoptera and its contribution to our knowledge of evolution. *Proc. 10th Intern. Cong. Entomology* **2**: 831–841.

KETTLEWELL, H. B. D., 1961a. Geographical melanism in the Lepidoptera of Shetland. *Heredity* **16**: 393–402.

KETTLEWELL, H. B. D., 1961b. Selection experiments on melanism in *Amathes glareosa* Esp. *Heredity* **16**: 415–434.

KETTLEWELL, H. B. D., 1965. Insect survival and selection for pattern. *Science* **148**: 1290–1296.

KETTLEWELL, H. B. D., and R. J. BERRY, 1961. The study of a cline. *Amathes glareosa* Esp. and its melanic *f. edda* Stand. (Lep) in Shetland. *Heredity* **16**: 403–414.

KETTLEWELL, H. B. D., and R. J. BERRY, 1969. Gene flow in a cline. *Amathes glareosa* Esp. and its melanic *f. edda* Stand. (Lep.) in Shetland. *Heredity* **24**: 1–14.

KEYFITZ, N., and W. FLIEGER, 1971. *Population, Facts and Methods of Demography.* W. H. Freeman and Company, Publishers, San Francisco.

KIMURA, M., 1960. Optimum mutation rate and degree of dominance as determined by the principle of minimum genetic load. *J. Genet.* **57**: 21–34.

KIMURA, M., 1964. Diffusion model in population genetics. *J. Appl. Prob.* **1**: 177–232.

KIMURA, M., and T. OHTA, 1971. *Theoretical Aspects of Population Genetics.* Princeton University Press, Princeton, N.J.

KIMURA, M., and G. H. WEISS, 1964. The stepping stone model of population structure and the decrease of genetic correlation with distance. *Genetics* **49**: 561–576.

KINGMAN, J. F. C., 1961a. A mathematical problem in population genetics. *Proc. Camb. Phil. Soc*, **57**: 574–582.

KINGMAN, J. F. C., 1961b. On an inequality in partial averages. *Quart. J. Math.* **12**: 78–80.

KOEHN, R. K., 1969. Esterase heterogeneity: Dynamics of a polymorphism. *Science* **163**: 943–944.

KOEHN, R. K., F. E. PEREZ, and R. B. MERITT, 1971. Esterase enzyme function and genetical structure of populations of the fresh water fish, *Notropis stramineus. Amer. Natur.* **105**: 51–69.

KOHN, A. J., 1959. The ecology of *Conus* in Hawaii. *Ecol. Mon.* **29**: 47–90.

KOHN, A. J., 1968. Microhabitats, abundance, and food of *Conus* on atoll reefs in the Maldive and Chagos Islands. *Ecology* **49**: 1046–1061.

KOJIMA, K., J. GILLESPIE, and Y. N. TOBARI, 1970. A profile of *Drosophila* species enzymes assayed by electrophoresis I. Number of alleles, heterozygosites and linkage disequilibrium in glucose-metabolizing systems and some other enzymes. *Biochem. Genet.* **4**: 627–637.

KOLMOGOROV, A. N., 1936. Sulla Teoria de Volterra della lotta per l'esisttenza. *Giorn. Instituto Ital. Attuari* **7**: 74–80.

LACK, D., 1947. *Darwin's Finches.* Cambridge University Press, New York.

LACK, D., 1968. *Ecological Adaptations for Breeding in Birds*. Methuen & Co. Ltd., London.

LAWLOR, L. B., 1978. A comment on randomly constructed model ecosystems. *Amer. Natur.* **112**: 445–447.

LAWLOR, L. B., and J. MAYNARD SMITH, 1976. The coevolution and stability of competing species. *Amer. Natur.* **110**: 79–99.

LAZELL, J., 1972. The *Anoles* (Saurea, Iguanidae) of the Lesser Antilles. *Bull. Mus. Comp. Zool.* Harvard University **143**: 1–115.

LEIGH, E. G., 1960. The mechanism of natural selection for the sex ratio. *Amer. Natur.* **94**: 373–377.

LEIGH, E. G., E. L. CHARNOV, and R. R. WARNER, 1976. Sex ratio, sex change, and natural selection. *Proc. Nat. Acad. Sci. (USA)* **73**: 3656–3660.

LEON, J. A., 1974. Selection in contexts of interspecific competition. *Amer. Natur.* **108**: 739–757.

LEVENE, H., 1953. Genetic equilibrium when more than one ecological niche is available. *Amer. Natur.* **87**: 311–313.

LEVIN, B. R., and W. L. KILMER, 1974. Interdemic selection and the evolution of altruism: A computer simulation study. *Evolution* **28**: 527–545.

LEVIN, S. A., 1970. Community equilibria and stability, and an extension of the competitive exclusion principle. *Amer. Natur.* **104**: 413–423.

LEVIN, S. A., 1974. Dispersion and population interaction. *Amer. Natur.* **108**: 207–228.

LEVIN, S. A., and R. T. PAINE, 1974. Disturbance, patch formation, and community structure. *Proc. Nat. Acad. Sci. (USA)* **71**: 2744–2747.

LEVIN, S. A., and L. A. SEGAL, 1976. Hypothesis for origin of planktonic patchiness. *Nature* **259**: 659.

LEVIN, S. A., and G. D. UDOVIC, 1977. A mathematical model of coevolving populations. *Amer. Natur.* **111**: 657–675.

LEVINS, R., 1967. Theory of fitness in a heterogeneous environment. VI. The adaptive significance of mutation. *Genetics* **56**: 163–178.

LEVINS, R., 1968. *Evolution in Changing Environments*. Princeton University Press, Princeton, N.J.

LEVINS, R., 1970. Extinction. In *Some Mathematical Questions in Biology*, M. Gerstenhaber, ed., Lecture on Mathematical Analysis of Fundamental Biological Phenomena. *Annals N.Y. Acad. Sci.* **231**: 123–138.

LEVINS, R., 1974. The qualitative analysis of partially specified systems. In Mathematical Analysis of Fundamental Biological Phenomena. *Annals N.Y. Acad. Sci.* **231**: 123–138.

LEVINS, R., and D. CULVER, 1971. Regional coexistence of species and competition between rare species. *Proc. Nat. Acad. Sci. (USA)* **68**: 1246–1248.

LEWONTIN, R. C., 1971. The effect of genetic linkage on the mean fitness of a population. *Proc. Nat. Acad. Sci. (USA)* **68**: 984–986.

LEWONTIN, R. C., 1974. *The Genetic Basis of Evolutionary Change*. Columbia University Press, New York.

LEWONTIN, R. C., and C. C. COCKERHAM, 1959. The goodness-of-fit for detecting natural selection in random mating populations. *Evol.* **13**: 561–564.

LEWONTIN, R. C., and J. L. HUBBY, 1966. A molecular approach to the study of genic heterozygosity in natural populations of *Drosophila pseudo-obscura. Genetics* **54**: 595–609.

LEWONTIN, R. C., and K. KOJIMA, 1960. The evolutionary dynamics of complex polymorphisms. *Evolution* **14**: 458–472.

LIMBAUGH, C., 1961. Cleaning symbiosis. *Sci. Amer.* **705**: 42–49.

LISTER, B. C., 1976. The nature of niche expansion in West Indian *Anolis* lizards II: Evolutionary components. *Evolution* **30**: 677–692.

LOTKA, A. J., 1945. Population analysis as a chapter in the mathematical theory of evolution, in *Essays on Growth and Form*, a testimonial volume presented to D'Arcy Wentworth Thompson. Oxford University Press, New York.

LUBCHENCO, J., 1978. Plant species diversity in a marine intertidal community: Importance of herbivore food preference and algae competitive abilities. *Amer. Natur.* **112**: 23–39.

LUDWIG, D., 1974. *Stochastic Population Theories*, Vol. 3, in *Lecture Notes in Biomathematics*. Springer-Verlag New York, Inc.

MACARTHUR, R. H., 1962. Some generalized theorems of natural selection. *Proc. Nat. Acad. Sci. (USA)* **231**: 123–138.

MACARTHUR, R. H., 1965. Ecological consequences of natural selection. In *Theoretical and Mathematical Biology*, T. H. Waterman and H. J. Morowitz, eds. John Wiley & Sons, Inc., New York.

MACARTHUR, R. H., 1968. The theory of the niche. In *Population Biology and Evolution*, R. C. Lewontin, ed. Syracuse University Press. Syracuse, N.Y., pp. 159–176.

MACARTHUR, R. H., 1970. Species packing and competitive equilibrium for many species. *Theor. Pop. Biol.* **1**: 1–11.

MACARTHUR, R. H., and R. LEVINS, 1967. The limiting similarity, convergence and divergence of coexisting species. *Amer. Natur.* **101**: 377–385.

MACARTHUR, R. H., and E. O. WILSON, 1963. An equilibrium theory of insular zoogeography. *Evolution* **17**: 373–387.

MACARTHUR, R. H., and E. O. WILSON, 1967. *The Theory of Island Biogeography*. Princeton University Press, Princeton, N.J.

MALÉCOT, G., 1969. *The Mathematics of Heredity*. W. H. Freeman and Company, Publishers. San Francisco.

MANDEL, S. P. H., 1959. The stability of a multiple allelic system. *Heredity*. **13**: 289–302.

MARSDEN, J. E., and M. MCCRACKEN, 1976. *The Hopf Bifurcation and Its Applications*. Springer-Verlag New York, Inc.

MARUYAMA, T., 1970. On the rate of decrease of heterozygosity in circular stepping stone models. *Theor. Pop. Biol.* **1**: 101–119.

MATESSI, C., and S. D. JAYAKAR, 1976a. Models of density-frequent dependent selection for the exploitation of resources I. Intraspecific competition. In *Population Genetics and Ecology*, S. Karlin and E. Nevo, eds. Academic Press, Inc., New York, pp. 707–721.

MATESSI, C., and S. D. JAYAKAR, 1976b. Conditions for the evolution of altruism under Darwinian selection. *Theor. Pop. Biol.* **9**: 360–387.

MAY, R. M., 1972. Limit cycles in predator-prey communities. *Science* **177**: 900–902.

MAY, R. M., 1973. *Stability and Complexity in Model Ecosystems*. Princeton University Press, Princeton, N.J.

MAY, R. M., 1974. Biological populations with non-overlapping generations: Stable points, stable cycles, and chaos. *Science* **186**: 645–647.

MAY, R. M., 1976. Simple mathematical models with very complicated dynamics. *Nature* **261**: 459–467.

MAY, R. M., and W. J. LEONARD, 1975. Nonlinear aspects of competition between three species. *SIAM J. Appl. Math.* **29**: 243–253.

MAY, R. M., and R. H. MACARTHUR, 1972. Niche overlap as a function of environmental variability. *Proc. Nat. Acad. Sci. (USA)* **69**: 1109–1113.

MAY, R. M., J. A. ENDLER, and R. E. MCMURTRIE, 1975. Gene frequency clines in the presence of selection opposed by gene flow. *Amer. Natur.* **109**: 659–676.

MAYNARD SMITH, J., 1964. Kin selection and group selection. *Nature* **201**: 1145–1147.

MAYNARD SMITH, J., 1965. The evolution of alarm calls. *Amer. Natur.* **99**: 59–63.

MAYNARD SMITH, J., 1968. Evolution in sexual and asexual populations. *Amer. Natur.* **102**: 469–473.

MAYNARD SMITH, J., 1970. Genetic polymorphism in a varied environment. *Amer. Natur.* **104**: 487–490.

MAYNARD SMITH, J., 1971a. What use is sex? *J. Theor. Biol.* **30**: 319–335.

MAYNARD SMITH, J., 1971b. The origin and maintenance of sex. In *Group Selection*, G. C. Williams, ed. Aldine Publishing Company, Chicago, pp. 163–175.

MAYNARD SMITH, J., 1974. *Models in Ecology*. Cambridge University Press, New York.

MAYNARD SMITH, J., 1978. *The Evolution of Sex*. Cambridge University Press, New York.

MAYR, E., 1965a. The nature of colonizations in birds. In *The Genetics of Colonizing Species*, H. B. Baker and G. L. Stebbins, eds. Academic Press, Inc. New York, pp. 29–47.

MAYR, E., 1965b. Avifauna: Turnover on islands. *Science* **150**: 1587–1588.

MENGE, B. A., 1972. Competition for food between two intertidal starfish species and its effect on body size and feeding. *Ecology* **53**: 635–644.

MERITT, R. B., 1972. Geographic distribution and enzymatic properties of lactate dehydrogenase in the fathead minnow, *Pimephales promelas*. *Amer. Natur.* **106**: 173–184.

MERTZ, D. B., 1975. Senescent decline in flour beetle strains selected for early adult fitness. *Physiol. Zool.* **48**: 1–23.

MILLER, R. S., 1967. Pattern and process in competition. *Adv. Ecol. Res.* **4**: 1–74.

MINORSKY, N., 1962. *Nonlinear Oscillations*. Reprinted 1974 by Robert E. Krieger Publishing Co., Huntington, N.Y.

MODE, C. J., 1958. A mathematical model for the co-evolution of obligate parasites and their hosts. *Evolution* **12**: 158–165.

MORAN, P. A. P., 1953. The statistical analysis of the Canadian lynx cycle I. Structure and prediction. *Aust. J. Zool.* **1**: 163–173.

MORAN, P. A. P., 1962. *The Statistical Processes of Evolutionary Theory*. The Clarendon Press, Oxford.

MORSE, P., and H. FESHBACH, 1953. *Methods of Theoretical Physics*, Parts I and II. McGraw-Hill Book Company, New York.

MULLER, H. J., 1932. Some genetic aspects of sex. *Amer. Natur.* **66**: 118–138.

MURDOCK, W. W., 1969. Switching in general predators: Experiments in predator specificity and stability of prey populations. *Ecol. Monogr.* **39**: 335–354.

MURDOCK, W. W., S. AVERY, and M. E. B. SMYTH, 1975. Switching in predatory fish. *Ecology* **56**: 1094–1105.

MURDOCK, W. W., and A. OATEN, 1975. Predation and population stability. *Adv. Ecol. Res.* **9**: 1–131.

MUZYCZBA, N., R. L. POLAND, and M. J. BESSMAN, 1972. Studies on the biochemical basis of spontaneous mutation I. A comparison of the DNA polymerases of mutator, antimutator and wild type strains of bacteriophage T4. *J. Biol. Chem.* **247**: 7116-7122.

NAGYLAKI, T., 1975. Conditions for the existence of clines. *Genetics* **80**: 595–615.

NEI, M., 1967. Modification of linkage intensity by natural selection. *Genetics* **57**: 625–626.

NEI, M., 1975. *Molecular Population Genetics and Evolution*. North-Holland Publishing Co., Amsterdam.

NEILL, W. E., 1974. The community matrix and interdependence of the competition coefficients. *Amer. Natur.* **108**: 399–408.

NEVO, E., and C. R. SHAW, 1972. Genetic variation in the subterranean mammal, *Spalax ehrenbergi. Biochem. Genet.* **7**: 235–241.

NORTON, H. T. J., 1928. Natural selection and mendelian variation. *Proc. Lond. Math. Soc.* **28**: 1–45.

NOTTEBOHM, F., 1969. The song of the Chengolo, *Zonotrichia capensis*, in Argentina: Description and evaluation of a system of dialects. *Condor.* **71**: 299–315.

NOTTEBOHM, F., 1975. Continental patterns of song variability in *Zonotrichia capensis*: Some possible ecological correlates. *Amer. Natur.* **109**: 605–624.

OATEN, A., 1977. Transit time and density-dependent predation on a patchily distributed prey. *Amer. Natur.* **3**: 1061–1075.

OATEN, A., and W. W. MURDOCK, 1975. Functional response and stability in predator-prey systems. *Amer. Natur.* **109**: 289–298.

O'DONALD, P., 1959. Possibility of assortative mating in the Arctic Skua. *Nature* **183**: 1210–1211.

O'DONALD, P., 1960. Inbreeding as a result of imprinting. *Heredity* **15**: 79–85.

O'DONALD, P., 1968. Models of the evolution of dominance. *Proc. Roy. Soc. Lond.* **B171**: 127–143.

O'DONALD, P., 1977. Theoretical aspects of sexual selection. *Theor. Pop. Biol.* **12**: 298–334.

OHTA, T., and M. KIMURA, 1969. Linkage disequilibrium due to random genetic drift. *Genet. Res.* **13**: 47–55.

OKUBO, 1974. Diffusion-induced instability in model ecosystems: Another possible explanation of patchiness. Tech. Rept. 86. Chesapeake Bay Inst., Johns Hopkins University, Baltimore, Md.

ORLOVE, M. J., 1975. A model of kin selection not invoking coefficients of relationship. *J. Theor. Biol.* **49**: 289–310.

OSTER, G., and Y. TAKAHASHI, 1974. Models for age-specific interactions in a periodic environment. *Ecol. Mon.* **44**: 483–501.

PAINE, R. T., 1966. Food web complexity and species diversity. *Amer. Natur.* **100**: 65–75.

PIANKA, E. R., 1973. The structure of lizard communities. *Ann. Rev. Ecol. and Syst.* **4**: 53–74.

PIANKA, E. R., 1976. Competition and niche theory. In *Theoretical Ecology, Principles and Applications*, R. M. May, ed. W. B. Saunders Company. Philadelphia, Pa., pp. 114–141.

PIELOU, E. C., 1969. *An Introduction to Mathematical Ecology.* Wiley-Interscience, John Wiley & Sons, Inc., New York.

PIELOU, E. C., 1972. Niche width and niche overlap: A method for measuring them. *Ecology* **53**: 687–692.

PIELOU, E. C., 1975. *Ecological Diversity.* Wiley-Interscience, John Wiley & Sons, Inc., New York.

PIMENTAL, D., 1968. Population regulation and genetic feedback. *Science* **159**: 1432–1437.

PIMENTAL, D., S. A. LEVIN, and D. OLSON, 1978. Coevolution and the stability of exploiter-victim systems. *Amer. Natur.* **112**: 119–125.

PIMENTEL, D., S. A. LEVIN, and A. B. SOANS, 1975. On the evolution of energy balance in some exploiter-victim systems. *Ecology* **56**: 381–390.

PLATT, T., and K. L. DENMAN, 1975. Spectral analysis in ecology. *Ann. Rev. Ecol. Syst.* **6**: 189–210.

POOLE, R. W., 1976. Stochastic difference equation predictors of population fluctuations. *Theor. Pop. Biol.* **9**: 25–45.

POOLE, R. W., 1978. A reply to Anderson. *Theor. Pop. Biol.* **13**: 190–196.

POWELL, T. M., et al., 1975. Spatial scales of current speed and phytoplankton biomass fluctuations in Lake Tahoe. *Science* **189**: 1088–1090.

PRAKASH, S., R. C. LEWONTIN, and J. L. HUBBY, 1969. A molecular approach to the study of genic heterozygosity in natural populations IV. Patterns of genic variation in central, marginal, and isolated populations of *Drosophila pseudoobscura*. *Genetics* **61**: 841–858.

PRESTON, F. W., 1962. The canonical distribution of commonness and rarity. *Ecology* **43**: 185–215, 410–432.

PROHOROV, Y. V., and Y. A. ROZANOV, 1969. *Probability Theory.* Springer-Verlag New York, Inc.

PROUT, T., 1965. The estimation of fitness from genotypic frequencies. *Evolution* **19**: 546–551.

PROUT, T., 1968. Sufficient conditions for multiple niche polymorphism. *Amer. Natur.* **102**: 493–496.

PROUT, T., 1969. The estimation of fitness from population data. *Genetics* **63**: 949–967.

PULLIAM, H. R., 1975. Coexistence of sparrows: A test of community theory. *Science* **189**: 474–476.

RAND, A. S., 1964. Ecological distribution of the anoline lizards of Puerto Rico. *Ecology* **45**: 745–752.

RAND, A. S., 1967. Ecological distribution of the anoline lizards around Kingston, Jamaica. *Breviora* **22**: 1–18.

RHOADES, M. M., 1941. The genetic control of mutability in maize. *Cold Spring Harbor Symp. Quant. Biol.* **9**: 138–144.

RICHARDSON, R. H., 1970. Models and analyses of dispersal patterns. In *Mathematical Topics in Population Genetics*, Ken-ichi Kojima, ed. Springer-Verlag New York, Inc.

RICHTER-DYN, N., and N. S. GOEL, 1972. On the extinction of a colonizing species. *Theor. Pop. Biol.* **3**: 406–433.

ROLLINS, R. C., 1963. The evolution and systematics of *Leavenworthia* (Cruciferae). *Cont. Gray Herb.* Harvard University **192**: 3–198.

RORRES, C., 1976. Stability of an age specific population with density dependent fertility. *Theor. Pop. Biol.* **10**: 26–46.

ROSENZWEIG, M. L., 1969. Why the prey curve has a hump. *Amer. Natur.* **103**: 81–87.

ROSENZWEIG, M. L., 1972. Stability of enriched aquatic ecosystems. *Science* **175**: 564–565.

ROSENZWEIG, M. L., 1973. Evolution of the predator isocline. *Evolution* **27**: 84–94.

ROTHSTEIN, S. I., 1973. The niche-variation model—Is it valid? *Amer. Natur.* **107**: 598–620.

ROUGHGARDEN, J., 1971. Density-dependent natural selection. *Ecology* **52**: 453–468.

ROUGHGARDEN, J., 1972. Evolution of niche width. *Amer. Natur.* **106**: 683–718.

ROUGHGARDEN, J., 1974a. Niche width: Biogeographic patterns among *Anolis* lizard populations. *Amer. Natur.* **108**: 429–442.

ROUGHGARDEN, J., 1974b. Species packing and the competition function with illustrations from coral reef fish. *Theor. Pop. Biol.* **5**: 163–186.

ROUGHGARDEN, J., 1974c. Population dynamics in a spatially varying environment: How population size "tracks" spatial variation in carrying capacity. *Amer. Natur.* **108**: 649–664.

ROUGHGARDEN, J., 1975a. A simple model for population dynamics in stochastic environments. *Amer. Natur.* **109**: 713–736.

ROUGHGARDEN, J., 1975b. Population dynamics in a stochastic environment: Spectral theory for the linearized N-species Lotka–Volterra competition equations. *Theor. Pop. Biol.* **7**: 1–12.

ROUGHGARDEN, J., 1975c. Evolution of marine symbiosis—A simple cost benefit model. *Ecology* **56**: 1201–1208.

ROUGHGARDEN, J., 1976. Resource partitioning among competing species—A coevolutionary approach. *Theor. Pop. Biol.* **9**: 388–424.

ROUGHGARDEN, J., 1977a. Coevolution in ecological systems: Results from "loop analysis" for purely density-dependent coevolution. In *Measuring Selection in Natural Populations*, F. B. Christiansen and T. M. Fenchel, eds., Vol. 19 in *Lecture Notes in Biomathematics*. Springer-Verlag New York, Inc., pp. 449–517.

ROUGHGARDEN, J., 1977b. Patchiness in the spatial distribution of a population caused by stochastic fluctuations in resources. *Oikos* **29**: 52–59.

ROUGHGARDEN, J., 1978a. Coevolution in ecological systems III. Coadaptation and equilibrium population size. In Invited Papers Presented to the Society for the Study of Evolution in Ithaca, 1977, P. Brussard, ed. To be published in *Lecture Notes in Biomathematics.* Springer-Verlag New York, Inc.

ROUGHGARDEN, J., 1978b. Influence of competition on patchiness in a random environment. *Theor. Pop. Biol.* **14**: 185–203.

ROUGHGARDEN, J., and M. FELDMAN, 1975. Species packing and predation pressure. *Ecology* **56**: 489–492.

ROUGHGARDEN, J., and E. FUENTES, 1977. The environmental determinants of size in solitary populations of West Indian Anolis lizards. *Oikos* **29**: 44–51.

ROUGHGARDEN, J. D., E. R. FUENTES, and G. GORMAN, 1979. Discrepancies in the parallel evolution of community structure in two-species communities of *Anolis* lizards: Data and a theoretical model. Unpublished ms.

ROYAMA, T., 1970. Factors governing the hunting behavior and selection of food by the Great Tit (*Parus major* L.). *J. Anim. Ecol.* **39**: 619–668.

SALE, P. F., 1977. Maintenance of high diversity in coral reef fish communities. *Amer. Natur.* **111**: 337–359.

SCHAFFER, W. M., 1977. Evolution, population dynamics, and stability: A comment. *Theor. Pop. Biol.* **11**: 326–329.

SCHAFFER, W. M., 1978. A note on the theory of reciprocal altruism. *Amer. Natur.* **112**: 250–253.

SCHOENER, T., 1967. The ecological significance of sexual dimorphism in size in the lizard *Anolis conspersus. Science* **155**: 474–477.

SCHOENER, T., 1968. The *Anolis* lizards of Bimini: Resource partitioning in a complex fauna. *Ecology* **49**: 704–726.

SCHOENER, T., 1969. Size patterns in West Indian *Anolis* lizards I. Size and species diversity. *Syst. Zool.* **18**: 386–401.

SCHOENER, T., 1973. Population growth regulated by intraspecific competition for energy or time. *Theor. Pop. Biol.* **4**: 56–84.

SCHOENER, T., 1974. Resource partitioning in ecological communities. *Science* **185**: 27–39.

SCHOENER, T., 1976. Alternatives to Lotka–Volterra competition: Models of intermediate complexity. *Theor. Pop. Biol.* **10**: 309–333.

SCHOENER, T., and G. GORMAN, 1968. Some niche differences in three Lesser-Antillean lizards of the genus *Anolis. Ecology* **49**: 819–830.

SCUDO, F. M., and S. KARLIN, 1969. Assortative mating based on phenotype. I. Two alleles with dominance. *Genetics* **63**: 479–498.

SELANDER, R. K., 1966. Sexual dimorphism and differential niche utilization in birds. *Condor* **68**: 113–151.

SELANDER, R. K., 1972. Sexual selection and dimorphism in birds. In *Sexual Selection and the Descent of Man (1871–1971),* B. G. Campbell, ed. Aldine Publishing Company, Chicago, pp. 180–230.

SELANDER, R. K., and W. E. JOHNSON, 1973. Genetic variation among vertebrate species. *Ann. Rev. Ecol. and Syst.* **4**: 75–91.

SETON, E. T., 1911. *The Arctic Prairies.* International University Press, New York.

SHAW, R. F., 1958. The theoretical genetics of the sex ratio. *Genetics* **93**: 149–163.

SIBLEY, C. G., and L. L. SHORT, 1959. Hybridization in the buntings, *Passerina*, of the Great Plains. *The Auk* **76**: 443–463.

SIMBERLOFF, D., 1972. Models in biogeography. In *Models in Paleobiology*, T. Schopf, ed. W. H. Freeman and Company, Publishers, San Francisco, pp. 160–191.

SIMBERLOFF, D. S., and E. O. WILSON, 1969. Experimental zoogeography of islands: The colonization of empty islands. *Ecology* **50**: 278–296.

SIMBERLOFF, D. S., and E. O. WILSON, 1970. Experimental zoogeography of islands: A two-year record of colonization. *Ecology* **51**: 934–937.

SINAI, Y. G., 1977. *Introduction to Ergodic Theory*. Princeton University Press, Princeton, N.J.

SINKO, J., and W. STREIFER, 1967. A new model for age-size structure of a population. *Ecology* **48**: 910–918.

SKELLAM, J. G., 1951. Random dispersal in theoretical populations. *Biometrika* **38**: 196–218.

SLATKIN, M., 1970. Selection and polygenic characters. *Proc. Nat. Acad. Sci. (USA)* **66**: 87–93.

SLATKIN, M., 1972. On treating the chromosome as the unit of selection. *Genetics* **72**: 157–168.

SLATKIN, M., 1973. Gene flow and selection in a cline. *Genetics* **75**: 733–756.

SLATKIN, M., 1974. Competition and regional coexistence. *Ecology* **55**: 128–134.

SLATKIN, M., 1978. On the equilibration of fitnesses by natural selection. *Amer. Natur.* **112**: 845–859.

SLATKIN, M., and R. LANDE, 1976. Niche width in a fluctuating environment-density independent model. *Amer. Natur.* **110**: 31–55.

SLOBODKIN, L. B., 1961. *Growth and Regulation of Animal Populations*. Holt, Rinehart and Winston, New York.

SMALE, S., 1976. On the differential equations of species in competition. *J. Math. Biol.* **3**: 5–7.

SMALE, S., and R. F. WILLIAMS, 1976. The qualitative analysis of a difference equation of population growth. *J. Math. Biol.* **3**: 1–4.

SMITH, N. G., 1968. The advantage of being parasitized. *Nature* **219**: 690–694.

SMITH-GILL, S. J., and D. E. GILL, 1978. Curvilinearities in the competition equations: An experiment with Ranid tadpoles. *Amer. Natur.* **112**: 557–570.

SMOUSE, P. E., 1976. The implications of density-dependent population growth for frequency- and density-dependent selection. *Amer. Natur.* **110**: 849–860.

SOKAL, R. R., and F. J. ROHLF, 1969. *Biometry*. W. H. Freeman and Company, Publishers, San Francisco.

SOLBRIG, O., 1971. The population biology of dandelions. *Amer. Scientist* **59**: 686–694.

SOLBRIG, O., 1972. Breeding system and genetic variation in *Leavenworthia*. *Evolution* **26**: 155–160.

SOULÉ, M. F., and B. R. STEWART, 1970. The "niche-variation" hypothesis: A test and alternatives. *Amer. Natur.* **104**: 85–97.

STADLER, L. J., 1946, 1948, 1949. Spontaneous mutation at the *R* locus in maize I. *Genetics* **31**: 377–394; II. *Amer. Natur.* **82**: 289–314; III. *Amer. Natur.* **83**: 5–30.

STEARNS, S. C., 1976. Life-history tactics: A review of the ideas. *Quart. Rev. Biol.* **51**: 3–47.

STEBBINS, G. L., 1950. *Variation and Evolution in Plants.* Columbia University Press, New York.

STEBBINS, G. L., 1971. *Chromosomal Evolution in Higher Plants.* Addison-Wesley Publishing Co., Inc., Reading, Mass.

STEELE, J. H., 1974a. *Structure of Marine Ecosystems.* Harvard University Press, Cambridge, Mass.

STEELE, J. H., 1974b. Spatial heterogeneity and population stability. *Nature* **243**: 83.

STEWART, F. M., 1971. Evolution of dimorphism in a predator-prey model. *Theor. Pop. Biol.* **2**: 493–506.

STROBECK, C., 1973. *N* species competition. *Ecology* **54**: 650–654.

STROBECK, C., 1974. Sufficient conditions for polymorphism with multiple niches and a general pattern of mating. *Amer. Natur.* **108**: 152–156.

Studies on *Cepaea*, 1968. *Phil. Trans.* **B253**: 383–593.

SUMNER, F. B., 1932. Genetic, distribution, and evolutionary studies of the subspecies of deer mice (*Peromyscus*). *Bibliogr. Genetica.* **9**: 1–116.

SUTHERLAND, J. P., 1974. Multiple stable points in natural communities. *Amer. Natur.* **108**: 859–873.

SVED, J. A., 1967. The stability of linked systems of loci with a small population size. *Genetics* **59**: 543–563.

SVED, J. A., 1971. An estimate of heterosis in *Drosophila melanogaster. Genet. Res.* **18**: 97–105.

SVED, J. A., and F. J. AYALA, 1970. A population cage test for heterosis in *Drosophila pseudo-obscura, Genetics* **66**: 97–113.

SVED, J. A., and D. MAYO, 1970. The evolution of dominance in *Topics in Mathematical Genetics*, K. Kojima, ed. Springer-Verlag New York, Inc.

TAYLOR, H. M., R. S. GOURLEY, C. E. LAWRENCE, and R. S. KAPLAN, 1974. Natural selection of life history attributes: An analytical approach. *Theor. Pop. Biol,* **5**: 104–122.

TEMPLETON, A. R., 1974. Density-dependent selection in parthenogenetic and self-mating populations. *Theor. Pop. Biol.* **5**: 229–250.

TIMOFEEFF-ROSSOVSKY, N. W., 1937. Experimentelle mutationsforschung in der vererbungslehre. Theodor Steinkopff, Dresden and Leipzig.

TINBERGEN, L., 1960. The natural control of insects in pinewoods I. Factors influencing the intensity of predation by songbirds. *Arch. Neerland. Zool.* **13**: 265–343.

TINKLE, D. W., H. M. WILBUR, and S. G. TILLEY, 1970. Evolutionary strategies in lizard reproduction. *Evolution* **24**: 55–74.

TRIVERS, R. L., 1971. The evolution of reciprocal altruism. *Quart. Rev. Biol.* **46**: 35–57.

TRIVERS, R. L., 1972. Parental investment and sexual selection. In *Sexual Selection and the Descent of Man 1871–1971*, B. Campbell, ed. Aldine Publishing Company, Chicago, pp. 136–179.

TRIVERS, R. L., 1974. Parent-offspring conflict. *Amer. Zool.* **14**: 249–264.

TULLOCK, G. 1971. The Coal Tit as a careful shopper. *Amer. Natur.* **105**: 77–80.

TURELLI, M., 1977. Random environments and stochastic calculi. *Theor. Pop. Biol.* **12**: 140–178.

TURELLI, M., 1978. A reexamination of stability in randomly varying versus deterministic environments with comments on the stochastic theory of limiting similarity. *Theor. Pop. Biol.* **13**: 244–267.

ULAM, S. M., and J. VON NEUMANN, 1947. *Bull. Amer. Math. Soc.* (abstr.) **53**: 1120.

VANCE, R., 1978. Predation and resource partitioning in one predator-two prey model communities. *Amer. Natur.* **112**: 797–813.

VANDERMEER, J., 1969. The competitive structure of communities: An experimental approach with Protozoa. *Ecology* **50**: 362–371.

VAN SICKLE, J., 1977. Analysis of a distributed-parameter population model based on physiological age. *J. Theor. Biol.* **64**: 571–586.

VAN VALEN, L., 1965. Morphological variation and the width of the ecological niche. *Amer. Natur.* **99**: 377–390.

VERWEY, J., 1930. Coral reef studies I. The symbiosis between damsel fishes and sea anemones in Batavia Bay. *Treubia* **12**: 305–366.

VON FOERSTER, H., 1959. Some remarks on changing populations. In *The Kinetics of Cellular Proliferation*, F. Stohlman, Jr., ed. Grune & Stratton, Inc., New York, pp. 382–407.

WAHLUND, S., 1928. Zuzammensetzung von populationen und korrelationserscheinungen vom standpunkt der vererbungslehre aus betrachtet. *Hereditas* **11**: 65–106.

WAKE, D. B., and J. F. LYNCH, 1976. The distribution, ecology, and evolutionary history of plethodontid salamanders in tropical America. *Sci. Bull. Nat. Hist. Mus. (Los Angeles Co.)* **25**: 1–65.

WALLACE, B., 1963. The elimination of an autosomal lethal from an experimental population of *Drosophila melanogaster*. *Amer. Natur.* **97**: 65–66.

WALLACE, B., 1968. *Topics in Population Genetics*. W. W. Norton & Company, Inc. New York.

WATT, W. B., 1977. Adaptation at specific loci. I. Natural selection on phosphoglucose isomerase of *Colias* butterflies: Biochemical and population aspects. *Genetics* **87**: 177–194.

WEIR, B. S., and C. C. COCKERHAM, 1973. Mixed selfing and random mating at two loci. *Genet. Res.* **21**: 247–262.

WEST EBERHARD, M. J., 1969. The social biology of polistine wasps. *Univ. Mich. Mus. Zool. Misc. Publ.* **140**: 1–101.

WEST EBERHARD, M. J., 1975. The evolution of social behavior by kin selection. *Quart. Rev. Biol.* **50**: 1–33.

WHITE, M. J. D., 1973. *Animal Cytology and Evolution*, 3rd. ed. Cambridge University Press, New York.

WHITTAKER, R. H., and P. P. FEENY, 1971. Allelochemics: Chemical interactions between species. *Science* **171**: 757–770.

WHITTAKER, R. H., and S. A. LEVIN, 1977. The role of mosaic phenomena in natural communities. *Theor. Pop. Biol* **12**: 117–139.

WIEBE, P. H., 1970. Small scale spatial distribution in oceanic zooplankton. *Limnol. Oceanogr.* **15**: 205–217.

WILBUR, H. M., 1972. Competition, predation and the structure of the *Ambystoma-Rana sylvatica* community. *Ecology* **53**: 3–21.

WILKES, H. G., 1977. The world's crop plant germplasm—An endangered resource. *Bull. Atomic Scientist* **33**: 8–14.

WILLIAMS, E. E., 1969. The ecology of colonization as seen in the zoogeography of Anoline lizards on small islands. *Quart. Rev. Biol.* **44**: 345–389.

WILLIAMS, E. E., 1972. Origin of faunas. Evolution of lizard congeners in a complex island fauna—A trial analysis. *Evol. Biol.* **6**: 47–89.

WILLIAMS, G. C., 1966. *Adaptation and Natural Selection: A Critique of Some Current Evolutionary Thought.* Princeton University Press, Princeton, N.J.

WILLIAMS, G. C., 1975. *Sex and Evolution.* Princeton University Press, Princeton, N.J.

WILLIAMS, G. C., and J. B. MITTON, 1973. Why reproduce sexually? *J. Theor. Biol.* **39**: 545–554.

WILLIAMS, G. C., and D. C. WILLIAMS, 1957. Natural selection of individually harmful social adaptations among sibs with special reference to social insects. *Evolution* **11**: 32–39.

WILLSON, M. F., 1969. Avian niche size and morphological variation. *Amer. Natur.* **103**: 531–542.

WILSON, A. C., S. S. CARLSON, and T. J. WHITE, 1977. Biochemical evolution. *Ann. Rev. Biochem.* **46**: 573–639.

WILSON, D. S., 1975a. A theory of group selection. *Proc. Nat. Acad. Sci. (USA)* **72**: 143–146.

WILSON, D. S., 1975b. The adequacy of body size as a niche difference. *Amer. Natur.* **109**: 769–784.

WILSON, E. O., 1971. *The Insect Societies.* Belknap Press of Harvard University Press, Cambridge, Mass.

WILSON, E. O. 1973. Group selection and its significance for ecology. *BioScience* **23**: 631–638.

WILSON, E. O., 1975. *Sociobiology, the New Synthesis.* Belknap Press of Harvard University Press, Cambridge, Mass.

WINGATE, D. B., 1965. Terrestrial herpetofauna of Bermuda. *Herpetologica* **21**: 202–218.

WORKMAN, P. L., 1964. The maintenance of heterozygosity by partial negative assortative mating. *Genetics* **50**: 1369–1382.

WRIGHT, S., 1929. Fisher's theory of dominance. *Amer. Natur.* **63**: 274–279.

WRIGHT, S., 1931. Evolution in mendelian populations. *Genetics* **16**: 97–159.

WRIGHT, S., 1945. The differential equation of the distribution of gene frequencies. *Proc. Nat. Acad. Sci. (USA)* **31**: 382–389.

WRIGHT, S., 1969. *Evolution and the Genetics of Populations*, Vol. 2. *The Theory of Gene Frequencies.* University of Chicago Press, Chicago.

YOSHIYAMA, R., and J. ROUGHGARDEN, 1977. Species packing in two dimensions. *Amer. Natur.* **111**: 107–121.

INDEX